TELE**HEALTH** AND MOBILE **HEALTH**

THE E-MEDICINE, E-HEALTH, M-HEALTH, TELEMEDICINE, AND TELEHEALTH HANDBOOK
VOLUME II

TELE**HEALTH** AND MOBILE **HEALTH**

Edited by
Halit Eren
John G. Webster

CRC Press
Taylor & Francis Group
Boca Raton London New York

CRC Press is an imprint of the
Taylor & Francis Group, an **informa** business

MATLAB® is a trademark of The MathWorks, Inc. and is used with permission. The MathWorks does not warrant the accuracy of the text or exercises in this book. This book's use or discussion of MATLAB® software or related products does not constitute endorsement or sponsorship by The MathWorks of a particular pedagogical approach or particular use of the MATLAB® software.

CRC Press
Taylor & Francis Group
6000 Broken Sound Parkway NW, Suite 300
Boca Raton, FL 33487-2742

First issued in paperback 2017

© 2016 by Taylor & Francis Group, LLC
CRC Press is an imprint of Taylor & Francis Group, an Informa business

No claim to original U.S. Government works

ISBN-13: 978-1-4822-3661-3 (hbk)
ISBN-13: 978-1-138-89349-8 (pbk)

Visit the Taylor & Francis Web site at
http://www.taylorandfrancis.com

and the CRC Press Web site at
http://www.crcpress.com

Contents

Section I   Medical Robotics, Telesurgery, and Image-Guided Surgery

Section II   Telenursing, Personalized Care, Patient Care, and eEmergency Systems

Section III   Networks and Databases, Informatics, Record Management, Education, and Training

Section IV   Business Opportunities, Management and Services, and Web Applications

Section V   Examples of Integrating Technologies: Virtual Systems, Image Processing, Biokinematics, Measurements, and VLSI

Preface

Introduction

The purpose of the *Telehealth and Mobile Health* handbook is to provide a reference that is both concise and useful for biomedical engineers in universities and medical device industries, scientists, designers, managers, research personnel, and students, as well as healthcare personnel, such as physicians, nurses, and technicians, who use technology over a distance. The handbook covers an extensive range of topics that comprise the subject of distance communication, from sensors on and within the body to electronic medical records. It serves the reference needs of a broad group of users—from advanced high school science students to healthcare and university professionals.

Recent development in digital technologies is paving the way for ever-increasing use of information technology and data-driven systems in medical and healthcare practices. Hence, this handbook describes how information and communication technologies, the Internet, wireless technologies and wireless networks, databases, and telemetry permit the transmission of information and control of information both within a medical center and between medical centers. Recent developments in sensors, wearable computing, and ubiquitous communications have provided medical experts and users with frameworks for gathering physiological data on a real-time basis over extended periods of time. Wearable sensor-based systems can transform the future of healthcare by enabling proactive personal health management and unobtrusive monitoring of a patient's health condition. Wireless body area networks permit a comfortable tank top with sensors to use wireless local area networks (e.g., Wi-Fi and Bluetooth) to continuously transmit to other systems such as smartphones, and then from any location, such as home, away from home, on the streets, or a nursing home, to a medical center for analysis of cardiac arrhythmias and ventilation. For example, a simple remotely located base unit can continually collect and locally integrate many incoming signals such as electrocardiography, oxygen saturation, heart rate, noninvasive blood pressure, temperature, and respiration, and provide the information required for detecting any possible emergency cases for the patients. The medical center can then accommodate all complementary and bulky systems, including telemedicine-enabled equipment such as intensive care units, intelligent analyzers, and automatic recorders plus a professionally managed database system supported by a professional service provider.

Today's technology allows clinical processes to be conducted at a distance; hence, it is an enabler, but in itself, the technology is not telemedicine. Telemedicine can be thought of as the tasks that the clinician carries out (such as observing, consulting, interpreting, and providing opinions), assisted by information and communication technologies, in circumstances where there is distance between the patient and the provider. Put succinctly, modern telemedicine is simply medicine at a distance.

This handbook also intends to bridge the gap between scientists, engineers, and medical professionals by creating synergy in the related fields of biomedical engineering, information and communication technologies, network operations, business opportunities, and dynamically evolving modern medical and healthcare practices. It includes how medical personnel use information and communication technologies, as well as sensors, techniques, hardware, and software. It gives information on wireless data transmission, networks, databases, processing systems, and automatic data acquisition, reduction, and analysis and their incorporation for diagnosis.

The chapters include descriptive information for professionals, students, and workers involved in eMedicine, telemedicine, telehealth, and mHealth. Equations in some chapters also assist biomedical engineers and healthcare personnel who seek to discover applications and solve diagnostic problems that arise in medical fields not in their specialty. All the chapters are written by experts in their fields and include specialized information needed by informed specialists who seek to find out advanced applications of the subject, evaluative opinions, and possible areas for future study.

Organization

The handbook is organized into two volumes and each volume in sections.
The sections in Volume 1 are

Section I: Integration of eMedicine, Telemedicine, eHealth, and mHealth

Section II: Wireless Technologies and Networks

Section III: Sensors, Devices, Implantables, and Signal Processing

Section IV: Implementation of eMedicine and Telemedicine

Section I contains information on the integration of modern eMedicine, telemedicine, eHealth, and telehealth. The interactions between these practices are explained and examples are given. Wireless technology, an essential part of telemedicine, is explained in Section II, with a particular emphasis on the fast deploying wireless body area networks. The state of the art on sensors, devices, and implantables is explained in Section III, while Section IV is dedicated to practical applications of all the information given in Sections I through III, ranging from telecardiology and teleradiology to teleoncology and acute care telemedicine.
The sections in Volume 2 are

Section I: Medical Robotics, Telesurgery, and Image-Guided Surgery

Section II: Telenursing, Personalized Care, Patient Care, and eEmergency Systems

Section III: Networks and Databases, Informatics, Record Management, Education, and Training

Section IV: Business Opportunities, Management and Services, and Web Applications

Section V: Examples of Integrating Technologies: Virtual Systems, Image Processing, Biokinematics, Measurements, and VLSI

We are all aware of telesurgery implementation as a routine process while the patient and the surgeon may be continents apart. This is explained in Section I, with emphasis on medical robotics and image guidance. Remote patient care, personalized care, and telenursing can be found in Section II. For an effective remote care, the use of networks, data management, record management, and the education and training aspects of personnel are given in Section III. Implementation of new technologies in eMedicine and eHealth bring many business, management, and service opportunities as explained in Section IV. Examples of emerging technologies, developing engineering, and scientific contributions are given in Section V. For example, the sound understanding of biokinematics has led to successful implementation of brain-controlled bionic human parts such as bionic arms and hands.

Locating Your Topic in the *Telehealth and Mobile Health* Handbook

Select your topic, skim the Contents, and peruse the chapter that describes your topic. Consider the alternative methods of distance communication with each of their advantages and disadvantages prior to selecting the most suitable method. For more detailed information, consult the Index, since certain principles of eMedicine, eHealth, mHealth, telemedicine, and telehealth may appear in more than one chapter.

MATLAB® is a registered trademark of The MathWorks, Inc. For product information, please contact:

The MathWorks, Inc.
3 Apple Hill Drive
Natick, MA 01760-2098 USA
Tel: 508 647 7000
Fax: 508-647-7001
E-mail: info@mathworks.com
Web: www.mathworks.com

Acknowledgments

We thank all 84 authors in this volume for sharing their expertise and sparing their valuable time to contribute to this volume. We also gratefully acknowledge the CRC Press team for their patience and tireless effort in putting everything together. We also thank all our readers in selecting this book to advance their knowledge and technical skills.

Editors

Halit Eren received BEng, MEng, and PhD degrees in 1973, 1975, and 1978, respectively, from the University of Sheffield, Sheffield, United Kingdom. He obtained an MBA degree from Curtin University, Perth, Australia, in 1999.

After his graduation, Dr. Eren worked in industry as an instrumentation engineer for two years. He held a position as assistant professor at Hacettepe University, Ankara, Turkey, in 1980–1981 and Middle East Technical University, Ankara, Turkey, in 1982. He has been at Curtin University since 1983, conducting research and teaching primarily in the areas of control systems, instrumentation, and engineering management. Currently, Dr. Eren holds an Adjunct Senior Research Fellow position at Curtin University.

Dr. Eren held an associate professor position at the Polytechnic University in Hong Kong in 2004, visiting professor position at the University of Wisconsin, Madison, Wisconsin, in 2013, and a visiting scholar position at the University of Sheffield, UK, in 2015. He is a senior member of the Institute of Electrical and Electronics Engineers, taking roles in various committees for organizing conferences and as a member of editorship in transactions. Dr. Eren has over 190 publications in conference proceedings, books, and transactions. He is the author of *Electronic Portable Instruments: Design and Applications* (2004) and *Wireless Sensors and Instruments: Networks, Design, and Applications* (Boca Raton: CRC Press, 2006). He coedited the fourth edition of *Instrument Engineers' Handbook, Volume 3: Process Software and Digital Networks*, in 2011 with Bela Liptak, and the second edition of the two-volume set *Measurement, Instrumentation, and Sensors Handbook* (Boca Raton: CRC Press, 2014) with John G. Webster. Dr. Eren is active in researching and publishing on wireless instrumentation, wireless sensor networks, intelligent sensors, automation and control systems, and large control systems.

John G. Webster received a BEE degree from Cornell University, Ithaca, New York, in 1953, and an MSEE and a PhD degree from the University of Rochester, Rochester, New York, in 1965 and 1967, respectively.

Dr. Webster is professor emeritus of biomedical engineering at the University of Wisconsin, Madison, Wisconsin. He was a highly cited researcher at King Abdulaziz University, Jeddah, Saudi Arabia. In the field of medical instrumentation, he teaches undergraduate and graduate courses and does research on an intracranial pressure monitor, electrocardiogram dry electrodes, and tactile vibrators.

Dr. Webster is author of *Transducers and Sensors*, an Institute of Electrical and Electronics Engineers/Educational Activities Board Individual Learning Program (Piscataway: IEEE, 1989). He is coauthor, with B. Jacobson, of *Medicine and Clinical Engineering* (Englewood Cliffs: Prentice-Hall, 1977); with R. Pallas-Areny, of *Sensors and Signal Conditioning*, second edition (New York: Wiley, 2001); and with R. Pallas-Areny, of *Analog Signal Conditioning* (New York: Wiley, 1999). He is editor of *Encyclopedia of Medical Devices and Instrumentation*, second edition (New York: Wiley, 2006); *Tactile Sensors for Robotics and Medicine* (New York: Wiley, 1988); *Electrical Impedance Tomography* (Bristol: Adam Hilger, 1990), *Teaching Design in Electrical Engineering* (Piscataway: Educational Activities Board, IEEE, 1990), *Prevention of Pressure Sores: Engineering and Clinical Aspects* (Bristol: Adam Hilger, 1991); *Design of Cardiac Pacemakers* (Piscataway: IEEE Press, 1995); *Design of Pulse Oximeters* (Bristol: IOP Publishing, 1997); *Encyclopedia of Electrical and Electronics Engineering* (New

York, Wiley, 1999); *Minimally Invasive Medical Technology* (Bristol: IOP Publishing, 2001); *Bioinstrumentation* (Hoboken: Wiley, 2004); *Medical Instrumentation: Application and Design*, fourth edition (Hoboken: Wiley, 2010); and *The Physiological Measurement Handbook* (Boca Raton: CRC Press, 2015). He is coeditor, with A. M. Cook, of *Clinical Engineering: Principles and Practices* (Englewood Cliffs: Prentice-Hall, 1979) and *Therapeutic Medical Devices: Application and Design* (Englewood Cliffs: Prentice-Hall, 1982); with W. J. Tompkins, of *Design of Microcomputer-Based Medical Instrumentation* (Englewood Cliffs: Prentice-Hall, 1981) and *Interfacing Sensors to the IBM PC* (Englewood Cliffs: Prentice Hall, 1988); with A. M. Cook, W. J. Tompkins, and G. C. Vanderheiden, of *Electronic Devices for Rehabilitation* (London: Chapman & Hall, 1985); and with H. Eren, of *Measurement, Instrumentation, and Sensors Handbook* (Boca Raton: CRC Press, 2014).

Dr. Webster has been a member of the Institute of Electrical and Electronics Engineers–Engineering in Medicine and Biology Society Administrative Committee and the National Institutes of Health Surgery and Bioengineering Study Section. He is a fellow of the Institute of Electrical and Electronics Engineers, the Instrument Society of America, the American Institute of Medical and Biological Engineering, the Biomedical Engineering Society, and the Institute of Physics. He is the recipient of the Institute of Electrical and Electronics Engineers Engineering in Medicine and Biology Career Achievement Award.

Contributors

Eric J. Addeo has more than 20 years of experience at the senior management level in the industrial research sector, where he has managed world-class applied research organizations at AT&T Bell Labs and Telcordia and most recently at Panasonic Labs in Princeton, New Jersey. He is the recipient of the Distinguished Alumni Medal for his management and leadership of industrial research organizations from the New Jersey Institute of Technology, Newark, New Jersey. He holds 10 patents and is the recipient of the New Jersey Inventors Hall of Fame Award in recognition of his seminal contributions to the development of current-generation cellular communications systems. He earned a bachelor and master of science degrees in electrical engineering from the New Jersey Institute of Technology. He has an earned PhD in electrical engineering from Stevens Institute of Technology, Hoboken, New Jersey. He is currently a professor in the College of Engineering and Information Sciences at DeVry University, North Brunswick, New Jersey.

Mahmud Ahsan has been developing web applications for over five years and mobile applications for the last four years. He recently completed his master's study at Multimedia University, Melaka Campus, Ayer Keroh, Malaysia, where he studied contextual information in personal health record systems. A research paper he coauthored won the Sydney Global Telehealth 2012 best paper award. Besides his studies, he publishes his mobile applications in both the Apple Store and Google Play Store. His portfolio of mobile applications can be found at http://ithinkdiff.net. He is currently working full time on mobile application development in both the iOS and Android platforms.

Ali Abdulwahab A. Al-Habsi is a researcher and laboratory member in the Health Informatics Laboratory at the Faculty of Information Science and Technology in Multimedia University, Melaka Campus, Ayer Keroh, Malaysia. He received his bachelor's degree in information technology, majoring in security, from Multimedia University in 2011. He received his MSc degree in Information Technology in 2015. His research interests are standard-compliant communication in healthcare, ethical hacking, security for medical information, and cognitive psychology. He is currently working as a part-time tae kwon do assistant instructor and looking forward to starting his PhD studies.

Arshad Ali is a prolific researcher and academician. He has over 100 publications to his credit in national and international conferences and refereed journals. He has been awarded a gold medal by the Pakistan Academy of Sciences and the Standing Committee on Scientific and Technological Cooperation for information technology research; the Presidents' Gold Medal for Best Researcher of the Year 2005 by the National University of Science and Technology, Islamabad, Pakistan; and a Distinguished Scientist of the Year 2006 Award by the Pakistan Academy of Sciences. Dr. Ali is also a recipient of President's Pride of Performance in information technology research from the government of Pakistan. He has also been awarded the NCR National IT Excellence Award in the information technology research and development category. Dr. Ali has also initiated research collaboration with the Center for European Nuclear Research, Switzerland, and earned the Associate Institute status of Compact Muon Solenoid–Center for European Nuclear Research for the

National University of Sciences and Technology. Dr. Ali formed a joint consortium with the University of the West of England, United Kingdom; Beijing Institute of Technology, China; University of Savoie, France; and National University of Science and Technology, Pakistan, and initiated various projects which attracted research funding from the European Union. He has also been part of the development team for a clinical decision-support system in partnership with Kyung Hee University, South Korea, and is an executive member of the Health Level 7 committee on the standardization of health information systems in Pakistan. Dr. Ali has also served as member of the International Advisory Board of the International ICFA Workshop on HEP Networking, Grids, and Digital Divide Issues for Global e-Science.

Emilia Ambrosini graduated cum laude in biomedical engineering in 2007 and obtained a PhD degree in bioengineering, cum laude, in 2011 from Politecnico di Milano, Milan, Italy. In 2009, she was a visiting PhD student at Technische Universität Berlin. Since 2011, she is a research fellow at the NeuroEngineering and Medical Robotics Laboratory, Department of Electronics, Information and Bioengineering, Politecnico di Milano (http://www.biomed .polimi.it/nearlab). She was involved in the European project Multimodal Neuroprosthesis for Daily Upper Limb Support (Specific Targeted Research Project 2010–2013) in the field of assistive and rehabilitative robotics and she currently collaborates with the Salvatore Maugeri Foundation in a project funded by the Italian Ministry of Health aimed at improving poststroke lower-limb rehabilitation. Her research interests are the design and clinical translation of novel methods based on electrical stimulations and robotic systems for neurorehabilitation and the assessment of functional improvements and cortical correlates by means of transcranial magnetic stimulation. She is a coauthor of about 16 papers in peer-reviewed international journals.

Ioannis Andreadis received a degree in electrical and computer engineering from the National Technical University of Athens, Athens, Greece, in 2006. Since 2007, he has been with the Biomedical Simulation and Imaging Laboratory, Department of Electrical and Computer Engineering, National Technical University of Athens, where he obtained his PhD degree in February 2014. He has contributed to a number of journal and conference papers. His main research interests include image processing and analysis, mainly concentrating on medical images, signal processing, artificial intelligence, and computer-aided diagnosis systems.

Nigel R. Armfield's primary research interest is in the development and formal evaluation of sustainable pediatric clinical telemedicine applications and services. He has particular interests in designing and conducting studies to assess the feasibility, efficacy, and clinical effectiveness of telemedicine for delivering neonatal and pediatric critical care at a distance. As a PhD candidate, Dr. Armfield developed and evaluated a novel real-time telemedicine application to support remote clinical consultation between a tertiary neonatal intensive care unit and four peripheral hospitals in Queensland. In addition, Dr. Armfield has a research interest in health geography, particularly in formally assessing and describing the effect of geography and regionalization on the availability, accessibility, and utilization of specialist health services for children. Dr. Armfield is involved in telemedicine service delivery at the Queensland Children's Hospital in Brisbane, Australia, and is coordinator for the Indigenous Health Screening Programme.

Anda Baharav holds degrees in physics and medicine from Tel Aviv University, Tel Aviv, Israel. She specialized in pediatrics and later in sleep medicine. In her works she has integrated clinical and physiological studies with signal processing at the Medical Physics Department of Tel Aviv University. Besides her clinical activity in academic medical institutions, Dr. Baharav has conducted research in the fields of autonomic nervous system, sleep physiology, and sleep disorders for more than 20 years. She mentored graduate and PhD students in the field and published in peer-reviewed journals. She is leading research on sleep and performance in young athletes at the Wingate Institute, the Israeli national sports institute. Those years led to results that have had significant value in sleep evaluation and simple home-based diagnosis of sleep disorders. She is the cofounder and chief scientist of a company that has the vision to make sleep diagnosis and sleep improvement widely available.

Giacinto Barresi is a PhD fellow (Robotics, Cognition and Interaction Technologies program of University of Genoa) at the Biomedical Robotics Laboratory of the Advanced Robotics Department, Istituto Italiano di Tecnologia, Genoa, Italy. He received his BSc and MSc degrees in experimental psychology from the University of Padua, Padua, Italy. His research background includes cognitive psychology and psychobiology, psychometrics, cognitive ergonomics, and user interface design and evaluation. He investigates human–machine interaction in biomedical (surgical and assistive) robotics in order to enhance the user's performance in tasks that require overcoming the cognitive and physical limitations of the human beings. He contributes to the European project Micro-Technologies and Systems for Robot-Assisted Laser Phonomicrosurgery.

Brett Bell received a PhD in biomedical engineering from the Purdue University, West Lafayette, Indiana, in 2009. He then joined the ARTORG Center for Biomedical Engineering at the University of Bern, Bern, Switzerland, to lead the robotic surgery group focused on microsurgical procedures on the lateral skull base. In 2015, Dr. Bell joined a company developing orthopedic robotic instruments.

Tushar Kanti Bera obtained his BE degree in electrical engineering from the North Bengal University, Siliguri, India, in 2000. He received his MTech degree in electrical engineering from the University of Calcutta, Kolkata, India, in 2003. He completed his PhD research work on electrical impedance tomography from the Department of Instrumentation and Applied Physics, Indian Institute of Science, Bangalore, India, in 2012 and he has been awarded a PhD degree in 2013. From August 2012 to April 2014 he worked as a postdoctoral researcher at the Department of Computational Science and Engineering, Yonsei University, Seoul, South Korea. Dr. Bera worked as an associate professor in the BMS College of Engineering (BMSCE), Bangalore, India, from June 2014 to January 2015, and is currently working as a Postdoctoral Fellow in the Composite and Heterogeneous Material Analysis and Simulation Laboratory, Department of Mechanical Engineering, King Abdullah University of Science and Technology, Thuwal, Saudi Arabia. His current research interests include electrical impedance tomography, medical imaging, biomedical instrumentation, inverse problems, image reconstruction, medical electronics, biosensors, bioelectrical impedance analysis, and impedance spectroscopy.

Elisa Beretta graduated cum laude in biomedical engineering at Politecnico di Milano, Milan, Italy, in 2011 with a thesis entitled "Hip Joint Center Localization with an Unscented Kalman Filter in Computer Assisted Orthopaedic Surgery applications." She is currently a PhD student in bioengineering at Politecnico di Milano, working at the NeuroEngineering and Medical Robotics Laboratory. Since 2012, she has been involved in the European project Active Constraints Technologies for Ill-defined or Volatile Environment and in the framework of Scuola Interpolitecnica di Dottorato. In 2013–2014, she was a visiting PhD student at the Mechatronics in Medicine Laboratory of Imperial College, London, United Kingdom, under the supervision of Professor Rodriguez y Baena, working on the development of adaptive control systems for cooperative robotic assistant in neurosurgery. Her research activities cover the fields of computer-assisted and robotic-assisted surgery. The focus of her research is the design and development of an advanced control platform for cooperative robotic surgery.

Liam Caffery is a senior research fellow and director of Telehealth Technology for the Centre for Online Health, The University of Queensland, Brisbane, Australia. He is an executive member of the Australasian Telehealth Society. Dr. Caffery is an associate investigator of the National Health and Medical Research Council's Centre for Research Excellence in Telehealth. He is actively involved in telehealth service delivery via his work programs, including RES-e-CARE, Health-e-Regions, Princess Alexandra Hospital Telehealth Centre, and Queensland Telepaediatric Service. He has an active research agenda in health services research and health informatics with a special interest in imaging informatics, indigenous health, and rural health.

Marco Caversaccio is the chief physician and the head of the Department of ENT Surgery at the Inselspital, Medical Faculty of the University of Bern, Bern, Switzerland. Following his studies on human medicine at the University of Geneva, Geneva, Switzerland, he finished his MD in 1993 on intracerebral hematomas. After various research and clinical fellowships, some at the Klinikum Rechts der Isar in Munich, Germany, and at the Imperial College and the Charing Cross Hospital (both in London, United Kingdom), he returned as a senior physician to the Inselspital in Bern. He became a private lecturer in 2004. Besides his clinical activities, Dr. Caversaccio participates in various research projects, some of them were funded by the Swiss National Science Foundation, such as the National Competence Center in Research for Computer Aided and Image Guided Medical Interventions or Nano-Tera. Since 2008, he has been codirecting the Bern Center for Computer Aided Surgery in the ARTORG Center for Biomedical Engineering. He is a well-known technology expert on the development and implementation of novel surgical techniques and systems in the area of rhinology and otology. He reviews for a number of scientific publications and conferences in those fields. Dr. Caversaccio is a member of the Swiss, German, and French societies for ear, nose, and throat medicine; the American Academy of Otolaryngology-Head and Neck Surgery; and the European Rhinological Society.

Panayiotis Constantinides is an associate professor of information systems and director of the MSc in Information Systems Management and Innovation program at the Warwick Business School, University of Warwick, Coventry, United Kingdom. Previously, he held positions at Lancaster University's Management School, Lancaster, United Kingdom, and the Judge Business School at the University of Cambridge, Cambridge, United Kingdom, where he also earned his PhD. Dr. Constantinides has carried out funded research projects on health information technology development and implementation in different national contexts. He has experience in applying coordination models to understand and provide guidelines for better management of healthcare service delivery and innovation. His work cuts across the fields of organization studies and information system research and has been published in leading journals in these fields, such as *Information Systems Research*, *MIS Quarterly*, and *Technological Forecasting and Social Change*. Dr. Constantinides is an associate editor for the International Conference of Information Systems and the annual Academy of Management Conference (Organizational Communication & Information Systems Division), among others.

Elena De Momi holds a PhD in bioengineering (2006) and an MSc in biomedical engineering (2002). Currently, she is an assistant professor at the Department of Electronics, Information and Bioengineering of Politecnico di Milano, Milan, Italy. She was a visiting PhD student at the Maurice E. Müller Research Center, University of Bern, Switzerland, and at the Institute for Computer Assisted Orthopaedic Surgery Laboratory and Carnegie Mellon University, both of which are in Pittsburgh, Pennsylvania, United States. She lectures in instrumentation and functional evaluation project management for the BSc in biomedical engineering program and in medical robotics for the PhD in bioengineering program. Since 2006, she has been involved in the NeuroEngineering and Medical Robotics Laboratory and has participated in several national and European grants in the field of computer-assisted surgery, keyhole robotic neurosurgery (FP7-ICT-2007-215190, ROBOCAST), and awake robotic neurosurgery (FP7-ICT-2009-6-270460, Active Constraints Technologies for Ill-defined or Volatile Environment, project manager). She is a principal investigator of the research network EUROSURGE (ICT 288233) for partner Politecnico di Milano. Dr. De Momi collaborates as an expert with the European Commission in the Seventh Framework Programme. She is part of the editorial board on medical robotics of the *International Journal of Advanced Robotic Systems*. Her research interests cover the fields of computer-aided surgery, medical robotics, biomechanics, and sensors.

Nikhil Deshpande is currently a postdoctoral researcher of the Biomedical Robotics group at the Istituto Italiano di Tecnologia, Genoa, Italy. He received his bachelor's degree in electrical engineering from the Government College of Engineering, University of Pune, India, in 2003 and his master's degree in integrated manufacturing systems engineering in 2007 and his PhD degree in electrical engineering in 2012, both from the North Carolina State University, Raleigh, North Carolina. His research interests are in the area of biomedical and surgical robotics, specifically developing novel surgeon–machine interfaces and smart surgical devices. He is a part of the European Union's Seventh Framework Programme–funded Micro-Technologies and Systems for Robot-Assisted Laser Phonomicrosurgery project, which is focused on robot-assisted systems for laser phonomicrosurgery. In the past, he led research activities into wireless sensor networks, autonomous robotics, and sensor integration and data fusion, which included distributed computation and using information from multiple sources. He is a member of the Institute of Electrical and Electronics Engineers.

William B. Dunbar received a BS degree in engineering science and mechanics from the Virginia Polytechnic Institute and State University, Blacksburg, Virginia, in 1997; an MS degree in applied mechanics and engineering science from the University of California, San Diego, California, in 1999; and a PhD degree in control and dynamical systems from the California Institute of Technology, Pasadena, California, in 2004. He is currently an associate professor at the Department of Computer Engineering, University of California–Santa Cruz, Santa Cruz, California. His research interests include applications of feedback control to problems in biophysics and sequencing with nanopores.

Sisira Edirippulige's main responsibilities at the Centre for Online Health, The University of Queensland, Brisbane, Australia, involved teaching and coordinating undergraduate and postgraduate courses in eHealthcare and coordinating continuing professional development courses in telehealth. He is active in research and development, specifically in the development, promotion, and integration of eHealth education within the healthcare sector with a strong conviction for training and education as the key component in promoting the use of eHealth. Dr. Edirippulige has been involved in designing and developing education and training on eHealth for different groups of health and related professionals. Before joining the University of Queensland, Dr. Edirippulige taught at Kobe Gakuin University, Kobe, Japan, and at the University of Auckland, Auckland, New Zealand. He has extensive experience in development studies, working in a number of countries, including Russia, Sri Lanka, South Africa, Japan, and New Zealand.

Giancarlo Ferrigno holds a PhD in bioengineering, and an MSc in electrical engineering. Currently, he is a full professor at Politecnico di Milano, Milan, Italy. He has also been the European Coordinator of three Seventh Framework Programme of the European Union projects on the topic of information and communication technology. Two of them, ROBOCAST (Specific Targeted Research Project 2008–2010) and Active Constraints Technologies for Ill-defined or Volatile Environment (Integrated Project 2011–2015), are in the field of surgical robotics. The project Multimodal Neuroprosthesis for Daily Upper Limb Support (Specific Targeted Research Project 2010–2013) is in the field of assistive and rehabilitative robotics. Ferrigno is the founder (2008) of the NeuroEngineering and Medical Robotics Laboratory of Politecnico di Milano's Department of Electronics, Information and Bioengineering and a lecturer on medical robotics. He is a coauthor of 20 papers (Institute for Scientific Information Web of Knowledge) on the field of robotics from 2011 to 2014 out of more than 190. He is working in the International Organization for Standardization's Joint Working Group 9, a standards group for surgical robots collateral standard, and has organized several workshops on topics on surgical robotics in the last three years.

Loris Fichera received his BSc and MSc degrees in computer engineering from the University of Catania, Italy, in 2008 and 2011, respectively. He is currently pursuing a PhD degree at the Department of Advanced Robotics of the Istituto Italiano di Tecnologia, Genoa, Italy. His research interests involve the development of assistive technologies for computer- and robot-assisted surgery, with particular focus on laser microsurgery.

Ann L. Fruhling, PhD, MBA, is a Mutual of Omaha Distinguished Professor of the College of Information Science and Technology, Peter Kiewit Institute, at the University of Nebraska at Omaha, Nebraska, and the founding director of the School of Interdisciplinary Informatics at the University of Nebraska at Omaha. The school is the academic home for the Bioinformatics, Biomedical Informatics, Information Assurance, and IT Innovation degree programs. She is also the director of the Consortium for Public Informatics. Dr. Fruhling's research focuses on evaluating and improving human–technology interaction efficiency and effectiveness. Her health informatics research has been funded by the National Institutes of Health, Centers for Disease Control and Prevention, Health Resources and Services Administration, Association of Public Health Laboratories, National Aeronautics and Space Administration, Department of Defense, International Business Machines, the Experimental Program to Stimulate Competitive Research, and the Nebraska Research Initiative. She has received over 45 grants and has been awarded more than $4.5 million in funding for health informatics-related research. Her research works have appeared in many journals, including *Applied Clinical Informatics, Journal Management Information Systems, Communications of the Association for Information Systems, Journal of Computer Information Systems, International Journal of Electronic Healthcare, International Journal of Medical Informatics, International Journal of Cooperative Information Systems, Health Systems, Journal of Electronic Commerce Research*, and *Journal of Information Technology Theory and Application*. She also has book chapters in *Value-Based Software Engineering, Patient-Centered E-Health*, and *Advances in Management Information Systems* and numerous conference papers.

Hamido Fujita is a professor at Iwate Prefectural University, Takizawa, Japan, as a director of Intelligent Software Systems. He is the editor in chief of *Knowledge-Based Systems* (Elsevier). He recently received a doctor honoris causa in 2013 and the title of honorary professor in 2011, both from from Obuda University, Budapest, Hungary. He received the title of distinguished honorary professor from the University of Technology, Sydney, Australia, in 2012. He is an adjunct professor at Stockholm University, Stockholm, Sweden; University of Technology, Sydney, Australia; National Taiwan Ocean University, Keelung, Taiwan; and others. He has supervised PhD students jointly with the University of Laval, Quebec, Canada; University of Technology, Sydney, Australia; University of Paris 1 Panthéon-Sorbonne, Paris, France; and University of Genoa, Genoa, Italy; and others. He is vice president of the International Society of Applied Intelligence. He heads a number of projects, including one on intelligent human–computer interaction, which is related to mental cloning as an intelligent user interface between human user and computers, and the SCOPE project on virtual doctor systems for medical applications.

Aimilia Gastounioti obtained a diploma in electrical and computer engineering in 2009 and a PhD degree in biomedical engineering in 2014 from the National Technical University of Athens, Athens, Greece. She is currently a postdoctoral researcher at the University of Pennsylvania, Philadelphia, Pennsylvania. Dr. Gastounioti has received scholarships from the Hellenic State Scholarships Foundation (2010–2014) and grants from the Institute of Computer and Communication Systems (2010 and 2012) for PhD studies. Her research interests include biomedical imaging, medical image analysis, and machine learning toward computer-aided clinical diagnosis and treatment selection. Dr. Gastounioti has authored/coauthored 23 papers in scientific journals and international conferences and one book chapter. She has also worked as a principal researcher in two national and two European funded projects on biomedical engineering. Since 2012, she has served as a scientific reviewer of the *Medical Image Analysis* (Elsevier), *Journal of Biomedical and Health Informatics* (Institute of Electrical and Electronics Engineers), *Scanning* (Wiley), *Computer Vision and Image Understanding* (Elsevier), and *Ultrasonics* (Elsevier) journals, as well as various international scientific conferences. Dr. Gastounioti is a member of the Institute of Electrical and Electronics Engineers; the Institute of Electrical and Electronics Engineers' Engineering in Medicine and Biology Society; the Institute of Electrical and Electronics Engineers' Ultrasonics, Ferroelectrics, and Frequency Control Society; and the Technical Chamber of Greece.

Kostas Giokas has a long career in consulting for several companies in the United Kingdom and Greece since 1994. He graduated in 2000 from the University of Westminster, London, United Kingdom, obtaining a BSc in business information technology. He received his MBA from the Open University, Milton Keynes, United Kingdom, in 2002. He then joined the Biomedical Engineering Laboratory of the National Technical University of Athens,

Athens, Greece, where he worked as a researcher in biomedical engineering, focusing on large network analysis, design, and deployment while consulting on national and European deployment projects. At the same time he has been part of the research and development team that undertakes European Union research projects in the Biomedical Engineering Laboratory. He has been involved in research proposals, leading more than 35 of them. He has been a team member/leader in more than 20 European and national research projects in the field of health informatics and has published 50 scientific papers. He is currently the leader of the Applied Informatics in mHealth Research Team.

Spyretta Golemati received a diploma in mechanical engineering from the National Technical University of Athens, Athens, Greece, in 1994, and MSc and PhD degrees in bio-engineering from the Imperial College of Science, Technology, and Medicine, University of London, London, United Kingdom, in 1995 and 2000, respectively. She then was a postdoctoral fellow in the Department of Electrical and Computer Engineering, National Technical University of Athens, Athens, Greece. She is currently a lecturer on biomedical engineering in the National and Kapodistrian University of Athens, Greece. Her research interests include ultrasound imaging and signal analysis, arterial biomechanics, and respiratory mechanics. She has coauthored 28 papers in international scientific journals, 8 book chapters, and 40 papers in international conference proceedings. She is an associate editor of Elsevier's *Ultrasonics* journal.

Florian Gosselin received an engineering degree in robotics from the Ecole Centrale de Nantes, Nantes, France, in 1995 and a PhD degree in mechanical engineering from the University of Poitiers, Poitiers, France, in 2000. He is now working as a research engineer and project manager at the Commissariat à l'Energie Atomique et aux Energies Alternatives–Laboratoire d'Intégration de Systèmes et des Technologies (CEA-LIST), Saclay, France. His research interests include haptics and robotics. He developed a high-fidelity master arm for abdominal telesurgery and supervised the design of a multimodal virtual-reality platform for the training of maxillofacial surgery skills. He also designed numerous other haptic interfaces (serial and parallel architectures, polyarticulated and tensed cable structures, rigid and compliant joints, and nonredundant and redundant actuation) and took part in their integration in state-of-the-art virtual-reality platforms with applications in computer-aided design, digital mock-up, cultural heritage, and digital arts. He is now working on the development of novel dexterous haptic devices and collaborative robots for surgery.

Kathleen Gray holds a PhD in eLearning in environmental science. She is a senior research fellow in health informatics in the Health and Biomedical Informatics Centre at the University of Melbourne, Melbourne, Australia, and the coordinator of the university's postgraduate programs in ehealth and biomedical informatics. She has over 100 peer reviewed publications and her research has been funded by the Australian Department of Health, Australian Government Office for Learning and Teaching, Australian Primary Health Care Research Institute, Australian Research Council, and Institute for a Broadband Enabled Society. Kathleen is active in research related to participatory health where patients' and consumers' use of the Internet health information and communication; influences of the Internet on health workforce professional learning and development needs; frameworks for understanding and evaluating the effects of the Internet on health service provision.

Donghai Guan received his PhD degree in computer science from Kyung Hee University, Seoul, South Korea in 2009. From 2009 to 2014, he was a postdoctor, research professor, and assistant professor successively at Kyung Hee University. Since November 2014, he has been an associate professor at the College of Computer Science and Technology, Nanjing University of Aeronautics and Astronautics, China. His research interests are machine learning, ubiquitous computing, and eHealth.

Deborah A. Helman is a professor of sales and marketing in the College of Business and Management at DeVry University, North Brunswick, New Jersey, and has earned a BA degree in history from Leicester University, an MPhil degree in marketing from Cranfield School of Management, and a PhD degree in commerce from the University of Birmingham. Dr. Helman's research interests include brands, Internet lifestyle, integrated marketing communications, adaption–innovation problem-solving styles of managers, eHealth, and user experience. She has experience as a researcher in the United States and the United Kingdom with involvement in a range of projects in the public and private sectors. Dr. Helman is co-managing editor of *DeVry University Journal of Scholarly Research* and is a member of the American Marketing Association and is an officer of the DeVry University Chapter of the American Marketing Association.

Guy T. Helman is a clinical research coordinator at the Department of Neurology of the Children's National Medical Center, Washington, DC. Prior to this, Mr. Helman earned his bachelor of science degree in biology at the George Washington University, Washington, DC. He has extensive research experience in the field of pediatric neurology, specifically in leukodystrophies and other neurogenetic conditions. He is an author of 15 peer-reviewed publications and 3 book chapters. In addition, Mr. Helman has been an active member of the Global Leukodystrophy Initiative, an international consortium made up of clinicians, researchers, and patient advocacy groups aimed at advancing the standard of care for leukodystrophy patients and prioritizing research needs.

Enrique Herrera-Viedma received BSc and PhD degrees in computer sciences, both from the University of Granada, Granada, Spain, in 1993 and 1996, respectively, and he is currently a professor at the Department of Computer Science and Artificial Intelligence at the University of Granada and the director of the Quality Evaluation and Information Retrieval Research Laboratory (SECABA). He is an associate editor of *IEEE Transactions on Systems, Man, and Cybernetics: Systems*; *KNOSYS*; *Journal of Intelligent & Fuzzy Systems*; and *INS* and a member of the editorial boards of *Fuzzy Sets and Systems*; *International Journal of Information Technology & Decision Making*, and *International Journal on Cybernetics & Informatics*. He has published extensively in leading international journals in his field, and several of his papers are classed as highly cited. He has also been consistently classed as one of the most cited scientists in his field and he is included in the list of highly cited researchers published by Thomson Reuters in June 2014. His h-index is 42 according to the Web of Science.

Kevin K. F. Hung is a lecturer at the School of Science and Technology of the Open University of Hong Kong, Hong Kong Special Administrative Region, China. His research interests include wearable devices, mobile health, eye tracking, pupillary dynamics, and modeling of biological systems. He is currently collaborating with the industry on the development of innovative healthcare products and mobile health systems. Before joining the Open University of Hong Kong, Dr. Hung was an assistant project manager at the Joint Research Centre for Biomedical Engineering, The Chinese University of Hong Kong, where he has coordinated several government-funded projects related to medical devices;

and an engineer at a medical device company, where he performed research and development of respiratory products.

Dimitra Iliopoulou received her degree in electrical and computer engineering from the National Technical University of Athens, Athens, Greece, in 2001. She received her PhD in 2009 (with the dissertation "A Smart Integrated System for the Consultation and Management of Diabetes Melitus") in the National Technical University of Athens's Biomedical Engineering Laboratory, where she has studied and applied informatics and telematic methods in a number of areas in the field of health services, in particular, areas related to diabetes. She has worked in information technology and telecoms in several companies (Multimedia A.E.—DOL, Bull ATS, Datamed SA, and Omnis M.). She also has experience in the field of health telecommunications and informatics as she has been active in several European and national research projects in regard to the application of information technology in health.

Sujitha Juliet is an assistant professor at the Department of Information Technology, Karunya University, Coimbatore, India. She received her bachelor of engineering degree, first class, from Bharathiar University, Coimbatore, India, in 2001 and her master of engineering degree in applied electronics, with distinction, from Anna University, Chennai, India, in 2003. She has received her PhD degree in information technology in 2014 from Karunya University. She has published eight papers in international journals, among which four are with very good impact factors. She has published nine papers in national and international conferences. Her principal research interests are near-lossless and lossless medical image compression, medical decision-support systems, and telemedicine networking.

Akihiro Kajiwara received a BS degree in electronic engineering from the University of Yamaguchi, Yamaguchi, Japan, in 1981 and MEng and PhD degrees in electrical engineering from Keio University, Minato, Japan, in 1989 and 1991, respectively. He was appointed as a professor at Ibaraki University, Ibaraki Prefecture, Japan, from 1996 to 2001. Since 2001, he has been a professor at the Department of Information and Media engineering of Kitakyushu University, Japan. He is also the vice president of the same university. Professor Kajiwara teaches undergraduate and postgraduate students in communication systems, telecommunications networks, and mobile communication systems. His research interests are wireless communications, microwave sensors, microwave and millimeter radio propagation, and ultrawideband radio.

Shamila Keyani has a master's degree in the field of Internet engineering from the University of East London, London, UK, and worked as a faculty member at the National University of Sciences and Technology, Islamabad, Pakistan, where she supervised a variety of industry projects and published numerous research papers. She has over seven years of working experience in academic and development sectors. Currently, she is serving as the director of development at UM Healthcare Trust and has cofounded a mobile-based Tele-healthcare project, "Jaroka," which aims to provide affordable and accessible health care in the rural and disaster-hit communities. Ms. Keyani is fluent in using information and communication technologies to address global health-related challenges. She has been awarded fellowships at Atlas Service Corps and Foundation of Youth Social Entrepreneurs for contributions in these nonprofit sectors.

Jungsuk Kim received a BS degree (magna cum laude) in electrical engineering from Sogang University, Seoul, South Korea, in 2003, an MS degree in electrical engineering from the University of Southern California, Los Angeles, California, in 2006, and a PhD degree from the University of California, Santa Cruz, California. After graduating in 2011, he worked as a postdoctoral scholar in computer engineering at the University of California, Santa Cruz, and as a senior engineer at Samsung Electronics. He is currently an assistant professor at the Department of Biomedical Engineering at Gachon University, Incheon, South Korea. His research interests include integrated circuit and system designs for nanopore sequencing technology and other biomedical applications, such as implantable prosthetic systems and neural–electronic interfaces.

Vasileios Kolias obtained a diploma in computer engineering and informatics from the University of Patras, Patras, Greece, in 2009 and an MSc degree in biomedical engineering from the University of Patras and National Technical University of Athens, Athens, Greece, in 2011. He is currently working toward a PhD degree in the Biomedical Simulations and Imaging Laboratory, School of Electrical and Computer Engineering, National Technical University of Athens. He has authored four research papers that were presented in international conferences, one journal paper, and two book chapters. He is participating as a research associate in the CAROTID project (09SYN-12-1054). His research interests are medical informatics, semantic web technologies, computer-aided design systems, clinical decision-support systems, and artificial intelligence in health.

Dimitris Koutsouris received his diploma in electrical engineering in 1978 (Greece), DEA in biomechanics in 1979 (France), doctorate in genie biologie medicale (France), and doctorat d'etat in biomedical engineering in 1984 (France). Since 1986, he has been a research associate at the University of Southern California, Los Angeles, California, and Paris Dèscartes University, Paris, France, and an associate professor at the School of Electrical and Computer Engineering of the National Technical University of Athens, Athens, Greece, where he is currently a professor at and the head of the Biomedical Engineering Laboratory. He has published over 150 research articles and book chapters and more than 350 peer-reviewed conference communications. He has been an elected president of the Hellenic Society of Biomedical Technology, Health Level 7 Hellas, and a chairman of the School of Electrical and Computer Engineering. Professor Koutsouris has been a principal investigator in more than 100 European and national research programs, especially in the field of telematics and informatics in healthcare. His work has received more than 1800 citations.

Efthyvoulos Kyriacou is an associate professor in the Department of Computer Science and Engineering of Frederick University, Lemesos, Cyprus. His research interests focus on eHealth systems, emergency telemedicine systems, medical imaging systems, and intelligent systems applications in medicine. He has published more than 120 journal and conference papers and invited book chapters and has one filed patent in these areas. He has been involved in numerous projects funded by the European Union, the National Research Foundation of Cyprus, the INTERREG, and other bodies. He was guest co-editor of special issues of the *IEEE Transactions on Information Technology in Biomedicine* and *Journal of Biomedical Signal Processing and Control*, Elsevier. He also served as an associate editor of the *IEEE Transactions on Information Technology in Biomedicine* from 2007 to 2010, he serves as a reviewer in many journals related to his research fields. He is a coeditor of the book

Ultrasound and Carotid Bifurcation Atherosclerosis (Springer, United Kingdom, 2012). He has been the Program cochair of Information Technology and Applications in Biomedicine 2009 and Bioinformatics & Bioengineering 2012 and has been in the program committee of many other scientific conferences. He is a senior member of the Institute of Electrical and Electronics Engineers and is currently the chairman of the Institute of Electrical and Electronics Engineers Cyprus Engineering in Medicine and Biology/Signal Processing chapter.

Kenneth Lai is currently a graduate student and research assistant at the Department of Electrical and Computer Engineering, Schulich School of Engineering, University of Calgary, Calgary, Canada. He has a BSc degree in electrical engineering from the University of Calgary. Lai is a member of the Biometric Technologies Laboratory at the University of Calgary. His research interests include biometrics, pattern recognition, computer vision, and image processing.

Moktar Lamari, PhD, is in the top 100 of the most quoted world scientists in the area of knowledge transfer and knowledge management (citation index—1562; Hirsch index—10). Dr. Lamari is presently a full professor at the School of Public Administration or École Nationale d'Administration Publique), Quebec University, Quebec City, Canada. He is also the head of the Centre de Recherche et d'Expertise en Evaluation (http://www.crexe .enap.ca). His ongoing research projects examine the impact of web 2.0 and digital tools on knowledge transfer and knowledge absorptive capacity. Dr. Lamari has won three awards: the Louis Brownlow Award in 2003 for the best article published by *Public Administration Review*; the Elsevier Award in 2002 for the best article published by *Technological Forecasting and Social Change*, an international journal; and the 2014 award for Outstanding World Research Leader, delivered by the International Multidisciplinary Research and the University of the Philippines Open University.

Sungyoung Lee received his PhD degree in computer science from Illinois Institute of Technology, Chicago, Illinois in 1991. He has been a professor at the Department of Computer Engineering, Kyung Hee University, Seoul, South Korea, since 1993. He is a founding director of the Ubiquitous Computing Laboratory and has been affiliated as a director of Neo Medical ubiquitous-Lifecare Research Center, Kyung Hee University, since 2006. His current research focuses on ubiquitous computing, cloud computing, cyberphysical systems, and eHealth.

Liping Liu is a professor of management and information systems at the University of Akron, Akron, Ohio. He received his bachelor of science in applied mathematics in 1986 from Huazhong University of Science and Technology, Wuhan, China, bachelor of engineering in river dynamics in 1987 from Wuhan University, Wuhan, China, master of engineering in systems engineering in 1991 from Huazhong University of Science and Technology, and PhD in business in 1995 from the University of Kansas, Lawrence, Kansas. His research interests are in the areas of uncertainty reasoning and decision making in artificial intelligence, electronic business, systems analysis and design, technology adoption, and data quality. Dr. Liu has published articles in *Decision Support Systems*; *European Journal of Operational Research*; *IEEE Transactions on System, Man, and Cybernetics*; *International Journal of Approximate Reasoning*; *Information and Management*; *Journal of Association for Information Systems*; *Journal of Optimization Theory and Applications*; *Journal of Risk and Uncertainty*; and others. His theories of coarse utilities and linear belief functions are currently taught at the nation's top PhD programs in accounting, computer science, economics, management, and psychology. Dr. Liu has served as a guest editor for the *International Journal of Intelligent Systems*, a coeditor of *Classic Works on Dempster-Shafer Theory of Belief Functions*, and on the editorial board of a few academic journals. He has served on the program committee or as a track chair of the Institute for Operations Research and the Management Sciences, Americas Conference on Information Systems, International Conferences on the Theory of Belief Functions, etc. Dr. Liu has strong practical and teaching interests in e-business systems design, development, and integration using advanced database management system, computer-aided software engineering, and rapid application development tools. His recent consulting experience includes designing and developing a patient record–management system, a payroll system, a course-management system, and an e-travel agent and providing corporate trainings on Oracle database administration, Oracle applications development, and object-oriented requirements analysis and modeling for large corporations.

Yiannis Makris has a PhD in medical and molecular genetics from King's College, London, United Kingdom, in cooperation with the Medical School of the University of Athens, Athens, Greece. He also holds a BSE in nursing (University of Athens), an MSc in molecular biology (University College London, London, United Kingdom), and a diploma in bioinformatics (Massachusetts Institute of Technology, Boston, Massachusetts). His PhD thesis included the identification and characterization of the genes responsible for the genetic disease orofacial clefting. He has been a scientific collaborator of the Electrical Engineering Department of Ethniko Metsovio Polytechnio (EMP) since 2007 and has actively participated in various European Union programs. He also participates in the preparation and teaching of the Introduction in Biomedical Technology and Measurements and Controls in Biomedical Technology lectures, as well as in the writing of a book on and teaching of the postgraduate course Introduction in Bioinformatics. Dr. Makris is the owner of CELLETHA LTD., a company that deals with novel biofuel processes. His scientific specialization and interests include bioinformatics, data mining, biomedical technology, microarrays, gene therapy, and medical information technology.

Leonardo S. Mattos is a research team leader and the head of the Biomedical Robotic Laboratory at the Istituto Italiano di Tecnologia in Genoa, Italy. His research background includes robotic microsurgery, micro-biomanipulation, systems integration, teleoperation, automation, and user interfaces. Leonardo received his BSc degree from the University of São Paulo, São Paulo, Brazil, and his MSc and PhD degrees in electrical engineering from the North Carolina State University, Raleigh, North Carolina, where he worked as a research assistant at the Center for Robotics and Intelligent Machines from 2002 until 2007. Mattos has been a researcher at the Istituto Italiano di Tecnologia's Department of Advanced Robotics since 2007. He is currently the principal investigator and coordinator of the European project Micro-Technologies and Systems for Robot-Assisted Laser Phonomicrosurgery, which is dedicated to the development of new tools and systems for robot-assisted laser microsurgeries.

Scott McGrath is a doctoral student in biomedical informatics at the University of Nebraska–Omaha, Nebraska. He was a graduate assistant for Dr. Ann Fruhling in the Public Health Informatics Research Laboratory. He has a Master of Science degree in biomedical informatics (bioinformatics) from the University of Nebraska–Omaha and a Bachelor of Science degree in biology from Linfield College, McMinnville, Oregon.

Jacey-Lynn Minoi is currently a senior lecturer in the Faculty of Computer Science and Information Technology and a research fellow at the Institute of Social Informatics and Technological Innovations, Universiti Malaysia Sarawak. Kuching, Malaysia. She received her PhD in computer engineering from Imperial College, London, United Kingdom, in 2009. She is also a member of the Association for Computing Machinery and the Institute of Electrical and Electronics Engineers' Computer Society. Her research interest lies in computational multivariate statistical methods for numerical data–analysis applications, such as biometric recognition, face analysis, medical imaging, surveillance, and security, but she also applies these technologies on social research, such as in the papers "Capturing Culture in a Role-playing Games using an Extended Technological Pedagogical Content Knowledge (TPACK) Approach," "Indigenous Technological Innovation in Malaysia: Reducing Vulnerability and Marginalization among Malaysia's Indigenous Peoples," and "Connected Village–Need Analysis and Sustainability."

Atif Mumtaz has over 18 years of experience in information communication technology across multiple domains and industries. He envisioned, designed, developed, and launched numerous companies and products. In 2004, he founded UM Healthcare Trust, a nonprofit organization that has treated over 200,000 patients free of charge in rural Pakistan. Under UM Healthcare, an innovative telehealth product (called Jaroka Tele-health) was configured, which has won numerous awards including the mBillionth Award for best mHealth application in South Asia. Atif takes pride in getting involved and providing services in local communities. He is an active member of the NUST University Corporate Advisory Council (for Social Sector), a member of the Pakistan Software Houses Association Central Executive Committee, and a member of the "PASHA Social Innovation Fund" with seed funding provided by Google. Atif won numerous awards, including Shell's Livewire Award of "Young Entrepreneur of the Year, 2004" and the DHL YES Award of "Young Social Entrepreneur of the Year" in 2007.

Ryohei Nakamura, PhD, is a research associate in the University of Kitakyushu, Kitakyushu, Japan. His research interests include communication network systems, wireless communications, radio propagation, and measurement engineering. He contributed in the publications of numerous papers on ultra-wideband radio propagations, wireless sensors, microwave sensors, and patient care monitoring systems.

Konstantina S. Nikita received a diploma in electrical engineering and a PhD degree from the National Technical University of Athens, Athens, Greece, as well as an MD degree from the Medical School, University of Athens. Since 2005 she has served as a professor at the School of Electrical and Computer Engineering of the National Technical University of Athens. She has authored or coauthored 155 papers in refereed international journals and 40 chapters in books and over 300 papers in international conference proceedings. She is an editor of four books in English and an author of two books in Greek. She holds two patents. She has been the technical manager of several European and national research and development projects. She has been honorary chair/chair of programs/organizing committees of several international conferences and she has served as keynote/invited speaker at international conferences, symposia, and workshops organized by the North Atlantic Treaty Organization, World Health Organization, International Commission on Non-Ionizing Radiation Protection, Institute of Electrical and Electronics Engineers, International Union of Radio Science, etc. She has been the advisor of 24 completed PhD theses, several of which have received various awards. Her current research interests include biological effects and medical applications of radio-frequency electromagnetic fields, biomedical telemetry, biomedical signal and image processing and analysis, simulation of physiological systems, and biomedical informatics. Dr. Nikita is an associate editor of the *IEEE Transactions on Biomedical Engineering*, the *IEEE Journal of Biomedical and Health Informatics*, Wiley's *Bioelectromagnetics*, the *Journal of Medical and Biological Engineering*, and *Computing* and a guest editor of several international journals. She has received various honors/awards, including the Bodossakis Foundation Academic Prize (2003) for exceptional achievements in theory and applications of information technology in medicine. She has been a member of the Board of Directors of the Atomic Energy Commission, the Hellenic National Academic Recognition and Information Center, and the Hellenic National Council of Research and Technology. She has also served as deputy head of the School of Electrical and Computer Engineering of the National Technical University of Athens. She is a member of the Hellenic National Ethics Committee, a founding fellow of the European Association of Medical and Biological Engineering and Science, and a member of the Technical Chamber of Greece and the Athens Medical Association. She is also the founding chair and ambassador of the Institute of Electrical and Electronics Engineers' Engineering in Medicine and Biology Society, Greece chapter, and vice-chair of the Institute of Electrical and Electronics Engineers Greece section.

Emidio Olivieri is a PhD fellow in the Biomedical Robotics Laboratory of the Advanced Robotics Department, Istituto Italiano di Tecnologia, Genoa, Italy. He is pursuing a PhD degree in robotics, cognition, and interaction technologies at the Università degli Studi di Genova, Genova, Italy. He received his BSc and MSc degrees in electronics engineering from the Università Politecnica delle Marche, Ancona, Italy, in 2010 and 2012, respectively. His research interests include haptics, augmented reality, human–robot interactions, and software engineering applied in the medical robotics area, in particular laser surgery. He contributes to the European project Micro-Technologies and Systems for Robot-Assisted Laser Phonomicrosurgery.

Tuna Orhanli is currently a research assistant at the Department of Electrical and Electronics Engineering of Hacettepe University, Ankara, Turkey. He received a BS degree in electrical and electronics engineering from Erciyes University, Kayseri, Turkey, in 2009. He completed his MSc degree in 2013 at the Electrical and Electronics Engineering Department of Hacettepe University in the field of design of smart above-knee prostheses. He is currently pursuing his PhD study under the supervision of Associate Professor Atila Yilmaz in the field of biomedical engineering, specifically the design and control of above-knee prostheses using a magnetorheological damper. He is a member of a research team for the project supported by the Scientific and Technological Research Council of Turkey. His research interests cover biomedical signal processing, design of embedded systems, and soft-computing methods.

Jesus Ortiz received a PhD in new automobile technologies from the University of Zaragoza, Zaragoza, Spain, in 2008. He worked as a researcher in the Department of Mechanical Engineering of the University of Zaragoza. In 2004 he was guest professor at the Ecole Nationale Supérieure d'Ingénieurs de Bourges, Bourges, France. Since 2006, he has been working at the Istituto Italiano di Tecnologia, Genoa, Italy, and he currently holds a technologist position at the Advanced Robotics Department, where he collaborates in two European projects and several internal projects. His principal research fields are motion bases, driving simulators, teleoperation, virtual reality, general-purpose computing on graphics processing units, computer vision, and medical robotics.

Anna Paidi is the chief executive officer of 1st Regional Healthcare Authority of Attica, Athens, Greece; the director of large-scale projects in the implementation of the Integrated Healthcare Information System of the Regional Healthcare Authority of Attica, cofunded by the European Union (operational program Information Society, Third Community Support Framework); a member of the organizing committee of the Unified Coding in Health in Greece; a member of the organizing committee of ehealthforum.gov.gr; a member of the Project Management Team of E-Governance; a member of the Working Group of Implementation of Diagnosis-Related Groups in Greece; the treasurer of the Greek Health Informatics Association; a society member of the International Medical Informatics Association; and a member of Health Level 7 Hellas. Anna Paidi worked as a research assistant at the Laboratory of Health Informatics (University of Athens, Athens, Greece) on European Union projects and an author of several publications and technical reports in scientific books and journals. Her main research interest is hospital information systems.

Andreas Panayides is a research associate at the Communications and Signal Processing Group at Imperial College, London, United Kingdom, and a visiting research assistant professor at the University of New Mexico, Albuquerque, New Mexico. Dr. Panayides received a BSc degree from the Department of Informatics and Telecommunications of the National and Kapodistrian University of Athens, Athens, Greece, in 2004, an MSc degree in computing and Internet systems from King's College, University of London, London in 2005, and a PhD degree in computer science from the University of Cyprus, Nicosia, Cyprus, in 2011. His research interests include image and medical video processing, dynamically reconfigurable medical video communications using multiobjective optimization, and mobile and pervasive computing for emergency crisis management and healthcare applications. He has published 7 journal papers, 4 book chapters, and 21 conference papers in related areas. He is also a research collaborator with the Department of Computer Science at the University of Cyprus, where he is a member of the Electronic Health (eHealth) Laboratory.

Diego Pardo received a BSc degree in electrical engineering (1996) from the Escuela Colombiana de Ingenieria, Bogota, Colombia. He obtained an MSc degree in electrical engineering (2002) from the Universidad de Los Andes, Bogotá, Colombia, and a PhD degree from the Polytechnic University of Catalonia, Barcelona, Spain, in 2009. He has conducted postdoctoral research in machine learning applied to robotics problems in the Barcelona Institute of Robotics (2011) and in the Italian Institute of Technology (2012). Since 2014, he has been with the Agile and Dexterous Robotics Laboratory at the Swiss Federal Institute of Technology Zurich.

Alessandra Pedrocchi received an MS degree in electrical engineering and a PhD degree in bioengineering from Politecnico di Milano, Milan, Italy, in 1997 and 2001, respectively. She is currently an assistant professor at the Department of Electronics, Informatics and Bioengineering of the Politecnico di Milano, where she teaches neuroengineering to graduate students and biomedical instrumentation to undergraduate students in the biomedical engineering program. She works at the NeuroEngineering and Medical Robotics Laboratory (http://www.nearlab.polimi.it) studying neurorobotics, bioartificial interfaces for in vitro neurons, and advanced technologies for neurorehabilitation. She has worked in the European Union's Multimodal Neuroprosthesis for Daily Upper Limb Support and REALNET projects and the Erasmus Mundus European MSc in advanced rehabilitation technologies program.

Veronica Penza received BS and MS degrees in biomedical engineering from Politecnico di Milano, Milan, Italy, in 2013. She is currently pursuing a PhD degree in bioengineering at the Istituto Italiano di Tecnologia, Genoa, Italy. Her main research topic is the development of an enhanced vision system to improve safety in single-port abdominal surgery. Previous research interests include bilateral telemanipulation systems, haptic feedback, and robotic neurosurgery. Veronica Penza was honored with the best video award at the Hamlyn Workshop on Augmented Reality and Surgical Vision in 2014.

Javier Pindter-Medina, since his childhood, has developed solid interests in science and knowledge about how things work. As an engineer, he made the decision in August 2012 to establish PindNET R&D, Mexico City, Mexico, with the intention of creating state-of-the-art electronic devices, mainly in the field of wearable technologies with the combination of biomedical engineering, to incorporate day-by-day the concept of mHealth. All of that is with the intention to contribute in the improvement of life quality and welfare of living beings, not only humans. Pindter-Medina understands that this may take a long path; nonetheless, he has the will to continue acquiring more and more knowledge in order to fulfill that journey.

Carmen C. Y. Poon graduated from the Engineering Science (Biomedical) Program of the University of Toronto, Ontario, Canada, and obtained her master's degree from a collaborative program offered by the Institute of Biomaterials and Biomedical Engineering and Department of Electrical and Computer Engineering, University of Toronto. She completed her PhD degree in electronic engineering (biomedical) at the Chinese University of Hong Kong, Hong Kong Special Administrative Region, China, where she is now a research assistant professor at the Department of Surgery. Dr. Poon is a senior member of the Institute of Electrical and Electronics Engineers, one of the three Asia-Pacific representatives of the Administrative Committee of the Institute of Electrical and Electronics Engineers' Engineering in Medicine and Biology Society (2014–2016), and the vice-chair of the Technical Committee of Wearable Sensors and Systems of the Society. Since 2009, she has served as the managing editor of *IEEE Transactions on Information Technology in Biomedicine*, which was retitled *IEEE Journal of Biomedical and Health Informatics* in 2013. She also served as a guest editor for a number of prestigious international journals on biomedical technology, including *IEEE Transactions on Biomedical Engineering*. Her research interests include wearable, ingestible, and implantable sensors and systems; body sensor networks; telemedicine and mobile health technologies; surgical and medical robots; public health informatics; and bioinformatics.

Marios Prasinos holds a BSc degree in computer science (University of Crete, Crete, Greece, 2010) and an MSc degree in engineering—economic systems (National Technical University of Athens, Athens, Greece, and University of Piraeus, Piraeus, Greece, 2013). During his postgraduate studies, Prasinos collaborated with the Biomedical Engineering Laboratory of the Electrical and Computer Engineering School of National Technical University of Athens. For his master's thesis, he designed and implemented an interactive software system for improving health compliance that includes web and mobile interfaces. Since 2014, Prasinos has been pursuing a PhD in computer science from City University London, London, United Kingdom. He is a member of the Center for Adaptive Systems and has received a bursary associated with the European Union project EMBalance. The aim of this research is to apply machine learning techniques and build an advanced model which will contribute to the prevention, diagnosis, and therapy of balance health problems.

Prasinos's research interests include health databases, health informatics, data mining, machine learning, and decision-support systems. Beyond research, Prasinos has worked in the software industry as a consultant and software engineer for more than two years.

Hammad Qureshi is an academic and a researcher in computer science. He is also an entrepreneur actively involved in technology start-ups. He holds a PhD degree in computer science and currently holds the post of assistant professor at the IT University-Punjab, Pakistan and College of Science-Zulfi, Majmaah University, Al Majmaah, Saudi Arabia. His main area of research interests are computer-assisted diagnosis and telehealthcare using machine learning, image processing, and pattern recognition. Dr. Qureshi has published over 25 academic and scientific articles in renowned journals and academic conferences. He has also published a book on automated cancer diagnosis. He is also a proud recipient of the mBillionth Award in 2010 for his work in telehealthcare and the University of Warwick's Institute of Advanced Study Early Career Fellowship for promising interdisciplinary work in automated healthcare.

Elijah Blessing Rajsingh is a professor and director of the School of Computer Science and Technology, Karunya University, Coimbatore, India. He received his master of engineering degree, with distinction, from the College of Engineering of Anna University, Chennai, India, where he also received a PhD degree in information and communication engineering in 2005. He has a strong research background in the areas of network security, mobile computing, wireless and ad hoc networks, and image processing. He is an associate editor of *International Journal of Computers and Applications* by Acta Press, Canada.

Sharmila Raman is a project manager at Intel Corporation located in Portland, Oregon. She was a graduate assistant for Dr. Ann Fruhling and worked on the STATPack research project. She graduated from the University of Nebraska at Omaha with a Master of Science degree in management information systems. She has a Bachelor of Engineering degree in electrical and electronics engineering from B.M.S. College of Engineering, Bengaluru, India.

Nathanael Sabbah is a young physician currently undergoing residency training in Radiology at the Rutgers University Hospital in Newark, New Jersey. He has completed his residency training in nuclear medicine at the Stanford University Medical Center, Stanford, California. A native of France, Dr. Sabbah received his medical degree from the Technion's Faculty of Medicine in Haifa, Israel, and subsequently completed his internship in surgery at the New York Presbyterian/Weill Cornell Medical Center in New York City, New York. Dr. Sabbah's clinical interests include the fields of both nuclear medicine and radiology, on which he has authored several peer-reviewed research publications.

Steven Samoil is currently a graduate student and research assistant at the Department of Electrical and Computer Engineering at the Schulich School of Engineering, University of Calgary, Calgary, Canada. He received a BSc degree in computer engineering from the University of Calgary and is currently working toward an MSc in electrical engineering. Samoil is a member of the Biometric Technologies Laboratory at the University of Calgary, where his research interests include gesture recognition, pattern recognition, computer vision, and image processing.

Fernando Martin Sanchez holds PhDs in informatics and in medicine. He is chair of Health Informatics at the University of Melbourne, Melbourne, Australia, and the director of the university's Health and Biomedical Informatics Centre. He has more than 100 peer-reviewed publications and his research has been funded by some 30 grants from the European Commission; the Spanish Ministries of Health, Science, and Defense; and the Australian National Health and Medical Research Council, the Institute for a Broadband Enabled Society, and Department of Health and Ageing. His research interests cover a wide range of topics related to the role of informatics in precision medicine (genomics and exposomics) and participatory health (social media, quantified-self, and mobile apps and devices).

N. Iwan Santoso has been working in research and development areas for more than 20 years. His expertise and research interests include embedded systems, intelligent control, data analysis, and decision support for medical, power, and industrial systems. He has MS and PhD degrees in electrical engineering from Louisiana State University, Baton Rouge, Louisiana. He served as research and development lead scientist, project/program director, and consultant in various companies. He is currently a professor and chair at DeVry University, North Brunswick, New Jersey, while continuing his function as a research and development consultant. He has authored numerous papers and was granted several global patents. He is a senior member of the Institute of Electrical and Electronics Engineers, the American Society for Engineering Education, and the Association for the Advancement of Medical Instrumentation and an active officer of the local Institute of Electrical and Electronics Engineers' Princeton Central Jersey Section.

H. Lee Seldon currently teaches near the Strait of Malacca in West Malaysia. He worked in Sarawak on the northwest edge of Borneo (2007–2010) after having served as a senior lecturer (of health informatics) at Monash University's Frankston campus in Frankston, Australia (1999–2007). In the 1980s and 1990s, Seldon worked as a senior research fellow associated with the Australian Bionic Ear Institute and as a telemedicine consultant for a company in Melbourne. He obtained his undergraduate degree in physics from Cambridge University, Cambridge, Massachusetts. He was a postgraduate student in biology in Freie University, West Berlin, Germany, then a student of medicine in University Koeln and ENT resident in University Hospital in Cologne, Germany. After earning his medical degree, he worked as an ear, nose, and throat resident at University Hospital in Cologne.

Anthony C. Smith is an associate professor at The University of Queensland and the deputy director of the Centre for Online Health, The University of Queensland, Brisbane, Australia. Dr. Smith has more than 14 years of research experience based on investigations of new telemedicine applications for the benefit of clinicians and patients in regional and remote areas of Queensland. Dr. Smith's specific research interests include the evaluation of feasibility, cost-effectiveness, and diagnostic accuracy of telemedicine applications in the context of pediatrics and child health. Current researches include the evaluation of wireless (robot) videoconference systems in pediatric wards; home telemedicine consultations for children with chronic health conditions; e-mail–based telemedicine support; and a community-based telemedicine health screening program for indigenous children. Dr. Smith has made a significant contribution to the literature on telepediatrics, publishing more than 90 papers in peer-reviewed journals, 3 edited books, and 12 book chapters on telemedicine-related topics.

Adrian Stoica has over 20 years of research and development experience in autonomous systems, developing novel adaptive, learning, and evolvable hardware techniques for applications ranging from measurement equipment to space avionics to robotics. He obtained his PhD degree in electrical engineering from Victoria University of Technology, Melbourne, Australia, in 1996, and MSc degree in electrical engineering from the Technical University of Iaşi, Iaşi, Romania, in 1986. Dr. Stoica joined the Jet Propulsion Laboratory of the National Aeronautics and Space Administration, Pasadena, California, in 1996, and led a variety of research projects as a principal scientist, a senior research scientist, and a group supervisor. He is currently a senior research scientist and supervisor of the Advanced Robotic Controls Group in the Mobility and Robotics Section at the Jet Propulsion Laboratory.

Nitish V. Thakor is a professor of biomedical engineering, electrical and computer engineering, and neurology at Johns Hopkins School of Medicine, Baltimore, Maryland, and directs the Neuroengineering Laboratory. He is also the director of the Singapore Institute for Neurotechnology at the National University of Singapore, Singapore. He received his undergraduate degree from the Indian Institute of Technology, Bombay, India (1974) and PhD degree from the University of Wisconsin, Madison, Wisconsin (1981). Dr. Thakor's technical expertise is in the areas of neural diagnostic instrumentation, neural microsystems, neural signal processing, optical imaging of the nervous system, neural control of prosthesis, and brain–machine interface. He is a coauthor of more than 260 refereed journal papers and is currently the editor-in-chief of the *Medical & Biological Engineering & Computing* journal. He was the editor-in-chief of *IEEE Transactions on Neural Systems and Rehabilitation Engineering* from 2005 to 2011. Dr. Thakor is a recipient of a Research Career Development Award from the National Institutes of Health and a Presidential Young Investigator Award from the National Science Foundation, and is a fellow of the American Institute of Medical and Biological Engineering, Institute of Electrical and Electronics Engineers; founding fellow of the Biomedical Engineering Society; and fellow of the International Federation of Medical and Biological Engineering. He is also a recipient of the Centennial Medal from the University of Wisconsin School of Engineering, an honorary membership from Alpha Eta Mu Beta Biomedical Engineering Student Honor Society, an award for technical excellence in neuroengineering from the Institute of Electrical and Electronics Engineers' Engineering in Medicine and Biology Society; Distinguished Alumnus Award from the Indian Institute of Technology, Bombay, India; and Centennial Medal from the University of Wisconsin, Madison School of Engineering.

David W. Walters is an honorary professor of logistics and supply-chain management at the Institute of Transport and Logistics Studies at the University of Sydney–St James Campus, Sydney, Australia. He has held senior academic posts at Oxford University (Templeton College), Oxford, United Kingdom; Cranfield School of Management, Cranfield, United Kingdom; the University of Western Sydney (Sydney Graduate School of Management), Sydney, Australia; Macquarie University, Sydney, Australia; and Sydney University, Sydney, Australia. He is an author of a series of business and marketing textbooks with a particular focus on value-chain management, commencing with *Operations Strategy* (Palgrave, 2002), then *Strategic Operations: A Value Chain Approach* (Palgrave, 2007), and *Managing in the Value Chain Network* (Prestige, 2012). Dr. Walters has published in professional journals and presented papers on value-chain management at conferences in Australia and internationally. He has extensive teaching experience across the globe and has acted as a consultant to numerous international companies, including CSR, Harrods, Laura Ashley, Kingfisher, Storehouse, British Oxygen Company, Marks and Spencer, and Tesco. Dr. Walters is a founding principal of the value-chain network.

Tomas E. Ward is vice president of engineering at Syntrogi Inc., a San Diego neurotechnology company which he joined in 2014. Prior to this he was a senior lecturer at the Department of Electronic Engineering, National University of Ireland Maynooth, Maynooth, Ireland (since 1999), where he led the Biomedical Engineering Research Group. Dr. Ward holds BE (electronic engineering), MEngSc (rehabilitation engineering), and PhD (biomedical engineering) degrees from University College Dublin, Ireland. His academic research includes the application of brain–computer interfaces for neurorehabilitation, particularly in stroke and signal processing for connected health. Dr. Ward serves in the Engineering Sciences Committee of the Royal Irish Academy and previously on the Irish Research Council. He has been a senior member of the Institute of Electrical and Electronics Engineers since 2011. Dr. Ward has authored more than 200 peer-reviewed publications and has supervised to completion 20 research students (12 PhD). He has licensed a range of technologies to industry since 2009 including sensor streaming technologies for eHealth, over-the-air programming, and mobile health applications.

Stefan Weber received a degree in electrical engineering and automation from the University of Ilmenau , Ilmenau, Germany, in 1998. He then joined the Robotics Laboratory of the University of Southern California, Los Angeles, California, as a Fulbright Fellow. He graduated with a PhD degree in augmented reality applications in medicine from the Humboldt University, Berlin, Germany, in 2004. From 2005 to 2008, he worked as a young scientist at the Technical University of Munich, Munich, Germany. From 2008 to 2012, he was an assistant professor of computer-assisted surgery and implantation technology at the University of Bern, Bern, Switzerland. In 2012, he became a full professor of image-guided therapy and the director of the ARTORG Center for Biomedical Engineering at the University of Bern. His research interests include fundamental and translational aspects of image-guided surgery, surgical robotics, and medical image analysis. Professor Weber is a member of the Cantonal Ethics Board in Bern.

Tom Williamson received his PhD in biomedical engineering at the ARTORG Center for Biomedical Engineering, University of Bern, Bern, Switzerland, after receiving his bachelor of engineering (biomedical) degree from Swinburne University, Melbourne, Australia, in 2006 and master of science (biomedical engineering) degree from the University of Bern in 2011. Ongoing projects include the utilization of novel sensor sources in the development of innovative control strategies in robot-assisted surgery and the development of intelligent instrumentation for the improvement of surgical outcomes.

Sinclair Wynchank, was educated at Oxford University, Oxford, United Kingdom, from which he gained MA (physics) and DPhil (nuclear physics) degrees. His postdoctoral studies at Columbia University, New York, New York, were followed by an assistant professorship at the City University of New York, New York until he joined the Physics Department of the University of Cape Town, Cape Town, South Africa. He obtained MB, ChB, and MD (nuclear medicine) degrees from the University of Cape Town and qualified as a CEng in London, United Kingdom. For two years, he was an associate professor in nuclear medicine at the University of Bordeaux, Talence, France. Thereafter, for 30 years, while a part-time member of the University of Cape Town's Faculty of Medicine, he also worked with the South African Medical Research Council (1983–present), Cape Town, South Africa, as the research institute's director, as well as other duties. His research has treated eHealth, clinical nuclear medicine, bioengineering, radiobiology, dosimetry, and the sociology/history of medicine. He is an author/coauthor of 132 published articles in these fields, has lectured extensively in 24 countries, had five visiting professorships and various consultancies, and was a president of the African Association of Nuclear Medicine.

Svetlana N. Yanushkevich is currently is a full professor at the Department of Electrical and Computer Engineering, Schulich School of Engineering, University of Calgary, Calgary, Canada. She has a PhD degree in electrical engineering from Belarusian State University of Informatics and Radioelectronics (1992) and a habilitated PhD degree in technical sciences from the Warsaw Institute of Technology, Warsaw, Poland (1999). Prior to joining the University of Calgary, she was an associate professor at the Technical University of Szczecin, Szczecin, Poland (now West Pomeranian University of Technology). Dr. Yanushkevich directs the Biometric Technologies Laboratory at the University of Calgary. Her research interests include biometrics and its applications to biomedical engineering, digital design (multivalued and probabilistic logic), computer vision, and machine learning.

Atila Yilmaz received a BS degree in electrical and electronics engineering from Hacettepe University, Ankara, Turkey, in 1986 and an MS degree in control and information technology from the University of Manchester, the Institute of Science and Technology–Control Systems Centre, Manchester, United Kingdom, in 1992. He received a PhD degree from the Department of Biomedical Engineering of the University of Sussex, Sussex, United Kingdom, in 1996. He joined the Department of Electrical and Electronics Engineering of Hacettepe University, Ankara, Turkey, in 1998 and then became an associate professor in 2009. He is vice-chair of the department at present. His current research interests include the development of prosthetic and orthotic devices, biomedical signal processing, neural/adaptive systems, and electric impedance tomography.

Basem F. Yousef is an associate professor at the Department of Mechanical Engineering of the United Arab Emirates University, Al-Ain, United Arab Emirates. Dr. Yousef received his BSc degree from the University of Jordan, Amman, Jordan, in 1993. He received his MSc and PhD degrees from the University of Western Ontario, London, Ontario, Canada, in 2001 and 2007, respectively. Prior to joining the United Arab Emirates University in August 2008, Dr. Yousef held the position of robotics system engineer at MAKO Surgical Corp., Fort Lauderdale, Florida. At MAKO, Dr. Yousef worked with a cross functional team of mechanical, electrical, and robotic engineers to develop a medical robotic system for knee surgery. He worked on developing formal test protocols and designs of complex electromechanical test equipment for design verification, reliability, and manufacturing at subassembly and system levels. His duties also included designing and conducting experiments on performance and reliability testing of medical robotic systems, diagnosing failures, performing root cause analysis, and suggesting design improvements. Dr. Yousef's research interests focus on developing smart robotic and mechatronic systems engineered to provide solutions for problems in industrial and healthcare fields.

Weiwei Yuan received her PhD degree in computer science from Kyung Hee University, Seoul, South Korea, in 2010. Since September 2010, she had been an assistant professor at Harbin Engineering University, China. Since November 2014, she has been an associate professor in the College of Computer Science and Technology, Nanjing University of Aeronautics and Astronautics, Nanjing, China. Her research interests are social computing, machine learning, and eHealth.

Saliha Ziam, PhD, is a professor at the school of business administration at the TÉLUQ, Quebec City, Canada. She completed her doctorate in business administration from Laval University (2010). Saliha Ziam has two master's degrees, one in public health from Laval University, Quebec City, Canada (2004), and another in health and social sciences from the Paris-Nord University, Paris, France (1995). Her research interests focus on innovation and knowledge transfer in the health sector, health assessment tools for decision support, and knowledge absorptive capacity strategies. Her recent works on knowledge transfer have been published in *International Review of Business Research Papers*; *Research in Higher Education Journal*; *Allergy, Asthma & Clinical Immunology*; and *Evidence & Policy*.

Section I

Medical Robotics, Telesurgery, and Image-Guided Surgery

1

Medical Robotics

**Giancarlo Ferrigno, Alessandra Pedrocchi, Elena De Momi,
Emilia Ambrosini, and Elisa Beretta**

CONTENTS

1.1 Introduction to Medical Robotics

What can be considered or not considered a medical robot is still under debate. The narrowest definition includes only those devices encompassing a robot or a robotic system which are used for a direct medical purpose, i.e., for surgical and radiation therapy and for rehabilitation, disregarding, for example, the assistive devices and the person carriers used in hospitals. By far, it is recognized that robots for surgical and radiation therapy belong to the *professional service robots*, while assistive and rehabilitation devices may be *robots for personal care* if a professional is not needed for their use. Groups of members of the International Standard Organization (ISO) and the International Electrotechnical Commission (IEC) are working to properly define *medical robots*. ISO has been traditionally responsible for industrial robot standards, while IEC has been largely involved with medical electrical equipment standards (e.g., the IEC 60601 family of standards). An ISO-IEC joint group, JWG9, has been working on these topics in the last five years and branched in the mid-2015 in two new groups: JWG35, medical robots for surgery, and JWG36, medical robots for rehabilitation.

1.1.1 Definitions and Standards

The international standard ISO 8373:2012, "Robots and Robotic Devices—Vocabulary," prepared by the Technical Committee (TC) ISO/TC 184 (Automation Systems and Integration) Subcommittee SC 2 (Robots and Robotic Devices), specifies the vocabulary used in relation with robots and robotic devices operating in both industrial and nonindustrial environments. It provides definitions and explanations of the most commonly used terms. According to this document, a *robot* is an "actuated mechanism programmable in two or more axes with a degree of autonomy, moving within its environment, to perform intended tasks." It "includes the control system and its interface with the operator." The classification of a robot into an *industrial robot* or a *service robot* is done according to its intended application. Service robots perform useful tasks for humans or equipment excluding industrial automation applications, which include, but are not limited to, manufacturing, inspection, packaging, and assembly. Among service robots, service robots for professional use include those used for a commercial task, usually operated by a properly trained operator, as, for example, cleaning robots for public places, delivery robots in offices or hospitals, fire-fighting robots, rehabilitation robots, and surgery robots in hospitals. In the case of medical robots, ISO 8373 defines the different role of the operator, i.e., a person designated to start, monitor, and stop the intended operation of a robot, the surgeon, for example, and the recipient or beneficiary, i.e., a person who interacts with a service robot to receive the benefit of its service, such as a patient receiving care from a medical robot.

1.1.2 Historical Perspective

The history of robots is quite recent in almost all the fields. In fact, notwithstanding the dreams of humans about slave automata dating back to the Middle Ages, like the golem was, and the drawings of automated warriors of Leonardo da Vinci, the word *robot* itself was coined in science fiction *R.U.R.* (*Rosumovi Univerzální Roboti*), written by the Czech writer Karel Čapek only in 1920.

The first real robots, Unimation's Programmable Universal Machine for Assembly (PUMA), entered automotive factories in 1961 (General Motors), and 25 years later, at the time when industrial robots started to be endowed with vision capacity, a PUMA 200 was used for the first robotic biopsy [1–2]. This project started at the Memorial Medical Center in Long Beach, California, United States, and was discontinued due to the safety rules set out by the manufacturer for the PUMA robots that were designed to operate when separated by a physical barrier (fence) from people, as occurs in industrial settings. The copresence of robots and humans in the operating room (OR) is still an issue nowadays and it is tackled with lightweight devices, body sensors, joint sensors, environmental surveillance cameras, modulation of the stiffness, etc. During the years, some ideas and inventions have worked and been expanded upon, while others have fallen by the wayside or become obsolete, mainly because of the difficulty of the market: high development cost (including preclinical studies and clinical trials), difficult actual demonstration of outcome improvement for the patient, hard safety standards, and patent monopoly.

Rehabilitation and assistive robotics stemmed from the common field of rehabilitation engineering. Within its subtopics, robotics has been applied to individuals with musculoskeletal and neurological diseases. The aims are mainly to recover or improve the motor control in congenital or acquired neurological diseases and assisting in strengthening and restoring functionality in the impaired muscle groups and the skeleton. Most of the work and development in rehabilitation robotics was initially oriented toward musculoskeletal diseases, although recently, the applications within the neurological field have been increasing based on recent insights of brain plasticity being enhanced by the use of robots. Rehabilitation engineering dates back to the development of functional prostheses, which can be considered the foundation of rehabilitation robotics, but the actual field of robotics in rehabilitation engineering officially started with powered human exoskeleton devices in the '60s. This research began in the United States and the former Yugoslavia, each with a different goal [3], i.e., to augment human capabilities on one hand and to help severely disabled people on the other. A research topic that is common among all medical robotics advancements is the enhancement of the human–machine interface, where it is needed either for executing a complex surgical procedure or for deciding how an activity of daily living (ADL) can be performed by a severely impaired subject. This is actually the critical part for it is strictly connected to the safety of the user. Safe and easy-to-use upper-limb and lower-limb robotic devices are available nowadays for neurological rehabilitation and assistance with interfaces ranging from residual electromyographic (EMG) activity to brain–computer interfaces (BCIs).

1.2 Surgical Robots

If we look at the current panorama of surgical robots, we see some major players sharing the market, which is mainly divided by applications, which drive, in turn, the requirements of the systems outlined in the following paragraphs.

1.2.1 General Requirements

Requirements for surgical robots are expressed in terms of safety provisions, which deeply involve risk analysis, and essential performance, both depending on the specific application. There are mainly two reasons for which the surgical robots, as in general medical equipment, should be carefully specified in terms of risk and effectiveness. The first is because a robot is a machine and, thus, can hurt the patient and the operators. The patient, in particular, can be severely or fatally hurt since often the robot is in close contact with the patient's inner organs. The second requirement is of paramount importance since the effectiveness of a therapy can make the difference between a healthy life and morbidity or death. Also in this case the essential performance specification depends on the clinical procedure. In the middle lies the autonomy of the system. Autonomy, in fact, allows effectiveness only when the decision time for a movement is too short to be approved by the operator, but, on the other hand, maneuvers that are wrong in a given situation can put the patient at risk. This kind of autonomy can be justified only by the benefit-to-risk ratio (as we will see for robots in radiation therapy). All these requirements vary with the application since the criticality of errors or misuse strongly depends on it. As an example, neurosurgery is more demanding in terms of accuracy than orthopedics: a 5 mm error hip joint center location can permanently damage important areas on the brain cortex, leading to a permanent functional loss, while no loss occurs in less than 1° of error in femur functional axis positioning.

Clause 4.2 of IEC 60601-1 general standard requires that manufacturers enforce a risk management system compliant with ISO 14971. The manufacturer must define what constitutes acceptable levels of risk and demonstrate that all risks associated with the device are acceptable in accordance with its own risk acceptance policy. Test data, clinical research, and scientific literature, used for risk analysis, must be placed in the risk management file. Clause 4.3 of IEC 60601-1 requires the manufacturer to identify essential performance by applying risk management. To determine which aspects of a device's performance are essential, it is first necessary to list all aspects of performance as potential hazards (i.e., sources of harm). For each function, the manufacturer must determine whether operation in excess of or below the specified level of performance would result in harm. The levels of performance at which harm would occur define the boundary of essential performance. If variation in a given performance characteristic does not result in injury, then it is not an essential performance. On the other hand, failure to provide essential performance by definition results in unacceptable risk. Essential performance must be guaranteed under specified fault conditions of any subsystem. Medical robots require (typical of the medical devices) the ability to move and interact with the patient's body and the operator(s). Potential harm could be due to pinching, impacts, etc., which can be also fatal if in direct contact with vital organs (e.g., brain injuries). In this respect, also essential performance in trajectory execution or fidelity in reproducing movements in teleoperation must be well controlled in the risk analysis.

1.2.2 Control

Several simple and combined control methods have been developed for robots that can be exported to the field of medical robotics. The following subsections describe the most used in the field.

1.2.2.1 Position Control

Position control is the most popular control mode in robotics. The Cartesian pose of the end effector is specified and the robot controller uses the actuators to reach it. A pose is a set of six independent variables that define the position of the reference frame's origin of the end effector and three angles that define the rotations around the world coordinate axes. Rotations can be represented in alternative ways with angle–axis or quaternion notation. In order to achieve such a pose, it should be turned into the joint space rotational, (i.e., the independent variable is a rotation, or prismatic, and, the independent variable is a translation). This is done through inverse kinematics that can be computed iteratively or analytically. For moving from one pose to another, at least the starting and the ending poses should be specified and the via points optionally.

Each transformation from a pose to another is of the type reported in Equation 1.1, where x, y, and z are the translations between the origins of the two reference frames $P1$ and $P2$ and the three-by-three submatrix R ($r_{i,j}$) is the rotation matrix between the two:

$$_{P1}T^{P2} = \begin{array}{cccc} r_{1,1} & r_{1,2} & r_{1,3} & x \\ r_{2,1} & r_{2,2} & r_{2,3} & y \\ r_{3,1} & r_{3,2} & r_{3,3} & z \\ 0 & 0 & 0 & 1 \end{array} \qquad (1.1)$$

Notice that the terms $r_{i,j}$ of the submatrix R are functions of the angles between the axes of two reference frames, so the actual independent variables are three and not nine. The notation of Equation 1.1 means that T is the tensor transforming poses from reference frame $P1$ to reference frame $P2$. Tensors can be inverted to reverse the transformation and chain multiplied to combine different transformations. In the multiplications the right upper index of the first matrix must equal the lower right index of the second.

For the purpose of defining the direct kinematics of a robotic link chain, a very convenient transformation is given by the Denavit–Hartenberg notation [4].

Interpolators compute a series of intermediate points, in space and time, compatible with the maximum torque or force available at the joint motors. For dynamic control, a precise mechanical model of the robot is needed in order to compute the torques or forces to be applied to actuators in order to follow a given trajectory. Sensors (typically optical encoders) at the joints allow closing the loop of the controller. Position control loop, usually provided by the manufacturer in a rather optimized way, can be exploited as the kernel for the synthesis of more complicated control schemes.

1.2.2.1.1 Tool Holders

Tool holders are commanded in their target position through a position control of the robot. They are used chiefly in image-guided surgery (IGS). In IGS a preoperative plan, usually defined on two-dimensional (2D) or three-dimensional (3D) medical images, is transferred to the surgical environment, thanks to a registration procedure, using tracking devices (spatial localizers). The plan can be also intraoperatively defined, on the basis of intraoperative measurements on the patient anatomy (e.g., in orthopedic surgery) or on the basis of intraoperative acquired images (e.g., in laparoscopy). In all the cases in which the surgical instrument must be precisely aligned with the preoperative plan (e.g., probes in stereotactic procedures in neurosurgery or cutting edges in orthopedics), robots can provide an accurate and repeatable alignment tool. Safety in these robots is guaranteed by continuous approval of the motion (quite slow) by the operator through the operation of a pedal.

Here we present several commercially available robots used for surgical tool positioning, which we classify on the basis of the used registration method. The registration defines the geometrical transformation between the preoperative plan and the intraoperative environment in which the task is defined.

The ROBODOC® system (Curexo Technology Corporation, Fremont, Canada) [5] and the MAKOplasty® (MAKO Surgical Corp., Fort Lauderdale, United States) [6], used in orthopedic arthroplasty procedures, are both registered to the intraoperative environment by touching pins screwed in the bones, visible in preoperative computed tomography (CT) images, with the robot tool center point (mechanical localization).

Intraoperative X-ray images are used for the target registration in the CyberKnife® system (Accuray Inc., Sunnyvale, California, United States), which is used for radiation therapy procedures. CyberKnife compensates also for patient/organ motions during the irradiation using real-time digital radiographies. This compensation cannot be confirmed by the operator since it happens too fast. This is an example of robot autonomy, which is allowed because not using it would be more harmful to the patient (wrong target) than the possible errors of the machine.

The neuromate® system (Renishaw Ltd., United Kingdom), used to position depth electrodes in the brain for stereoelectroencephalography (SEEG), is registered to the intraoperative target (the patient's head) using intraoperative fluoroscopic images. The same approach is used by SpineAssist (Mazor, Cesarea, Israel), a parallel miniaturized platform for pedicle screw insertion. This class of systems is registered by images. The ROSA™ (Medtech, France) system, used in neurosurgical endoscopy, SEEG, and biopsy, acquires the patient head surface by using a laser scanner and surface-matching techniques during the registration procedure.

1.2.2.1.2 *Steering Needles and Intraluminal Devices*

One particular application of robotic devices in surgery is the capability of operating with flexible probes in tissues or cavities. The advantage of using steerable probes in tissues, for biopsies or drug release, is to reach deeply seated targets hidden by sensible areas. There are several approaches to the steering in the bulk of tissues. When a needle with an asymmetric tip is inserted into a solid tissue, the shape of the tip and its interaction with the medium creates an imbalance in lateral force. If the tissue does not deform significantly, the resultant steering force causes the needle tip to follow a circular arc, and if the needle is flexible relative to the medium, the rest of the needle shaft will follow the same path [7]. Similarly, in another work [8], the authors devised and tested a means of generating curvilinear trajectories by spinning a needle with a variable duty cycle (i.e., in on–off fashion). The technique can be performed using image guidance and trajectories can be adjusted intraoperatively via the joystick. Cadaver testing demonstrated the efficacy of the needle-steering system and its precision down to 2 ± 1 mm. The bevel tip steers against tissue resistance only. This assumption does not hold in cerebral parenchyma, so a different approach has recently been proposed by Ko and Rodriguez [9], which uses a multisection flexible probe in which only one section, out of four, moves forward and uses the other three as a stable guide that does not push against the tissue. Extrusion of the section regulates the trajectory curvature. The control of these systems is based on a nonholonomic model, resembling the trajectory of a planar bicycle with limited steering angle. Another family of steering probes has been designed for moving in cavities by exploiting multi–degrees-of-freedom (DoF) snake robots [10], which use a hybrid tendon–micromotor actuation scheme coupled via an internal ring gear and concentric prebent nitinol tubes [11], which twist when extruded from the parent container tube. A different approach is crawling motion over the organs, as in the HeartLander [12], a miniature mobile robot designed to navigate over the epicardium of the beating heart for minimally invasive therapy (Figure 1.1). The HeartLander is an inchworm-like robot crawler that uses suction to grip the epicardial surface of the heart and the drive-wire actuation of the tandem body sections to generate motion. The tethered design of the crawler allows for the transmission of relatively high suction and drive-wire forces from external instrumentation. Each body section is $5.5 \times 8 \times 8$ mm^3, giving the crawler a length of 16 mm when fully contracted.

1.2.2.2 *Shared Control*

Surgical robots can also be used in a shared-control mode with the surgeon and be in direct contact with the tissue. When the robot has to interact with the so-called environment (i.e., hard or soft tissues or user interaction), it is necessary that the robot controller

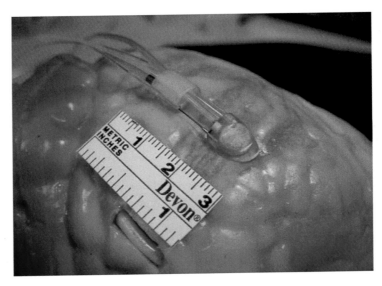

FIGURE 1.1
HeartLander, a miniature mobile robot designed to navigate over the epicardium of the beating heart for minimally invasive therapy. (Courtesy of Nicholas Patronik.)

provides the force needed to overcome the resistance of the environment or to comply with it, assuring system stability. Different control schemata have been implemented to achieve specific desired dynamic behaviors of the manipulator. Active interaction control strategies can be divided into "direct force control," i.e., the desired motion and desired contact force and moment need to be specified in a consistent way with respect to the geometrical constraints imposed by the environment, and "indirect force control," i.e., force control is achieved via motion control without requiring in principle the explicit closure of a force feedback loop and, thus, measurements of the contact force [13].

1.2.2.2.1 Admittance/Impedance Control

In the indirect force control schema, the mechanical impedance/admittance of a manipulator is set with the following relationship between effort and motion about a nominal end-effector trajectory x_0:

$$\mathbf{f} = \mathbf{K}(\mathbf{x} - \mathbf{x}_0) + \mathbf{B}(\dot{\mathbf{x}} - \dot{\mathbf{x}}_0) + \mathbf{M}(\ddot{\mathbf{x}} - \ddot{\mathbf{x}}_0) \tag{1.2}$$

where $\mathbf{f}$ is a vector of forces and torques acting on the manipulator due to contact with the environment; $\mathbf{x}$, $\dot{\mathbf{x}}$, and $\ddot{\mathbf{x}}$ are the linear and angular motions of the end effector and their derivatives; and $\mathbf{K}$, $\mathbf{B}$, and $\mathbf{M}$ are the stiffness, damping, and inertia matrices.

Since $\mathbf{f}$ is identically zero if $\mathbf{x} - \mathbf{x}_0$ is zero during the time, the trajectory $\mathbf{x}_0$ is interpreted as the noncontact end-effector trajectory [14]. In absence of interaction, this control scheme makes the end effector follow the desired motion; in the presence of contact with the environment, a compliant dynamic behavior is imposed on the end effector, according to the stiffness, damping, and inertia matrices. These parameters can be selected to correspond to various manipulation task objectives [15]. We define control "in impedance causality" (torque-based impedance control; Figure 1.2) if it reacts to motion deviation by generating forces, while we define it as a "controlled in admittance causality" (position-based

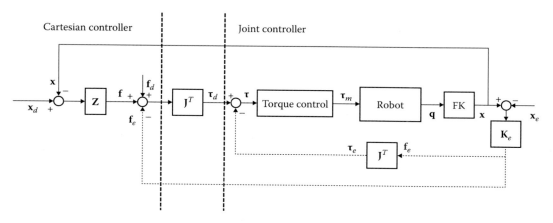

FIGURE 1.2

Torque-based impedance controller. $\mathbf{x}$ is the robot actual pose in the task space computed from the actual joint configuration $\mathbf{q}$ with the forward kinematics (FK) block; $\mathbf{J}$ is the robot Jacobian; $\mathbf{x}_d$ is the desired pose in the task space; $\mathbf{x}_e$ is the equilibrium pose of the environment; $\mathbf{K}_e$ is the net stiffness of the sensor and of the environment; $\mathbf{f}_e$ and $\mathbf{\tau}_e$ are the external environment forces expressed in the task space and in the joint space, respectively; $\mathbf{f}_d$ is the desired force vector; $\mathbf{\tau}_d$ is the desired torque vector computed from the force equilibrium; $\mathbf{\tau}$ is the torque input vector of the inner torque control loop; and $\mathbf{\tau}_m$ is the commanded motor torque vector. The command force $\mathbf{f}$ is defined as $\mathbf{f} = \mathbf{Z}(\mathbf{x}_d - \mathbf{x})$, where $\mathbf{Z}$ is the impedance matrix. When the environmental forces are available (dotted lines), the measurements are used to decouple the dynamic of the system.

impedance control; Figure 1.3) if it reacts to interaction forces by imposing a deviation from the desired motion (where the robot speed is the controlled variable). The first approach is better suited to keep contact forces small and for application where manipulation gravity loads are small and the motion is slow. The precise dynamic manipulator model needs to be available and the joint inner loop must be provided at high rate.

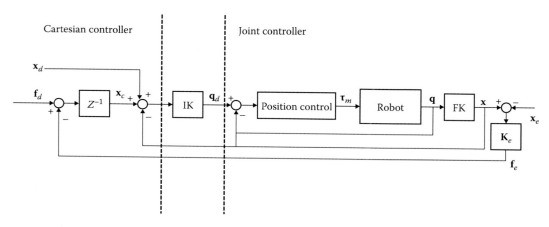

FIGURE 1.3

Position-based impedance controller. $\mathbf{x}$ is the robot actual pose in the task space computed from the actual joint configuration $\mathbf{q}$ with the FK block, $\mathbf{x}_d$ is the desired pose in the task space, $\mathbf{x}_e$ is the equilibrium pose of the environment, $\mathbf{K}_e$ is the net stiffness of the sensor and of the environment, $\mathbf{f}_e$ is the external environment forces expressed in the task space, $\mathbf{f}_d$ is the desired force vector, $\mathbf{q}_d$ is the desired joint configuration computed with the inverse kinematic (IK) block, and $\mathbf{\tau}_m$ is the commanded motor torque vector. The command trajectory $\mathbf{x}_c$ is defined as $s.\mathbf{x}_c = Z^{-1}(\mathbf{f}_d - \mathbf{f}_e)$, where Z^{-1} is the admittance matrix and s is the argument of the Laplace transform.

Conversely (Figure 1.3), the second approach focuses more on desired force tracking control [16]. It suffers from the inability to provide very "soft" impedance, while it is fitting in cases where stiff joint position control is required, e.g., for high-accuracy positioning in specific Cartesian direction [14].

1.2.2.2.2 Active Constraints

Another human–machine interaction class is the active constraint, also called virtual fixture, virtual wall, and electronic tunnel. The concept of active constraints was first introduced by Rosemberg [17] as an overlay of abstract sensory information on a work space in order to improve human performance in direct and remotely manipulated surgical tasks in order to increase the system safety. Recently, Bowyer et al. [18] presented a classification of different types of constraints in regional/guidance, attractive/repulsive, and unilateral/bilateral constraints. In case of regional and repulsive constraints, the system control prevents the robot to overcome predefined space regions either if the robot is manually driven (i.e., the controller nullifies the motion that violates a constraint applying a force to the user) or if the robot is remotely operated (only motion components which do not violate a constraint are allowed by the controller).

There are several constraint enforcement methods (see Figure 1.4 for their classification). In the simpler implementation, the effect of the constraint (i.e., the force $\mathbf{f}_c$) is computed as a function of the proximity of the end effector from the closest point on the constraint and of the velocity of the robot tool center point approaching, such as

$$\mathbf{f}_c = k_p(\mathbf{p}_{\text{const}} - \mathbf{p}_{\text{ee}}) \tag{1.3}$$

where $\mathbf{f}_c$ is the constraint force vector, k_p is the proportional gain, and $\mathbf{p}_{\text{const}}$ and $\mathbf{p}_{\text{ee}}$ are the closest points on the constraint and the tool center point position [18].

In this case the constraint works as a spring connecting the two points, so with positive k_p the force is attractive, enforcing guidance, and with k_p negative the force is repulsive, enforcing a repulsive constraint. Regional constraints require the use of nonlinear functions. When the constraint geometry varies in time due to physiological or voluntary movement, this is commonly referred as "dynamic active constraints" [19].

1.2.2.3 Cooperative Control

In cooperative-controlled (or hands-on) systems, the surgeon and the robot share the control of the surgical system and this is particularly suited where high positional accuracy is required (e.g., retinal surgery) or when both manual and robotic-assisted operations are performed in different surgical work-flow steps (e.g., in orthopedic surgery). In the first category we can find the steady-hand eye robot [20], which is an actively and cooperatively controlled robot assistant designed for retinal microsurgery. Cooperative control allows the surgeon to have full control of the robot, with his hand movements dictating the displacement of the robot. Also, the micron device was designed as a fully handheld instrument by MacLachlan et al. [21] and Yang et al. [22]. This handheld robot (see Figure 1.5) can actively compensate in real time for the motion due to hand tremor.

In the second category we can find back-drivable serial arms that can be manually positioned by surgeons. Transparency, which quantifies the ability of a robot to follow human movements without requiring any human-perceptible resistive force drive, is one of the major issues in the field of human–robot interaction for assistance in manipulation tasks [23]. Cooperatively controlled robotic systems are used in robotic surgery, e.g., the

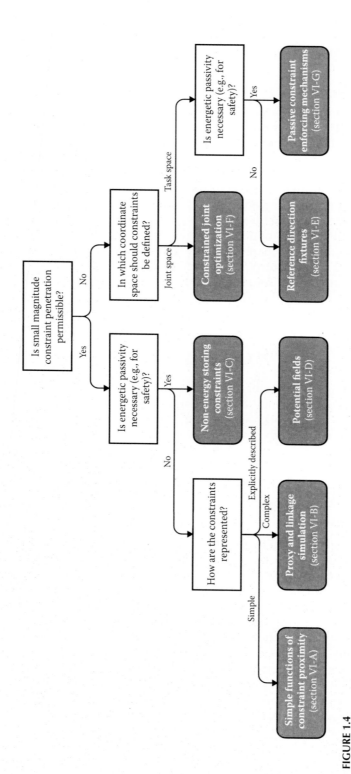

FIGURE 1.4

Simplified decision tree for constraint enforcement methods according to the working scenario requirements. (From Bowyer, S. A. et al., *IEEE Transactions on Robotics,* 30, 138–157, 2013. © 2013 IEEE. With permission.)

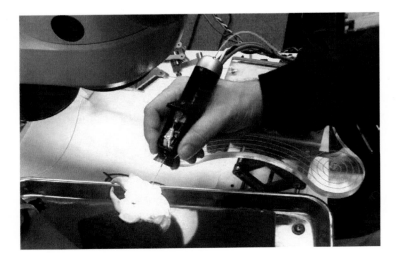

FIGURE 1.5
Micron: six-degree-of-freedom handheld tremor-canceling microsurgical instrument. (Courtesy of Robert MacLachlan.)

RIO system (MAKO Surgical, now owned by Stryker Corp.) and ROBODOC (Curexo Technology) for orthopedic surgery, the ROSA system (Medtech, Montpellier, France) for neurosurgery, and needle-insertion devices [24].

1.2.2.4 Teleoperation

Teleoperation has been used to control robots in hostile (nuclear plants and disaster relief) or remote environments (space, underwater, and aerial). In medical robotics, teleoperation achieves the following goals:

- Motion scaling (joined with magnification of visual field)
- Tremor filtering
- Ergonomic position of the operator, even for tools working in uncomfortable positions
- Expansion of hand–wrist range of movement

Teleoperation encompasses a master–slave architecture in which the motion/force impressed on the master side is reflected to the slave side. The master side can just be a position reference, but it can also reflect forces experienced at the slave side, possibly scaled up–down, through the use of haptics.

1.2.2.4.1 Haptic Devices

Haptics is a tactile/kinesthetic feedback technology that provides the sense of touch, giving information on the material properties of an object (such as stiffness, texture, weight, size, orientation, and curvature). The haptic stimulation can be tactile (achieving cutaneous feedback through mechanoreceptors stimulated by passive pressure or by current injection) and kinesthetic (achieving force feedback revolving around muscle receptors' engagement through an active touch).

The computational task in haptic rendering is to generate feedback information that is relevant to particular applications. Whether this information should refer to forces, displacements, or a combination of these and their derivatives is still the object of debate. Tactile sensitivity depends on size, density, frequency range, nerve fiber branching, and type of stimulation (skin motion or sustained pressure). The force exerted must be greater than 0.06 to 2 N/cm² [25]. For simulating tactile sensations, a haptic display must

- Maintain active pressure for the user to feel a hard surface after initial contact;
- Maintain a slight positive reaction against the skin after initial contact for soft surfaces (without active pressure or relative motion); and
- Provide some relative motion between the haptic surface and the skin to accurately display texture.

Kinesthesia allows the perception of limb movement and position together with the perception of force. This sensory perception originates primarily from muscle mechanoreceptors. The differential threshold for force perception averages 7%–10% over a force range of 0.5–200 N [26]. Discrimination deteriorates for forces smaller than 0.5 N, with the threshold increasing to 15%–27%. Forces as small as 0.14–0.2 N can still be distinguished.

A proper haptic interface depends on the particular surgical task to be accomplished. There are stylus-type interfaces in which the surgeon grips a scalpel-like protrusion or wearable glove-type interfaces. The haptic device is also characterized by its work space (range of motion that is mechanically allowable by its structural design which should match the human limb), size, encoder resolution, and maximum force. The mechanism should have low inertia, high stiffness, and good kinematic conditioning through the work space: high transmission ratios and kinematic singularities. The latter can be avoided, maximizing manipulability, the mechanism isotropy, and the force output in the worst direction [27].

Rotary optical quadrature encoders are typically used as position sensors on the joints of haptic devices. They are integrated with rotary motors, which serve as actuators. Force sensors are used in haptic devices as the operator input to an admittance-controlled device or as a mechanism for canceling device friction in an impedance-controlled device.

1.2.2.4.2 Master–Slave Control

In the telerobotic system the master side is operated by the human operator and the master movements are replicated on the slave side. If the slave is under position control and if the master and slave kinematics are similar, the kinematic coupling between master and slave is achieved with a straightforward mapping transformation, once a perfect coupling is achieved.

In surgical applications, the master and the slave kinematics are often dissimilar, especially because the master is interfaced to the human user, while the slave has to operate in, e.g., a mini-invasive environment. In case that they also have different dimensions, the work space has to be mapped and scaled, consequently.

Also, the user observes the operating scene on a separate monitor, which replicates the scene observed by an on-field video camera (often 3D). A mapping procedure is, therefore, needed: the slave position and orientation should be measured relative to the camera, while the master position and orientation should be measured relative to the user's view.

In the position-position control architecture, a proportional-derivative (PD) controller is usually implemented:

$$\mathbf{f}_m = -\mathbf{K}_m(\mathbf{x}_m - \mathbf{x}_{md}) - \mathbf{B}_m\left(\dot{\mathbf{x}}_m - \dot{\mathbf{x}}_{md}\right) \tag{1.4}$$

$$\mathbf{f}_s = -\mathbf{K}_s(\mathbf{x}_s - \mathbf{x}_{sd}) - \mathbf{B}_s\left(\dot{\mathbf{x}}_s - \dot{\mathbf{x}}_{sd}\right) \tag{1.5}$$

where $\mathbf{f}_m$ and $\mathbf{f}_s$ are the master and the slave forces, respectively; $\mathbf{x}_m$ and $\mathbf{x}_s$ are the master and the slave positions, respectively; $\mathbf{x}_{md}$ and $\mathbf{x}_{sd}$ are the master and the slave desired positions, respectively; $\dot{\mathbf{x}}_m$ and $\dot{\mathbf{x}}_s$ are the master and slave velocities, respectively; and $\dot{\mathbf{x}}_{md}$ and $\dot{\mathbf{x}}_{sd}$ are the master and slave desired velocities, respectively. $\mathbf{K}_m$ and $\mathbf{K}_s$ and $\mathbf{B}_m$ and $\mathbf{B}_s$ are the position and velocity gains, respectively.

In the position–force architecture, the force measured at the slave side is fed back to the user. In such architecture, a force sensor is fixed at the tip of the slave robot and, therefore,

$$\mathbf{f}_m = \mathbf{f}_{\text{sensor}} \tag{1.6}$$

$$\mathbf{f}_s = -\mathbf{K}_s(\mathbf{x}_s - \mathbf{x}_{sd}) - \mathbf{B}_s\left(\dot{\mathbf{x}}_s - \dot{\mathbf{x}}_{sd}\right) \tag{1.7}$$

where $\mathbf{f}_{\text{sensor}}$ is the sensor-measured force.

Transparency describes how close the user-perceived impedance comes to recreating the true environment impedance. However, this architecture is less stable; there might be lags in slave motion tracking or delay in communication or in the control loop can be very high; e.g., a small motion command can turn into a large force if the slave is pressing against a stiff environment.

A common tool that avoids some stability issues is the concept of passivity, which provides a sufficient (not necessary) condition for stability: a system is stable if energy is dissipated instead of generated; i.e., the output energy from the system is limited by the initial and accumulated energy (i.e., the integral of power over time, where power is computed as velocity times the applied force) in the system. Without the human operator, telerobotic systems are passive if they interact with a passive environment. It can be demonstrated that position-position architecture is inherently passive [28].

A well-known commercial system using master–slave architecture for laparoscopy is the da Vinci® (Intuitive Surgical, Inc., Sunnyvale, California, United States). The master haptic and visual station is shown in Figure 1.6 together with a single-port three-arm slave. Currently the da Vinci has no force feedback on the master.

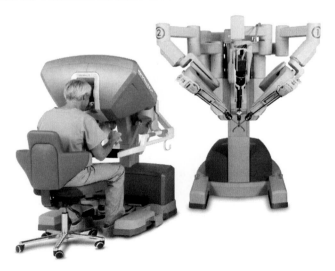

FIGURE 1.6
The Intuitive da Vinci master and slave stations (single port). (Courtesy of Intuitive Surgical, Inc., Sunnyvale, California.)

1.2.2.4.3 Simulators for Training

Surgical simulators are being developed to teach both the behavioral and procedural aspects of medicine and surgery. Procedural simulators stress the cognitive reasoning that goes into successful completion of a surgical intervention, often incorporating physiological response and anatomic findings that can influence a surgeon's intraoperative decision. They can also improve the fidelity of behavioral simulations of tasks and emergencies and provide an objective assessment of trainee performance [29].

Airline pilots train for commercial jet flight on huge, expensive, high-fidelity simulators that recreate the attitude of the plane in flight in response to pilot maneuvers. These high-complexity simulated environments improve the fidelity of behavioral simulations of tasks and emergencies. Simultaneously, simulator performance provides an objective assessment of trainee performance [29]. In the United States, the Accreditation Council for Graduate Medical Education (ACGME) currently requires general surgery training programs to maintain simulation laboratories and many U.S. medical schools are constructing dedicated simulation centers for teaching both behavioral and procedural aspects of medicine and surgery [30]. Procedural simulators stress the cognitive reasoning that goes into the successful completion of a surgical intervention, often incorporating physiological response and anatomic findings that can influence a surgeon's intraoperative decisions. Many players are active in this dynamic market, such as, amongst others, CAE Healthcare, Immersion, Mentice AB, Simbionix, VirtaMed, Surgical Science, and GMV Healthcare. Simulators are designed for a large variation in procedures including epidural procedures, needle biopsy, endoscopy, endovascular procedures, laparoscopic interventions, and colonoscopy.

In the field of neurosurgery, the Dextroscope (Bracco, Princeton, New Jersey) simulator is designed to support surgical evaluation and decision making for different types of interventions, including intracranial aneurysm clipping and temporal bone surgery [31]. The first neurosurgery haptic simulators have focused on navigation for ventriculostomy catheter placement [32]. The NeuroTouch system [33] allows performing soft-tissue manipulation such as tumor debulking and electrocautery. The interface mimics the binocular microscope and provides haptic feedback. Bleeding and even brain pulsation are simulated [34]. Azarnoush et al. [35] presented a pilot study with innovative metrics to assess neurosurgeons' performance using the NeuroTouch platform with simulated brain tumors.

Several different simulators have been developed, including those for ventriculostomy, neuroendoscopic procedures, and spinal surgery, with evidence for improved performance in a range of procedures. Feedback from users has generally been favorable. However, study quality was found to be poor overall, with many studies hampered by nonrandomized design, presenting normal rather than abnormal anatomy, lack of control groups and long-term follow-up, poor study reporting, lack of evidence of improved simulator performance translating into clinical benefit, and poor reliability and validity evidence [36].

1.2.3 Recent Developments

In the field of neurosurgery, the most important commercialization attempt is currently being done by IMRIS (Canada), which is trying to put the neuroArm system [37–38] in the market. The research in the field was started in 2001 by the University of Calgary and MacDonald, Dettwiler and Associates Ltd., which defined the project requirements for a teleoperated system for robotic neurosurgery magnetic resonance imaging (MRI) compatible and agreed on the overall feasibility of the project. The primary neurosurgeon requirement is to have access to sophisticated imaging data, such as intraoperative MRI, without interrupting the surgical procedure. After the manufacture and testing of the first neuroArm prototype, a company was established, neuroArm Surgical, to hold the neuroArm intellectual property (IP) (i.e., to create value for the product through IP protection, paving the way for future commercialization). The system is composed of two MR-compatible robotic arms, seven DoFs, and each titanium joint is equipped with two absolute encoders. Each end effector is equipped with six-DoF force/torque sensors (Nano17, ATI). The system has an open-loop control configuration with piezoelectric motors (1 m/s–200 mm/s of velocity range). By 2013, the system was tested on 35 brain tumor cases.

In other fields, such as laparoscopy, the trend is toward single-port devices to reduce invasiveness and toward natural orifice transluminal endoscopic surgery (NOTES) [39] to completely avoid scars. Other groups are working on devices that can be assembled inside cavities (e.g., stomach or abdomen), where they perform their job, then disassembling and exiting the body, as in the case of Assembling Reconfigurable Endoluminal Surgical System (ARES) project [40]. The ARES robot was designed to self-assemble inside the body after patients swallows up to 15 parts. In using a modular approach, each of these plays roles in image control, communications, structural functions, and diagnostics while forming the structure needed to carry out a particular operation. Each module is around 15 mm in diameter and 36 mm in length, and each represents a single pill to be ingested by the patient. A series of other European Union (EU)–funded project are currently researching the cutting edge of the robotic surgery. Links to the following projects are available on the Active Constraints Technologies for Ill-defined or Volatile Environments (ACTIVE) project website [41]: Accurate Robot Assistant (AccuRobAs), Array of Robots Augmenting the Kinematics of Endoluminal Surgery (ARAKNES), Cognitive Autonomous Catheter Operating in Dynamic Environments (CASCADE), Colonic Disease Investigation by Robotic Hydro-colonoscopy (CODIR), European Robotic Surgery (EuRoSurge), Intelligent Surgical Robotics (I-SUR), Nano-Actuators and Nano-Sensors for Medical Applications (NANOMA), Remote Medical Diagnostician (ReMeDi), Robot and Sensors Integration as Guidance for Enhanced Computer Assisted Surgery and Therapy (ROBOCAST), Patient Safety in Robotic Surgery (SAFROS), Smart Catheterization (SCATh), Stiffness Controllable Flexible and Learnable Manipulator for Surgical Operations (STIFF-FLOP), Micro-Technologies and Systems for Robot-Assisted Laser Phonomicrosurgery (μRALP), and Versatile Endoscopic Capsule for Gastrointestinal Tumor Recognition and Therapy (VECTOR).

1.3 Rehabilitation Robots

1.3.1 Introduction: Why Robots in Rehabilitation?

The ability to deliver high-dosage and high-intensity training is the most relevant advantage of using robots in rehabilitation. Research into rehabilitation robotics has been growing rapidly, and the number of therapeutic rehabilitation robots has increased dramatically during the last two decades [42]. Stroke is a major cause of chronic impaired motor functions and may affect many activities of daily living. Physical and mental training helps in improving the recovery of poststroke patients and reducing their disability. Thus, there exists an urgent need for new inpatient and outpatient rehabilitation and training strategies that match the specific needs. Stroke in the field of therapeutic rehabilitation is generally assumed as the target pathology because of its high prevalence and the good perspectives of recovery, making it the most relevant target for the industrial market. Neuroplasticity is the basic mechanism underlying improvements in functional outcomes after stroke and most movement disorders. Therefore, one of the most important goals of rehabilitation is the effective use of neuroplasticity for functional recovery, leveraging over the intrinsic capability of the nervous system to learn. The principles of therapeutic rehabilitation are goal setting, high-intensity practice, and task-specific training. Overall, high-dose intensive training and repetitive practice of specific functional tasks are the crucial ingredients for neuromotor recovery. Robotic training allows people to practice a task more intensively by making the task safe, allows participants to progress in task difficulty and to achieve the desired movements, and, thus, may serve to motivate repetitive, intensive practice by reconnecting intention to action. These requirements make stroke rehabilitation a labor-intensive process. Robotic therapy needs to control the robots so that the exercises induce and facilitate motor plasticity, in order to improve motor recovery. However, how this goal can best be achieved is still matter of discussion and of scientific research [43].

1.3.2 The Mechanical Design: Exoskeleton versus End-Effector Robots—Some Examples

Two categories of robotic devices are used for rehabilitation: end-effector-type devices and exoskeleton-type devices. End-effector devices apply mechanical forces to the distal segments of the limbs. In the case of gait end-effector robots, the patient's feet are placed on footplates, whose trajectories simulate the stance and swing phases during gait training. They have the advantage of easy setup but suffer from limited control of the proximal joints of the limb, which could result in abnormal movement patterns. On the other hand, exoskeleton-type robotic devices have robot axes aligned with the anatomical axes of the subject. These robots provide direct control of individual joints, which can minimize abnormal posture or movement. Their construction is more complex and more expensive. Gait exoskeletons are outfitted with programmable drives or passive elements, which move the knees and hips during the gait phases over predefined trajectories, set over physiological profiles. However, since the gait pattern that occurs within a robotic orthosis limits the legs and pelvis movements (exoskeleton-assisted approach), it might lead to changes in naturally occurring muscle activation patterns. Examples of end-effector devices assisting gait are the G-EO System [44], the LokoHelp [45], the HapticWalker [46], and the Gait Trainer GT I [47]; end-effector devices for upper limbs are the MIT-Manus (InMotion ARM) and the Reo Go (Figure 1.7). Examples of exoskeletons for gait training are the Lower Extremity Powered Exoskeleton (LOPES) [48] and the Lokomat [49], whereas examples of exoskeletons for upper limbs are the Armeo and the ARMin and for the hand is the Gloreha (Figure 1.8).

1.3.3 The Problem of Control

Control algorithms represent the core of the robot and aim at implementing strategies to provoke plasticity. Marchal-Crespo and Reinkensmeyer [43] have defined four categories of robotic devices depending on the type of control algorithms: assisting, challenge based,

(a) (b) (c)

FIGURE 1.7
Examples of end-effector-based system for rehabilitation: (a) InMotion ARM (Interactive Motion Technologies), (b) InMotion HAND (Interactive Motion Technologies), and (c) Gait Trainer GT I (Reha-Stim). (Courtesy of Interactive Motion Technologies, Watertown, Massachusetts, and Reha-Stim, Berlin, Germany.)

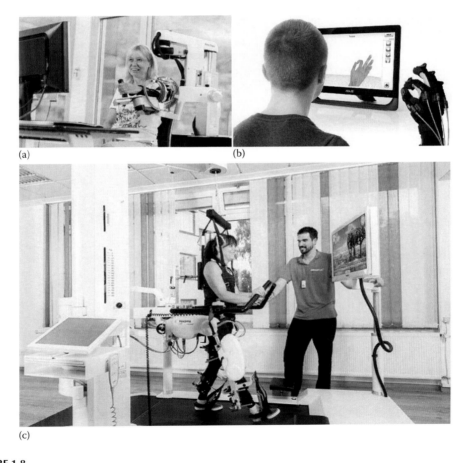

(a) (b)

(c)

FIGURE 1.8
Examples of exoskeleton type devices for rehabilitation: (a) Armeo®Power (picture: Hocoma, Switzerland), (b) Gloreha (picture: Idrogenet, Italy), and (c) Lokomat (picture: Hocoma, Switzerland).

haptic solutions simulating normal tasks, and coaching. Clearly, in some cases multiple strategies could be combined and used in a complementary fashion.

Assistive controllers mimic the "active-assist" exercises performed by rehabilitation therapists and help the patients to move their affected limbs along desired trajectories. The basic concept of assistive controllers is that when the participant moves along the desired trajectory (or inside a deadband around it), the robot just follows the patient in a transparent way, whereas if the patient deviates from the dead band, the robot creates a restoring force, which is generated by mechanical impedance. The rationale for assistive control for motor plasticity is that physically demonstrating the desired movement may help a participant learn how to achieve the pattern; indeed it provides a novel somato-sensory stimulation that helps induce brain plasticity. Further, creating a reference pattern of sensory input may facilitate the motor system in relearning the correspondent sequence of motor output. Repetition of this normal pattern might reinforce it, improving unassisted motor performance. On the other hand, there is also a history of motor control research that suggests that physically guiding a movement may actually decrease motor learning for some tasks [50]. The reason is that the dynamics of the assisted task is not the real target movement, since assistance usually cancels at least part of the effort (for example, to reduce the weight of the limb or of the body to be supported). Guiding the movement also reduces the burden on the learner's motor system to discover the principles necessary to perform the task successfully, and sometimes prevents the subject from exploring strategies, different from the physiological one, which could best suit their status. Guiding movement also appears in some cases to cause people to decrease physical effort during motor training [51]. These findings have been termed "the slacking hypothesis": a robotic device could potentially decrease recovery if it encourages slacking, i.e., a decrease in motor output, effort, energy consumption, and/or attention during training. In order to avoid this attitude, a commonly stated goal in active assist exercise is to provide assistance as needed, which means to assist the participant only as much as is needed to accomplish the task. Exemplary strategies to encourage participants' effort are self-initiated movements or to trigger the assistance only when the subject achieves a force or a velocity threshold, making the robot compliant, or including a forgetting factor in the robotic assistance [52–53]. Providing weight counterbalance to a limb is another assistance strategy that has been developed [54]. In this context, also surface electromyographic signals (sEMGs) have been proposed to trigger the assistance, so to be able to recognize the subject's intention also when the weakness is still very high and does not produce a visible and measurable motion. Alternatively, to assure a complete alignment between the subject's intention and the assisting forces, "proportional myoelectric control" for the arm movements or for walking has been developed [55–56]. With this approach the robot compensates for weakness, generating a force proportional to the EMG signal needed to accomplish the task. However, there are some limitations to the use of EMG signals: sensitivity to electrode placements, muscles crosstalk, etc., which result in the need to recalibrate the EMG parameters for every individual and for each experimental session. In addition, in case of complex tasks, such as upper-limb reaching, the number of EMG channels to fully control the tasks would be very high and the redundancy between muscles makes the control system very challenging. For this reason, only solutions with a reduced number of DoFs and simple tasks have been exploited, making the learned task farther from the natural one, but anyway allowing a good relearning of the control of the limb.

The assistance-as-needed approach is often based on the use of adaptable control parameters, which have the potential advantage that the assistance can be automatically tuned to the participant's individual changing needs, both throughout the movement and over the course of rehabilitation [52,57–58].

Challenge-based controllers can be described as a continuum of the assistive controllers, passing from a compensation strategy to a resistive approach, making the movement tasks more difficult or challenging. Examples include controllers that provide resistance to the participant's limb movements during exercise [59–60], require specific patterns of force generation, or increase the size of movement errors (error-amplification strategies) [61]. The rationale focuses on the idea that bigger errors can be easily captured and more easily corrected as well as that more ample tasks are best controllable, then resizing the level of force is the second stage of learning but the muscular pattern is the same. Haptic simulation refers to the practice of ADL movements in a virtual environment. Haptic simulation has flexibility, convenience, and safety advantages compared to practice in a physical environment [62]. Finally, coaching controllers deal with robots that provide directions to exercises but do not really interfere with the patient's execution of the task.

1.3.4 Impact on Clinical Practice and First Evidence-Based Studies of Rehabilitation Robotics

Conclusive scientific evidence about the benefits of using robot training in rehabilitation has not been reached yet, but some robust studies have been published so far. Since the kind of treatments, the target population, the assisted motion, and the robot-control algorithms are very different from study to study, the reported results cannot be summarized into unique and clear guidelines on the best treatments. Overall, the results are either that robotic training is comparable to conventional training or that it achieves greater or faster improvements. The economic reasons that support the use of robotic training to assure high-dosage treatments to the population that is ever growing in number and disability in the so-called aging society make the use of robotic training a potentially interesting solution in most of the rehabilitation hospitals in Western countries. Nonetheless, it is worthy to provide a brief review over the studies so far published on the clinical impact of rehabilitation robotics.

Considering end-effector gait-training devices, two studies conducted on patients with chronic stroke reported comparable effects on gait function between robot-assisted therapy and conventional gait training [63–64]. On the contrary, five trials which enrolled patients with subacute stroke demonstrated that robot-assisted therapy in combination with conventional physiotherapy produced greater improvement in gait function than conventional gait training alone [65–69]. About exoskeleton devices, results in literature are contrasting. Hidler et al. [70] concluded that the diversity of conventional gait-training interventions appeared to be more effective than robot-assisted gait training for improving walking ability. However, other reports documented similar or superior effects of robot-assisted therapy in combination with conventional physiotherapy versus conventional therapy alone on gait recovery, especially in patients with subacute stroke [71–72]. In 2009, a study recruiting a larger number of participants concluded that locomotor therapy by using robot devices in combination with regular physiotherapy produced promising effects on gait function in patients with subacute stroke in comparison with regular physiotherapy alone [73]. Therefore, robot-assisted therapy with exoskeleton devices may not be able to replace conventional physiotherapy for improving gait function in patients with stroke but rather is recommended for use in combination with conventional physiotherapy, preferably in the subacute stage. However, there is insufficient research on the additional effect of robot-assisted therapy on gait function in the chronic stage of stroke.

Regarding upper-limb robotic training, a Cochrane review including 19 trials (enrolling a total of 666 participants, poststroke patients) was published in 2012 [74]. Their main results were that robot-assisted arm training did improve activities of daily living (standardized mean difference [SMD] 0.43, 95% confidence interval [CI] 0.11 to 0.75, $P = 0.009$, $I^2 = 67\%$) as well as arm function (SMD 0.45, 95% CI 0.20 to 0.69, $P = 0.0004$, $I^2 = 45\%$), but arm muscle strength did not improve (SMD 0.48, 95% CI −0.06 to 1.03, $P = 0.08$, $I^2 = 79\%$). Electromechanical and robot-assisted arm training did not increase the risk of patients dropping out (risk difference [RD] 0.00, 95% CI −0.04 to 0.04, $P = 0.82$, $I^2 = 0.0\%$), and adverse events were rare. However, the results must be interpreted with caution because there were variations between the trials in the duration and amount of training, type of treatment, and the characteristics of the patients. The most relevant study in terms of population size in the literature was conducted by Lo et al. and involved 127 chronic stroke patients [75]. They observed that robot-assisted therapy and conventional therapy produced similar amounts of improvement after 12 weeks of treatment. However, after 36 weeks of therapy, the robot-assisted therapy achieved greater motor improvement than did conventional therapy.

1.3.5 Perspectives and Challenges

Overall, robotic technology has entered the clinical practice in rehabilitation; the major reason is that in an aging society the labor-intensive practice required for rehabilitation needs a one-to-one approach in the conventional therapy approach and this strongly limits the capability to deliver high-dosage and high-intensity treatments, which are absolutely recognized as the major positive elements for the recovery process. The possibility to continuously refine the robotic control algorithms so as to manage best the exercises in order to facilitate neuroplasticity and favor the process of relearning of the nervous system is the most relevant challenge in the design of robotic devices. In this view, one very interesting perspective that has started to be explored is the combination between robotic assistance and neuromuscular electrical stimulation. In induced artificial muscle contraction, possibly the residual capability of the patients and robotic assistance cooperate with each other. Once these elements are harmonized, the task becomes safe, immersive, goal oriented, and under the volitional control of the subject, providing not only a sensory normative reference but also a muscular reference pattern to the brain of the patients, combining the best blend for motor recovery. Unfortunately, hybrid systems are still not easy to set up and operate, and the long procedures are still the major barrier which prevents translation to clinical practice [76–78].

The other major perspective of robotic treatments in therapeutic rehabilitation is in the direction of home-based treatments with remote control. Of course, it is well known that especially poststroke patients would benefit from intensive training throughout their life and especially during the first months after the brain lesion. However, economical sustainability limits the duration of hospital admission and also the outpatient treatments are limited in time and often reduced by the consequent burden on the families. The possibility of training family members for home rehabilitation is a major challenge for the future. The use of robotic technology in this perspective would assure the possibility of safe and controlled training with limited supervision, possibly without any formal caregivers, depending on the patient's status. Some research is currently ongoing at the border between assistive technology and therapeutic robotic devices, but still few examples are available and mostly at the research prototype level [79].

1.4 Assistive Robots

1.4.1 Introduction

The International Classification of Functioning, Disability and Health defines assistive technology as "any product, instrument, equipment or technology adapted or specially designed for improving the functioning of a disabled person." The recent advances in the robotics field have also influenced assistive technology. Several robots that assist individuals with disabilities have been developed and some of them are becoming commercially available. This trend is fostered by the rapid aging of the population, which results in a larger number of elderly and chronically impaired subjects, who require social and health-care services, as well as in a smaller number of informal caregivers (i.e., relatives) and workers to financially sustain health and social services.

Assistive robots (ARs) are becoming more affordable, functional, and aesthetically pleasing systems; they enhance mobility, independence in ADLs, vocational tasks, communication and cognitive abilities, and social integration and participation, improving self-image and life satisfaction [80]. However, for a system to be accepted and utilized it is crucial to meet the unique needs of each single user [81]. For this reason, on one hand the users' perspective has been increasingly included in the technology design, and on the other hand it is important for the healthcare providers to be aware of the available systems and to avoid improper device prescription.

ARs can be divided into the following [80]:

- *Physical assistance robots* (PARs), i.e., contact assistive robots that provide physical or mechanical assistance in order to help the user perform specific tasks
- *Socially assistive robots* (SARs), i.e., contactless assistive robots that are perceived as social entities which communicate with the user, providing physiological and cognitive support

Since this chapter deals with medical robotics, in what follows we will focus only on PARs.

1.4.2 Physical Assistance Robots

PARs are designed to increase independence and function in physical tasks. PARs can be classified into the following:

- *Mobility robotic aids*, such as powered and/or smart wheelchairs, lower-limb exoskeletons, and active orthoses and prostheses, which improve the mobility level of the user
- *Robots for ADL support*, which include systems that support the user in performing upper-limb functional tasks (e.g., wheelchair-mounted robot arms, upper-limb prostheses, and exoskeletons), feeders, and vocational robotic workstations

1.4.3 Mobility Aids

Independent mobility is an important aspect of self-esteem and quality of life [82]. Many individuals with disabilities can take advantage of manual or powered wheelchairs, but some of them require more advanced systems to regain independent mobility. These individuals might benefit from smart wheelchairs, defined as "either a standard powered wheelchair to which a computer and a collection of sensors have been added or a mobile robot base to which a seat has been attached" [82]. An alternative to the traditional input method associated with the powered wheelchair (e.g., joysticks), smart wheelchairs implement more sophisticated user interfaces, such as voice recognition [83–84] or brain–computer interface [85–86]. They integrate different sensors, such as sonar, infrared sensors, laser range finders, or cameras, based on which obstacle avoidance algorithms are designed. Finally, they implement advanced control algorithms: they can work either in an automatic manner, i.e., the user provides the final destination and the wheelchair plans and executes a path to the target location, or in a cooperative manner, i.e., the user plans and controls the navigation and the wheelchair assists to assure obstacle avoidance. An example of a smart wheelchair is the Let Unleashed Robots Crawl the House (LURCH) system designed at Politecnico di Milano (Figure 1.9) [87–88]. In LURCH, the user can choose among several autonomy levels, ranging from simple obstacle avoidance to full autonomy, and different interfaces: a classical joystick, a touch screen, an EMG interface, and a BCI.

Other examples of mobility robotic aids are lower-limb exoskeletons or active orthoses. An *exoskeleton* is defined as "an active mechanical device that is essentially anthropomorphic in nature, is 'worn' by an operator and fits closely to his or her body, and works in concert with the operator's movements" [3,89]. As previously reported, research in powered exoskeletons began in the late 1960s with a twofold objective: (i) to augment the capabilities of able-bodied subjects (e.g., for military purposes) and (ii) to assist persons with disabilities. Focusing on the second objective, one of the first examples of powered

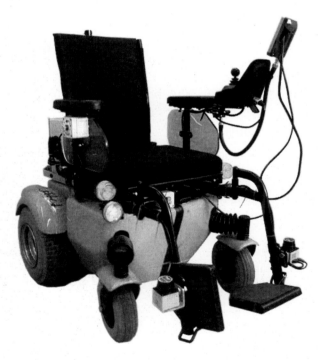

FIGURE 1.9
The LURCH prototype with mounted sensors. (Courtesy of Andrea Bonarini and Matteo Matteucci.)

exoskeletons was developed by Vukobratovic et al. in the 1970s [90]. This system incorporated pneumatic actuators for flexion/extension of hip, knee, and ankle; an actuated abduction/adduction joint at the hip for greater stability on the frontal plane; and a corset which enclosed the entire chest of the patient for providing trunk support. Favoring the clinical use of exoskeletons, portability, which requires energetic autonomy, reduced weight, and low encumbrance, is crucial. Nowadays, there are some commercially available lower-limb exoskeletons. ReWalk from ReWalk Robotics Ltd., for example, provides powered hip and knee motion to enable individuals with spinal cord injury to stand up, walk, and climb stairs (Figure 1.10a) [91]. The system is powered by a backpack battery and includes a motion–tilt sensor which detects shifts in body weight and balance and propels the machine forward when appropriate. Another example of commercially available exoskeleton is Ekso™ from Ekso Bionics™ (Figure 1.10b) [92]. Lower-limb exoskeletons were primarily thought of as rehabilitative devices for paraplegics, but some of them (i.e., ReWalk Personal [93]) are now becoming available for everyday use, offering some practical benefits which are difficult to be achieved in a wheelchair, such as speaking to colleagues face to face or doing exercises.

Lower-limb-powered prostheses are another example of mobility robotic aids. In the past 30 years, the rapid advances in prosthetic technology resulted in a number of devices that improve the functional mobility and quality of life in individuals with lower-limb amputations [94]. Focusing on above-knee amputation, knee prosthesis design varies from the very simple single-axis knee to high-tech microprocessor-controlled prosthetic knees, such as the C-Leg from Ottobock [95]. One of the most important advancements in the

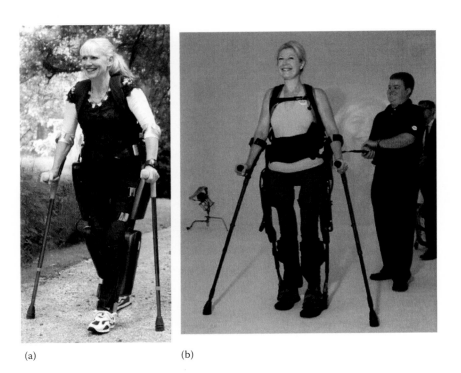

(a)   (b)

FIGURE 1.10
Examples of lower-limb-powered exoskeletons: (a) ReWalk (picture: ReWalk Robotics Inc) and (b) Ekso (picture: Ekso Bionics).

knee prosthesis design is the control of the swing phase. Initially, this control was assured using a friction-based system in which an adjustable friction cell was pressed against the knee axle. Later, fluid swing-phase control mechanisms, either pneumatic or hydraulic, were introduced. These mechanisms are more flexible since they function over a range of gait speeds [96].

1.4.4 Activity of Daily Living Support

Many robotic feeding systems are commercially available and used in clinical and home settings to support the user while eating. Some examples are the Mealtime Partners Assistive Dining Device [97], which is equipped with rotating bowls, a mechanical spoon, and a positioning arm; the Neater Eater [98], which consists of a two-DoF arm and one dish; and the SECOM My Spoon system [99], which consists of a five-DoF manipulator, a gripper, and a meal tray.

Manipulator robotic aids assist disabled people in performing activities of daily life. Manus, currently produced by Exact Dynamics as iARM, is one of the most widespread wheelchair-mounted robotic arms [100]. It has six DoFs plus a gripper as end effector (Figure 1.11a). The wheelchair joystick and an extra switch are used as user interface and the robot can be operated in joint mode (each joint is controlled separately) or in Cartesian control mode (controlling the position of the end effector). Although the device is very helpful, the control system has been found very challenging. To increase the number of

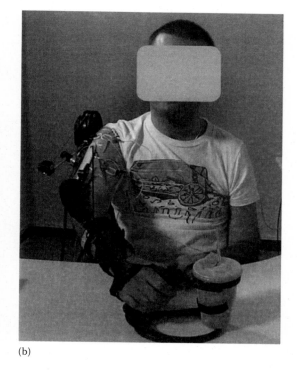

(a)   (b)

FIGURE 1.11
Examples of manipulation robots that support daily life activities: (a) iARM (picture: Exact Dynamics) and (b) the passive exoskeleton developed in the Multimodal Neuroprosthesis for Daily Upper Limb Support (MUNDUS) project. Surface electrodes for stimulation of the biceps and shoulder muscles are also shown.

potential users, several research groups have been adding sensors (e.g., cameras and sonar sensors) and developing advanced control algorithms, such as computer vision algorithm or path-planning algorithm [80].

Another example of wheelchair-mounted robotic arm is the KAIST (Korea Advanced Institute of Science and Technology) Rehabilitation Engineering Service System II (KARES II) [101]. KARES II is a six-DoF robotic arm and can support 12 predefined tasks, such as eating, drinking, shaving, opening/closing doors, picking up objects, or turning switches on and off. It provides various kinds of human–robot interfaces, such as eye–mouse; shoulder/head interface; and EMG signal-based control, which can be selected according to the level of disability. Among wheelchair-mounted manipulators, JACO is a lightweight robotic arm developed by Kinova, which consists of six interlinked carbon fiber segments joined to a three-fingered hand [102]. It is characterized by seven DoFs which allow six movements in three-dimensional space, six movements of JACO's wrist (abduction, adduction, flexion, extension, pronation, and supination), and opening and closing of the three fingers. A three-axis joystick is used for control. To provide an efficient, hands-free control modality, some researchers have recently developed an upper-limb gesture-recognition system combined with object tracking and face-recognition systems as substitute for the joystick [103]. Among the different user interfaces, one of the most attractive is the possibility of using neural interface systems to translate neuronal activity directly into control signals for assistive devices. After experiments performed on monkeys [104], in a paper published in *Nature*, the authors showed the ability of two people with long-standing tetraplegia to perform three-dimensional reach and grasp movements with a robotic arm controlled by brain activities [105].

To support upper-limb functions, passive or powered exoskeletons have also been proposed. Passive exoskeletons use spring and/or damper elements to compensate for the arm's weight, assisting the users in performing functional tasks with their own residual muscle activity [106–107]. In the European project MUNDUS, a lightweight spring-loaded robotic exoskeleton arm has been integrated with functional electrical stimulation to provide additional support in the case of users with weak or no residual volitional muscle activity (Figure 1.11b) [79]. An alternative solution to providing gravity compensation consists of the application of electrorheological fluids as controlled resistive torque elements to allow coordinated joint damping or locking/unlocking of the DoF [108]. A number of powered exoskeletons for the upper extremity have been also developed (e.g., MIT-MANUS [109], ArmeoPower from Hocoma [110], and ARMin [111]). However, these systems are heavy, cumbersome, and expensive, thereby being mainly used as stationary rehabilitative systems in a clinical environment rather than as assistive devices at home. Surveys on the usage of assistive robotic arms have shown that the users prefer mobile devices capable of functioning in a variety of unstructured environments, instead of fixed workstations [112]. For this reason, wheelchair robotic arms are usually preferred.

Other examples of robots for ADL support are powered prostheses for helping upper-limb amputees to perform ADLs. The majority of the robotic prosthetic arms are myocontrol systems in which the electrical activity naturally generated by a contracting muscle in the residual part of the limb is used to control the operation of the artificial limb [113]. Although myoelectric prostheses were first designed in the late 1940s [114] and have been clinically implemented since the 1960s [115], they still hardly substitute for the functionality of a missing arm or hand [116]. The prosthetic hardware has advanced more than the control system. Devices with multiple DoFs and equipped with different sensors have been developed [117–119]. However, patients are usually able to control only one joint or function at a given time and the difficulty in controlling prosthesis is widely considered one of the key reasons for rejection [120]. Ideally, a prosthesis should reproduce the bidirectional link between the user's nervous system and the environment by exploiting the neural pathways that persist after amputation. The possibility to restore the sensory feedback is mainly crucial for hand prostheses. Transversal intrafascicular multichannel electrodes connected to artificial hand sensors have been used to restore touch sensation in a person with hand amputation [121]. However, the system was tested on a single subject, and more extensive evaluations are needed to prove its functioning.

Finally, among robots for ADL support we also include vocational robotic workstations, such as the Desktop Vocational Assistant Robot (DeVAR), which is a voice-controlled robotic arm mounted on an overhead track system to keep the desk surface available for job-related objects and equipment. This system was designed to allow people with high-cervical spinal cord injuries to function more independently in a work environment [122]. Some other examples of robotic workstations can be found but they have been used with success only in a limited number of cases due to their bulkiness, high cost, and reduced number of supported tasks.

1.4.5 Future Perspectives

Research in the AR field has been growing rapidly. Although sophisticated prototypes that have multiple DoFs and exploit advanced user interfaces have been proposed in research laboratories, the commercially available systems are much simpler and the number of users who benefit from ARs in their daily life is still very low. Some issues need to be tackled to promote the use of ARs. First of all, human safety during the interaction with a robot needs to be improved [123]. Mechanical or electrical constraints (e.g., speed or force limits) or back-drivable joints can be integrated into the system to enhance the level of safety. To increase user acceptance, the control needs to be simplified and a natural interaction should be assured. The recent advances in neural interface systems could play a crucial role to achieve this aim. Another desirable feature is the possibility of customizing the system to the needs of each individual user. Some users, in the absence of commercially available systems, have developed their own device to meet their specific needs, underlying the importance of personalization. In the design process, it is important to learn from the users' experiences [124]. Independent operation has been defined as one of the most important design criteria [81]. ARs need to promote participation within families, communities, and society. In the selection of the ideal assistive system, the users balance the stigma they perceive to be associated with a device against the function and autonomy achieved with that device. Based on these considerations, a new definition of *assistive technology* has been suggested: "devices or adaptations that serve as an interface between people with disabilities and their unique environments, used to engage in activities of choice, and to promote self-perceived participation" [124]. This new definition might guide the design of future ARs.

In parallel with the continuous development of innovative robotic systems, in order to promote their use it is also crucial to provide an extensive evaluation of efficacy, safety, reliability, and users' and families' acceptance of the developed systems. The MUNDUS system was evaluated on six end users affected by neuromotor disorders (three spinal cord injuries, one multiple sclerosis, and two Friedreich's ataxia), and a high level of usability, user satisfaction, and motor performance were observed [125]. In the future, more trials conducted in natural environments are needed. Last but not least, to increase accessibility both investment and maintenance costs need to be reduced.

References

1. Kwoh YS, Reed IS, Chen JY, Shao H, Truong TK, and Jonckheere EA, A new computerized tomographic aided robotics stereotactic system, *Robot Age*, vol. 7, pp. 17–21, 1985.
2. Kwoh YS, Hou J, Jonckheere EA, and Hayati S, A robot with improved absolute positioning accuracy for CT guided stereotactic brain surgery, *IEEE Transactions on Biomedical Engineering*, vol. 35:2, pp. 153–160, 1988.
3. Dollar AM and Herr H, Lower extremity exoskeletons and active orthoses: Challenges and state-of-the-art, *IEEE Transactions on Robotics*, vol. 24:1, pp. 144–158, 2008.
4. Denavit J and Hartenberg RS, A kinematic notation for lower-pair mechanisms based on matrices, *Transactions of the ASME, Journal of Applied Mechanics*, vol. 23, pp. 215–221, 1955.
5. Taylor RH, Joskowicz L, Williamson B, Guéziec A, Kalvin A, Kazanzides P, Van Vorhis R, Yao J, Kumar R, Bzostek A, Sahay A, Börner M, and Lahmer A, Computer-integrated revision total hip replacement surgery: Concept and preliminary results, *Medical Image Analysis*, vol. 3:3, pp. 301–319, 1999.
6. Conditt MA and Roche MW, Minimally invasive robotic-arm-guided unicompartmental knee arthroplasty, *Journal of Bone and Joint Surgery, American volume*, vol. 91:suppl. 1, pp. 63–68, 2009.
7. Reed KB, Majewicz A, Kallem V, Alterovitz R, Goldberg K, Cowan NJ, and Okamura AM, Robot-assisted needle steering, *IEEE Robotics and Automation Magazine*, vol. 18, pp. 35–46, 2011.
8. Engh JA, Minhas DS, Kondziolka D, and Riviere CN, Percutaneous intracerebral navigation by duty-cycled spinning of flexible bevel-tipped needles, *Neurosurgery*, vol. 67:4, pp. 1117–1122, 2010.
9. Ko SY and Rodriguez y Baena F, Toward a miniaturized needle steering system with path planning for obstacle avoidance, *IEEE Transactions on Biomedical Engineering*, vol. 60, pp. 910–917, 2013.
10. Noonan DP, Vitiello V, Shang J, Payne CJ, and Yang GZ, A modular, mechatronic joint design for a flexible access platform for MIS, *2011 IEEE/RSJ International Conference on Intelligent Robots and Systems*, September 25–30, 2011, San Francisco, California, United States, pp. 949–954, 2011.
11. Rucker DC, Jones BA, and Webster RJ III, A geometrically exact model for externally loaded concentric tube continuum robots, *IEEE Transactions on Robotics*, vol. 26:5, pp. 769–780, 2010.
12. Patronik NA, Ota T, Zenati MA, and Riviere CN, Synchronization of epicardial crawling robot with heartbeat and respiration for improved safety and efficiency of locomotion, *International Journal of Medical Robotics and Computer Assisted Surgery*, vol. 8:1, pp. 97–106, 2012.
13. Hogan N, Impedance control: An approach to manipulation: Parts I–III, *ASME Journal of Dynamic Systems, Measurement, and Control*, vol. 107, pp. 1–24, 1985.
14. Lawrence DA, Impedance control stability properties in common implementation, *Proceedings of the IEEE International Conference on Robotics and Automation (ICRA)*, pp. 1185–1190, 1988.
15. Villani L and De Schutter J, Force Control, in *Springer Handbook of Robotics*, Siciliano B and Kathib O, eds., pp. 161–185, Springer, 2008.
16. Zeng G and Hemami A, An overview of robot force control, *Robotica*, vol. 15:5, pp. 473–482, 1997.
17. Rosemberg LB, Virtual fixtures: Perceptual tools for telerobotic manipulation, *IEEE Virtual Reality Annual International Symposium* 1993, pp. 76–82, 1993.
18. Bowyer SA, Davies BL, and Rodriguez y Baena F, Active constraints/virtual fixtures: A survey, *IEEE Transactions on Robotics*, vol. 30:1, pp. 138–157, 2013.
19. Gibo TL, Verner LN, Yuh DD, and Okamura AM, Design considerations and human–machine performance of moving virtual fixtures, *IEEE International Conference on Robotics and Automation*, pp. 671–676, 2009.
20. Gonenc B, Balicki MA, Handa J, Gehlbach P, Riviere CN, Taylor RH, and Iordachita I, Preliminary evaluation of a micro-force sensing handheld robot for vitreoretinal surgery, *2012 IEEE/RSJ International Conference on Intelligent Robots and Systems*, pp. 4125–4130, October 7–12, 2012, doi: 10.1109/IROS.2012.6385715, 2012.

21. MacLachlan RA, Becker BC, Tabarés JC, Podnar GW, Lobes LA Jr, and Riviere CN, Micron: An actively stabilized handheld tool for microsurgery, *IEEE Transactions on Robotics*, vol. 28:1, pp. 195–212, 2012.

22. Yang S, MacLachlan RA, and Riviere CN, Manipulator design and operation for a six-degree-of-freedom handheld tremor-cancelling microsurgical instrument, *IEEE/ASME Transactions on Mechatronics*, vol. 20:2, pp. 761–772, 2015.

23. Jarrassé N, Paik J, Pasqui V, and Morel G, How can human motion prediction increase transparency? *Proceedings of IEEE International Conference on Robotics and Automation*, pp. 2134–2139, 2008.

24. De Lorenzo D, Koseki Y, De Momi E, Chinzei K, and Okamura AM, Coaxial needle insertion assistant with enhanced force feedback, *IEEE Transactions on Biomedical Engineering*, vol. 60:2, pp. 379–389, 2013.

25. Hale KS and Stanney KM, Deriving haptic design guidelines from human physiological, psychophysical, and neurological foundations, *IEEE Computer Graphics and Applications*, vol. 24:2, pp. 3–39, 2004.

26. Jones LA, *Kinesthetic Sensing, Human and Machine Haptics*, MIT Press, 2000.

27. Hannaford B and Okamura AM, Haptics, in *Springer Handbook of Robotics*, Siciliano B and Kathib O, eds., pp. 719–739, Springer, 2008.

28. Niemeyer G, Preusche C, and Hirzinger G, Telerobotics, in *Springer Handbook of Robotics*, Siciliano B and Kathib O, eds., pp. 741–757, Springer, 2008.

29. Selden NR, Origitano TC, Hadjipanayis C, and Byrne R, Model-based simulation for early neurosurgical learners, *Neurosurgery*, vol. 73:4, pp. S15–S24, 2013.

30. Danzer E, Dumon K, Kolb G et al., What is the cost associated with the implementation and maintenance of an ACS/APDS-based surgical skills curriculum? *Journal of Surgical Education*, vol. 68:6, pp. 519–525, 2011.

31. Kockro RA and Hwang PY, Virtual temporal bone: An interactive 3-dimensional learning aid for cranial base surgery, *Neurosurgery*, vol. 64:5, pp. 216–229, 2009.

32. Lemole GMJ, Banerjee PP, Luciano C, Neckrysh S, and Charbel FT, Virtual reality in neurosurgical education: Part-task ventriculostomy simulation with dynamic visual and haptic feedback, *Neurosurgery*, vol. 61:1, pp. 142–149, 2007.

33. Delorme S, Laroche D, Diraddo R, and Del Maestro FR, Neurotouch: A physics-based virtual simulator for cranial neurosurgery training, *Neurosurgery*, vol. 71, pp. S32–S42, 2012.

34. Chan S, Conti F, Salisbury K, and Blevins NH, Virtual reality simulation in neurosurgery: Technologies and evolution, *Neurosurgery*, vol. 72:suppl. 1, pp. A154–A164, 2013.

35. Azarnoush H, Alzhrani G, Winkler-Schwartz A, Alotaibi F, Gelinas-Phaneuf N, Pazos V, Choudhury N, Fares J, DiRaddo R, and Del Maestro RF, Neurosurgical virtual reality simulation metrics to assess psychomotor skills during brain tumor resection, *International Journal of Computer Assisted Radiology and Surgery*, vol. 10:5, pp. 603–618, 2015.

36. Kirkman MA, Ahmed M, Albert AF, Wilson MH, Nandi D, and Sevdalis N, The use of simulation in neurosurgical education and training, *Journal of Neurosurgery*, vol. 20, pp. 1–19, 2014.

37. Sutherland GR, Latour I, and Greer AD, Integrating an image-guided robot with intraoperative MRI: A review of the design and construction of neuroArm, *IEEE Engineering in Medicine and Biology Magazine*, vol. 7:3, pp. 59–65, 2008.

38. Sutherland GR, Wolfsberger S, Lama S, and Zareinia K, The evolution of neuroArm, *Neurosurgery*, vol. 72:suppl. 1, pp. 27–32, 2013.

39. Swanström LL, Khajanchee Y, and Abbas MA, Natural orifice transluminal endoscopic surgery: The future of gastrointestinal surgery, *The Permanente Journal*, vol. 12:2, pp. 42–47, 2008.

40. http://ec.europa.eu/research/fp6/nest/pdf/projects/ares.pdf, accessed on August 29, 2014.

41. http://www.active-fp7.eu/index.php/related-eu-projects/39-relprojects, accessed on August 29, 2014.

42. Chang WH and Kim YH, Robot-assisted therapy in stroke rehabilitation, *Journal of Stroke*, vol. 15:3, pp. 174–81, 2013.

43. Marchal-Crespo L and Reinkensmeyer DJ, Review of control strategies for robotic movement training after neurologic injury, *Journal of NeuroEngineering and Rehabilitation*, vol. 6:20, 2009.

44. Hesse S, Waldner A, and Tomelleri C, Innovative gait robot for the repetitive practice of floor walking and stair climbing up and down in stroke patients, *Journal of NeuroEngineering and Rehabilitation*, vol. 7:30, 2010.

45. Freivogel S, Schmalohr D, and Mehrholz J, Improved walking ability and reduced therapeutic stress with an electromechanical gait device, *Journal of Rehabilitation Medicine*, vol. 41, pp. 734–739, 2009.

46. Schmidt H, Hesse S, Bernhardt R, and Krüger J, HapticWalker—A novel haptic foot device, *ACM Transactions on Applied Perception*, vol. 2, pp. 166–180, 2005.

47. Hesse S, Sarkodie-Gyan T, and Uhlenbrock D, Development of an advanced mechanised gait trainer, controlling movement of the centre of mass, for restoring gait in non-ambulant subjects, *Biomedizinische Technik (Berlin)*, vol. 44, pp. 194–201, 1999.

48. Veneman J, Kruidhof R, van der Helm FCT, and van der Kooy H, Design of a series elastic- and Bowden cable-based actuation system for use as torque-actuator in exoskeleton-type training robots, *Proceedings of the ICOOR*, Chicago, United States, June 28–30, 2005.

49. Colombo G, Joerg M, Schreier R, and Dietz V, Treadmill training of paraplegic patients using a robotic orthosis, *Journal of Rehabilitation and Research Development*, vol. 37, pp. 693–700, 2000.

50. Marchal-Crespo L and Reinkensmeyer DJ, Haptic guidance can enhance motor learning of a steering tasks, *Journal of Motor Behaviour*, vol. 40:6, pp. 545–557, 2008.

51. Israel JF, Campbell DD, Kahn JH, and Hornby TG, Metabolic costs and muscle activity patterns during robotic- and therapist-assisted treadmill walking in individuals with incomplete spinal cord injury, *Physical Therapy*, vol. 86:11, pp. 1466–1478, 2006.

52. Duschau-Wicke A, von Zitzewitz J, Caprez A, Lünburger L, and Riener R, Path control: A method for patient-cooperative robot-aided gait rehabilitation, *IEEE Transactions on Neural Systems and Rehabilitation Engineering*, vol. 18:1, pp. 38–48, 2010.

53. Cai LL, Fong AJ, Otoshi CK, Liang Y, Burdick JW, Roy RR, and Edgerton VR, Implications of assist-as-needed robotic step training after a complete spinal cord injury on intrinsic strategies of motor learning, *Journal of Neuroscience*, vol. 26:41, pp. 10564–10568, 2006.

54. Sanchez RJ, Liu J, Rao S, Shah P, Smith R, Cramer SC, Bobrow JE, and Reinkensmeyer DJ, Automating arm movement training following severe stroke: Functional exercises with quantitative feedback in a gravity-reduced environment, *IEEE Transactions on Neural and Rehabilitation Engineering*, vol. 14:3, pp. 378–389, 2006.

55. Song R, Tong KY, Hu X, and Li L, Assistive control system using continuous myoelectric signal in robot-aided arm training for patients after stroke, *IEEE Transactions on Neural Systems and Rehabilitation Engineering*, vol. 16:4, pp. 371–379, 2008.

56. Ferris DP, Czerniecki JM, and Hannaford B, An ankle-foot orthosis powered by artificial pneumatic muscles, *Journal of Applied Biomechanics*, vol. 21:2, pp. 189–197, 2005.

57. Krebs HI, Palazzolo JJ, Dipietro L, Ferraro M, Krol J, Rannekleiv K, Volpe BT, and Hogan N, Rehabilitation robotics: Performance based progressive robot-assisted therapy, *Autonomous Robots*, vol. 15, pp. 7–20, 2003.

58. Wolbrecht ET, Chan V, Reinkensmeyer D, and Bobrow JE, Optimizing compliant, model-based robotic assistance to promote neurorehabilitation, *IEEE Transactions on Neural Systems and Rehabilitation Engineering*, vol. 16:3, pp. 286–297, 2008.

59. Lum PS, Burgar CG, Shor PC, Majmundar M, and Van der Loos M, Robot assisted movement training compared with conventional therapy techniques for the rehabilitation of upper-limb motor function after stroke, *Archives of Physical Medicine and Rehabilitation*, vol. 83:7, pp. 952–959, 2002.

60. Casellato C, Pedrocchi A, Zorzi G, Rizzi G, Ferrigno G, and Nardocci N, Error-enhancing robot therapy to induce motor control improvement in childhood onset primary dystonia, *Journal NeuroEngineering and Rehabilitation*, vol. 9:46, 2012.

61. Emken JL and Reinkensmeyer DJ, Robot-enhanced motor learning: Accelerating internal model formation during locomotion by transient dynamic amplification, *IEEE Transactions on Neural Systems and Rehabilitation Engineering*, vol. 13, pp. 33–39, 2005.

62. Casellato C, Pedrocchi A, Zorzi G, Vernisse L, Ferrigno G, and Nardocci N, EMG-based visual-haptic biofeedback: A tool to improve motor control in children with primary dystonia, *IEEE Transactions on Neural Systems and Rehabilitation Engineering*, vol. 21:3, pp. 474–80, 2013.

63. Peurala SH, Tarkka IM, Pitkanen K, and Sivenius J, The effectiveness of body weight-supported gait training and floor walking in patients with chronic stroke, *Archives of Physical Medical Rehabilitation*, vol. 86, pp. 1557–1564, 2005.

64. Dias D, Laíns J, Pereira A, Nunes R, Caldas J, Amaral C, Pires S, Costa A, Alves P, Moreira M, Garrido N, and Loureiro L, Can we improve gait skills in chronic hemiplegics? A randomised control trial with gait trainer, *Europa Medicophysica*, vol. 43, pp. 499–504, 2007.

65. Morone G, Bragoni M, Iosa M, De Angelis D, Venturiero V, Coiro P, Pratesi L, and Paolucci S, Who may benefit from robotic-assisted gait training? A randomized clinical trial in patients with subacute stroke, *Neurorehabilitation Neural Repair*, vol. 25, pp. 636–644, 2011.

66. Peurala SH, Airaksinen O, Huuskonen P, Jakala P, Juhakoski M, Sandell K, Tarkka IM, and Sivenius J, Effects of intensive therapy using gait trainer or floor walking exercises early after stroke, *Journal of Rehabilitation Medicine*, vol. 41, pp. 166–173, 2009.

67. Pohl M, Werner C, Holzgraefe M, Kroczek G, Mehrholz J, Wingendorf I, Hoölig G, Koch R, and Hesse S, Repetitive locomotor training and physiotherapy improve walking and basic activities of daily living after stroke: A single-blind, randomized multicentre trial (DEutsche GAngtrainerStudie, DEGAS), *Clinical Rehabilitation*, vol. 21, pp. 17–27, 2007.

68. Tong RK, Ng MF, and Li LS. Effectiveness of gait training using an electromechanical gait trainer, with and without functional electric stimulation, in subacute stroke: A randomized controlled trial, *Archives of Physical Medicine and Rehabilitation*, vol. 87, pp. 1298–1304, 2006.

69. Werner C, Von Frankenberg S, Treig T, Konrad M, and Hesse S, Treadmill training with partial body weight support and an electromechanical gait trainer for restoration of gait in subacute stroke patients: A randomized crossover study, *Stroke*, vol. 33, pp. 2895–2901, 2002.

70. Hidler J, Nichols D, Pelliccio M, Brady K, Campbell DD, Kahn JH, and Hornby TG, Multicenter randomized clinical trial evaluating the effectiveness of the Lokomat in subacute stroke, *Neurorehabilitation Neural Repair*, vol. 23, pp. 5–13, 2009.

71. Chang WH, Kim MS, Huh JP, Lee PK, and Kim YH, Effects of robot-assisted gait training on cardiopulmonary fitness in subacute stroke patients: A randomized controlled study, *Neurorehabilitation Neural Repair*, vol. 26, pp. 318–324, 2012.

72. Westlake KP and Patten C, Pilot study of Lokomat versus manual-assisted treadmill training for locomotor recovery post-stroke, *Journal of NeuroEngineering Rehabilitation*, vol. 6:18, 2009.

73. Schwartz I, Sajin A, Fisher I, Neeb M, Shochina M, Katz-Leurer M, and Meiner Z, The effectiveness of locomotor therapy using robotic-assisted gait training in subacute stroke patients: A randomized controlled trial, *PM&R*, vol. 1, pp. 516–523, 2009.

74. Mehrholz J, Hadrich A, Platz T, Kugler J, and Pohl M, Electromechanical and robot-assisted arm training for improving generic activities of daily living, arm function, and arm muscle strength after stroke, *Cochrane Database of Systematic Reviews*, no. 6:CD006876, 2012.

75. Lo AC, Guarino PD, Richards LG, Haselkorn JK, Wittenberg GF, Federman DG, Ringer RJ, Wagner TH, Krebs HI, Volpe BT, Bever CT Jr, Bravata DM, Duncan PW, Corn BH, Maffucci AD, Nadeau SE, Conroy SS, Powell JM, Huang GD, and Peduzzi P, Robot-assisted therapy for long-term upper-limb impairment after stroke, *New England Journal of Medicine*, vol. 362, pp. 1772–1783, 2010.

76. Braz GP, Russold M, and Davis GM. Functional electrical stimulation control of standing and stepping after spinal cord injury: A review of technical characteristics. *Neuromodulation*, vol. 12:3, pp. 180–190, July 2009.

77. Ambrosini E, Ferrante S, Schauer T, Ferrigno G, Molteni F, and Pedrocchi A, An automatic identification procedure to promote the use of FES-cycling training for hemiparetic patients, *Journal of Healthcare Engineering*, vol. 5:3, pp 275–292, 2014.

78. Ambrosini E, Ferrante S, Schauer T, Klauer C, Gaffuri M, Ferrigno G, and Pedrocchi A, A myocontrolled neuroprosthesis integrated with a passive exoskeleton to support upper limb activities, *Journal of Electromyography and Kinesiology*, vol. 24:2, pp. 307–317, 2014.

79. Pedrocchi A, Ferrante S, Ambrosini E et al., MUNDUS project: MUltimodal Neuroprosthesis for daily upper limb support, *Journal of NeuroEngineering and Rehabilitation*, vol. 10:1, p. 66, 2013.

80. Brose SW, Weber DJ, Salatin BA, Grindle GG, Wang H, Vazquez JJ, and Cooper RA, The role of assistive robotics in the lives of persons with disability, *American Journal of Physical Medicine & Rehabilitation*, vol. 89:6, pp. 509–521, 2010.

81. Collinger JL, Boninger ML, Bruns TM, Curley K, Wang W, and Weber DJ, Functional priorities, assistive technology, and brain-computer interfaces after spinal cord injury, *Journal of Rehabilitation Research and Development*, vol. 50:2, pp. 145–160, 2013.

82. Simpson RC, Smart wheelchairs: A literature review, *Journal of Rehabilitation Research and Development*, vol. 42:4, pp. 423–436, 2005.

83. Levine SP, Bell DA, Jaros LA, Simpson RC, Koren Y, and Borenstein J, The NavChair assistive wheelchair navigation system, *IEEE Transactions on Rehabilitation Engineering*, vol. 7:4, pp. 443–451, 1999.

84. Peixoto N, Nik HG, and Charkhkar H, Voice controlled wheelchairs: Fine control by humming, *Computer Methods and Programs in Biomedicine*, vol. 112:1, pp. 156–165, 2013.

85. Carlson T and Millan JdR, Brain-controlled wheelchairs: A robotic architecture, *IEEE Robotics and Automation Magazine*, vol. 20:1, pp. 65–73, 2013.

86. Lamti HA, Ben Khelifa MM, Gorce P, and Alimi AM, A brain and gaze-controlled wheelchair, *Computer Methods in Biomechanics and Biomedical Engineering*, vol. 16:suppl. 1, pp. 128–129, 2013.

87. Ferrigno G, Baroni G, Casolo F, De Momi E, Gini G, Matteucci M, and Pedrocchi A, Medical robotics, *IEEE Pulse*, vol. 2:3, pp. 55–61, 2011.

88. Bonarini A, Ceriani S, Fontana G, and Matteucci M, On the development of a multi-modal autonomous wheelchair, *HandBook of Research on ICTs for Healthcare and Social Services*, pp. 727–748, IGI Global, 2013.

89. Verl A, Albu-Schaffer A, Brock O, and Raatz A, eds., *Soft Robotics-Transferring Theory and Application*, Springer-Verlag, Berlin, 2015.

90. Vukobratovic M, Hristic D, and Stojiljkovic Z, Development of active anthropomorphic exoskeletons, *Medical and Biological Engineering*, vol. 12:1, pp. 66–80, 1974.

91. Fineberg DB, Asselin P, Harel NY, Agranova-Breyter I, Kornfeld SD, Bauman WA, and Spungen AM, Vertical ground reaction force-based analysis of powered exoskeleton-assisted walking in persons with motor-complete paraplegia, *Journal of Spinal Cord Medicine*, vol. 36:4, pp. 313–321, 2013.

92. http://www.eksobionics.com/ekso, accessed on August 29, 2014.

93. http://rewalk.com/products/rewalk-personal/, accessed on August 29, 2014.

94. Laferrier JZ and Gailey R, Advances in lower-limb prosthetic technology. *Physical Medicine & Rehabilitation Clinics of North America*, vol. 21:1, pp. 87–110, 2010.

95. Bellmann M, Schmalz T, Ludwigs E, and Blumentritt S, Immediate effects of a new microprocessor-controlled prosthetic knee joint: A comparative biomechanical evaluation, *Archives of Physical Medicine and Rehabilitation*, vol. 93:3, pp. 541–549, 2012.

96. Tang PC, Ravji K, Key JJ, Mahler DB, Blume PA, and Sumpio B, Let them walk! Current prosthesis options for leg and foot amputees, *Journal of the American College of Surgeons*, vol. 206:3, pp. 548–560, 2008.

97. http://mealtimepartners.com/, accessed on August 29, 2014.

98. http://www.neater.co.uk/, accessed on August 29, 2014.

99. http://www.secom.co.jp/english/myspoon/index.html, accessed on August 29, 2014.

100. Driessen BJ, Evers HG, and van Woerden JA, MANUS—A wheelchair-mounted rehabilitation robot, *Proceedings of the Institution of Mechanical Engineers, Part H*, vol. 215:3, pp. 285–290, 2001.

101. Zeungnam B, Myung-Jin C, Pyung-Hun C, Dong-Soo K, Dae-Jin K, Jeong-Su H, Jae-Hean K, Do-Hyung K, Hyung-Soon P, Sang-Hoon K, Kyoobin L, and Soo-Chul L, Integration of a rehabilitation robotic system (KARES II) with human-friendly man–machine interaction units, *Autonomous Robots*, vol. 16:2, pp. 165–191, 2004.

102. Maheu V, Frappier J, Archambault PS, and Routhier F, Evaluation of the JACO robotic arm: Clinico-economic study for powered wheelchair users with upper-extremity disabilities, *IEEE International Conference on Rehabilitation Robotics*, Zurich, Switzerland, June 29–July 1, 2011.

103. Jiang H, Wachs JP, and Duerstock BS, Integrated vision-based robotic arm interface for operators with upper limb mobility impairments, *IEEE International Conference on Rehabilitation Robotics*, Seattle, Washington, DC, United States, June 24–26, 2013.

104. Velliste M, Perel S, Spalding MC, Whitford AS, and Schwartz AB, Cortical control of a prosthetic arm for self-feeding, *Nature*, vol. 453, pp. 1098–1101, 2008.

105. Hochberg LR, Bacher D, Jarosiewicz B, Masse NY, Simeral JD, Vogel J, Haddadin S, Liu J, Cash SS, van der Smagt P, and Donoghue JP, Reach and grasp by people with tetraplegia using a neurally controlled robotic arm, *Nature*, vol. 485:7398, pp. 372–375, 2012.

106. Iwamuro BT, Cruz EG, Connelly LL, Fischer HC, and Kamper DG, Effect of a gravity-compensating orthosis on reaching after stroke: Evaluation of the Therapy Assistant WREX, *Archives of Physical Medicine and Rehabilitation*, vol. 89, pp. 2121–2128, 2008.

107. Rahman T, Sample W, Jayakumar S, King MM, Wee JY, Seliktar R, Alexander M, Scavina M, and Clark A, Passive exoskeletons for assisting limb movement, *Journal of Rehabilitation Research & Development*, vol. 43, pp. 583–590, 2006.

108. Nikitczuk J, Weinberg B, and Mavroidis C, Control of electro-rheological fluid based resistive torque elements for use in active rehabilitation devices, *Smart Materials and Structures*, vol. 16, pp. 418–428, 2007.

109. Krebs HI, Hogan N, Aisen ML, and Volpe BT, Robot-aided neurorehabilitation, *IEEE Transactions on Rehabilitation Engineering*, vol. 6:1, pp. 75–87, 1998.

110. http://www.hocoma.com/products/armeo/, accessed on August 29, 2014.

111. Nef T, Mihelj M, Riener R, ARMin: A robot for patient-cooperative arm therapy, *Medical & Biological Engineering & Computing*, vol. 45, pp. 887–900, 2007.

112. Stanger CA, Anglin C, Harwin WA, and Romilly DP, Devices for assisting manipulation: A summary of user task priorities, *IEEE Transactions on Rehabilitation Engineering*, vol. 2:4, pp. 256–265, 1994.

113. Muzumdar A, ed., *Powered Upper Limb Prostheses: Control, Implementation and Clinical Application*, Springer, 2004.

114. Reiter R, Eine neu elecktrokunstand, *Grenzgebeiter der Medicin*, vol. 1, pp. 133–135, 1948.

115. Parker P, Englehart K, and Hudgins B, Control of Powered Upper Limb Prostheses, in *Electromyography: Physiology, Engineering, and Noninvasive Applications*, Merletti R and Parker P, eds., pp. 453–475, Wiley-IEEE, 2004.

116. Ortiz-Catalan M, Brånemark R, Håkansson B, and Delbeke J, On the viability of implantable electrodes for the natural control of artificial limbs: Review and discussion, *BioMedical Engineering OnLine*, vol. 11, p. 33, 2012.

117. Carrozza MC, Cappiello G, Micera S, Edin BB, Beccai L, and Cipriani C, Design of a cybernetic hand for perception and action, *Biological Cybernetics*, vol. 95, pp. 629–644, 2006.

118. Johannes MS, Bigelow JD, Burck JM, Harshbarger SD, Kozlowski MV, and Van Doren T, An overview of the developmental process for the modular prosthetic limb, *Johns Hopkins APL Technical Digest*, vol. 30, pp. 207–216, 2011.

119. Connolly C, Prosthetic hands from Touch Bionics, *Industrial Robot*, vol. 35, 290–293, 2008.

120. Biddiss E, Chau T, Upper-limb prosthetics: Critical factors in device abandonment, *American Journal of Physical Medicine & Rehabilitation*, vol. 86, pp. 977–987, 2007.

121. Raspopovic S, Capogrosso M, Petrini FM, Bonizzato M, Rigosa J, Di Pino G, Carpaneto J, Controzzi M, Boretius T, Fernandez E, Granata G, Oddo CM, Citi L, Ciancio AL, Cipriani C, Carrozza MC, Jensen W, Guglielmelli E, Stieglitz T, Rossini PM, and Micera S, Restoring natural sensory feedback in real-time bidirectional hand prostheses, *Science Translational Medicine*, vol. 6, p. 222ra19, 2014.

122. Taylor B, Cupo ME, and Sheredos SJ, Workstation robotics: A pilot study of a Desktop Vocational Assistant Robot, *American Journal of Occupational Therapy*, vol. 47:11, pp. 1009–1013, 1993.

123. Groothuis SS, Stramigioli S, and Carloni R, Lending a helping hand: Toward novel assistive robotic arms, *IEEE Robotics & Automation Magazine*, vol. 20:1, pp. 20–29, 2013.
124. Ripat JD and Woodgate RL, The role of assistive technology in self-perceived participation, *International Journal of Rehabilitation Research*, vol. 35, pp. 170–177, 2012.
125. Ambrosini E, Ferrante S, Rossini M, Molteni F, Gföhler M, Reichenfelser W, Duschau-Wicke A, Ferrigno G, and Pedrocchi A, Functional and usability assessment of a robotic exoskeleton arm to support activities of daily life, *Robotica*, July 22, 2014 [ePub ahead of print].

2

Modern Devices for Telesurgery

Florian Gosselin

CONTENTS

2.1 Introduction and History

Telesurgery, or remote surgery, consists of teleoperating slave robots performing a surgical operation on a patient situated at a remote site. The surgeon grasps and controls the movements of small handles attached to purposely designed master input devices. These displacements are reproduced by the slave robots, giving the surgeon direct control of the surgery while not being present at the operating site.

This technology has the potential to give equal access to advanced surgery procedures to people living in areas away from large medical facilities. One can, for example, easily imagine slave robots installed in small-town offices in less densely populated areas or in third-world countries and remotely controlled by expert surgeons in large city hospitals. Medical trucks equipped with fully operational operating rooms (ORs) integrating slave robots could also be a solution to project surgical capabilities wherever they are needed without the requirement to have surgeons on site. This provides the capability to operate in difficult or even impossible-to-access sites like war zones or spatial environments. Indeed telesurgery was largely financed by the United States National Aeronautics and Space Administration (NASA) with the aim to allow surgical care of astronauts and by the Defense Advanced

Research Projects Agency (DARPA) for the purpose of remotely operating on injured soldiers. This research inspired the development of two commercial telesurgery systems.

The first one is the Zeus® Robotic Surgical System (ZRSS) (Figure 2.1) [1] developed by Computer Motion, Goleta, California. It integrates a voice-controlled arm supporting the endoscope called Automated Endoscopic System for Optimal Positioning (AESOP) and two teleoperated robotic arms carrying surgical tools. The ZRSS was approved by the United States Food and Drug Administration (FDA) in 2001. It was used on September 7, 2001, to perform Operation Lindbergh, the first intercontinental complete telesurgery procedure ever made on a human patient. The surgeon, Professor Jacques Marescaux, successfully performed a cholecystectomy on a 68-year-old patient located in an operating room in Strasbourg, France, using a remote console situated in New York, United States, 7000 km away from the operating site [2–3].

The second robot on the market is the da Vinci, developed by Intuitive Surgical, Inc., Sunnyvale, California (Figure 2.2) [4]. It was also initially composed of three robotic arms carrying the endoscope and the surgery tools. A fourth arm was added later to allow controlling a third tool. These arms, with design similar to the Black Falcon developed at the Massachusetts Institute of Technology (MIT) [5], are successively teleoperated by the surgeons. The robots controlling the tools mimic the hand movements while the endoscope follows the mean of both hands' movements in translation and their difference in rotation (as if the camera was fixed on a stick grasped at its tips). The da Vinci was approved by the FDA in 2000 for minimally invasive surgery (MIS). In 2003, Computer Motion and Intuitive Surgical merged and the da Vinci Surgical System remained as the sole FDA-approved robotic platform on the market. Several versions of the system have since then been approved for various surgical procedures including general, urologic, gynecologic,

FIGURE 2.1
Zeus® Surgical System. © 2014 Intuitive Surgical, Inc., Sunnyvale, California, used with permission.

FIGURE 2.2
da Vinci® Si™ Surgical System. © 2014 Intuitive Surgical, Inc., Sunnyvale, California, used with permission.

transoral, cardiac, thoracic, and pediatric surgeries. Despite its relatively high price, about 3000 units were sold and approximately two million patients worldwide have benefitted from da Vinci surgery (http://www.intuitivesurgical.com).

The da Vinci currently has a quasi-monopolistic situation in MIS, even if competitors try to emerge, like the MIRO Lab system developed by the Deutsches Zentrum für Luft und Raumfahrt (DLR) (German Aerospace Research Center; Oberpfaffenhofen, Germany) for both MIS and open surgery (Figure 2.3) [6–7] and the Surgeon's Operating Force-feedback Interface Eindhoven (Sofie) robot developed at the Technical University of Eindhoven [8].

Note that despite the extensive media exposure offered by iconic operations like Operation Lindbergh, telesurgery systems are still nowadays mostly used in a more familiar environment where the patient is lying on an operating table just beside the control station of the surgeon. The capability to remotely operate on a patient anywhere in the world is not required in practice for day-to-day procedures. The critical aspect of long-distance telesurgery is communication delays. Surgeons need real-time visual feedback of their movements. This is easily obtained over short distances but it becomes difficult for longer ones. As an example, Operation Lindbergh required a complete team of France Télécom engineers to ensure small communication delays (about 155 ms, well below the maximum acceptable of 500 ms).

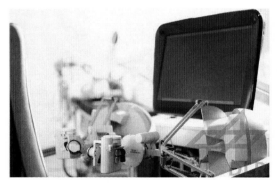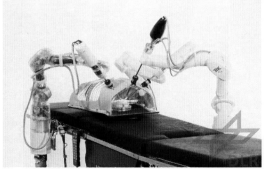

FIGURE 2.3
MIRO Lab system. (Courtesy of DLR, Oberpfaffenhofen, Germany.)

Having the surgeon on site is much more comfortable. He or she can, in particular, easily terminate the procedure with conventional laparoscopic tools in case of system failure or even convert to open surgery. Indeed, the main interest of robotic telesurgery systems is not to allow remotely operating on a patient far away (which is still rare) but, more pragmatically, to overcome the limitations of manual surgery, especially in endoscopy and microsurgery.

MIS is performed with long instruments inserted through small incisions equipped with trocars while the body is inflated with gas to improve vision and accessibility. A video monitor is used to display high-definition images of the operating site captured by an endoscope (a camera mounted at the tip of a long shaft inserted through another trocar). Compared to open surgery, MIS allows smaller incisions, thus resulting in reduced pain and shorter recovery time. However, it introduces several shortcomings. As surgeon's hands and surgical tools are on both sides of the trocars, which act as fulcrums, hands and tools move in opposite directions. Moreover, their movements are restricted to four degrees of freedom (DoFs) as the tool can translate along any direction but can rotate only along its extension axis. Last but not least, the surgeon loses force and tactile feedback and can only rely on visual cues to estimate the forces and torques applied to the tissues. These constraints strongly limit dexterity, all the more so as the hand–eye coordination is lost as the surgeon looks at the video monitor instead of his hands. Manipulating long instruments with arms raised is also tiring and uncomfortable. The introduction of remotely operated robots equipped with intracorporal dexterous wrists allows canceling all these drawbacks. The robots have at least seven DoFs and can grasp tools and move in any direction in translation and rotation. The surgeon is comfortably installed on a master console equipped with ergonomic input devices. His movements are directly mimicked by the robots, in seven DoFs and without the mirror effect. Dexterity is comparable to open surgery and natural hand–eye coordination can be recovered with a colocalized three-dimensional (3D) visual feedback. The surgeon is virtually immersed in the patient's body as if he or she had direct access to internal organs. In principle, such systems should even be capable of restoring force and tactile feedback. However, this capability is still restricted to laboratory equipment due to the unavailability of precise force sensors that can be either sterilized after use or produced at sufficiently low price for single use. As a consequence, force feedback is not implemented on actual commercial robots.

Regarding microsurgery, modern microscopes and miniaturized tools have offered the capability to see and operate on tiny structures of the human body. However, this challenges the surgeon's ability to make precise movements and apply forces at the microscale. Telemicrosurgery robotic systems can scale down hand movements, filter tremors, and go beyond these limitations. Eye surgery is a good example. As in MIS, surgeons make use of long and thin instruments inserted through the eye envelope, resulting in inverted hand and tool movements and nonergonomic postures. They face the additional challenge of operating structures down to 10 µm, well below human accuracy, which is about 125 µm. The resulting forces are also well below the human detection limit. As a consequence, some operations are almost impossible to perform manually. Teleoperated robots like the vitrotretinal eye surgery robot developed by the Technical University of Eindhoven [9] overcome these limitations. Visual feedback is provided on a compact video monitor integrated in an ergonomic and intuitive master console. Tremors are filtered and the master movements are scaled down on the slave side, allowing reaching the required accuracy. Finally, the master arm provides force feedback based on sensors integrated on the slave robot having sensitivity below the human threshold. Neurosurgery is another example. The requirements are still the same, with the exception of the mirror effect, which is not present in this case as the skull is drilled to get a direct access to the brain. As a consequence, robots like neuroArm, developed at the University of Calgary (Figure 2.4), are usable for both microsurgery and stereotactic brain surgery [10–11]. Thanks to the use of titanium and polyether ether ketone (PEEK) manufactured parts and ultrasonic piezoelectric motors, the robots are MRI compatible. The robots are controlled via a multimodal master station featuring visual, audio, and haptic feedback.

Note that along with market-intended developments, research has also been made on open platforms. A notable initiative of this kind is the collaborative development of the Raven and Raven II surgical robotic systems [12–13].

To further reduce pain, recovery time, and aesthetic outcomes associated with MIS, two other approaches were more recently proposed, namely, single-port (or single-incision) laparoscopic surgery (SPLS) and natural orifice transluminal surgery (NOTES).

SPLS consists of inserting the endoscope, air pressure, and instruments through a single larger incision, typically in the range of 20 to 25 mm, instead of several 5 to 15 mm cuts. To keep a similar arrangement of working space and tools, the instruments cross at the level of the trocar and have a curved shape so that they arrive from both sides as in MIS. Apart from a reduction in the instruments' size, such systems are comparable to MIS and single-port telesurgery platforms already available on the market like, for example, the Single-Site da Vinci Surgery System. It is, however, restricted to abdominal surgery.

With both MIS and SPLS, the surgical site is accessed from the outside of the body. On the contrary, with NOTES, the instruments are inserted through natural orifices (mouth, vagina, or rectum). A long and flexible endoscope platform with a diameter typically between 14 and 20 mm is used, the surgery tools being either fixed on flexible arms at the tip of the platform or inserted through two to four working channels a few millimeters wide. This technique totally avoids pain and scarring associated with abdominal incisions. However, current manually operated NOTES endoscope platforms suffer from limited forces and insufficient stiffness. They are not stable enough to expose the organs and operate comfortably and are used only by few medical teams [14]. Despite their theoretical potential to be the next revolution in surgery, they still require more time and resources to perform and their real benefit is controversial [15]. The future is probably robotized NOTES instruments remotely controlled with dexterous master arms. Such systems are currently under development; see, for example, Reynoso et al.'s [16] and Zhou et al.'s [17] works for the state of the art. These systems have not, however, entered operating rooms to date and will not be discussed in detail in this chapter.

(a)

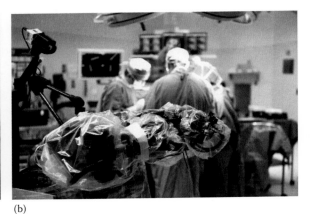
(b)

FIGURE 2.4
neuroArm master station (a) and slave robot (b). (Courtesy of the University of Calgary, Calgary, Canada.)

2.2 Main Components and Functionalities of a Robotic Telesurgery System

2.2.1 General Overview

Figure 2.5 illustrates the main components of a telesurgery system. The surgery is performed by slave robots (1) remotely controlled from a master station (2). A cabinet (3) is often used for carrying the robot's controllers and additional equipment. Finally, high-performance communication means (4) transfer information within the whole system. Further details on these components will be given below.

2.2.2 Slave Surgery Robots

Slave robots (1) are usually composed of several components. First, a positioning stage (1a) allows for an optimal placement and orientation of the robot relative to the patient. This structure is often of selective compliance articulated robot arm (SCARA) type, with a vertical sliding axis and three vertical pivot joints allowing it to move on a horizontal plane. This solution is very safe as the weight of the robot can be easily compensated for. A simple counterbalancing mass is sufficient in the z axis, while gravity has no influence on the pivot joints. The optimal placement of the robots, taking into account the constraints imposed by the presence of the other robots, equipment, and surgery team, can be experience based or obtained using path-planning software. The second and third stages are the external robot structure (1b) and the instrument (1c). Their roles are, respectively, to move the instrument to the operating site and to orient it using an internal wrist whose rotation axes are close to the tool tip. In endoscopic surgery, the robot (1b) must respect the additional constraint of rotating around the fixed point imposed by the trocar. This can be obtained by several means. The Zeus robot, for example, is not motorized in orientation and the instrument automatically tilts to point at the trocar when it is moved [1]. This

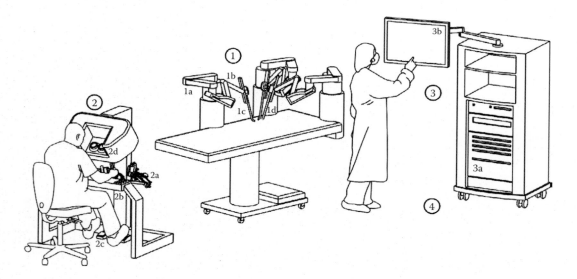

FIGURE 2.5

Main components of a telesurgery system. 1: slave robots; 2: master station; 3: nurses' cart; 4: communications means.

solution is very simple and convenient. However, the movements of the tool tip are not completely controlled as they are also a function of the patient-dependent behavior of the abdominal wall through which the trocar is inserted. To avoid this drawback, all joints must be actuated. In this case, the fixed trocar constraint can be solved using remote-center-of-rotation (RCR) architectures, for example, pivot joints which intersect at the trocar as used on the Raven robot [12] or double parallelograms as implemented on da Vinci [4–5] (see Figure 2.6). Such solutions are intrinsically safe as the RCR remains mechanically fixed. However, the robot's placement must be very precise in order to ensure that the fixed point and trocar coincide. Moreover, such architectures are quite bulky and have limited amplitudes of movements. While well suited for abdominal and chest surgeries, they do not appear optimal for some other operations. Another solution consists of using multipurpose conventional serial robots and programming their controllers so that the trocars remain fixed when they are used in MIS. This way, a more compact and slender design can be obtained, as illustrated by the MIRO Lab System [6]. Such robots are also more versatile. They are equally suitable for MIS and open surgery, simply replacing the complex robotized MIS instruments with simple tools. This advantage comes, however, at the price of a higher complexity. Indeed, while three actuated DoFs are sufficient to move the instruments of purely MIS robots as the tool's orientation is not controlled by the robot itself but by the internal actuated wrist, such multipurpose robots must have at least six actuated

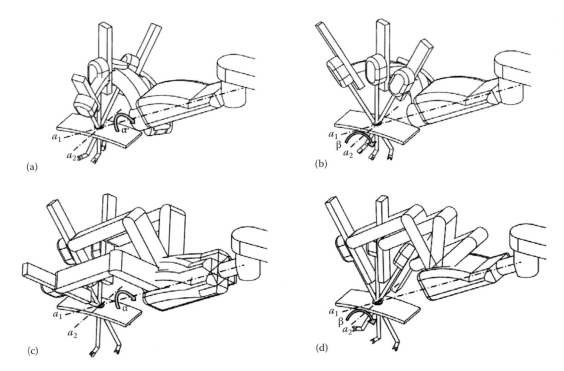

(a)

(b)

(c)

(d)

FIGURE 2.6
Two different solutions for implementing an RCR and solving the fixed trocar constraint. *Top*: spherical architecture with intersecting axes. (a) rotation β along axis a_1 is simply achieved by moving the first link of the structure; (b) rotation β along axis a_2 is obtained when the angle between the two links is modified. *Bottom*: double-parallelogram architecture (hidden by the robot's housing). (c) α is generated by a rotation of the whole robot along its first joint; (d) β is obtained by moving the double parallelogram, and the coordinated movement of its different links maintains the trocar fixed.

DoFs to allow controlling open-surgery passive tools in both position and orientation. As an internal wrist is still required in endoscopic surgery, this results in a redundant architecture when using them for MIS.

The instrument (1c) is composed of four parts: an interface with the robot, a long shaft going through the trocar, an internal wrist allowing rotation of the tool around any axis, and the tool itself. The instrument is generally passive, the actuators being physically placed on the robot and their movements transmitted via small clutches integrated in the interface. This way, the instrument's sterilization is easier. The shaft and wrist come in different diameters, typically within 5 and 10 mm. The largest wrists are composed of small pivot joints actuated with cables and pulleys. Smaller ones are often manufactured with multiple coupled joints, resulting in a more flexible structure. The tools themselves are various and comparable to those encountered in conventional MIS, including graspers, needle drivers, clip appliers, cautery instruments, scissors, scalpels, and retractors.

The robot and instrument must be designed in order to reach the desired accuracy. The latter depends on the operation. For abdominal telesurgery, 100 μm seems reasonable, while less than 10 μm is required for vitroretinal eye surgery [9]. This calls for precise manufacturing, play-free mounting and assembly, and high-resolution position sensors. Note that in teleoperation, the human is always in the loop and can adjust the position of the tool based on visual feedback. Resolution is, thus, more important than absolute accuracy.

Another requirement is the possibility to easily move the positioning stage and the robot during the installation phase. They must resist as little as possible to the movement of the surgical team. This calls for a frictionless and low-inertia design. The same holds for the instrument, which must be easily extracted from the trocar in the case of a change of tool or malfunction. The whole system must be as mechanically transparent as possible. There are two alternatives, therefore: cable-based back-drivable mechanical design as encountered on most nuclear teleoperation systems and haptic interfaces [18–22] and implemented on the first version of da Vinci or lightweight robots equipped with force sensors, inertia and friction being partially cancelled by dedicated compensation feedback loops.

For typical operations, at least three robots (1b) and their positioning stage are required. Two of them carry instruments, while the last is used to control the endoscope (1d). As the endoscope does not require an internal wrist, its actuation system is simplified. In the most advanced systems, an additional robot is added, with a total of three instruments plus the endoscope.

These robots can be independently fixed on the operating table as illustrated in Figure 2.5 in order to minimize their footprint. This is the solution proposed in the MIRO Lab system. Each robot is light and can be moved from operating room to operating room and installed easily. They can also be fixed on a single structure like the da Vinci patient-side cart. This solution is more bulky but the robots are more precisely positioned with respect to each other. They are also easier to remove in case of conversion following a system failure.

2.2.3 Master Control Station

To control these robots, the surgeon is comfortably installed at a master station (2). He or she grasps and controls the displacements of the input devices (2a), whose movements are mimicked by the slave robots operating the patient. The master arms (2a) have generally seven DoFs, three DoFs in translation to pilot the displacements of the slave robots, three DoFs in rotation to control wrist orientations, and one grasping DoF for dual-jaw tools like forceps or scissors. To allow for stable movements, an armrest (2b) is often provided close to the input devices' handles. The surgeon is also provided with additional commands like foot pedals (2c) or switches and buttons integrated on the armrest or on the devices' handles. They allow controlling, for example, electric scalpels. Finally, a visual feedback from the surgery site is provided in a video display (2d). On the da Vinci's console, a binocular-type visual interface provides high-definition 3D images from inside the patient's body. This configuration is very immersive and the surgeon's hands are naturally located in his line of sight. However, he or she is quite isolated from the other members of the medical team. On the contrary, most other systems like Zeus and the MIRO Lab system propose more conventional nonimmersive consoles. While visuohaptic colocation is lost, the surgeon is more aware of its environment and can more easily look at complementary monitors and interact with the team.

As the surgeon only has two hands, he or she cannot manually control more than two slave robots at a time. He or she has to switch, thus, between the different robots to successively teleoperate the instruments and endoscope. Another solution is to use a voice-controlled endoscope as implemented on the AESOP arm of the Zeus Robotic Surgical System. Of course, the endoscope can also be manipulated by a second surgeon sitting at a second master station, which can also be used to control a fourth arm holding a third instrument, typically a retractor or a clip applier used in conjunction with the instruments of the main surgeon. Finally, the endoscope can automatically follow the object of interest using a visual servoing function.

2.2.4 Additional Equipment and Communication Means

An additional cabinet (3) is usually provided. It can be used to carry the robots and master station's controllers (3a), as well as other medical equipment. An external video display (3b) is also generally present. It provides the nurses, the anesthetist, and other people in the medical team with a visualization of the progress of the procedure.

Finally, rapid, efficient, and secured communication means (4) (not illustrated in Figure 2.5) are implemented. As previously explained, the communication delay has to be well below the maximum acceptable of 500 ms to allow efficient teleoperation of the slave robots in the presence of visual feedback only. If a force feedback is implemented (only on research systems to date), it has to be in the order of a few milliseconds to a few tenths of milliseconds to avoid compromising the robots' stability. In case of larger delays, the control gains have to be decreased, which dramatically diminishes the sense of presence and teleoperation efficiency. Another solution is to couple the master arms with a virtual model of the slave robots and patient. This model can be used to generate the force feedback and control the slave robots. Implementing a physiologically correct virtual model of the patient is, however, very difficult and such a solution is not proposed on commercially available systems.

2.2.5 Main Functionalities

2.2.5.1 Master–Slave Teleoperation

The slave robots are remotely controlled with the input devices of the master station. As previously explained, commercial systems to date do not allow force feedback and a simple position-control loop is used to pilot the system so that the position and orientation of the tool tip mimic those of the master arm's handle. In order to favor ergonomics, the handle and tool axes are generally kept aligned during the surgery. Therefore, the master arm's joints are actuated and the motors are used when the unit is switched on to reach the slave arm's orientation.

Force feedback is a strong surgeons' demand. Without it, they have to rely on visual cues to infer forces from tissue displacements. However, despite huge research efforts in the last decade, this functionality is still not available in commercial systems. The key missing component is a precise force/torque sensor which can be sterilized or produced at very low price. While the latter requirement is beyond the current state of the art, miniature sensors exist and are already implemented on research telesurgery systems like the MIRO Lab system [23]. Such robots implement force feedback. Both classical passive proportional derivative position-position control loops and position-force schemes can be used in surgery [5,24–25]. The former has the advantage of being passive but suffers from limited force sensitivity. The latter allows improving the system transparency in free space and detecting very tiny forces in contact. It is, however, unstable when the robots are in contact with a stiff environment, which is fortunately not the case when interacting with soft tissues and organs.

2.2.5.2 Motion (and Force) Scaling

A gain between the master and slave robots' positions can be implemented in the system controller, allowing an amplification of the handle's motions compared to the tool. When the surgeon moves his hands, the tool makes smaller movements, allowing to reach subhuman precision and accuracy. This is especially important in telemicrosurgery. It is worth noting that if the motion scaling is too large, the surgeon will not be able to reach distant points with the slave arm. A clutching function is made available to solve this issue. When the surgeon has reached the limit of the input device's work space, he or she can declutch the tool from the handle, go back to the center of the master arm's working volume, and reengage the master slave coupling to go further.

Contrary to positions, the ratio between the orientations of the handle and tool is always kept equal to 1. Amplifying rotations would also be feasible but it would misalign the hands' and tools' movements, breaking the hand–eye coordination and reducing the system's intuitiveness and ergonomics. For the same reasons, the clutch is not available in rotation, and the master arm is actuated so as to keep the handle in fixed orientation when the clutch is activated in translation.

With force feedback systems, a second gain can also be implemented between the forces sensed by the slave robots and those sent back to the user. Position and force amplification, however, change the perceived behavior of the environment.

2.2.5.3 Tremor Cancellation

A low-pass filter can be inserted between the master arm's measured positions and the orders sent to the slave arm. This way, the surgeon's hand tremors can be rejected and the slave robots have smooth and very precise movements.

2.2.5.4 Shared Control

Within a robotic telesurgery platform, the control of the slave robots can be shared between the surgeon and the system. An example is found in beating-heart surgery. The robots carrying the tools and endoscope can automatically compensate for heart movements, so that it appears fixed to the surgeon who can operate as if the anastomosis site were maintained with mechanical stabilizers.

Control of the robots can also be shared in time, either between several surgeons, each one installed at a master station and controlling the robots during specific tasks, or between a surgeon and the system. One could, for example, imagine automatic suturing or knot tying after the surgeon has positioned the tissues in correct position.

2.2.5.5 Augmented Haptic Feedback

Thanks to the advancement of measurement techniques and computer science, it is easier to model or scan and digitize the surgical environment, including the OR table, robots, equipment, and surgery team. The continuous progress in imaging technologies also allows obtaining 3D models of the patients. After proper calibration and alignment of the environment and patient models, a complete 3D model of the surgery site can be generated. Such a model can be used for optimal placement of the robots and planning of the surgery. In particular, optimal trajectories along which the user can be haptically guided or sensitive areas which have to be avoided can be defined. The complete procedure can even be rehearsed on the virtual model before it is realized in the real OR. It has, however, to be ensured that the virtual model and real patient are perfectly aligned. Functional MRI as used in neurosurgery is a solution, therefore, but robots cannot integrate magnetic field–sensitive materials in this case. This is a huge additional constraint and in practice only few robots fit this requirement.

2.3 Optimal Design of an Advanced Input Device for Telesurgery

We will now focus on the design of the control station master arms. We will rely, therefore, on the author's previous works [26–29].

Teleoperation appeared in the fifties in the nuclear industry from the need to manipulate radioactive materials. Purely mechanical systems were first developed, followed by bilaterally coupled robots. The use of this technology has since extended to other environments which are inaccessible due to their hostile nature (e.g., space and subsea), to their scale (minimally invasive surgery and micro- and nanomanipulation), or to their virtual nature (digital environments and virtual reality) [30–31].

2.3.1 Design Guidelines

A consensus has emerged on the characteristics a telerobotic system should exhibit. Whether considering virtual reality, telesurgery, or heavy arms remote handling, it must be "transparent." The operator must feel that he or she is performing the task directly in the remote environment. To identify the associated design drivers, two system states are usually distinguished. First, in free space, he or she must feel free, with as low as possible resistance from the system. The robot must, in particular, have a large and singularity-free work space, low inertia, and low friction. Second, when the slave arm hits an obstacle, the operator must clearly feel this contact and the touch sensation must remain realistic until contact is released. This calls for a sufficient force capacity, a high bandwidth, and a large stiffness. The same holds for the master arm (also called haptic interface in virtual reality).

Almost all existing input devices were designed in order to answer these general requirements. These specifications are, however, contradictory as, for example, a stiff robot required to realistically simulate hard environments is usually made of large diameter links which are relatively heavy and resist operator's accelerations in free space. Similarly, using big actuators to allow rendering large forces impedes the user's movements in free space as they introduce friction. A compromise has to be found, which depends on the application. For dismantling nuclear facilities, for example, work space and force are of primary importance, while for telesurgery, precision and sensitivity are essential. As a consequence, very different master arms and haptic interfaces have been developed over time [32–35]. They differ in types, shape, dimensions, and performance. To fit with a specific application, these requirements must be refined.

Regarding abdominal telesurgery, the design drivers are the following:

- Bimanual manipulation: Most surgery tasks are bimanual in nature, e.g., knot tying. To perform them, the surgeon has to control two slave robots simultaneously. The control station must, thus, be equipped with two master arms.

- Dexterous manipulation: The goal of minimally invasive robotic surgery (MIRS) is to restore similar motion capabilities and dexterity as in open surgery despite the presence of the trocar. This requires seven-DoF slave robots able to move and rotate in any direction and grasp surgery tools (e.g., needles). Their movements are directly mimicking those of the master arms which must, thus, also have seven DoFs. In order to maximize dexterity and precision, a precision grasp (of stylus or pinch–grasp type) and an elbow support are recommended. Also, the rotations and translations must be decoupled. Regarding their amplitude, they can be considered separately. As explained before, movements can be scaled in translation and a clutch can be used to expand the master arm's work space. As a consequence, the slave's movements are not really limited by those of the input device, which can be designed solely based on ergonomic considerations. The comfortable work space associated with a precision grasp and an elbow support is about 200 mm in every direction [36]. It will be used here as a design driver. On the contrary, rotations cannot be extended using scaling factors or clutch. They have to cover the surgeon's hand work space, which is about 140° to 160° around any axis in this configuration [36].

- Realistic haptic feedback: Even if commercial input devices do not allow force feedback based on slave arms measured efforts, they should in the future. As a consequence, it is worth anticipating this capability. In such case, the input device should be actuated to allow force feedback in the range of efforts associated with typical surgery procedures, i.e., from 0 to 15 N [37]. Note that the master arm's actuators are also useful to implement virtually generated augmented haptic feedback (e.g., virtual guides, avoidance areas, and interactions with a virtual model of the patient). In this case, the forces should remain within a noticeable yet comfortable range, which is about 10 N for a precision grasp [36]. Force resolution has also to be optimized to allow feeling slight contacts against the organs, the stiffness maximized and the mass minimized. This optimization concerns not only force, mass, and friction amplitudes but also their isotropy and homogeneity over the whole work space. Anisotropic behavior would result in preferred directions of motion and has to be avoided, as well as inhomogeneous performance, as it would result in various haptic feelings depending on the position of the handle, even if the slave robot performs the same task. Both would perturb the surgeon.

- Safe manipulation: In order to avoid the system falling under its own weight in case the handle is inadvertently released, the input device has to be statically balanced. A mechanical balancing system using counterweights or springs is preferred over actively compensating for the weight of the links with the actuators as it remains efficient in case of electric failure. Passive or active weight cancellation also avoids requiring the user to support the weight of the device and minimizes fatigue.

2.3.2 Design Methodology

In order to reach previous specifications, the input device has to be carefully designed. As previously mentioned, the optimized criteria are not independent. Also, they vary nonlinearly as a function of the design parameters whose variations can be continuous (e.g., link lengths and angles between joints) or discrete (e.g., actuator type and sensor resolution) and are constrained by physical limits. Globally, we face a constrained multicriterion nonlinear coupled optimization problem. Advanced optimization methodologies can be used to solve such a problem. However, the result highly depends on the weight given to the different criteria, this weight being quite arbitrary. Of course Pareto fronts can be used to show the influence of each parameter. However, ensuring a sufficiently large number of combinations to understand the system behavior leads to a large number of optimizations. The time required increases and the advantage compared to studying all configurations decreases. We rather prefer choosing classical well-understood robot architectures and trying to optimize each criterion at a time to better control the optimization process. The disadvantage of this method is that is does not apply well to more complex original architectures.

The first criterion is the work space of the robot. It is defined as the set of configurations the robot can reach. It can be expressed in the joint space as a set of n-dimensional vectors with n being the number of actuated joints or as a set of six-dimensional vectors in the Cartesian space (seven dimensions are necessary if the handle opening is taken into account, but in practice, this motion is usually independent from the robot configuration, i.e., position and orientation, and can be studied separately). The robot's work space depends on the robot's architecture and joints' arrangement, on the links' length and shape, and on external constraints like collision avoidance with other console equipment. Two alternative approaches can be used for its study. First, we can span the joint space and plot for each joint position q the associated configuration X computed using the direct geometric model of the robot, $X = f(q)$. This solution has the advantage of determining the exact limits of the operational work space. However, the geometric model being nonlinear, the Cartesian configurations of the robot are not evenly distributed. To avoid this drawback, the second solution consists of scanning the Cartesian space and checking the valid configurations by using the inverse geometric model of the robot, $q = g(X)$. In this case, the points are evenly distributed in the work space but its limits are only approached. For a precise estimation, small steps have to be used. This model is used to tune the size of the robot until its work space encompasses the specified volume in six dimensions.

The second optimization criterion is the force capacity, defined as the minimum amount of force the robot can apply in any direction. To study this parameter, instead of using the manipulability and force ellipsoids, defined as the operational forces produced by 1 N·m motor torques, we propose to use force-dimensioning ellipsoids, defined as the torques necessary to generate 1 N forces and/or 1 N·m torques in the operational space [36]. Calling

J_{mot} and G_{mot} respectively the direct and inverse Jacobian matrices allowing to express the n-dimensional joint speeds $\dot{q}_{mot}$ and torques τ_{mot} expressed at the motor level as a function of the six-dimensional operational speeds $V = \dot{X}$ and efforts F by using the equations $\dot{X} = J_{mot} \cdot \dot{q}_{mot}$, $\dot{q}_{mot} = G_{mot} \cdot \dot{X}$, $\tau_{mot} = J_{mot}^T \cdot F$, and $F = G_{mot}^T \cdot \tau_{mot}$, these ellipsoids can be defined using the following formulas:

$$F^2 = 1, \tag{2.1}$$

$$\tau_{mot}^T \cdot \left(G_{mot} \cdot G_{mot}^T\right) \cdot \tau_{mot} = 1. \tag{2.2}$$

Equation 2.2 is the equation for an ellipsoid in the joint torque space. It represents the torques which are necessary on the robot's joint axes to apply 1 N and/or 1 N·m efforts on the handle. In practice, forces and torques are studied separately and the results are added. When studying forces, we first write that the torques applied on the handle are null and express the three motor torques as a function of the others. Replacing these torques in Equation 2.2 gives three forces as a function of only three actuators (the same is done in case of a redundant robot, using the force-closure equations to exclude redundant actuators). Doing this with different sets of actuators allows dimensioning each of them. Conversely, when studying torques, we first write that the forces applied on the handle are null and express three motor torques as a function of the others. Equation 2.2 gives in this case three torques as functions of three motors. As illustrated in Figure 2.7 in a two-dimensional (2D) case, the motor torques are simply obtained as the bounding boxes of those ellipsoids. For a given desired operational force F_d and torque τ_d, the results are simply multiplied by F_d and τ_d and added.

Equation 2.2 and its pure force and torque derivatives allow computing the required actuation torques in each valid configuration. Final dimensioning is obtained as the worst case over the useful work space of the robot. Note that these torques depend on both the actuators and the reducers. It is, thus, possible to adjust either the motors or the reduction ratios or both to obtain the desired values. The choice depends on several factors. Larger electric actuators, as usually used on haptic interfaces, are more efficient than smaller ones. They require smaller reducers, which produce less friction than larger ones. However, such combination requires higher-definition encoders to reach the same accuracy as those of small motors and large reducers. Control stiffness and mass also are largely impacted by this choice as will be explained below.

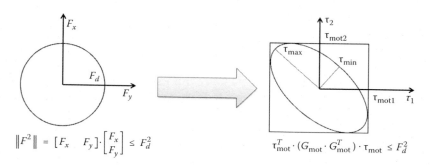

FIGURE 2.7
Force-dimensioning ellipsoids in two dimensions.

Note that force-dimensioning ellipsoids, as well as manipulability ones, also allow computing the local isotropy index of the device as $I_{Floc} = \tau_{max}/\tau_{min}$. A global index over the work space can also be computed, taking into account either its mean value or the worst case. The robot's architecture, dimensions, and actuators can be adjusted to reach a desired force/torque capacity and a predefined isotropy. Also, as the total work space of the robot is usually larger than the specified useful one, the position of the latter relative to the base of the robot can be optimized, i.e., placed where the force/torque capacity is maximum and/or where it is the most isotropic (the same holds for apparent stiffness and mass).

The third design driver taken into account is the master arm's apparent stiffness. Based on the hypothesis that each joint is basically composed of an actuation unit, a transmission, and a link, the robot behaves as these three components arranged in series. Denoting K_{mot}, K_{trans}, and K_{link} as their respective stiffnesses, the global apparent stiffness K_{rob} equals

$$K_{mot} = 1/(1/K_{mot} + 1/K_{trans} + 1/K_{link}). \tag{2.3}$$

K_{trans} and K_{link} depend on the transmissions design and links shapes and can be optimized using finite-element models of the robot. K_{mot} is called the motor electric equivalent stiffness. It is defined as the minimum static gain in any direction deduced in the operational space from the maximum stable static gain K_{max} of the motor's control loops. These gains can be written as

$$\tau_{mot} = K_{max} \cdot dq_{mot}, \tag{2.4}$$

$$F = K_{mot} \cdot dX = G_{mot}^T \cdot K_{max} \cdot G_{mot} \cdot dX. \tag{2.5}$$

To study the K_{mot} parameter, we make use of the apparent-stiffness ellipsoid, defined as the operational forces produced by a normalized 1 m displacement. It can be defined using the following formulas:

$$\|dX^2\| = 1, \tag{2.6}$$

$$F^T \cdot \left(K_{mot} \cdot K_{mot}^T \right)^{-1} \cdot F = 1. \tag{2.7}$$

Equation 2.7 is the equation for an ellipsoid which can be used to compute the maximum and minimum actuators' equivalent stiffnesses (see Figure 2.8 for a 2D example). It can be used to compute the minimum apparent stiffness in any direction in a given position or all over the work space as well as the local stiffness isotropy index $I_{Kloc} = K_{max}/K_{min}$ or its global derivative. The design of the robot can be adjusted until the required values are obtained. The parameters with the largest influence on the electric equivalent stiffness are the actuator's behavior, the encoder's resolution, and the control-loop frequency, which limit the maximum stable static gain of the motors, as well as the reduction ratio R, which allows increasing the apparent stiffness by the factor R^2. Note that different actuator–reducer combinations with different electric equivalent stiffnesses can be chosen to reach the same desired force capacity as the force is a function of R, while the stiffness varies as R^2. On the other hand, it is demonstrated by Townsend [21] that it is preferable to implement the reducers as close as possible to the actuated joints.

Finally, the fourth design driver is the apparent mass of the robot. It is defined as the maximum mass sensed by the surgeon when accelerating the handle in free space. It is

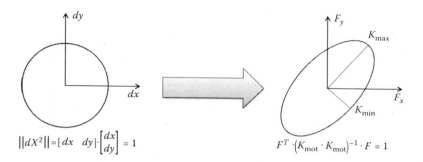

FIGURE 2.8

Apparent actuators' equivalent stiffness ellipsoids in two dimensions.

computed using a simplified dynamic model of the system. With the hypothesis of low speeds (a surgeon moves slowly to preserve precision and accuracy), the centrifugal and Coriolis terms vanish, as well as the speed term, when deriving the inverse kinematic model $\dot{q}_{art} = G_{art} \cdot \dot{X}$ expressed at the articular level ($\dot{q}_{art}$ is the n-dimensional joint speed vector and G_{art} is the inverse Jacobian matrix at that level). Assuming also that the system is statically equilibrated, the gravity forces are also null and the simplified dynamic model can be written at joint level as $\tau_{art} = A_{art}(q_{art}) \cdot \ddot{q}_{art}$, with τ_{art} and $\ddot{q}_{art}$ as the n-dimensional vectors of the joint torques and accelerations expressed at the articular level and $A_{art}(q_{art})$ the inertia matrix of the robot's links. We can write

$$F = M\dot{V} = G_{art}^T A_{art}(q_{art}) G_{art} \dot{V}. \tag{2.8}$$

This equation can be used to compute the apparent-mass ellipsoid, defined as the operational forces produced by a normalized 1 m/s² acceleration. This ellipsoid can be defined by the formulas

$$\| \dot{V}^2 \| = 1, \tag{2.9}$$

$$F^T(MM^T)^{-1} F = 1. \tag{2.10}$$

Equation 2.10 is similar to Equation 2.7 and allows computing the maximum apparent mass in all directions in every configuration and over the work space, as well as the local and global inertial isotropy indices. The link's dimensions and shapes can be optimized to reach the desired value, with the constraint, however, of remaining within an acceptable stiffness.

Note that the actuator's rotors also contribute to the apparent mass of the robot. To estimate their contribution, the same reasoning can be applied, with all equations expressed, however, at the motor level. It can be proven that the apparent inertia of the actuators increases as a function of R^2. As a consequence, the inertia of systems using large reducers, for example, harmonic drives, is relatively important (often larger than the links' apparent inertia) and requires force-sensing and force-control closed loops to reduce it. On the contrary, the inertia of actuation units using low-reduction-ratio cable capstan reducers is usually very limited (much smaller than the link's inertia).

As explained before, these models are successively used to optimize the robot architecture and dimensioning step by step.

2.3.3 Application to the Design of a Telesurgery Master Arm

Different architectures can be envisaged for the design of telesurgery master arms. The da Vinci master station, for example, makes use of eight-axis serial robots [38–39] as shown in Figure 2.2. Serial robots have the advantage of a large range of motions, especially in orientation. Moreover, force feedback can be easily and efficiently obtained in translation. However, this is not the case in orientation. Indeed, the actuators are the heaviest components of haptic interfaces. In order to limit the apparent inertia felt by the surgeon, they have either to be placed close to the base or to be very light. The former solution calls for complex transmissions which introduce friction and reduce the transparency, while the latter does not allow rendering large torques. As the da Vinci does not allow force feedback, its actuators are used only to equilibrate the handle's weight and to align it with the surgery tools when the system is switched on. Small actuators are sufficient in this case but they would not be optimal for the implementation of torque feedback.

The master arms of the MIRO Lab system make use of a different architecture. They are composed of a Delta parallel stage for the translations and a serial wrist for the rotations. Two versions were successively used: first an omega.7 haptic interface with a passive wrist, then a sigma.7 device with seven actuated axes (both from Force Dimension, Nyon, Switzerland). This hybrid architecture suffers from the same drawbacks as serial robots. While the Delta is very light and transparent, it has to carry a bulky and relatively heavy actuated wrist. In order to remain within an acceptable volume and weight, the torque feedback is limited and, even so, the two master arms cannot be placed side by side very close to each other.

In order to limit the weight of the wrist, we proposed to use a parallel robot instead. As fully parallel robots (i.e., robots with the same number of substructures and DoFs) suffer from a limited range of motions in orientation and complex control models, they will be discarded. We will make use of partially parallel robots instead, composed of a limited number of substructures having several actuated DoFs each. Note that a lot of haptic interfaces make use of a pivot joint and a parallelogram in series. This simple three-DoF architecture efficiently allows implementing force feedback in translation. The actuators can be placed close to the base and the architecture is very light and powerful. At the same time, it remains thin and compact. It will be used here as an elementary substructure. To obtain six-DoF robots from this, three solutions can be envisaged: (1) using three such substructures arranged in parallel, each of them having two or three actuated joints (we will call such architectures 3 × 2 and 3 × 3, respectively); (2) using two such robots in parallel, each with three motors, and implement a sixth axis in series (also called 2 × 3 + 1; see details below); and (3) using only one such architecture with a serial wrist (1 × 6). In order to rank these alternatives, we first computed their homogeneous orientation work space (HOW). HOW is defined in each position in the work space as the orientation that can be obtained simultaneously in every direction (i.e., a HOW of n degrees means that the handle can span a cone with a half top angle of n degrees, the handle being rotated around its own axis of any angle between ±n degrees). This calculation was performed for normalized 1 m long structures, with a handle of 15 cm, taking into account the joints' motions limits and collisions between the robot's links. The results illustrated in Figure 2.9 show that the HOW of 3 × 2 parallel robots is limited to about 40° to 50° in a large part of their work space. It reaches 70° to 80° for 2 × 3 + 1 robots (several configurations were tested, with parallel and opposed substructures) and 90° for serial structures. These values can be considered as their useful HOWs. We can conclude from this study that 3 × 2/3 × 3 parallel robots are not suitable for dexterous telesurgery. Their range of motion

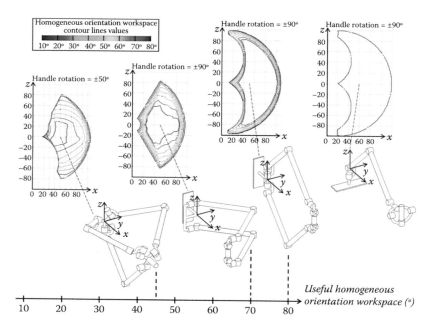

FIGURE 2.9
Maps of the homogeneous orientation work spaces of different robot architectures using pivot and parallelogram substructures on the *x-z* plane.

in orientation is insufficient. Serial 1 × 6 robots reach much larger orientations but suffer from the aforementioned limitations. In between these extreme solutions, hybrid 2 × 3 + 1 architectures appear as an interesting alternative. They have a large work space, comparable to those of serial robots, especially in orientation. Yet they require actuating only one DoF away from the basis. Different configurations can be envisaged. Having opposite substructures limits their collisions and increases the work space. However, it would be practically impossible to integrate such robots into a master station without colliding with the surgeons' knees. As a consequence, the 2 × 3 + 1 architecture with parallel branches was finally selected.

The robot was then dimensioned to fit with the design drivers presented in Subsection 2.3.1. In order to allow taking into account the specifications of the selected components and the design constraints in the procedure explained in Subsection 2.3.2, the computer-aided design (CAD) of the robot was conducted in parallel with its dimensioning and optimization. The final design and dimensions are illustrated in Figure 2.10. This redundant and parallel robot links a fixed base to a mobile platform via two five-DoF robots, each composed of the aforementioned three-DoF actuated structure and a passive universal joint. An extra serial DoF is added on this platform to allow handle rotation. A total of six operational DoFs with force feedback are obtained using seven motors. Note that, traditionally, the first wrist axis of pivot and parallelogram haptic interfaces is implemented along the forearm (third moving link). This very simple solution allows optimally integrating the first wrist motor inside the forearm, thus resulting in a thin and slender design. In this case, however, the robot is in singular configuration as soon as the handle aligns with the forearm. For some positions of the work space, it happens for small handle orientations, well below the specified angle of 70° to 80°. The same holds here. In order to reject these singularities outside the useful work space, additional links were integrated

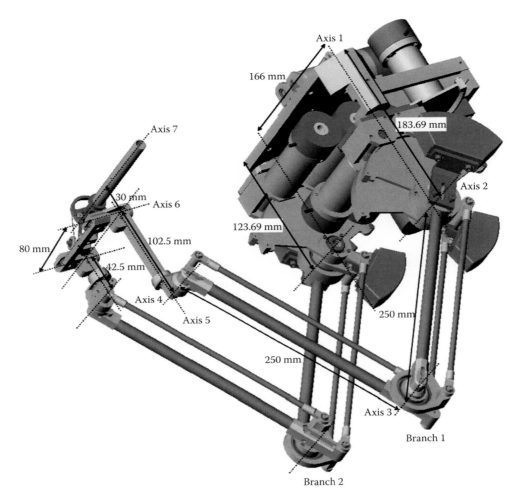

Axis 1

166 mm

183.69 mm

Axis 7

Axis 2

30 mm

Axis 6

123.69 mm

80 mm

102.5 mm

42.5 mm

250 mm

Axis 4

Axis 5

250 mm

Axis 3

Branch 1

Branch 2

FIGURE 2.10
CAD and dimensions of the first hybrid-architecture telesurgery master prototype arm with handle actuator fixed on the base and remotely actuated handle.

between the serial three-DoF robots and the U joints. Their orientation is kept constant, thanks to double parallelograms running in parallel with the arm and forearm.

The arm and forearm are 250 mm long. The mobile platform height is 80 mm. The handle is shifted 30 mm away from axis 6 in order to allow comfortable finger positioning around the handle without collisions with the robot structure. The three-DoF robots are shifted 60 mm relative to each other in order to allow integrating the motors of the two substructures on top of each other and optimize the master arm's compactness. The actuators are high-performance ironless direct-current (DC) motors from Maxon Motor (Sachseln, Switzerland). Maxon RE 35 actuators are used on the three-DoF substructures. They are associated with high-efficiency cable capstan reducers ($R = 19.15$ on first axis and $R = 15.45$ on second and third axes). The same actuator was first used on the handle axis, this motor being fixed on the base and its torque transmitted to the handle with cables and pulleys. This solution, however, introduced too much friction and was finally replaced with a smaller Maxon RE 16 motor and a small cable capstan reducer directly integrated into the mobile

FIGURE 2.11
Second prototype of the hybrid-architecture telesurgery master arm with handle actuator integrated on the mobile platform.

platform. Counterweights are used to ensure the static balancing of the robot. It is worth mentioning that for static balancing, small weights placed away from the axes or heavier masses placed at their close proximity are equivalent. The latter, however, is much better for the dynamic performances of the robot. Small counterweights made of DENAL, a very dense tungsten alloy with density above 18, were used here. Further details of the robot's performances can be found in Gosselin et al.'s work [27], as well as details on its controller.

Two prototype arms were manufactured. The first one is similar to the design in Figure 2.10. The second one, illustrated in Figure 2.11 and used for a virtual intervention by Dr. F. Taha from the University Hospital of Amiens, France, makes use of slightly reinforced links and simplified mechanical parts. Centering pins are integrated on the fixed basis for initial calibration at start-up. This architecture is remarkably simple to implement while having both a large work space, similar to that of serial robots, and a high transparency, similar to parallel structures.

Note that this hybrid architecture has applications also outside abdominal telesurgery. The requirement to have a large work space and a high transparency is also found in virtual reality-based surgical training. In this field, a similar robot was developed and successfully used for the training of the basic skills involved in maxilla facial surgery [28–29].

2.4 Conclusion

In this chapter, we first gave a historic overview of telesurgery. We introduced the limitations of manual surgery and the advantages offered by teleoperated surgical robots. We presented representative existing systems commercially available or developed in research laboratories. We concluded with an opening to SPLS and NOTES, which could be the next steps for less invasive and more efficient surgeries.

In Section 2.2, we made a detailed presentation of the different components of a telesurgery system: the slave robots composed of a positioning stage, an external structure, and a dexterous internal wrist carrying the surgery tools; the master station, which integrates the master arms used to intuitively and efficiently control the slave robots and a video monitor; and the additional equipment and communication means. This section ends with a presentation of the main functionalities of telesurgery robots: position control, position and force scaling, tremor filtering, shared control, and augmented haptic feedback.

Finally, in Section 2.3, we focused on the design of telesurgery master arms. After presenting the design guidelines associated with abdominal telesurgery, we introduced modeling and dimensioning tools allowing to conceive a robot so that it has the specified work space, force capacity, stiffness, and inertia. We also briefly explained how these tools can be successively used to refine the design step by step. We finally briefly presented how this methodology was applied in the design of a telesurgery master arm. This robot proves very interesting and a similar design was also successfully used for surgical training in virtual reality.

References

1. Ghodoussi, M., Butner, S.E., and Wang, Y., Robotic surgery—The transatlantic case, *Proc. IEEE Int. Conf. on Robotics and Automation*, pp. 1882–1888, 2002.
2. Wall, J., and Marescaux, J., History of telesurgery, in *Telemicrosurgery*, Liverneaux, L. et al., Eds., Springer-Verlag, Paris, France, Ch. 2, pp. 15–18, 2013.
3. Marescaux, J., Leroy, J., Gagner, M., Rubino, F., Mutter, D., Vix, M., Butner, S.E., and Smith, M.K., Transatlantic robot-assisted telesurgery, *Nature*, 413, pp. 379–380, 2001.
4. Guthart, G.S., and Salisbury, J.K., The Intuitive telesurgery system: Overview and application, *Proc. IEEE Int. Conf. on Robotics and Automation*, pp. 618–621, 2000.
5. Madhani, A.J., Design of teleoperated surgical instruments for minimally invasive surgery, PhD diss., Massachusetts Institute of Technology, Cambridge, Massachusetts, 1998.
6. Hagn, U., Nickl, M., Jorg, S., Passig, G., Bahls, T., Nothhelfer, A., Hacker, F., Le-Tien, L., Albu-Schaffer, A., Konietschke, R., Grebenstein, M., Warpup, R., Haslinger, R., Frommberger, M., and Hirzinger, G., The DLR MIRO: A versatile lightweight robot for surgical applications, *Industrial Robot: An Int. J.*, 35(4), pp. 324–336, 2008.
7. Konietschke, R., Hagn, U., Nickl, M., Jorg, S., Tobergte, A., Passig, G., Seibold, U., Le-Tien, L., Kubler, B., Groger, M., Frohlich, F., Rink, C., Albu-Schaffer, A., Grebenstein, M., Ortmaier, T., and Hirzinger, G., The DLR MiroSurge—A robotic system for surgery, *Proc. IEEE Int. Conf. on Robotics and Automation*, pp. 1589–1590, 2009.
8. Eindhoven University of Technology, Better surgery with new surgical robot with force feedback, *ScienceDaily*, 2010.
9. Meenink, H.C.M., Hendrix, R., Naus, G.J.L., Beelen, M.J., Nijmeijer, H., Steinbuch, M., van Oosterhout, E.J.G.M., and de Smet, M.D., Robot-assisted vitreoretinal surgery, in *Medical robotics: Minimally invasive surgery*, Gomes, P., Ed., Woodhead Publishing, Cambridge, England, pp. 185–209, 2012.
10. Sutherland, G.R., Latour, I., and Greer, A.D., Integrating an image-guided robot with intraoperative MRI, *IEEE Engineering in Medicine and Biology Mag.*, 27(3), pp. 59–65, 2008.
11. Pandya, S., Motkoski, J.W., Serrano-Almeida, C., Greer, A.D., Latour, I., and Sutherland, G.R., Advancing neurosurgery with image-guided robotics, *J. Neurosurg.*, 111, pp. 1141–1149, 2009.
12. Lum, M.J.H., Friedman, D.C.W., Sankaranarayanan, G., King, H., Fodero II, K., Leuschke, R., Hannaford, B., Rosen, J., and Sinanan, M.N., The Raven: Design and validation of a telesurgery system, *Int. J. Robotics Research*, 28(9), pp. 1183–1197, 2009.

13. Hannaford, B., Rosen, J., Friedman D.W., King, H., Roan, P., Cheng, L., Glozman, D., Ma, J., Kosari, S.N., and White, L., Raven-II: An open platform for surgical robotics research, *IEEE Trans. on Biomedical Engineering*, 60(4), pp. 954–959, 2013.

14. Dallemagne, B., and Marescaux, J., NOTES: Past, present and future, *Asian J. Endosc. Surg.*, 3, pp. 115–121, 2010.

15. Boni, L., Dionigi, G., and Rovera, F., Natural orifices transluminal endoscopic surgery (NOTES) and other allied "ultra" minimally invasive procedures: Are we losing the plot?, *Surg. Endosc.*, 23, pp. 927–929, 2009.

16. Reynoso, J., Meyer, A., Unnirevi, J., and Oleynikov, D., Robotics for minimally invasive surgery (MIS) and natural orifice transluminal endoscopic surgery (NOTES), in *Medical robotics: Minimally invasive surgery*, Gomes, P., Ed., Woodhead Publishing, Cambridge, England, pp. 210–222, 2012.

17. Zhou, Y., Ren, H., Meng, M.Q.H., Tse, Z.T.H., and Yu, H., Robotics in natural orifice transluminal endoscopic surgery, *J. Mechanics in Medicine and Biology*, 13(2), 2013.

18. Vertut, J., Charles, J., Coiffet, P., and Petit, M., Advance of the new MA 23 force reflecting manipulator system, *Second CISM/IFTOMM International Symp. on the Theory and Practice of Robots and Manipulators*, Warsaw, Poland, pp. 307–322, 1977.

19. Vertut, J., and Coiffet, P., *Les robots—Tome 3A: Téléopération, évolution des technologies*, Hermes Publishing, Paris, France, 1984.

20. Massie, T.H., and Salisbury, J.K., The PHANToM haptic interface: A device for probing virtual objects, *Proc. ASME Haptic Interfaces for Virtual Environment and Teleoperator Systems in Dynamic Systems and Control*, pp. 295–301, 1994.

21. Townsend, W.T., The effect of transmission design on force-controlled manipulator performance, PhD diss., Massachusetts Institute of Technology, Cambridge, Massachusetts, 1998.

22. Hayward, V., Gregorio, P., Astley, O., Greenish, S., Doyon, M., Lessard, L. Mac Dougall, J., Sinclair, I., Boelen, S., Chen, X., Demers, J.-P., Poulin, J., Benguigui, I., Almey, N., Makuc, B., and Zhang, X., Freedom 7: A high fidelity seven axis haptic device with application to surgical training, *Proc. Int. Symp. Experimental Robotics*, pp. 445–456, 1997.

23. Thielmann, S., Seibold, U., Haslinger, R., Passig, G., Bahls, T., Jorg, S., Nickl, M., Nothhelfer, A., Hagn, U., and Hirzinger, G., MICA—A new generation of versatile instruments in robotic surgery, *Proc. IEEE/RSJ Int. Conf. on Intelligent Robots and Systems*, pp. 871–878, 2010.

24. Sallé, D., Gosselin, F., Bidaud, P., and Gravez, P., Analysis of haptic feedback performances in telesurgery robotic systems, *Proc. IEEE Int. Workshop on Robot and Human Communication*, pp. 618–623, 2001.

25. Tobergte, A., Passig, G., Kuebler, B., Seibold, U., Hagn, U.A., Fröhlich, F.A., Konietschke, R., Jörg, S., Nickl, M., Thielmann, S., Haslinger, R., Groeger, M., Nothhelfer, A., Le-Tien, L., Gruber, R., Albu-Schäffer, A., and Hirzinger, G., MiroSurge—Advanced user interaction modalities in minimally invasive robotic surgery, *Presence*, 19(5), pp. 400–414, 2010.

26. Friconneau, J.P., Karouia, M., Gosselin, F., Gravez, P., Bonnet, N., and Leprince, P., Force feedback master arms, from telerobotics to robotics surgery training, *Proc. Computer Assisted Radiology and Surgery*, pp. 31–36, 2002.

27. Gosselin, F., Bidard, C., and Brisset, J., Design of a high fidelity haptic device for telesurgery, *Proc. Int. Conf. on Robotics and Automation*, pp. 206–211, 2005.

28. Gosselin, F., Ferlay, F., Bouchigny, S., Mégard, C., and Taha, F., Specification and design of a new haptic interface for maxillo facial surgery, *Proc. Int. Conf. on Robotics and Automation*, pp. 737–744, 2011.

29. Gosselin, F., Bouchigny, S., Mégard, C., Taha, F., Delcampe, P., and D'Hauthuille, C., Haptic systems for training sensori-motor skills, a use case in surgery, *Robotics and Autonomous Systems*, 61(4), pp. 380–389, 2013.

30. Sheridan, T.B., Telerobotics, *Automatica*, 25(4), pp. 487–507, 1989.

31. Hokayem, P.F., and Spong, M.W., Bilateral teleoperation: An historical survey, *Automatica*, 42, pp. 2035–2057, 2006.

32. Köhler, G.W., *Manipulator type book*, Thiemig Taschenbücher, München, Germany, 1981.

33. Hayward, V., Astley, O.R., Cruz-Hernandez, M., Grant, D., and Robles-De-La-Torre, G., Haptic interfaces and devices, *Sensor Review*, 24(1), pp. 16–29, 2004.
34. Gosselin, F., Andriot, C., and Fuchs, P., Hardware devices of force feedback interfaces, in *Virtual reality: Concepts and technologies*, Fuchs, P., Moreau, G., and Guitton, P., Eds., CRC Press, Boca Raton, Florida, Ch. 8, pp. 137–178, 2011.
35. Gosselin, F., Guidelines for the design of multi-finger haptic interfaces for the hand, *Proc. CISM-IFToMM RoManSy Symp.*, pp. 167–174, 2012.
36. Gosselin, F., Développement d'outils d'aide à la conception d'organes de commande pour la téléopération à retour d'effort, PhD diss., University of Poitiers, Poitiers, France, 2000.
37. Toledo, L., Analyse des actions elémentaires en chirurgie endoscopique: Applications au développement d'un instrument basé sur le concept du poignet articulé, Master's thesis, University of Paris 5, Paris, France, 1995.
38. Niemeyer, G., Kuchenbecker, K.J., Bonneau, R., Mitra, P., Reid, A.M., Fiene, J., and Weldon, G., THUMP: An immersive haptic console for surgical simulation and training, *Proc. Medicine Meets Virtual Reality*, pp. 272–274, 2004.
39. Tobergte, A., Helmer, P., Hagn, U., Rouiller, P., Thielmann, S., Grange, S., Albu-Schäffer, A., Conti, F., and Hirzinger, G., The sigma.7 haptic interface for MiroSurge: A new bi-manual surgical console, *Proc. Int. Conf. on Intelligent Robots and Systems*, pp. 3023–3030, 2011.

3

Microsurgery Systems

Leonardo S. Mattos, Diego Pardo, Emidio Olivieri, Giacinto Barresi,
Jesus Ortiz, Loris Fichera, Nikhil Deshpande, and Veronica Penza

CONTENTS

3.1 Introduction

Microsurgery is a term used to classify delicate surgical procedures that require the use of an operating microscope. Still, early microsurgical techniques have been developed based on direct observation of the surgical site, without the assistance of optical magnification systems. This was the case, for example, of otologic operations such as fenestration and stapes mobilization performed in the late 19th century.

The application of microscopes to surgery started in the 1920s. This enabled not only the improvement of early microsurgical techniques but also the development of a large range of new methods, especially in vascular surgery. With microscopes, the surgical connection (anastomosis) of diminutive blood vessels and nerves became realistic, enabling the creation of procedures for replantation and transplantation of tissues and organs. Since then, vascular surgery has been in continuous progress, driving the advancement of microsurgical techniques and technologies.

The history of microsurgery is, nevertheless, somehow controversial as a large number of specialties have incorporated microsurgical techniques into surgical practice over the past two centuries. As a result, different areas report different surgeons as "the father of microsurgery." Even so, it is known that techniques of vascular ligature have been described in the 1500s, and there is general agreement that the first use of a microscope in surgery was by Carl-Olof Siggesson Nylén in 1921 [1].

Currently, microsurgeries are carried out under high-magnification operating microscopes that offer magnificent three-dimensional views of the surgical site. In addition, dedicated microsurgical tools and suture materials are now available for these operations. However, microsurgeries are still performed by the dexterous hands of highly capable clinicians, who go through extensive training periods to acquire the specialized skills necessary for realizing successful microoperations.

The new frontier in microsurgery is the surgical robot or, more precisely, robot-assisted microsurgery. As this chapter will show, robotics can provide increased dexterity, controllability, and visualization capabilities to surgeons, allowing the execution of previously highly difficult or unfeasible procedures or even the pioneering of new techniques. Robotics can also eliminate the need for microscopes and their associated requirement for direct line of sight to the surgical site, thus enabling minimally invasive microsurgical procedures. They also offer great potential to increase the precision, safety, and quality of microsurgeries, much more so than for other types of (larger) surgeries.

Robot-assisted microsurgery is an emerging discipline, but it is rapidly establishing itself through the clinical use of the surgical robot da Vinci and the development of other robots and systems dedicated to this area. New technologies include, for example, assistive systems for safety supervision, intraoperative planning, augmented visualization, and force-controlled operations.

This chapter introduces the reader to the wonderful world of microsurgery systems, starting from an overview of major specialty areas that rely on microsurgical techniques and of existing tools used for these operations. Existing and under-development microsurgical robotic systems are also presented, leading to an analysis of current major challenges associated with the development of such technologies. Finally, concluding remarks about the future of microsurgery systems are offered.

3.2 Clinical Applications

Microsurgery systems and techniques are used in a large range of surgical specialties, from otolaryngology and ophthalmology to plastic and general surgery. Overall, microsurgical procedures are technically demanding, requiring high degrees of surgical skills and high levels of dexterity from surgeons. A large percentage of microsurgeries involve the anastomosis of small blood vessels and nerves, while other important applications involve precise excision of tissue from delicate organs, typically to treat benign or malignant lesions. Significant clinical applications are illustrated in Figure 3.1 and described below.

3.2.1 Pediatric and Fetal Surgery

Microsurgery techniques are required during many pediatric and fetal surgeries due to the small nature of the patients. The diminutive size and fragility of structures in infants and fetuses are ideally suited to the skills of microsurgeons. Significant examples of pre-natal microsurgeries include atrial septostomy (opening of the interatrial septum in the heart), valvuloplasty (opening of the aortic or pulmonary heart valves), the treatment of congenital diaphragmatic hernias, and the treatment of twin–twin transfusion syndromes.

3.2.2 Ophthalmology

Similarly as with the case above, the size and delicacy of eye structures require the use of microsurgical techniques when operating on this organ. Significant procedures here include vitreoretinal surgery, cataract surgery, and retinal vein cannulation.

3.2.3 Otolaryngology

Historically, otolaryngology was the first medical specialty to use microsurgical techniques. The need in this case also comes from the diminutive size of structures in the ear, nose, and throat (ENT) and from their high importance to the life and well-being of the patients. Typical examples in this case include the excision of benign and malignant lesions, stapedectomy operations for treatment of otosclerosis, and the implantation of hearing aids.

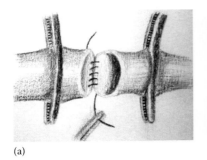
(a)

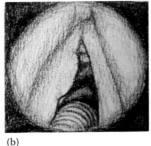
(b)

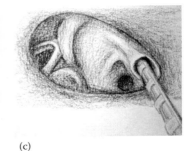
(c)

FIGURE 3.1
Examples of clinical applications of microsurgery systems: (a) vascular anastomosis, (b) excision of vocal cord carcinoma, and (c) cochlear implantation.

3.2.4 Plastic Surgery

This is likely the field that makes the most extensive use of microsurgical techniques. In addition to cosmetic surgeries, plastic surgeries involve highly important and lifesaving procedures such as replantations and reconstructive procedures, which rely on the anastomosis of diminutive blood vessels and nerves. These procedures are most often performed to correct functional impairments caused by traumatic injuries, but may also be done to recreate normal appearance after devastating illnesses or traumas. Common reconstructive surgeries include hand surgery, laceration repair and replantation of limbs, cleft lip and palate repair, ear reconstruction, and breast reconstruction after a mastectomy.

3.2.5 Nerve Surgery

Microsurgical techniques are fundamental for the treatment of abnormal nerve conditions such as nerve entrapments, nerve ruptures, and brachial plexus injuries. These operations involve delicate nerve repairs through direct anastomosis or the implantation of nerve grafts.

3.2.6 Urology

Within this specialty, microsurgery is used intensely in the treatment of male infertility and chronic testicular or groin pain. One common procedure in this area is microscopic vasectomy reversal, for which the use of microscopes and microsurgical techniques have brought greater patency and fertility rates.

3.3 Microsurgery Systems in Clinical Use

The key element in current microsurgery systems is the microscope. This device is required to provide magnified visualization of the surgical site to the surgeon, allowing the execution of the highly delicate microsurgeries. As mentioned previously, microscopes started to be used in surgeries in the 1920s and since then they have been greatly improved to offer higher magnifications, wider field of view (FOV), and high-quality undistorted views of the surgical site from relatively long focal distances. Major manufacturers of surgical microscopes include Carl Zeiss, Leica Microsystems, and several other smaller companies like ATMOS, Haag-Streit, Takagi, and Alcon. Most of these companies offer different types of microscopes specifically developed for different microsurgical specialties. Figure 3.2 shows examples of microscopes for throat, eye, and ear microsurgery.

Dedicated sets of microsurgical tools are also available to each different surgical specialty. These allow the surgeon to manipulate delicate tissue with precision and realize accurate cuts, ablations, injections, or challenging anastomosis. Microsurgical tools are often handheld. They complete the traditional surgical system as schematized in Figure 3.3. Examples of state-of-the-art tools for eye microsurgery and for laryngeal operations are shown in Figure 3.4. Differences between the tool sets are evident and expected as each type of microsurgery has very particular requirements.

Another type of tool that is being increasingly used in microsurgery is the surgical laser. Lasers allow the execution of precise incisions and ablations on both soft and hard

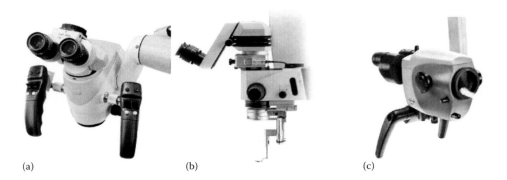

(a)   (b)   (c)

FIGURE 3.2
Examples of operating microscopes: (a) Zeiss Sensera for laryngeal surgeries, (b) Leica M620 for ophthalmic surgery, and (c) ATMOS iView microscope for ear surgery. (Courtesy of Carl Zeiss, Jena, Germany; Leica Microsystems, Wetzlar, Germany; and ATMOS, Lenzkirch, Germany.)

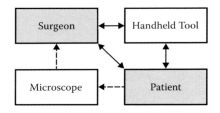

FIGURE 3.3
Components of traditional microsurgery systems: surgeon, patient, microscope, and handheld microsurgery tools. The solid arrows indicate direct interaction, while the dashed arrows indicate limited interaction (visual information flow only).

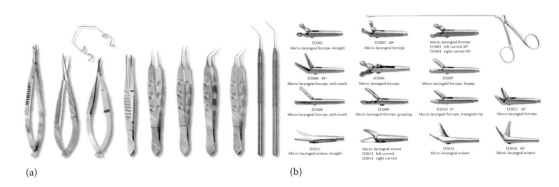

(a)   (b)

FIGURE 3.4
Precision tools for (a) eye microsurgery and (b) laryngeal microsurgery. (Courtesy of Beaver-Visitec International, Waltham, Massachusetts, and Orient Medical, Hangzhou, China.)

tissues. In addition, they have demonstrated to be a viable solution for tissue soldering [2], with potential application for several types of tissues, including cornea [3], liver [4], dura, trachea, urinary bladder, and blood vessels [5]. Currently, lasers are a major tool in both laryngology and ophthalmology. A popular example is the laser-assisted in situ keratomileusis (LASIK) procedure for eyesight correction. As other microsurgical tools, lasers are selected and customized for specific surgical applications. The reason here is that laser–tissue interaction characteristics, such as absorption coefficient, penetration depth, and scattering, are highly dependent on the laser light wavelength and the target tissue properties. For example, excimer lasers (193 nm wavelength) are used in eye surgery, while CO_2 lasers (10.6 µm wavelength) are the preferred type in laryngeal microsurgeries.

The use of lasers in microsurgeries requires additional specialized systems for controlling the aiming and motions of the laser beam. In LASIK this control is highly automated through motorized mirror systems. Conversely, in other types of microsurgery it depends totally on the dexterity of the surgeon, who manually guides the laser. Nevertheless, tools and devices do exist to facilitate the work of the surgeon. In the case of optical fiber–coupled lasers, the fiber can be attached to handheld tools such as those shown in Figure 3.5, which assist the surgeon in bringing the laser energy to the surgical site and in controlling it. In these systems, due to the divergence of the laser coming out of the fiber, the effect of the laser on tissue depends on the distance between the fiber tip and the tissue. Lasers normally have a cutting effect when the fiber is in contact with the tissue, but present ablative or coagulative effects as the distance between the fiber and the tissue increases.

Lasers in microsurgery can also be delivered through air. The devices in this case are called free-beam laser systems and typically rely on movable mirrors to bring the laser to the surgical site in a controllable fashion. Fine laser control is achieved through the use of a laser micromanipulator device. Traditionally, this is a hand-controlled mechanical system that allows the surgeon to precisely control the motions of a mirror and, thus, the aiming of the laser. Several laser micromanipulator systems are currently commercially available for microsurgeries, including systems from the companies Lumenis, KLS Martin, and DEKA. One example is shown in Figure 3.6. In addition to the controllable mirror, these systems include focusing optics with long focal distance, which are specifically designed for the intended surgical application. For example, the focusing distance in ear microsurgery is typically 200 mm, while in laryngeal microsurgery this distance is normally 400 mm.

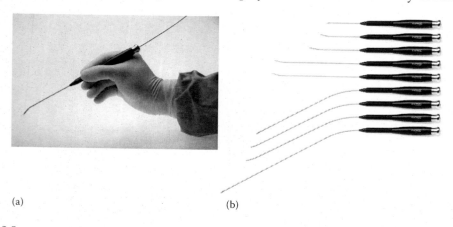

(a)   (b)

FIGURE 3.5

The OmniGuide handpieces for laser microsurgery: (a) demonstration of the usage of handpieces, and (b) various handpiece products for different surgical procedures. (Courtesy of OmniGuide Inc., Cambridge, Massachusetts.)

FIGURE 3.6
The Micro Point 2 R™ laser micromanipulator from KLS Martin. (Courtesy of the KLS Martin Group, Freiburg, Germany.)

The quality of laser microsurgeries has seen great improvements with the development of pulsed laser systems and scanning laser micromanipulators [6]. These technologies allow improved control over the laser incision depth and minimize thermal damage to surrounding tissue, which translates into increased surgical precision, minimal formation of scar tissue, and reduced healing time. Examples of such systems include the AcuBlade™ (Lumenis), the SoftScan Plus R™ (KLS Martin), and the Hi-Scan Surgical™ (DEKA).

Scanning laser micromanipulators are one of the pioneering computer-assisted systems to be used clinically in microsurgeries. Even though they do not actually enhance the surgical user interface (since the microscope and a traditional laser micromanipulator are still required), they prove such technologies can enhance the surgeon's capabilities, facilitating, and improving surgical procedures. The next section will present more advanced examples of computer- and robot-assisted systems specifically developed or adapted for microsurgeries.

3.4 Robot-Assisted Microsurgery Systems

Robotics has played a major role in the success and quality of microsurgeries. Many companies and research groups concentrated toward the development of new surgical tools and systems to augment the capabilities of microsurgeons. These efforts include the creation of handheld robotic tools and teleoperated robotic systems, which makes the operating setup more ergonomic and intuitive, increases the precision and controllability of surgical instruments, and eliminates the need for microscopes and their associated requirement for direct line of sight to the surgical field. The incorporation of robotics in microsurgery modifies the overall surgical system configuration as presented in Figure 3.7.

Robot-assisted systems offer vast potential to improve microsurgeries through both new hardware and software tools. The insertion of such systems in the surgical setup allows for the augmentation of the surgeon's actions (augmented actuation) and of the feedback he/she receives from the surgical field (augmented feedback). In fact, robotics transforms the traditional surgical system, offering the chance to add assistive features to any of the identified blocks in Figure 3.7. For example, software algorithms can improve the identification of diseases through video processing [7]; mechatronic systems can improve tissue manipulation by providing motion scaling and hand tremor filtration [8]; sensors and haptic devices can allow the surgeon to feel fine interaction forces [9]; and virtual objects

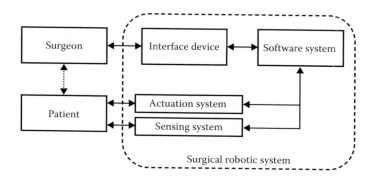

FIGURE 3.7
Robot-assisted surgical system components. The dotted arrow indicates that the surgeon may not have the possibility to directly interact with the patient during surgery.

can be defined to protect delicate structures or even offer guidance to improve surgical precision [10].

Due to the availability of the Intuitive Surgical's da Vinci robot in hospitals around the globe, the clinical use of such a system for microsurgeries is steadily growing. This robotic system, shown in Figure 3.8, can be defined as a highly advanced teleoperated manipulation system, which is located on the side of patient and controlled by the surgeon from a remote console. It offers an immersive 3D operating environment from which the surgeon can observe magnified views of the surgical site and precisely control small articulated surgical instruments. This is accomplished through an intuitive interface that translates the motions of the surgeon's hands into scaled-down movements of the robotic instruments, resulting in significantly improved surgical accuracy in small operating spaces. Even though this robotic system was originally developed for general laparoscopy, microsurgeons have been demonstrating that it can also bring great benefits to a growing number of clinical procedures in urology and in vascular, nerve, and plastic surgery.

The da Vinci system presents, however, several drawbacks related to its application in microsurgeries. These include the large size of the system and the associated requirement for a large operating room; the high acquisition and operative costs (associated with

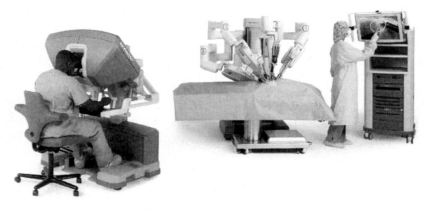

FIGURE 3.8
The da Vinci surgical system. (Courtesy of Intuitive Surgical Inc., Sunnyvale, California.)

disposables, tool life, etc.); the need for specifically trained operating room personnel; and the fact that the surgeon operates from a remote and nonsterile location, which can be problematic in case of emergencies. These issues and also parallel efforts toward the development of specialized robotic systems for specific microsurgery applications have resulted in the creation of several other systems over the last two decades. Significant examples are presented below.

One of the first examples of robot-assisted microsurgery systems was created in 1994 at the National Aeronautics and Space Administration's (NASA) Jet Propulsion Laboratory (JPL) in collaboration with the company MicroDexterity Systems. The system, shown in Figure 3.9, was named robot-assisted microsurgery (RAMS) and consisted of a six-DoF master–slave telemanipulator with programmable controls [11]. It was intended to be a general purpose system for delicate microsurgeries in the brain, eye, ear, nose, etc., offering force reflection and scaling capabilities, miniature forceps for manipulation, and dual-arm coordination. However, the technology has never reached the market for clinical use.

Another early example of robot-assisted microsurgery system is the steady-hand robot created by Taylor et al. at the Johns Hopkins University (United States) in 1999 [12]. This system introduced the concept of cooperative-controlled robot assistant, which allows the surgical tool to be held simultaneously by both the operator's hand and a specially designed actively controlled robot arm. The system was created for microsurgery augmentation and found important applications in retinal microsurgery. Researchers continue to improve the system design to bring it to clinical use (see Figure 3.10) [13]. In addition, the control principle has been translated to other microsurgery applications, such as in ENT surgery [14].

Yet another example of ophthalmic microsurgical robot conceptualized and developed during the 1990s, and which is still focus of research and development efforts, is the Micron system [8,15]. This system, shown in Figure 3.11, is a fully handheld actively stabilized tool for microsurgery. It improves the surgeon's accuracy during precision manipulation tasks by removing involuntary hand motions such as tremor.

Significant examples of robot-assisted microsurgery systems have also been developed for otologic applications. Within these, the robotic microdrilling tool for cochleostomy developed by Brett et al. [16] is an additional example of a great microsurgery system from the 1990s that is still trying to reach the clinical market. This system is able to create

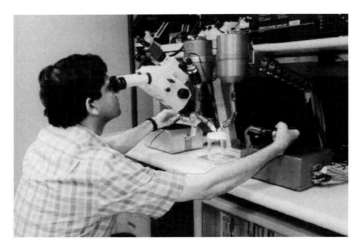

FIGURE 3.9
RAMS. (Courtesy of NASA/JPL–California Institute of Technology, La Cañada Flintridge, California.)

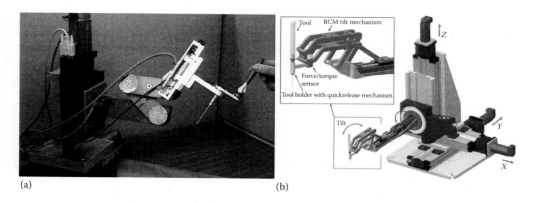

(a)   (b)

FIGURE 3.10
Steady-hand eye robot evolution: (a) original version from 1999 and (b) Eye Robot 2.1 design from 2012. (Courtesy of the Computer Integrated Interventional Systems Laboratory, Johns Hopkins University, Baltimore, Maryland.)

FIGURE 3.11
Micron: an actively stabilized handheld tool for microsurgery. (Courtesy of the Robotics Institute, Carnegie Mellon University, Pittsburgh, Pennsylvania.)

a fenestration on the outer bone tissue of the cochlea without damaging the underlying endosteal membrane by detecting changes in the drilling environment, a capability that is highly valuable in hearing aid implantation. The level of precision and difficulties associated with this specific application has also inspired Salzmann et al. to develop an image-guided robot for precise bone drilling during otologic surgery [17]. These two systems are presented in Figure 3.12.

Within the laryngology area, significant examples of robot-assisted microsurgery systems include the work of Giallo [18] and Mattos et al. [19], who focused on improving laser phonomicrosurgeries, i.e., the highly delicate laser-based operations performed on vocal cords. Giallo sought to create an improved surgeon interface based on the motorization of a commercial laser micromanipulator and the use of an electronic joystick for aiming control [18]. Mattos et al., on the other hand, proposed a new motorized laser

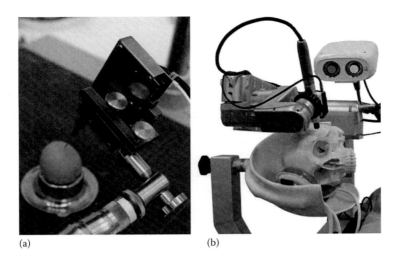

(a)   (b)

FIGURE 3.12
Robotic systems for otologic microsurgeries: (a) smart surgical drill from Brunel University (United Kingdom) and (b) miniature robot for hearing-aid implantation from the University of Bern (Switzerland).

micromanipulator and the implementation of the virtual scalpel concept, which allows surgeons to perform operations by using a stylus and a graphics tablet on which a live video of the surgical site is displayed [19] (see Figure 3.13). The success of this latter system associated with its robotic capabilities, such as intraoperative surgical planning and fast and precise laser scanning, attracted attention to the area and resulted in the formation of the European research consortium Micro-Technologies and Systems for Robot-Assisted Laser Phonomicrosurgery (μRALP) [20], which became the first microsurgery robotics program funded by the European Union. μRALP is an ongoing research project expected to greatly improve laryngeal operations in terms of precision, controllability, safety, and efficiency through the development of novel technologies for endoscopic laser microsurgery. In this process, it is facing many of the challenges described in the next section.

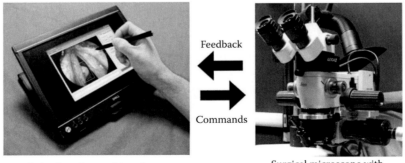

Feedback

Commands

Stylus-based control interface   Surgical microscope with motorized laser micromanipulator

FIGURE 3.13
The Virtual Scalpel robot-assisted laser microsurgery system. (Courtesy of the Department of Advanced Robotics, Istituto Italiano di Tecnologia, Genova, Italy.)

3.5 Current Challenges for Next-Generation Microsurgery Systems

The future of microsurgery is undoubtedly connected to robotic systems. As the examples presented in the previous section demonstrate, robotics has a great potential to revolutionize microsurgeries by offering technological solutions to major challenges associated with these procedures, which currently include access, exposure, visualization, and control. Access relates to problems in safely reaching and bringing tools to the surgical area. Exposure relates to the problem of having enough space to manipulate and operate on the diseased tissue. Visualization relates to difficulties in observing the surgical field and identifying the diseased tissue. Finally, control relates to issues in accurately controlling the surgical instruments during the operation.

Robotics can address all of these clinical issues and offer significant assistance to surgeons, enabling them to perform more precise, more effective, safer, or even currently impossible microsurgical procedures. However, several technological challenges have yet to be surmounted before robotic systems can achieve their full potential and become required clinical tools in microsurgery operating rooms. This section presents a summary of these challenges and describes recent research progress toward their solution.

3.5.1 Miniaturization

The shift toward miniaturization is seen as one of the most important trends in surgical devices. Similar models of development from other industries, such as semiconductor devices and electronics, have laid the foundation for this trend. Smaller and lighter cellular phones, laptop computers, and digital cameras, among others, serve the ever-growing demand for double the functionality at half the size and cost. *Miniaturization* is seen as the core technology of the *third* industrial revolution in the 21st century [21]. Richard Feynman coined the term *miniaturization* in the context of manufacturing in his seminal lecture [22]. He drew the parallel between biological cells, which store information as well as maneuver and make substances, and miniature manufactured devices that could be of similar scale and serve that purpose. Some of the existing technologies that lie at the heart of this revolution include (i) single-chip, multipurpose processing and sensing, (ii) micro- and nanoelectromechanical systems for sensing and actuation, and (iii) miniaturized, high-capacity memory chips. For the surgical devices, miniaturization lies at the convergence of the evolutions in the semiconductor, electronics, and manufacturing technologies and the requirements of modern surgery. Consequently, every part of the surgical device is a candidate for miniaturization.

Miniaturized surgical equipment can bring significant advantages to microsurgical procedures, including improved access to the surgical site, greater precision and accuracy, and faster actuations. However, the actualization of such systems poses stiff challenges to designers, especially in terms of implementation and control. Brief descriptions of the challenges are presented below.

3.5.1.1 Materials and Robustness

A key step to enabling miniaturized surgical devices is the development of robust yet flexible milli- and micrometer scale designs. As is noted by Hsu [21], similar micromachining technology as that used in the fabrication of silicon-based integrated circuits (ICs) is used in the manufacture of microelectromechanical systems. Surgical devices are complex

3D structures with moving parts and it is a considerable challenge to replicate the larger surgical devices at a small scale in a cost-effective manner, with the correct mechanical properties. The lack of standards in design, material selection, and packaging compounds the challenge manifold.

A shift toward utilizing physicochemical processes, newer materials, and techniques is inevitable [23]. Processes such as wafer bonding [24], stereolithography [25], and self-assembly [26] can enable complex 3D structure fabrication. Smart composite microstructures [27] of carbon, silicon, polymer, etc., can provide for robust structures with flexible joints. For tools, metals such as stainless steel, platinum–tantalum, and nickel–titanium or natural materials such as gelatin and collagen can be used.

3.5.1.2 Maneuverability

Any surgical device is characterized by its dexterity and maneuverability at the surgical site. As the devices are miniaturized, they can reach more areas of the human anatomy. Yet in targeting the devices to a surgical site, their entry, reach, stabilization, operation, and eventual retrieval are major challenges.

For tethered devices, continuing research by Simaan [28], Bergelas and Dupont [29], and Burgner et al. [30], among others, have not only brought the challenges into greater focus but also demonstrated the utility of robotically controlled miniature snakelike and concentric tube catheters in improved surgical access and device movement. Planning and executing stable trajectories for the catheters and end effectors is still an open challenge.

Tetherless devices hold great promise due to their compact nature, the reduced number of components, and the simplicity of introducing them in the body. Iddan et al. [31] and Gettman and Swain [32] explore the challenges of guidance and motion of the capsules to reach target sites.

3.5.1.3 Sensing

In several robot-assisted procedures, the surgeon is no longer required to directly handle tissue or peer through incisions in the skin to visualize the surgical site. The robotic devices provide the tools to accomplish such tasks. Yet, maintaining the tactile and visual feedback for the surgeons remains a significant challenge. Force/Tactile feedback aids precise manipulation, grasping, or palpation of soft organs. It also improves the transparency in master/slave teleoperation of robotic devices [33]. Additionally, biophysical information of the surgical site such as tissue density and temperature can assist the surgeon in improving surgical actions and outcomes.

Thanks to the enormous growth in the cell-phone industry, high-resolution miniature cameras with system-on-chip technology have become ubiquitous [34]. The presentation of 3D visualization to the surgeon is critical for a holistic view of the surgical site. For tactile feedback, researchers have investigated six- and three-DoF miniature force sensors [35–36]. It is important to translate the signals from these sensors into intuitive feedback for the surgeons. For biosensors, nanotechnology should play a vital role in the fabrication. This requires the investigation of different types of physical principles and the development of a method to produce these devices. Most miniature catheters are sub–10 mm [30]. Adapting these different sensing technologies to the tip of the catheters or miniature capsule endoscopes, optimizing their location at the surgical site for maximum sensing capability, and translating their signals for maximum utility by the surgeons are some of the open challenges in miniaturization technologies.

3.5.1.4 Actuation

Surgical actions generally involve gripping, manipulating, dissecting, and suturing tissue. The mechanical components that execute these actions, such as grippers, tweezers, cutters, forceps, and needles, have to be actuated in a manner that will allow optimal force exertion without causing trauma or damage at the surgical site. The actuation mechanisms face critical challenges in miniaturization with the size of the moving parts dictated by the type of surgery, the anatomical part involved, and the consequent forces and torques required. The miniaturization challenges include (i) the design of the actuation mechanisms; (ii) the transmission of the actuation forces to the surgical site; (iii) the size, number of components, and materials which can satisfy the constraints and requirements of the surgery; and (iv) the maximum force and torque the actuators can output.

3.5.2 Microsurgical Tools

The creation and utilization of tools that are small enough to pass through the various conduits in the body is an emerging area of medical innovation in the microsurgery domain. It involves the fabrication of miniaturized devices having physical, chemical, mechanical, and electronic functionalities that can be controlled or autonomously trigger based on the application. Broadly, the tools can be divided into two categories: (i) sensing tools that measure the surgical site and (ii) actuation tools that manipulate and affect the site.

3.5.2.1 Sensing Tools

As discussed above, miniaturized surgical instruments bring the advantage of accessibility. But as a consequence, the surgeons have to completely rely on the dexterity inherent to the tools, which reduces (or completely eliminates) their ability to directly visualize or to have tactile feedback of the surgical site. The challenges lie in reintroducing these feedbacks for the surgeons, using similarly miniaturized sensing devices.

3.5.2.1.1 Force/Tactile Sensing

Surgical devices with force sensing capability can provide surgeons with force information that (i) helps "feel" the tissue at the site, (ii) guides through the surgical maneuvers, and consequently (iii) improves the safety and efficacy of the procedures. Following the widespread adoption of laparoscopic surgery [37], researchers have carried out extensive developments of force sensors. The technologies adopted for developing miniaturized force sensors need to satisfy size and sterilization constraints, as well as cover the critical force range in tissues. A review of the research reveals that most minimally invasive surgery (MIS) devices require sub–10 mm sensors and need to cover a force range of up to 40 mN while allowing sub-millinewton resolutions. Achieving these requirements are currently major challenges associated with the development of sensing tools applicable to microsurgery.

3.5.2.1.2 Imaging

The quality of the surgical site visualization directly impacts the quality of surgical outcomes. Recent trends toward miniaturization in medical instruments highlight the functionality of miniature cameras. Commonly denoting a total diameter of less than 3 mm, miniature cameras bring new challenges together with their inherent benefits [38]. These are intrinsically related to the camera technology and to parameters such as focal length, field of view, size, frame rate, and resolution.

3.5.2.1.2.1 Fiber-Based Cameras This imaging method involves the use of distal-tip optical lenses connected to fiber-optic bundles containing thousands of individual fibers. In this case the imaging sensor is located at the proximal end, facilitating the capture of best-quality images. Fiber-based cameras can easily go down to sub–1 mm size. For example, Jacobs et al. [39] successfully demonstrated the use of a LaDuScope (PolyDiagnost, Pfaffenhofen an der Ilm, Germany), a 0.55 mm camera with 70° FOV, a 75 mm focal length, and up to 6000 pixel resolution. The Richard Wolf [40] company also has multiple rigid and flexible endoscopes using this technology. Yet these cameras suffer from shortcomings like (i) "blank spaces" between the bundled fibers, (ii) high cost, (iii) fragility of the fibers, (iv) limited radius of curvature, and (v) limited field of view and angle of view. To overcome some of these issues, researchers are pursuing novel algorithm-based solutions that use a single fiber and different light patterns that can allow the reconstruction of the observed field from reflected patterns [41–42].

3.5.2.1.2.2 CCD/CMOS-Based Cameras Having the camera sensor and lens at the distal tip itself overcomes some of the above issues. This has been made possible by the recent development of miniaturized charge-coupled device (CCD)/complementary metal–oxide–semiconductor (CMOS) technology driven mainly by the cell-phone and smartphone markets. Several companies have developed and now offer medical imaging products using miniature cameras. These include Olympus, Awaiba, Medigus, Misumi, etc. The remaining challenges related to the use of this technology in new microsurgical systems are currently centered on integration, packaging, and sterilization issues. In addition, the wish for smaller cameras with higher resolution continues to be a significant challenge to miniature camera manufacturers.

3.5.2.2 Actuation Tools

As mentioned earlier, the problem of navigating microsurgical tools to the surgical site continues to be a major challenge associated with their development. Researchers have been exploring different technologies to solve this issue, including planar and spatial linkages [43] and serial articulated wrists [44], as well as snakelike devices [45–46].

Another important challenge in this case is the surgical requirement for the manipulation of tissue using different tools, such as grippers, cutters, and forceps. This need for multifunctionality is aggravated by the desire for small tools that facilitate surgical site access and by the need to provide enough force to manipulate tissue. Researchers are trying to solve these issues through the investigation of novel processes for microforcep fabrication and actuation [47–49]. In addition, alternative ways of powering microsurgical tools are being sought, such as intrabody biochemical processes [50]. Companies such as Alcon, Medtronic, and Sklar are active in this domain.

Finally, exposure of the surgical site can be highly challenging during some microsurgery applications, especially when these are performed in narrow and confined spaces, such as the case of minimally invasive brachial plexus operations. Consequently, the capability to appropriately expose the surgical area is also major challenge in the development of future microsurgical tools.

3.5.3 Visualization Methods and Systems

The definition and implementation of appropriate visualization methods and systems continue to be a challenge associated with the development of new microsurgery systems. As described earlier, in microsurgery the surgeon typically visualizes the surgical scene through a microscope. However, the surgeon does not necessarily need to use the microscope binoculars directly, since most of the microscopes are suitable for mounting mono or stereo cameras on them. The use of cameras on a microscope allows decoupling the surgeon from it, enabling a reconfiguration of the surgical system setup and improvements to its ergonomics. In addition, the use of electronic displays facilitates the incorporation of augmented-reality (AR) techniques to the surgical system, bringing advantages such as the possibility of adding extra information to the surgeon on the video streaming.

In endoscopic microsurgery, the microscope has already been substituted by an endoscope with cameras mounted on its tip or, externally, at a distal point. In these cases, the visualization system is equivalent to a microscope equipped with cameras. Consequently, here also the visualization system is critical. The main aspects to take into account are the following:

- **Stereo:** It is important to have a system with stereo capabilities, since the depth perception helps the surgeon better understand the clinical situation and leads to improved performance. However, the visualization has to be properly designed to avoid user discomfort; i.e., the left and right images must be synchronized and properly aligned or corrected [51].
- **Quality:** The image quality has to be comparable with the one of the direct visualization. Low image quality can lead to surgical errors or surgeon discomfort.
- **Frame rate:** The frame rate is the number of images displayed per unit of time. The effect of the frame rate is linked to controllability and surgical performance. The range of acceptable frame rates is placed between 10 and 30 frames/s. Users can adapt to lower frame rates, but that compromises performance and comfort [52].
- **Delays:** The delays correspond to the time between the acquisitions and displays of an image. Users can easily adapt to delays below 50 ms in teleoperated systems. Higher delays impact user performance and cause discomfort [52].

Unfortunately, quality, frame rate, and delay are linked together. A high-quality image requires more bandwidth for transmission. If this capability is limited, it can cause reductions in frame rate and an increase in transmission delay. Therefore, it is important to find a good trade-off between these three parameters.

3.5.3.1 Visualization Devices

The creation of an ergonomic and effective operative setup is the first challenge to be overcome when developing new microsurgery systems. This involves careful selection and configuration of visualization devices. Several kinds of stereoscopic visualization systems can be used to display video coming from the microscope or endoscope cameras. Table 3.1 presents a summary of the advantages and disadvantages of the following devices:

- **Head-mounted display (HMD):** An HMD is usually a helmet or a pair of glasses featuring a small optic display in front of each eye. Helmets are usually heavier but better cover the visual field, increasing the immersion. In contrast, glasses are lighter but provide less immersion.

TABLE 3.1

Comparison of Existing Technologies of Stereoscopic Visualization Devices

Device	Advantages	Disadvantages
Head-mounted display	• High mobility	• Wearing heavy helmet/glasses • Lower image quality than that of a screen • Too immersive (device covers most of the visual field)
Virtual binoculars	• Similar visualization as the classical microscope • No need for wearing glasses	• Lower image quality than that of a screen • Lower mobility
Shutter glasses + monitor	• Several people can see the scene in three dimensions • Good image quality	• Wearing glasses • Lower image brightness (each image left/right is shown half of the time)
Autostereoscopic display	• Several people can see the scene in three dimensions • No need for wearing glasses	• The correct visualization depends on the position of the viewer relative to the display

- **Virtual binoculars (VBs):** This system is technically equivalent to the HMD, but it is mounted on a fixed position and is meant to be used as the classical microscope binoculars. There are commercial microscope systems that include their own VBs, but the concept can be applied to any microscope equipped with cameras or even to endoscope systems.

- **Shutter glasses + 3D monitor:** A 3D monitor is a screen that is able to display images at a high frequency, alternating left and right images. The shutter glasses can, synchronously with the monitor refresh rate, occlude selectively each eye. This way the user sees different images with the left and right eyes, creating the stereoscopic effect. The shutter glasses can be wireless and light and have good autonomy.

- **Autostereoscopic display:** This kind of monitor displays simultaneously the left and the right images but, thanks to the construction of the screen, each image can be seen only with one eye. The correct visualization of the images and the stereoscopic effect work only at specific viewer positions [53].

3.5.3.2 Augmented Reality

Another challenging area related to visualization systems for surgery is augmented reality. The use of AR techniques can significantly improve the visual perception of the surgical area. Such techniques can be used to combine video streaming with computer-generated images, allowing the addition of relevant and useful information directly in the surgeon's field of view. However, the correct combination of different information in a consistent image is a complex task, especially when the scene being observed is highly deformable such as in soft tissue microsurgery. Thus, a correct tracking of features and estimation of tissue deformations is mandatory for an efficient surgical AR system.

Examples of important potential uses of AR in surgery include the following:

- **Highlighting of critical tissue:** This can offer significant help to surgeons. For example, cancerous tissue can be highlighted to assist in the identification and definition of surgical margins. Or, on the contrary, AR can be used to highlight and protect delicate tissue such as blood vessels and nerves. The detection and identification of the critical tissues can be done in a preoperation phase or during operation using image-processing techniques. These techniques are currently under research and are hardly used in real surgical scenarios.

- **Display of assistive information:** AR can also be used to render operation information, such as cutting paths or ablation areas planned graphically. This is the case, for example, of the computer-assisted laser microsurgery system presented by Mattos and Caldwell [54]. In this system, laser trajectories defined intraoperatively enable subsequent automatic laser control and precise execution of the planned action [54].

3.5.4 Haptic Feedback

Providing force and tactile feedback to surgeons is a major goal in the surgical robotics area. This type of information is expected to bring safety and precision to telesurgery and also to enable the use of common diagnosis methods such as palpation and thermal inspection, which have became impossible with the use of surgical robots. Due to the delicacy of tissues typically involve in microsurgeries, haptic feedback has the potential to significantly improve these procedures and, thus, constitute a major challenge area in this context.

The term *haptic* comes from the Greek verb ἅπτεσθαι (*haptesthai*), which means "to touch." It refers to tactile interaction and can be extended to include the perception of objects and the self in space. As a perceptual system, haptics relies on the person's cutaneous and kinesthetic subsystems [55]:

- **Cutaneous subsystem:** This subsystem consists of mechanoreceptors, thermoreceptors, and chemoreceptors embedded in the skin. It is responsible for detecting *static* characteristics of touched items, such as roughness and smoothness; shape and orientation; stickiness; compliance; and temperature.
- **Kinesthetic subsystem:** This subsystem uses mechanoreceptors of skeletal striated muscles, tendons, and joints. It is responsible for acquiring *dynamic* information from limbs, which are highly important in manipulations. These include position, orientation, and movements in space.

By merging information acquired from these two subsystems, the human brain can reconstruct more complex data, such as weight and inertial characteristics, or the shape of objects. This capability has fundamental importance in everyday life [56] and, thus, also for surgical procedures [57].

The challenges involved in adding haptic feedback to robot-assisted microsurgery systems currently lay on both the integration of sensors into the microsurgical tools and the creation of appropriate interface devices capable of rendering haptic information to the users. Current research and technological development efforts are expected to bring significant improvements to these areas in the near future.

3.5.5 Control Interfaces and Ergonomics

Because microsurgery targets a site that is beyond the typical direct perceptual and manual capabilities of humans, a critical aspect of this field is the usage of technologies that mediate and assist the microsurgeon's activity during the visualization of the target, the control of surgical tools, and the management of the multisensory feedback derived from tool–target interactions, all of which occur within a specific space: the control interface.

In particular, microsurgery technologies should offer a control interface able to match the skills, limits, and needs of the surgeon in order to increase his/her performance and the safety of the procedures. Considering the multidimensional (perceptual, cognitive, affective, and motor) processes underlying any human activity, it is important to adopt an interdisciplinary human-centered perspective to represent a specific activity and to design and evaluate the control interfaces that support it. Such an approach, embracing constructs and methodologies of psychology, biology, medicine, and engineering, is offered by the discipline of ergonomics.

Stone and McCloy [58] analyzed the approach of ergonomics as science of the interaction between humans and their working environment in medical and surgical fields. The authors highlighted different factors that should be investigated to optimize interface system performance and to maximize user well-being and operational effectiveness, including anthropometric and biomechanic aspects of human body (which are related to gestures and posture of the surgeon), and cognitive processes, whose limitations (e.g., limited attention and mental workload) become critical during complex tasks requiring high levels of attention and control over time. Such cognitive processes are also affected by cultural stereotypes and expectations about the equipment and the operation. Furthermore, they are also affected by the level of knowledge of the user about surgical equipment, medical practices, and clinical conditions (including emergency).

In this context, surgeon training is a challenge for new microsurgery systems. This should be integrated in the design of any control interface that aims at assisting directly the user's cognitive skills. Moreover, the control interface should be intuitive and require little training to be used effectively. This leads to an approach to equipment design that considers the flexibility of human cognition and its limitations in information processing [59], which requires investigating how the new tool will affect the surgeon and how the surgeon will utilize the new tool.

The subjective nature of constructs that define the study of behaviors during complex tasks requires specific methodologies of analysis and evaluation. All constructs are interrelated and they emphasize different task-relevant aspects of human activity, like *mental workload*: the amount of cognitive resources allocated and employed during a task [60]. For instance, the construct of *usability* is defined as "the extent to which a product can be used by specified users to achieve specified goals with effectiveness, efficiency, and satisfaction in a specified context of use" (ISO standard 9241-11, 1998). It depends on the capability of performing the task (effectiveness) with the least use of resources (efficiency) and the gratification experienced by the user (satisfaction). Other dimensions are also crucial in training [61]: *learnability* (ease of learning the task during the first trials); *memorability* (ease of restoring the previous level of operation outcome after a period of inactivity); and *errors* (number and severity of errors, ease of coping, and recovering from errors).

The evaluation of a control interface according to the described ergonomic constructs can be performed based on different methodologies designed to measure user behavior and system performance as a function of the tools used [62], the task, and the context. Subjective methods considering the user's point of view include *focus groups* [63] and, in particular, *questionnaires* like the system usability scale (SUS) [64] for usability and the NASA Task Load Index (NASA-TLX) [65] and surgery task load index (SURG-TLX) [66] for mental workload. Also "thinking aloud" during the interaction trial can offer information about the user experience [67].

The interface evaluation can be done based on quantitative performance data. Such data may be obtained experimentally using *imaging-based metrics* of surgery-like tasks [68] or *biometric variables* about user state and workload (e.g., eye tracking [69], psychophysiological changes [70], and motion analysis [71]).

Furthermore, qualitative methods like video analysis of behavior [72] can also be used. These consider the direct observation of user's behavior during the actual task in order to find the meaning of certain situated actions and, for instance, the reasons for a breakdown during the interaction.

The data collected using these subjective and objective measures can be integrated, generating a comprehensive assessment of the microsurgical system control interface. For example, Mattos et al. [73] compared two different systems for laser phonomicrosurgery according to usability questionnaires and imaging-based metrics as measures of performance in surgery-like tasks. The authors also compared the self-evaluated time and the actual time spent to accomplish the tasks in order to obtain an estimation of mental workload [74]. Using this same methodology, Barresi et al. [75] obtained data proving the benefits of the novel microsurgery control interface in terms of reduced mental workload.

Considering this user-centered design perspective, surgical robotic systems currently employ control interface solutions that are based on a master–slave relationship [76]. This means that the surgeon sits at a control console, manipulating the master system and visualizing operations on a video display. Systems of this type typically lack haptic feedback to the surgeon, who operates from a nonsterile location away from the operating table. This can be a critical issue in case of emergency, so novel control interface designs are focusing on bringing the surgeon back to the patient's side.

3.5.6 Surgical Planning

The term *surgical planning* can comprehensively describe the process of defining and fitting specific surgical actions to the actual operating condition to increase the operation accuracy and safety [77]. Planning is extremely important for any kind of surgery but becomes critical for many types of microsurgeries due to the required precision and the delicacy of tissues involved.

Preoperative surgical planning typically starts with a collection of information about the patient, including 2D or 3D multimodal images, which can be later combined with general information about human anatomy to produce patient-specific anatomical models. In the case of MIS, careful preoperative planning helps the surgeon to individualize the best entry points and robot configuration for specific procedures on the specific patient. In general, the benefits of preoperative planning for robotic surgery include [78] the following:

- Minimization of setup time by predefining the intraoperative placement of the surgical robot and the procedure for approaching the surgical target.
- Maximization of the intraoperative movable range of the robot, which can have significant impact on the success of the operation.
- Minimization of risks and "surprises" during surgery since the surgeon can study the case and start the procedure knowing exactly what has to be done.
- Education: Preoperative planning can serve as an educational tool for less experienced surgeons, especially if it is associated with a virtual training system.

Once patient data are obtained, the main subsequent steps during surgical planning include preoperative reconstruction and intraoperative registration. These are two areas posing significant challenges to the creation of intuitive and clinically usable surgical planning systems.

3.5.6.1 Preoperative Reconstruction

The main approaches used to visualize 3D patient anatomy using data acquired through computed tomography (CT) or magnetic resonance imaging (MRI) are (i) direct volume rendering from raw data and (ii) surface rendering from segmented data.

Volume rendering is of great clinical interest when dealing with contrasted malformations (e.g., in the vascular system or bones) or with pathologies visible through enhanced contrast, such as vessel diseases (aneurism, atheroma, and calcification) or tumors [79]. However, current technology is still limited, making volume rendering unfeasible in a number of the clinical situations. For example, organs cannot be visualized independently if they have the same gray level in the acquired images, making independent volume computation a very hard problem. Consequently, volume rendering is still not usable for advanced surgical planning in these cases. On the other hand, volume rendering presents a great benefit: It allows 3D visualization without preprocessing; i.e., no delineation is required to generate useful results for diagnosis and surgical planning in many other cases.

The visualization of 3D patient anatomy based on surface rendering requires first the delineation of the structure of interest. This can be performed semiautomated or automated [80]. The subsequent step consists of rendering the 3D structures based on a mesh generation process that provides the surface of the delineated structures. This preprocessing is the main drawback of the technique due to the possible difficulty and duration of the initial delineation step. Moreover, automated segmentation is in practice not guaranteed in clinical routine due to the large variability in medical image quality. This means that the segmentation quality has to be always checked by an expert.

3.5.6.2 Intraoperative Registration

The process of medical data registration consists of the identification of the position of the preoperative information in the frame of the intraoperative display [81]. Intraoperative information on anatomical structures can be recovered with different run-time imaging modalities [82]. An example is intraoperative ultrasound (US), which is noninvasive and can allow the identification of structures of interest. However, US images present disadvantages in terms of quality: Besides noise and echoic reflections, such images may be distorted due to varying velocities of sonic propagation in different kinds of tissue [83]. Another intraoperative imaging option is open MRI [84]. Unfortunately, such systems can be affected by the surgical instruments, which can cause electromagnetic interference and create artifacts in the imaging data. Tomographic intraoperative imaging modalities are yet another option. These can potentially provide anatomically coregistered information about the 3D shape and morphology of soft tissue, but their deployment in operating theaters is still a significant challenge. Currently, the most practical method of recovering the patient anatomy intraoperatively is through optical techniques using stereo cameras.

Intraoperative registration techniques can be classified as manual, semiautomatic, and automatic [85]. Manual registration provides simultaneous visualization of the patient and the preoperative 3D model, allowing the user to manually modify the model position, scale, and orientation to achieve proper superposition over the patient image. The use of tissue landmarks is generally helpful in this process. These are often anatomical structures such as ribs, iliac crests, and clavicles or boundaries between different tissues.

Semiautomatic registration methods are also based on landmarks. In this case, these are identified on both the preoperative 3D model and the real patient. Once landmarks are identified on the model, the user is typically requested to point at the corresponding landmarks visible on the patient by using a tracked pointer. The algorithm then computes the registration parameters to properly superimpose the landmarks and, thus, the 3D model. In comparison with the manual method, semiautomatic registration is more complex and requires more time. However, it results in higher registration precision, especially when dealing with rigid structures such as bones.

In order to achieve fully automatic intraoperative registration, two main issues have to be considered: (i) physiological organ or patient motion and (ii) surgical instrument's interactions with the organs. Moreover, in case the registration has to be accomplished on video from a moving camera or display (typically an endoscopic camera or a head-mounted display), the system is required to track the position and viewpoint of these devices. In the operating room, 3D model registration has to be done using intraoperative sensing, which typically involves the use of 3D localization systems, X-ray or ultrasound images, and the surgical robot (if one is used).

Intraoperative registration accuracy is a major concern when the registration results are used for surgical guidance or safety supervision. This is particularly the case for microsurgeries, which require high precision due to the small nature of the operations. Discrepancies in the registration process can generate erroneous guidance information and lead to dangerous conditions. However, the creation of a high-accuracy automatic system for intraoperative registration during microsurgery is yet an open challenge. This problem is especially hard since most microsurgeries are performed on soft tissue, requiring real-time tracking of tissue deformations.

3.5.7 Safety

The introduction of robotics in medicine and surgery has been very slow compared to other fields like manufacturing processes or exploratory missions where robots can perform specific, delicate, precise, or even dangerous tasks. Safety is a major reason for this delay. This section discusses the safety aspects that make this combination of task (surgery) and tool (robot) so complicated.

Robot actions on human bodies are associated with high risk for health, even if the task has nothing to do with medicine or surgery. The usual approach to guarantee safety during robotic tasks is to keep human beings as far as possible from the robot motion. This restriction has inspired new research areas to develop hardware [86] and control algorithms [87] that allow robots to interact more safely with humans.

Surgical applications seem to be the worst-case scenario for the use of robots. Not only the task, by definition, implies close proximity to a human being where the actions of the robot are meant to have a consequence, but also the patient has no opportunity to perceive and, thus, react to robot actions. Furthermore, safety is a real and very important concern for surgery. Potential risks associated with any new technology to be used in operating rooms are meticulously evaluated before its approval.

The intersection of these restrictions leaves very little room for developing robotic systems useful and safe enough for surgical applications. Nevertheless, this is possible and robotic technologies can indeed offer interesting and safe solutions to most drawbacks and limitations of current surgical procedures.

The important fact to note here is that, differently from other fields where safety specifications for robot operation are global, the required safety for surgical robots depends on the specific surgical procedure it is intended for. This is compatible with the safety protocols applied during conventional surgeries, where risks are explicitly listed and evaluated on a case-by-case basis. In this sense, and as a reference point, it is useful to compare the benefits and risks introduced by a surgical robot to those from conventional procedures.

3.5.8 Autonomous Behaviors

Procedures involving microscopic tool–tissue interaction pose significant challenges to clinicians. These are associated with the complexity of doing actions at miniscule scales, which require precise control of small movements and forces. In addition, further challenges are posed by the fact that during microsurgery the surgeon might not visualize the surgical site directly. Inadequate perception may affect the performance of clinicians, hindering the execution of even potentially simple tasks.

To compensate for these issues, surgical devices may offer a level of assistance that goes beyond the mere execution of the surgeon's commands. This includes the automation of certain tasks. For instance, the device may be able to perform suturing in an autonomous fashion.

Generally speaking, autonomy is regarded as the capability of a person or system to perform a task with little or no direction from another entity [88]. Figure 3.14 shows a typical operational loop during surgery. In this case, the surgeon is in direct control of the surgical device, which operates on the patient. Based on his/her perception, the surgeon decides the actions to be performed. This is the case of traditional microsurgery and also of robot-assisted surgery performed with systems such as the da Vinci robot.

Autonomy does not just involve performing a sequence of tasks; it also involves the capability to perceive the environment (the patient) and adapt its behavior and actions based on this perception (see Figure 3.15). Over the last decade, several autonomous solutions have been investigated in the scope of surgical robots, especially in the field of microsurgical devices because of the high precision required [89–94]. Not all autonomous surgical robots offer the same level of assistance, which may vary significantly.

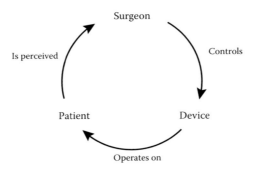

FIGURE 3.14
Traditional operation loop.

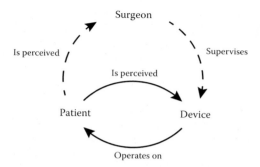

FIGURE 3.15
Operation loop with a device capable of performing tasks autonomously.

The level of autonomy of a surgical device can be evaluated by the amount of user interaction required to accomplish a task. Highly autonomous devices necessitate little guidance from the surgeon and are able to operate without human intervention. At the other end of the spectrum, the devices are completely under the control of the surgeon, who provides motion commands through a master control interface.

A classification of surgical robots based on the amount of user interaction has been recently proposed by Moustris et al. [88]. Based on the framework originally proposed by Niemeyer et al. [95], three categories were defined: (i) manual control, (ii) shared control, and (iii) supervisory control. Robots belonging to the first category are under direct control of the surgeon and replicate his/her gestures. Robots in the second category share the control of the actuators with the surgeon. This is useful for the implementation of safety features such as virtual fixtures to protect delicate areas. Finally, robots belonging to the third category are capable of performing an entire task autonomously, with the surgeon simply supervising the operation.

3.6 Conclusion

Microsurgery systems are being revolutionized by robotic technologies. Handheld robotic tools and teleoperated robotic systems have already started to transform the way these highly delicate and demanding surgeries are performed. The new technologies bring improved precision, safety, efficiency, and quality to microsurgery by augmenting the surgeon's sensing and actuation capabilities. Nevertheless, this revolution is still in its nascent phase. Many challenges have yet to be overcome until robotics can achieve their full potential in microsurgery and make robot-assisted systems become required tools in operating rooms.

Major challenges associated with the progress of microsurgery systems have been discussed in this chapter, including miniaturization and the development of novel control interfaces. Miniaturization will allow surgical tools to reach and treat more parts of the human anatomy, but progress from the current state will require multidisciplinary research and development efforts involving engineering, chemistry, biology, optical physics, and medical expertise. New microsurgery tools are expected to be created from a mix of nano- and microtechnologies. This will entail major shifts in terms of materials, manufacturing processes, assembly, and packaging of micromedical devices. Biodegradable materials, hybrid structures of metals and natural materials, and tissue-engineered materials are potential options for the future. Additionally, the emergence of graphene as a strong and hard, yet thin and flexible material, promises to propel the miniaturization of everything from tools to actuators to supersensitive sensors [96].

The future of microsurgery systems also involves an expanded exploitation of alternative energy sources such as lasers, ultrasound, and electromagnetics, which can bring significant improvements to both imaging and treatment during surgery. Lasers, for example, have the potential to miniaturize surgical actions to the cellular level or even eliminate the need for microsutures by realizing tissue welding. In any case, the success of new microsurgery technologies in the operating room will be contingent on the ergonomics and usability of their control interfaces, so progress in this area also faces challenges related to human factors. This corroborates the assertion that a multidisciplinary approach is required to create new clinically significant microsurgery systems.

Finally, surgical planning, augmented reality, and surgical robot autonomy are bound to enable surgeons to achieve superhuman performance in microsurgery tasks. Once critical challenges related to computer vision are overcome, these capabilities will greatly contribute to improvements in surgical precision, safety, and quality. The possibility of merging multimodal images of diseases and anatomical structures to the intraoperative view, for example, will assist surgeons to completely eradicate malignancies with higher safety and minimal damage to healthy tissue. Moreover, the autonomous execution of specific tasks and surgeon-defined surgical plans by robotic systems will help minimize operative time and further improve microsurgery accuracy, safety, and quality to ultimate levels.

References

1. S. Tamai, "The history of microsurgery," in *Experimental and Clinical Reconstructive Microsurgery*, S. Tamai, M. Usui, and T. Yoshizu, eds., pp. 3–24, Springer, Japan, 2003.
2. L. S. Bass and M. R. Treat, "Laser tissue welding: A comprehensive review of current and future," *Lasers Surg. Med.*, 17: 315–349, 1995.
3. A. Barak, T. Ma-Naim, M. Belkin, and A. Katzir, "Temperature-controlled CO_2 laser tissue welding of ocular tissues," in *BiOS '97, Part of Photonics West*, pp. 103–105, International Society for Optics and Photonics, 1997.
4. Y. Wadia, H. Xie, and M. Kajitani, "Liver repair and hemorrhage control by using laser soldering of liquid albumin in a porcine model," *Lasers Surg. Med.*, 27: 319–328, 2000.
5. A. Katzir, "LASERS—The future of suture," *SPIE Professional*, open-access article, July 2010, Available: https://spie.org/x41031.xml, accessed on August 12, 2014.
6. M. Remacle, G. Lawson, M. Nollevaux, and M. Delos, "Current state of scanning micromanipulator applications with the carbon dioxide laser," *Ann. Otol., Rhinol. & Laryngol.*, 117(4): 239–244, April 2008.
7. S. A. Karkanis, D. K. Iakovidis, D. E. Maroulis, D. A. Karras, and M. Tzivras, "Computer-aided tumor detection in endoscopic video using color wavelet features," *IEEE Trans. Information Technol. Biomed.*, 7(3): 141–152, September 2003.
8. C. Riviere and N. V. Thakor, "Modeling and canceling tremor in human–machine interfaces," *IEEE Eng. Med. Biol. Magazine*, 15(3): 29–36, June 1996.
9. A. M. Okamura, "Methods for haptic feedback in teleoperated robot-assisted surgery," *Ind. Rob.*, 31(6): 499–508, December 2004.

10. A. Kapoor, M. Li, and R. H. Taylor, "Spatial motion constraints for robot assisted suturing using virtual fixtures," *Med. Image Comput. Comput. Assist. Interv.*, 8(part 2): 89–96, 2005.
11. P. S. Schenker et al., "Development of a telemanipulator for dexterity enhanced microsurgery," *Second Annu. Int. Symp. Med. Robotics Comput. Assisted Surg. (MRCAS)*, pp. 81–88, 1995.
12. R. Taylor et al., "A steady-hand robotic system for microsurgical augmentation," *Int. J. Robot. Res.*, 18: 1201–1210, December 1999.
13. X. He, D. Roppenecker, D. Gierlach, M. Balicki, K. Olds, P. Gehlbach, J. Handa, R. Taylor, and I. Iordachita, "Toward clinically applicable steady-hand eye robot for vitreoretinal surgery," *Proc. ASME IMECE*, paper no. IMECE 2012-88384, pp. 145–153, 2012.
14. C. He, K. Olds, I. Iordachita, and R. H. Taylor, "A new ENT microsurgery robot: Error analysis and implementation," *IEEE Int. Conf. Robotics Automation (ICRA)*, May 2013.
15. R. MacLachlan et al., "Micron: An actively stabilized handheld tool for microsurgery," *IEEE Trans. Robotics*, 28(1): 195–212, February 2012.
16. P. Brett, L. Reyes, and J. Blanshard, "An automatic technique for micro-drilling a stapedotomy in the flexible stapes footplate," *Proc. IMechE, Part H*, 209(4): 255–262, 1995.
17. J. Salzmann, G. Zheng, N. Gerber, C. Stieger, A. Arnold, U. Rohrer, L. Nolte, M. Caversaccio, and S. Weber, "Development of a miniature robot for hearing aid implantation," *IEEE/RSJ Int. Conf. Intelligent Robots Syst. (IROS 2009)*, pp. 2149–2154, St. Louis, United States, 2009.
18. J. F. Giallo, "A medical robotic system for laser phonomicrosurgery," PhD Thesis, North Carolina State University, 2008.
19. L. S. Mattos, G. Dagnino, G. Becattini, M. Dellepiane, and D. Caldwell, "A virtual scalpel system for computer-assisted laser microsurgery," *Proc. IEEE Int. Conf. Intelligent Robots Syst., IROS 2011*, September 2011.
20. μRALP—Micro-technologies and systems for robot-assisted laser phonomicrosurgery, Available: http://www.microralp.eu, accessed on August 12, 2014.
21. T.-R. Hsu, "Miniaturization—A paradigm shift in advanced manufacturing and education," in *Intl. Conf. Advanced Manufacturing Technol. Educ. 21st century*, Plenary Speech, Taiwan, pp. 10–15, August 2002.
22. R. Feynman, "There's plenty of room at the bottom: An invitation to enter a new physics," First presented at the American Physical Society at California Institute of Technology, December 29, 1959. Published in Engineering and Science, California Institute of Technology, February 1960.
23. R. Fernandes and D. H. Gracias, "Toward a miniaturized mechanical surgeon," *Materials Today*, 12(10): 14–20, 2009.
24. N. Miki, X. Zhang, R. Khanna, A. A. Ayón, D. Ward, and S. M. Spearing, "Multi-stack silicon-direct wafer bonding for 3D MEMS manufacturing," *Sensors Actuators A: Physical*, 103(1–2): 194–201, 2003.
25. V. K. Varadan and V. V. Varadan, "Micro stereo lithography and fabrication of 3D MEMS and their applications," *Proc. SPIE—Int. Soc. Opt. Eng.*, pp. 9–20, 2001.
26. H. O. Jacobs, A. R. Tao, A. Schwartz, D. H. Gracias, and G. M. Whitesides, "Fabrication of a cylindrical display by patterned assembly," *Science*, 296(5566): 323–325, 2002.
27. R. J. Wood, S. Avadhanula, R. Sahai, E. Steltz, and R. S. Fearing, "Microrobot design using fiber reinforced composites," *J. Mech. Design*, 130(5): 052304, 2008.
28. N. Simaan, "Snake-like units using flexible backbones and actuation redundancy for enhanced miniaturization," *IEEE Intl. Conf. Robotics Automation*, Spain, pp. 3012–3017, 2005.
29. C. Bergelas and P. E. Dupont, "Planning stable paths for concentric tube robots," *IEEE/RSJ Intl. Conf. Intelligent Robots Syst.*, Japan, pp. 3077–3082, 2013.
30. J. Burgner, P. J. Swaney, T. L. Bruns, M. S. Clark, D. C. Rucker, E. C. Burdette, and R. J. Webster, "An autoclavable steerable cannula manual deployment device: Design and accuracy analysis," *J. Med. Devices*, 6(4): 041007, 2012.
31. G. Iddan, G. Meron, A. Glukhovsky, and P. Swain, "Wireless capsule endoscopy," *Nature*, 405(6785): 417–418, 2000.
32. M. T. Gettman and P. Swain, "Initial experimental evaluation of wireless capsule endoscopes in the bladder: Implications for capsule cystoscopy," *Eur. Urol.*, 55(5): 1207–1212, 2009.

33. M. Tavakoli, R. P. Patel, and M. Moallem, "Bilateral control of a teleoperator for soft tissue palpation: Design and experiments," *IEEE Intl. Conf. Robotics Automation, United States*, pp. 3280–3285, 2006.

34. "NanEye," Awaiba LDA, Available: http://www.awaiba.com/en/products/medical-image-sensors/, accessed on April 16, 2014.

35. P. Estevez, J. M. Bank, M. Porta, J. Wei, P. M. Sarro, M. Tichem, and U. Staufer, "6 DOF force and torque sensor for micro-manipulation applications," *Sensors Actuators A: Physical*, 186: 86–93, 2012.

36. X. He, J. Handa, P. Gehlbach, R. Taylor, and I. Iordachita, "A submillimetric 3-DOF force sensing instrument with integrated fiber Bragg grating for retinal microsurgery," *IEEE Trans. Biomed. Eng.*, 61(2): 522–534, 2014.

37. F. Dubois, G. Berthelot, and H. Levard, "Laparoscopic cholecystectomy: Historic perspective and personal experience," *Surg. Laporoscopy Endoscopy*, 1(1): 52–57, 1991.

38. T. Kolatt, "Miniature cameras—The big picture," Available: http://www.smallestvideocameras.com/professional-information/103-miniature-cameras-the-big-picture, accessed on August 12, 2014.

39. V. R. Jacobs, M. Kiechle, B. Plattner, T. Fischer, and S. Paepke, "Breast ductoscopy with a 0.55-mm mini-endoscope for direct visualization of intraductal lesions," *J. Minimally Invasive Gynaecol.*, 12 (4): 359–364, 2005.

40. Richard Wolf GmbH. Available: http://www.richard-wolf.com/en/company/endoscopy/endoscope.html, accessed on April 16, 2014.

41. D. Yelin, I. Rizvi, W. M. White, J. T. Motz, T. Hasan, B. E. Bouma, and G. J. Tearney, "Three-dimensional miniature endoscopy," *Nature*, 443: 765, 2006.

42. A. Myers, "Stanford engineers develop high-resolution endoscope as thin as a human hair," Available: http://news.stanford.edu/news/2013/march/high-resolution-endoscope-031413.html, accessed on August 12, 2014.

43. S. Ma, I. Kobayashi, S. Hirose, and K. Yokoshima, "Control of a multi-joint manipulator, 'MorayArm'," *IEEE/ASME Trans. Mechatronics*, 7(3): 304–317, 2002.

44. D. Asai, S. Katopo, J. Arata, S. I. Warisawa, M. Mitsuishi, A. Morita, S. Sora, T. Kirino, and R. Mochizuki, "Micro-neurosurgical system in the deep surgical field," *Proc. Intl. Conf. Medical Image Comput. Comput.-Assisted Intervention*, pp. 33–40, 2004.

45. N. Simaan, "Snake-like units using flexible backbones and actuation redundancy for enhanced miniaturization," *IEEE Intl. Conf. Robotics Automation*, Spain, pp. 3012–3017, 2005.

46. J. Burgner, P. J. Swaney, T. L. Bruns, M. S. Clark, D. C. Rucker, E. C. Burdette, and R. J. Webster, "An autoclavable steerable cannula manual deployment device: Design and accuracy analysis," *J. Med. Devices*, 6(4): 041007, 2012.

47. A. Menciassi, A. Eisinberg, G. Scalari, C. Anticoli, M. Carrozza, and P. Dario, "Force feedback-based microinstrument for measuring tissue properties and pulse in microsurgery," *Proc. IEEE Intl. Conf. Intelligent Robots Syst.*, pp. 626–631, 2001.

48. U. Seibold, B. Kubler, and G. Hirzinger, "Prototype of instrument for minimally invasive surgery with 6-axis force sensing capability," *Proc. IEEE Intl. Conf. Intelligent Robots Syst.*, pp. 496–501, 2005.

49. B. Bell, S. Stankowski, B. Moser, V. Oliva, C. Stieger, L.-P. Nolte, M. Caversaccio, and S. Weber, "Integrating optical fiber force sensors into microforceps for ORL microsurgery," *Proc. 33rd IEEE Int. Conf. Eng. Med. Biol. Soc.*, pp. 1848–1851, 2010.

50. T. G. Leong, C. L. Randall, B. R. Benson, N. Bassik, G. M. Stern, and D. H. Gracias, "Tetherless thermobiochemically actuated microgrippers," *Proc. Nat. Acad. Sci. U. S. A.*, 106(3): 703–708, 2009.

51. F. L. Kooi and A. Toet, "Visual comfort of binocular and 3D displays," *Displays*, 25(2): 99–108, 2004.

52. C. Ware and R. Balakrishnan, "Reaching for objects in VR displays: Lag and frame rate," *ACM Trans. Comput.-Human Interaction (TOCHI)*, 1(4): 331–356, 1994.

53. A. Neil, "Autostereoscopic 3D displays," *Computer*, 8: 32–36, 2005.

54. L. Mattos and D. Caldwell, "Safe teleoperation based on flexible intraoperative planning for robot-assisted laser microsurgery," *Proc. 2012 Int. Conf. IEEE Eng. Med. Biol. Soc.*, EMBC 2012, August 2012.

55. S. J. Lederman and R. L. Klatzky "Haptic perception: A tutorial," *Attention, Perception, & Psychophysics*, 71(7): 1439–1459, 2009.

56. G. Robles-De-La-Torre, "The importance of the sense of touch in virtual and real environments," *IEEE Multimedia*, 13(3): 24–30, 2006.

57. G. Tholey, J. P. Desai, and A. E. Castellanos, "Force feedback plays a significant role in minimally invasive surgery: Results and analysis," *Ann. Surg.*, 241(1): 102, 2005.

58. R. Stone and R. McCloy, "Ergonomics in medicine and surgery," *Br. Med. J.*, 328(7448): 1115–1118, 2004.

59. E. Hollnagel, "Cognitive ergonomics: It's all in the mind," *Ergonomics*, 40(10): 1170–1182, 1997.

60. T. E. Nygren, "Psychometric properties of subjective workload measurement techniques: Implications for their use in the assessment of perceived mental workload," *Human Factors: J. Human Factors Ergonomics Soc.*, 33(1): 17–33, 1991.

61. M. Van Welie, G. C. Van Der Veer, and A. Eliëns, "Breaking down usability," *Proc. INTERACT*, 99: 613–620, September 1999.

62. J. C. Bastien and D. L. Scapin, "Evaluating a user interface with ergonomic criteria," *Int. J. Human-Comput. Interaction*, 7(2): 105–121, 1995.

63. S. Caplan, "Using focus group methodology for ergonomic design," *Ergonomics*, 33(5): 527–533, 1990.

64. J. Brooke, "SUS: A 'quick and dirty' usability scale," in *Usability Evaluation in Industry*, P. W. Jordan, B. Thomas, I. L. McClelland, and B. Weerdmeester, eds., CRC Press, 1995.

65. S. Hart and L. Staveland, "Development of NASA-TLX (Task Load Index): Results of empirical and theoretical research," in *Human Mental Workload*, ser. *Advances in Psychology*, P. A. Hancock and N. Meshkati, eds., Elsevier, 1988.

66. M. R. Wilson, J. M. Poolton, N. Malhotra, K. Ngo, E. Bright, and R. S. Masters, "Development and validation of a surgical workload measure: The surgery task load index (SURG-TLX)," *World J. Surg.*, 35(9): 1961–1969, 2011.

67. A. W. Kushniruk and V. L. Patel, "Cognitive computer-based video analysis: Its application in assessing the usability of medical systems," *MEDINFO*, 8: 1566–1569, 1994.

68. N. Deshpande, L. S. Mattos, G. Barresi, A. Brogni, G. Dagnino, L. Guastini, G. Peretti, and D. Caldwell, "Imaging based metrics for performance assessment in laser phonomicrosurgery," *Proc. 2013 IEEE Int. Conf. Robotics Automation* (ICRA 2013), Karlsruhe, Germany, May 2013.

69. X. Jiang, B. Zheng, G. Tien, and M. S. Atkins, "Pupil response to precision in surgical task execution," *Medicine Meets Virtual Reality (MMVR)*, February 2013.

70. E. Haapalainen, S. Kim, J. F. Forlizzi, and A. K. Dey, "Psycho-physiological measures for assessing cognitive load," *Proc. 12th ACM Int. Conf. Ubiquitous Comput.*, Association for Computing Machinery, pp. 301–310, September 2010.

71. G. Lee, T. Lee, D. Dexter, R. Klein, and A. Park, "Methodological infrastructure in surgical ergonomics: A review of tasks, models, and measurement systems," *Surg. Innovation*, 14(3): 153–167, 2007.

72. J. V. Laws and P. J. Barber, "Video analysis in cognitive ergonomics: A methodological perspective," *Ergonomics*, 32(11): 1303–1318, 1989.

73. L. Mattos, N. Deshpande, G. Barresi, L. Guastini, and G. Peretti, "A novel computerized surgeon–machine interface for robot-assisted laser phonomicrosurgery," *Laryngoscope*, doi: 10.1002/lary.24566, December 2013.

74. M. Lind and H. Sundvall, "Time estimation as a measure of mental workload," in *Engineering Psychology and Cognitive Ergonomics*, D. Harris, ed., pp. 359–365, Springer, 2007.

75. G. Barresi, N. Deshpande, L. Mattos, A. Brogni, L. Guastini, G. Peretti, and D. G. Caldwell, "Comparative usability and performance evaluation of surgeon interfaces in laser phonomicrosurgery," *Proc. 2013 IEEE Int. Conf. Intelligent Robots Syst.* (IROS 2013), doi:10.1109 /IROS.2013.6696871, Tokyo, Japan, November 2013.

76. A. Simorov, R. S. Otte, C. M. Kopietz, and D. Oleynikov, "Review of surgical robotics user interface: What is the best way to control robotic surgery?," *Surg. Endoscopy*, 26(8): 2117–2125, 2012.

77. G. P. Moustris et al., "Evolution of autonomous and semi-autonomous robotic surgical systems: A review of the literature," *Int. J. Med. Robotics Comput. Assisted Surg.*, 7(4): 375–392, 2011.

78. M. Hayashibe et al., "Preoperative planning system for surgical robotics setup with kinematics and haptics," *Int. J. Med. Robotics Comput. Assisted Surg.*, 1(2): 76–85, 2005.

79. P. S. Calhoun et al., "Three-dimensional volume rendering of spiral CT data: Theory and method," *Radiographics*, 19(3): 745–764, 1999.

80. C. Koehl, L. Soler, and J. Marescaux, "PACS-based interface for 3D anatomical structure visualization and surgical planning," *Med. Imaging 2002*, International Society for Optics and Photonics, 2002.

81. P. Markelj et al., "A review of 3D/2D registration methods for image-guided interventions," *Med. Image Analysis*, 16(3): 642–661, 2012.

82. M. Baumhauer et al., "Navigation in endoscopic soft tissue surgery: Perspectives and limitations," *J. Endourol.*, 22(4): 751–766, 2008.

83. H. H. Holm and B. Skjoldbye, "Interventional ultrasound," *Ultrasound Med. Biol.*, 22(7): 773–789, 1996.

84. A. Lauro et al., "Laparoscopic and general surgery guided by open interventional magnetic resonance," *Minerva Chirurgica*, 59(5): 507–516, 2004.

85. S. Nicolau et al., "Augmented reality in laparoscopic surgical oncology," *Surg. Oncol.*, 20(3): 189–201, 2011.

86. A. Albu-Schaeffer, O. Eiberger, M. Grebenstein, S. Haddadin, C. Ott, T. Wimboeck, S. Wolf, and G. Hirzinger, "Soft robotics: From torque feed-back controlled lightweight robots to intrinsically compliant systems," *IEEE Robotics Automation Magazine*, 15(3): 20–30, 2008.

87. L. Zollo et al., "Compliance control for an anthropomorphic robot with elastic joints: Theory and experiments," *J. Dynamic Syst., Measurement, Control*, 127(3): 321–328, 2005.

88. G. P. Moustris, S. C. Hiridis, K. M. Deliparaschos, and K. M. Konstantinidis, "Evolution of autonomous and semi-autonomous robotic surgical systems: A review of the literature," *Int. J. Med. Robotics Comput. Assisted Surg.*, 7(4): 375–392, December 2011.

89. S. G. Yuen, D. T. Kettler, P. M. Novotny, R. D. Plowes, and R. D. Howe, "Robotic motion compensation for beating heart intracardiac surgery," *Int. J. Robotics Res.*, "Special Issue on Medical Robotics," 28(10): 1355–1372, October, 2009.

90. X. Du et al., "Robustness analysis of a smart surgical drill for cochleostomy," *Int. J. Med. Robotics Comput. Assisted Surg.*, 9(1): 119–126, March 2013.

91. T. Ortmaier, M. Groger, D. H. Boehm, V. Falk, and G. Hirzinger, "Motion estimation in beating heart surgery," *IEEE Trans. Biomed. Eng.*, 52(10): 1729–1740, October, 2005.

92. A. Hussong, T. S. Rau, T. Ortmaier, B. Heimann, T. Lenarz, and O. Majdani, "An automated insertion tool for cochlear implants: Another step towards atraumatic cochlear implant surgery," *Int. J. Comput. Assisted Radiol. Surg.*, 5(2): 163–171, March 2010.

93. G. Zong, Y. Hu, D. Li, and X. Sun, "Visually servoed suturing for robotic microsurgical keratoplasty," *2006 IEEE/RSJ Int. Conf. Intelligent Robots Syst. (IROS 2006)*, pp. 2358–2363, October 2006.

94. L. Fichera, D. Pardo, P. Illiano, D. G. Caldwell, and L. S. Mattos, "Feed forward incision control for laser microsurgery of soft tissues," *Proc. 2015 IEEE Int. Conf. Robotics Automation (ICRA)*, pp. 1235–1240, May, 2015.

95. G. Niemeyer, C. Preusche, and G. Hirzinger, "Telerobotics," *Springer Handbook of Robotics*, pp. 741–757, Springer, Berlin, 2008.

96. S. Ismael and J. Lewis, "Graphene: Will it be the future material in orthopaedic and trauma surgery?," *Bone Joint J.*, 95-B(Supp. 27), 2013.

4

Image-Guided Microsurgery

Tom Williamson, Marco Caversaccio, Stefan Weber, and Brett Bell

CONTENTS

4.1 Introduction

Microsurgical procedures require surgeons to work at the limits of human dexterity and perception. A microsurgical procedure can generally be defined as any surgical technique which requires the surgeon to perform actions or observe structures at the micrometer level (i.e., less than 1 mm). Microsurgery often makes use of an operating microscope, as well as a variety of other specialized instrumentation allowing increased perception and dexterity, which requires years of training and experience to master the techniques needed for safe and effective surgery and telesurgery in remote applications. Subsequently, the field is ideally suited for the use of assistive technologies such as computer assistance or image guidance. In general, image guidance may allow improved surgeon confidence, spatial orientation, and anatomical understanding. Some microsurgery systems such as da Vinci and Raven mimic the motions of the user, allowing motion scaling, remote center compensation, and tremor reduction in remote applications, as will be discussed in this chapter.

4.1.1 What Is Image Guidance?

Any surgical procedure in which medical imaging data are used to assist the surgeon in the performance of that procedure may be considered, at some level, as image guided. From cases in which a patient's broken arm is X-rayed and the image is examined by the physician in order to provide information about the location and extent of the injury, to highly delicate neurosurgical procedures involving multiple imaging modalities, computer-assisted planning, and intraoperative navigation or robotics. The development of image-guided surgery (IGS) as a whole is closely tied to the history and progression of medical imaging techniques. Image-guided surgery can be thought to have started with the discovery of the X-ray; however, the modern age of image guidance as we now know it was born with the advent of computed tomography techniques. By allowing the reconstruction of the three-dimensional patient anatomy, it became possible to more accurately and intuitively guide the surgeon to the desired location within the body; this was initially accomplished through the use of stereotactic frames, the predecessor to modern surgical navigation techniques.

The first reported mention of surgical navigation in a general context was around 1989; when Schlöndorff et al. presented a system in which the position of a tool was displayed relative to the anatomy on a computer screen within the operating room (OR) [1], thereby setting the scene for all subsequent work. It is worth noting that while early image-guidance systems utilized concepts identical to those used today, they differed significantly in the details of hardware and software interfaces. Note, however, that work utilizing image-guided surgical robotics came about even earlier (circa 1985 [2]), and that image guidance through stereotactic frames earlier again.

4.1.2 Why Image Guidance?

In the subsequent sections a number of examples of the use of image guidance in the context of microsurgery will be examined. First, however, a detailed consideration of the advantages of image-guided surgery will be presented along with an analysis of image-guided system components. Microsurgery requires extremely high levels of dexterity, perception, spatial coordination, anatomical knowledge, and clinical experience. This requires years of training and, no matter how experienced the surgeon is, one slip of the scalpel can have disastrous results for the patient. It is in this context that image guidance can be useful; by using the technology now available, image guidance aims to make microsurgical procedures both easier to perform and safer for the patient. These technologies also have the potential to make new surgical techniques possible or allow the treatment of previously untreatable conditions. Used inappropriately, however, image guidance also has the potential to make things worse.

What is important in an image-guided surgical system? One of the primary requirements is a system well adapted to the surgical procedure for which it is being used. This includes appropriate accuracy, intraoperative feedback, OR integration, and a variety of other factors. In terms of accuracy, as the procedures typically defined as microsurgery involve work at scales below 1 mm, this can be defined as a pseudo–upper limit for the accuracy; however, one must consider the exact procedure being performed as the accuracy requirements may be even more severe (or lax). There are a number of potential error sources which may limit the overall accuracy of navigation. The spatial resolution and accuracy of imaging data, the aforementioned registration process, and the accuracy of

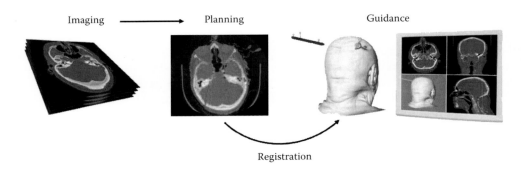

FIGURE 4.1
A typical image-guided surgery work flow. Medical image data are obtained preoperatively, registered to the patient intraoperatively, and then utilized to guide the surgeon to a target within the body while avoiding critical structures.

the tracking system used directly affect achievable accuracy. Other factors such as bending of nonrigid tools or movement of organs within the body may also play a significant role.

How information is fed back to the surgeon is also vital. Confusing or contradictory display of information can discourage use or, worse, lead to improper use. The same applies if the use of image guidance increases parameters such as operative time or invasiveness. In these cases the benefits provided by the use of the system must be large enough to overcome these drawbacks. Image-guided surgical systems are expensive, and clear benefits should be demonstrated; early adopters may be convinced to use the systems, but the success of image guidance is predicated on the benefits to the patient.

Limiting the evaluation of image guidance to the context of microsurgery, the range of techniques and procedures is still vast. Therefore, it may be instructive to consider the typical components of image guidance, as shown in Figure 4.1, individually before considering a variety of examples in fields representative of microsurgery, such as otorhinolaryngology (ORL) and neurosurgery.

4.2 Image Guidance Components and Workflow

4.2.1 Image Acquisition

There are two actions required for all types of image guidance: the acquisition of medical imaging data and the presentation of these data to the surgeon. Image acquisition can be preoperative (prior to the surgery) or intraoperative (during the surgery); the images can be functional (provide physiologic information) or structural/anatomical (provide information only about the patient's anatomy), and images can be obtained using one of a variety of imaging modalities or constructed from a combination of them. The imaging modality (or modalities) used can have an important impact on the subsequent image-guided work flow. Every imaging method has advantages and disadvantages. Some modalities may be appropriate for general surgery but not within a microsurgical context; we will consider each modality in turn in the sections below.

CT is the imaging modality most commonly used for image guidance and particularly within the context of microsurgical procedures. CT has a number of advantages: high spatial accuracy, high resolution, good image quality, lack of noise, and widespread availability of systems; but it is not without a number of issues, the most limiting of which is the use of ionizing radiation, which generally precludes the use of CT as a real-time intraoperative modality. The relatively poor soft-tissue contrast also disqualifies CT for use in certain procedures or in certain anatomical regions. Fluoroscopy is similar to CT in that it requires ionizing radiation for image formation; however, it differs significantly in the details. In general, the dose is lower but less information is obtained (i.e., a single 2D slice), and imaging can be completed in real time. Similar issues with soft-tissue contrast are also present, although the use of contrast agents, for example, in the case of cardiovascular imaging, may alleviate these problems.

Cone-beam computed tomography (CBCT) (also flat-panel CT, or digital volume tomography) uses a conical beam during image acquisition, reconstructing the patient anatomy from the acquired volumes (Figure 4.2). Only a single rotation of the gantry is required to fully reconstruct the anatomy; however, the increased complexity of the reconstruction process can lead to potential increases in imaging artifacts [3–4]. Additionally, while CT provides absolute information about the density of structures through the Hounsfield scale, this does not apply to CBCT images as the intensity is dependent on the distance from the X-ray source. CBCT is used extensively in dental procedures, and that it has

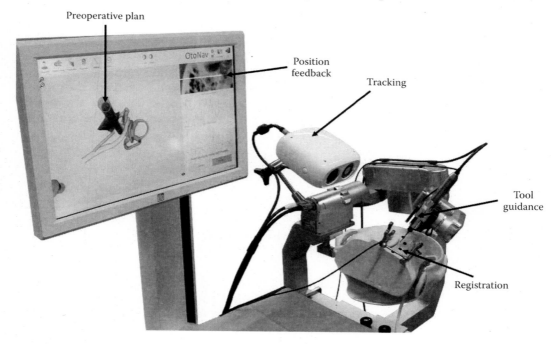

FIGURE 4.2
An image-guided robotic system for minimally invasive otologic microsurgery. Cone-beam CT data are acquired preoperatively and a case-specific plan is created through the segmentation of vital structures and the definition of a drilling trajectory. Registration of the plan to the patient is achieved through the use of bone-anchored fiducial screws, which are detected semiautomatically both in the image and on the patient. A lightweight five-degree-of-freedom robotic arm is used to guide the surgical tools through the utilization of high-accuracy optical tracking. Feedback about the progress of the procedure is provided to the surgical team on a monitor outside the surgical field.

various technological advances means that it is gaining popularity in other areas, particularly in ear, nose, and throat (ENT) surgery. Cone-beam CT may offer higher resolution than that available with standard CT.

Ultrasound has a number of advantages; it is relatively cheap, portable, safe, and simple to use and imaging can be completed in real time. The high-frequency sound waves used can effectively penetrate soft tissue; however, bone or gas pockets generally shield the anatomy within or behind them from imaging. The spatial resolution of ultrasound is limited by the frequency used. Higher frequency can increase resolution but at the cost of penetration depth as this is inversely related to the frequency. As many microsurgical procedures involve structures attached to or embedded within bone, ultrasound is mostly utilized intraoperatively and typically in combination with other, high-resolution modalities.

MRI remains the newest of the major imaging modalities. It functions through the use of powerful magnetic fields, which interact with the hydrogen atoms within the body, aligning the articles with the field. Particles are then knocked out of alignment by electromagnetic pulses. Eventually, particles are realigned with the field at a rate dependent on the particular tissue type. MRI does not require ionizing radiation, provides excellent soft-tissue contrast, and uses a variety of protocols to achieve specific imaging needs. Drawbacks include lower resolution than that available with CT and long scanning and reconstruction times. The large magnetic fields involved also means that using MRI as an intraoperative modality with standard (metal) surgical tools is impossible. The small bore of the machine is a further problem which can drastically reduce the surgical access in cases requiring intraoperative imaging. Specialized intraoperative MRI machines exist, typically with field strengths on the lower end of the spectrum and specialized designs to allow for increased surgical access, usually at the expense of image quality. Examples include "double-donut"-type superconducting machines [7–8] or the Odin PoleStar system [9], a semiportable, dedicated neuroimaging machine. MRI is used extensively in microsurgical procedures within neurosurgery and is often used in combination with CT in order to provide high-resolution guidance information within the bony structures of the skull, as well as more detailed information with respect to the borders of tumors or other soft-tissue pathologies. Examples of images obtained with each of the major structural imaging modalities are shown in Figure 4.3.

Functional imaging modalities such as positron emission tomography (PET) or single-photon emission computed tomography (SPECT) differ from the other modalities in that they provide very little structural or anatomical information but enable functional information to be collected from the tissue. Subsequently, they are often combined with other,

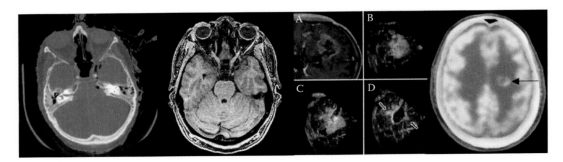

FIGURE 4.3
A variety of imaging modalities may be utilized for image guidance; each modality has individual advantages and disadvantages. *Left to right:* CT, MRI, ultrasound [5], and fluoroscopy [6] images of the head and the brain.

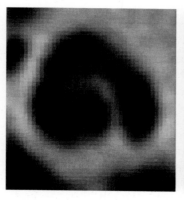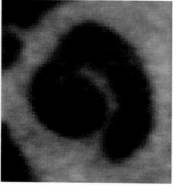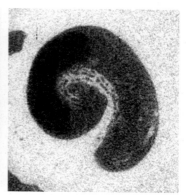

FIGURE 4.4
Left to right: images of the cochlea utilizing standard CT (0.5 mm resolution), high-resolution cone-beam CT (0.15 mm resolution), and nonclinical micro-CT (0.015 mm resolution). As microsurgery occurs at the micrometer scale, the resolution and quality of the acquired images is critical for accurate segmentation and planning.

anatomical, imaging modalities when used for image guidance (for example, PET-CT) [10] within a microsurgical context.

When acquiring images for microsurgical procedures, a number of specific requirements must be met. The resolution of the images must be high enough that small structures, potentially smaller than 1 mm, can be perceived. There must also be sufficient contrast between tissue types to arrive at a proper diagnosis and/or define a target. Accurate spatial reconstruction is also vital for effective guidance to targets within the body (Figure 4.4).

4.2.2 Surgical Planning

The concept of surgical navigation is an important one in modern image guidance. The commonly applied analogy is that of global positioning system (GPS), in which the surgeon is guided to the correct surgical site, or away from delicate structures, through the combination of image data, a surgical plan, and tracking of surgical tools. The imaging data, representing the map in our GPS analogy, can be obtained using any appropriate imaging technique. After images are obtained, the surgeon can plan a route to the target.

The planning phase can be completed pre- or intraoperatively and typically involves the segmentation of anatomical structures and the definition of a target position and/ or avoidance regions. Segmentation is the delineation of specific anatomical structures within image data. The simplest method, manual segmentation, requires the user, typically a surgeon, radiologist, or other qualified person, to trace the borders of the structure on each of the available slices or images. This can be time consuming, tedious, and prone to errors in selection or interpretation. As such, a significant body of research is focused on the development of automatic or semiautomatic segmentation methods. For example, statistical or active shape-modeling techniques rely on the acquisition of large amounts of data from which the statistical variation within the desired population can be extracted and utilized for segmentation of new, unknown data sets. Semiautomatic methods may require initialization from the user, for example, clicking a point within the region to be segmented. Subsequently, algorithms such as active contours and level-set

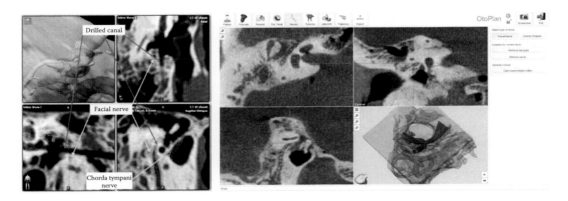

FIGURE 4.5
Left: The iPlan 2.6 surgical planning system from Brainlab allows segmentation and planning of generic image-guided interventions [11]. *Right:* The custom otologic planning software OtoPlan from the University of Bern, Switzerland, is specialized for segmentation of structures in the lateral skull base and allows planning of minimally invasive microsurgical procedures.

algorithms are initialized with this point. Comprehensive discussions of medical image segmentation methods are available in Pham et al.'s [12] and Heimann and Meinzer's [13] papers. The planning process for microsurgery does not differ significantly from that of other types of surgery as only the scale of image data is different; however, the selection of suitable imaging modalities is critical such that structures can be successfully segmented and targets defined (Figure 4.5).

4.2.3 Registration

In order to successfully execute an image-guided procedure, the surgeon requires feedback about the tool position. Before this can be done, a process known as registration is required. Registration involves the transfer of anatomical or planning data from images to the patient in the OR. That is, this process represents the calculation of the 3D transformation between the image data and the patient on the operating room table.

A variety of techniques exist, but all result in a transformation of 3D coordinates in the image to their equivalent positions on the patient. We can divide these methods into three separate categories based on the process used: point based, template based, and surface based.

Point-based registration represents the matching of corresponding points, whether artificial or anatomical, in the image and on the patient. The accuracy of these methods depends only on the ability of the system (with or without user error) to accurately identify identical points in both coordinate systems. This error is commonly referred to as the fiducial localization error, or FLE. As perfect point selection is impossible; matching of the two point sets implies finding the best fit, i.e., by minimizing the difference between points. This optimization process results in a parameter known as fiducial registration error (FRE), which is essentially the remaining error after the matching. In general, assuming no systematic error is present; the registration accuracy will improve with increased numbers of points. FRE (error of matching process) is the most commonly reported measurement of accuracy for image-guided systems. This parameter, however, has been shown to have no mathematical or physical relationship to the targeting or navigation error at other points [14]. The mathematics behind point-based registration methods has been studied in depth, with seminal works from Fitzpatrick et al. [15] and Maurer et al. [16] and more recent works including those from Wiles et al. [17] and Moghari et al. [18].

Surface-based registration involves the matching of surfaces segmented within the image and the same surfaces measured intraoperatively on the patient. The accuracy of surface-based registration is highly dependent on a number of factors and is also susceptible to the same error sources as those present in point-based matching. Of primary concern is the accurate detection of the surface points, while the rigidity of the surface and the ability to digitize the "true" surface, i.e., the same surface as extracted from the imaging data, also play a role. This ability may be confounded by the surface being covered. As an example, the surface of the bone can be directly digitized with a pointer; however, the skin, fat, muscle, and fascia must be removed in order to reach the bone. An overview of algorithmic considerations in surface-based registration is provided by Audette et al. [19].

Finally, image-guidance systems which utilize template-based registration methods are essentially self-registering once attached. Template-based methods involve the creation of a patient-specific template which fits directly onto the patient anatomy; these rely on accurate segmentation in the image, accurate manufacturing, and accurate intraoperative positioning [20–21]. Modifications of the techniques utilized during the use of a stereotactic frame have also been proposed. Frame-based methods require a known coordinate system to be imaged along with the patient; noninvasive methods may involve the use of a bite block to which the frame is then attached in a known configuration. They can be cumbersome and may suffer from loss of accuracy due to movement if not anchored to bone or the fit between pieces is not perfect [22].

It is important to note that the registration process typically results in a large percentage of the error encountered during navigation; as the achievement of high accuracy is vital to the successful completion of image-guided microsurgical procedures, the registration procedure used is vital. The gold standard, the method capable of providing the highest possible accuracy, is represented by the use of bone-anchored fiducial markers which can be located with high-level accuracy both in the image and by the surgeon on the patient [23–24].

If intraoperative imaging is used in combination with preoperative imaging, the registration of the two data sets may be necessary (Figure 4.6). Image-to-image registration can be a challenging process, particularly in cases of registration between imaging modalities or moving, deforming, or otherwise dynamic anatomical structures. It is possible to register not only between imaging modalities of the same dimensionality (i.e., 3D-3D) but

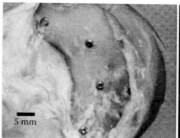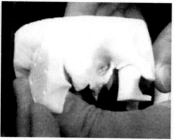

FIGURE 4.6
A variety of methods can be used for registering the patient to pre- or intraoperative imaging data including point-based methods using (left) bone-implanted fiducials [25], (center) surface-based methods [26], or (right) template-based methods [20].

also between lower- and higher-dimensional modalities. Examples include the registration processes by Zollei et al. [27], in which intraoperative, two-dimensional fluoroscopy data are registered to preoperative CT data, and by Rietzel and Chen [28], where the registration of moving anatomy in CT is considered. In a microsurgical context, this may be necessary in neurosurgical procedures in which brain shift plays a significant role and intraoperative imaging is available.

For interested readers, a comprehensive overview of registration methods and procedures was provided by Maintz and Viergever in 1998 [29]; information regarding more recent developments can be found in Zitova and Flusser's [30], Eggers et al.'s [31], and Hill et al.'s papers [32], or detailed information regarding image–image techniques in Oliveira and Tavares's paper [33].

4.2.4 Tracking

During navigated procedures it is necessary to track the tools, i.e., determine the position of the tools in space, by some method. There are a number of possibilities, each with a variety of advantages and disadvantages. Articulated-arm technologies, or mechanical tracking systems, utilize a passive mechanical arm, extracting positions from joint encoder values and the kinematics of the arm. Mechanical tracking was used in some of the earliest image-guided navigation systems [34]; however, it has mostly been replaced by other tracking modalities. While these systems may demonstrate high levels of accuracy and good repeatability, they tend to be bulky and cumbersome. Furthermore, each tracked object must be attached to its own individual mechanical arm in order to track it, and the movements of the surgeon are limited by the range of the mechanism.

Optical tracking systems localize surgical tools through the simultaneous acquisition of images by at least two cameras, fixed in a known configuration; the 3D coordinates can then be extracted by triangulation. The tracked features are used to create a coordinate system and the instrument must be calibrated with respect to this coordinate frame in order to effectively navigate the tool. Optical tracking systems require a direct line of sight between the tracking system and the tracked tools, and the tools must be modified such that they can be tracked, typically through the attachment of a marker frame containing at least three trackable points or other features such as edges, corners, or defined patterns. Due to the relatively high application accuracy, optical tracking is the most commonly used tracking method in microsurgical procedures; one should, however, be aware that accuracy can be reduced by a number of factors. Errors in the detection of individual features lead to errors in the definition of the coordinate system and are magnified when tracking a position at the tip of a long tool. Errors may also be introduced during the calibration process. A range of optical tracking systems are available with varying accuracies and work spaces; these are listed in Table 4.1.

In cases where very long tools are required, or line of sight is unavailable, alternative means of tracking are available, including electromagnetic tracking. Electromagnetic tracking involves the generation of a magnetic field and the tracking of small sensor coils to determine the position and orientation of a tool [35–36]. Electromagnetic tracking has the advantage of not requiring direct line of sight between the tools and field generator, meaning that tools can be tracked directly at the tip. However, it can be susceptible to interference from metal tools within the magnetic field (although significant progress has been made in this regard) and is less technically accurate than optical tracking.

If intraoperative imaging is used, this can also be used to track the position of surgical instruments. For example, the utilization of ultrasound for tool tracking has been reported

TABLE 4.1

Reported Accuracies of Commonly Used Tracking Systems[a]

	Manufacturer	Type	Measurement Range (mm)	Accuracy (mm)
Polaris Vicra	NDI	Optical	557–1336	0.25
Optotrak	NDI	Optical	1500–7000	0.1
accuTrack 500	Atracsys	Optical	154–2000	0.19–0.25
MicronTracker	Claron	Optical	300–2900	0.2
CamBar B1	AXIOS 3D	Optical	150–300	<0.05
Aurora	NDI	Electromagnetic	50–500	0.48

[a] Note that the majority of modern image-guided systems utilize optical tracking. Electromagnetic tracking systems are also sometimes utilized; however, these systems typically have lower accuracy than optical systems.

in the literature [37]. There also exist task-specific methods for localizing tools; one example is in the context of minimally invasive cochlear implantation, in which a tunnel is drilled from the surface of the skull to the inner ear. This method of localization combines observed drilling force data with the inhomogeneous nature of the mastoid bone through which the drill passes; the tool position is estimated by comparing observed forces with the variable bone density throughout the region [38].

In general, standard surgical tools can be used for navigated procedures with some degree of modification. Optically or electromagnetically tracked tools require attachment of a tracking marker, while tools tracked using intraoperative imaging can generally be used without modification, although alteration to simplify tracking may be possible. Note that in practice almost all modern navigation systems utilize either optical or electromagnetic tracking; the first example of the use of an optical tracking system in the context of image-guided surgery was in 1994, in a work by Galloway et al. [39], and the arrangement of image-guided surgical systems has not changed significantly since this date.

4.2.5 Instrumentation and Instrument Guidance

During image guidance, the surgical team must carry out a preoperative plan using feedback from the image guidance system; the simplest case involves the freehand use of tracked surgical tools by the user. If a stereotactic frame of reference attachment was fixed prior to imaging, this may be used to guide the tool position. Robotic assistance may also be used to guide tools. Robots have been present in the OR environment since the mid-1980s; many surgical robotic systems have been proposed and constructed but relatively few have reached a wide level of acceptance in a clinical context; the reasons for this are varied, but high costs relative to the achieved benefits play a large role as well as more general concerns such as safety.

The first surgical robotic systems were designed for specific applications, many within the field of orthopedics. The Acrobot system was developed to assist the surgeon during the cutting of bones during total-knee-replacement procedures; active constraint control is used to ensure that only the predefined cuts can be made and milling regions reached. The surgeon maintains control of the manipulator, guiding the arm manually [40]. The ROBODOC system was also developed within an orthopedic context, for the milling of a cavity in the femur for the precise attachment and fitting of hip prostheses. The surgeon is required to manually move the robotic arm to three implanted fiducials for registration, after which the robot will complete the desired cuts under the supervision of the surgical team [41–42]. In the context of neurosurgery, the neuromate was designed specifically for microsurgical procedures on the brain and can be used in a number of configurations (frame based or frameless) [43] and for a variety of tasks, including the placement of electrodes for the treatment of functional neurosurgical procedures, including movement disorders [44]. Similarly, the Pathfinder system can be used in a similar manner to a stereotactic frame, indicating and constraining a biopsy needle or electrode to a preoperatively planned path [45–46]. Image-guided robotics has also been the subject of significant research and development in other areas, including otorhinolaryngology (the OTOBOT system [25]) and spinal surgery (SpineAssist [47]).

To date, the most popular (and commonly used) robotic system is the da Vinci (Intuitive Surgical, Inc., Sunnyvale, California, United States). A master–slave manipulator, the da Vinci mimics the motions of the user, allowing motion scaling, remote center compensation, and tremor reduction. Similar systems include the Raven, designed for remote or telesurgery [48], or MRI-compatible robotic systems designed for assistance during neurosurgical interventions [49], illustrated in Figure 4.7. These systems can theoretically be utilized in any field, although the advantages (most commonly decreased invasiveness) may not outweigh the drawbacks (cost, increased operating time, and the additional training and staff required) for microsurgical tasks. Manipulators of this type are not strictly image guided in a traditional sense; however, the surgeon relies on endoscopic images to guide the robotic tools and modifications to include preoperative imaging data in an augmented-reality context have been proposed and investigated [51].

Other novel arrangements for robotic systems have been developed; some examples include self-assembling robotic systems, snake and concentric tube robots, magnetically actuated manipulators, and "microrobots." Concentric tube robots allow increased surgical access without increasing invasiveness by utilizing precurved concentric tubes; the

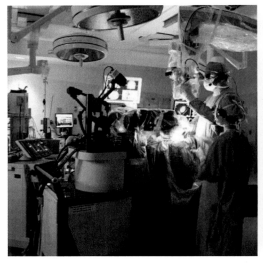 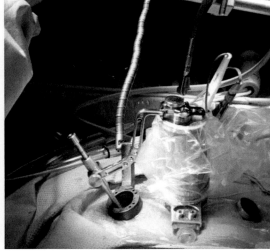

FIGURE 4.7
Left: neuroArm is an MRI-compatible robotic system for neurosurgery [50]. *Right:* SpineAssist is designed to assist with accurate minimally invasive placement of pedicle screws on the spine [47].

translation and rotation of individual components allows the manipulator to form specific shapes and reach previously inaccessible areas [52]. Snake robots have similar goals; however, these are nature inspired, having been created as flexible (or at least highly mobile) robots for increased access within the body. Examples include the robots in Webster et al.'s [53], Cianchetti et al.'s [54], and Simaan et al.'s works [55]. Self-assembling systems, for example, those by Harada et al. [56] and Nagy et al. [57], are introduced into the body in individual components, which are then connected into some predefined structure in order to perform tasks within the body. The use of external magnetic guidance and control has been studied in detail, for a variety of manipulator arrangements. Di Natali et al. [58] presented a novel master–slave–type manipulator which utilized external magnets to control three manipulators (including one camera) within the inflated abdomen; Kummer et al. [59] presented control of microrobots in the eye by using external magnetic fields. The majority of these developments have taken place outside the realm of microsurgery; however, the potential for their usage within a microsurgical context is great. A number of comprehensive reviews of surgical robotic technology are available for the interested reader; these include those by Cleary and Nguyen [60], Camarillo et al. [61], and Haidegger et al. [62].

4.2.6 Information Presentation

Presentation of imaging data can be as simple as an X-ray slice on a light box or computer screen or as complicated as intraoperative 3D displays, projections directly onto the patient, physical models created by rapid prototyping techniques, or the navigation of surgical tools relative to the patient anatomy.

A variety of novel image-presentation methods have been proposed; for example, heads-up displays, as in the works of Rolland and Fuchs [63] or Salb et al. [64], in which a semi-transparent display is mounted directly onto the surgeon's head. The projection of images directly onto the patient anatomy has also been the subject of recent research; examples include ceiling-mounted projection [65], in which the patient is positioned underneath the projector and the internal anatomy is displayed directly on the skin. Handheld projection has also been explored; this allows hidden anatomy or targeting information to be displayed directly on the patient [66].

Some surgical tasks do not have direct line of sight to the target organ or area, particularly laparoscopic and endoscopic procedures, and in these procedures the surgeon already relies on imaging information from the scope presented on the screen. The combination of preoperative imaging, planning, and navigation information with these images through the use of augmented-reality techniques have also been investigated [67–68]; however, suitable methods for combining this three- and two-dimensional information and displaying hidden structures in an intuitive manner are still under investigation. The da Vinci robotic interface allows the three-dimensional information to be perceived through stereoendoscopy and viewing; this has been shown to improve the performance of surgical tasks, such as suturing, when compared to standard two-dimensional displays [69]. The use of three-dimensional displays in other contexts has not, however, been shown to provide any significant benefit [70].

Recent research has been completed into how the addition of navigation information affects the performance of surgical tasks, including whether the cognitive load of the surgeon is increased [71]. The use of projection or head-mounted displays may help to solve diversion-of-sight issues, other problems such as parallax (changes in the perception of the location of deep structures when displayed on the skin as the viewer changes location) may be introduced [72]. Further cognitive load may be introduced in cases where conflicting information may be presented to the surgeon, or a confusing display. Methods for visualizing tracking or registration errors have also been explored [73]; however, it is currently unknown exactly how useful this information is. Overall, relatively little research has focused on user interface and interaction factors within the context of microsurgery. However, as image-guided technology evolves, acceptance increases and the possibility for use within new procedures presents itself; these factors are likely to become more important and the subject of future research.

4.3 Image Guidance by Surgical Domain

4.3.1 Image Guidance in Otorhinolaryngology

The field of ORL has proven to be a pioneer in both microsurgery and image-guided surgery. The structures of the head and neck, often embedded in or attached to bone, provide an ideal rigid environment for the development of these systems; however, the density or arrangement of vital structures within the surgical region of interest may impose daunting accuracy restrictions. Microsurgical techniques were revolutionized by the invention of the operating microscope, and ORL surgeons were among the first to adopt the technology. By allowing increased perception of structures or pathologies, a new range of procedures became possible. Image guidance has become standard practice within ORL surgery; the American Association of Otolaryngology—Head and Neck Surgery endorse this technology and recommends its use, particularly for cases of revision or abnormal anatomy [74].

One representative example with respect to the progress in microsurgical navigation technologies is that of cochlear implantation. Though the first attempts at utilizing image guidance for minimally invasive access to the inner ear for cochlear implantation were not until 2004 [75], advancement has proceeded rapidly over the last decade. Cochlear implantation is a highly delicate procedure. The surgeon is required to mill away a large portion of the mastoid behind the ear in order to visualize a number of vital structures which lie between the surface and the inner ear. While the general use of traditional image guidance may allow young or inexperienced surgeons to complete the procedure with greater confidence, recent developments have demonstrated the potential for image guidance to be transformative. Direct cochlear access (DCA) or minimally invasive cochlear implantation (MICI) is a proposed modification to the cochlear implantation procedure which reduces invasiveness by removing the need for a mastoidectomy by replacing the milled cavity with a tunnel drilled directly from the surface of the temporal bone to a target on the cochlea (see Figure 4.8). The major challenge in the successful completion of a DCA procedure is in the achievement of a suitable level of accuracy. The drill trajectory must pass through a region known as the facial recess, bounded by the external auditory canal, facial nerve (responsible for the enervation of the ipsilateral facial muscles), and chorda tympani (responsible for the sense of taste), which typically has a width of approximately 2.5 mm (although this varies throughout the population). To effectively pass an electrode through the tunnel, a drill diameter of at least 1.5 mm is required, leaving less than 0.5 mm between the instrument and the anatomy. Initial approaches to the procedure involved standard navigation [75]; however, it was concluded that the navigation techniques and technology available at the time were simply not accurate enough to safely implement the technique.

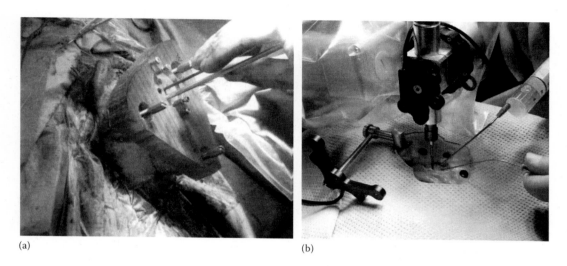

(a)   (b)

FIGURE 4.8
Two approaches to minimally invasive cochlear implantation [76]: (a) patient-specific microstereotactic frame and (b) custom-built high-accuracy robotic system.

Two approaches were proposed to overcome the issues presented by standard navigations (see Figure 4.9). The first involved the use of patient-specific templates or frames [76]. Based on the principles of the stereotactic frame, this approach involved the construction of a patient-specific drill guide specific to the patient anatomy and the locations of several preimplanted attachment posts. The constructed template is mounted to the posts, and the drill constrained to the planned path. The use of these frames is currently undergoing clinical evaluation [77]. The second approach involves the use of robotics; early work involved the modification of industrial robots [78]; however, suitable clinical accuracy could not be achieved and methods for integration into the operating room remained unsolved.

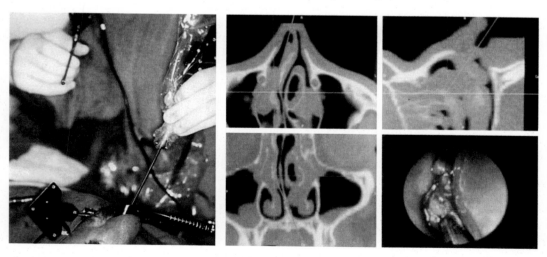

FIGURE 4.9
Image guidance during functional endoscopic sinus surgery [90]. (From Caversaccio, M. et al. "Frameless computer-aided surgery system for revision endoscopic sinus surgery." *Otolaryngology—Head and Neck Surgery* 122.6 (2000): pp. 808–813.)

Custom serial [79] and parallel [80–81] manipulators were also developed and evaluated. MICI presents an excellent case study in terms of what is required to move successfully from image-guided surgery to image-guided microsurgery (IGµS); Table 4.2 presents the approximate improvements required at each stage of the image-guided work flow.

Imaging quality and resolution are vital to the successful completion of microsurgical procedures as they determine how accurately critical structures can be segmented, as well as playing a role in the definition of a clinically suitable target. In the case of MICI, imaging resolution can be reduced, without increases in radiation dose, through the use of high-resolution cone-beam CT, allowing for improved segmentation and visualization. The registration process is responsible for a large part of the error in image-guided systems. It is, therefore, vital that the process used in microsurgical procedures be as accurate and robust as possible. Accurate registration relies on the accurate detection of corresponding features in both the patient and the image; bone-anchored fiducials provide the highest accuracy of available registration methods. Semiautomatic methods can be used to remove user bias or error in point selection at both stages of the registration procedure, as described by Gerber et al. [23], in which the mean target registration error was reduced to approximately 0.1 mm. Tracking errors can be reduced through the use of high-accuracy tracking systems such as the CamBar B1 (AXIOS 3D GmbH, Oldenburg, Germany). Other errors may be introduced during the guidance phase, and the use of robotics can help to alleviate user-specific errors during alignment or positioning of instruments. The minimization of other errors, for example, due to backlash or the bending of drills or burrs, relies on the careful design and manufacturing of tools specifically for the required task. The robotic system presented by Bell et al. [25] utilizes error-minimization concepts at each stage of the image-guided procedure as described above, achieving an end-to-end application accuracy of 0.15 ± 0.08 mm at a target at the surface of the cochlea during in vitro trials.

Other recent work in otology and skull base microsurgery has included the integration of robotics into other stages of the cochlear implantation procedure. For example, work by Brett et al. [82] utilized a small robotic device for the creation of an atraumatic cochleostomy, in which the bony outer surface of the cochlear could be milled away but the endosteum left intact. Robotics has also been utilized for assistance in standard mastoidectomy, for example, the da Vinci, as in the work of Liu et al. [83], or a custom parallel frame, as in the work of Dillon et al. [84]. Robotics has also been utilized within other otologic procedures; Miroir et al. [85] described a robotic system designed for use within middle-ear surgery. This master–slave manipulator allowed increased visualization of the target site and improved tool stability during otosclerosis surgery.

Functional endoscopic sinus surgery (FESS) has become the standard surgical approach for specific rhinologic procedures, particularly the treatment of chronic sinusitis and nasal polyps (see Figure 4.9). These procedures originally required the opening of the skin and

TABLE 4.2

Estimated Accuracy Possible at Each Stage of the Image-Guided Work Flow and Estimated Improvements Required for Image-Guided Microsurgery

	Imaging Voxel (mm)	Registration TRE[a] (mm)	Tracking RMS[b] (mm)	Guidance (mm)	Overall (mm)
IGS	0.5	1.0–2.0	0.25	0.5–1.0	2.0–3.0
IGµS	~0.15	~0.1–0.2	~0.05	~0.05	~0.2–0.3

[a] Target registration error.
[b] Root-mean-square.

were completed using a surgical microscope; however, seminal works from Messerklinger [86], Stammberger and Posawetz [87], and Kennedy in 1985 [88] have played a large role in the wider adoption of the endoscopic procedure. FESS has been shown to be as effective as open procedures in the treatment of the pathologies for which it is designed [89]. The major difficulties with the technique are due to the reduced access; difficulties with spatial orientation may occur if landmarks cannot be identified, particularly in cases of abnormal anatomy or surgeon inexperience. Research suggests that complication rates may be reduced and that FESS may be useful in particular anatomical cases or for training and assistance of inexperienced surgeons [91]. Many ORL surgeons now have extensive experience utilizing these navigation techniques, for example, those by Koele et al. [92] and Caversaccio et al. [93], within a rhinologic context.

Laryngeal microsurgical procedures require an array of specialized instruments and expertise. Classical image guidance or navigation is not widely used in this field; however, a variety of recent works have been completed, aimed at reducing complications and improving patient outcomes. As an example, the current standard technique during phonomicrosurgery (microsurgery relating to the quality of the voice) involves the use of a surgical laser (typically CO_2) in combination with a surgical microscope. A low-powered beam of visible light coincident to the high-powered beam is used for feedback during guidance, completed manually by the surgeon using a joystick. This has been shown to be error prone and could be potentially improved by the use of robotics [94]. The use of robotics within the throat has also been considered in other procedures, for example, laryngeal or hypopharyngeal cancer as in the works by Park et al. [95] and by Hockstein et al. [96], in which the da Vinci system was used. Robotic surgery in these cases allowed improved movement precision and improved visualization of the surgical site when compared to standard surgery utilizing an operating microscope or laryngoscope.

4.3.2 Image Guidance in Neurosurgery

Neurosurgery presents a number of unique challenges due to the enclosed nature of the brain within the skull and the importance of the tissue between the access point and the target, as well as the fragility of this same tissue; subsequently, the accurate localization of tools and anatomy within the skull is both extremely difficult and of vital importance [97]. These factors led to the development of the stereotactic frame, the spiritual predecessor to intraoperative navigation (see Figure 4.10). Stereotactic frames are inherently invasive devices, and while they allow the achievement of relatively high levels of accuracy, this invasiveness and the additional time required for anesthetization and fixation, as well as their relative limitations in terms of tool guidance, whereby only linear trajectories are possible, led to the development of less invasive, more universal devices and methods. Stereotactic frames also have the disadvantage of requiring manual adjustment and alignment; this process can be time consuming and prone to human error.

Computer assistance and robotic support has been used to assist neurosurgeons almost since the inception of the technology. Early work includes that by Kwoh et al. [2], representing the earliest integration of robotics within the operating room, (a Unimation Puma 200 industrial robot was utilized as a drill guide replacing the stereotactic frame) and Galloway et al. [97], whereby the surgeon was able to orient themselves accurately within the skull through the use of a custom articulated arm. Both of these examples utilize the basic required image-guidance work flow and principles of imaging, planning, registration, tracking, and guidance that more recent examples rely on. The technology is now well established and a vast body of literature exists on its use within specific situations, for

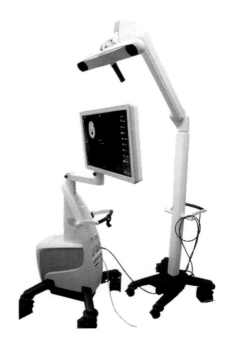

FIGURE 4.10
Left to right: The stereotactic frame and a modern frameless stereotactic navigation system from Brainlab, with optical tracking [31].

example, within the context of deep-brain stimulation, or more generally within neurosurgery [98].

Image guidance can support the surgeon in aligning drills or other cutting tools in order to gain access to the brain through the skull, as in the case of a craniotomy; the definition and achievement of a suitable and safe craniotomy location is of vital importance such that critical functional areas of the brain are avoided. Once access to the brain has been achieved, image guidance can further help the surgeon to orient themselves within the skull. The accuracy of the image guidance in this case is compromised by the shifting of the brain once the skull has been opened, due to changes in intracranial pressure, and by manipulation of the organ [99]. The location of identifiable points on the cortex may shift by as much as 24 mm when compared to their preoperative locations [100]. A variety of methods have been utilized to compensate for this movement [101–102], but accuracy can be guaranteed only through the use of intraoperative imaging [103]. Specialized intraoperative MRI machines can be utilized for this task [104], as can ultrasound, which may be combined with preoperative MRI [105].

The placement of electrodes into the deep-lying structures of the brain allows the treatment of debilitating neurological disorders such as Parkinson's disease [106] or epilepsy [107]. Accurate targeting of the desired location in these cases is vital in order to successfully treat the disease. Stereotactic frames [108] or frameless navigation techniques [109] can be used. Brain shift is typically less of an issue in these procedures; however, movement of tissue has been observed and quantified, as, for example, by Miyagi et al. [110]. As the exact areas to be target may not be visible on preoperative imaging data, recent research has expanded to include the integration of data from previous interventions. A map of previously successful stimulation locations (from previous patients) can be warped to new imaging data, providing the surgical team with further information regarding the approximate location that should be targeted [111]. This integration of additional data may signal a shift in the future toward "information-guided interventions," as opposed to purely image-guided ones. Robotic systems such as the Neuromate have also been used extensively for these types of procedures [44].

4.3.3 Image Guidance in Ophthalmic Surgery

Ophthalmic surgery has a long history and presents some of the most demanding constraints in terms of perception and dexterity, complicated by the deformable, fragile, and constantly moving target structures within the eye. Ophthalmologists and ophthalmic surgeons have a variety of technologies at their disposal which they can utilize in order to perform procedures at the micrometer level. Ophthalmic specific imaging modalities such as slit lamps, stereophotography, or confocal scanning laser ophthalmoscopy, are available and used extensively. These are often combined with standard medical imaging modalities such as MRI or CT. Modified versions of standard imaging modalities, particularly ultrasound (B-mode and A-mode ultrasound and ultrasound biomicroscopy), are also used [112]. Optical coherence tomography is utilized extensively for noninvasive imaging, particularly of the retina. A variety of ophthalmic procedures utilize image guidance; examples include laser or traditional cataract surgery, laser treatment of retinal pathologies [113], orbital surgery [114], and the treatment of intraocular tumors through the use of external beam radiotherapy [115].

The use of robotics within the context of ophthalmic surgery is not widespread; however, great possibilities exist and a great deal of research is underway. Early work included the development of a novel six-degree-of-freedom manipulator for the precise placement of needles in the retina [116] and a master–slave manipulator including force feedback, which could also be used as a surgical simulator [117]. An overview of other work was covered in two 2010 reviews of the subject by Jeganathan et al. [118] and Fine et al. [119]. Examples include the steady-hand robot [120], which aims to reduce the natural tremor of a surgeon's hands, improving fine motor control during delicate ophthalmic (or other microsurgical) procedures. Similarly, work at Carnegie Melon University has included the development of a handheld device for the cancellation of hand tremor [121]. The da Vinci system has also been utilized within an ophthalmic context in order to repair corneal lacerations [122], while at the opposite end of the size spectrum, miniature microrobots were proposed for drug delivery within the retina [123]. These microrobots are steered through the use of external magnetic fields, with a robotic system known as the OctoMag, used to generate complex magnetic field patterns, allowing the miniature device to move through five degrees of freedom [59].

4.3.4 Image Guidance in Other Surgeries

The above represent some of the major microsurgical fields in which image guidance is commonly, or routinely, used. These are not the only examples, however. As the technology improves, the opportunity exists for more and more surgical procedures to become a form of "microsurgery," relying on actions performed at the micrometer level.

Within the context of craniomaxillo facial surgery, the use of surgical planning for procedures is well established, allowing the surgeon to plan the cuts required in advance [124], select and shape implants and fixation plates, and predict the final outcome of the procedure, including aesthetic factors [125]. While these procedures can be, and typically are, performed using traditional tools and techniques, recent developments include the use of high-power laser devices for high-accuracy cutting of bone [126–127]. On the spine, the implantation of pedicle screws into the vertebrae requires accuracies of better than 1 mm in order to safely avoid penetration of nearby neural structures. Image guidance and robotics have been used extensively to assist with this task. Examples include the SpineAssist [128] and Neuroglide [129] surgical robots, as well as a variety of literature describing the use of standard navigation techniques [130].

Except in the case of ophthalmology, the accuracy of microsurgical navigation as presented above is dependent on (and takes advantage of) the rigidity of bones; it is assumed that all transformations are rigid and that structures do not deform. These constraints hold (to different degrees) in the majority of cases that we have discussed, and this allows high application accuracy to be achieved through the use of image-guided systems; the systems can subsequently demonstrate some degree of usefulness in microsurgical situations. These assumptions cannot be made in cases where soft tissue is not constrained by bone, that is, cases in which significant motion or deformation of the target organ may occur. A subset of these issues was discussed in the context of neurosurgery, with the shifting of the brain due to changes in pressure or handling; however, movements or deformations can be significantly larger elsewhere in the body. The movement of the soft-tissue structures of the abdomen due to breathing, body position, handling, or a variety of other factors complicates every stage of the image-guidance procedure. Imaging artifacts due to breathing movements are common, planning is complicated as structures may be difficult to segment or correctly identify, the registration process may be compromised due to difficulties in identifying landmarks or surfaces, or the matching process may fail due to deformation. Tracking of tools may be difficult; long tools may be required to reach structures, affecting accuracy; targets may shift within the body; and tools, particularly needles, may bend. For these reasons, the application accuracy of image-guided systems is typically significantly reduced in nonrigid environments, and navigation is not as widely used within these fields. Despite these issues, however, significant progress has been made.

4.4 Conclusions

This chapter has presented an overview of image guidance within the context of microsurgical procedures. While image guidance has existed for a number of years, significant progress has been made since the introduction of CT and the stereotactic frame, particularly through improvements in imaging technology and computing power. Microsurgical procedures are performed at the limits of human dexterity and perception and rely on assistive technologies such as the operating microscope. Image guidance may be considered as simply another tool available to the surgeon. The technology is now standard for many microsurgical procedures, chiefly in the fields of neurosurgery and otolaryngology. By combining medical imaging information with surgical experience and expertise, the technology has the potential to make procedures less invasive and more precise, as well as to allow entirely new procedures such as minimally invasive cochlear implantation. Future progress in the field relies on continuing partnerships between medical and research or engineering experts to solve complex medical or surgical problems and ensure improved outcomes for the patient.

References

1. Schlöndorff, G. et al. "CAS (computer assisted surgery): A new procedure in head and neck surgery." *HNO* 37.5 (1989): p. 187.
2. Kwoh, Y. S. et al. "A new computerized tomographic-aided robotic stereotaxis system." *Robotics Age* 7.6 (1985): pp. 17–22.
3. Boas, F. E., and Fleischmann, D. "CT artifacts: Causes and reduction techniques." *Imaging in Medicine* 4.2 (2012): pp. 229–240.
4. Jaju, P. P. et al. "Artefacts in cone beam CT." *Open Journal of Stomatology* 3.5 (2013).
5. Unsgaard, G. et al. "Neuronavigation by intraoperative three-dimensional ultrasound: Initial experience during brain tumor resection." *Neurosurgery* 50.4 (2002): pp. 804–812.
6. Eröss, L. et al. "Neuronavigation and fluoroscopy-assisted subdural strip electrode positioning: A simple method to increase intraoperative accuracy of strip localization in epilepsy surgery: Technical note." *Journal of Neurosurgery* 110.2 (2009): pp. 327–331.
7. Schenck, J. F. et al. "Superconducting open-configuration MR imaging system for image-guided therapy." *Radiology* 195.3 (1995): pp. 805–814.
8. Black, P. M. L. et al. "Development and implementation of intraoperative magnetic resonance imaging and its neurosurgical applications." *Neurosurgery* 41.4 (1997): pp. 831–845.
9. Schulder, M., Azmi, H., and Biswal, B. "Functional magnetic resonance imaging in a low-field intraoperative scanner." *Stereotactic and Functional Neurosurgery* 80.1–4 (2004): pp. 125–131.
10. Townsend, D. W. "Dual-modality imaging: Combining anatomy and function." *Journal of Nuclear Medicine* 49.6 (2008): pp. 938–955.
11. Majdani, O. et al. "A robot-guided minimally invasive approach for cochlear implant surgery: Preliminary results of a temporal bone study." *International Journal of Computer Assisted Radiology and Surgery* 4.5 (2009), pp. 475–486.
12. Pham, D. L., Xu, C., and Prince, J. L. "Current methods in medical image segmentation 1." *Annual Review of Biomedical Engineering* 2.1 (2000): pp. 315–337.
13. Heimann, T., and Meinzer, H.-P. "Statistical shape models for 3D medical image segmentation: A review." *Medical Image Analysis* 13.4 (2009): pp. 543–563.
14. Fitzpatrick, J. M. "Fiducial registration error and target registration error are uncorrelated." *SPIE Medical Imaging*. International Society for Optics and Photonics, 2009.
15. Fitzpatrick, J. M., West, J. B., and Maurer Jr., C. R. "Predicting error in rigid-body point-based registration." *IEEE Transactions on Medical Imaging* 17.5 (1998): pp. 694–702.
16. Maurer, C. R. et al. "The accuracy of image-guided neurosurgery using implantable fiducial markers." *Computer Assisted Radiology*. Springer-Verlag, Berlin, 1995.
17. Wiles, A. D. et al. "A statistical model for point-based target registration error with anisotropic fiducial localizer error." *IEEE Transactions on Medical Imaging* 27.3 (2008): pp. 378–390.
18. Moghari, M. H., Ma, B., and Abolmaesumi, P. "A theoretical comparison of different target registration error estimators." *Medical Image Computing and Computer-Assisted Intervention—MICCAI 2008*. Springer, Berlin, Heidelberg, 2008, pp. 1032–1040.
19. Audette, M. A., Ferrie, F. P., and Peters, T. M. "An algorithmic overview of surface registration techniques for medical imaging." *Medical Image Analysis* 4.3 (2000): pp. 201–217.
20. Matsumoto, N. et al. "A minimally invasive registration method using Surface Template-Assisted Marker Positioning (STAMP) for image-guided otologic surgery." *Otolaryngology—Head and Neck Surgery* 140.1 (2009): pp. 96–102.
21. Matsumoto, N. et al. "Cochlear implantation assisted by noninvasive image guidance." *Otology & Neurotology* 33.8 (2012): pp. 1333–1338.
22. Fitzpatrick, J. M. "Error prediction for probes guided by means of fixtures." *SPIE Medical Imaging*. International Society for Optics and Photonics, 2012.

23. Gerber, N. et al. "High accuracy patient-to-image registration for the facilitation of image guided robotic microsurgery on the head," *IEEE Transactions on Biomedical Engineering*, 60.4 (January 2013): pp. 960–968.

24. Widmann, G., Stoffner, R., and Bale, R. "Errors and error management in image-guided craniomaxillofacial surgery." *Oral Surgery, Oral Medicine, Oral Pathology, Oral Radiology, and Endodontology* 107.5 (2009): pp. 701–715.

25. Bell, B. et al. "In vitro accuracy evaluation of image-guided robot system for direct cochlear access." *Otology & Neurotology* 34.7 (2013): pp. 1284–1290.

26. Amstutz, C. et al. "A-mode ultrasound-based registration in computer-aided surgery of the skull." *Archives of Otolaryngology—Head & Neck Surgery* 129.12 (2003): pp. 1310–1316.

27. Zollei, L. et al. "2D-3D rigid registration of X-ray fluoroscopy and CT images using mutual information and sparsely sampled histogram estimators." *Proceedings of the 2001 IEEE Computer Society Conference on Computer Vision and Pattern Recognition, 2001 (CVPR 2001)*, Vol. 2. IEEE, 2001.

28. Rietzel, E., and Chen, G. T. "Deformable registration of 4D computed tomography data." *Medical Physics* 33.11 (2006): pp. 4423–4430.

29. Maintz, J. B., and Viergever, M. A. "A survey of medical image registration." *Medical Image Analysis* 2.1 (1998): pp. 1–36.

30. Zitova, B., and Flusser, J. "Image registration methods: A survey." *Image and Vision Computing* 21.11 (2003): pp. 977–1000.

31. Eggers, G., Mühling, J., and Marmulla, R. "Image-to-patient registration techniques in head surgery." *International Journal of Oral and Maxillofacial Surgery* 35.12 (2006): pp. 1081–1095.

32. Hill, D. L. G. et al. "Medical image registration." *Physics in Medicine and Biology* 46.3 (2001): R1.

33. Oliveira, F. P. M., and Tavares, J. M. R. S. "Medical image registration: A review." *Computer Methods in Biomechanics and Biomedical Engineering* ahead-of-print (2012): pp. 1–21.

34. Zinreich, S. J. et al. "Frameless stereotaxic integration of CT imaging data: Accuracy and initial applications." *Radiology* 188.3 (1993): pp. 735–742.

35. Milne, A. D. et al. "Accuracy of an electromagnetic tracking device: A study of the optimal operating range and metal interference." *Journal of Biomechanics* 29.6 (1996): pp. 791–793.

36. Yaniv, Z. et al. "Electromagnetic tracking in the clinical environment." *Medical Physics* 36.3 (2009): pp. 876–892.

37. Gobbi, D. G., Comeau, R. M., and Peters, T. M. "Ultrasound probe tracking for real-time ultrasound/MRI overlay and visualization of brain shift." *Medical Image Computing and Computer-Assisted Intervention–MICCAI '99*. Springer, Berlin, Heidelberg, 1999.

38. Williamson, T. M. et al. "Estimation of tool pose based on force–density correlation during robotic drilling." *IEEE Transactions on Biomedical Engineering* 60.4 (2013): pp. 969–976.

39. Galloway Jr., R. L. et al. "Optical localization for interactive, image-guided neurosurgery." *Medical Imaging 1994*. International Society for Optics and Photonics, 1994.

40. Jakopec, M. et al. "The hands-on orthopaedic robot 'Acrobot': Early clinical trials of total knee replacement surgery." *IEEE Transactions on Robotics and Automation* 19.5 (2003): pp. 902–911.

41. Paul, H. A. et al. "Development of a surgical robot for cementless total hip arthroplasty." *Clinical Orthopaedics and Related Research* 285 (1992): pp. 57–66.

42. Taylor, R. H. et al. "An image-directed robotic system for precise orthopaedic surgery." *IEEE Transactions on Robotics and Automation* 10.3 (1994): pp. 261–275.

43. Li, Q. H. et al. "The application accuracy of the NeuroMate robot—A quantitative comparison with frameless and frame-based surgical localization systems." *Computer Aided Surgery* 7.2 (2002): pp. 90–98.

44. Varma, T. R. K., and Eldridge, P. "Use of the NeuroMate stereotactic robot in a frameless mode for functional neurosurgery." *The International Journal of Medical Robotics and Computer Assisted Surgery* 2.2 (2006): pp. 107–113.

45. Eljamel, M. S. "Robotic application in epilepsy surgery." *The International Journal of Medical Robotics and Computer Assisted Surgery* 2.3 (2006): pp. 233–237.

46. Deacon, G. et al. "The Pathfinder image-guided surgical robot." *Proceedings of the Institution of Mechanical Engineers, Part H: Journal of Engineering in Medicine* 224.5 (2010): pp. 691–713.

47. Sukovich, W. et al. "Miniature robotic guidance for pedicle screw placement in posterior spinal fusion: Early clinical experience with the SpineAssist®." *The International Journal of Medical Robotics and Computer Assisted Surgery* 2.2 (2006): pp. 114–122.

48. Lum, M. J. H. et al. "Objective assessment of telesurgical robot systems: Telerobotic FLS." *Studies in Health Technology and Informatics* 132 (2008): 263.

49. Hempel, E. et al. "An MRI-compatible surgical robot for precise radiological interventions." *Computer Aided Surgery* 8.4 (2003): pp. 180–191.

50. Sutherland, G. R. et al. "The evolution of neuroArm." *Neurosurgery* 72 (2013): pp. A27–A32.

51. Buchs, N. C. et al. "Augmented environments for the targeting of hepatic lesions during image-guided robotic liver surgery." *Journal of Surgical Research* 184.2 (2013): pp. 825–831.

52. Dupont, P. E. et al. "Design and control of concentric-tube robots." *IEEE Transactions on Robotics* 26.2 (2010): pp. 209–225.

53. Webster, R. J., Okamura, A. M., and Cowan, N. J. "Toward active cannulas: Miniature snake-like surgical robots." *2006 IEEE/RSJ International Conference on Intelligent Robots and Systems*. IEEE, 2006.

54. Cianchetti, M. et al. "STIFF-FLOP surgical manipulator: Mechanical design and experimental characterization of the single module." *2013 IEEE/RSJ International Conference on Intelligent Robots and Systems (IROS)*. IEEE, 2013.

55. Simaan, N., Taylor, R., and Flint, P. "High dexterity snake-like robotic slaves for minimally invasive telesurgery of the upper airway." *Medical Image Computing and Computer-Assisted Intervention—MICCAI 2004*. Springer, Berlin, Heidelberg, 2004, pp. 17–24.

56. Harada, K. et al. "Wireless reconfigurable modules for robotic endoluminal surgery." *IEEE International Conference on Robotics and Automation, 2009 (ICRA '09)*. IEEE, 2009.

57. Nagy, Z. et al. "Experimental investigation of magnetic self-assembly for swallowable modular robots." *IEEE/RSJ International Conference on Intelligent Robots and Systems, 2008 (IROS 2008)*. IEEE, 2008.

58. Di Natali, C. et al. "Trans-abdominal active magnetic linkage for robotic surgery: Concept definition and model assessment." *2012 IEEE International Conference on Robotics and Automation (ICRA)*. IEEE, 2012.

59. Kummer, M. P. et al. "OctoMag: An electromagnetic system for 5-DOF wireless micromanipulation." *IEEE Transactions on Robotics* 26.6 (2010): pp. 1006–1017.

60. Cleary, K., and Nguyen, C. "State of the art in surgical robotics: Clinical applications and technology challenges." *Computer Aided Surgery* 6.6 (2001): pp. 312–328.

61. Camarillo, D. B., Krummel, T. M., and Salisbury Jr, J. K. "Robotic technology in surgery: Past, present, and future." *The American Journal of Surgery* 188.4 (2004): pp. 2–15.

62. Haidegger, T. et al. "Future trends in robotic neurosurgery." *14th Nordic-Baltic Conference on Biomedical Engineering and Medical Physics*. Springer, Berlin, Heidelberg, 2008.

63. Rolland, J. P., and Fuchs, H. "Optical versus video see-through head-mounted displays in medical visualization." *Presence: Teleoperators and Virtual Environments* 9.3 (2000): pp. 287–309.

64. Salb, T. et al. "Intraoperative presentation of surgical planning and simulation results using a stereoscopic see-through head-mounted display." *Electronic Imaging*. International Society for Optics and Photonics, 2000.

65. Sugimoto, M. et al. "Image overlay navigation by markerless surface registration in gastrointestinal, hepatobiliary and pancreatic surgery." *Journal of Hepato-Biliary-Pancreatic Sciences* 17.5 (2010): pp. 629–636.

66. Gavaghan, K. A. et al. "A portable image overlay projection device for computer-aided open liver surgery." *IEEE Transactions on Biomedical Engineering* 58.6 (2011): pp. 1855–1864.

67. Buchs, N. C. et al. "Augmented environments for the targeting of hepatic lesions during image-guided robotic liver surgery." *Journal of Surgical Research* 184.2 (2013): pp. 825–831.

68. Fusaglia, M. et al. "Endoscopic image overlay for the targeting of hidden anatomy in laparoscopic visceral surgery." *Augmented Environments for Computer-Assisted Interventions*. Springer, Berlin, Heidelberg, 2013, pp. 9–21.

69. Munz, Y. et al. "The benefits of stereoscopic vision in robotic-assisted performance on bench models." *Surgical Endoscopy and Other Interventional Techniques* 18.4 (2004): pp. 611–616.

70. Herron, D. M. et al. "The 3-D monitor and head-mounted display." *Surgical Endoscopy* 13.8 (1999): pp. 751–755.

71. Strau, G. et al. "Evaluation of a navigation system for ENT with surgical efficiency criteria." *The Laryngoscope* 116.4 (2006): pp. 564–572.

72. Gavaghan, K. et al. "Evaluation of a portable image overlay projector for the visualisation of surgical navigation data: Phantom studies." *International Journal of Computer Assisted Radiology and Surgery* 7.4 (2012): pp. 547–556.

73. Simpson, A. L., Ma, B., Vasarhelyi, E. M., Borschneck, D. P., Ellis, R. E., and Stewart, A. J. "Computation and visualization of uncertainty in surgical navigation." *The International Journal of Medical Robotics and Computer Assisted Surgery* 10.3 (2013): pp. 332–343.

74. "Intra-Operative Use of Computer Aided Surgery" *American Academy of Otolaryngology—Head and Neck Surgery*. Web. 2 March 2014.

75. Schipper, J. et al. "Navigation as a quality management tool in cochlear implant surgery." *The Journal of Laryngology & Otology* 118.10 (2004): pp. 764–770.

76. Labadie, R. F. et al. "Customized, rapid-production microstereotactic table for surgical targeting: Description of concept and in vitro validation." *International Journal of Computer Assisted Radiology and Surgery* 4.3 (2009): pp. 273–280.

77. Labadie, R. F. et al. "Minimally invasive image-guided cochlear implantation surgery: First report of clinical implementation." *The Laryngoscope* 124.8 (2014): pp. 1915–1922.

78. Eilers, H. et al. "Navigated, robot assisted drilling of a minimally invasive cochlear access." *IEEE International Conference on Mechatronics, 2009 (ICM 2009)*. IEEE, 2009.

79. Bell, B. et al. "A self-developed and constructed robot for minimally invasive cochlear implantation." *Acta Oto-laryngologica* 132.4 (2012): pp. 355–360.

80. Kobler, J.-P. et al. "Design and analysis of a head-mounted parallel kinematic device for skull surgery." *International Journal of Computer Assisted Radiology and Surgery* 7.1 (2012): pp. 137–149.

81. Kratchman, L. B. et al. "Design of a bone-attached parallel robot for percutaneous cochlear implantation." *IEEE Transactions on Biomedical Engineering* 58.10 (2011): pp. 2904–2910.

82. Brett, P. N. et al. "A surgical robot for cochleostomy." *Conf. Proc. IEEE Eng. Med. Biol. Soc.* (2007): pp. 1229–1232.

83. Liu, W. P. et al. "Cadaveric feasibility study of da Vinci Si–assisted cochlear implant with augmented visual navigation for otologic surgery." *JAMA Otolaryngology—Head & Neck Surgery* (2014).

84. Dillon, N. P. et al. "Preliminary testing of a compact bone-attached robot for otologic surgery." *SPIE Medical Imaging*. International Society for Optics and Photonics, 2014.

85. Miroir, M. et al. "RobOtol: From design to evaluation of a robot for middle ear surgery." *2010 IEEE/RSJ International Conference on Intelligent Robots and Systems (IROS)*. IEEE, 2010.

86. Messerklinger, W. "Technics and possibilities of nasal endoscopy." *HNO* 20.5 (1972): p. 133.

87. Stammberger, H., and Posawetz, W. "Functional endoscopic sinus surgery." *European Archives of Otorhinolaryngology* 247.2 (1990): pp. 63–76.

88. Kennedy, D. W. "Functional endoscopic sinus surgery: Technique." *Archives of otolaryngology* 111.10 (1985): pp. 643–649.

89. Tschopp, K. P., and Thomaser, E. G. "Outcome of functional endonasal sinus surgery with and without CT-navigation." *Rhinology* 46.2 (2008): pp. 116–120.

90. Caversaccio, M. et al. "Frameless computer-aided surgery system for revision endoscopic sinus surgery." *Otolaryngology—Head and Neck Surgery* 122.6 (2000): pp. 808–813.

91. Ragab, S. M., Lund, V. J., and Scadding, G. "Evaluation of the medical and surgical treatment of chronic rhinosinusitis: A prospective, randomised, controlled trial." *The Laryngoscope* 114.5 (2004): pp. 923–930.

92. Koele, W. et al. "Image guided surgery of paranasal sinuses and anterior skull base—Five years' experience with the InstaTrak®-System." *Rhinology* 40.1 (2002): pp. 1–9.

93. Caversaccio, M. et al. "Impact of a self-developed planning and self-constructed naviga-tion system on skull base surgery: 10 years' experience." *Acta Oto-laryngologica* 127.4 (2007): pp. 403–407.

94. Mattos, L. S. et al. "Design and control of a robotic system for assistive laser phonomicro-surgery." *2010 Annual International Conference of the IEEE Engineering in Medicine and Biology Society (EMBC).* IEEE, 2010.

95. Park, Y. M. et al. "Transoral robotic surgery (TORS) in laryngeal and hypopharyngeal cancer." *Journal of Laparoendoscopic & Advanced Surgical Techniques* 19.3 (2009): pp. 361–368.

96. Hockstein, N. G. et al. "Robot-assisted pharyngeal and laryngeal microsurgery: Results of robotic cadaver dissections." *The Laryngoscope* 115.6 (2005): pp. 1003–1008.

97. Galloway Jr, R. L., Maciunas, R. J., and Edwards, C. A. "Interactive image-guided neurosur-gery." *IEEE Transactions on Biomedical Engineering* 39.12 (1992): pp. 1226–1231.

98. Grunert, P. et al. "Computer-aided navigation in neurosurgery." *Neurosurgical Review* 26.2 (2003): pp. 73–99.

99. Letteboer, M. M. J. et al. "Brain shift estimation in image-guided neurosurgery using 3-D ultrasound." *IEEE Transactions on Biomedical Engineering* 52.2 (2005): pp. 268–276.

100. Hastreiter, P. et al. "Strategies for brain shift evaluation." *Medical Image Analysis* 8.4 (2004): pp. 447–464.

101. Audette, M. A. et al. "An integrated range-sensing, segmentation and registration frame-work for the characterization of intrasurgical brain deformations in image-guided surgery." *Computer Vision and Image Understanding* 89.2 (2003): pp. 226–251.

102. Ding, S. et al. "Estimation of intra-operative brain shift using a tracked laser range scanner." *29th Annual International Conference of the IEEE Engineering in Medicine and Biology Society, 2007 (EMBS 2007).* IEEE, 2007.

103. Nabavi, A. et al. "Serial intraoperative magnetic resonance imaging of brain shift." *Neurosurgery* 48.4 (2001): pp. 787–798.

104. Schulder, M., and Carmel, P. W. "Intraoperative magnetic resonance imaging: Impact on brain tumor surgery." *Cancer Control* 10.2 (2003): pp. 115–124.

105. Gobbi, D. G., Comeau, R. M., and Peters, T. M. "Ultrasound probe tracking for real-time ultra-sound/MRI overlay and visualization of brain shift." *Medical Image Computing and Computer-Assisted Intervention—MICCAI '99.* Springer, Berlin, Heidelberg, 1999.

106. Obeso, J. A. et al. "Surgery for Parkinson's disease." *Journal of Neurology, Neurosurgery, and Psychiatry* 62.1 (1997): p. 2.

107. Mehta, A. D. et al. "Frameless stereotactic placement of depth electrodes in epilepsy surgery." *Journal of Neurosurgery* 102.6 (2005): pp. 1040–1045.

108. Fiegele, T. et al. "Accuracy of stereotactic electrode placement in deep brain stimulation by intraoperative computed tomography." *Parkinsonism & Related Disorders* 14.8 (2008): pp. 595–599.

109. D'Haese, P.-F. et al. "Computer-aided placement of deep brain stimulators: From planning to intraoperative guidance." *IEEE Transactions on Medical Imaging* 24.11 (2005): pp. 1469–1478.

110. Miyagi, Y., Shima, F., and Sasaki, T. "Brain shift: An error factor during implantation of deep brain stimulation electrodes." (2007).

111. Finnis, K. W. et al. "Three-dimensional database of subcortical electrophysiology for image-guided stereotactic functional neurosurgery." *IEEE Transactions on Medical Imaging* 22.1 (2003): pp. 93–104.

112. "Ophthalmic Services Guidance—Ophthalmic Imaging," *Royal College of Ophthalmologists,* Web. February 2009.

113. Broehan, A. M. et al. "Real-time multimodal retinal image registration for a computer-assisted laser photocoagulation system." *IEEE Transactions on Biomedical Engineering* 58.10 (2011): pp. 2816–2824.

114. Klimek, L., Wenzel, M., and Mösges, R. "Computer-assisted orbital surgery." *Ophthalmic Surgery* 24.6 (1993): pp. 411–417.

115. Rüegsegger, M. B. et al. "Statistical modeling of the eye for multimodal treatment planning for external beam radiation therapy of intraocular tumors." *International Journal of Radiation Oncology, Biology, Physics* 84.4 (2012): pp. e541–e547.

116. Jensen, P. S. et al. "Toward robot-assisted vascular microsurgery in the retina." *Graefe's Archive for Clinical and Experimental Ophthalmology* 235.11 (1997): pp. 696–701.
117. Hunter, I. W. et al. "Ophthalmic microsurgical robot and associated virtual environment." *Computers in Biology and Medicine* 25.2 (1995): pp. 173–182.
118. Jeganathan, V. S. E., and Shah, S. "Robotic technology in ophthalmic surgery." *Current Opinion in Ophthalmology* 21.1 (2010): pp. 75–80.
119. Fine, H. F. et al. "Robot-assisted ophthalmic surgery." *Canadian Journal of Ophthalmology/Journal Canadien d'Ophtalmologie* 45.6 (2010): pp. 581–584.
120. Taylor, R. et al. "A steady-hand robotic system for microsurgical augmentation." *The International Journal of Robotics Research* 18.12 (1999): pp. 1201–1210.
121. Riviere, C. N., Ang, W. T., and Khosla, P. K. "Toward active tremor canceling in handheld microsurgical instruments." *IEEE Transactions on Robotics and Automation* 19.5 (2003): pp. 793–800.
122. Bourla, D. H. et al. "Feasibility study of intraocular robotic surgery with the da Vinci surgical system." *Retina* 28.1 (2008): pp. 154–158.
123. Dogangil, G. et al. "Toward targeted retinal drug delivery with wireless magnetic microrobots." *IEEE/RSJ International Conference on Intelligent Robots and Systems, 2008 (IROS 2008).* IEEE, 2008.
124. Westermark, A., Zachow, S., and Eppley, B. L. "Three-dimensional osteotomy planning in maxillofacial surgery including soft tissue prediction." *Journal of Craniofacial Surgery* 16.1 (2005): pp. 100–104.
125. Bianchi, A. et al. "Facial soft tissue esthetic predictions: Validation in craniomaxillofacial surgery with cone beam computed tomography data." *Journal of Oral and Maxillofacial Surgery* 68.7 (2010): pp. 1471–1479.
126. Bruno, A. E., Jürgens, P., and Zeilhofer, H.-F. "Carlo-computer assisted and robot guided laserosteotome." U.S. Patent Application 13/497,520.
127. Kuttenberger, J. J. et al. "Computer-guided CO2-laser osteotomy of the sheep tibia: Technical prerequisites and first results." *Photomedicine and Laser Surgery* 26.2 (2008): pp. 129–136.
128. Devito, D. P. et al. "Clinical acceptance and accuracy assessment of spinal implants guided with SpineAssist surgical robot: Retrospective study." *Spine* 35.24 (2010): pp. 2109–2115.
129. Kostrzewski, S. et al. "Robotic system for cervical spine surgery." *The International Journal of Medical Robotics and Computer Assisted Surgery* 8.2 (2012): pp. 184–190.
130. Merloz, Ph. et al. "Pedicle screw placement using image guided techniques." *Clinical Orthopaedics and Related Research* 354 (1998): pp. 39–48.

Section II

Telenursing, Personalized Care, Patient Care, and eEmergency Systems

5

eHealth and Telenursing

Sinclair Wynchank and Nathanael Sabbah

CONTENTS

ABSTRACT   Telenursing, that part of eHealth relating to nursing, is the use of information and communication technology (ICT) to transmit data relevant to any aspect of nursing activity, encompassing many activities, such as triage, teleconsultations, home care, education, and research. After a definition and explanation of telenursing, a short history of the field is given. The recent expansion of nursing responsibilities worldwide has been accompanied by the steady growth of nurses' power to manage patients, exemplified by the nurse practitioner, and all this is shown to be strongly supported by telenursing. It is cost effective and its applications have rapidly increased in number and variety over the last two decades, in parallel with the increasing power and decreasing price of the necessary ICT equipment. An account of relevant ICT, examples of the wide range of telenursing's clinical activities, and its important role in education, especially in developing countries, are presented. Ethical considerations arising from telenursing are outlined. It is concluded that use of telenursing has increased rapidly because it is a most powerful and cost-effective modality, significantly improving the healthcare provided by the nursing professionals.

KEY WORDS: *Telenursing, developed countries, developing countries, distance learning.*

5.1 Introduction

Nursing is generally held to be a compassionate vocation and its members predominantly care for the sick and disabled and nurture normal development in the young. But it is also a profession that enthusiastically and efficiently embraces modern technology, as exemplified by the explosive increase in telenursing activities worldwide in the last two decades. In this way technological advances have allowed traditional nursing activities to be greatly enhanced and extended. Florence Nightingale stated that "what nursing has to do … , is to put the patient in the best condition for nature to act upon him" [1]. Today's nurse can do so much more than her 19th-century prototype, for technology, such as eHealth, extends the range and depth of her professional activities to attain such a "best condition." (Since the great majority of nursing professionals are female, we use *she* and *her* to refer to a nursing professional in this chapter.)

eHealth has many definitions and one of the most recent and comprehensive, provided by the European Union in February 2014, is that

> eHealth is a broad definition for a variety of technologies and services, from electronic health records to telecare systems that enable patients to be treated remotely while staying in the comfort of their own homes. But while the implementation and application of eHealth systems may vary, the overriding goal is fundamentally the same: using information and communication technology (ICT) to provide better care more efficiently at a lower cost [2].

Telenursing is the subset of eHealth (often exchangeable with telehealth) applied to nursing. It has existed for at least as long as the telephone, but in the last 20 years or so its activity has greatly increased. Nursing's emphases in developed and developing countries differ, so telenursing emphasis varies, too, between them, as elaborated below. The broad range of services and technologies encompassed by telenursing includes remote consultations; surveillance of self-care by patients in their residence; transfer of relevant data, such as that for patient consultation and research, statistical, or other purposes; and provision of availability of electronic libraries and databases. It is particularly important in developing countries, through facilitation of distance learning.

Nurses have a crucial role in healthcare, yet there is currently a great lack of nurses and other healthcare workers. The World Health Organization (WHO) has recently noted that "one third of the world's population lacks access to healthcare because of gaps in the health workforce … and Burundi, for example, has a deficit of 33 health workers per 10,000 population, which leaves 95% of its population without access to healthcare" [3]. Since the costs of training more nurses are a main drawback in the developing world, the savings resulting from use of e-learning (distance learning using ICT) for nurses, and others, can be an important tool for their education.

The WHO observes that worldwide the responsibilities of nurses have been increasing steadily in the last few decades, as their range of activities has been broadening. This is exemplified by the concept of nurse practitioners. They are described by the International Council of Nurses as "registered nurses who have acquired the knowledge base, decision-making skills, and clinical competencies for expanded practice beyond that of a registered nurse." The term *nurse practitioner* was introduced into the United States in 1965, and there, nurse practitioners are now a familiar part of the healthcare system, especially for managing primary healthcare (PHC) problems. About 60% of Americans have used their services, with an overwhelming number (82%) of their patients satisfied or highly satisfied [4]. (In contrast there are few reports from developing countries of patients' opinions of healthcare received from nurses.) As a typical example in Africa, and elsewhere, nurses now frequently provide services for those infected by the human immunodeficiency virus (HIV), which were previously exclusively administered by physicians [5]. Also in developing nations, PHC clinics directed by nurses, without doctors, have the greatest role in provision of public healthcare, whether in urban or nonurban settings. As the spectrum of nursing practice broadens, nursing education must also increase in breadth and depth. Since distance learning is an efficient and cost-effective way of providing nursing education, it has been greatly emphasized in developing countries to the extent that it has been recently noted that "there are few sustained telemedicine services in Africa with the exception of tele-education" [6]. This crucial role is treated more fully below.

Telephonic telenursing consultations often form an important part of nurse practitioners' activities. They should be structured to provide maximum benefit, by including the following: after assessment of the caller, advice and referral should be supplied as required (and this is often the core of the interaction); the caller should be always supported and strengthened and in addition taught (as appropriate) with necessary facilitation of the caller's prior learning [7]. Telenursing consultations can frequently substitute for on-site home visits, [8] and estimation indicates 45% of them can thereby be obviated [9].

With the constant increase in availability, capacity, and speed of ICT, accompanied by its steadily decreasing costs, the role of telenursing in all settings and its benefits have seen an explosive increase during the last two decades. This is confirmed by the rise in number of relevant publications (from the first, dated 1974, up to the time of writing, March 2014). In the 20 years to 1993 there was an average of 0.7 publications per year on all aspects of telenursing listed in the U.S. National Library of Medicine's database. In the 5-year intervals since 1993, the numbers of such publications have risen monotonically from 195 to 878, hence, strongly indicating rapidly growing attention to telenursing and the steady widening of its applications. Incidentally, of this total of 2236 publications, only about 1% directly concerned telenursing in developing countries and all of them were published since 2003.

So it is clear that there has been a great increase in interest in telenursing in the last 20 or so years, but most of this interest has been concentrated in developed countries. Therefore, it is now appropriate to examine the place of telenursing in the activities of the nursing profession, as exercised in all types of nation.

5.2 How Telenursing Came About

The telephone was very much used, soon after its introduction in the late 19th century, by medical doctors, and it continues its pivotal role in all medical communications. Medical radio communications began soon after the First World War and by 1930 they were available in some remote parts of Alaska and Australia [10]. In 1935 an Italian organization was founded to provide free medical assistance to ships at sea, without any healthcare facilities aboard, and this organization still flourishes [11]. But modern telemedical communications were introduced in the 1960s and 1970s as a result of the United States' military and space programs [12]. Initial telemetering of physiological data from spacecraft and space suits evolved into 15 active telemedical projects in 1975, which were supported by the U.S. NASA [13]. From these beginnings telenursing has developed into many varied activities, serving every type of community imaginable. Today some of its most important activities include triage following a telephone call describing a health problem; case management and home care for disability and chronic or incapacitating conditions; monitoring of the early postsurgical course; wound care; and telemonitoring.

5.3 Nursing's Applications of Information and Communication Technology

5.3.1 Computerized Decision-Support Systems

A telenurse service, such as for triage or teleconsultation, depends on the knowledge and experience of the telenurse, which can be enhanced by complementary use of a computerized decision-support system, of which several are now available. Initially this software was to aid triage, when acute conditions required assessment but its use has been extended to other forms of teleconsultation [14]. The use of support systems is intended to improve both the safety and the quality of the consultation, but the resulting patient care should still be centered on the patient, even while being evidence based. So it is essential to ensure a strong component of human contact. Some telenurses have reported such computerized aids to their teleconsultations as providing inhibition as well as assistance to their work. Yet overall there was strong preference for the computerized aid to be made available to them [15].

5.3.2 Databases

A database devoted to nursing can provide varied and most useful applications. The worldwide lack of nurses is associated with insufficient funds for adequate education and training of nurses and is compounded by the problem of nurse migration. However, appropriate use of a nursing database can permit a more effective dispersion of the remaining nursing workforce. Such databases, according to the WHO, should contain comparable and reliable statistics for all health workforce personnel and also details of relevant infrastructure, medications, accessibility of diagnostic procedures, acceptable measures to prevent nosocomial infections, and, therefore, a comprehensive view of nursing service availability and readiness [16]. In an attempt to understand and determine possible means of improving this comprehensive view, the African Health Regulatory Collaborative for Nurses and Midwives, was set up in 2011, on the initiative of the Centers for Disease Control and Prevention in the United States. It addresses mismatches between the knowledge and competencies of nurses in over 35 sub-Saharan nations, keeping in mind the needs of the populations they serve. A centralized and readily accessible database is an important part of this project [17]. Rarely in developing nations have such national nursing databases been set up, but there are examples from Africa. In Kenya the project started in 2002 [18] and is still functioning strongly. It has impacted effectively on a range of improvements in personnel management and workforce planning and policy [19]. In Uganda a similar project was initiated in 2005 and it too has proved valuable in strategic healthcare planning [20]. It is well known that outcomes after trauma are in general significantly worse in developing countries than in the well-resourced nations. There are many reasons for this and to ameliorate the situation will require much funding (e.g., to improve staff knowledge and numbers, medical facilities, transport, and other infrastructure). However, one relatively simple and inexpensive aid is availability of an appropriate trauma database. In a pilot survey project, conducted in Ethiopia, it was found that trauma mortality was higher than predicted, so that the appropriate database would result in better prognosis, leading to better understanding and care for trauma patients [21].

5.3.3 Telephony

Common applications of telenursing are triage and teleconsultations, most often using a telephone. Videoconferencing is much less frequently available. For optimal benefits to the caller, this requires appropriate training of the telenurses, who normally receive postgraduate training. One of the earliest such courses of study was founded in Queensland, Australia, in 2006. It has a preceptorship program run in conjunction with other training and appropriately some of the instruction can be received using distance learning. Those who qualify can work as telenurses in a Queensland telenursing service, which accepts phone calls 24/7. Assessment of the caller's problem is aided by a computerized clinical decision-support system. Action resulting from the phone call ranges from providing self-care advice to requesting an ambulance service. This service, called 13 HEALTH, offers more than triage. It can provide a variety of health and well-being information, for example, which may be required during an epidemic of measles, influenza, etc. This single point of contact is particularly useful for all types of health assessment and information (V. Chalmers, private communication, 2014). A telenursing training program was set up at the Edith Cowan University, in Perth, Australia, in 2007, which resulted in the awarding of a postgraduate certificate. Those eligible for the course required possession of a nursing degree and a 3 years' full-time professional nursing experience. All of the course materials were available online. However, it was found that there was insufficient demand for such formal telenursing education and the course was discontinued in 2012, after 12 nurses had completed these studies (B. Ewens, private communication, 2014).

5.3.4 Videoconferencing

Mainly because of cost, videoconferencing is much less used in telenursing than telephony, but one enterprising application took place in the United Kingdom, as part of the extensive National Health Service's (NHS) telenursing service called NHS Direct. NHS Direct was established in 2000, offering mainly telephonic interactions with patients. But it will be superseded by a new service in 2014, called NHS 111 [22]. In Birmingham, United Kingdom, the pilot telenursing videoconference service used an ordinary, domestic, digital television (TV) to provide a one-way videoconference link. The caller saw the nurse and also could see diagrams and videos, if necessary, to provide information to help solve the problem in hand. One surprising result was that such consultations were shorter on average than consultations using a telephone. However, it never progressed beyond being a pilot scheme [23].

5.3.5 mHealth

Mobile telephone usage throughout the world is now widespread. Currently in very many developing countries, and especially Africa, telephone owners probably will possess only mobile devices (and never own fixed line telephones). This results from a recent, very rapid increase in inexpensive mobile telephone handset availability and the necessary infrastructure. In many regions the world's lowest socioeconomic levels coexist with satisfactory mobile phone signals [24]. Healthcare using mobile ICT devices (called mHealth) is commonplace everywhere and in developed countries it is becoming ubiquitous, especially for home care surveillance. In parallel, mHealth is also now increasingly used in developing nations and, for example, in 2011 the first published report on mobile phone teledermatology came from Egypt [25]. There are also mHealth trials underway in other disciplines. mHealth cannot easily send highly detailed images, but it is appropriate for delivery of reports, etc. mHealth applications often use the short message service. Particularly successful mHealth applications include reminding patients to take chronic medication (e.g., for HIV infection or multidrug-resistant tuberculosis [TB] [26–27]) and to remember to attend clinic (and other healthcare) appointments [28]. Similarly mHealth has been shown to be effective for assessment of acute wound care need and may consequently reduce the number of visits required to a hospital's emergency department [29].

5.3.6 Telenursing Services

The telenurse's function is well described as follows: "the primary responsibility of telenurses is to assess callers' symptoms and provide advice with the guiding principle that the patient receives the right level of care at the right time." From the earliest telenursing services in parts of Australia, New Zealand, Norway, the United Kingdom, and the United States, where their residents have enjoyed access to telenursing advice and triage for symptoms and problems, there has been extension to other countries [30]. Currently, there are many nations with some telenursing service available, including Australia, Brazil, Canada, Finland, India, New Zealand, Norway, Philippines, Sweden, the United Kingdom, and the United States. Africa also has telenursing activity, with some individual projects described elsewhere, but not generally in the form of a formal telenursing service [31]. In countries with multiple regions, each of which has a different licensing authority for the practice of nursing, there are resulting limitations on telenurses' activity. The United States is such a country, but some of its states are members of a multistate licensing authority, which broadens the region in which telenurses can operate [32]. This agreement is called the Multistate Nursing Licensure Compact and it was founded in 2000, with 4 member states [33]. In 2014 this number had grown to 24 states. For such countries as the United States, with varying (state) jurisdictions within a single nation, it has been recommended that two types of nursing databases be used. One should be based on the state and the other a standardized nationwide database. In this way, a comprehensive depository for nurse staffing data would be available for researchers, as well as providing for administrative and other analytical needs [34].

5.4 Telenursing's Healthcare Applications

5.4.1 Triage

One of the first and most extensively applied telenursing activities was triage, for this is a most efficient procedure. It much reduces uncertainty and time wasted in traveling to an inappropriate healthcare establishment. However, triage is not necessarily a straightforward procedure. If a computerized decision-support system is used by the triage nurse, there can be resulting errors. These have been studied in a large Swedish telenursing network and 41% were due to problems with accessibility for patients after they had been directed to appropriate healthcare services. The next most important source of error (25%) was incorrect assessment. So those aspects deserve special attention in training triage nurses [35]. There can also be some more subtle issues involved. One Swedish study of patient satisfaction after telenursing triage concentrated on patients who received a less urgent level of healthcare than they expected, and this resulted in inefficiency, for there were subsequent visits seeking unnecessarily high levels of care. After investigation it was found that those telenurses who had initially suggested a less urgent course of action than the patient desired were independently found to have lower "ability to listen and take note of the callers' health problems." Although this finding seems blindingly obvious, there was waste of time and facilities. Hence, telenursing triage training should anticipate and aim to prevent this situation [36].

5.4.2 Maternity and Pediatrics

A simple and obvious, but valuable, benefit of the availability of a telenursing link is the reassurance that it can bring. Such benefits were evident in observations of first-time mothers and their reactions to the onset of labor. There were clearly reduced levels of anxiety about "getting to the hospital on time" and more confidence that they could remain at home longer [37]. A Cochrane review has shown that telenursing links can also provide significant benefits for antenatal care. It is clear that "reduced programs of antenatal care are associated with an increase in perinatal mortality." So when the planned number of antenatal visits is low, for reasons of inadequate resources, etc., they should not be reduced further unless there is close monitoring of fetal and neonatal outcome [38].

Primary healthcare in pediatrics benefits in many ways from telenursing, especially with counseling. Telephonic counseling for lactation in Malaysia has resulted in an increase in exclusive breast-feeding in the first month of life, but with less success for the 6 months postpartum [39]. This is important for, as a Cochrane review has shown, such exclusive breast-feeding results in less morbidity from gastrointestinal infection, compared with partial breast-feeding only. Such morbidity is regrettably very common in developing countries. In all types of communities, wealthy and poor, there have been no deficits in growth for such infants exclusively breast-fed for their first 6 months [40]. Another serious problem for infants in developing countries is neonatal sepsis. Community-based approaches to reduce its effects have been found to be very effective in a joint study by the WHO and UNICEF (United Nations Children's Fund) and telenursing is an approach that was recommended [41]. Although in recent years childhood mortality has decreased in most parts of the world, neonatal mortality has not, and a neonate dies every 10 s, with 99% of such deaths in poorly resourced nations [42]. So the WHO has instituted a pilot educational program promoting essential neonatal care in Zambia and Brazil. This included a self-directed, computer-controlled course and it achieved the same success, with greater cost-effectiveness, as that of a conventional course taught by instructors. These findings are of great importance for all nations with limited resources [43]. Unsuitable foods, complementary to breast milk, are a principal cause of infant malnutrition in developing countries. However, for the nurse or other advisor to have a list of appropriate foods and feeding schemes to pass on to the mother is not enough to solve the problem. Advice to the mother must contain more than a list of dos and don'ts about what foods to give and how much. In particular, the instructions to the mother must be integrated with locally available foods and sensitive to the mother's culture. If this is done appropriately, then it has been shown that "educational intervention can effectively improve complementary feeding practices and child nutrition and growth" [44]. Preparation for such education by the nurse in the infant clinic, or elsewhere, is often not included in the nursing curriculum, so tele-education using these principles can be applied with a significant beneficial effect.

In Africa a foremost cause of pediatric mortality and morbidity is malaria and about 90% of all life-threatening malaria occurs in children. Appropriate management starts with improved quality of care at the earliest levels of referral and it was found in a joint Norwegian-Malawian project that a clinical audit was able to lead to an assessment of appropriate clinical care, which, in turn, would lead to means to improve the overall management. Telenursing can provide practical means of disseminating these findings [45]. Adequate knowledge about pediatric resuscitation was found to be lacking in Vietnam. To remedy this, some Vietnamese healthcare workers, with a wide range of professional backgrounds, received appropriate training in Australia, which took into account the clinical practices and disease patterns current in Vietnam. To disseminate their newly obtained skills and knowledge and to include updates in their resuscitation practice, tele-education is a most suitable means of achieving this [46]. Overall the most frequent pediatric problems that telenurses must deal with are ear conditions, fever, wounds, and rashes. When complex pediatric situations are dealt with by telenurses, there is more effective assistance if enhanced electronic health records are readily available, which is currently possible only in technologically advanced nations [47]. This is also true for adult patients.

5.4.3 Posthospitalization

It is increasingly common now for surgical patients to be sent home sooner after the procedure than was the case about a decade ago and telenursing is playing an important role in their follow-up. The most effective monitoring phone calls are made in the first 12–24 h [48]. There is a current lack of intensivists in the United States and consequently about 55,000 lives are lost annually. A practical and cost-effective interim measure to remedy this situation is to set up a tele–intensive care unit (tele-ICU), which has been shown to provide better service than an ICU lacking a full-time intensivist. This requires increased nursing participation and so is a novel form of telenursing [49].

There are quantitative assessments of telenursing practice. Patients with congestive heart failure, who were checked with an interactive voice-response system and received daily self-management tips based on self-reporting of symptoms, were found to require less than half the number of readmissions to hospital in the month following their discharge. In addition to being cost effective, it improved adherence to the treatment regime [50]. Another advantage of telenursing for such patients is that it has been shown to result in more effective recognition of comorbidities [51]. So overall their quality of life is improved with telenursing. But this benefit is greater for male patients [52].

5.4.4 Home Care

Home care is an important part of telenursing, for it is cost effective. A telenurse can "visit" more patients in a day than making conventional home visits in person and also the distances between the nurse's base and patients' homes in telenursing are inconsequential. Telephonic transitional care after discharge for high-risk conditions resulted in an average of $1225 saved per patient in a U.S. study and also there were fewer rehospitalizations [53]. Similar findings resulted from a study of telenursing for home care of patients with chronic heart disease. It was perceived to be useful by all parties and easy in application, which together resulted in its general acceptance as a worthwhile form of management [54]. For chronic obstructive pulmonary disease, telenursing has been shown to reduce numbers of hospital admissions and their durations, with increased patient satisfaction [55]. When possible, palliative care is best performed at home and telenursing allows improved symptom management and fewer visits to the emergency department, especially if an after-hours telenursing service is available. Also it provides enhanced support for the family [56]. For patients in a hospice there are also clear cost–benefit advantages from telenursing. It reduces the number of healthcare visits to the hospice by over 60% [57]. In addition, general satisfaction for such a service has been expressed by both patients and their caregivers [58]. Telenurses working with family doctors also bring about improved patient care, adherence to medication regimes, and less use of other healthcare facilities [59].

However, telenursing is not a panacea that can be applied in all situations and this must be kept in mind, for teleconsultations can have drawbacks. Some studies [60–61] show that telenurses' patients have misinterpreted the advice they received, so continuous interventions to maintain quality control are necessary to reduce such miscommunication. When hypertensive patients were managed by telenurses (and some teledoctors), the results in one study provided only moderate benefits overall, although those with poor blood pressure control at baseline benefited most [62]. One common application of telenursing has been in the home care of wound healing. When outcome measures, such as time to heal, resultant costs, and length of hospital stay, and wound status and progress in its size, were considered, it is found that telenursing provided no clear advantages [63]. But other aspects of telenursing for wound care were less negative. Both nurses and patients acquire knowledge about wound care and the telenursing contacts have a calming effect. Also the participating nurses appreciated its "humane component" [64]. Telenursing should sometimes augment, not totally replace, nursing visits, as was found when investigating heart failure patients. The combination clearly improved patient outcomes [65]. Only a few patients in the largest Netherlands telenursing program were able to substitute the telenursing for all regular home care visits, so this must be taken into account when visits in person are being reduced with the onset of telenursing [66].

5.4.5 Chronic Illnesses

The WHO has identified chronic illness as the world's major health burden, with women and rural dwellers its most vulnerable sufferers [67]. Some representative applications, from the wide variety of chronic conditions that can benefit from telenursing, will be outlined and they indicate the successes of this cost-effective approach, which reduces the need for in-person home visits and/or clinic attendance. One of the ways of combating the deleterious effects of chronic illness was a woman-to-woman program in five North Central states of the United States, with nurse monitor facilitators participating. The women suffered from conditions such as cancers, multiple sclerosis, diabetes mellitus, and rheumatic illnesses. The nurses indicated which resources were appropriate, responded to queries, and attempted to augment the participating women's ability to solve problems [68–69]. Later an online nonsynchronous telenursing link further aided their healthcare and also improved social support, reducing loneliness and enhancing their ability to adapt to chronic illness, self-manage more effectively, and be less susceptible to depression. This was especially evident in those suffering from breast cancer [70]. Those with complex endocrine problems are especially likely to benefit from telenursing associated with home care, for such contacts facilitate symptom management, obtaining medication and patient education, so much so that 81.2% of patients in one survey agreed that telenursing services provided high, or very high, levels of satisfaction [71]. Fibromyalgia patients are mainly female and their home care aims to ensure that they cope with their anxiety, chronic and widespread pain, and the consequent desire to minimize activity. When a telenursing intervention using cognitive behavioral therapy was compared with web-based access to information, the former was found to provide more support, increase patient motivation, and expand consciousness. This can be interpreted as the superiority of the personal means of contact in this example of a chronic illness [72]. Telenursing has also been shown to be effective in chronic conditions that are episodic. For ulcerative colitis, which requires medication during both active and quiescent periods, telenursing was shown to improve adherence, for it was believed to be more effective in paying attention to the patients' emotional and cognitive reactions [73]. There are clear advantages for the application of telenursing counseling for sufferers of chronic headache too. When compared with standard care, there was a greater improvement in the quality of life and better adherence to treatment regimes, reduced pain intensity, and fewer visits to general practitioners and to casualty departments. But the symptoms reduced only in the last 5 weeks of a 3-month study [74].

5.4.6 Human Immunodeficiency Virus/Acquired Immunodeficiency Syndrome

Developing countries, especially those in southern Africa, are much more severely affected by HIV/acquired immunodeficiency syndrome (AIDS) than other nations. One of the most effective means of ensuring a lifestyle that avoids HIV infection is voluntary counseling and testing (VCT) [75]. It can reduce risk of acquiring the virus by teaching appropriate sexual behavior modification and has amply confirmed its importance as a crucial HIV prevention strategy [76]. Nurses are at the forefront of such activity [77] and telenursing can support their efforts very effectively in many different ways. But in South Africa, with the world's largest number of HIV/AIDS sufferers, the application of VCT remains low [78]. Nurses there struggle to initiate VCT in part because their training was inadequate [79]. Traditional methods of instruction can be effective and have been shown in a pilot project to bring the level of understanding of rural clinic staff up to that of urban HIV specialists after a short training course to disseminate VCT expertise [80]. To increase its effectiveness, such proven success must be greatly extended, with scaling up community VCT on the large scale necessary [81]. Telenursing is the most effective and cost-efficient way to do so. All sorts of persons likely to contract an HIV infection, as well as those already infected, can benefit from extensive and readily available VCT. This has been demonstrated for young adults, for whom peer-based education is a valuable approach to ensure success [82]. For males, an outreach program using the Internet and encouraging a visit to existing VCT clinics has proved successful [75]. One way in which nurses can ensure that patients are enrolled in a HIV care program, long before they become overtly ill and, therefore, much more difficult to manage, is to enroll them in a home-based VCT program. This effective approach has been used in western Kenya, resulting from a cooperative program with the United States [83]. In addition to prevention of infection, VCT can lead to its early diagnosis. This is critical for it can ensure that those who are HIV positive have earlier access to antiretroviral therapy and effective care services. So with the establishment of a VCT program, there are advantages which accrue beyond its immediate effects. Since TB often coexists with HIV positivity, similar procedures can ensure adherence to both anti-TB and antiviral medication regimes. With the involvement of the VCT clinic and appropriate messages to the patients, it has been established that there are improved TB treatment completion rates and better prevention of infection by multidrug-resistant TB. In many regions the full potential of an integrated TB-HIV service has not been achieved, for the two infections are too often treated separately. Such an integrated attack, with messaging, has a greater impact, quite apart from being less costly [84].

5.4.7 Mental Health

In many situations it has been shown that mental health patients may benefit from telemedical links, and the availability of a telenursing service comprises an important component. Examples follow. One application is the nursing part of a distributed care team for suicidal patients, who may feel isolated and with whom trust must be built. The participating nurses often use their telelinks to deal with otherwise hard-to-contact patients [85]. Often outpatient care is preferable for psychiatric patients, and telenursing is well suited to supplement it. An application that has been successful was to increase motivation to attain cessation of tobacco consumption for those with the posttraumatic stress disorder. Home-based care with telenursing was the approach used [86]. Quantitative improvement has also been reported for Medicaid patients with "serious and persistent mental illness," who received telenursing phone calls (a mean of 3.5 calls over 9 months). As a result there were fewer visits to hospitals and emergency departments and improved adherence to antipsychotic medication regimes, in all a most cost-effective outcome. The participating telenurses were trained using "cognitive behavioral and motivational interviewing techniques" [87].

5.4.8 Geriatrics

When there is dementia in geriatrics, whose caregivers are family members, it is necessary to recognize, and if possible preempt, patient behaviors which may lead to distress for those around the patients. A telenursing project allowed identification of such potential triggers, whether prompted by communication or environmental factors. Increased caregiver understanding of the situation often allowed modification of the triggers and increased confidence in managing such events. An additional benefit was the telenurses' recognition of otherwise undiagnosed medical conditions [88]. Geriatric patients, who experience bypass surgery, tend to have delayed recovery and worse outcomes compared with those who are younger if they had poor preoperation functioning. A telenursing adjunct during their early recovery period succeeded in improving the patients' general functioning and physical activity, with a consequent reduction in their postoperative morbidity and mortality [89].

5.5 Nurse Migration

Nurse migration can have serious consequences, especially for the typically less wealthy nations, from which the nurses usually emigrate. Motives for this emigration, often to the United States, the United Kingdom, Canada, and Australia [90], are usually economic but there are also push factors (which may be professional, social, personal, and political) and the situation can be complex [91]. Negative consequences of nurse migration are readily understood, for the migrating nurses' country of origin's health system is often weakened and can even be considered to be close to collapse [92]. However, some positive features of migration's impact are less well appreciated. It generates billions of dollars in remittances to the nations supplying the emigrants. Also health workers not infrequently return, bringing significant skills and expertise back to their home countries [92]. Paradoxically such migration may improve nursing standards in the nurses' country of origin, as has been observed in the case of Mexico as a result of its trade treaty with the United States and Canada, which are both destinations for Mexican migrant nurses [93]. However, using telenursing to supplement clinical support and increase distance learning opportunities, preferably with formal qualifications resulting from success in such studies, can reduce some of the nurse migrants' push factors [94].

5.6 Telenursing and Distance Education

5.6.1 Bases of e-Learning

Nursing education has benefited from telenursing in many different ways. Classical educational methods require the teacher and pupil to be present at the same time and at the same place for the transfer and acquisition of knowledge to take place. e-learning removes these two conditions, so decreasing costs (for example, by eliminating needs for teachers and/or pupils to travel to the same location), increasing convenience of all parties, and removing limits on the numbers of participating students, although they must have access to appropriate ICT equipment. Computer-controlled instruction can allow self-paced learning and it can also dispense with the need for a teacher's presence or interaction. All these benefits are well known and apply both to telenursing education and to all other types of distance learning. In spite of all these distinct advantages, often there is sometimes resistance to the adoption of new methods in health education, especially in replacing formal lectures, which normally comprise the greater part of the nursing course. Such resistance is often a stubborn adherence to traditional educational structures [95]. However, e-learning is not a total replacement for classical learning techniques, for it can often exist profitably in combination with traditional nurse training methods and classroom instruction [96]. Nursing education levels are currently far from uniform, even in a relatively wealthy region such as Europe. There they range between a nursing diploma and the full academic pathway, including bachelors' and masters' degrees and doctorates. As a result a call for full statistical analyses of European nursing education has been made in conjunction with establishing an appropriate, accessible database [97].

It has been long recognized that distance education methods could facilitate some nursing training. Since nurses are acknowledged as being "at the forefront of healthcare in disaster-stricken areas" [98], there is an e-learning program that is part of a master of science course in disaster relief nursing. It is based in a multicultural society and so this program emphasizes transcultural and transnational nursing. Videoconferencing has been shown to be a powerful and cost-beneficial means of training clinical nursing skills, since clinical experts are often not widely dispersed. When compared with conventional face-to-face methods, such training is more cost effective but also has been found to permit more "dialog, reflection, and synthesis" for the participants in one survey [99], when trainees were most enthusiastic about extending its application.

5.6.2 Educating Laypersons

An important part of nursing activity is to educate laypersons on appropriate ways of improving their health status and preventing morbidity. Such instruction ranges over very many topics, including appropriate infant feeding, measures to prevent infant dehydration, means to reduce spread of sexually transmitted diseases, and care for those who are approaching death with only palliation possible. However, the onus for general wellness is not only the responsibility of healthcare workers. The lay public plays an important role in their own lifestyle and in other ways. As a result, the nursing profession has a crucial role in educating the general population. The concept of health literacy for the public has "evolved from a history of defining, redefining, and quantifying the functional literacy needs of the adult population" [100]. It is believed to be increasingly needed, for otherwise there may be a detrimental effect on health and healthcare. One such successful example of nurses educating the general public and certain health workers resulted from the presentation, in the waiting areas of some South African public and private PHC and other clinics, of appropriate videos concerning means to avoid HIV infection. Viewing of the videos was followed up with explanatory talks by nurses and assessment. The participating nurses received the necessary training by telenursing education [101]. Another example of patient benefit that is more directly clinically involved was for self-care patients with heart failure, who were shown to benefit from a nurse educational program involving telenursing and supplemented by literature and home visits [102].

The use of mHealth technology, in particular mobile phones, has been shown to empower not only nurses and other healthcare givers but also the patients and to allow much more effective communication between them [103]. In Latin America an informal network of peer care was established, in addition to the telenursing contacts. The informal caregivers also provided much mutual support [104]. This is a good example of the use of social networking as an adjunct to telenursing in self-care, which is particularly valuable for the elderly and persons with limited access to formal health services through geographic or cultural barriers. When a telenursing facility is available, the resulting improvement in efficiency for a healthcare service has been well established. For example, after a PHC consultation normally there are follow-up visits and such visits are significantly reduced in number if telenursing is available [105].

5.6.3 Formal Instruction

Telenursing can employ formal instruction courses as part of its e-learning function. Many such courses are currently available, worldwide, and two typical examples are provided by the University of Cape Town, in South Africa. They teach palliative care, use distance learning except for 2 contact teaching weeks, and lead to either a postgraduate diploma or a master of philosophy degree [106]. Nursing plays a major role in palliative care. In the last 13 years since the courses became available, 312 students have graduated, with 23 failing to complete the courses. Most pupils are from South Africa, but 59 have come from other countries, mainly African, for in Africa formal studies in this field are very rare. The curriculum takes into account local socioeconomic and cultural conditions and stresses the participation of the family (both nuclear and nonnuclear), for this is especially relevant to childhood bereavement and HIV/AIDS, currently of major importance throughout much of the continent (E. Gwyther, private communication, 2014). Other tele-education to train telenurses has been described in Subsection 5.3.3.

5.6.4 Web-Based Education

A much used and effective way of extending e-learning, to any location with access to the Internet, is to establish a web-based program and telenursing has frequently adopted this method. Many examples of such programs exist, such as the palliative medicine courses mentioned above. Further representative examples follow. Some of this method's striking advantages, not already mentioned, include the learner's ability to examine the course materials whenever this is required, using any connection to the Internet and without any limitation to repeated access. Many web designs and instructional methods can be employed and these must be carefully evaluated before use in a course [107]. But it is likely that improved educational outcomes are associated with inclusion of feedback, practice exercises, repetition, and other interactions between the participating parties [108].

An extensive web-based distance learning network in pediatric hematology–oncology, for nurses and others in developing countries, especially in Latin America, has been set up by the International Outreach Program, of the St. Jude Children's Research Hospital, Memphis, Tennessee, United States [109]. Such dissemination of expertise is necessary, for about three times as many pediatric cancers achieve remission in developed countries compared with those in less advanced nations and over 80% of all pediatric cancers occur in children from poor nations, where survival can be under 10% and it can be the leading cause of childhood death in the age range 5–15 years [110]. After nursing training provided by the outreach program and other interactions, there was a clearly enhanced outcome for childhood acute lymphoblastic leukemia and other subsequent successes [111]. Since it is clear that neonatal and perinatal deaths are greatly reduced with appropriate training of traditional birth attendants, several such instruction programs have been set up [112]. One of the most extensive and successful programs is the work of the WHO in a wide range of developing countries. It emphasizes improvement of the ability of birth attendants to cope with obstetric and neonatal complications. This program depends strongly on an effective web-based network. It has been shown to be successful in reducing maternal mortality, the fifth Millennium Development Goal [113]. There are over half a million deaths annually worldwide resulting from obstetric complications. Most are preventable and occur in developing countries [114]. It is important to teach the limits of practice, for frequently there is an unclear line of accountability. So the midwife must not feel an obligation to perform clinical procedures which go beyond her training or level of competence. In a study of post–myocardial revascularization patients, telephonic follow-up included an assessment of the patient's relevant knowledge, an offer to address concerns, and monitoring of anxiety levels and mood symptoms, together with encouragement to make appropriate changes to lifestyle and behavior. The phone calls were adapted to each patient's needs and were found to be effective in ensuring satisfactory progress. So the telenurse needs more than the relevant medical knowledge of a patient's condition and this adjunct understanding should be part of the telenurse's training [115]. Telenursing also has proved its value in more mundane situations. In a large Brazilian city a PHC training program involving 148 basic health units was shown to be so successful in providing up-to-date information and tools for PHC practice and widening the scope of the care offered that it was extended to 900 other towns and cities throughout Brazil [116].

Web-based programs can also be of great utility in well-resourced nations, as shown in one project with nurses leading the course. It was intended to reduce potential dangers for patients with high-risk vascular conditions. It succeeded and they became more aware of their capability to manage a variety of situations [117]. Similarly to teach advanced pediatric life-support skills, a web-based program [118] has shown signal success in improving relevant knowledge and abilities. Teaching about ICU nursing has some specific difficulties, for in the ICU, students are often overwhelmed by the physical milieu, especially its technology, its many and varied sounds, and the unpredictable, sudden accelerations of activity. As a result, the students are often shunted aside, especially when there is a patient emergency or instability. The development of a tele-ICU has proved a valuable method of teaching about ICU workings to nursing students, for it allows them to participate in complex situations when the "patient's" situation becomes critical, yet always in perfect safety [49].

In summary, the place of web-based nursing education has been shown in a systematic review to have "encouraging effects in improving both participants' knowledge and skills performance, and in enhancing self-efficacy in performing nursing skills, with a high satisfaction rate expressed by participants" [119]. For many nursing fields, web-based educational programs provide benefits for those nurses who possess and maintain appropriate computer skills. Such fields include critical care [120–121] and palliative care practiced by rural nurses [122]. Note that appropriate websites can also benefit patients by making available relevant information to them. A recent example has shown, from a survey, that this is particularly effective for cardiac patients, although the more elderly patients were less likely to use the Internet [123].

5.6.5 Recent Trends

Recent trends in nursing education are emphasizing patient-centered, team-based care, and, where practical, interprofessional education [124]. Multicultural understanding is steadily becoming more relevant to nursing in many societies, especially in most of the rich nations, which have become more culturally mixed during the last half century or so, because of immigrants from many different ethnic backgrounds. Also there is a growing tendency for nurse migration bringing in nurses with a cultural background which may differ from that of the indigenous population. The nursing profession must be aware of all these differences and its education should reflect them. "Many countries are becoming culturally diverse, but healthcare systems and nursing education often remain monocultural and focused on the norms and needs of the majority culture." This is especially true for developed nations [125]. Nurses should be aware of the needs of all persons living in multicultural societies, with appropriate sensitivity expressed in their professional activities and should express "cultural humility, rather than cultural competency" [126]. Appropriate inclusions into the nursing curriculum initially will be taught by relatively few specialized instructors and so this teaching is well suited to later much wider dissemination by telenursing education techniques.

5.7 Telenursing and Ethical Questions

Telenursing ethics extend general nursing ethics, about which much has been written. Telephony plays a central role in telenursing and can result in new and important ethical questions. For example, the telenurse may not be entirely sure of the caller's identity, which can cause considerable difficulties, concerning documentation, etc. [127]. Multicultural problems are becoming increasingly frequent. They can affect discussion of some sensitive issues, when the telenurse and caller have different cultural backgrounds or genders. It is essential to respect such cultural differences, even if, in the opinion of the telenurse, differences in belief and practice are contributing to the problem under discussion. Neither participant should consequently feel that his/her autonomy is under threat. It has been argued that all of these questions offer greater challenges for telenurses when compared with conventional nurse–patient interactions [127]. When the telenurse is expected to adhere to fixed protocols, this may be at variance with another course of action felt by the telenurse to be preferable for one of several reasons. These may include previous relevant experience, or a lack of resources, especially in poor countries. Male and female telenurses may react differently to ethical problems, as has been shown in a preliminary study, because of possible gender differences in moral reasoning processes. (This is the use of logic to determine whether a specific personal situation is right or wrong.) Mothers using a telenursing service for pediatric problems are much more likely to receive self-care advice, while fathers tend to be referred to other healthcare services. This implies that telenurses may need more gender competency, which can be an apparent lack in their current training [128]. Gender plays an important role in the work of telenurses in other respects too, for most are female and they can stereotype their dialogs "to fit better with female ways of communicating." During their training clinical supervision is required to avoid this [129]. Also female telenurses must cope when "handling covert or overt power messages based on male superiority" [129]. So for ethical telenursing, it is essential that the telenurses' gender competence is of a high standard.

At all times the telenurse must be aware of the concepts of equity of access, informed consent, and privacy and this is especially important for elderly patients [130]. Telenurses' training should, therefore, include presentation of appropriate ethical questions, focus on the ethical dilemmas of telenursing, and allow discussion of their resolution [127]. Telenursing malpractice results principally from the nurse paying inadequate attention to the caller, who may present with the commonest symptoms, which are chest and abdominal pains. To avoid future malpractice, the holding of work group discussions, where practical, is considered effective. Also, all third-party communications are held to be very risky for a telenurse and callers making recurring contacts require repeated evaluation of their need for telecare. But telenursing malpractice is exceedingly uncommon. For example, during the years 2003–2010 there was a total of only 33 malpractice claims made against a large Swedish telenursing network (in stark contrast to a total of about 4.5 million calls made to the service in 2010 alone). These claims have been subsequently analyzed and discussed [131].

5.8 Discussion

Whenever a telenursing service is to be set up, a careful evaluation of the situation is crucial, for not all situations and patients will benefit, as has been indicated above. It is necessary to have a survey of the pretelenursing situation and another (or more than one) after the service has been underway for at least about 6 months. A comparison of costs incurred, satisfaction/dissatisfaction of all parties concerned, numbers of persons served, avoidance of unnecessary travel, etc., should be investigated. But it is essential not to emphasize costs above all else. If the telenursing service is overall more costly than the system it replaced, then this is not necessarily regression, for there are two important further considerations in the evaluation. Even if introduction of telenursing resulted in greater costs, do the consequent benefits outweigh such additional costs? The necessary quantification required to answer this question is not an easy matter, but there are relevant techniques of health economics which can be applied. The other important question relevant to the evaluation asks, Is the potential for expansion of the service and associated telehealth activity likely to provide greatly increased benefits? Under most circumstances it is clear that telenursing is cost beneficial. In 1998 there were a million and a half home health visits daily in the United States. If they were made by a registered nurse, their average cost was about $90 and she can perform only five or six such visits in a day. But even then it was appreciated that about 70% of such visits could be replaced or supplemented by telenursing and because of travel time saved, about three or four times as many telenursing visits could be made daily by a telenurse [132].

A benefit of teleconsultations is that after referring several patients with similar conditions, the referring nurse practitioner frequently has acquired the expertise to recognize and manage the same type of problem without further referral. This has been exemplified in a South African teledermatological pilot project. Many patients involved were HIV positive, for well over 90% of those infected with HIV have dermatological manifestations, whose severity mirrors the progress of the infection. After telereferral the nurse practitioners receive a diagnosis and suggestions for management. When several such referrals had been made, the senders' ability to recognize the condition rapidly increases and with it their confidence in being able to manage future similar patients without any teleconsultation. So it became quickly evident that many fewer HIV-positive patients with skin problems were being referred and that they were being competently managed by nurse practitioners in the sending clinic [133]. There are also many other benefits to telenursing links, elaborated elsewhere in this chapter. A fundamental principle is that such "clinical" telenursing links can normally transmit bidirectional communications and are only in clinical use for a small part of the day. Therefore, they can be also be used for other purposes, principally distance learning.

Deficiencies in the extensive United Kingdom NHS Direct telenursing service will lead to its replacement by the NHS 111 service in 2014. The original problems, which led to the establishment of NHS Direct in 1998, were experienced by those who required help for urgent and emergency situations. Before 1998 there was uncertainty about the available services, especially lack of understanding about the most appropriate service to access, for there were a variety of possible contacts relevant to the same medical situation. Although NHS Direct was founded to resolve this, the problems of access continued. Hence, NHS 111 was devised to replace it in 2014. It offers a telephonic facility for all sorts of urgent care, which can include primary care during the evening and night, problems normally requiring ambulance services, and even provides health advice and information. All these are integrated and a nonnurse responds, with a nurse's immediate assessment, if appropriate. This assessment may be immediate or may require a return phone call. This integration of all necessary services allows immediate connection of the caller, or an appointment to be made, as required. When other telenursing services are planned or set up, this experience should be borne in mind.

A worldwide lack of healthcare personnel has been recognized for several decades, especially in the poorest nations. For nursing this has given rise to several associated problems which have relevance to telenursing. In the 47 nations of sub-Saharan Africa, for each 10,000 of the population there are 11 nurses/midwives, while in Europe the equivalent number is 78 nurses/midwives [134]. The International Labor Organization has studied this lack and states that "30 to 36% of the global population has no access to healthcare when in need" [135]. In general for developing countries, fewer funds are available for healthcare and training its personnel. So various telenursing programs to facilitate nurse training have proved their cost-effectiveness and they are being extended. Another means of overcoming the lack of healthcare personnel in developing countries is to involve private medical establishments in the education and training of nurses. Programs to do so have been mooted and pilot schemes are under way in a range of developing countries, including India, Kenya, South Africa, and Thailand [136]. But when there is a rapid expansion of nurses, it is often difficult to find suitable teachers and those available are then usually overworked. In such circumstances tele-education can greatly alleviate these difficulties.

Telenursing activity reflects availability of telemedical resources and also the nursing needs of the region. Presentation of many conditions is frequently much later in developing countries and the prevalence of common diseases very different. This is true for most poorly resourced nations, although the incidence of typically Western diseases of lifestyle there is steadily increasing. In developing countries most health problems result from communicable diseases (60%–80% in Africa), such as TB, HIV/AIDS, malaria, pneumonias, and various viral infections, which are often preventable [137]. So the relevant education of the community by nurses in PHC clinics is a priority. This can come about after appropriate distance learning by nurses. Currently the predominant application of telenursing in developing countries should be education [6]. Because it has been so successful to date, this is likely to continue for some years. However, the basic education of nurses has not kept pace with this trend. For example, in South Africa the basic nursing curriculum contains no material on ICT, no practical instruction on computers or computing, and no reference to telenursing, eNursing, etc. However, this situation will change for the syllabus to be used in 2015 (D. J. Sibiya, private communication). Thus, it is clear that all South African telenurses must receive a considerable amount of instruction beyond their basic nursing education and this is also likely to be true in most developing countries.

Overall a very wide range of applications of telenursing is in use in all parts of the world and applications have increased rapidly in the last decade or so, because of a combination of technical improvements, reduced costs of equipment, and cost-effectiveness of telenursing itself. This trend shows no sign of abating.

5.9 Conclusions

After an explosive increase in telenursing activity of the last two decades or so, its value is recognized in all types of nation for two main reasons. It reduces wastage of time, effort, and funds, for both the patient and telenurse, so it is cost effective and can improve the quality of a nursing service. Also it facilitates education at all levels. Therefore, health planning often works toward an extension of telenursing. The important economic benefits are due in part to the steadily increasing sophistication and technical capacity of both the telenursing equipment and its associated software. In parallel there has been a crucial monotonic reduction in hardware costs. Telenursing practice in many ways reflects general nursing activity and so it differs between developed and developing nations because of contrasts in their access to all forms of resources, infrastructure levels, disease prevalences, and healthcare personnel numbers. In resource-poor nations there has been a greater emphasis on tele-education than on home care, telemonitoring, and a full telenursing service, which are all more typical of richer nations. Many noneducational telenursing programs in poor nations resulted from collaborations with wealthy countries and international organizations, but after recognition of such programs' merit, there is increasing initiation of purely local projects. It is believed that the equipment trends, which were instrumental in aiding the increasing application of telenursing, are highly likely to continue for at least a decade. This bodes very well for the future of telenursing in all types of society and, hence, for the whole nursing profession.

Abbreviations

AIDS	acquired immunodeficiency syndrome
HIV	human immunodeficiency virus
ICT	information and computer technology
ICU	intensive care unit
NHS	National Health Service
PHC	primary healthcare
TB	tuberculosis
UNICEF	United Nations Children's Fund
VCT	voluntary counseling and testing
WHO	World Health Organization

Nomenclature

e-learning	distance learning using ICT
mHealth	healthcare using mobile ICT devices, such as mobile telephones
telenursing	the subset of eHealth applied to nursing

Acknowledgments

The authors are happy to acknowledge informative discussions and valuable information received from Professor E. Gwyther, Dr. V. Chalmers, D. J. Sibiya, and B. Ewens.

References

1. https://www.goodreads.com/author/quotes/63031.Florence_Nightingale. Accessed on March 17, 2014.
2. Anon. The European Union's definition of eHealth, http://ec.europa.eu/information_society /newsroom/cf/dae/itemdetail.cfm?type=379&typeName=Research%20Result&item_id=8170 Accessed on March 17, 2014.
3. Scheil-Adlunga, X., Health workforce benchmarks for universal health coverage and sustainable development, *Bull World Health Organ* 2013; 91:888–889, 2013.
4. Brown, D. J., Consumer perspectives on nurse practitioners and independent practice, *J Am Acad Nurse Pract* Oct; 19(10):523–529, 2007.
5. McCarthy, C. F. et al., Nursing and midwifery regulatory reform in east, central, and southern Africa: A survey of key stakeholders, *Hum Resour Health* Jun 25; 11(1):29, 2013.
6. Mars, M., Telemedicine and advances in urban and rural healthcare delivery in Africa, *Prog Cardiovasc Dis* Nov–Dec; 56(3):326–335, 2013.
7. Kaminsky, E., Rosenqvis, U., and Holmström, I., Telenurses' understanding of work: Detective or educator? *J Adv Nurs* Feb; 65(2):382–90, 2009.
8. Allen, A. et al., An analysis of the suitability of home health visits for telemedicine, *J Telemed Telecare* 5(2):90–96, 1999.
9. Wootton, R. et al., A joint US-UK study of home telenursing, *J Telemed Telecare* 4(Suppl 1):83–85, 1998.
10. Zundel, K. M., Telemedicine: History, applications, and impact on librarianship, *Bull Med Libr Assoc* 84(1), 71–79, 1996.
11. Amenta, F., Dauri, A., and Rizzo, N., Organization and activities of the International Radio Medical Centre (CIRM), *J Telemed Telecare* 2(3):125–131, 1996.
12. Anon., Telemedicine: Opportunities and developments in Member States: Report on the second global survey on eHealth 2009, *Global Observatory for eHealth Series, Volume 2*, WHO, Geneva, Switzerland, 2010. http://www.who.int/goe/publications/goe_telemedicine_2010.pdf Accessed on March 17, 2014.
13. Basher, R., and Lovett, J., Assessment of telemedicine: Results of the initial experience, *Aviation Space Environmental Medicine* 48(1):65–70, 1977.
14. Holmström, I., Decision aid software programs in telenursing: Not used as intended? Experiences of Swedish telenurses, *Nurs Health Sci* Mar; 9(1):23–28, 2007.
15. Ernesäter, A., Holmström, I., and Engström, M., Telenurses' experiences of working with computerized decision support: Supporting, inhibiting and quality improving, *J Adv Nurs* May; 65(5):1074–1083, 2009.
16. Boerma, T., and Siyam, A., Health workforce indicators: Let's get real, *Bull World Health Organ* 91:886, 2013.
17. McCarthy, C. F., and Riley, P. L., The African health profession regulatory collaborative for nurses and midwives, *Hum Resour Health* Aug 29; 10(1):26, 2012.
18. Riley, P. L. et al., Developing a nursing database system in Kenya, *Health Serv Res* Jun; 42(3 Pt 2):1389–1405, 2007.
19. Waters, K. P., Kenya's health workforce information system: A model of impact on strategic human resources policy, planning and management, *Int J Med Inform* Sep; 82(9):895–902, 2013.
20. Spero, J. C., McQuide, P. A., and Matte, R., Tracking and monitoring the health workforce: A new human resources information system (HRIS) in Uganda, *Hum Resour Health* Feb 17; 9:6, 2011.
21. Mengistu, Z., and Azaj, A., Trauma severities scores and their prediction of outcome among trauma patients in two hospitals of Addis Ababa, Ethiopia, *Ethiop Med* Jul; 50(3):231–237, 2012.
22. Turner, J. et al., Impact of the urgent care telephone service NHS 111 pilot sites: A controlled before and after study *BMJ Open* Nov 14; 3(11):e003451. doi:10.1136/bmjopen-2013-003451, 2013.

23. Gann, B., NHS Direct Online: A multi-channel eHealth service, *Stud Health Technol Inform*, 100:164–168, 2004.

24. Byass, P., and D'Ambruoso, L., Cellular telephone networks in developing countries, *Lancet* Feb 23; 371(9613):650, 2008.

25. Tran, K. et al., Mobile teledermatology in the developing world: Implications of a feasibility study on 30 Egyptian patients with common skin diseases, *J Am Acad Dermatol* Feb; 64(2):302–309, 2011.

26. Horvath, T., Azman, H., Kennedy, G. E., and Rutherford, G. W., Mobile phone text messaging for promoting adherence to antiretroviral therapy in patients with HIV infection, *Cochrane Database Syst Rev* Mar 14; 3:CD009756, 2012.

27. Cook, P. F., McCabe, M. M., Emiliozzi, S., and Pointer, L., Telephone nurse counseling improves HIV medication adherence: An effectiveness study, *J Assoc Nurses AIDS Care* Jul–Aug; 20(4):316–325, 2009.

28. Car, J. et al., Mobile phone messaging reminders for attendance at healthcare appointments, *Cochrane Database Syst Rev* Jul 11; 7:CD007458, 2012.

29. Sikka, N. et al., The use of mobile phones for acute wound care: Attitudes and opinions of emergency department patients, *J Health Commun* 17(Suppl 1):37–42, 2012.

30. St George, I., Cullen, M., Gardiner, L., and Karabatsos, G., Universal telenursing triage in Australia and New Zealand—A new primary health service, *Aust Fam Physician* Jun; 37(6):476–479, 2008.

31. Wynchank, S., and Fortuin, J., Telenursing in Africa, in *Telenursing*, Khumar, S., and Snooks, H. (ed), Springer-Verlag, New York, United States, 2011.

32. Simpson, R. L., State-based licensure: Are we regulating away the promise of telemedicine? *Nurs Adm Q* Oct–Dec; 32(4):346–348, 2008.

33. Litchfield, S. M., Update on the nurse licensure compact, *AAOHN J* Jul; 58(7):277–279, 2010.

34. Unruh, L. et al., Can state databases be used to develop a national, standardized hospital nurse staffing database? *West J Nurs Res* Feb; 31(1):66–88, 2009.

35. Ernesäter, A. et al., Incident reporting in nurse-led national telephone triage in Sweden: The reported errors reveal a pattern that needs to be broken, *J Telemed Telecare* 16(5):243–247, 2010.

36. Rahmqvist, M., Ernesäter, A., and Holmström, I., Triage and patient satisfaction among callers in Swedish computer-supported telephone advice nursing, *J Telemed Telecare* 17(7):397–402, 2011.

37. Green, J. M. et al., Converting policy into care: Women's satisfaction with the early labour telephone component of the All Wales Clinical Pathway for Normal Labour, *J Adv Nurs* Oct; 68(10):2218–2228, 2012.

38. Dowswell, T. et al., Alternative versus standard packages of antenatal care for low-risk pregnancy, *Cochrane Database Syst Rev* Oct 6; (10):CD000934, 2010.

39. Tahir, N. M., and Al-Sadat, N., Does telephone lactation counselling improve breastfeeding practices? A randomised controlled trial, *Int J Nurs Stud* Jan; 50(1):16–25, 2013.

40. Kramer, M. S., and Kakuma, R., Optimal duration of exclusive breastfeeding, *Cochrane Database Syst Rev* Aug 15; 8:CD003517, 2012.

41. Anon., Expert Consultation on Community-Based Approaches for Neonatal Sepsis Management, *WHO Meeting Report*, 26–28 September 2007, London, United Kingdom, 2007.

42. Størdal, K., Neonatal mortality—The key to reduced neonatal mortality? *Tidsskr Nor Laegeforen* Nov 5; 129(21):2270–2273, 2009.

43. McClure, E. M. et al., Evaluation of the educational impact of the WHO Essential Newborn Care course in Zambia, *Acta Paediatr* Aug; 96(8):1135–1138, 2007.

44. Shi, L., and Zhang, J., Recent evidence of the effectiveness of educational interventions for improving complementary feeding practices in developing countries, *J Trop Pediatr* Apr; 57(2):91–98, 2011.

45. Diep, P. P., Lien, L., and Hofman, J., A criteria-based clinical audit on the case-management of children presenting with malaria at Mangochi District Hospital, Malawi, *World Hosp Health Serv* 43(2):21–29, 2007.

46. Young, S. et al., Teaching paediatric resuscitation skills in a developing country: Introduction of the Advanced Paediatric Life Support course into Vietnam, *Emerg Med Australas* Jun; 20(3):271–275, 2008.
47. Looman, W. S. et al., Meaningful use of data in care coordination by the advanced practice RN: The TeleFamilies project, *Comput Inform Nurs* Dec; 30(12):649–654, 2012.
48. Flanagan, J., Postoperative telephone calls: Timing is everything, *AORN J* Jul; 90(1):41–51, 2009.
49. Nielsen, M., and Saracino, J., Telemedicine in the intensive care unit, *Crit Care Nurs Clin North Am* Sep; 24(3):491–500, 2012.
50. Austin, L. S., Landis, C. O., and Hanger, Jr., K. H., Extending the continuum of care in congestive heart failure: An interactive technology self-management solution, *J Nurs Adm* Sep; 42(9):442–446, 2012.
51. Radhakrishnan, K. et al., Association of comorbidities with homecare service utilization of patients with heart failure while receiving telehealth, *J Cardiovasc Nurs* May–Jun; 28(3):216–227, 2013.
52. Iavazzo, F., and Cocchia, P., Quality of life in people with heart failure: Role of telenursing, *Prof Inferm* Oct–Dec; 64(4):207–212, 2011.
53. Kind, A. J. et al., Low-cost transitional care with nurse managers making mostly phone contact with patients cut rehospitalization at a VA hospital, *Health Aff (Millwood)* Dec; 31(12):2659–2668, 2012.
54. Or, C. K. et al., Factors affecting homecare patients' acceptance of a web based interactive self-management technology, *J Am Med Inform Assoc* Jan–Feb; 18(1):51–59, 2011.
55. Annandale, J., and Lewis, K. E., Can telehealth help patients with COPD? *Nurs Times* Apr 19–May 2; 107(15–16):12–14, 2011.
56. Roberts, D., Tayler, C., MacCormack, D., and Barwich, D., Telenursing in hospice palliative care, *Can Nurse* May; 103(5):24–27, 2007.
57. McDermott, R., Telenursing can reduce costs and improve access for rural patients, *Oncol Nurs Forum* Jan; 32(1):16, 2005.
58. Doolittle, G. C., Yaezel, A., Otto, F., and Clemens, C., Hospice care using home-based telemedicine systems, *J Telemed Telecare* 4(Suppl 1):58–59, 1998.
59. Berkley, R., Bauer, S. A., and Rowland, C., How telehealth can increase the effectiveness of chronic heart failure management, *Nurs Times* Jul 6–12; 106(26):14–15, 2010.
60. Leclerc, B. S. et al., Callers' ability to understand advice received from a telephone health-line service: Comparison of self-reported and registered data, *Health Serv Res* Apr; 38(2):697–710, 2003.
61. Leclerc, B. S. et al., Recommendations for appeal to another resource in a health case brought by nurses in the Info-Health Service CLSC: Convergence between utilization declarations and informants' data, *Can J Public Health* Jan–Feb; 94(1):74–78, 2003.
62. Bosworth, H. B. et al., Home blood pressure management and improved blood pressure control: Results from a randomized controlled trial, *Arch Intern Med* Jul 11; 171(13):1173–1180, 2011.
63. Lauderdale, M. E., Feasibility study of homecare wound management using telemedicine, *Adv Skin Wound Care* Aug; 22(8):358–364, 2009.
64. Jönsson, A. M., and Willman, A., Implementation of telenursing within home healthcare, *Telemed J E Health* Dec; 14(10):1057–1062, 2008.
65. Browning, S. V. et al., Telehealth monitoring: A smart investment for homecare patients with heart failure? *Home Healthc Nurse* Jun; 29(6):368–374, 2011.
66. Van Offenbeek, M. A., and Boonstra, A., Does telehomeconsultation lead to substitution of home visits? Analysis and implications of a telehomecare program, *Stud Health Technol Inform* 157:148–153, 2010.
67. Cudney, S. et al., Chronically ill rural women: Self-identified management problems and solutions, *Chronic Illn* Mar; 1(1):49–60, 2005.
68. Cudney, S., and Weinert, C., Computer-based support groups: Nursing in cyberspace, *Comput Nurs* Jan–Feb; 18(1):35–43, 2000.
69. Cudney, S., Winters, C., Weinert, C., and Anderson, K., Social support in cyberspace: Lessons learned, *Rehabil Nurs* Jan–Feb; 30(1):25–28, 2005.

70. Weinert, C., Cudney, S., and Hill, W. G., Rural women, technology, and self-management of chronic illness, *Can J Nurs Res* Sep; 40(3):114–134, 2008.

71. Vinson, M. H. et al., Design, implementation, and evaluation of population-specific telehealth nursing services, *Nurs Econ* Sep–Oct; 29(5):265–272, 2011.

72. Jelin, E., Granum, V., and Eide, H., Experiences of a web-based nursing intervention—Interviews with women with chronic musculoskeletal pain, *Pain Manag Nurs* Mar; 13(1):2–10, 2012.

73. Cook, P. F. et al., Telephone nurse counseling for medication adherence in ulcerative colitis: A preliminary study, *Patient Educ Couns* Nov; 81(2):182–186, 2010.

74. Cicolini, G. et al., Effectiveness of the telephonic-case-management for treatment of headache: A pilot study, *Prof Inferm* Jul–Sep; 64(3):173–178, 2011.

75. Zou, H. et al., Internet-facilitated, voluntary counseling and testing (VCT) clinic-based HIV testing among men who have sex with men in China, *PLoS One* 8(2):e51919, 2013.

76. Fonner, V. A. et al., Voluntary counselling and testing (VCT) for changing HIV-related risk behaviour in developing countries, *Cochrane Database Syst Rev* Sep 12; 9:CD001224, 2012.

77. Moore, J. et al., Challenges and successes of HIV voluntary counselling and testing programme in antenatal clinics in greater Kingston, Jamaica, *West Indian Med J.* Jun; 57(3):269–273, 2008.

78. Mall, S. et al., Changing patterns in HIV/AIDS stigma and uptake of voluntary counselling and testing services: The results of two consecutive community surveys conducted in the Western Cape, South Africa, *AIDS Care* 25(2):194–201, 2013.

79. Leon, N., Lewin, S., and Mathews, C., Implementing a provider-initiated testing and counselling (PITC) intervention in Cape Town, South Africa: A process evaluation using the normalisation process model, *Implement Sci* Aug 26; 8:97, 2013.

80. Capital Ka Cyrilliciriazova, T. K. et al., Evaluation of the effectiveness of HIV voluntary counselling and testing trainings for clinicians in the Odessa region of Ukraine, *AIDS Behav* Jan; 18(Suppl 1):S89–S95, 2014.

81. Tromp, N. et al., Cost effectiveness of scaling up voluntary counselling and testing in West-Java, Indonesia, *Acta Med Indones* Jan; 45(1):17–25, 2013.

82. Ngo, A. D., Ha, T. H., Rule, J., and Dang, C. V., Peer-based education and the integration of HIV and sexual and reproductive health services for young people in Vietnam: Evidence from a project evaluation, *PLoS One* Nov 28; 8(11):e80951, 2013.

83. Wachira, J. et al., What is the impact of home-based HIV counseling and testing on the clinical status of newly enrolled adults in a large HIV care program in Western Kenya? *Clin Infect Dis* Jan 15; 54(2):275–281, 2012.

84. Perumal, R., Padayatchi, N., and Stiefvater, E., The whole is greater than the sum of the parts: Recognising missed opportunities for an optimal response to the rapidly maturing TB-HIV co-epidemic in South Africa, *BMC Public Health* Jul 16; 9:243, 2009.

85. Ackerman, B., Pyne, J. M., and Fortney, J. C., Challenges associated with being an off-site depression care manager, *J Psychosoc Nurs Ment Health Serv* Apr; 47(4):43–49, 2009.

86. Battaglia, C. et al., Building a tobacco cessation telehealth care management program for veterans with posttraumatic stress disorder, *J Am Psychiatr Nurses Assoc* Mar–Apr; 19(2):78–91, 2013.

87. Cook, P. F., Emiliozzi, S., Waters C., and El Hajj D., Effects of telephone counseling on antipsychotic adherence and emergency department utilization, *Am J Manag Care* Dec; 14(12):841–846, 2008.

88. Gitlin, L. N. et al., Targeting and managing behavioral symptoms in individuals with dementia: A randomized trial of a nonpharmacological intervention, *J Am Geriatr Soc* Aug; 58(8):1465–1474, 2010.

89. Barnason, S., Zimmerman, L., Schulz, P., and Tu, C., Influence of an early recovery telehealth intervention on physical activity and functioning after coronary artery bypass surgery among older adults with high disease burden, *Heart Lung* Nov–Dec; 38(6):459–468, 2009.

90. Zubaran, C., The international migration of healthcare professionals, *Australas Psychiatry* Dec; 20(6):512–517, 2012.

91. Dywili, S., Bonner, A., and O'Brien, L., Why do nurses migrate?—A review of recent literature, *J Nurs Manag* Apr; 21(3):511–520, 2013.

92. Anon., Migration of health workers, Fact Sheet No 301, *WHO, Geneva, Switzerland,* 2010.

93. Squires, A., The North American Free Trade Agreement (NAFTA) and Mexican nursing, *Health Policy Plan* Mar; 26(2):124–132, 2011.

94. Ogilvie, L. et al., The exodus of health professionals from sub-Saharan Africa: Balancing human rights and societal needs in the twenty-first century, *Nurs Inq* Jun; 14(2):114–124, 2007.

95. Burg, G., and French, L. E., The age of Gutenberg is over: A consideration of medical education— Past, present and future, *Hautarzt* Apr; 63(Suppl 1):38–44, 2012.

96. Masic, I., E-learning as new method of medical education, *Acta Inform Med* 16(2):102–117, 2008.

97. Lahtinen, P., Leino-Kilpi, H., and Salminen, L., Nursing education in the European higher education area—Variations in implementation, *Nurse Educ Today;* 34(6):1040–1047, 2013.

98. Davies K., Deeny P., and Raikkonen M., A transcultural ethos underpinning curriculum development: A master's programme in disaster relief nursing, *J Transcult Nurs* Oct; 14(4):349–357, 2003.

99. Grady, J. L., The Virtual Clinical Practicum: An innovative telehealth model for clinical nursing education, *Nurs Educ Perspect* May–Jun; 32(3):189–194, 2011.

100. Berkman, N. D., Davis, T. C., and McCormack, L., Health literacy: What is it? *J Health Commun* 15(Suppl 2):9–19, 2010.

101. Deverell, A. et al., An evaluation of mindset health: Using ICT to facilitate an innovative training methodology for healthcare providers in South Africa, *11th Conference of the International Society for Telemedicine and eHealth,* Cape Town, 2006.

102. Rodríguez-Gázquez, M. de L. et al., Effectiveness of an educational program in nursing in the self-care of patients with heart failure: Randomized controlled trial, *Rev Lat Am Enfermagem* Mar–Apr; 20(2):296–306, 2012.

103. Blake, H., Mobile technology: Streamlining practice and improving care, *Br J Community Nurs* Sep; 18(9):430, 432, 2013.

104. Sapag, J. C., Lange, I., Campos, S., and Piette, J. D., Innovative care and self-care strategies for people with chronic diseases in Latin America, *Rev Panam Salud Publica* Jan; 27(1):1–9, 2010.

105. Huibers, L., Follow-up after telephone consultations at out-of-hours primary care, *J Am Board Fam Med* Jul–Aug; 26(4):373–379, 2013.

106. Gwyther, L., and Rawlinson, F., Palliative medicine teaching program at the University of Cape Town: Integrating palliative care principles into practice, *J Pain Symptom Manage* May; 33(5):558–562, 2007.

107. Cook, D. A. et al., What do we mean by web-based learning? A systematic review of the variability of interventions, *Med Educ* Aug; 44(8):765–774, 2010.

108. Cook, D. A. et al., Instructional design variations in Internet-based learning for health professions education: A systematic review and meta-analysis, *Acad Med* May; 85(5):909–922, 2010.

109. Wilimas, J. A., and Ribeiro, R. C., Pediatric hematology-oncology outreach for developing countries, *Hematol Oncol Clin North Am* Aug; 15(4):775–787, 2001.

110. Day, S. W., Dycus, P. M., Chismark, E. A., and McKeon, L., Quality assessment of pediatric oncology nursing care in a Central American country: Findings, recommendations, and preliminary outcomes, *Pediatr Nurs* Sep–Oct; 34(5):367–373, 2008.

111. Howard, S. C. et al., Establishment of a pediatric oncology program and outcomes of childhood acute lymphoblastic leukemia in a resource-poor area, *JAMA* May 26; 291(20):2471–2475, 2004.

112. Wilson, A. et al., Effectiveness of strategies incorporating training and support of traditional birth attendants on perinatal and maternal mortality: Meta-analysis, *BMJ* Dec 1; 343:d7102, 2011.

113. Harvey, S. A. et al., Are skilled birth attendants really skilled? A measurement method, some disturbing results and a potential way forward, *Bull World Health Organ* Oct; 85:783–790, 2007.

114. Islam, M., The Safe Motherhood Initiative and beyond, *Bull World Health Organ,* 85: 735, 2007.

115. Furuya, R. K. et al., Original research: Telephone follow-up for patients after myocardial revascularization: A systematic review, *Am J Nurs* May; 113(5):28–31, 2013.

116. dos Santos Ade, F. et al., Telehealth in primary healthcare: An analysis of Belo Horizonte's experience, *Telemed J E Health* Jan–Feb; 17(1):25–29, 2011.

117. Goessens, B. M. et al., A pilot-study to identify the feasibility of an Internet-based coaching programme for changing the vascular risk profile of high-risk patients, *Patient Educ Couns* Oct; 73(1):67–72, 2008.

118. Cheng, A. et al., Evolution of the Pediatric Advanced Life Support course: Enhanced learning with a new debriefing tool and Web-based module for Pediatric Advanced Life Support instructors, *Pediatr Crit Care Med* Sep; 13(5):589–595, 2012.

119. Du, S. et al., Web-based distance learning for nurse education: A systematic review, *Int Nurs Rev* Jun; 60(2):167–177, 2013.

120. Wolbrink, T. A., and Burns, J. P., Internet-based learning and applications for critical care medicine, *J Intensive Care Med* Sep–Oct; 27(5):322–332, 2012.

121. Kleinpell, R. et al., Web-based resources for critical care education, *Crit Care Med* Mar; 39(3):541–553, 2011.

122. Phillips, J. L., Piza, M., and Ingham, J., Continuing professional development programmes for rural nurses involved in palliative care delivery: An integrative review, *Nurse Educ Today* May; 32(4):385–392, 2012.

123. Jones, J. et al., Delivering healthcare information: Cardiac patients' access, usage, perceptions of usefulness and website content preferences, *Telemed J E Health* Mar; 20(3):223–228, 2014.

124. Schmitt, M. H. et al., The coming of age for interprofessional education and practice. *Am J Med* Apr; 126(4):284–288, 2013.

125. Ruddock, H. C., and Turner, de S., Developing cultural sensitivity: Nursing students' experiences of a study abroad programme, *J Adv Nurs* Aug; 59(4):361–369, 2007.

126. Levi, A., The ethics of nursing student international clinical experiences, *J Obstet Gynecol Neonatal Nurs* Jan–Feb; 38(1):94–99, 2009.

127. Holmström, I., and Höglund, A. T., The faceless encounter: Ethical dilemmas in telephone nursing, *J Clin Nurs* Oct; 16(10):1865–1871, 2007.

128. Kaminsky, E., Carlsson, M., Höglund, A. T., and Holmström, I., Paediatric health calls to Swedish telenurses: A descriptive study of content and outcome, *J Telemed Telecare* 16(8):454–457, 2010.

129. Höglund, A. T., and Holmström, I., "It's easier to talk to a woman": Aspects of gender in Swedish telenursing, *J Clin Nurs* Nov; 17(22):2979–2986, 2008.

130. Demiris, G., Doorenbos, A. Z., and Towle, C., Ethical considerations regarding the use of technology for older adults: The case of telehealth, *Res Gerontol Nurs* Apr; 2(2):128–136, 2009.

131. Ernesäter, A., Winblad, U., Engström, M., and Holmström, I. K., Malpractice claims regarding calls to Swedish telephone advice nursing: What went wrong and why? *J Telemed Telecare* Oct; 18(7):379–383, 2012.

132. Wheeler, T., Strategies for delivering tele-home care-provider profiles, *Telemed Today* Aug; 6(4):37–40, 1998.

133. Colven, R., Shim, M. H., Brock, D., and Todd, G., Dermatological diagnostic acumen improves with use of a simple telemedicine system for underserved areas of South Africa, *Telemed J E Health* Jun; 17(5):363–369, 2011.

134. Naicker, S., Eastwood, J. B., Plange-Rhule, J., and Tutt, R. C., Shortage of healthcare workers in sub-Saharan Africa: A nephrological perspective, *Clin Nephrol* Nov; 74(Suppl 1):S129–S133, 2010.

135. Scheil-Adlung, X., and Bonnet, F., Beyond legal coverage: Assessing the performance of social health protection, *Int Soc Secur Rev* Mar; 64:21–38, 2011.

136. Reynolds, J. et al., A literature review: The role of the private sector in the production of nurses in India, Kenya, South Africa and Thailand, *Hum Resour Health* Apr 12; 11(1):14, 2013.

137. Crisp, N., Gawanas, B., and Sharp, I., Task force for scaling up education and training for health workers, *Lancet* Feb 23; 371(9613):689–691, 2008.

6

mHealth: Intelligent Closed-Loop Solutions for Personalized Healthcare

Carmen C. Y. Poon and Kevin K. F. Hung

CONTENTS

6.1 Introduction

Mobile health (mHealth) is an emerging concept which refers to the combinational use of mobile computing, communication technologies, and medical sensors for healthcare [1]. In addition to the passive provision of health-related information via mobile platforms, its scope also extends to more active applications such as disease management, medical prognosis, diagnosis, and even treatment. A typical mHealth system is characterized by its features of mobility, wireless connectivity, location independence, and timely response—which together provide seamless and personalized healthcare.

According to census statistics, aging of the world's population is pervasive, profound, and enduring. The world's elderly population is projected to be almost two billion by the year 2050 [2]. The situation is most severe in developed countries and is accompanied by the prevalence of chronic diseases, which are becoming the leading causes of death and disability. In response to the current and foreseen burden on the healthcare system, there is growing interest in the use of mHealth for chronic and geriatric care. In a World Health

Organization (WHO) conference, mHealth was addressed as a "marriage of technologies that can improve the quality of patient monitoring while reducing overall healthcare costs," and was recognized as "a way to help lower a patient's health risk, raise a patient's comfort level, reduce the number of checkups and time spent in the hospital, and enhance the effectiveness of pharmaceutical field tests" [3]. In addition to fulfilling the needs of the world's demographic shift, mHealth has also been proven beneficial in providing services to remote and underserved areas, as well as in prehospital emergency.

6.2 Historical Overview of mHealth

Although mHealth may utilize the newest mobile and wireless technologies, the concept itself is not completely new. Instead, it is a gradual evolution from traditional telemedicine in response to technological advancement and increasing demand for diversified modes of healthcare.

6.2.1 Evolution from Telemedicine to mHealth

The telepsychiatry program at the Nebraska Psychiatric Institute in the 1950s is commonly cited as the first institution to practice telemedicine. The institute used two-way closed-circuit televisions (CCTVs) for consultation, neurological examination, speech therapy, and staff training [4]. Numerous telemedicine projects have since emerged in the United States and Canada. Most were focused on remote consultation and teleradiology using teleconferencing equipment in hospitals. In the 1960s, a small step toward mobility was achieved in the Space Technology Applied to Rural Papago Advanced Health Care project [5] and Applications Technology Satellite-6 satellite biomedical demonstration [6], in which mobile medical examination rooms with satellite communication facility were deployed in villages of Indian reservations.

The widespread use of personal computers (PCs) in the 1980s, followed by the emergence of the Internet in the 1990s, has enabled delivery of healthcare to homes, as could be seen from the launch of personalized medical devices with data-transmitting functions. The term *telehealth* was used to describe this enlarged scope, while telemedicine was considered as its hospital-oriented subset. A decade later, reports suggested that telemedicine and telehealth were becoming part of a broader scope called electronic health (eHealth), which further comprises elements such as electronic patient records, business transactions, health insurance, and staff training [7].

As soon as mobile phones were launched into the market, there was overwhelming interest in their usage in eHealth. Different names were suggested for this specialized group of applications. Starting in the early 2000s, some terms that have emerged were *unwired eMed*, *mobile eHealth*, and *wireless telemedicine*. However, the generally accepted term is *mHealth*, which is defined as "mobile computing, medical sensor, and communication technologies for healthcare" [1].

6.2.2 Initial mHealth Applications

Before smartphones became widely available in the market, initial mHealth applications were realized with notebook computers connected to wireless modems. Such trials date back to as early as 1995, when Yamamoto [8] demonstrated the feasibility with mobile computing and cellular communication. Yamamoto transmitted scanned CT and X-ray scan images from a PC to a notebook computer with pocket modems and earlier versions of networking protocols [8]. Years later, Yamamoto [9], Yamamoto and Williams [10], and Singh et al. [11] performed similar tests viewing 12-lead electrocardiogram (ECG) recordings on pocket computers. In the late 1990s, the emergence of personal digital assistant (PDA) phones has further enhanced the mobility of mHealth applications. As an example, Reponen et al. demonstrated transmitting CT scans (as shown in Figure 6.1) from a file server via transmission control protocol (TCP)/Internet protocol (IP) to PDA phones over the Global System for Mobile Communication (GSM) network [12–13]. In general, it took several minutes to download one file. The group concluded that handheld devices at that time were suitable for reading of the most common emergency CT findings.

The above examples involved transfer of static files between mobile terminals only. An early attempt of using wireless Internet for interactive mHealth was the wireless application protocol (WAP)-based patient-monitoring system developed by Hung and Zhang in the early 2000s [14]. Their system utilized WAP-enabled devices for general inquiry and patient monitoring. On a mobile phone, authorized users could interactively view and analyze patients' data, blood pressure (BP) readings, and ECG. The applications, written in wireless markup language (WML), WMLScript, Perl, and Java servlets, resided in a remote server; while a database server was set up to store the contents and records. WML mainly supported text, user input, basic navigation mechanisms, and simple images in wireless bitmap (WBMP) format. To further demonstrate the concept, the group built a wireless subsystem for recording and storing indoor ambulatory ECG, as shown in Figure 6.2.

For testing, a WAP-enabled phone was used at GSM 1800 MHz by circuit-switched data (CSD) to connect to the server through a WAP gateway, which was provided by a mobile phone service provider in Hong Kong. Data were successfully retrieved from the database and displayed on the WAP phone. The time required for establishing a connection to the server was about 10 to 12 s. Starting from the time of request for information at the phone, the time required for query, dynamic generation of WML and WBMP, and refresh of information at the device ranged from 3 to 5 s. The study showed that WAP devices were

FIGURE 6.1
Display of head CT image on a PDA phone. (From K. Hung and Y. T. Zhang, *IEEE Trans. Inform. Technol. Biomed.*, 7, 2, 101–107, © 2003 IEEE.)

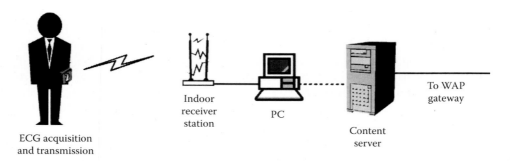

FIGURE 6.2
A wireless ambulatory ECG-monitoring subsystem as part of a WAP-based patient-monitoring system.

feasible for mHealth, but limited only to store-and-forward, client/server, and low-bandwidth operations. Figure 6.3 shows screenshots of a mobile phone displaying the information during the trials.

6.2.3 Recent mHealth Applications

Due to the low data rate, earlier forms of mHealth were restricted to store-and-forward operation. However, a milestone was reached when third-generation mobile telecommunications (3G) became widely available in the mid-2000s. High-speed downlink packet access (HSDPA), an enhancement of wideband code-division multiple access (WCDMA), can offer a peak downlink data rate of up to 14 Mbps.

Because such systems can provide sufficiently high bandwidth for streaming data, it is recognized as a key enabler of modern mHealth. Various 3G-based mHealth projects have targeted emergency scenarios. An example is the study by Viruete Navarro et al. [15]. As shown in Figure 6.4, the group has assembled a system using off-the-shelf medical devices, free software, and commercial 3G Universal Mobile Telecommunications System (UMTS) data services. The system was capable of real-time transmission of multiple biosignals, simultaneous videoconferencing, and other non-real-time medical services.

A new dimension in the evolution of mHealth is wearable sensors which target long-term and continuous patient monitoring. An early example was the advanced care and alert portable telemedical monitor (AMON), which was a wrist-worn system for high-risk cardiac and respiratory patients [16]. The device was integrated with blood oxygen saturation

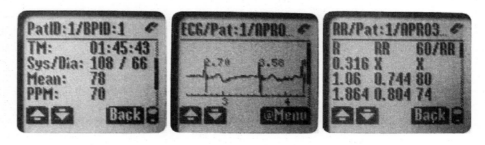

FIGURE 6.3
Screenshots of a WAP phone accessing monitored physiological parameters. *Left to right:* Displays of BP readings, ECG, and ECG analysis data. (From K. Hung and Y. T. Zhang, *IEEE Trans. Inform. Technol. Biomed.*, 7, 2, 101–107, © 2003 IEEE.)

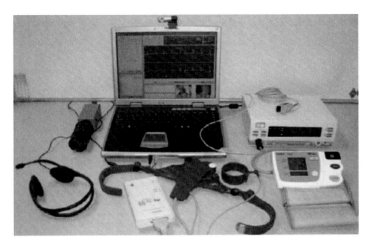

FIGURE 6.4
3G-based mHealth system. (From E. A. Viruete Navarro et al., "Performance of a 3G-based mobile telemedicine system," *3rd IEEE Consumer Communications and Networking Conference, 2006: CCNC 2006* (Volume: 2), pp. 1023–1027, 2006.)

level SpO_2 by pulse oximetry, ECG, BP, movement, and temperature sensors, which captured and sent the information to a telemedical center for continuous monitoring.

In the past decade, an increasing number of mHealth systems appeared in the form of clothing. For example, the LifeShirt from VivoMetrics is embedded with inductive plethysmographic sensors, accelerometer, and ECG electrodes [17]. Data are recorded in a small belt-worn PDA where they are stored or sent to a care center via mobile network. Another group integrated biosensors in the fabric of a shirt and developed their prototype which contains dry ECG electrodes, shock/fall sensor, breath rate sensor, temperature sensors, global positioning system (GPS) receiver, and wireless module [18].

The wearable healthcare system (WEALTHY), developed by Paradiso et al. [19], was based on wearable sensors, electrodes, and connections in fabric forms. Piezoresistive fabric sensors, which behaved as strain gauges, were built from Lycra fabric coated with carbon-loaded rubber and electroconductive yarn, while contact electrodes were fabricated with stainless-steel yarns. The system could simultaneously acquire ECG, respiration, activity, and temperature signals, which were transmitted to a monitoring center via GPRS link. The h-Shirt, developed by Zhang et al. [20], monitored continuous BP based on pulse-transit time. Using conductive fabrics as ECG electrodes and medium for signal transmission, ECG was captured from the two wrists, with reference to an electrode placed on the forearm to avoid noise induced by respiration, while a photoreflective sensor on a fingertip captured photoplethysmogram (PPG) signals. Using fasten snap buttons as connection terminals, the signals were fed into a wristwatch for real-time signal processing and BP display. As shown in Figure 6.5, the system has been further developed into an armband that can be comfortably worn on a subject for measuring blood pressure during nighttime without using the cuff [21]. Data can be transmitted to a mobile phone in real time. Another group has also reported the measurement of cuffless differential blood pressure by using smartphones [22].

Meanwhile, numerous medical device companies are adding communication capabilities into their existing products. Examples are BP meters, blood glucose meters, heart-rate (HR) monitors, thermometers, and spirometers with universal serial bus (USB), Bluetooth, and/or Wi-Fi connectivity. In a typical usage scenario, the mobile phone or Wi-Fi access

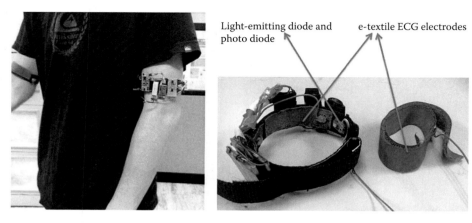

Light-emitting diode and photo diode   e-textile ECG electrodes

FIGURE 6.5
A wearable armband for overnight and cuffless blood pressure measurement.

point would act as a patient-side hub, which relays the stored readings from the medical device to a remote center for disease management. Along with this generation of devices are wearable physiological sensors in alternative forms.

For example, the finger-ring sensor developed by Rhee et al. measured photoplethysmogram (PPG) to extract blood oxygen saturation level (SaO_2) and HR information [23]. The ring was designed for resistance to motion artifacts. Another group developed a ring sensor as part of a multiparameter-monitoring system. They have also developed application-specific integrated circuits (ASICs) for such purpose. The near-infrared (NIR) HR measurement ASIC used the current-steering technique and very low light-emitting diode (LED) pulsing duty cycle to achieve very low cutoff frequency and low power consumption [24].

As wearable sensors are becoming prevalent in mHealth systems, many have suggested connecting them with a body sensor network (BSN) for efficient use of resources [25]. The concept of BSN was introduced in 2002 for interconnection of a variety of wearable implantable sensing devices on an individual [26]. In cases where several sensors placed at different body parts are required, it is necessary to set up BSNs for resource-use optimization to satisfy the stringent constraints in the sensing devices and to enhance local control so that the system would be adaptive to the changing environments. Wireless body sensor network (WBSN) is formed by modules (nodes) of sensing unit, processor, wireless transceiver, and battery. It ensures accurate capture of data from the sensing unit to which it is connected, carries data preprocessing, and transmits the data wirelessly to a patient terminal or local station. Data are then further processed and fused before being sent to the central server. WBSN has been demonstrated in heart-rate variability studies, ECG monitoring, and rehabilitation. Personalized WBSN is an elementary unit in mHealth systems and is designed for seamless integration in homes, workplaces, and hospital environments.

6.3 Mobile Apps for mHealth

With the high penetration of mobile phones and mobile data service, mobile applications (apps) have become an indispensable part of daily routines. Over 80 billion mobile apps were downloaded in the year 2013 alone, accumulating to over 165 billion downloads in total [27]. Specifically, mHealth apps are playing a significant role in shaping the modes of healthcare provision. Users include healthcare professionals, consumers, and patients. According to industry estimates, 50% of the more than three billion smart device users will have downloaded an mHealth app by 2017; and the corresponding revenue is forecasted to reach US$26 billion by that time [28].

Some have suggested three phases to describe the development trends of mHealth apps. According to reports, the market has just completed the *initial trial phase* and started the *commercialization phase*, which is featured by a large increase in the number of solutions, new business models, and products which mainly target private, health-interested people, patients, and corporations. The main market driver in this phase is smartphone user penetration. The *integrated phase* is expected to start years later, when it becomes more common to integrate mHealth apps into traditional healthcare systems [28].

6.3.1 Overview of mHealth Apps

Unlike the traditional medical device market, most revenue generated from mHealth is not from purchases of the apps but rather from the sales of other related services and hardware. Operating under new business models, the apps often serve as platforms to sell the related products. In addition to appearing in online apps stores, more mHealth apps have made their way to traditional health distributive channels. While the first batch of mHealth apps mainly targeted the health and fitness domain, the second generation will focus on chronic care. MHealth apps can generally be divided into the categories listed in Table 6.1. Most of them are native apps instead of web-based applications.

Despite the encouraging figures, some major market barriers have to be tackled before the widespread use of mHealth apps. For example, there is still need for more supporting data about the effectiveness of mHealth. Many people are also concerned about the conflicting health system priorities and the lack of regulations for mHealth [28–29].

6.3.2 Regulation of mHealth Apps

Foreseeing the challenges posed by mHealth apps, in July 2011 the U.S. Food and Drug Administration (FDA) drafted a document which outlined the types of mobile apps they

TABLE 6.1

Categories and Examples of mHealth Apps

Category	Examples
Remote health monitoring	Monitoring of HR/ECG
Disease management	Measurement and logging of BP/blood glucose/reduced forced expiratory volume in 1 second (FEV1)
Interface/extension of existing medical devices	BP meter with display and storage on phone; thermometer communicating with a tablet via Bluetooth
Assistance in diagnosis	Teleradiology, telecardiology, and review of lab test results
Sports and fitness	Measurement and logging of HR and physical activities
Healthy living	Diet plans and exercise log
Reminders and alerts	Appointment/medication reminders
Productivity for healthcare providers	Scheduling
References	Medical guides

intended to regulate. The scope included a small subset of medical apps that could pose risks to patients if they do not function as intended. Upon receiving more than 130 comments from the public, the agency issued the final guidance in September 2013 [30]. It aims to give apps developers a clear road map to determine whether their products fall within scope of FDA's oversight. According to the document, the agency will focus on medical mobile apps that meet the definition of device in the Federal Food, Drug, and Cosmetic Act and

- Are intended to be used as an accessory to a medical device regulated by the FDA; or
- Transform a mobile platform into a medical device regulated by the FDA.

For example, apps that allow doctors to perform diagnosis by viewing medical images from a picture archiving and communication system (PACS) on a smartphone and apps that turn mobile tablets into ECG recorders to detect abnormalities would fall within the scope. The document has also listed out examples of (i) apps that are not considered as medical devices, (ii) mobile apps that the agency intends to exercise enforcement discretion (those which pose minimal risk to consumers), and (iii) mobile medical apps that the FDA will regulate. Apps that need to undergo review by FDA will be assessed with the same regulatory standards and risk-based approach that are applied to other medical devices. For example, apps are also categorized as class I, II, or III devices depending on the risks they may pose. By the end of 2013, FDA has already cleared about 100 mobile medical apps. As new mHealth apps will continue to emerge, it is expected that joint efforts worldwide will be needed to further develop and harmonize the standards and regulations.

6.4 Cloud Computing

The recent development in cloud computing has revolutionized the information technology (IT) industry. With cloud computing, software designers can now deploy their products as a service over the Internet, without the need to invest large capital outlays in hardware and human expense to support them. Moreover, provisioning the popularity of a service, and, thus, predicting the sales return, is not essential for launching the software in cloud computing, which allows an extremely flexible way to allocate resources. As a result, cloud computing has been recently adopted by many different service providers, including those in the healthcare domain.

6.4.1 Definitions

Cloud computing refers to both the applications delivered as services over the Internet and the hardware and system software in the data centers that provide those services [31]. The computing community usually named the hardware and software of a data center *cloud*, which is by definition "a type of parallel and distributed system consisting of a collection of interconnected and virtualized computers that are dynamically provisioned and presented as one or more unified computing resource(s) based on service-level agreements established through negotiation between the service provider and consumers" [32]. A cloud is *public* when it is made available to the general community in a pay-for-use

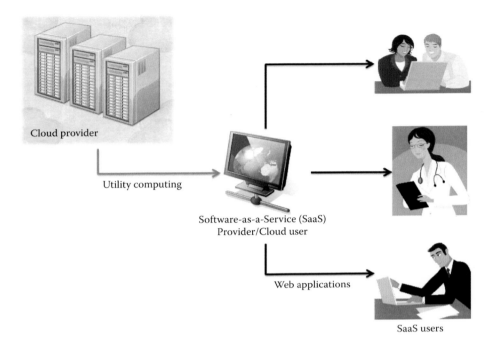

FIGURE 6.6
Users and providers of cloud computing in the healthcare domain. (Courtesy of Microsoft™ clip arts, Microsoft Corp., Redmond, Washington.)

manner, while the service being sold is utility computing. On the other hand, *private clouds* usually refer to internal data centers of a business or other organizations that are inaccessible to the general public.

In the healthcare domain, this technological advancement has led to novel ways of how healthcare services can be delivered, e.g., real-time monitoring and analysis systems that are scalable and economical for people who require frequent monitoring of their health can now be designed [33]. Cloud computing also serves as a powerful tool to close the loop for health management in mHealth. It allows active health management outside the hospitals, in addition to collecting health data passively. Figure 6.6 shows the basic relationships of users and providers of cloud computing in the healthcare domain. Note that users of cloud computing in the healthcare domain can be both patients and healthcare providers.

6.4.2 Selected Applications

With the support of mobile devices, cloud services are particularly useful for capturing acute and transient disease symptoms as well as monitoring chronic patients. For example, an autonomic cloud environment that collects and analyses the electrocardiogram has been proposed [33]. The data were disseminated to a cloud-based information repository and analyzed by software services hosted in the cloud. Cloud computing has also been proposed as a clinical decision-support system for the accurate assessment of pain, which is a highly subjective measure and sometimes difficult to be evaluated by the healthcare professionals during the short visits of their patients [34]. Mobile computing and cloud services are also useful for emergency healthcare, where emergency care processes were supported by an integrated computer system with a cloud-based architecture that cross-links institutional healthcare systems [35]. The system allowed authorized users to access emergency case information in standardized format and exchange operational data with different hospitals in order to triage and select the most appropriate ambulances and hospitals for each specific case [35]. Another potential application of mHealth and cloud services is to report vital signs of outpatients after cancer treatment [36]. For these patients, physicians usually have to keep track of their health-related quality of life (HRQoL). For such a situation, a prototype on cloud computing services that can be accessed by a mobile application can extend a standard clinical trial informatics model and allows medical practitioners to efficiently monitor discharged patients' HRQoL and vital signals.

Not only that cloud computing is revolutionizing the way healthcare can be delivered, but it is also reforming healthcare education, to both the professionals and the general public. A recent report by Cisco reveals that global mobile data traffic grew 81% in 2013 and it is anticipated that by the end of 2014, the number of mobile-connected devices will exceed the number of people on earth [37]. The prevalence and handiness of these mobile devices led to new learning models, where mobile devices and cloud learning platform such as Google+ can be used as innovative tools to save manpower and reduce errors while enhancing medical practitioners' professional knowledge and skills [38]. Cloud computing facilitates a personalized, diverse, and virtual learning environment. When this virtual healthcare education platform becomes mature, it can potentially be opened to the general public and thereby further enhance the future *participatory* health model.

6.5 Closed-Loop Solutions for Personalized Health Interventions

The advancement of mobile and cloud computing technologies will no doubt change how one can self-monitor one's own health and well-being. Nevertheless, in order to bring about higher clinical impact, it is important to have intelligent closed-loop solutions where personalized health interventions can take place to change patient management schemes. In this regard, the aforementioned technologies will have crucial roles; however, there are still enormous technical challenges.

6.5.1 Challenges in Sensor Design and Fabrication

For some applications, miniature sensors that can robustly and accurately measure the desired physiological parameters in real-life dynamic situations are still urgently needed. For example, a model-based predictive algorithm for blood glucose control in type 1 diabetic patients has been developed for over a decade [39]. Nevertheless, robust closed-loop systems for insulin delivery are not yet widely available. A main limiting factor in this application is the glucose monitors in presently available monitors fail to demonstrate satisfactory characteristics in terms of reliability and/or accuracy [40]. In a recent study [41], the loop was closed manually; i.e., glucose measurements were fed every 15 min into a control algorithm calculating rate of insulin infusion, and a nurse adjusted the insulin pump accordingly. The study found that closed-loop systems could reduce risk of nocturnal hypoglycemia in children and adolescents with type 1 diabetes [41]. Similarly, although measurement of blood pressure by the cuffless pulse-transit-time approach was found to be promising during nighttime, studies also reported that the current method still has difficulties in tracking blood pressure accurately in dynamic situations, e.g., during exercises or vascular drug administration [42–44].

Current wearable sensors are developed from rigid printed circuit boards and integrated circuit chips. Although these developments show promising results in controlled laboratory settings, a number of practical issues must be solved before they can be widely deployed in daily applications [21]. First, most of these sensors do not conform and adhere well to the curvilinear body surface and, therefore, will result in huge motion artifacts in the collected signals when the subject moves. Second, most of these systems are still as large as if they were to be used by "healthy" subjects rather than patients. In order to enhance user compliance, these systems must be miniaturized and designed such that they are completely unobtrusive.

Fortunately, sensing technologies have greatly advanced in the last two decades. A new direction in future physiological and behavioral sensor design is to utilize flexible and stretchable electronic technologies to partly address the aforementioned issues. Sensors which are fabricated on flexible substrate such as polydimethylsiloxane (PDMS) or polyethylene naphthalate, for example, those shown in Figure 6.7, have the potential to be made ultrathin and stretchable [45]. They can, therefore, adhere well to the irregular body surface unobtrusively and, even when the subjects move, signals could be robustly obtained.

The next generation of flexible and stretchable electronics are emerging research areas to solve the disadvantages of rigid electronics. At present, due to the limitation of material strength and fabrication process, these fabrication technologies are mostly applied on single-device integrated circuits and sensors but not on complicated electronic components such as microprocessors and instrumentation amplifiers. As an intermediate step before fully flexible and stretchable systems are available, solutions for lightweight and unobtrusive physiological sensors would be to use copper/polyamide films as flexible printed circuit boards with conventional surface-mount electronic components.

6.5.2 Challenges in Mining and Managing Big Health Data

With the advancement of sensing and communication infrastructure, health information that can be collected from a single patient has dramatically increased. For example, fetal heart-rate patterns at 20 to 24 weeks can now be recorded and analyzed, revealing unique patterns with accelerations and decelerations, as well as higher baseline variability [46]. Nevertheless, such devices have not penetrated into clinical practice, due to a lack of research into "intelligent" analysis methods that are sufficiently robust to support

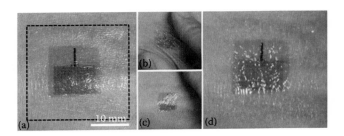

FIGURE 6.7
The fabricated serpentine-shape electrode after transfer onto the skin of the human subject (a) with PDMS as adhesive tape (black dashed line pane); (b) under compressing; (c) being stretched; and (d) after stretching. (From B. H. K. Leung et al., "Mobile health: Design of flexible and stretchable electrophysiological sensors for wearable systems," *Proceedings of the 11th International Conference on Wearable and Implantable Body Sensor Networks*, Zurich, Switzerland, 2014.)

large-scale deployment [47]. Current systems typically have large false-alarm rates, and are unable to handle artifacts in the collected signals diligently. Thus, in order for mHealth monitoring to be widely adopted within the actual clinical environment, further study of these mHealth systems with intelligent algorithms must be carried out at scale to prove they are capable of improving patient outcomes via personalized healthcare. Recently, a principled, probabilistic approach set within a Gaussian process framework has been proposed for vital-signs monitoring of patients [47]. The study adopted a patient-specific approach to the construction of Gaussian process, allowing the personalization of the posterior distribution to those physiological data acquired from each patient. Based on the framework, a distribution of values of the physiological data for any test points can be reliably estimated, coping with periods where data are incomplete due to sensor artifacts and communication failure and, thus, reducing the number of false alarms [47].

In addition, closed-loop solutions that help physicians to make clinical decisions are still missing in some applications. For example, a wireless capsule endoscope has been launched in the market for over a decade. The technology is extremely useful in helping physicians to explore the small bowel, which is difficult to examine using a traditional flexible endoscope. With this technology, data collected from a single patient can be over 8 h. Thus, a significant workload is needed for physicians and healthcare providers who utilize this technology. Moreover, the current wireless capsule technologies provide only a passive function of viewing the gastrointestinal tract without therapeutic treatment schemes. If pathological conditions have been identified, the physicians will need to use the conventional flexible endoscope to provide treatment. These issues have limited the wide adoption of the technology in a clinical setting and, thus, lead to the developments of novel systems that can combine with mHealth technologies to provide intelligent closed-loop solutions to these problems, as shown in Figure 6.8.

There are several technical challenges in this kind of system. First, miniaturized actuation modules of swallowable size must be designed. Second, intelligent algorithms must be developed to enable the real-time triggering of the therapeutic treatment when needed. Third, robust transmission of massive data from inside the body to the body surface must be available, which requires specially designed antennas that can penetrate well through

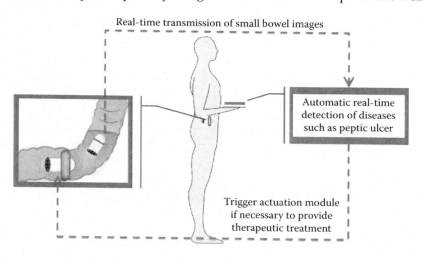

FIGURE 6.8
Illustration of potential closed-loop system for real-time treatment of digestive diseases using mHealth technologies with a future novel wireless capsule endoscope with actuation mechanisms for therapeutic treatment.

biological tissues. When these issues can be fully addressed, the systems are anticipated to bring forth new clinical usages that mark mHealth in a new stage of development, where in addition to monitoring functionality, therapeutic treatments can also be available.

6.6 Conclusions

After decades of development, telemedicine and mHealth have evolved into a new era, where a variety of applications can now be made available beyond teleconsultation and monitoring of health signals. Nevertheless, challenges remain before these technologies can be widely adopted in clinical and healthcare settings. Technically, flexible and stretchable electronics are anticipated to change the next generation of wearable and implantable sensors and systems. Mobile apps and cloud communication infrastructures will continue to assist the penetration of health services such that they will become an indispensable part of daily routines. Novel intelligent algorithms to deal with the enormous amount of information in a timely and robust manner must be derived. Moreover, regulation policies are urgently needed to assure that these advancements in technologies can truly be beneficial to public health.

Abbreviations and Nomenclature

ASIC	application-specific integrated circuit
BP	blood pressure
BSN	body sensor network
CCTV	closed-circuit television
CSD	circuit-switched data
ECG	electrocardiogram
FDA	Food and Drug Administration
HR	heart rate
HRQoL	health-related quality of life
HSDPA	high-speed downlink packet access
IT	information technology
mHealth	mobile health
NIR	near infrared
PACS	picture archiving and communication system
PC	personal computer
PDMS	polydimethylsiloxane
PPG	photoplethysmogram
SaO$_2$	blood oxygen saturation level
WBMP	wireless bitmap
WBSN	wireless body sensor network
WCDMA	wideband code-division multiple access
WEALTHY	wearable healthcare system
WML	wireless markup language

Acknowledgments

This work was supported in part by the Hong Kong Innovation and Technology Commission (ITS/159/11 and ITS/197/12).

References

1. R. S. Istepanian, E. Jovanov, and Y. Zhang, "Guest editorial introduction to the special section on m-health: Beyond seamless mobility and global wireless health-care connectivity," *IEEE Transactions on Information Technology in Biomedicine*, vol. 8, no. 4, pp. 405–414, 2004.
2. World Health Organization, "Global Health and Aging Report," National Institute on Aging and National Institutes of Health. NIH Publication, 2011.
3. World Health Organization Regional Office for the Eastern Mediterranean, "Address by Dr. Hussein A. Gezairy," in Fifth Regional Conference on E-Health and Related Applications, Cairo, Egypt, 2006. http://applications.emro.who.int/docs/message_2011_july_egypt_cairo_13720.pdf
4. M. J. Field, *Telemedicine: A guide to assessing telecommunications for health care*. National Academies Press, 1996.
5. G. Freiburger, M. Holcomb, and D. Piper, "The STARPAHC collection: Part of an archive of the history of telemedicine," *Journal of Telemedicine and Telecare*, vol. 13, no. 5, pp. 221–223, 2007.
6. D. Foote, D. Parker, and H. Hudson, "Telemedicine in Alaska: The ATS-6 satellite biomedical demonstration," *Stanford University Report*, vol. 1, 1976.
7. J. Mitchell, "From telehealth to e-health: The unstoppable rise of e-health." Canberra: National Office for the Information Economy, Department of Communications, Information Technology and the Arts, 1999.
8. L. G. Yamamoto, "Wireless teleradiology and fax using cellular phones and notebook PCs for instant access to consultants," *The American Journal of Emergency Medicine*, vol. 13, no. 2, pp. 184–187, 1995.
9. P. M. Ti, L. G. Yamamoto, and L. K. Shirai, "Instant telemedicine ECG consultations with cardiologist using pocket wireless computers," *American Journal of Emergency Medicine*, vol. 19, supplement 4, pp. 248–249, May 2001.
10. L. G. Yamamoto and D. R. Williams, "A demonstration of instant pocket wireless CT teleradiology to facilitate stat neurosurgical consultation and future telemedicine implications," *American Journal of Emergency Medicine*, vol. 18, no. 4, pp. 423–426, July 2000.
11. P. M. T. Singh, L. G. Yamamoto, and L. K. Shirai, "Instant telemedicine ECG consultation with cardiologists using pocket wireless computers," *American Journal of Emergency Medicine*, vol. 19, no. 3, pp. 248–249, May 2001.
12. J. Reponen, E. Ilkko, L. Jyrkinen et al., "Initial experience with a wireless personal digital assistant as a teleradiology terminal for reporting emergency computerized tomography scans," *Journal of Telemedicine and Telecare*, vol. 6, no. 1, pp. 45–49, 2000.
13. J. Reponen, E. Ilkko, L. Jyrkinen et al., "Digital wireless radiology consultations with a portable computer," *Journal of Telemedicine and Telecare*, vol. 4, no. 4, pp. 201–205, 1998.
14. K. Hung and Y. T. Zhang, "Implementation of a WAP-based telemedicine system for patient monitoring," *IEEE Transactions on Information Technology in Biomedicine*, vol. 7, no. 2, pp. 101–107, June 2003.
15. E. A. Viruete Navarro, J. Ruiz Mas, J. Fernandez Navajas et al., "Performance of a 3G-based mobile telemedicine system," *3rd IEEE Consumer Communications and Networking Conference, 2006: CCNC 2006* (Volume: 2), pp. 1023–1027, 2006.

16. U. Anliker, J. A. Ward, P. Lukowicz et al., "AMON: A wearable multiparameter medical monitoring and alert system," *IEEE Transactions on Information Technology in Biomedicine*, vol. 8, no. 4, pp. 415–427, 2004.

17. K. J. Heilman and S. W. Porges, "Accuracy of the LifeShirt® (Vivometrics) in the detection of cardiac rhythms," *Biological Psychology*, vol. 75, no. 3, pp. 300–305, July 2007.

18. N. Noury, A. Dittmar, C. Corroy et al., "A smart cloth for ambulatory telemonitoring of physiological parameters and activity: The VTAMN project," *Proceedings 6th International Workshop on Enterprise Networking and Computing in Healthcare Industry, 2004: HEALTHCOM 2004*, pp. 155–160, 2004.

19. R. Paradiso, G. Loriga, and N. Taccini, "A wearable health care system based on knitted integrated sensors," *IEEE Transactions on Information Technology in Biomedicine*, vol. 9, no. 3, pp. 337–344, 2005.

20. Y.-T. Zhang, C. C. Poon, C.-H. Chan et al., "A health-shirt using e-textile materials for the continuous and cuffless monitoring of arterial blood pressure," *3rd IEEE/EMBS International Summer School on Medical Devices and Biosensors, 2006*, pp. 86–89.

21. Y.-L. Zheng, B. P. Yan, Y.-T. Zhang et al., "An armband wearable device for overnight and cuffless blood pressure measurement," *IEEE Transactions on Biomedical Engineering*, vol. 61, no. 7, pp. 2179–2186, Jul 2014.

22. V. Chandrasekaran, R. Dantu, S. Jonnada et al., "Cuffless differential blood pressure estimation using smart phones," *IEEE Transactions on Biomedical Engineering*, vol. 60, no. 4, pp. 1080–1089, 2013.

23. S. Rhee, B.-H. Yang, and H. H. Asada, "Artifact-resistant power-efficient design of finger-ring plethysmographic sensors," *IEEE Transactions on Biomedical Engineering*, vol. 48, no. 7, pp. 795–805, 2001.

24. A. Wong, K.-P. Pun, Y.-T. Zhang et al., "A near-infrared heart rate measurement IC with very low cutoff frequency using current steering technique," *IEEE Transactions on Circuits and Systems I: Regular Papers*, vol. 52, no. 12, pp. 2642–2647, 2005.

25. C. C. Y. Poon, Y. T. Zhang, and S. D. Bao, "A novel biometrics method to secure wireless body area sensor networks for telemedicine and m-health," *IEEE Communications Magazine*, vol. 44, no. 4, pp. 73–81, 2006.

26. G.-Z. Yang and M. Yacoub, "Body Sensor Networks," Spring-Verlag, London, 2006.

27. Portio-Research, "Mobile Applications Futures 2013–2017: Analysis and Growth Forecasts for the Worldwide Mobile Applications Market," 2013.

28. Research2Guidance, "Mobile Health Trends and Figures 2013–2017," July 2013.

29. World Health Organization, "mHealth: New Horizons for Health through Mobile Technologies: Second Global Survey on eHealth," 2011.

30. U.S. FDA, "Mobile Medical Applications: Guidance for Industry and Food and Drug Administration Staff," 2014.

31. M. Armbrust, A. Fox, R. Griffith et al., "A view of cloud computing," *Communications of the ACM*, vol. 53, no. 4, pp. 50–58, Apr 2010.

32. R. Buyya, C. S. Yeo, S. Venugopal et al., "Cloud computing and emerging IT platforms: Vision, hype, and reality for delivering computing as the 5th utility," *Future Generation Computer Systems—The International Journal of Grid Computing and Escience*, vol. 25, no. 6, pp. 599–616, June 2009.

33. S. Pandey, W. Voorsluys, S. Niu et al., "An autonomic cloud environment for hosting ECG data analysis services," *Future Generation Computer Systems—The International Journal of Grid Computing and Escience*, vol. 28, no. 1, pp. 147–154, Jan 2012.

34. N. Pombo, P. Araújo, and J. Viana, "Knowledge discovery in clinical decision support systems for pain management: A systematic review," *Artificial Intelligence in Medicine*, vol. 60, no. 1, pp. 1–11, 2014.

35. M. Poulymenopoulou, F. Malamateniou, and G. Vassilacopoulos, "Emergency healthcare process automation using mobile computing and cloud services," *Journal of Medical Systems*, vol. 36, no. 5, pp. 3233–3241, 2012.

36. C. Cheng, T. H. Stokes, and M. D. Wang, "CaREMOTE: The design of a cancer reporting and monitoring telemedicine system for domestic care," *33rd Annual International Conference Proceedings of the IEEE Engineering in Medicine and Biology Society*, Atlanta, GA, pp. 3168–3171, 2011.

37. Cisco, "Cisco Visual Networking Index: Global Mobile Data Traffic Forecast Update, 2013–2018," 2014.

38. T. T. Wu and T. W. Sung, "Public health practice course using Google Plus," *CIN—Computers Informatics Nursing*, vol. 32, no. 3, pp. 144–152, 2014.

39. R. S. Parker, F. J. Doyle, and N. A. Peppas, "A model-based algorithm for blood glucose control in type I diabetic patients," *IEEE Transactions on Biomedical Engineering*, vol. 46, no. 2, pp. 148–157, Feb 1999.

40. R. Hovorka, "Continuous glucose monitoring and closed-loop systems," *Diabetic Medicine*, vol. 23, no. 1, pp. 1–12, Jan 2006.

41. R. Hovorka, J. M. Allen, D. Elleri et al., "Manual closed-loop insulin delivery in children and adolescents with type 1 diabetes: A phase 2 randomised crossover trial," *Lancet*, vol. 375, no. 9716, pp. 743–751, Feb 2010.

42. R. Payne, C. Symeonides, D. Webb et al., "Pulse transit time measured from the ECG: An unreliable marker of beat-to-beat blood pressure," *Journal of Applied Physiology*, vol. 100, no. 1, pp. 136–141, 2006.

43. B. McCarthy, C. Vaughan, B. O'Flynn et al., "An examination of calibration intervals required for accurately tracking blood pressure using pulse transit time algorithms," *Journal of Human Hypertension*, vol. 27, no. 12, pp. 744–750, Dec 2013.

44. Q. Liu, B. P. Yan, C. Yu et al., "Attenuation of systolic blood pressure and pulse transit time hysteresis during exercise and recovery in cardiovascular patients," *IEEE Transactions on Biomedical Engineering*, vol. 61, no. 2, pp. 346–352, Feb 2014.

45. B. H. K. Leung, N. Luo, J. Ding, B. P. K. Wong, J. L. Xu, N. Zhao, and C. C. Y. Poon, "Mobile health: Design of flexible and stretchable electrophysiological sensors for wearable systems," *Proceedings of the 11th International Conference on Wearable and Implantable Body Sensor Networks*, Zurich, Switzerland, 2014.

46. F. Hofmeyr, C. A. Groenewald, D. G. Nel et al., "Fetal heart rate patterns at 20 to 24 weeks gestation as recorded by fetal electrocardiography," *Journal of Maternal-Fetal & Neonatal Medicine*, vol. 27, no. 7, pp. 714–718, Apr 2014.

47. L. Clifton, D. A. Clifton, M. A. F. Pimentel et al., "Gaussian processes for personalized e-health monitoring with wearable sensors," *IEEE Transactions on Biomedical Engineering*, vol. 60, no. 1, pp. 193–197, Jan 2013.

7

Patient Care Sensing and Monitoring Systems

Akihiro Kajiwara and Ryohei Nakamura

CONTENTS

7.1 Introduction

Wireless sensors which have more advantages relative to wired sensors are playing an important role in improving the quality of life in daily indoor space. The applications include home security systems for improving the safety of our home and vital-signs monitoring systems for improving our health. The sensor systems can remotely detect the echo reflected from a person and then estimate the state and/or vital signs such as respiration rate and heart rate, for example. Several types of wireless sensor device have thus far been developed such as video camera, passive infrared (IR), and microwave. The video camera is unacceptable for some applications from a privacy protection point of view, although it can monitor a wide area of the indoor space. Passive IR sensors are generally developed to be dedicated to a relatively small area such as the home entrance and toilet area. Microwave sensors such as the Doppler and the frequency-modulated continuous-wave (FM-CW) sensor, which are capable of penetrating a variety of nonmetallic materials such as inner walls, provide a wider coverage area when compared with video cameras and IR but are susceptible to interference from other wireless radio systems, thereby causing false alarms or false detection. Ultrawideband impulse radio (UWB-IR) has lately attracted considerable attention in short-range remote-sensor applications since it offers high ranging accuracy, multipath reduction, and environmental friendliness due to the very low energy emission [1–4]. However, it requires a very high-speed analog-to-digital conversion (ADC) and high-level processors in order to synchronize and detect the received nanosecond pulse. It may also cause interference with existing or future wireless systems using the same or nearby bands because it occupies a bandwidth wider than 500 MHz. Therefore, the ultrawideband (UWB) device operated in the lower band of 850 MHz to 3.14 GHz is required to implement the detect-and-avoid (DAA) technique in many countries such as Europe, Korea, and Japan [5].

To solve these problems, the use of the stepped-frequency modulation (FM) scheme has been suggested for the UWB sensor. The scheme has the following main advantages relative to UWB-IR [6–10]:

- Lower-speed ADC and lower-level processor: It transmits a series of bursts of narrowband pulses where each burst is a sequence consisting of many pulses shifted in frequency from pulse to pulse with a fixed frequency step. Each received narrowband pulse is phase detected and then combined into the large effective bandwidth (sequentially over many pulses). Therefore, the hardware requirement is less stringent relative to that of UWB-IR. The detector bandwidth is smaller, resulting in lower noise bandwidth and higher signal-to-noise (SN) ratio when compared with UWB-IR.

- Inherent DAA function: It is capable of coexisting with other narrowband wireless systems operating in the same frequency range. It detects interference radio potential by searching and is then designed to have some spectrum hole (nonactivated within a portion of the wide radio spectrum) to prevent any conflict and to coexist with the other narrowband wireless systems [11]. For example, the frequency band of interference can be detected by the phase detector, while the spectrum hole is adaptively assigned according to the interference band.

We have fabricated the sensor setup where the algorithm detecting and avoiding the interference mentioned above has also been developed. Note that the ADC speed is 10 kilosamples per second (kS/s) for the 2 GHz bandwidth. Measurements were conducted for some scenarios and the results are presented. From the results, the scheme is shown to be useful as a wireless sensor and can also coexist with other wireless systems operating in an overlaid frequency band.

7.2 Stepped-Frequency Modulation Ultrawideband Scheme

7.2.1 Ultrawideband Impulse Radio Sensor

UWB-IR uses nanosecond pulses spreading out all over a continuous wide bandwidth as shown in Figure 7.1a. It offers many advantages such as low-cost implementation, low transmission power, ranging, multipath immunity, and low interference. Due to the ultrashort duration pulses, subcentimeter ranging is possible, thereby resulting in the correct identification of the complex shaped target. It does not require up-and-down conversion; thus, it may reduce the implementation cost and low power consumption. In spite of the advantages, however, there are several engineering issues which need to be considered. Due to the very short pulses, accurate synchronization and detection may be difficult. Interference with existing or future radio systems using the same band is one of the difficult issues to be solved. Designing wideband radio-frequency (RF) components is also a big challenge which includes high-speed ADCs.

Now, the received one-dimensional signal, which is referred to as range profile, is generally presented by multiple impulses with gains $\{\beta_k\}$ and propagation delays $\{\tau_k\}$, where k

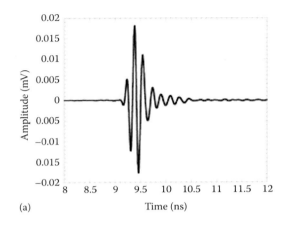

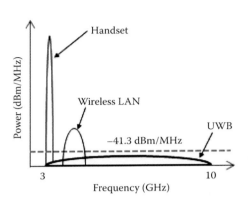

(a)

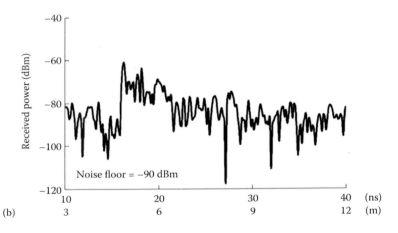

(b)

FIGURE 7.1
UWB-IR: (a) waveform and the spectrum and (b) example of power range profile (1 GHz bandwidth).

is the impulse index. Suppose for a nanosecond pulse of $s(t)$, the range profile $y(\tau, t)$ is the time convolution of $s(t)$ and the impulse echo response $\sum \beta_k s(t - \tau_k)$ as follows:

$$y(\tau, t) = \sum_k \beta_k s(t - \tau_k). \tag{7.1}$$

Figure 7.1b shows an example of received power range profile for a bandwidth of 1 GHz in a room (corresponding to 1 ns pulse).

7.2.2 Stepped-Frequency Modulation Ultrawideband Sensor

A different approach to UWB, which solves the above problems, is the use of stepped-FM pulse sequences instead of transmitting a nanosecond short pulse. We have fabricated an experimental setup of a stepped-FM UWB sensor. The block diagram is illustrated in Figure 7.2a. Waveforms at each stage and the external photo are given in Figure 7.2b and c,

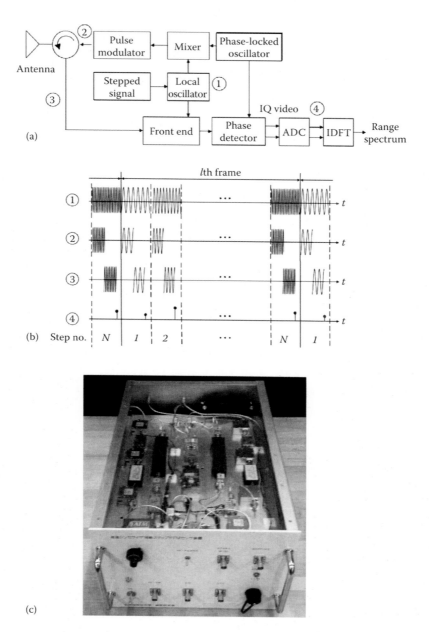

FIGURE 7.2
Stepped-FM UWB sensors: (a) block diagram, (b) signal waveforms, and (c) external photo of the setup.

respectively. The sensor transmits a series of bursts of narrowband pulses, where each burst is a sequence consisting of N narrowband pulses shifted in frequency from pulse to pulse with a fixed frequency step Δf [4]. The received echo from an object is phase detected with the transmitted stepped signal (homodyne detection) and is then in-phase and quadrature (I-Q) sampled by a relaxed speed of ADC.

Then, the nth complex sample R_n is given by

$$R_n = A_n \exp(-j\theta_n), \tag{7.2}$$

$$\theta_n = 2\pi[f_c + (n-1)\Delta f] \cdot \frac{2d}{c}, \tag{7.3}$$

where A_n is the amplitude of nth pulse (A_n can be approximated by A for a stationary object), f_c is the fundamental frequency, and c is the velocity of light.

Next, each complex sample is applied to the inverse discrete Fourier transformation (IDFT) device in order to obtain an N-element synthetic range profile, which is called range spectrum. The N-element range spectrum is given by

$$R(\phi) = \left| \sum_{n=1}^{N} R_n \cdot \exp\left[j\frac{2\pi}{N}(n-1)\cdot\phi \right] \right|$$

$$= N\cdot A \cdot \left| \frac{\sin c\left[\pi\left(\phi - N\Delta f \frac{2d}{c} \right) \right]}{\sin c\left[\frac{\pi}{N}\left(\phi - N\Delta f \frac{2d}{c} \right) \right]} \right|, \tag{7.4}$$

$$\phi = \frac{2dN\Delta f}{c}. \tag{7.5}$$

It is clear that the range resolution ΔR is approximately $1/N\Delta f$. For example, suppose $\Delta f = 34.5$ MHz and $N = 30$, the resolution is approximately 30 cm, which is equivalent to a UWB-IR with 1 GHz. Hence, it does not require high-speed ADC devices and a high-level processor at the receiver side. Note that the unambiguous range (maximum detectable range) R_{max} is given by $c/2\Delta f$.

An example of the range spectrum is shown in Figure 7.3, where $\Delta f = 34.5$ MHz and $N = 30$ and zero padding was also used for the IDFT operation. The phase detector should be generally ADC sampled by 10 times of stepped pulse repetition rate or more. For example, when the pulse repletion rate of the setup is 1 kHz, the used ADC is 10 kS/s. Therefore, the hardware requirements become less stringent when compared with UWB-IR.

7.2.3 Detect-and-Avoid and Spectrum Hole Technique

The stepped-FM UWB sensor can detect radio interference by monitoring the phase detector output in passive mode (transmitter power off/receiver on) and is then designed to have some spectrum hole (nonactivated within a portion of the wide radio spectrum) at the transmitter in order to prevent any conflict and to coexist with the other narrow-band wireless systems. Note that the transmit radio consists of independent narrow-band pulses with a different frequency. As such, it provides the flexibility to support

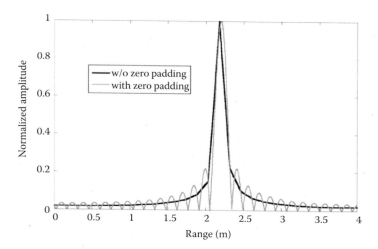

FIGURE 7.3

Example of range spectrum with zero padding (Δf = 34.5 MHz and N = 30).

regulatory measures in different areas of the world and ease concerns about interference. This subsection describes an interference avoidance technique called spectrum hole.

Figure 7.4a shows the power spectrum for Δf = 34.5 MHz and N = 30, where some stepped pulses from 3.655 to 3.724 GHz are not transmitted corresponding to the spectrum hole of 6.6% (simulation result). The PSD of the hole is seen to be suppressed less than –13 dB. Consider the Federal Communications Commission (FCC) regulation of –41.3 dBm/MHz; the PSD is suppressed to less than –55 dBm. Therefore, it does not interfere with existing radios. Figure 7.4b also shows the power spectrum against three narrowband interferences where we assumed three spectrum holes of 3.310–3.379, 3.621–3.655, and 3.897–3.966 GHz (corresponding to 16.6%). Figure 7.5 shows the range spectra for 6.6% and 16.6% where a zero padding was used for the IDFT processing (1024 points).

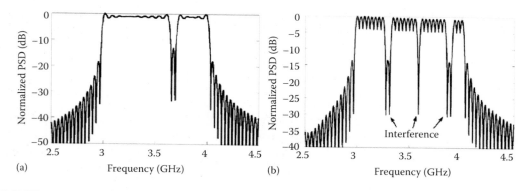

FIGURE 7.4

Power spectra (in terms of power spectral density [PSD]) of transmit signal with spectrum hole: (a) spectrum hole of 6.6% and (b) spectrum hole of 16.6%.

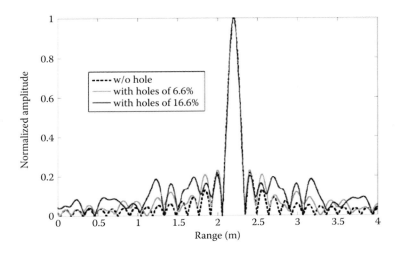

FIGURE 7.5
Range spectra with and without spectrum hole.

7.3 Detect-and-Avoid Technique

The UWB device operating in the 3.1–4.2 GHz band is required to implement a DAA (with narrowband signal detection and avoidance function) that allows it to detect an active wireless system operating in the same frequency range. The stepped-FM UWB sensor has inherently a DAA function unlike UWB-IR.

The flowchart of the suggested DAA algorithm is shown in Figure 7.6. Prior to transmitting the UWB radio, the receiver is activated (passive mode) and detects some narrowband

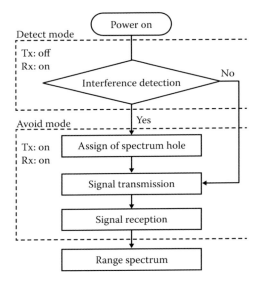

FIGURE 7.6
Flowchart of DAA algorithm.

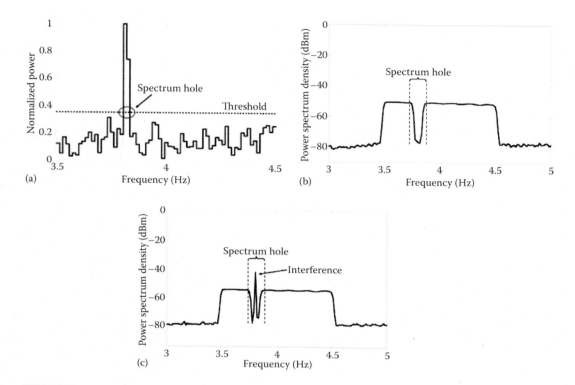

FIGURE 7.7
DAA technique: (a) phase detector output, (b) range spectrum of transmit signal, and (c) range spectrum of received signal.

radios such as communications and broadcast signals. Figure 7.7a shows the output of the phase detector where a narrowband radio of 3.84 to 3.85 GHz is assumed. In the measurement, the radio was generated by an Agilent E8254A signal generator. It is seen that the radio is detected corresponding to the step number of 23 and 24 where $\Delta f = 14.5$ MHz and $N = 70$. Based on the result, the transmitter is then activated. Figure 7.7b shows the range spectrum where seven stepped pulses from 3.76 to 3.85 GHz are not transmitted. Figure 7.7c shows the transmit signal together with the interference in the frequency domain. It is seen that the UWB radio can coexist with the radio.

7.4 Patient Care Sensing and Monitoring System

A patient care sensing and monitoring system has been developed using the stepped-FM UWB sensor. The measurements were conducted and the results are presented.

7.4.1 Sensing and Monitoring Algorithm

The requirement for monitoring the state of elderly persons in care facilities and hospitals is increasing year by year since the increase in accidents involving elderly persons is of

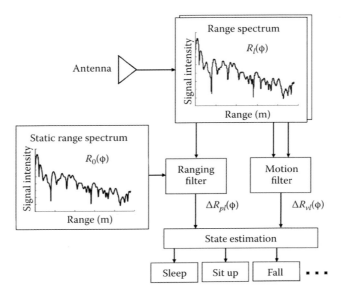

FIGURE 7.8
Flowchart of monitoring algorithm.

great concern. When an elderly person attempts to leave the bed alone, for example, it has been reported that fall accidents occur frequently. Thereby, it is important to monitor the state of the person. The flowchart is shown in Figure 7.8, where the sensor detects various states or state in a room. The received range spectrum consists of echoes from various obstructions such as the bed, person, and walls, where the state should be estimated by employing a "ranging filter" and a "motion filter."

The ranging filter $\Delta R_{pl}(\phi)$ is to detect the range from the sensor to a person, which is given by

$$\Delta R_{pl}(\phi) = |R_l(\phi) - R_0(\phi)|, \tag{7.6}$$

where $R_0(\phi)$ is a reference range spectrum which represents the range spectrum in a static room without a person and $R_l(\phi)$ is the range spectrum of the lth frame.

For example, when a person moves, the range spectrum would also fluctuate. Also, some motion in a room is expected to be detected by the motion filter $\Delta R_{vl}(\phi)$, which is given by

$$\Delta R_{vl}(\phi) = |R_l(\phi) - R_{l-1}(\phi)|. \tag{7.7}$$

Note that $R_l(\phi)$ and $R_{l-1}(\phi)$ represent the range spectra of the lth and $(l - 1)$th frames, respectively.

The motion filter is to detect some motion in a room, while the ranging filter is to estimate the range of a person. The trajectory of a moving person can also be estimated; thereby, the state should be estimated such as "walk in room" and "fall." It is also possible to observe his or her state in bed such as "tossing about in bed" and "sitting up in bed" by using the motion filter without invasion of privacy. Therefore, the state of a person that can be detected includes "out of room," "static," "walk in room," "sleep in bed," "tossing about in bed," "sitting up in bed," and "fall" in this algorithm.

7.4.2 Measurement Results

The measurements were conducted in a care room, shown in Figure 7.9, and the results are presented. The specification is shown in Table 7.1. The algorithm has been developed for six states of out of room ("out" for short), walk in room ("walk"), sleep in bed ("sleep"), tossing about in bed ("toss about"), sitting up in bed ("sit up"), and fall. Figure 7.10 represents an example of the measured results where the solid line is the sensor's estimated state, while

FIGURE 7.9
Scene for measurement environment.

TABLE 7.1

Measurement Specification

Type	Specification
Transmit power	−12 dBm
Frequency	3.5–4.5 GHz
Stepped bandwidth Δf	14.5 MHz
Number of steps N	70
Frame period	0.1 s
ADC device	70 kS/s
Antenna	Horn (9.88 dBi)

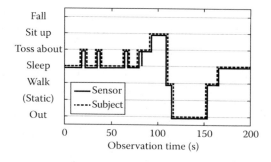

FIGURE 7.10
An example of measured results.

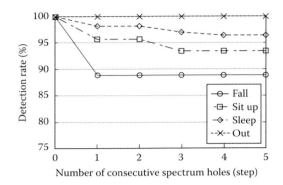

FIGURE 7.11
Detection rate as a function of spectrum hole number.

TABLE 7.2

Detection Rate

State	Detection Rate
Fall	100%
Sit up	100%
Sleep	100%
Out	100%
Toss about	98.80%
Walk	100%

the dashed line is the actual state identified by the video camera, as shown in Figure 7.9. The measurements were conducted for different subjects and the detection rate was investigated. Next, it is important to investigate the detection performance for a spectrum hole. The results of fall, sit up, sleep, and out are shown in Figure 7.11. Table 7.2 summarizes the results. A detection rate of more than 88% is seen to be attained for a spectrum hole of 10%. Note that the rate is approximately 100% without a spectrum hole.

7.5 Conclusions

The increase in accidents involving elderly patients is becoming a great concern and the requirement for monitoring their activity and state is especially increasing in care facilities. This chapter presents a bed state–monitoring sensor of elderly patients that uses a stepped-FM UWB scheme and the performance has been investigated for each spectrum by measurements in care facilities. The sensor provides the following advantages:

- Lower-speed ADC and lower-level processor: It transmits a series of bursts of narrowband pulses where each burst is a sequence consisting of many pulses shifted in frequency from pulse to pulse with a fixed frequency step. Each received narrowband pulse is phase detected and then combined with the large effective bandwidth (sequentially over many pulses). Therefore, the hardware requirement is less stringent relative to UWB-IR.

- Inherent DAA function: It is capable of coexisting with other narrowband wireless systems operating in an overlaid frequency band. It can be designed to have any spectrum hole which does not cause interference with other wireless systems and medical equipment. For example, the location of the spectrum hole can be adaptively assigned according to the interference band detected by the DAA.

From the results, the scheme has been found to be useful and can also coexist with other wireless systems operating in the overlaid frequency band. Also it has been shown that various states of the patient can be detected including sleeping in bed, sitting up in bed, fall, walk in room, static, and going out and in at the door.

References

1. I. Immoreev, S. Samkov, and T. Tao, "Short-distance ultra wideband radars," *IEEE Aerospace Electronic Syst. Magazine*, Vol. 20, No. 6, pp. 9–14, June 2005.
2. K. Ota, Y. Ota, M. Otsu, and A. Kajiwara, "Elderly-care motion sensor using UWB-IR," *Proc. IEEE Sensors Appl. Symp. 2011*, San Antonio, pp. 159–162, Feb. 2011.
3. J. Taylor, *Introduction to ultra-wideband radar systems*, CRC Press, Boca Raton, 1995.
4. J. Lansford, "UWB radio issues and opportunities," USC UWB Symposium, April 2006.
5. Electronic Communications Committee (ECC), ECC Report 120, "Technical requirements for UWB DAA (Detect and Avoid) devices to ensure the protection of radiolocation services in the bands 3.1–3.4 GHz and 8.5–9 GHz and BWA terminals in the band 3.4–4.2 GHz," Kristiansand, June 2008.
6. D. Wehner, *High resolution radar*, Second edition, pp. 197–237, Artech House, Boston, 1995.
7. M. Otsu, R. Nakamura, and A. Kajiwara, "Remote respiration monitoring sensor using stepped-FM," *Proc. IEEE Sensors Appl. Symp. 2011*, pp. 159–162, Feb. 2011.
8. A. Kajiwara and H. Yamaguchi, "Clutter suppression characteristics of stepped-FM radar with MUSIC algorithm," *Trans. IEICE* (in Japanese), Vol. J84-B, No. 10, pp. 1848–1856, Oct. 2001.
9. M. Otsu, R. Nakamura, and A. Kajiwara, "Elderly-care monitoring sensor using stepped-FM UWB Scheme," *Proc. IEEE Sensor Appl. Symp.* pp. 151–154, Feb. 2012.
10. Y. Ota, R. Nakamura, and A. Kajiwara, "UWB stepped-FM sensor for home security," *Proc. IEEE Sensor Appl. Symp.* pp. 126–129, Feb. 2012.
11. R. Nakamura and A. Kajiwara, "Empirical study of stepped-FM UWB microwave sensor," *Proc. IEEE Radio and Wireless Symposium (RWS 2010)*, pp. 363–366, Jan. 2011.

8

Mobile Health Sleep Technologies

Anda Baharav

CONTENTS

8.1 Introduction

8.1.1 Background about Sleep

Sleep is ubiquitous and vital. Roughly, humans spend about a third of the time sleeping, more during the neonatal period, childhood, and adolescence, and then around 8 h a day during adult life. The need remains quite constant along the adult life span; however, the ability to get enough sleep may deteriorate with time [1].

Although the functions of sleep are not completely understood, there is increasing scientific evidence indicating that sleep has a role in memory development and learning and in the normal function of the immune and cardiovascular systems, as well as in motor and cognitive performance. Abnormal and insufficient sleep have both an immediate and long-term impact on human health, well-being [2], and mood [3–4].

Sleep is a very complex, cyclical process depending on the balance between two processes: process S representing the sleep drive and process C originating in the circadian behavior of the internal clock. Thus, normally the sleep drive is minimal when waking up (alertness is high) and the sleepiness accumulates with the time of the day, facilitating the next night's sleep episode. Process C is ruling over the sleep–wake cycle, with a main sleep episode during nighttime or dark time, and being awake and active during daytime when there is light; the internal clock has a period a bit longer than 24 h; thus, without any external influence most human beings have a tendency to go to sleep a bit later each night and wake up a bit later the next morning. External cues, light–dark cycle, work demands, and food intake have a role in synchronizing the sleep–wake cycle to exactly 24 h. Normally, during the late evening–early night hours, the sleep drive is high, the biological clock pushes toward sleep. Additional factors interfere with sleep, mainly motivation and individual needs. Most human beings need around 8 h of sleep a day, yet they sleep less because in the modern society artificial light permits us to extend activities during dark hours. Motivation overcomes the sleep drive, up to a certain point; the result is chronic sleep deprivation, and the price is daytime sleepiness and reduced cognitive and physical performance. The consequences are serious health problems including cardiovascular morbidity, being overweight, and workplace and traffic accidents.

8.1.2 Sleep Problems and Their Implications

Falling asleep implicates a disconnection from the external world, a kind of cyclical unconsciousness, reversible and ruled by internal control systems. One has to be ready and relaxed in order to let go and fall asleep. When hyperarousal, anxiety, or stress is present, it interferes with the natural disconnection process and a complaint of difficulty falling asleep or maintaining sleep appears. The result is insomnia, a prevalent sleep disorder of poor and nonrefreshing sleep. Its consequences are similar to those of insufficient sleep.

Many sleep problems are the result of a misalignment between the internal clock and occupational or societal obligations. Sleep disorders related to the biological clock are known as circadian sleep disorders and include delayed sleep phase (characterized by difficulties falling asleep and waking up at normal hours, yet when the timings are late sleep is normal), irregular sleep, advanced sleep phase syndrome, shift work, and jet lag.

Insufficient sleep, insomnia, and circadian rhythm disorders have a physiological root with added behavioral and psychological components that lead to poor sleep, sleep loss, negative effect of daytime efficiency, and health deterioration.

There are also medical sleep disorders, such as obstructive sleep apnea (OSA) and additional sleep related breathing disorders and snoring, as well as other medical sleep problems including narcolepsy and hypersomnias of other origin and sleep disorders secondary to other medical problems, such as pain or neurological issues or heart failure, that influence sleep duration and quality.

Since sleep is a vital human function, sleep problems in general are important contributors to general health, quality of life, and daytime optimal performance.

Sleep problems are prevalent, yet they remain mostly not diagnosed and untreated. The reasons for this are the fact that sleep in general is the "hidden," inactive part of our life and is sometimes considered as a waste of time; regular sleep diagnosis is not comfortable and is mostly limited to one night only in settings that differ to various extents from the regular sleep conditions of a patient, and the costs are high.

The vast majority of sleep disorders, including circadian rhythm disorders, insomnia, and insufficient sleep, need a diagnostic procedure that includes multiple nights, making the regular diagnostic tools inefficient. Even obstructive sleep apnea can vary in its severity from night to night in direct relation with personal behavior such as sleep deprivation and alcohol consumption. Thus, proper diagnosis also demands sleep in the home environment and several nights of testing. All other sleep disorders require extensive physiological signals testing in a special laboratory that is most efficient as inpatient, attended setting. This sophisticated and expensive procedure is indicated in about 2% of all sleep disorders. The remaining situations may and should be diagnosed in the natural sleep environment, in own bed at home.

8.2 Sleep and Technology

8.2.1 The Role of Technology in General and Mobile Technology in Particular in Inducing Sleep Disorders

In the dawn of human history, sleep was confined to dark hours after sunset and until sunrise. Activity was limited to natural light time. Industrial development brought artificial light and the need for a 24 h active society with shift work, long transmeridian travel, and more recently media and mobile communication. Thus, within a matter of about a century and a half, the Western human society skipped from a rest–activity pattern depending on the natural environment, with activity and sleep hours rigidly divided between daytime light period and dark nights, to an "always-on" society. This is an abrupt behavioral change and has not been matched by an evolutionary change in sleep needs.

The Bedroom Poll 2013 by the American National Sleep Foundation [5] (http://sleepfoun dation.org/sites/default/files/RPT495a.pdf) reveals that the presence of electronic devices in the bedroom is pervasive. Importantly, these electronic devices are found in both the parents' bedrooms and the children's bedrooms: 89% of adults and 75% of children have at least one electronic device in their bedrooms; television was the most common (over 60% of parents and 45% of children had a TV set in their bedrooms). Moreover, multiple electronic devices in the bedroom are highly prevalent (68% of parents and 51% of children have two or more devices in their bedroom at night). Interestingly, smartphones and tablets also have a significant presence (45% in parents' bedrooms, 30% in children's), with mobile technology being left on at night in around a third of the bedrooms; about the same amount of adults and children leave the TV set on as well at night. Leaving electronics on at night can be particularly disruptive to sleep.

8.2.2 Why Mobile Interface Is Most Suitable for Sleep

Mobile technology has become ubiquitous, at least in the Western developed countries. Smartphones have extensive computing and memory capabilities and they became a personal assistant and companion. People use them not only for communication, text, or voice, but also as digital agendas, watches, and alarm clocks. They are widely used to track activities and health parameters, and their use for sleep became a natural outcome.

The challenge is how to use them to improve sleep instead of being an addictive sleep destructor.

Mobile technology can be used as a one-way communication channel allowing for regular medical sleep tests to be performed in the home environment. To achieve this, a patient gets a regular polygraph delivered to his home along with hook-up instructions, performs tests for a few nights in a row, and, using his smartphone, uploads recorded tests for scoring and interpretation. The diagnostic equipment is then returned and the test results can be delivered to the referring physician and to the patient by using either mobile or internet technology. Some examples for this medical sleep diagnostics are the following:

- *Verizon–NovaSom cooperation:* The AccuSom, NovaSom's second-generation home sleep test, is a smartphone-sized portable monitor system that tracks five physiological signals (respiration airflow, respiration effort, oxygen saturation, heart rate [HR], and snoring) necessary for sleep-related breathing disorders diagnosis, mainly OSA diagnosis. Once a physician submits a prescription for a home sleep test, the AccuSom is mailed directly to the patient to be self-administered in the comfort and privacy of the home environment. AccuSom uses voice prompts, which can be programmed in any language, to walk the patient through the entire testing procedure. Upon completion of the test, the sleep data are wirelessly transmitted via the Verizon network to NovaSom's online cloud-based sleep management platform, where they are accessed and analyzed by an interpreting physician. The AccuSom is capable of multinight testing, which is proven to identify more patients with positive OSA diagnosis than a one-night sleep lab test.

- *CleveMed* has a similar service using an eight-channel lightweight sleep test device, the *SleepView,* and a web portal to securely access and interpret patient sleep data.

- *NiteWatch by LifeWatch Inc.* also offers a sleep testing service as a physician-prescribed test for unattended sleep monitoring of patients with suspected obstructive sleep apnea in their home private bedroom. The NiteWatch sleep testing system utilizes a very user-friendly kit (available channels include respiration effort, oxygen saturation, electrocardiogram, heart rate, snoring, airflow, and body position).

The mobile technology allows for short patient wait times and fast and comprehensive reports for the physician.

Mobile technology can also be used as an aid to self-sleep evaluation, as an initial step in understanding one's own sleep needs, pattern, quality, and structure in connection with subjective complaints regarding difficulties related to sleep and daytime performance, fatigue, or sleepiness.

8.3 Methods for Evaluating Sleep

From the dawn of history, sleep represented a reason for concern and caused philosophers to think about it being a mysterious necessity, poets to write about it fulfilling essential emotions, and doctors to enquire about bodily and mind changes that occur during sleep.

Hippocrates wrote, "With regard to sleep—as is usual with us in health, the patient should wake during the day and sleep during the night. If this rule be anywise altered it is so far worse: but there will be little harm provided he sleep in the morning for the third part of the day; such sleep as takes place after this time is more unfavorable; but the worst of all is to get no sleep either night or day; for it follows from this symptom that the insomnolency is connected with sorrow and pains, or that he is about to become delirious" [6]. Sleep is a vital function, since without sleep, like without nutrients, water, or air, life cannot continue for a long time.

Many imaginative theories about sleep appeared and phased out; yet modern sleep science was born when observation and analysis of the brain's electrical activity became available noninvasively for researchers, during the fourth decade of the last century.

Humans need sleep, enjoy healthy sleep, and are worried when they are not able to fall asleep when they intend to, when they are not able to return to sound sleep after some natural sleep interruption, and when they wake up too early, or the opposite, when they are not able to wake up at a planned time needed for their daily activities. They perceive intuitively that sleep was good or nonrefreshing, yet they are not able to correlate between their subjective sleep evaluation and objective measures. Falling asleep is a process of gradual disconnection, both motor and sensory, from the environment, in a way a loss of consciousness that, unlike coma, is cyclical and reversible by itself. The very moment of falling asleep (sleep onset) can be defined in different ways, depending on the way it is measured. Thus, if during the very early first instants of disconnection from the environment, a human subject is asked to report about his/her state, the spontaneous answer to the question, "are you sleeping?" would be, "certainly not"; at this very moment an outside observer would say, "this person is not moving, eyes are closed, respiration is very regular, and he/she does not seem to perceive we are here or react to noise, or light, or other stimuli—this person is asleep." A measurement of the muscle activity would report a lower level than during wakefulness; a measurement of eye movements would report slow, rolling eye wandering; the brain activity would be of mixed frequency, low amplitude, and nonsynchronized, with some very specific patterns; heart rate will be quite variable, and so will be respiration. Different organs and systems reset their level of activity while falling asleep. Different physiological functions gradually or abruptly change their level of activity sometime during the transition from wakefulness to sleep. However, sleep onset has an arbitrary definition based on three physiological variables, namely, (i) electrical activity of the brain based on surface measurements at predefined locations on the scalp, (ii) eye movements, and (iii) muscle tone derived from surface electrodes on the chin muscles. There is no proof that all physiological phenomena that accompany sleep onset coincide with this arbitrarily defined event. Thus, we know that movements stop earlier, whereas conscious disconnection from the environment occurs a bit later generally. Thus, when all these complex factors are taken into account, it is understandable that different scales for sleep measures are available, and the outcomes depend on the specific metric used.

8.3.1 Subjective Information and Questionnaires

Any sleep evaluation begins with acquiring basic verbal/written information from a person regarding personal health, sleep habits and beliefs, mood, sleep environment, individual satisfaction with sleep and daytime function, fatigue and sleepiness, circadian preferences, and more. Population-validated questionnaires regarding all these variables are usually administered during individual visits with sleep or healthcare personnel. Obviously, individual administration of these questionnaires is time consuming, requires special appointments and expert attention, and is followed by additional workup, as dictated by the results obtained when manual analysis of this general information is performed. Automatic analysis of most of the above data is easy to perform, and based on this, further individual evaluation can be prescribed by a physician, or ways to deal with sleep issues may be suggested. Mobile technology suits administration of questionnaire-based sleep evaluation and some applications are available (Sleepio and SleepBot). These applications are based on extensive expert knowledge that has been validated; some involve payment, and some are free. Many of the sleep problems can be evaluated this way, but some cannot. Obviously, solutions for the uncovered sleep issues are needed, and some are offered (Sleepio), while others are based on self-help (SleepBot). Large-scale usage or prospective validation is needed to acknowledge usability and effectiveness of these available tools.

8.3.2 Diaries

Sleep diaries have served as a vital aid for sleep experts when they needed to evaluate and treat their poor sleeper patients. Diaries can be easily digitized and used as a self-measuring tool. Mobile technology can allow continuous and easy to manage automatic sleep diaries (SleepBot). There is no doubt that keeping a diary is a first step in dealing with a problem. Nutrition or fitness diaries help many in dealing with dieting and weight loss plans, as well as with keeping a fitness program, or preparing for a bike or running event. Mobile technology offers many such opportunities (Fitbit and Jawbone).

8.3.3 Gold-Standard Polysomnography

Sleep medicine is a relatively young branch of medicine, initiated during the second half of the last century. It implicates understanding of the mutual influences between wakefulness and sleep, normal sleep for healthy durations on a regular basis, and good health and normal physical and cognitive performance and mood while awake. Gold-standard sleep diagnostic studies require a number of physiological variables to be acquired before, during, and after the main sleep episode. Those include one to three electrical brain activity channels, two eye movement channels, and one to three muscle activity channels, in addition to electrocardiogram, oxygen saturation, two respiratory effort channels, one to two airflow channels, body position, and snoring. Due to this multitude of signals, the test process is not very comfortable for the patient, is work consuming, and requires sleep in an attended laboratory setting. Obviously, the procedure is expensive and limited mostly to a single night in a sleep laboratory. The scoring of the results is based on standard, yet arbitrary rules deriving from the history of the technology of analog polygraphs, which used paper as a monitoring and storage device. No reliable automatic analysis device to score sleep structure and quality has been developed and validated. Gold-standard sleep studies are manually scored by trained personnel according to arbitrary set rules [7–8]. Interscorer reliability is limited to 75%–80% [9]. An additional limitation of the procedure derives from the fact that it is restricted to one night, in a not natural and known sleep environment; moreover, a single not typical sleep night does not accurately represent a regular night's sleep in familiar bed. There is extensive evidence that sleep differs from night to night and depends on the sleep environment.

8.3.4 Electroencephalography

Electroencephalography (EEG) signals represent the electrical brain activity and can be measured with scalp electrodes that are based on conductive technology. The signal-to-noise ratio is limited and a good signal usually requires special electrodes and an electrical contact substance to be used and thorough application. The use of most diagnostic sleep EEG devices is limited by the need to have a trained technician to apply the electrodes in order to obtain a good quality signal. Theoretical background on supporting the use of dry electrodes to monitor EEG exists. In particular, a consumer device (Zeo) based on this technology was on the market for a number of years. The last version had mobile sleep-monitoring capabilities and included sleep structure evaluation and some generic comparison between measured sleep time and structure and normal values. Zeo was pioneering a mobile sleep technology, giving consumers the opportunity to get objective insight into their own sleep across multiple nights. Sadly, the company closed its activities due to economic reasons. The limitations of Zeo resided in the lack of validation of the device based on dry EEG electrodes. Moreover, Zeo supplied the user plenty of exciting information, but offered no detailed personal advice on how to proceed in order to solve some personal sleep issues. An additional limitation was that while Zeo had a cool wearable device, this device was not very comfortable to wear while awake or asleep.

8.3.5 Heart-Rate Variability

The possibility of performing home sleep diagnostics without sacrificing clinically relevant information regarding sleep efficiency and sleep architecture means that one can reach a larger proportion of undiagnosed sufferers of sleep apnea. Sleep apnea is estimated to affect 4%–9% of the population, yet 75% remain undiagnosed and untreated [2]. A multitude of physiological parameters can be measured during sleep and wakefulness, and some of them are used to allow standard description of different sleep–wake states. As mentioned above, the gold-standard sleep disorder diagnostic method is overnight polysomnography, which requires recordings of electroencephalogram, electromyogram, electrooculogram, respiratory effort, and oronasal airflow, as well as electrocardiogram, oxygen saturation, limb movements, and additional variables in an attended setting. This testing process is uncomfortable, and since it requires manual scoring, it is costly. There is a great need for patient-friendly, accurate, cost-effective home testing.

The physiological interconnections between the central nervous system and the autonomic nervous system, specifically autonomic cardiovascular control (at the sinus node level), allow the uncovering of information concerning sleep architecture based on noninvasive analysis of the heart-rate variability (HRV) as detected from the electrocardiogram [10]. Moreover, morphologic changes in the cardiac electrical complex occur during respiratory cycles and body position shifts. The subtle changes in the electrical signal can now be quantified to allow obtaining a sound respiratory signal and body position information. Power spectral analysis of instantaneous heart-rate fluctuations reveals three components, high frequency, low frequency, and very low frequency, which are correlates of autonomic nervous system function. Novel techniques of time-frequency decomposition of these fluctuations allow quantitative evaluation of transient physiological phenomena as they occur during sleep or wakefulness. These components display differential profiles in the different sleep stages, permitting classification of sleep stages from the electrocardiogram.

Based on these simple and well-known facts, HypnoCore–SleepRate has developed a sophisticated algorithm which permits conducting a sleep study based on a single electro-cardiography (ECG) channel that can be automatically scored to obtain information on sleep architecture and efficiency, arousals, autonomic nervous function, and respiratory function during sleep. Thus, the HC1000P algorithm provides a powerful new tool for the diagnosis of sleep disorders in general and sleep apnea in particular [11]. The performance of this new software has been clinically validated, showing that it is ready for practical use. The new procedure for sleep testing in the home environment is easy to implement at low cost. Thus, mass diagnosis of frequent sleep disorders such as insomnia and sleep apnea may be easily performed to allow suitable treatment. Since access to ECG devices is limited and expensive, a consumer version of the software has been developed to allow sleep evaluation based on instantaneous accurate heart-rate signal provided by off-the-shelf heart-rate monitors, such as popular fitness belts (Polar, Suunto, Wahoo, and more). Thus, the technology becomes sensor agnostic and can be readily provided via mobile devices. The advantages are low-cost sleep evaluation with an off-the-shelf familiar device and the possibility of evaluating many nights in a row in the home environment. There are some limitations related to battery life, inconvenience of wearing a fitness belt at night time, and sometime connectivity with the mobile device. The advantages are that any new instantaneous HR monitors, including contactless ones, can be readily integrated into the system. In addition, the same technology can be used for both daytime performance and stress and nighttime sleep and stress.

8.3.6 Movement Actigraphy

A motion-based method of analyzing sleep has been widely used to assess sleep in children and to evaluate circadian rhythm disorders [12]. The first devices were wearable, watch-like, wrist worn, and validated for sleep–wake discrimination; and some similar devices are used now in medical sleep evaluation (Actiwatch by Mini-Mitter–Respironics and Actigraph by Ambulatory Monitoring). The main limitations of those devices is an overestimation of sleep time and underestimation of sleep onset than either subjectively reported by insomniacs or measured by gold-standard polysomnography. Although there are some actigraphs equipped with software that offer sleep structure (deep sleep, light sleep, and even rapid eye movement [REM] sleep by some), there is no real physiological basis for this and there is no validated evidence that sleep stages can be detected by motion loggers' analysis. The technique has been adopted by companies that support self-quantification, including fitness and sleep (Fitbit and Jawbone). Smartphones have built-in gyrometers to allow flipping their screens when the phone changes its position. These built-in technologies can be used as a motion-based sleep–wake detector when the smartphone is placed on the bed during sleep time. This, however, implies sleeping with a device in very close proximity to the body and has limited validity: although it may detect movement/awakenings, it cannot detect sleep structure. Many apps are available (Sleep Cycle, MotionX, and more) supporting this feature.

8.3.7 Behavioral: Audio–Video Monitoring

Noises and light in the bedroom can disturb sleep and they may be easily detected using the microphone and light meter or cameras of smartphones. Snore sound is one of the signals usually acquired in a sleep laboratory when a sleep-related breathing disorder is suspected. There is evidence that when a microphone is calibrated and placed at a specified distance from a snorer, the sound analysis can be used to detect OSA. Snore detection is feasible with a smartphone, and there are applications that detect and measure snoring in the bedroom (SnoreLab, Quit Snoring, and SleepRate). Note that it is difficult to identify the source of the snoring when there is a bed partner or a pet sharing the bedroom.

Respiratory pattern changes from wakefulness to almost wakefulness and during the different stages of sleep. Thus, at least theoretically, breathing sound analysis using smartphones can be used to evaluate sleep and sleep structure.

8.4 Adding Treatment

Sleep evaluation and sleep education, including tips regarding healthy sleep practice and sleep hygiene, constitute the first step in dealing with a sleep issue, be it voluntary-motivational sleep restriction, poor sleep related to environmental or occupational or family and social factors (sleeping in the proximity of a train station or an airport, shift work, frequent transmeridian travel, a new baby in the family, or extensive use of media and social media), insomnia, sleep wake timing that is not in line with professional or social duties, or an intrinsic sleep disorder (such as OSA or periodic limb movement of sleep). Some people can use the evaluation results and the education received in order to try and solve some of the problems uncovered when using a sleep-tracking mobile application. By continuing tracking, they can understand the effect of their actions on their sleep and adjusting further their sleep-related behavior to get better sleep. This represents, in a way, a self-contained biofeedback process mediated by mobile technology.

In a next step, mobile technology can integrate objectively measured variables regarding sleep, such as movements, short awakenings, sleep depth, or relaxation/stress, with subjective information such as desired wake-up time. This integration may allow a smart alarm to wake up a person at the physiologically most suitable moment around the desired wake-up time. Many sleep trackers on the market have this feature, yet they operate on different basic objective information; thus, the wake-up time depends on the physiological information obtained (movement, heart-rate fluctuations, EEG, or a combination of those).

Effective treatments with long-lasting effects for insomnia are based on cognitive behavioral therapy for insomnia (CBTI) methods that integrate intake by a trained therapist including basic questionnaires, sleep diaries, and sometimes actigraphy, and continue with sleep education and behavioral changes to improve sleep. These are accompanied by cognitive tasks, including relaxation. A course of CBTI requires four to six therapy sessions over 4–6 weeks. The availability of trained therapists is limited and the costs are high, leaving most insomniacs without an effective solution for their burning problem. The need for cost-effective sleep improvement programs led to computerized CBTI administered via the web (Sleepio), which proved to be quite efficient.

In spite of the fact that many sleep trackers based on mobile technology are made available, the need to provide dedicated personalized individual therapy for insomnia and other causes of poor sleep has become apparent only lately. Sleep tracking allows one to get some objective information regarding sleep quality and duration and sleep habits. This knowledge, in correlation with people's complaints, provides a good basis for cognitive behavioral therapy that enables adaptation of existing CBTI methods and protocols to a mobile platform. There are many advantages: (1) changes and instructions can be administered more frequently than with regular therapist-led CBTI; (2) reminders and notifications inherent to mobile technology can improve compliance with treatment; (3) costs are low; (4) the treatment can be reused from time to time to prevent relapse; and (5) objective evaluation of sleep patterns and sleep improvement or deterioration in large populations becomes possible.

Mobile technology can further allow screening for sleep disorders such as OSA. The screening procedures may use available signals (heart rate, respiratory movement or sound, oxygen saturation, and snore analysis). Screening procedures would allow detecting many who are undiagnosed (75% of OSA cases remain undiagnosed) and offering treatment. Effective treatment has been proven to improve health, decrease morbidity and mortality related to OSA, lower healthcare costs, and improve quality of life.

8.5 Players on the Market

The need to evaluate sleep in its natural environment, the epidemic of sleep deprivation that hits modern society, the high prevalence of sleep problems, and the significant consequences on personal health, performance, and well-being all call for action. New technologies derived from the medical sleep procedures and the availability of mobile technologies facilitate implementation of sleep evaluation for the end user, the consumer who shops for understanding, evaluating, and solving issues connected with personal health and well-being. There are many players out there. Some of the most significant are listed in Table 8.1.

Actigraphic sleep evaluation overestimates sleep time and mainly underestimates sleep latency [13]; thus, its usefulness for insomnia may be limited and this fact should be taken into consideration. This method works well for sleep–wake circadian issues, such as delayed sleep or advanced sleep, without symptoms of insomnia. Methods based on EEG and HR variability seem to work very well for sleep evaluation of normal and of poor sleep, including insomnia.

The only EEG-based mobile sleep tracker for the consumer was Zeo and it is no longer available. It is not comfortable to wear at night and does not have a treatment program for improving sleep.

Accurate instantaneous HR is needed in order to assess sleep and sleep structure and fragmentation as well as stress at night. Presently all wrist-worn HR monitors do not provide instantaneous HR signal, but one hopes that this limitation will be overcome soon to be one of the available improved technologies. An example of graphic nightly data processed by SleepRate appears in Figure 8.1.

Sleep improvement treatment is available only in a couple of applications (Sleepio and SleepRate) based on validated CBTI protocols which have been modified to fit automated administration using mobile technology. SleepRate integrates also snoring and environmental noises along with objective sleep evaluation based on HR variability analysis. HRV allows also assessment of sleep fragmentation, which correlates well with OSA. This may help those whose insomnia is aggravated by the coexistence of OSA to get diagnosed and treated for both conditions: OSA in a medical dedicated setting and insomnia using the personalized sleep improvement plan based on CBTI. Figure 8.2 represents baseline data obtained for a few nights in order to generate a personal sleep assessment and a custom-made, individual sleep improvement plan based on CBTI protocols.

TABLE 8.1

Mobile Sleep

Product	Device	Signal Analyzed	Results	Sleep Interventions	Mobile System	Link
MyZeo	Headband	EEG	Detailed sleep structure	Smart alarm, sleep education	iOS	http://www.myzeo.com/
Jawbone	Wristband	Motion	Daytime activity, food diary, sleep quality, structure	Smart alarm	iOS, Android	https://jawbone.com/up
BodyMedia	Armband	Motion, ECG, HR	Weight, activity, sleep monitoring	None	None	http://www.bodymedia.com/ (acquired by Jawbone)
Basis	Wristwatch	Motion, HR derived from PPG, temperature and perspiration	Activity, sleep	None	Android, iOS	http://www.mybasis.com/ (acquired by Intel)
Lark	Armband	Motion	Sleep	Smart alarm	iOS	http://lark.com/
Fitbit	Wristband, clip	Motion	Activity, nutrition, sleep	None	iOS, Android	http://www.fitbit.com/
SleepRate	Instantaneous HR, sensor agnostic, Polar H7	HR, sound	Sleep structure, quality, stress, snoring, environmental noise, awakenings	Personalized CBTI, relaxation, smart alarm, send to expert if suspected sleep disorder	iOS	http://www.sleeprate.com/
Sleep Cycle	Smartphone	Motion	Sleep structure	Smart alarm	iOS, Android	http://www.sleepcycle.com/
Sleepio	None	Questionnaires, diaries	Sleep quality, subjective	CBTI	Desktop, iOS	http://www.sleepio.com/

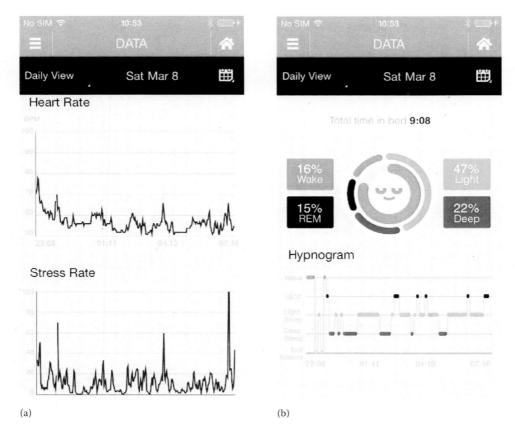

(a)   (b)

FIGURE 8.1

Example of a nightly graphic report by SleepRate. (a) *Top:* Heart rate as a function of time during the night. The decrease in mean HR at the beginning of the night is followed by subsequent fluctuations at much lower levels. The few peaks occur with awakening at night. *Bottom:* Stress at night calculated based on the time-frequency decomposition of the instantaneous HR fluctuations. Peaks may be related to awakenings. (b) Sleep macrostructure at night: the percentage of time spent in each sleep stage and hypnogram representing the way sleep stages change across the night's sleep.

8.6 Advantages

The high prevalence of poor or insufficient sleep makes the phenomenon perceived as normative, and in spite of documented correlation between sleep and health, performance, cognition, memory, weight gain, and mood, people do not complain much about sleep issues. Many get sleeping pills on a regular or intermittent basis, and others self-medicate; some try CBTI, but most suffer quietly. The availability of sleep assessment and personal treatment in the natural sleep environment and at a reasonable cost offers the opportunity to achieve an overall better life.

Sleep data including sleep habits, times, structure, and duration from large populations allow us to reach a better understanding of existing problems and improve treatment protocols for the benefit of individuals and society overall.

(a) (b)

FIGURE 8.2

Example of some of the sleep data during a week of evaluation that make possible a detailed individualized sleep improvement plan. (a) Measured sleep efficiency (the percentage of time asleep out of the total time in bed during the night) and the subjective sleep rating; (b) bedtime and wake-up time.

8.7 Next Steps

Insufficient sleep and insomnia have been identified as contributing to significant cardio-vascular morbidity. The epidemic of obesity, diabetes, and metabolic syndrome may be attributed, at least partially, to sleep loss. Traffic accidents due to sleepiness are directly connected to sleep deprivation. OSA has been pointed out as causally related to hyper-tension. Depression gets better after insomnia is successfully treated with CBTI. Sleep disorders drive healthcare costs up. These findings represent a call to action: make sleep evaluation and improvement a mission of modern society. A first step is happening now: spread the news to educate the population. The next one is at its dawn: offer affordable solutions to detect and improve sleep. Mobile technology has great potential to serve an important role in the fulfillment of this mission.

Standard metrics and suitable treatments should be developed and validated. As sleep evaluation and sleep treatment for sleep problems becomes a readily available commodity, there is a need to secure quality and privacy standards, validate procedures, and decide where and when regulatory rules are needed.

The smartphone has already become an indispensable personal companion. We should learn to use it efficiently and shut off incoming communication of any kind when we intend to sleep and stop sending messages at the same time. Thus, we allow the natural disconnection from the environment that accompanies sleep to occur. In a further step, we should use our smartphones to measure our sleep and sleep environment features, and when we wake up renew the connection and get our sleep metrics and some indications on how to behave or what to modify in order to sleep better. We are now reconnected and our personal companion, Dr. Smartphone, helps us adhere to the behavioral plans.

Abbreviations

CBT   cognitive behavioral therapy
CBTI   cognitive behavioral therapy for insomnia
HR   heart rate
HRV   heart-rate variability
OSA   obstructive sleep apnea

References

1. Dijk, D.-J., Sleep of Aging Women and Men: Back to Basics, editorial, *Sleep* 29, pp. 12–13, 2006.
2. Laurel, F., Young, T., Palta, M., and Fryback, D.G., Sleep-disordered breathing and self-reported general health status in the Wisconsin Sleep Cohort Study, *Sleep* 29, pp. 701–706, 2006.
3. Dinges, D.F., Pack, F., Williams, K., Gillen, A.K., Powell, W.J. et al., Cumulative sleepiness, mood disturbance, and psychomotor vigilance decrements during a week of sleep restricted to 4–5 hours per night, *Sleep* 20, pp. 267–277, 1997.
4. Neckelmann, D., Mykletun, A., and Dahl, A.A., Chronic insomnia as a risk factor for developing anxiety and depression, *Sleep*, 30, pp. 873–880, 2007.
5. Poll 2013 by the American National Sleep Foundation [http://sleepfoundation.org/sites/default/files/RPT495a.pdf] accessed on January 22, 2015.
6. Adams, C.D., editor, *The Book of Prognostics*, in *The Genuine Works of Hippocrates*, New York, Dover, 1868.
7. Reschaffen, A., and Kales, A., editors, *A Manual of Standardized Terminology, Techniques and Scoring System for Sleep Stages of Human Subjects*. Los Angeles, Brain Information System, 1968.
8. The AASM Manual for the Scoring of Sleep and Associated Events: Rules, Terminology and Technical Specifications, American Academy of Sleep Medicine, Westchester, 2007.
9. Danker-Hopfe, H., Kuntz, D., Gruber, G., Klosch, G., Lorenzo, J.L. et al., Interrater reliability between scorers from eight European sleep laboratories in subjects with different sleep disorders, *J. Sleep Res.* 13, pp. 63–69, 2004.
10. Baharav, A., Kotagal, S., Gibbons, V., Rubin, B.K., Pratt, G. et al., Fluctuations in autonomic nervous activity during sleep displayed by power spectrum analysis of heart rate variability, *Neurology* 45, pp. 1183–1187, 1995.
11. Decker, J.M., Eyal, S., Shinar, Z., Fuxman, Y., Cahan, C. et al., Validation of ECG-derived sleep architecture and ventilation in sleep apnea and chronic fatigue syndrome, *Sleep Breath* 14, pp. 233–239, 2010.
12. Morgenthaler, T., Alessi, C., Friedman, L., Owens, J., Kapur, V. et al., Practice parameters for the use of actigraphy in the assessment of sleep and sleep disorders: An update for 2007 standards of practice committee; American Academy of Sleep Medicine, *Sleep* 30, pp. 519–529, 2007.
13. Tryon, W.W., Issues of validity in actigraphic sleep assessment, *Sleep* 27, pp. 158–165, 2004.

9

Cardiovascular Disease Management via Electronic Health

Aimilia Gastounioti, Spyretta Golemati, Ioannis Andreadis,
Vasileios Kolias, and Konstantina S. Nikita

CONTENTS

9.1 Introduction

Cardiovascular disease (CVD) is a group of disorders of the heart and blood vessels leading often to serious events, including heart attacks and strokes. Heart attacks and strokes are usually acute events and are mainly caused by a blockage that prevents blood from flowing to the heart or brain. The most common reason for this is atherosclerosis, i.e., a buildup of fatty deposits on the inner walls of the blood vessels that supply the heart or brain. CVD remains the number one cause of death globally [1] and is projected to remain the single leading cause of death [2]. If appropriate action is not taken, by 2030, an estimated 23.3 million people will die from CVD every year, mainly from heart attacks and strokes. In addition to this, CVD bears a substantial financial burden, especially in low- and middle-income countries. Over the period 2011–2025, the cumulative lost output in low- and middle-income countries associated with CVD is estimated to be US$3.76 trillion [3].

To address the high morbidity and mortality rates of CVD, efficient management of cases of variable severity is crucial [4]. This includes (a) primary CVD prevention, which focuses on reducing the population risk factor burden, toward reducing the overall incidence of disease in the community; (b) early and valid identification of subjects at increased risk, for whom timely treatment is important; and (c) appropriate treatment of acute events. Efficient management of all these three phases of care is challenged by a number of factors, including (1) lack of multisectoral action to support reduction of behavioral risk factors and their determinants, (2) weak public health and healthcare system capacity for forging an accelerated national response, and (3) inefficient use of limited resources [5].

The enhanced use of technological advancements in the last decades has greatly aided all phases of CVD care and is even more promising for further improving them. Technological advancements have initially allowed the development and clinical use of sophisticated devices, able to measure valuable physiological parameters [6] and to image anatomical structures [7] in real time using different physical properties. The measurements obtained by such devices are particularly informative of the normal (or abnormal) anatomy and function of the interrogated organs and, as a result, have greatly facilitated clinical decision making.

Going one step further, the efficient exploitation of the huge amount of the derived information remains a great challenge, which can be addressed by the use of digital technologies. Initially, digital technology has been used to archive, store, and manage patient data through electronic health records (EHRs). The recent rapid adoption of EHRs has the potential to improve the management of CVD by removing variability and assuring at least consideration of guideline-recommended care and appropriate use criteria, thus leading to improved cardiac outcomes at all phases of care [8]. An additional advantage of EHRs is that they allow patient involvement in the management of their own health [9].

More advanced applications of digital technology include (a) computer-aided diagnosis (CAD) [10–11] and (b) transfer of patient data through various communication routes [12]. CAD has been shown to influence clinical decision making and, more importantly, to improve the decision quality, by enabling the physician to take a systematic approach toward diagnosis and treatment. In addition to this, wireless communication can be applied in a multitude of innovative ways to improve the speed and efficiency of patient care. To maximize the value of digital technology to clinical and public health, semantic interoperability, i.e., the ability of a given system to operate efficiently with a set of data originating from a different system, is necessary [13].

The purpose of this chapter is to provide an overview of the state-of-the-art digital health, or eHealth, technologies in terms of management of CVD. To this end, in the sections that follow, basic principles and recent advances in the areas of CAD and eHealth technologies are outlined, including telehealth and mobile and web-based applications. The use of the semantic web and related ontologies, toward improved interoperability, is also discussed.

9.2 Computer-Aided Diagnosis

CAD is a broad concept that integrates signal processing, artificial intelligence, and statistics into computerized techniques that assist health professionals in their decision-making processes. CAD methodologies seek to automatically extract objective and quantitative information from medical signals or images and perform as a supplement tool to assist physicians in their diagnostic task. Their role is mainly focused on increasing the knowledge of normal and diseased anatomy, in order to contribute to the early detection and diagnosis of various diseases and improve physicians' efficiency [14].

There seems to exist a great variety of CAD systems reported in the literature for several diseases. All these systems are based on the automatic computation and classification of features extracted from images or other biological signals. For example, CAD systems have become a part of the routine clinical work for detection and diagnosis of breast cancer or lung cancer on radiographic images at many screening sites and hospitals in the United States [15]. Based on the analysis of mammograms and chest radiographs, CAD systems have been proposed in the past years with the aim to support radiologists in the discrimination of benign and malignant lesions and provide a reliable second opinion for the subsequent patient management [16]. Moreover, various CAD methodologies analyzing different types of images have been proposed for the early detection of Alzheimer's disease [17]. Similarly, numerous approaches based on signal processing of voice recordings have been introduced to diagnose speech disorders and Parkinson's disease [18].

9.2.1 Analysis of Cardiovascular Signals and Images

Early detection of heart diseases can prolong life and enhance the quality of living through appropriate treatment. Therefore, for several years, the development of CAD methodologies for CVD has received great attention from the biomedical engineering community [15]. The analysis of the ECG signal is located at the center of the specific growth. ECG is a biological signal of vital importance, since it provides cardiologists with useful information about the rhythm and functioning of the heart, showing the electrical impulses over time. The state of the heart is generally reflected in the shape of ECG waveform and heart rate. A typical structure of the ECG signal is shown in Figure 9.1.

There are several aspects of the ECG trace, which are particularly helpful in diagnosing anomalies. Indeed, the QRS complex is the most crucial step in automatic electrocardiogram analysis, such as arrhythmia detection and classification of ECG diagnosis and heart-rate variability studies. Various factors such as the morphology and regularity of the signal, the heart rate, the existence of wave segments, and relative amplitudes contain important information on the pumping action of various chambers of the heart. Therefore, a CAD system oriented toward ECG analysis is able to evaluate these specific parameters and provide objective recommendation toward the identification of the abnormality.

Apart from the ECG signal, cardiovascular imaging may be recommended to support a diagnosis of a heart condition. There exist several different imaging techniques such as echocardiography, computed tomography, angiography, and MRI. Cardiovascular imaging is followed to obtain images of the beating heart and get insights of its structure and function. The specific images help the physicians either to diagnose anomalies or to determine

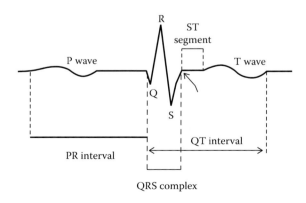

FIGURE 9.1
Diagrammatic representation of the basic electrocardiographic deflection. (From Ingole M et al., *International Journal of Engineering Sciences & Research Technology* 2014; 3(1):39–44.)

the treatment plans. Through the CAD inspection of such medical images, useful information may be obtained to identify patients susceptible to heart abnormalities. Proper image-analysis methods are required by the CAD system, in order to provide a reliable second opinion to the physician and enhance consequently the diagnostic process. The morphology of the myocardium and its texture and the existence of a scarred myocardial tissue after infarction and its relative location inside the myocardium are factors of high importance for abnormalities, such as arrhythmias, and play a valuable role in the CAD pipeline for CVD.

Vascular imaging also holds a prominent position in CAD for atherosclerosis, with different imaging modalities being most suitable for different vessels [7]: computed tomography for evaluating atherosclerosis in the coronary artery, MRI for detecting abnormal vessel boundaries and structural components, intravascular ultrasound for assessing ambiguous lesions in the left main coronary artery and at bifurcation sites, and duplex ultrasound for assessing risk in carotid atherosclerosis. In particular, the development of CAD for atherosclerotic carotid arteries by using ultrasound image analysis has been recently placed among the grand challenges in the field of life sciences and medicine, which has motivated related research on upgrading the potential of this low-cost imaging examination into a powerful tool for objective and personalized clinical assessment of patients with carotid atherosclerosis [11].

9.2.2 Generating a Diagnostic Decision

An image-based CAD flowchart is usually divided into image preprocessing, selection of the region of interest (ROI), feature extraction, feature selection, and classification (Figure 9.2). Similarly, ECG analysis includes data acquisition, preprocessing, feature extraction and selection, and finally classification. After the signal's acquisition, the denoising of the initial signal is usually a prerequisite step in order to smooth the signal. The pretreatment of the ECG is usually completed based on proper application of high-pass and low-pass filters, eliminating that way the noise and achieving precise waveforms so as to detect easily the signal's peaks. After applying the preprocessing, the CAD system has to extract objective features through the analysis of the obtained ECG waveform.

From a review of literature, there seem to exist various feature-extraction methods for the analysis of the ECG signal [20]. These methods mainly include time-domain methods,

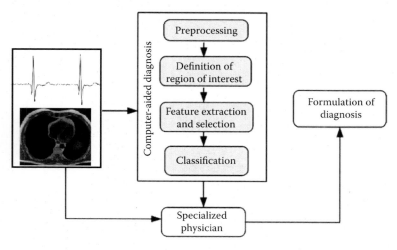

FIGURE 9.2
Schematic diagram illustrating the CAD flowchart.

frequency-domain methods, time-frequency domain analysis, wavelet transform's application, statistical representation, Lyapunov exponents, and Hermite coefficients. Because the ECG is a continuous time signal, various features may be computed, such as temporal features (distances between fiducial points), amplitude features (peaks), and angle features (formed within the waveform). Time intervals have also been considered to extract corresponding features, while high-order statistics (cumulants of second, third, and fourth orders) have been evaluated to extract statistical features of the QRS complexes of ECG. Transforming the signal into the frequency domain may also provide valuable information for its analysis. The Fourier transform has proven to be a useful tool in observing the changes in QRS complex, while wavelet transform has also been considered, as it represents the signal in different translations and scales. The coefficients obtained from the decomposition process are considered the filtered signal in the sub-bands and may provide features that can efficiently represent the characteristics of the original signal in different details.

As in the case of the ECG signal, there are common trends concerning the feature-extraction phase when images are considered as input in the CAD framework. The corresponding images have to be automatically analyzed in order to extract information on cardiac function and morphology. For example, in cases of patients having suffered from myocardial infarction, cardiac magnetic resonance (CMR) images may be used to identify patients susceptible to fatal arrhythmias. Several feature-extraction techniques, including shape and texture analysis, can be applied on the images in order to extract significant features for the identification of arrhythmia. Shape analysis has been proposed in order to estimate the size and the morphology of the scar [21], while texture analysis (first-order statistics and co-occurrence matrices) have been investigated to quantify the spatial distribution pattern of gray levels in both scarred and nonscarred areas of the myocardium [22]. Consequently, it has been shown that the use of texture-analysis techniques plays a crucial role toward the classification of patients at high or low risk of arrhythmia. This brief overview of the literature reveals clearly that research in the field of automatic ECG classification has reached a good level of maturation, since various feature-extraction methods have been investigated. Despite the fact that the CAD analysis of cardiac images is not yet as extended as in the case of ECG signal, there are still many methodologies under consideration mainly focused on the size, morphology, function, and tissue characteristics.

The application of such feature-extraction methodologies for both sources may lead to a relatively great number of features that can feed the classification phase of the CAD system. However, despite the existence of many considered features, their discriminative power varies: while some of them may have important discriminative ability, others may be redundant or even irrelevant to the classification task. As a result, the first task toward the implementation of the final classification phase concerns the use of feature-selection methods, in order to locate a satisfactory subset able to reduce the computational cost and improve the classification performance. In general, there are two main approaches to feature selection: the first approach is called filter approach and contains methods that are generally independent of the employed classifier [23]. In filter methods, features are selected based on a metric, which usually measures their discriminative power. The second approach is called wrapper and includes methods that wrap the target classifier to find a good feature subset using a nonexhaustive search strategy. The first approach is simpler and has the advantage of being independent of the selection of classifier, but it is not able to count for dependencies among features. On the contrary, the wrapper methods take under consideration different combinations of features and manage to locate a satisfactory subset. Many methods have been proposed for feature selection or reduction, such as info gain ratio, recursive feature elimination, genetic algorithms, sequential forward selection and the related sequential backward elimination, and linear discriminant analysis.

Finally, we focus on the last task in the CAD flowchart, which concerns the prediction of a final estimation. In the majority of the studies reported above, the role of the CAD system is to provide for the corresponding problem at hand a discrete value as its final output, e.g., benign or malignant or existence of arrhythmia or not [16]. Hence, the final phase of a CAD system's flowchart consists of a two-class classification problem, where methodologies from the machine learning scientific field are required. Proper methodologies and classification algorithms have to be implemented in order to model a supervised learning problem, which is defined as the prediction of the corresponding discrete value for any valid input after training a learner by using examples of input and target output pairs. A great variety of classifiers have been applied in the state-of-the-art CAD approaches to solving this problem [24]. Artificial neural networks (ANNs) and support vector machines are the most commonly used types of classifiers for discrimination between instances of two different groups. Other simple classifiers, such as the *k*-nearest neighbor classifier, the Bayesian classifier, and classification trees, are also widely used, while a relatively recent classification scheme, the random forests classifier, has already become popular enough in the machine learning community in general.

9.3 Telehealth Systems

CVDs have always been at the forefront of what is now termed *telehealth*, including remote patient monitoring, teleconsultation, and home care. Transtelephonic monitoring of pacemakers was introduced in the 1970s and of implantable cardioverter defibrillators in the early 1990s. Today, telehealth relies on wireless medical telemetry, which provides real-time measurements from wearable/implantable medical sensors [25]. Antenna components embedded in the sensor nodes make it possible for the generated measurements to be transmitted wirelessly to a body-worn or closely located hub device. The hub device, in turn, receives the data generated from the various sensor nodes on the body and may process the data locally and/or transmit it wirelessly for centralized processing, display, and storage (Figure 9.3). Among the numerous advantages of wireless medical telemetry are the ease of use, reduced risk of infection, reduced patient discomfort, and enhanced mobility; yet there are also many requirements to be considered, such as radio-frequency (RF) radiation safety, reliability, biocompatibility, and energy-aware communication.

Telehealth has revolutionized the field of CVD management in several ways. Its primary role is to collect real-time physiological (oxygen levels, cardiac rhythm, blood pressure, wake–sleep patterns, etc.), medical (clinical profile, biochemical indices, and cardiovascular images), and behavioral data (dietary intake, exercise patterns, and medication) to be evaluated remotely by cardiovascular experts (Figure 9.4). In more advanced telehealth systems, the collected data are analyzed with CAD software, providing an alert of potential medical emergency (e.g., increase in blood pressure and abnormal ECG measurement), so that a precautionary action can be taken.

The potential of telehealth technology in CVD management has been investigated in several small- and large-scale studies on heart failure. Specifically, meta-analysis of a number of small-scale studies on patients with chronic heart failure (CHF) reported considerable clinical benefit from telemonitoring when compared with the usual care provided at

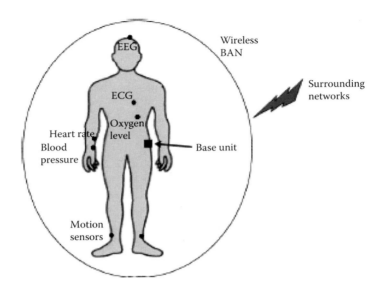

FIGURE 9.3
Illustration of wireless physiological measurements. (From Hao Y and Foster R, *Physiological Measurement*. 2008; 29(11):R27–R56.)

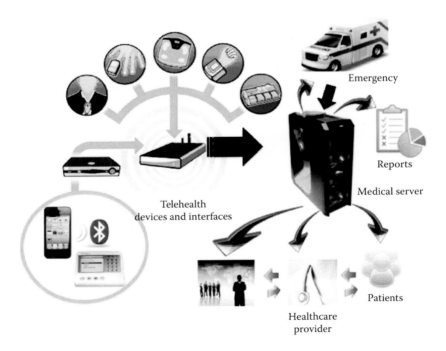

FIGURE 9.4
Schematic presentation of telehealth network and services.

that time [28]. More recent trials, such as Telemedical Interventional Monitoring in Heart Failure Study (TIM-HF) in Germany [29] and Telemonitoring to Improve Heart Failure Outcomes (Tele-HF) in the United States [30], investigated the role of telehealth on a larger scale. In TIM-HF and Tele-HF, 710 and 1653 CHF patients were randomly assigned (1:1) to telemonitoring or usual care, respectively. The primary end point was readmission for any reason or death, while secondary end points included hospitalization for heart failure, number of days in the hospital, and number of hospitalizations. Both studies suggested that the specific type of technology used is crucial, as are the characteristics of the patients, to ensure that the technology used matches the need of the patient group.

Furthermore, related studies have shown that remotely monitored biosignals facilitate early detection of arrhythmias and better identification of patients at increased risk of deterioration and of technical problems with medical devices [31–32]. Moreover, TEMEO, a validated ECG telemonitoring system, was used recently to follow up with patients with cryptogenic stroke or transient ischemic attack (TIA) in the previous 3 months [33]. According to the results, ECG telemonitoring resulted in detection of atrial fibrillation, a significant factor in stroke prevention, in 25% of patients. Further to the above, a small-size Bluetooth device by Alive Technologies (Alive Bluetooth Heart and Activity Monitor) offers self-monitoring of the heart rate and activity, providing ECGs, blood oximeters, and blood glucose meters, and targets early diagnosis of chronic CVDs, including atrial fibrillation, congestive heart failure, and sleep apnea.

9.4 Mobile Applications

A number of mobile health systems show promise for addressing various aspects of self-management of CVD. A wearable and mobile system to remotely manage heart failure was proposed by Villalba et al. [34]. The system includes (a) sensors for measuring ECG, respiration, and resting and exercise activity, which are embedded on a piece of clothing or a bed garment; (b) commercially available weighing scales and a blood pressure cuff; and (c) a personal digital assistant, which collects all measurements via Bluetooth, processes them, and provides advice for the daily healthcare of patients.

MHealth has been shown to improve medication adherence in subjects at increased cardiovascular risk [35]. Specifically, a medication adherence management system was suggested, comprising (a) an electronic medication blister, to track objectively the dosage and timing of medication intake; (b) a mobile-phone-based data gateway, to read and process the data in (a); and (c) a remote telemonitoring service, to receive the data sent from the mobile phone and analyze aspects, including timing and number of pills taken. Processed data were used to remind the patients automatically via short message service (SMS) messages [35]. A text messaging program, namely, Text4Heart, was recently presented by Dale et al. in the context of a randomized controlled trial, in an effort to improve self-management of coronary heart disease [36]. The program involves sending predefined text messages to the mobile phones of participants, with advice and information about illness perceptions and lifestyle changes.

The effect of home monitoring using mobile phone technology on the outcome of acute worsening of heart failure was investigated in the framework of the Mobile Telemonitoring in Heart Failure Patients (MOBITEL) study [37]. Random allocation of 120 patients into groups receiving (group A) and not receiving (group B) telemedical surveillance in combination with the conventional medication treatment was done. Patients of group A transmitted daily vital signals and medication dosage via mobile-phone-based terminals to physicians. There was a 54% reduction in risk for readmission or death among group A, while with respect to group B, patients of group A who were hospitalized for deteriorating heart failure had a significantly shorter length of stay.

Several other trials have also reported favorable results of mHealth in terms of improved medical profile and quality of life, enhanced clinical management, and reduced length of hospitalization [38–39]. Specifically, mHealth technology was used in patients with cardio-vascular disease to (a) send a text message on vitals and signals and receive instructions by physicians [38], (b) transmit alerts regarding atrial fibrillation and tachycardia activity for patients with implantable cardioverter-defibrillator [40], and (c) monitor physical activity and diet and receive motivational and educational text messages [39–41].

Moreover, there are an increasing number of mobile applications which are available for smartphones. These applications contribute to cardiovascular disease management by measuring and recording vital signs (iBP Blood Pressure, Blood Pressure Monitor, BP Monitor, and HeartWise), monitoring heart activity (Instant Heart Rate, Cardiograph, and Digifit), evaluating risk (Wellframe, Heart Age, Heart Risk Calculator, and Stroke Riskometer), improving lifestyle (Cholesterol Manager and Cardiac Assist), and assisting patient education (Cardiovascular System Pro, Cardiovascular Medicine Focus Apps, In Case Emergency: Medi Alert, and Heart Failure Trials).

9.5 Web-Based Telemedicine

The large-scale use of web technology has had a significant impact on the global healthcare sector. In particular, the role of web technology in eHealth has been revealed in web-based platforms, patient web portals, and social media, which are fast becoming a valuable cost-effective and easily accessible solution for both healthcare providers and online patients.

An indicative example is the ITS My HealthFile telemedicine system by InSight Telehealth, an interactive web-based system for cardiovascular disease management and healthcare delivery [42]. The ITS My HealthFile was used to investigate the impact of telemedicine on rural and urban underserved populations; 465 subjects at risk of developing cardiovascular disease were randomly allocated into a telemedicine-supported nurse management program (group 1) or usual care (group 2) [43]. After 1 year of surveillance, a significant reduction in risk was observed for group 1 (19.1%) compared to the usual care group (8.1%). Besides the favorable impact of telemedicine on CVD management, the related research also revealed that telemedicine can increase the accessibility of health services to those who are not in close proximity to a healthcare provider and to those with limited ability to obtain primary care services.

Additionally, CAROTID is a recently proposed web-based platform for optimal management of patients with carotid atherosclerosis [44]. CAROTID addresses this clinical challenge via two interconnected modules, which offer (a) objective selection of high-risk patients, using a multifaceted description of the disease consisting of ultrasound imaging and biochemical and clinical markers, and (b) effective storage and retrieval of patient data to facilitate frequent follow-ups and direct comparisons with related cases. CAROTID was evaluated in a real clinical environment in terms of CAD performance, end-user satisfaction, and time spent on different functionalities and the results were particularly encouraging for the potential of the platform in assisting the clinical practice for carotid atherosclerosis.

Patient web portals, which integrate medical resources and patient health records, have grown in both significance and popularity in recent years, with a number of major health organizations developing portals for CVDs (HeartHub, CardioSmart, Heart Information Center, HealthView, etc.). Patient web portals offer the possibility of truly patient-centered care through robust mechanisms for patient participation in the management of (chronic) cardiovascular disorders; health providers and patients can communicate via the web portal, thereby enabling meaningful participation by the patient as an equal partner in the care plan and its implementation. Moreover, they offer unique opportunities for patient education and empowerment across the broader continuum of care, including prevention, treatment, and wellness. By using targeted information resources and risk-assessment tools, patient web portals increase satisfaction with care, improve access to health information, enhance patient–provider communication, and eventually result in better overall cardiovascular disease management and patient outcomes.

The HeartHub by the American Heart Association [45] is a rich source of information on CVD; it also offers useful tools, namely, Heart360, for tracking health information and sharing results with physicians, and My Life Check, for patient engagement through a simple lifestyle assessment. The CardioSmart by the American College of Cardiology [46] provides information resources on multiple conditions of heart disease, educational videos, and interactive tools of risk assessment, as well as peer-to-peer connections between patients and caregivers. The Heart Information Center by the Texas Heart Institute [47] is dedicated to providing educational information on prevention, diagnosis, and treatment of cardiovascular disease. Moreover, real-time consultation by professionals is available via the Ask a Texas Heart Institute Doctor module. HealthView by the Duke Heart Center [48] offers access to tools and applications for helping in management of major cardiovascular diseases. HealthView supports home monitoring for the collection of vital biosignals and medical data of the patient. Furthermore, patients can review prescriptions, laboratory results, image data, and procedure reports.

Social media have also gained tremendous popularity over the past few years, encompassing a wide range of online forums, blogs, collaborative websites (wikis), social networking sites, and virtual worlds. This trend has also affected the area of CVD management, with patients using social media to (a) gather information about their condition, (b) identify other individuals with similar health conditions and share clinical information and receive support, and (c) identify sources of education regarding their disease. As an example, Inspire is a social networking platform designed by WomenHeart (an association of women with heart disease) and it allows sharing information, medical resources, networking, and support opportunities [49].

9.6 Semantic Interoperability and Ontologies

Diagnosis and management of CVD relies on a variety of medical data, such as ECG, ultrasound images, MRI images, CT images, biochemical exams, clinical profiles, and other data, such as patient history. The data that are mentioned above contain a wealth of information that is very useful for the physician but, due to their nonstructured format and the fact that the sources are heterogeneous, their retrieval is neither fast nor accurate. Therefore, semantic interoperability is important. Semantic interoperability can be broadly defined as the ability of one system to receive information from another system and reliably apply its rules against the information received [50]. This definition represents a well-established, consensus-based view from the international health information exchange community for shared messaging (syntax) and meaning (semantics) between health systems.

In an attempt to create a structured form of data and to enhance semantic interoperability, the scientific community has proposed the use of Semantic Web technologies and ontologies. Tim Berners-Lee defines the semantic web as a web of data that can be processed directly and indirectly from the machines. Ontology is the central element of the semantic web ontology. The definition of *ontology* in computer science is that ontology is a formal, explicit specification of a shared conceptualization [51]. This definition implies that ontology provides a vocabulary which can be used to model a domain, that is, the types of concepts and objects that exist and the relationships between them. There are two types of ontologies, namely, the reference or domain ontologies and the application ontologies. Reference ontologies represent knowledge about a particular part of the world in a way that is independent from specific objectives, through a theory of the domain [52]. On the contrary, application ontologies are designed in order to perform specific tasks and are narrower than the reference ontologies.

Many reference ontologies include fragments that represent the cardiovascular domain. The Foundational Model of Anatomy [53] has concepts that model the anatomic parts of the cardiovascular domain; the 10th revision of the International Statistical Classification of Diseases and Related Health Problems (ICD-10) [54] has terms that represent the diseases of the cardiovascular system; and the Systematized Nomenclature of Medicine—Clinical Terms (SNOMED CT) [55] contains clinical findings for CVDs. The above ontologies describe some parts of the cardiovascular domain, but they are too generic and do not combine all the elements that are essential for a doctor to make a diagnosis. Additionally to the above, ontologies cannot be used in order to estimate risk factors of the CVDs. Therefore, new ontologies have been developed in order to overcome the previous issues.

The ontology that was created by Soguero-Ruiz et al. [56] is focused on heart-rate turbulence (HRT). The ontology is designed to create CAD tools based on EHR data and ECG. To achieve this, the designers used SNOMED CT concepts and extended them to model the HRT domain. The goal of this ontology is to create connections between biosignals, their representation, and their anatomical, electrophysiological, and clinical concepts. Also, the use of SNOMED CT and EHRs offers enhanced capabilities for interoperability and allows the sharing of scientific knowledge of the domain. A similar approach has been followed in HEARTFAID [57] and Model Morphisms (MoMo) [58]. Both of these ontologies utilize the EHR data by modeling them with the use of SNOMED CT, but they also create a framework for integrating and utilizing data from mobile devices and sensors. Also the ontologies of these platforms use the semantic web rule language (SWRL) [59]. Another approach for modeling the risk factors of CVD is with the use of fuzzy ontologies [60]. With the use of fuzzy ontologies, patients with incomplete data can be mapped to similar patients based on the cluster to which they belong.

9.7 Future Trends

Electronic health technologies have had a substantial impact on the management of CVD. Despite some limitations and barriers to achieving the full potential of digital technology, it has been shown that the results of its applications in health have been predominantly positive [61]. Existing constraints are but the major challenges for the future achievements in this continuously evolving interdisciplinary field.

In particular, CAD has provided a large variety of useful tools for facilitating clinical decision making. Extended clinical trials remain to be performed so as to corroborate the usefulness and applicability of these methods in the clinical practice and, more importantly, to assess their value in promoting CVD management as well as public health in general. Among other applications, CAD tools may be useful in combination with portable systems, which allow not only performance of examinations outside the conventional clinical environment but also performance of these examinations by nonspecialized scientists [62]. Because these devices allow only a limited range of measurements to be performed, compared to conventional devices, their usefulness may be enhanced if their outputs are combined with an advanced data processing and analysis scheme.

Telecommunication technologies have allowed reliable transfer of a wide range of patient data between different locations, thus transforming the care of the cardiac patient from a hospital-centric to a patient-centric approach. Allowing healthcare providers to monitor the patients' cardiac status in near real time allows for timely, personalized interventions and maximizes efficiencies while reducing costs. It seems that management of CVD is moving toward a concept of self-management of CVD.

The use of semantic technologies in the cardiovascular domain can provide extended capabilities in sharing and retrieving information. Due to this the diagnosis procedures can be faster and also knowledge that is hidden in vast pools of data can be exploited. The ontologies can be used to combine heterogeneous data from many resources in a formal way and also enhance the interoperability between the sources. With the integration of SWRL rules and fuzzy ontologies, monitoring systems for CVD patients with extended capabilities can be created, providing personalized healthcare services.

9.8 Conclusions

The widespread use of electronic health technologies is the dawn of a new era where evidence-based guidelines can be seamlessly translated to patient care and where patients are actively involved in their own health. The large-scale use of web technology has already had a significant impact on the global healthcare sector, enabling web-based platforms to become cost-effective and easily accessible solutions for both healthcare providers and patients. As transformative as this will be, it is important to recognize that we are currently experiencing only the very earliest potential of eHealth in improving CVD outcomes. Enormous efforts are still required in order to exploit the great advantages that electronic health technologies may offer, as new challenges have to be faced and unexpected limitations should be overcome.

However, despite the fact that we are still in the beginning of this new era, research efforts reported in the literature have generated encouraging results toward the management of CVD. Considerable efforts have been already performed toward the automated diagnosis of such diseases, aiming to prolong life and enhance the quality of living through appropriate treatment. Various methodologies for signal and image analysis have been investigated, in order to develop CAD methodologies able to support the diagnostic process. The encouraging preliminary results reported form the basis of transforming the current diagnostic decision-making process, as it seems that there is a potential to adopt these methods in daily clinical practice and become a part of clinical work.

Nevertheless, one great challenge for the research community is to integrate the aforementioned novel CAD methods into telehealth systems [63]. The wireless medical telemetry that has been developed during recent years aims to ensure home care and continuous remote monitoring of the patient [25]. Toward the same direction, mHealth technologies through mobile applications and smartphones have to be exploited to contribute to CVD. These eHealth technologies offer continuous measuring and recording of vital signs, as well as monitoring of heart activity. Therefore, the automated collection of physiological, behavioral, and medical data of the patient through eHealth or mHealth applications has to be combined with CAD methodologies in order to ensure timely precautionary actions can be taken. In additional to that, the vast amount of these data has be structured in a formal way with the use of semantic web technologies in order to have better exploitation of the data and also to make reusable the information derived from them for diagnostic and research purposes.

It is obvious that the concurrent use of CAD and telehealth/mobile technologies addresses the health management of CVD diseases in a holistic approach, placing the patient in the center of this procedure and putting increased emphasis on patient empowerment, secondary prevention, and self-management of individual conditions. The proper combination of these technologies transforms the diagnostic process and mainly benefits patients that have no access to traditional delivery of health services, either because of distance or due to lack of local specialist clinicians able to deliver service.

References

1. World Health Organization. Cardiovascular Diseases [cited April 16, 2014]. Available from: http://www.who.int/topics/cardiovascular_diseases/en/.
2. Mathers CD and Loncar D. Projections of global mortality and burden of disease from 2002 to 2030. *PLoS Medicine.* 2006; 3(11):e442.
3. World Health Organization and World Economic Forum. From Burden to "Best Buys": Reducing the Economic Impact of Non-Communicable Diseases in Low- and Middle-Income Countries Geneva, Switzerland: World Health Organization and World Economic Forum; 2011 [cited April 15, 2014]. Available from: http://www.who.int/nmh/publications/best_buys _summary/en/.
4. Cooney MT, Cooney HC, Dudina A, and Graham IM. Total cardiovascular disease risk assessment: A review. *Current Opinion in Cardiology.* 2011; 26(5):429–437.
5. Mendis S and Chestnov O. The global burden of cardiovascular diseases: A challenge to improve. *Current Cardiology Reports.* 2014; 16(5):486.
6. Huebner T, Goernig M, Schuepbach M, Sanz E, Pilgram R, Seeck A et al. Electrocardiologic and related methods of non-invasive detection and risk stratification in myocardial ischemia: State of the art and perspectives. *German Medical Science: GMS E-journal.* 2010; 8:Doc27.
7. Nikita KS. Atherosclerosis: The evolving role of vascular image analysis. *Computerized Medical Imaging and Graphics.* 2013; 37(1):1–3.
8. Roumia M and Steinhubl S. Improving cardiovascular outcomes using electronic health records. *Current Cardiology Reports.* 2014; 16(2):451.
9. Blumenthal D and Glaser JP. Information technology comes to medicine. *The New England Journal of Medicine.* 2007; 356(24):2527–2534.

10. Anchala R, Pinto MP, Shroufi A, Chowdhury R, Sanderson J, Johnson L et al. The role of Decision Support System (DSS) in prevention of cardiovascular disease: A systematic review and meta-analysis. *PloS One.* 2012; 7(10):e47064.

11. Golemati S, Gastounioti A, and Nikita KS. Toward novel noninvasive and low-cost markers for predicting strokes in asymptomatic carotid atherosclerosis: The role of ultrasound image analysis. *IEEE Transactions on Bio-medical Engineering.* 2013; 60(3):652–658.

12. Nikita KS, Lin JC, Fotiadis DI, and Arredondo MT. Editorial, "Special Issue on Mobile and Wireless Technologies for Healthcare Delivery." *IEEE Transactions on Biomedical Engineering,* 2012; 59(11):3083–3089.

13. Dixon BE, Vreeman DJ, and Grannis SJ. The long road to semantic interoperability in support of public health: Experiences from two states. *Journal of Biomedical Informatics.* 2014.

14. Giger ML, Chan HP, and Boone J. Anniversary paper: History and status of CAD and quantitative image analysis: The role of medical physics and AAPM. *Medical Physics.* 2008; 35(12):5799–5820.

15. Doi K. Computer-aided diagnosis in medical imaging: Historical review, current status and future potential. *Computerized Medical Imaging and Graphics.* 2007; 31(4–5):198–211.

16. Elter M and Horsch A. CADx of mammographic masses and clustered microcalcifications: A review. *Medical Physics.* 2009; 36(6):2052–2068.

17. Ramirez J, Gorriz JM, Segovia F, Chaves R, Salas-Gonzalez D, Lopez M et al. Computer aided diagnosis system for the Alzheimer's disease based on partial least squares and random forest SPECT image classification. *Neuroscience Letters.* 2010; 472(2):99–103.

18. Tsanas A, Little MA, McSharry PE, Spielman J, and Ramig LO. Novel speech signal processing algorithms for high-accuracy classification of Parkinson's disease. *IEEE Transactions on Bio-medical Engineering.* 2012; 59(5):1264–1271.

19. Ingole M, Alaspure SV, and Ingole DT. Electrocardiogram (ECG) signals feature extraction and classification using various signal analysis techniques. *International Journal of Engineering Sciences & Research Technology* 2014; 3(1):39–44.

20. Wang J, Chiang WC, Hsu YL, and Yang YTC. ECG arrhythmia classification using a probabilistic neural network with a feature reduction method. *Neurocomputing* 2013; 116:38–45.

21. Bello D, Fieno DS, Kim RJ, Pereles FS, Passman R, Song G et al. Infarct morphology identifies patients with substrate for sustained ventricular tachycardia. *Journal of the American College of Cardiology.* 2005; 45(7):1104–1108.

22. Roes SD, Borleffs CJ, van der Geest RJ, Westenberg JJ, Marsan NA, Kaandorp TA et al. Infarct tissue heterogeneity assessed with contrast-enhanced MRI predicts spontaneous ventricular arrhythmia in patients with ischemic cardiomyopathy and implantable cardioverter-defibrillator. *Circulation Cardiovascular Imaging.* 2009; 2(3):183–190.

23. Saeys Y, Inza I, and Larranaga P. A review of feature selection techniques in bioinformatics. *Bioinformatics.* 2007; 23(19):2507–2517.

24. Melgani F and Bazi Y. Classification of electrocardiogram signals with support vector machines and particle swarm optimization. *IEEE Transactions on Information Technology in Biomedicine.* 2008; 12(5):667–677.

25. *Handbook of Biomedical Telemetry.* Nikita KS, editor. Wiley–IEEE Press; 2014.

26. Hao Y and Foster R. Wireless body sensor networks for health-monitoring applications. *Physiological Measurement.* 2008; 29(11):R27–R56.

27. Otto C, Milenkovic A, Sanders C, and Jovanov E. System architecture of a wireless body area sensor network for ubiquitous health monitoring. *Journal of Mobile Multimedia.* 2006; 1(4):307–326.

28. Inglis SC, Clark RA, McAlister FA, Ball J, Lewinter C, Cullington D et al. Structured telephone support or telemonitoring programmes for patients with chronic heart failure. *The Cochrane Database of Systematic Reviews.* 2010(8):CD007228.

29. Koehler F, Winkler S, Schieber M, Sechtem U, Stangl K, Bohm M et al. Impact of remote telemedical management on mortality and hospitalizations in ambulatory patients with chronic heart failure: The telemedical interventional monitoring in heart failure study. *Circulation.* 2011; 123(17):1873–1880.

30. Chaudhry SI, Mattera JA, Curtis JP, Spertus JA, Herrin J, Lin Z et al. Telemonitoring in patients with heart failure. *The New England Journal of Medicine.* 2010; 363(24):2301–2309.
31. Bourge RC, Abraham WT, Adamson PB, Aaron MF, Aranda JM, Jr., Magalski A et al. Randomized controlled trial of an implantable continuous hemodynamic monitor in patients with advanced heart failure: The COMPASS-HF study. *Journal of the American College of Cardiology.* 2008; 51(11):1073–1079.
32. Abraham WT, Adamson PB, Bourge RC, Aaron MF, Costanzo MR, Stevenson LW et al. Wireless pulmonary artery haemodynamic monitoring in chronic heart failure: A randomized controlled trial. *Lancet.* 2011; 377(9766):658–666.
33. Simova I, Mateev H, Katova T, Haralanov L, and Dimitrov N. Telemonitoring boosts atrial fibrillation detection in cryptogenic stroke patients—Preliminary findings. *Cardiology and Angiology: An International Journal.* 2013; 1(2):47–57.
34. Villalba E, Salvi D, Ottaviano M, Peinado I, Arredondo MT, and Akay A. Wearable and mobile system to manage remotely heart failure. *IEEE Transactions on Information Technology in Biomedicine.* 2009; 13(6):990–996.
35. Brath H, Morak J, Kastenbauer T, Modre-Osprian R, Strohner-Kastenbauer H, Schwarz M et al. Mobile health (mHealth) based medication adherence measurement—A pilot trial using electronic blisters in diabetes patients. *British Journal of Clinical Pharmacology.* 2013; 76(Suppl 1):47–55.
36. Dale LP, Whittaker R, Jiang Y, Stewart R, Rolleston A, and Maddison R. Improving coronary heart disease self-management using mobile technologies (Text4Heart): A randomised controlled trial protocol. *Trials.* 2014; 15:71.
37. Scherr D, Kastner P, Kollmann A, Hallas A, Auer J, Krappinger H et al. Effect of home-based telemonitoring using mobile phone technology on the outcome of heart failure patients after an episode of acute decompensation: Randomized controlled trial. *Journal of Medical Internet Research.* 2009; 11(3):e34.
38. Seto E, Leonard KJ, Cafazzo JA, Barnsley J, Masino C, and Ross HJ. Mobile phone-based telemonitoring for heart failure management: A randomized controlled trial. *Journal of Medical Internet Research.* 2012; 14(1):e31.
39. Walters D, Varnfield M, Karunanithi M, Ding H, Honeyman E, Arnold D et al. Technology based home-care model improves outcomes of uptake, adherence and health in cardiac rehabilitation. *Heart Lung and Circulation.* 2012; 21(Suppl 1): S315.
40. Crossley GH, Boyle A, Vitense H, Chang Y, and Mead RH. The Connect (Clinical Evaluation of Remote Notification to Reduce Time to Clinical Decision) Trial: The value of wireless remote monitoring with automatic clinician alerts. *Journal of the American College of Cardiology.* 2011; 57(10):1181–1189.
41. Worringham C, Rojek A, and Stewart I. Development and feasibility of a smartphone, ECG and GPS based system for remotely monitoring exercise in cardiac rehabilitation. *PloS One.* 2011; 6(2):e14669.
42. Santamore WP, Homko CJ, Kashem A, McConnell TR, Menapace FJ, and Bove AA. Accuracy of blood pressure measurements transmitted through a telemedicine system in underserved populations. *Telemedicine Journal and E-health.* 2008; 14(4):333–338.
43. Bove AA, Santamore WP, Homko C, Kashem A, Cross R, McConnell TR et al. Reducing cardiovascular disease risk in medically underserved urban and rural communities. *American Heart Journal.* 2011; 161(2):351–359.
44. Gastounioti A, Kolias V, Golemati S, Tsiaparas NN, Matsakou A, Stoitsis JS et al. CAROTID— A web-based platform for optimal personalized management of atherosclerotic patients. *Computer Methods and Programs in Biomedicine.* 2014; 114(2):183–193.
45. HeartHub for patients [cited April 15, 2014]. Available from: http://www.hearthub.org.
46. CardioSmart [cited April 15, 2014]. Available from: https://www.cardiosmart.org/.
47. Texas Heart Institute [cited April 15, 2014]. Available from: http://www.texasheart.org/HIC /his.cfm.
48. HealthView [cited January 22, 2015]. Available from: https://healthview.dukehealth.org/wps /portal.

49. WomanHeart [cited January 22, 2015]. Available from: https://www.inspire.com/groups/womenheart/.

50. Dolin RH and Alschuler L. Approaching semantic interoperability in Health Level Seven. *Journal of the American Medical Informatics Association.* 2011; 18(1):99–103.

51. Gruber TR. A translation approach to portable ontology specifications. *Knowledge Acquisition.* 1993; 5(2):199–220.

52. Burgun A. Desiderata for domain reference ontologies in biomedicine. *Journal of Biomedical Informatics.* 2006; 39(3):307–313.

53. Rosse C and Mejino JL, Jr. The foundational model of anatomy ontology, in *Anatomy Ontologies for Bioinformatics.* Springer; 2008. pp. 59–117.

54. Möller M, Sintek M, Biedert R, Ernst P, Dengel A, and Sonntag D, editors. *Representing the International Classification of Diseases Version 10 in OWL.* KEOD; 2010.

55. Stearns MQ, Price C, Spackman KA, and Wang AY, editors. SNOMED Clinical Terms: Overview of the development process and project status. *Proceedings of the AMIA Symposium.* American Medical Informatics Association; 2001.

56. Soguero-Ruiz C, Lechuga-Suarez L, Mora-Jimenez I, Ramos-Lopez J, Barquero-Perez O, Garcia-Alberola A et al. Ontology for heart rate turbulence domain from the conceptual model of SNOMED-CT. *IEEE Transactions on Biomedical Engineering,* 2013; 60(7):1825–1833.

57. Gamberger D, Prcela M, Jovic A, Smuc T, Parati G, Valentini M et al. *Medical Knowledge Representation within Heartfaid Platform.* HEALTHINF (1); 2008.

58. Hervás R, Fontecha J, Ausín D, Castanedo F, Bravo J, and López-de-Ipiña D. Mobile monitoring and reasoning methods to prevent cardiovascular diseases. *Sensors.* 2013; 13(5):6524–6541.

59. Lezcano L, Sicilia M-A, and Rodríguez-Solano C. Integrating reasoning and clinical archetypes using OWL ontologies and SWRL rules. *Journal of Biomedical Informatics.* 2011; 44(2):343–353.

60. Parry D and MacRae J. Fuzzy ontologies for cardiovascular risk prediction—A research approach. *2013 IEEE International Conference on Fuzzy Systems (FUZZ);* July 7–10, 2013.

61. Buntin MB, Burke MF, Hoaglin MC, and Blumenthal D. The benefits of health information technology: A review of the recent literature shows predominantly positive results. *Health Affairs.* 2011; 30(3):464–471.

62. Tofield A. The use of pocket size imaging devices: A position statement by the European Association of Echocardiography. *European Heart Journal.* 2011; 32(4):385–386.

63. Zhang YT, Zheng YL, Lin WH, Zhang HY, and Zhou XL. Challenges and opportunities in cardiovascular health informatics. *IEEE Transactions on Bio-medical Engineering.* 2013; 60(3):633–642.

10

mHealth eEmergency Systems

Efthyvoulos Kyriacou, Andreas Panayides, and Panayiotis Constantinides

CONTENTS

10.1 Introduction

mHealth eEmergency systems can be defined as "emerging mobile communications and network technologies for emergency healthcare support" [1]. This concept represents the evolution of traditional eHealth systems from desktop platforms and wired connections to the use of more compact devices and wireless connections in emergency healthcare support systems. The emerging development of mHealth systems in the last decade was made possible due to the recent advances in wireless and network technologies, linked with recent advances in nanotechnologies, compact biosensors, wearable devices and clothing, and pervasive and ubiquitous computing systems. These advances will have a powerful impact on some of the existing healthcare services and will reshape the work flow and practices in the delivery of these services [1].

This chapter reviews recent eEmergency systems, including the wireless technologies used, as well as the data transmitted (electronic patient record, biosignals, medical images and video, and others).

Wireless telemedicine systems and services are expected to enhance traditional emergency care provision not only within the emergency department but also in a variety of prehospital emergency care situations where geographically remote consultation and monitoring can be implemented [2]. A timely and effective way of handling emergency cases can prove essential for a patient's recovery or even for a patient's survival. Especially in cases of serious injuries of the head, the spinal cord, and internal organs, the way of transporting and generally the way of providing care are crucial for the future of the patient. Furthermore, during cardiac disease cases, much can be done today to stop a heart attack or resuscitate a victim of sudden cardiac death (SCD). Time is the enemy in the acute treatment of a heart attack or SCD. The first 60 min (the golden hour) are the most critical regarding the long-term patient outcome. Therefore, the ability to remotely monitor the patient and guide the paramedical staff in their management of the patient can be crucial. This paper provides an overview of the main systems and technological components used in mHealth eEmergency systems.

The structure of the chapter is as follows: Section 10.2 presents the eEmergency systems enabling technologies covering the wireless transmission technologies, mobile computing technologies, biosignals, medical imaging, and video. In Section 10.3, protocols and processes for eEmergency management and response are covered, followed by Section 10.4, which covers a review of mobile health eEmergency systems.

10.2 Enabling Technologies

10.2.1 Wireless Transmission Technologies

The unparalleled growth of mHealth systems and services over the past two decades is tightly coupled with associated advances in wireless networks infrastructure and video-compression standards (see Subsection 10.2.5). Especially for bandwidth demanding medical image and video communication systems, milestone advances in mobile cellular networks reflect on breakthrough achievements in mHealth systems and services. The digital era in mobile cellular networks was introduced by the Global System for Mobile Communications (GSM), signifying the transition from analog first-generation (1G) to digital second-generation (2G) mobile telecommunications. The general packet radio service (GPRS) and enhanced data rates for GSM evolution (EDGE) 2.5G technologies facilitated the data-transfer rates that allowed mHealth systems to integrate medical image and video transmission, in addition to biomedical signals. Toward this direction, third-generation mobile telecommunications (3G) systems (Universal Mobile Telecommunications System [UMTS]) set the foundations of establishing telemedicine systems in standard clinical practice, exploiting extended coverage for a virtually always-on service provision. Today's 3.5G and near-4G (fourth-generation mobile telecommunications) wireless networks, namely, high-speed downlink packet access (HSDPA), high-speed uplink packet access (HSUPA), high-speed packet access (HSPA), HSPA+, mobile WiMAX, and Long-Term Evolution (LTE), allow the deployment of responsive mHealth and eEmergency systems, minimizing delays and extending upload data-transfer rates that can accommodate high–diagnostic quality medical video that can rival the quality of in-hospital examinations. Ongoing deployment of 4G wireless networks and emerging fifth-generation mobile telecommunications (5G) systems theoretical upload data-transfer rates comparable to wired infrastructure will expedite the adoption of such systems and services in standard clinical practice.

The afore-described wireless networks facilitate incremental data-transfer rates while minimizing end-to-end delay. Evolving wireless communications networks' theoretical upload data rates range from 50 kbps to 86 Mbps. In practice, typical upload data rates are significantly lower. More specifically, typical upload data rate ranges are (i) for GPRS, 30–50 kbps; (ii) for EDGE, 80–160 kbps; (iii) for evolved EDGE, 150–300 kbps; (iv) for UMTS, 200–300 kbps; (v) for HSPA, 500 kbps–2 Mbps; (vi) for HSPA+, 1–4 Mbps; and (vii) for LTE, 6–13 Mbps [3].

10.2.2 Mobile Computing Platforms

The introduction of portable devices, like tablets, smartphones, small-size laptops, and single-board computers, enables eEmergency system application developers to create systems that are efficient with respect to computing power and functionality, consuming less power, and are compact and smaller in size. Such systems have already appeared in the last decade and will certainly continue to appear in the coming years.

In a recent study, where mobile and fixed computer use by doctors and nurses on hospital wards was investigated, it was found that the choice of device was related to clinical role, nature of the clinical task, and degree of mobility required [4]. Nurses' work and clinical tasks performed by doctors during ward rounds require highly mobile computer devices, and they showed a strong preference for generic computers mobile on wheels (including laptops) over all other devices. Tablet personal computers (PCs) were selected by doctors for only a small proportion of clinical tasks. Note that even when using mobile devices, clinicians completed a very low proportion of observed tasks at the bedside [4].

A systematic review of personal digital assistant (PDA) usage surveys by healthcare providers was carried out by Garritty and El Emam [5]. It was documented that younger physicians and residents and those working in large and hospital-based practices are more likely to use a PDA. Moreover, it appeared that professional PDA use in healthcare settings involved more administrative and organizational tasks than those related to patient care. Garritty and El Emam concluded that physicians are likely accustomed to using a PDA; however, there is still a need to evaluate the effectiveness and efficiency PDA-based applications [5].

10.2.3 Biosignals

The biosignals usually collected in an eEmergency system include the following: ECG signals, 3, 7, or 12 leads, depending on the monitor used; oxygen saturation (SpO_2); capnography values (CO_2); heart rate (HR); noninvasive blood pressure (NIBP); invasive blood pressure (IP); temperature (Temp); respiration (Resp); and their corresponding trends and alarms based on preset settings. ECG signals are sampled at an average rate of 200 samples/s at 10 bits/sample (at least) (depending on the monitor used), thus resulting in a generation of at about 2000 bits/s per ECG channel. SpO_2 and CO_2 waveforms are sampled at a rate of 100 samples/s by 10 bit/sample, thus resulting in a generation of 1000 bits/s for one channel. Trends for SpO_2, HR, NIBP, BP, Temp, Resp, and monitor data are updated with a refresh rate of one per second. This adds a small fraction of data to be transmitted approximately up to 200 bits/s.

The collection of biosignals [6–9], such as ECG signals, until now was performed using expensive devices which could be handled and supported only by medical personnel. More recently, the collection of biosignals, such as ECG signals, can be performed by very small devices. These are not always devices on their own but they might connect to a smartphone, tablet, laptop, or PC in order to display or send the signals and usually have Bluetooth or GPRS connectivity to wirelessly transfer the signals. They might be wearable, have the shape and weight of a necklace, etc. These devices will enable the use of wireless telemedicine systems almost anywhere and at less cost. Such devices can be used for home care purposes much more easily than the standard medical devices. Examples have already being presented by major companies like Samsung Electronics, where the idea of open-source wearable products (Simband) is presented as a future tool trying to attract collaborators and researchers from all over the world [10].

Biosignal compression is very desirable for eEmergency applications or long-term recordings. Efficient compression algorithms achieve a reduction in the number of bits required to describe a biosignal that could facilitate the data transmission. However, this must be done with great care not affecting the diagnostic loss of information. According to Hadjileontiadis [11], although new emerging data-compression techniques with very promising results are seen in the recent years, some problems have not been entirely addressed. In particular, the following items are still under consideration: there is a lack of (i) widely adopted compression quality assurance criteria and standards, (ii) widely available benchmark biosignal databases, and (iii) interoperability of data acquisition and processing equipment and exchange of compressed data between databases from different research groups and/or manufacturers because of their incompatibility.

The standard communications protocol for computer-assisted electrocardiography (SCP-ECG) (International Organization for Standardization [ISO]/Institute of Electrical and Electronics Engineers [IEEE] 11073 standards), health level 7 (HL7) annotated electrocardiogram (aECG), and digital imaging and communication in medicine (DICOM)–ECG are three widely used open standards facilitating the exchange of ECG signals. These three open standards are supported by the ECG tool kit, recently published by van Ettinger et al. (2008) [12]. In this paper it is documented that SCP-ECG gives the smallest size of file, requiring the smallest bandwidth; however, it is complex to implement. HL7 aECG supports extensible markup language (XML) text files (and it is easy to be compressed by gzip or any other compression utility) and is the preferred standard for exchanging and comparing ECGs required by the FDA for drug trials. DICOM allows integration of ECGs with other imaging modalities; however, very few DICOM viewers support the proper ECG display. Moreover, the need to develop eHealth systems integrating the ISO/IEEE 11073 SCP-ECG standard still exists, and the EN13606 electronic health record (EHR) standard, thus facilitating an end-to-end standard-based solution, as it has recently been demonstrated by Martínez et al. [13].

10.2.4 Transmission of Digital Images

The use of digital images in medicine has benefited from the formation of the DICOM committee [14]. The committee was formed in 1983 by the American College of Radiology (ACR) and the National Electrical Manufacturers Association (NEMA). For still images, DICOM has adopted various JPEG variants such as lossless JPEG (JPEG-LS) [15] and JPEG 2000 [16].

We first consider lossless image-compression methods that provide for exact reconstruction of the input images. Lossless image-compression eliminates the need for diagnostic validation of compression artifacts [17–18]. Unfortunately, lossless methods provide limited compression ratios, usually ranging between 2 and 3.7 [19]. Lossy image-compression methods can provide much better compression ratios. However, the use of lossy image compression requires a careful evaluation of the effect of compression artifacts on diagnostic performance [20]. While not directly relevant to diagnostic performance, lossy image compression of general images attempts to be perceptually lossless. Here, a compressed image is termed *perceptually lossless* if an (average) human observer cannot differentiate it from its uncompressed version. In general, optimal performance requires the study of the impact of lossy compression in different clinical scenarios.

A general lossy compression approach that can be directly applied to medical images is to use diagnostic regions of interest (ROIs). Here, the parts of an image that are of diagnostic interest will see little or no compression. On the other hand, the parts that are not of diagnostic interest can be compressed significantly. For example, if the region of interest covers about 20% of the entire image, average compression ratios of about 15:1 have been reported using JPEG-LS, while an average compression ratio of only 2.58 was achieved when using the entire image as the region of interest [21].

Similar to perceptually lossless compression for general images, another approach is to use lossy compression that does not allow clinicians to differentiate between the compressed image from the uncompressed one. Clearly, if the uncompressed image cannot be identified, then the (clinical) visual inspection of the compressed images should not impact the diagnosis [21]. This technique leads to near-lossless techniques where the uncompressed image differs from the original in only a small number of levels (± 1 and ± 2 out of possible 4096 levels). For comparison, JPEG-LS in lossless mode provides for an average compression ratio of 2.58 that improves to 3.83 in the near-lossless mode (± 1 levels) [21]. In addition, for a region of interest that covers 20% of the image region, the average compression ratio improves from 15.1 to 22.0 [21].

10.2.5 Transmission of Digital Video

The successful and efficient deployment of medical video communication systems relies on timely integration of video coding standards technologies and adaptation to the underlying wireless channel's capacity, as already highlighted in Subsection 10.2.1. Given the fact that wireless networks impose a limit on the available upload data-transfer rates, exploiting the trade-off between compression efficiency and clinical video quality is of primary importance. In contrast to wireless communication of conventional video and multimedia applications, clinical video quality cannot be compromised, as the latter would translate to misdiagnosis. Consequently, mHealth medical video communication systems need to be diagnostically driven [21]. The objective of diagnostically driven systems is to maximize the clinical capacity of the communicated video by adapting the employed algorithms for video encoding, wireless communication, and both objective and clinical video quality assessments on the underlying medical video modality. While sharing common concepts and principles, diagnostically driven approaches are often medical video modality specific.

In terms of video resolutions facilitated by video coding standards, the H.261 standard [22] (first video coding standard—early 1990s) supported the quarter common intermediate format (QCIF) (176×144) and the common intermediate format (CIF) (352×288) video resolutions. Subsequently, the H.262 standard [23] (released in 1995) extended video resolutions support up to 4CIF (720×576) and 16CIF (1408×1152). Then, the H.263 standard [24] (introduced in 1996) provided for improved quality at lower bit rates while also allowing lower, sub-QCIF (128×96) video resolution encoding. The aforementioned video coding standards were employed in early medical video communications systems of limited clinical capacity. This was partly attributed to the inability to encode medical video at video resolutions and frame rates (to match the available bit rate of wireless channels) that would not compromise diagnostic quality and clinical motion, respectively. The H.264/advanced video coding (AVC) standard [25] dominated mHealth systems and services of the past decade (released in 2003). The H.264/AVC standard, linked with 3G and 3.5G wireless networks, enabled efficient and timely encoding to match the available data rates and conform to real-time transmission requirements. In addition to coding efficiency and error-resilience coding tools, H.264/AVC introduced the network abstraction layer (NAL), a novel concept enabling network-friendly adaptation of the encoded content to candidate heterogeneous networks and/or storage devices (and cloud infrastructure), a significant feature toward H.264/AVC success [26]. As documented in Subsection 10.2.1, high-resolution and high–frame rate medical video communication is feasible using H.264/AVC over 3.5G wireless channels, leading to a plethora of mHealth systems and services of high diagnostic value, being a milestone for mHealth systems and services. However, in contradiction to initial projections and enthusiasm, the adoption of wireless telemedicine systems in standard clinical practice remains limited. The new high efficiency video coding (HEVC) standard [27] specifically designed for beyond high-definition video communications that provides 50% bit rate gains for comparable visual quality compared to the H.264/AVC standard [28–29]. Together with 4G and beyond (5G) wireless networks deployment, the HEVC standard is expected to play a decisive role toward wider adoption of such systems and services. To achieve this, mHealth medical video communication systems that can transmit medical video at the clinically acquired resolution and frame rate that can be robustly transmitted in low delay without compromising clinical quality are envisioned.

10.3 Protocols and Processes for eEmergency Management and Response

10.3.1 Emergency Management and Response: The Challenge of Coordination

Research on medical informatics jointly with research in computer science has led to the development of so-called smart devices, third-generation wireless connectivity, and positioning technologies, all of which have application in emergency management and response because they are location aware; i.e., they combine timely clinical information with accurate geographic information. These technologies are being evaluated to improve and enhance patient care and tracking; foster greater provider safety; enhance incident management at the scene and coordination of emergency medical services and hospital resources; and greatly enhance informatics support at the scene and at receiving emergency departments and hospitals.

Despite these advances, however, many of the logistical problems faced in emergencies are not caused by shortages of medical and technological resources but rather by failures to coordinate their distribution [30], with significant possibilities for error and destruction, as occurred at the New York's World Trade Center command post on September 11, 2001 [31–32].

To address the challenge of coordination, emergency dispatch centers around the world are increasingly using some form of priority dispatch protocol when handling emergency calls. The key objective behind these protocols is to ensure that each incident is appropriately responded to within time limitations and standard procedures until professional help arrives [33].

Over the last 30 years or so, organizations such as the European Emergency Number Association (http://www.eena.org) and the National Academies of Emergency Dispatch in North America (http://www.emergencydispatch.org) have established standards for the development of priority dispatch protocols for helping the coordination of medical and technological resources toward the effective management and response to emergency incidents. The most popular system of priority dispatch protocols is the Medical Priority Dispatch System (MPDS), developed by the National Academies of Emergency Dispatch in collaboration with the National Association of Emergency Medical Services Physicians, the American Society for Testing and Materials, the American College of Emergency Physicians, the U.S. Department of Transportation, the National Institutes of Health, and the American Medical Association. The original set of protocols contained 29 sets of two 8×5 in^2 (21×13 cm^2) cards. Each caller complaint was listed in alphabetical order, as they are today, and reflected either a symptom (e.g., abdominal pain, burns, or cardiac/respiratory arrest) or an incident (e.g., electrocutions, drowning, or traffic injury accident). The core card contained three color-coded areas: key questions, prearrival instructions, and dispatch priorities. The MPDS has since gone through 18 revisions to reflect advances in medicine, such as the addition of compressions first cardiopulmonary resuscitation (CPR), and it shares the stage with protocol designed for police and fire dispatch centers. The MPDS now contains 34 "chief complaint" protocols, case entry and exit information, call termination scripts, and additional verbatim instruction protocols for ambulance and emergency dispatch, cardiopulmonary resuscitation, childbirth assistance, tracheostomy airway and breathing, and the Heimlich maneuver. The MPDS is now used by more than 3000 emergency dispatch centers and in 23 countries worldwide, including Great Britain, Ireland, Germany, Italy, Azerbaijan, New Zealand, and Australia. Priority Dispatch Corporation is licensed to design and publish the MPDS and associated support products (http://www .prioritydispatch.net). Figure 10.1 provides an illustration of the set of case entry questions employed in MPDS that lead to questions for identification of a convulsion/fitting incident (Figure 10.2) and to a set of prearrival instructions (Figure 10.3).

As evident from the table in Figure 10.1, the process starts with the dispatcher asking a set of so-called entry questions to establish the exact location of the incident, the phone number of the caller, the exact type of problem, the age of the victim, and whether they are conscious and breathing. These questions are then followed by more specific questions according to the answers received, through which the dispatcher identifies the type of chief complaint—in this case a convulsions/fitting incident (see Figure 10.2). Depending on the answers to questions such as, "Does s/he have a history of heart problems?" and "Is s/he an epileptic or ever had a fit before?," the dispatcher assigns a severity score to the call according to different "determinant descriptors," which allow him/her to categorize the call into one of four priority dispatch levels from A to D. Level A would indicate that the victim is in a stable, nonemergency condition, thus not requiring immediate assistance, especially if units are required for other incidents. Level B would indicate a response but still not immediate, whereas levels C and D would indicate the need for immediate responses with an advanced life-support (ALS) ambulance (see Figure 10.2). After having triaged the call and dispatched an ambulance, the dispatcher would continue to give instructions to the caller until the ambulance arrived at the scene of the incident. Depending on answers to questions such as, "Is s/he breathing?," the dispatcher would give instructions to check the airway, to check breathing, and to administer cardiopulmonary resuscitation (see Figure 10.3, steps 1–9).

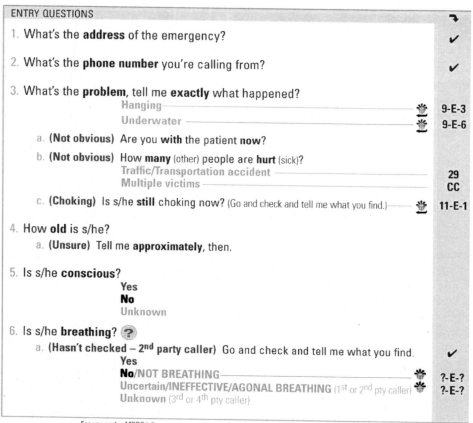

FIGURE 10.1
Case entry protocol for priority medical dispatch. (Courtesy of Priority Dispatch Corp., Salt Lake City, Utah.)

10.3.2 Computer-Aided Medical Dispatch Systems

Since the early 1990s, there has been a consistent effort to implement the MPDS and similar priority dispatch protocols through computer-based systems in an effort to automate processes and further minimize human error rates [34]. We have carried out a literature review of empirical case studies illustrating the advantages and disadvantages of computer-aided dispatch systems using priority dispatch protocols.

The search was initially carried out in the database ScienceDirect by using the term *computer-aided medical dispatch systems* across all search fields. We limited our search to the years 2000–2010 to focus on computer and telecommunication developments of the last decade. The search resulted in 169 articles. Many of these articles were found in subject areas not related to emergency care so the results were filtered according to the following subjects: computer science + decision sciences + medicine and dentistry + nursing and health professions. This search resulted in 84 articles. Then we further filtered down our results to 38 articles, by keeping only articles from journal titles directly related to emergency care, including *Resuscitation* (13), *Prehospital Emergency Care* (9), *Air Medical*

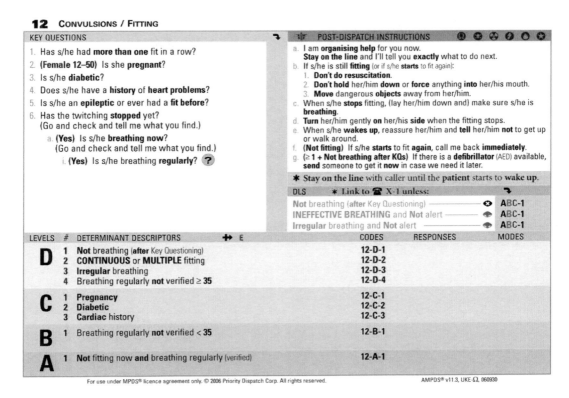

FIGURE 10.2
Chief complaint protocol for convulsions/fitting. (Courtesy of Priority Dispatch Corp., Salt Lake City, Utah.)

Journal (4), *Annals of Emergency Medicine* (4), *International Journal of Medical Informatics* (4), *American Journal of Preventive Medicine* (1), *Disaster Management & Response* (1), *Journal of the American Medical Informatics Association* (1), and *The Journal of Emergency Medicine* (1). To triangulate our filtered results, we carried out a further search in the databases PubMed and Cumulative Index to Nursing and Allied Health Literature (via EBSCOhost) using the same term *computer-aided medical dispatch systems* and came up with five more articles. From these combined results, we excluded review and editorial articles that did not report on empirical research findings on the direct or indirect impact of computer-aided medical dispatch systems on the outcome of emergency care provided by emergency teams, including the success rate of dispatchers' decisions on the severity of emergency calls. Table 10.1 lists the 10 most representative articles found in our literature review.

The key findings reported in these empirical case studies show that, despite the evident advantages of computer-aided dispatch systems utilizing priority dispatch protocols, effective emergency management and response cannot be minimized in ritualistic behavior through "blind" protocol following [45–46]. Protocols should be used as standardized, guidance tools, but emergency professionals should not rely on them to complete their tasks. This is because it is impossible to create work-flow scenarios that will adequately handle every type of call a dispatcher will take. Further, the criteria in these work-flow guides are not based on a strict yes-or-no answer and the questions are listed so that

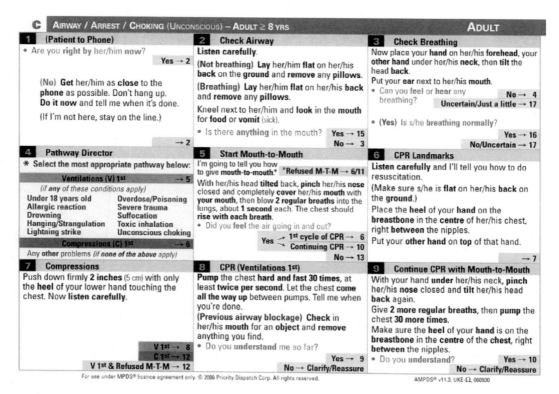

FIGURE 10.3
Ambulance prearrival instruction protocol. (Courtesy of Priority Dispatch Corp., Salt Lake City, Utah.)

dispatchers can easily guide a caller into follow-up questions but not necessarily to make a final diagnosis. More importantly, the effective management of emergency incidents is very much dependent on team characteristics such as how long have emergency team members spent on the same team, whether team members share the same levels of knowledge and expertise in dealing with different incidents, whether team members trust one another to complete a task successfully, and whether a task is an easy routine or involves a more complex scenario such as a mass accident [47–48].

10.4 mHealth eEmergency Systems

Technological applications for emergency healthcare support appeared in the literature more than 100 years ago [49] where Einthoven demonstrated a telemedicine application by connecting his lab with a university hospital in Holland at a distance of 1.5 km. Since then, many studies have been presented. Recent advances in mobile communications have also impacted mHealth eEmergency systems. In this section we present a literature review of studies published in journals related to mobile systems for emergency healthcare support (eEmergency and mHealth systems) that appeared since 2000.

TABLE 10.1

Empirical Case Studies on the Impact of Computer-Aided Dispatch Systems in Emergency Management and Response

Authors, Year	Country of Study	Key Objective	Key Findings
Garza et al. 2003 [35]	United States	Analyze the accuracy of dispatchers in predicting cardiac arrest through computer-aided dispatch protocols and assess the effect of the caller party on dispatcher accuracy	A higher level of medical training may improve dispatch accuracy for predicting cardiac arrest. The type of calling party influenced the dispatcher-assigned condition.
Dale et al. 2003 [36]	United Kingdom	Investigate the potential impact for ambulance services of telephone assessment and computer-aided triage for nonserious emergency calls as classified by ambulance service call takers	Telephone assessment and computer-aided triage can identify nonserious calls, which could have a significant impact on emergency ambulance dispatch rates. Nurses were more likely than paramedics to assess calls as requiring an alternative response to emergency ambulance dispatch, but the extent to which this relates to aspects of training and professional perspective is unclear.
Michael and Sporer 2005 [37]	United States	Evaluate a group of computer-aided dispatch protocols defined as requiring ALS intervention	There was variation in clinical practice toward ALS intervention due to the more precautionary approach to care found in this computer-aided dispatch system.
Deakin et al. 2006 [38]	United Kingdom	Examine patients with acute coronary syndrome (ACS) to identify whether a computer-aided dispatch system enabled dispatchers to allocate an appropriate emergency response	The system was found not to be an appropriate tool designed for clinical diagnosis, and its extension into this field does not enable accurate identification of patients with ACS. However, the system can be used to guide the appropriate level of clinical response to different emergency incidents.
Flynn et al. 2006 [39]	Australia	Undertake a sensitivity/specificity analysis to determine the ability of a computer-aided dispatch system to detect cardiac arrest	The system correctly identified 76.7% of cardiac arrest cases, but the number of false negatives suggests that there is room for improvement to maximize chances for survival in out-of-hospital cardiac arrest.
Reilly 2006 [40]	United States	Assess the relationship between dispatches of cardiac nature through a computer-aided dispatch system and the actual clinical diagnosis as determined by an emergency department physician	The system may overtriage emergency medical responses to cardiac emergencies. This can result in the only ALS unit in the community being unavailable in certain situations. Future studies should be conducted to determine what level of overtriage is appropriate when using such systems.

(Continued)

TABLE 10.1 (CONTINUED)

Empirical Case Studies on the Impact of Computer-Aided Dispatch Systems in Emergency Management and Response

Authors, Year	Country of Study	Key Objective	Key Findings
Feldman et al. 2006 [41]	Canada	Determine the relationship between a computer-aided dispatch system and an out-of-hospital patient acuity scale	The system exhibits at least moderate sensitivity and specificity for detecting high acuity of illness or injury. This performance analysis may be used to identify target protocols for future improvements.
Clawson et al. 2007 [42]	United Kingdom	Establish the accuracy of the emergency medical dispatcher's decisions to override the automated computer-aided dispatch system's triage recommendations based on at-scene paramedic-applied transport acuity determinations and cardiac arrest findings	Automated, protocol-based call taking is more accurate and consistent than the subjective, anecdotal, or experience-based determinations made by individual emergency medical dispatchers.
Buck et al. 2009 [43]	United States	Assess the diagnostic accuracy of the current national protocol guiding dispatcher questioning of emergency callers to identify stroke	Dispatcher recognition of stroke calls using a computer-aided dispatch system algorithm is suboptimal, with failure to identify more than half of stroke patients as likely stroke. Revisions to the current national dispatcher structured interview and symptom identification algorithm for stroke may facilitate more accurate recognition of stroke by emergency medical dispatchers.
Johnson and Sporer 2010 [44]	United States	Evaluate the number of emergency dispatches per cardiac arrest in cardiac arrest and non-cardiac arrest determinants found in a computer-aided dispatch system	The system was designed to detect cardiac arrest with high sensitivity, leading to a significant degree of mistriage. The number of dispatches for each cardiac arrest may be a useful way to quantify the degree of mistriage and optimize emergency dispatch.

10.4.1 Overview

The MEDLINE and IEEE Xplore databases were searched with the following key words: *wireless telemedicine emergency, wireless telemedicine ambulance, wireless telemedicine disaster, wireless ambulance, wireless disaster,* and *wireless emergency.* The number of journal papers found to be published under these categories is around 220. Out of these a total of 40 applications were selected and are briefly summarized in Tables 10.2 through 10.4. We tried to select systems that cover the whole spectrum of medical informatics applications for emergency cases that have been published in the last decade.

These papers are grouped under the following eEmergency areas as illustrated in Tables 10.2 through 10.4: ambulance systems (see Table 10.2), rural health center systems and in-hospital systems (see Table 10.3), and civilian systems (see Table 10.4). The column headings in these tables are coded as follows: ECG; Other Biosignals (biosignals like SpO_2, CO_2, heart rate, blood pressure, temperature, and respiration); Images (incident/patient scenery [SCN], X-ray, computed tomography imaging [CT], and magnetic resonance imaging [MRI]); video (SCN and ultrasound [US]); and Communication Link (GSM/GPRS/3G [WT], satellite [SAT], wireless local area network [WLAN], Bluetooth [BLUET], and sensor networks [SENSN]).

As shown in Tables 10.2 through 10.4, most of the mHealth eEmergency systems fall under the category of ambulance systems (see Table 10.2). These systems exploit the wireless telephone connectivity GSM/GPRS/3G. Almost all of the systems supported ECG transmission and other biosignals, whereas a few recent studies supported the ultrasound video transmission. Systems tabulated in Table 10.3 for eEmergency systems for rural health centers and in hospital cover mainly the transmission of medical images, including X-rays, CTs, and MRIs; two applications supported ultrasound video transmission.

Civilian eEmergency systems cover mainly the transmission of ECG in emergency cases (see Table 10.4). The system introduced by Virgin Atlantic Airways in 2006 [88] supports the monitoring of a passenger's blood pressure, pulse rate, temperature, ECG, blood oxygen, and carbon dioxide levels in emergency cases in the aircraft via the satellite communications link. Etihad Airlines has recently also announced the installation of an eEmergency system for monitoring the condition of passengers who display signs of illness that might require immediate medical attention [89].

10.4.2 Case Studies

Two case studies were selected and presented here in order to show examples of the evolution of medical informatics used for emergency healthcare support.

10.4.2.1 Case Study 1: Emergency Telemedicine— The AMBULANCE and Emergency 112 Projects

The availability of prompt and expert medical care can meaningfully improve healthcare services at understaffed rural or remote areas. The provision of effective emergency telemedicine and home monitoring solutions are the major fields of interest of Ambulance HC1001 and Emergency 112 HC4027 projects that were partially funded by the European Commission/Directorate General (DG) XIII Telematics Application Programme.

The aim of the AMBULANCE [56] project was the development of a portable emergency telemedicine device that supports real-time transmission of critical biosignals as well as still images of the patient using initially the GSM link and then the GPRS and 3G links.

TABLE 10.2

Overview of Selected Applications of mHealth eEmergency Systems

Authors	Year	ECG	Other Biosignals	Images	Video	Communication Link	Comments
			Data Transmitted				
Karlsten and Sjoqvist [50]	2000	✓				WT	Triage support
Xiao et al. [51]	2000	✓	✓		SCN	WT	Ambulance neurological examination support
Anantharaman and Han [52]	2001	✓	✓			WT	Prehospital support
Rodríguez et al. [53]	2001	✓	✓			WT	Cardiac arrest treatment
Istepanian et al. [54–55]	2001	✓	✓	SCN		WT	Transmission of ECG data and still images for emergency use; compression of ECG using a wavelet compression method
Pavlopoulos et al. [56]	1998	✓	✓			WT	Portable teleconsultation medical device
Chiarugi et al. [57]	2003	✓				WT	Transmission of 12-lead ECG in order to support ambulance and rural health center emergencies (HYGEIAnet)
Garrett et al. [58]	2003				US	WT	Echocardiogram transmission in cardiac emergency from an ambulance in transit to a tertiary care facility
Kyriacou et al. [59]	2003	✓	✓	SCN		WT	Wireless transmission of biosignals and images from a moving ambulance vehicle to a central hospital
Chu and Ganz [60]	2004	✓		SCN	SCN	WT	Trauma care through transmission of patient's video, medical images, and ECG
Clarke [61]	2004	✓				WT	Wireless connection to sensors and transmission of data from an ambulance
Clemmensen et al. [62]	2005	✓				WT	Transmission of ECG signals directly to a cardiologist's PDA to improve time to reperfusion

(Continued)

TABLE 10.2 (CONTINUED)

Overview of Selected Applications of mHealth eEmergency Systems

Authors	Year	Data Transmitted				Communication Link	Comments
		ECG	Other Biosignals	Images	Video		
Campbell et al. [63]	2005	✓				WT	Wireless transmission of ECG from the emergency scenery to the department and then through a WLAN to the on-call cardiologist who is carrying a PDA
Giovas et al. [64]	2006	✓				WT	Wireless transmission of 12-lead ECG from a moving ambulance vehicle to a central hospital
Sillesen et al. [65]	2006	✓				WT	Transmission of ECG signals to a cardiologist's PDA in order to improve time for PCI treatment
Kontaxakis et al. [66]	2006				US	WT	Tele-echography system and 3D US
Garawi et al. [67]	2006				US	WT	Teleoperated robotic system for mobile tele-echography (OTELO project)
Tsapatsoulis et al. [68]	2007				US	WT	Low-bit rate US video coding based on the d-ROI
Martini and Hewage [69]	2010				US	WT	Cardiac US video transmission using d-ROI H.264/AVC coding
Panayides et al. [70]	2011				US	WT	Diagnostically relevant (d-ROI) and resilient H.264/AVC coding of atherosclerotic plaque US video over wireless channels
Cavero et al. [71]	2012				US	WT	Communication protocol for cardiac US transmission over wireless networks
Panayides et al. [72]	2013				US	WT	Open-source telemedicine platform for medical video communications

Note: d-ROI: diagnostic region of interest; OTELO: mObile Tele-Echography using an ultra Light rObot; SCN: incident/patient scenery; US: ultrasound; WT: GSM/ GPRS/3G.

TABLE 10.3

Selected mHealth eEmergency Rural Health Center Systems and In-Hospital Systems

Authors	Year	Data Transmitted				Communication Link	Comments
		ECG	Other Biosignals	Images	Video		
Rural Health Centers							
Strode et al. [73]	2003				US	SAT	Examination of trauma using focused abdominal sonography (military)
Chiarugi et al. [57] Kouroubali et al. [74]	2003 2005	✓				WT	Transmission of 12-lead ECG to support ambulance and rural health centers emergencies (HYGEIAnet)
Kyriacou et al. [75]	2001	✓	✓	SCN		WT	Wireless transmission of biosignals and images from a moving ambulance vehicle to a central hospital
Garawi et al. [67] Vieyres et al. [76] Canero et al. [77]	2006 2006 2005				US	WT, SAT	Teleoperated robotic system for mobile tele-echography (OTELO project)
Hospital Systems							
Reponen et al. [78]	2000			CT		WT	Transmission of CT scans using GSM and PDAs; images transmitted to a neuroradiologist for a preliminary consultation
Oguchi et al. [79]	2001			CT		WT	Use of a personal handy phone system to transmit CT images using a web-based application
Voskarides et al. [80] Hadjinicolaou et al. [81]	2003 2009			X-ray		WT	Transmission of X-ray images in emergency orthopedics
Hall et al. [82]	2003					CT, WLAN	Wireless access to electronic patient record
Pagani et al. [83]	2003			CT		WLAN	Web-based transmission of cranial CT images
Lorincz et al. [84]	2004	✓				SENSN	Sensor networks for emergency response; system tested using two vital-signs monitors
Campbell et al. [63]	2005	✓				WLAN	Wireless transmission of ECG from the emergency scenery to the department and then through a WLAN to the on-call cardiologist who is carrying a PDA
Kim et al. [85]	2005			CT, MRI		WLAN	Transmission of CT and MRI images through a PDA and wireless high-bandwidth net to neurosurgeons
Kim et al. [86]	2009				✓	WLAN	Transmission of video and audio to consult on the treatment of acute stroke patients

Note: OTELO: mObile Tele-Echography using an ultra Light rObot; SCN: incident/patient scenery; SENSN: sensor networks; US: ultrasound; WT: GSM/GPRS/3G.

TABLE 10.4

Selected mHealth eEmergency Civilian Systems

Authors	Year	Data Transmitted				Communication Link	Comments
		ECG	Other Biosignals	Images	Video		
Salvador et al. [87]	2005	✓				WT	Transmission of ECG and other parameters to support patients with chronic heart diseases during an emergency case.
Virgin Atlantic Airways [88]	2006	✓	✓	SCN		SAT	The Tempus 2000 device will be used for monitoring a passenger's blood pressure, pulse rate, temperature, ECG, blood oxygen, and carbon dioxide levels in emergency cases in the aircraft.
Etihad Airways [89]	2010	✓	✓	SCN		SAT	Tempus IC will be installed on long-haul aircraft flights for monitoring the condition of passengers who display signs of illness that might require immediate medical attention.
Maki et al. [90]	2004	✓				SENSN	Wireless monitoring of sensors on persons that need continuous monitoring; when an emergency occurs, the specialized personnel hears a sound alarm or receives a notification through mobile phone.
Palmer et al. [91]	2005		✓			WLAN	Wireless blood pulse oximeter system for mass casualty events designed to operate in Wi-Fi hot spots; the system is capable of tracking hundreds of patients; suitable for disaster monitoring.
Lenert et al. [92]	2005	✓				WLAN	Medical care during mass casualty events, transmission of signals, and alerts monitor.
Nakamura et al. [93]	2003					WLAN	Wireless emergency telemedicine LAN with over 30 km distance coverage; used in the Japanese Alps for monitoring mountain climbers in emergency cases.
Lee et al. [94]	2007		✓			WT	Patient continuous monitoring/alert in case of emergency; signals are transmitted through GSM and acquisition to device is through Bluetooth.
Chin-Teng et al. [95]	2010	✓				BLUTH	A wearable system for detecting atrial fibrillation that uses expert systems.

Note: BLUTH: Bluetooth; SCN: incident/patient scenery; SENSN: sensor networks; US: ultrasound; WT: GSM/GPRS/3G.

This device can be used by paramedics or nonspecialized personnel that handle emergency cases, in order to get directions from expert physicians. The system comprises two different modules: (i) the mobile unit, which is located in an ambulance vehicle near the patient, and (ii) the consultation unit, which is located at the hospital site and can be used by the experts in order to give directions. The system allows telediagnosis, long-distance support, and teleconsultation of mobile healthcare providers by experts located at an emergency coordination center or a specialized hospital.

Emergency 112 [96–97], which was the extension of the AMBULANCE project, aimed at the development of an integrated portable medical device for emergency telemedicine. The system enables the transmission of critical biosignals (ECG, BP, HR, SpO_2, and temperature) and still images of the patient, from the emergency site to an emergency call center, thus enabling physicians to direct prehospital care in a more efficient way, improving patient outcomes, and reducing mortality. The system was designed in order to operate over several communication links such as satellite, GSM/GPRS/3G/HSPA, plain old telephone service (POTS), asymmetric digital subscriber line (ADSL), and integrated services digital network (ISDN). In Emergency 112, emphasis was given on maximizing the system's future potential application, through the utilization of several communication links (both fixed and wireless), as well as through an increase in the overall system's usability, focusing on advanced user interface and ergonomics. The system comprises two different modules: (i) The patient unit, which is the unit located near the patient, can operate automatically over several communication means and has several operating features (depending on the case used). (ii) The physician's unit, which is the unit located near the expert doctor, can operate over several communication links (depending on the place where the expert doctor is located). A snapshot from the transmitted biosignals between the two units can be seen in Figure 10.4. This system was further expanded through an Interreg III B Archimed project called Intermed. The current version of the portable unit of the system, entitled Abaris, is shown in Figure 10.5. According to Greek mythology, Abaris

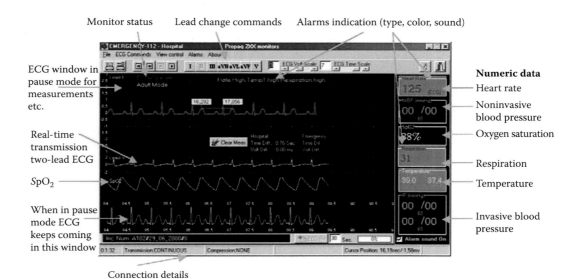

FIGURE 10.4
Biosignals transmitted between the telemedicine and base unit.

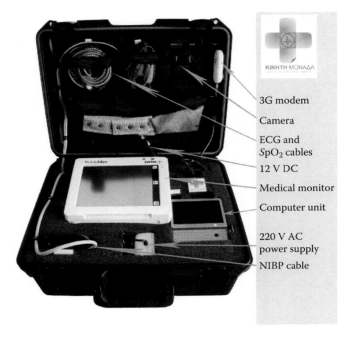

3G modem

Camera

ECG and
SpO$_2$ cables

12 V DC

Medical monitor

Computer unit

220 V AC
power supply

NIBP cable

FIGURE 10.5
Portable version of telemedicine unit.

was an ancient magician who had the gift of traveling from place to place with his arrow and curing people. The new system consists of a Welch Allyn monitor that is connected to a netbook that has the software and the control of the unit. The communication link is achieved via a Vodafone 3G modem.

The final system was used in emergency healthcare provision via different scenarios: in an ambulance vehicle, in rural health centers, in navigating ships, and others.

10.4.2.2 Case Study 2: Diagnostically Robust Ultrasound Video Transmission over Emerging Wireless Networks

The objective of the Diagnostically Robust Ultrasound Video Transmission over Emerging Wireless Networks (DRIVEN) project is to develop a unifying framework for the diagnostically resilient communication of medical video over existing 3.5G and currently deployed 4G wireless networks, enabling confident remote diagnosis. The goal is to develop new and integrate state-of-the-art methods in video coding, wireless networks, and medical video quality assessment into an adaptive framework. It is anticipated that the developed system will aid in establishing mHealth medical video communications systems in standard clinical practice.

Figure 10.6 depicts diagnostically driven medical video communication systems architecture. Preprocessing steps range from medical video denoising (removing speckle noise induced by ultrasound equipment during the acquisition process) to medical video segmentation (identifying the clinically important video regions) and spatiotemporal subsampling. Diagnostically relevant and resilient encoding adapts the encoding process to the

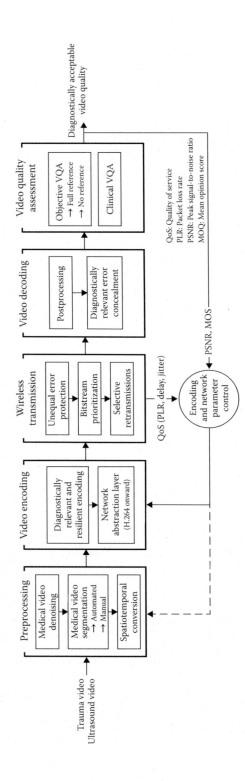

FIGURE 10.6
Diagnostically driven mHealth medical video communication systems architecture. (From A.S. Panayides et al., *IEEE Signal Proc. Mag.*, vol. 30, no. 6, pp. 163–172, Nov. 2013. doi:10.1109/MSP.2013.2276512.)

underlying medical video modality. A widely used method also adopted by the DRIVEN project is variable quality encoding, where diagnostically important video regions (d-ROIs) are encoded in finer quality than nonimportant (clinically wise) regions. Likewise, a d-ROI is protected more strongly using error-resilience mechanisms. In terms of wireless transmission, diagnostically driven approaches for reliable communication include bitstream prioritization compared to less demanding applications, and unequal error protection and selective retransmission of the d-ROI (where applicable). At the medical experts' (remote) side, postprocessing and diagnostically relevant decoding precedes objective and clinical video quality assessment.

While the DRIVEN framework can support all medical video modalities, including trauma and emergency scenery video, with minor modifications, developed algorithms focus on cardiovascular remote diagnosis applications, more specifically, in the transmission of videos of symptomatic atherosclerotic plaque (associated with critical stroke events) and abdominal aortic aneurysm (AAA) ultrasound, which can cause life-threatening internal bleeding. In both cases, timely interventions, especially for the elderly and patients with mobility problems, are supported. Early diagnosis and mass population screening of patients in isolated and rural areas, especially in developing countries, is expected to reduce patients' risks to more general cardiovascular diseases. Additional application scenarios range from emergency telematics to second opinion provision and medical education. The objective is to establish how specific diagnostic decisions can be safely made based on the communicated medical video.

High-resolution, low-delay, and high–frame rate medical video communication over 3.5G mobile WiMAX and HSPA wireless networks using diagnostically driven techniques is demonstrated by the DRIVEN platform in Panayides et al.'s works [72,98]. The diagnostically driven approach for atherosclerotic plaque ultrasound video encoding adopted in the DRIVEN project includes identifying the clinically sensitive video regions by using automated segmentation algorithms (alternatively, the d-ROI can be outlined by the relevant medical expert). The d-ROIs are, in turn, encoded in higher quality than nondiagnostically important regions and protected more strongly during transmission by using redundant slices (an H.264/AVC error-resilience technique). Moreover, intraencoded frames that are inserted approximately every second in the coded bitstream aim at limiting error propagation between cardiac cycles, hence maximizing the video's clinical robustness. The experimental evaluation demonstrates that significant reductions in bit rate demands are achieved using the proposed diagnostically relevant encoding methods, with bit rate gains extending up to 42% for 4CIF resolution [98], while also providing for transmission in noisy channels [70].

Ongoing work includes integrating the new HEVC standard in the DRIVEN framework and investigating efficient ultrasound video despeckling algorithms prior to encoding and transmission [99–101]. Initial results show that both approaches provide noteworthy bit rate gains that can be used toward encoding and communication of ultrasound video at the acquired resolution and frame rates. The latter will allow for a comparable experience of the standard, in-hospital ultrasound examination to be materialized for remote diagnosis purposes. As depicted in Figure 10.7 for a data set consisting of 12 atherosclerotic plaque ultrasound videos with a video resolution of 560 × 416 at 50 frames/s, HEVC achieves higher PSNR and SSIM objective ratings than its predecessor, the H.264/AVC standard, at lower bit rates.

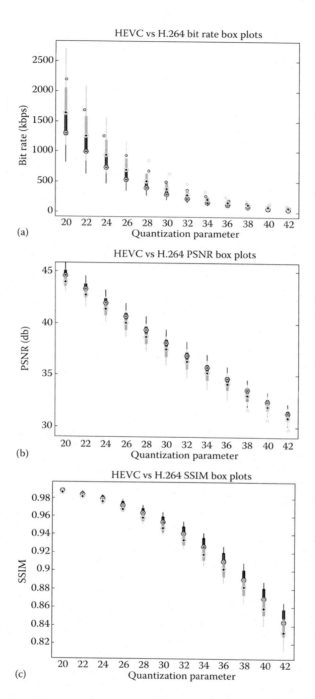

FIGURE 10.7
Box plots of HEVC (gray) versus H.264/AVC (dark) that demonstrate (a) bit rate and image quality, (b) in terms of peak signal-to-noise ratio (PSNR) and (c) structural similarity (SSIM), as a function of the quantization parameter (QP).

References

1. R.H. Istepanian, S. Laxminarayan, and C.S. Pattichis, Eds., *M-Health: Emerging Mobile Health Systems*, Springer, New York, 2006.
2. E. Kyriacou, M.S. Pattichis, C.S. Pattichis, A. Panayides, and A. Pitsillides, "m-Health e-Emergency systems: Current status and future directions," *IEEE Antenn. Propag. Mag.*, vol. 49, no. 1, pp. 216–231, 2007.
3. Rysavy Research, LLC, "Mobile Broadband Explosion: The 3GPP Wireless Evolution," Aug. 2013, http://www.4gamericas.org/.
4. P. Andersen, A.M. Lindgaard, M. Prgomet, N. Creswick, and J.L. Westbrook, "Mobile and fixed computer use by doctors and nurses on hospital wards: Multi-method study on the relationships between clinician role, clinical task, and device choice," *J. Med. Internet Res.*, vol. 11, no. 3, p. e32, Aug. 4, 2009.
5. C. Garrity and K. El Emam, "Who's using PDAs? Estimates of PDA use by health care providers: A systematic review of surveys," *J. Med. Internet Res.*, vol. 8, no. 2, p. e7, Apr.–Jun. 2006. Published online May 12, 2006. doi:10.2196/jmir.8.2.e7.
6. F. Axisa, C. Gehin, G. Delhomme, C. Collet, O. Robin, and A. Dittmar, "Wrist ambulatory monitoring system and smart glove for real time emotional, sensorial and physiological analysis," in *Proc. of the 26th Annual Int. Conf. of the IEEE EMBS*, San Francisco, California, pp. 2161–2164, 2004.
7. M. Bolaños, H. Nazeran, I. Gonzalez, R. Parra, and C. Martinez, "A PDA-based electrocardiogram/blood pressure telemonitor for telemedicine," in *Proc. of the 26th Annual Int. Conf. of the IEEE EMBS*, San Francisco, California, pp. 2169–2172, 2004.
8. Wealthy Project: Wearable Health Care System, IST 2001–3778, Commission of the European Communities, http://www.wealthyist.com, last accessed on Jan. 22, 2015.
9. E. Jovanov and D. Raskovic, "Wireless intelligent sensors," in *M-Health: Emerging Mobile Health Systems*, R.H. Istepanian, S. Laxminarayan, and C.S. Pattichis, Eds. Springer, New York, 2006, pp. 33–49.
10. Samsung Digital Health site, http://www.samsung.com/us/globalinnovation/innovation_areas/#digital-health, May 2014.
11. L.J. Hadjileontiadis, "Biosignals and compression standards," in *M-Health: Emerging Mobile Health Systems*, R.H. Istepanian, S. Laxminarayan, and C.S. Pattichis, Eds. Springer, New York, 2006, pp. 277–292.
12. M.J.B. van Ettinger, J.A. Lipton, M.C.J. de Wijs, N. der Putten, and S.P. Nelwan, "An open source ECG toolkit with DICOM," *Comput. Cardiol.*, vol. 35, pp. 441–444, 2008. Online: http://www.open-ecg-project.org, Aug. 2010.
13. I. Martínez, J. Escayola, J. Trigo, P. Muñoz, J. García, M. Martínez-Espronceda, and L. Serrano, "Implementation Guidelines for an End-to-End Standard-Based Platform for Personal Health," in *2009 Fourth International Multi-Conference on Computing in the Global Information Technology (ICCGI)*, pp. 123–131, 2009.
14. National Electrical Manufacturers Association (NEMA), Digital Imaging and Communications in Medicine (DICOM) Publication PS 3, Rosslyn, Virginia, U.S., 1996.
15. ISO/IEC 14995-1 Final Draft International Standard (FDIS), "Information Technology—Lossless and Near-Lossless Coding of Continuous-Tone Still Images: Baseline," JPEG-LS Standard, Part 1, Mar. 1999.
16. M. Boliek, C. Christopoulos, and E. Majani, Eds., "JPEG2000 Part I Final Draft International Standard" (ISO/IEC FDISI5444-1), ISO/IEC JTC1/SC29/WGINI855, Aug. 18, 2000.
17. S. Wong, L. Zaremba, D. Gooden, and H.K. Huang, "Radiologic image compression—A review," *Proc. IEEE*, vol. 83, no. 2, pp. 194–219, 1995.
18. P.W. Jones and M. Rabbani, "JPEG compression in medical imaging," Ch. 5 in *Handbook of Medical Imaging, Vol. 3: Display and PACS*. Y. Kim and S.C. Horii, Eds. SPIE Press, Bellingham, 2000, pp. 221–275.

19. J. Kivijarvi et al., "A comparison of lossless compression methods for medical images," *Comput. Med. Imag. Graphics*, vol. 22, pp. 323–339, 1998.

20. H. MacMahon et al., "Data compression: Effect on diagnostic accuracy in digital chest radiography," *Radiology*, vol. 178, pp. 175–179, 1991.

21. A.S. Panayides, M.S. Pattichis, and C.S. Pattichis, "Mobile-health systems use diagnostically driven medical video technologies," *IEEE Signal Proc. Mag.*, vol. 30, no. 6, pp. 163–172, Nov. 2013. doi:10.1109/MSP.2013.2276512.

22. ITU-T, "Video Codec for Audiovisual Services ar px64 kbit/s," ITU-T Recommendation H.261, Nov. 1990.

23. ITU-T, "Information Technology—Generic Coding of Moving Pictures and Associated Audio Information: Video," ITU-T Recommendation H.262, Jul. 1995.

24. ITU-T, "Video Coding for Low Bit Rate Communication," ITU-T Recommendation H.263, Nov. 1995.

25. ITU-T, "Advanced Video Coding for Generic Audiovisual Services," ITU-T and ISO/IEC 14496-10 Recommendation H.264 (MPEG4-AVC), May 2003.

26. T. Wiegand, G.J. Sullivan, G. Bjontegaard, and A. Luthra, "Overview of the H.264/AVC video coding standard," *IEEE Trans. Circ. Syst. Vid. Tech.*, vol. 13, p. 560–576, Jul. 2003.

27. ITU-T, "H.265: High Efficiency Video Coding," ITU-T Recommendation H.265, Jun. 2013.

28. J.-R. Ohm, G.J. Sullivan, H. Schwarz, T K. Tan, and T. Wiegand, "Comparison of the coding efficiency of video coding standards—Including high efficiency video coding (HEVC)," *IEEE Trans. Circ. Syst. Video Tech.*, vol. 22, no. 12, pp. 1669–1684, Dec. 2012.

29. G.J. Sullivan, J.-R. Ohm, W.-J. Han, and T. Wiegand, "Overview of the high efficiency video coding (HEVC) standard," *IEEE Trans. Circ. Syst. Video Tech.*, vol. 22, no. 12, pp. 1649–1668, Dec. 2012.

30. E. Aufderheide, "Disaster planning, part II: Disaster problems, issues, and challenges identified in the research literature," *Emerg. Med. Clin. North Am.*, vol. 14, pp. 453–480, 1996.

31. Improving NYPD Emergency Preparedness and Response. Post-9/11 Report of the NYPD. McKinsey & Company, New York, Aug 2002.

32. Post-9/11 Report of the Fire Department of New York (FDNY). McKinsey & Company, New York, Aug. 2002.

33. B.S. Zachariah and P.E. Pepe, "The development of emergency medical dispatch in the USA: A historical perspective," *Eur. J. Emerg. Med.*, vol. 2, no. 3, pp. 109–112, 1995.

34. B.S. Zachariah, P.E. Pepe, and P.A. Curka, "How to monitor the effectiveness of an emergency medical dispatch system: The Houston model," *Eur. J. Emerg. Med.*, vol. 2, no. 3, pp. 123–127, 1995.

35. A.G. Garza et al., "The accuracy of predicting cardiac arrest by emergency medical services dispatchers: The calling party effect," *Acad. Emerg. Med.*, vol. 10, no. 9, pp. 955–960, 2003.

36. J. Dale et al., "Computer assisted assessment and advice for 'non-serious' 999 ambulance service callers: The potential impact on ambulance dispatch," *Emerg. Med. J.*, vol. 20, no. 2, pp. 178–183, Mar. 2003.

37. G.E. Michael and K. Sporer, "Validation of low-acuity emergency medical services dispatch codes," *Prehosp. Emerg. Care*, vol. 9, no. 4, pp. 429–433, 2005.

38. C.D. Deakin et al., "Does telephone triage of emergency (999) calls using advanced medical priority dispatch (AMPDS) with Department of Health (DH) call prioritisation effectively identify patients with an acute coronary syndrome? An audit of 42 657 emergency calls to Hampshire Ambulance Service NHS Trust," *Emerg. Med. J.*, vol. 23, pp. 232–235, 2006.

39. J. Flynn et al., "Sensitivity and specificity of the medical priority dispatch system in detecting cardiac arrest emergency calls in Melbourne," *Prehosp. Disaster Med.*, vol. 21, no. 2, pp. 72–76, 2006.

40. L.J. Reilly, "Accuracy of a priority medical dispatch system in dispatching cardiac emergencies in a suburban community," *Prehosp. Disaster Med.*, vol. 21, no. 2, pp. 77–81, Mar.–Apr. 2006.

41. M.J. Feldman et al., "Comparison of the medical priority dispatch system to an out-of-hospital patient acuity score," *Acad. Emerg. Med.*, vol. 13, no. 9, pp. 954–960, 2006.

42. J. Clawson et al., "Accuracy of emergency medical dispatchers' subjective ability to identify when higher dispatch levels are warranted over a Medical Priority Dispatch System automated protocol's recommended coding based on paramedic outcome data," *Emerg. Med. J.*, vol. 24, pp. 560–563, 2007.

43. B. Buck et al., "Dispatcher recognition of stroke using the National Academy Medical Priority Dispatch System," *Stroke*, vol. 40, pp. 2027–2030, 2009.

44. N. Johnson and K. Sporer, "How many emergency dispatches occurred per cardiac arrest," *Resuscitation*, vol. 81, no. 11, pp. 1499–1504, Nov. 2010.

45. T.E. Drabek and D.A. McEntire, "Emergent phenomena and the sociology of disaster: Lessons, trends and opportunities from the research literature," *Disaster Prev. Manag.*, vol. 12, no. 2, pp. 97–112, 2003.

46. K.J. Tierney, M.K. Lindell, and R.W. Perry, *Facing the Unexpected: Disaster Preparedness and Response in the United States*. Joseph Henry Press, Washington, D.C., 2001.

47. A. Majchrzak, S.L. Jarvenpaa, and A.B. Hollingshead, "Coordinating expertise among emergent groups responding to disasters," *Organ. Sci.*, vol. 18, no. 1, pp. 147–161, 2007.

48. P. Constantinides, A. Kouroubali, and M. Barrett, "Transacting expertise in emergency management and response," in *Proc. of the International Conference of Information Systems*, Paris, France, Dec. 8–10, 2008.

49. W. Einthoven, "Le telecardiogramme," *Archives Internationales Physiologie*, vol. IV, pp. 132–164, 1906.

50. R. Karlsten and B.A. Sjoqvist, "Telemedicine and decision support in emergency ambulances in Uppsala," *J. Telemed. Telecare*, vol. 6, no. 1, pp. 1–7, 2000.

51. Y. Xiao, D. Gagliano, M. LaMonte, P. Hu, W. Gaasch, and R. Gunawadane, "Design and evaluation of a real-time mobile telemedicine system for ambulance transport," *J. High Speed Netw.*, vol. 9, no. 1, pp. 47–56, 2000.

52. V. Anantharaman and L.S. Han, "Hospital and emergency ambulance link: Using IT to enhance emergency pre-hospital care," *Int. J. Med. Inform.*, vol. 61, nos. 2–3, pp. 147–161, May 2001.

53. A. Rodríguez, J.L. Villalar, M.T. Arredondo, M.F. Cabrera, and F. Del Pozo, "Transmission trials with a support system for the treatment of cardiac arrest outside hospital," *J. Telemed. Telecare*, vol. 7, suppl. 1, pp. 60–62, Sept. 2001.

54. R.S. Istepanian, L.J. Hadjileontiadis, and S.M. Panas, "ECG data compression using wavelets and higher order statistics methods," *IEEE Trans. Inf. Technol. Biomed.*, vol. 5, no. 2, pp. 108–115, Jun. 2001.

55. R.S Istepanian, E. Kyriacou, S. Pavlopoulos, and D. Koutsouris, "Effect of wavelet compression on data transmission in a multipurpose wireless telemedicine system," *J. Telemed. Telecare*, vol. 7, suppl. 1, pp. 14–16, 2001.

56. S. Pavlopoulos, E. Kyriacou, A. Berler, S. Dembeyiotis, and D. Koutsouris "A novel emergency telemedicine system based on wireless communication technology—AMBULANCE," *IEEE Trans. Inform. Tech. Biomed.—Special Issue on Emerging Health Telematics Applications in Europe*, vol. 2, no. 4, pp. 261–267, 1998.

57. F. Chiarugi et al., "Continuous ECG monitoring in the management of pre-hospital health emergencies," *Comput. Cardiol.*, pp. 205–208, 2003.

58. P.D. Garrett et al., "Feasibility of real-time echocardiographic evaluation during patient transport," *J. Am. Soc. Echocardiogr.*, vol. 16, no. 3, pp. 197–201, 2003.

59. E. Kyriacou et al., "Multi-purpose healthcare telemedicine systems with mobile communication link support," *BioMedical Engineering OnLine*, http://www.biomedical-engineering-online.com, vol. 2, no. 7, 2003.

60. Y. Chu and A. Ganz, "A mobile teletrauma system using 3G networks," *IEEE Trans. Inf. Technol. Biomed.*, vol. 8, no. 4, pp. 456–462, 2004.

61. M. Clarke, "A reference architecture for telemonitoring," *Stud. Health Technol. Inform.*, vol. 103, pp. 381–384, 2004.

62. P. Clemmensen et al., "Diversion of ST-elevation myocardial infarction patients for primary angioplasty based on wireless prehospital 12-lead electrocardiographic transmission directly to the cardiologist's handheld computer: A progress report," *J. Electrocardiol.*, vol. 38, no. 4, pp. 194–198, 2005.

63. P.T. Campbell et al., "Prehospital triage of acute myocardial infarction: Wireless transmission of electrocardiograms to the on-call cardiologist via a handheld computer," *J. Electrocardiol.*, vol. 38, no. 4, pp. 300–309, 2005.

64. P. Giovas et al., "Medical aspects of prehospital cardiac telecare," in *M-Health: Emerging Mobile Health Systems*, R.H. Istepanian, S. Laxminarayan, and C.S. Pattichis, Eds. Springer, New York, 2006, pp. 389–400.

65. M. Sillesen et al., "Telemedicine in the transmission of prehospitalisation ECGs of patients with suspected acute myocardial infarction," *Ugeskr Laeger*, vol. 168, no. 11, pp. 1133–1136, 2006.

66. G. Kontaxakis, G. Sakas, and S. Walter, "Mobile tele-echography systems—TELEINVIVO: A case study," in *M-Health: Emerging Mobile Health Systems*, R.H. Istepanian, S. Laxminarayan, and C.S. Pattichis, Eds. Springer, New York, 2006, pp. 445–460.

67. S. Garawi, R.S.H. Istepanian, and M.A. Abu-Rgheff, "3G wireless communications for mobile robotic tele-ultrasonography systems," *IEEE Commun. Mag.*, vol. 44, no. 4, pp. 91–96, 2006.

68. N. Tsapatsoulis, C. Loizou, and C. Pattichis, "Region of interest video coding for low bit-rate transmission of carotid ultrasound videos over 3G wireless networks," in *Proc. of IEEE EMBC '07*, Lyon, France, Aug. 23–26, 2007.

69. M.G. Martini and C.T.E.R. Hewage, "Flexible macroblock ordering for context-aware ultrasound video transmission over mobile WiMAX," *Int. J. Telemed. Appl.*, vol. 2010, article ID 127519, 14 pages, 2010.

70. A. Panayides, M.S. Pattichis, C.S. Pattichis, C.P. Loizou, M. Pantziaris, and A. Pitsillides, "Atherosclerotic plaque ultrasound video encoding, wireless transmission, and quality assessment using H.264," *IEEE Trans. Inform. Tech. Biomed.*, vol. 15, no. 3, pp. 387–397, May 2011.

71. E. Cavero, A. Alesanco, and J. Garcia, "Enhanced protocol for real time transmission of echocardiograms over wireless channels," *IEEE Trans. Biomed. Eng.*, vol. 59, no. 11, pp. 3212–3220, Nov. 2012.

72. A. Panayides, I. Eleftheriou, and M. Pantziaris, "Open-source telemedicine platform for wireless medical video communication," *Int. J. Telemed. Appl.*, vol. 2013, article ID 457491, 12 pages, doi:10.1155/2013/457491, 2013.

73. C.A. Strode et al., "Wireless and satellite transmission of prehospital focused abdominal sonography for trauma," *Prehosp. Emerg. Care*, vol. 7, no. 3, pp. 375–379, 2003.

74. A. Kouroubali, D. Vourvahakis, and M. Tsiknakis, "Innovative practices in the emergency medical services in Crete," in *Proc. of the 10th International Symposium on Health Information Management Research—iSHIMR 2005*. P.D. Bamidis et al., Eds. pp. 166–175, 2005.

75. E. Kyriacou, S. Pavlopoulos, D. Koutsouris, A. Andreou, C. Pattichis, and C. Schizas, "Multipurpose health care telemedicine system," *Proceedings of the 23rd Annual International Conference of the IEEE/EMBS*, Istanbul, Turkey, 2001.

76. P. Vieyres et al., "A tele-operated robotic system for mobile tele-echography: The OTELO project," in *M-Health: Emerging Mobile Health Systems*, R.H. Istepanian, S. Laxminarayan, and C.S. Pattichis, Eds. Springer, New York, pp. 461–473, 2006.

77. C. Canero et al., "Mobile tele-echography: User interface design," *IEEE Trans. Inf. Tech. in Biomed.*, vol. 9, no. 1, pp. 44–49, 2005.

78. J. Reponen, E. Ilkko, L. Jyrkinen, O. Tervonen, J. Niinimäki, V. Karhula, and A. Koivula, "Initial experience with a wireless personal digital assistant as a teleradiology terminal for reporting emergency computerized tomography scans," *J. Telemed. Telecare*, vol. 6, no. 1, pp. 45–49, 2000.

79. K. Oguchi et al., "Preliminary experience of wireless teleradiology system using personal handyphone system," *Nippon Igaku Hoshasen Gakkai Zasshi*, vol. 61, no. 12, pp. 686–687, Oct. 2001.

80. S.Ch. Voskarides, C.S. Pattichis, R. Istepanian, C. Michaelides, and C.N. Schizas, "Practical evaluation of GPRS use in a telemedicine system in Cyprus," in *CD-ROM Proceedings of the 4th Int. IEEE EMBS Special Topic Conference on Information Technology Applications in Biomedicine*, Birmingham, U.K., 4 pages, 2003.

81. M.G. Hadjinicolaou, R. Nilavalan, T. Itagaki, S.Ch. Voskarides, C.S. Pattichis, and A.N. Schizas, Emergency teleorthopaedics m-health system for wireless communication links, *IET Commun.*, vol. 3, pp. 1284–1296, 2009.

82. E.S. Hall, D.K. Vawdrey, C.D. Knutson, and J.K. Archibald, "Enabling remote access to personal electronic medical records," *IEEE Eng. Med. Biol. Mag.*, vol. 22, no. 3, pp. 133–139, 2003.

83. L. Pagani et al., "A portable diagnostic workstation based on a Webpad: Implementation and evaluation," *J. Telemed. Telecare*, vol. 9, no. 4, pp. 225–229, 2003.

84. K. Lorincz et al., "Sensor networks for emergency response: Challenges and opportunities," *IEEE Pervasive Computing*, vol. 3, no. 4, pp. 16–23, Oct.–Dec. 2004.

85. D.K. Kim, S.K. Yoo, and S.H. Kim, "Instant wireless transmission of radiological images using a personal digital assistant phone for emergency teleconsultation," *J. Telemed. Telecare*, vol. 11, no. 2, pp. S58–S61, 2005.

86. D.K. Kim et al., "A mobile telemedicine system for remote consultation in cases of acute stroke," *J. Telemed. Telecare*, vol. 15, pp. 102–107, 2009.

87. C.H. Salvador et al., "Airmed-cardio: A GSM and Internet services-based system for out-of-hospital follow-up of cardiac patients," *IEEE Trans. Inf. Technol. Biomed.*, vol. 9, no. 1, pp. 73–85, 2005.

88. "Virgin to upgrade telemedicine across fleet," E-Health Insider, http://www.e-health-insider .com/news/item.cfm?ID=1925, Announced: Jun. 6, 2006.

89. "Etihad to Install Onboard Health Monitor System on Long—Haul Fleet," http://www.air linesanddestinations.com/airlines/etihad-to-install-onboard-health-monitor-system-on-long -haul-fleet/, Jul. 30, 2010.

90. H. Maki et al., "A welfare facility resident care support system," *Biomed. Sci. Instrum.*, vol. 40, pp. 480–483, 2004.

91. D.A. Palmer, R. Rao, and L.A. Lenert, "An 802.11 wireless blood pulse-oximetry system for medical response to disasters," in *Proc. AMIA Annual Symposium*, p. 1072, 2005.

92. L.A. Lenert, D.A. Palmer, T.C. Chan, and R. Rao, "An intelligent 802.11 triage tag for medical response to disasters," in *Proc. AMIA Annual Symposium*, pp. 440–444, 2005.

93. M. Nakamura, Y. Yang, S. Kubota, H. Shimizu, Y. Miura, K. Wasaki, Y. Shidama, and M. Takizawa, "Network system for alpine ambulance using long distance wireless LAN and CATV LAN," *Igaku Butsuri*, vol. 23, no. 1 pp. 30–39, 2003.

94. R.G. Lee et al., "A mobile care system with alert mechanism," *IEEE Trans Inf. Tech. Biom.*, vol. 11, no. 5, pp. 507–517, 2007.

95. L. Ching-Teng et al., "An intelligent telecardiology system using a wearable and wireless ECG to detect atrial fibrillation," *IEEE Trans. Inf. Technol. Biomed.*, vol. 14, no. 3, pp. 726–733, May 2010.

96. C. Antoniades, A. Kouppis, S. Pavlopoulos, E. Kyriakou, A. Kyprianou, A.S. Andreou, C. Pattichis, and C. Schizas, "A novel telemedicine system for the handling of emergency cases," *Proc. of the 5th World Conference on Injury Prevention and Control*, World Health Organization (WHO), New Delhi, India, 2000.

97. E. Kyriacou, S. Pavlopoulos, A. Bourka, A. Berler, and D. Koutsouris, "Telemedicine in Emergency Care," *Proc. of the VI Int. Conf. on Medical Physics, Patras 99*, pp. 293–298, Patra, Greece, Sep. 1999.

98. A. Panayides, Z. Antoniou, Y. Mylonas, M.S. Pattichis, A. Pitsillides, and C.S. Pattichis, "High-resolution, low-delay, and error-resilient medical ultrasound video communication using H.264/AVC over mobile WiMAX networks," *IEEE J. Biomed. Health Informat.*, vol. 17, no. 3, pp. 619–628, May 2013.

99. A. Panayides, Z. Antoniou, M.S. Pattichis, C.S. Pattichis, and A.G. Constantinides, "High effi-ciency video coding for ultrasound video communication in m-health systems," *2012 Annual International Conf. of the IEEE Engineering in Medicine and Biology Society (EMBC)*, pp. 2170–2173, Aug. 28–Sep. 1, 2012.

100. A. Panayides, M.S. Pattichis, and C.S. Pattichis, "HEVC encoding for reproducible medical ultrasound video diagnosis," in *Proc of Asilomar Conference on Signals, Systems, and Computers 2013*, pp. 1117–1121, California, U.S., Nov. 3–6, 2013.

101. A.S. Panayides, M.S. Pattichis, C.P. Loizou, M. Pantziaris, A.G. Constantinides, and C.S. Pattichis, "An effective ultrasound video communication system using despeckle filtering and HEVC," *IEEE J. Biomed. Health Informat.*, vol. 19, no. 2, pp. 668–676. doi: 10.1109/JBHI.2014.2329572. Epub ahead of print.

Section III

Networks and Databases, Informatics, Record Management, Education, and Training

11

Global and Local Health Information, Databases, and Networks

**Kostas Giokas, Yiannis Makris, Anna Paidi, Marios Prasinos,
Dimitra Iliopoulou, and Dimitris Koutsouris**

CONTENTS

11.1 Introduction

In today's world, people stay in touch with each other by using different communication methods and have access to information via smartphones or conventional Internet networks. This ease in communication capabilities and access to all types of information flow from individual users to local and global health systems. Any loss of communication can have profoundly adverse effects on our lives and our ecosystems, thus exerting stress on our health, education, and livelihood. Also, many IT systems (including health systems) in the market today are not designed to be interoperable. This imposes an additional problem in the overall policy design of effective healthcare delivery.

In this chapter, we discuss health data that are collected and are accessed on local and global levels. Health data on the local level includes all sources that generate, collect, and record health data citywide, region-wide, and nationwide.

11.2 Local Health Data

Healthcare data are sensitive and complex, are produced by many sources, and have to be quality controlled, fresh, timely, and readily retrieved or should be used for any purpose. Additionally, healthcare data volumes are and can be very large.

The European Committee for Standardization or Comité Européen de Normalisation (CEN)/Technical Committee (TC) 251 gives the following definition for datum: a *datum* is any single observation or fact. A medical datum generally can be regarded as the value of a specific parameter at a specific time [1].

Health data are obtained or acquired from patients, either by listening, by looking, or via devices that record signals or images. Healthcare professionals generate and record data as a result of their observations or from the use of various devices. The data healthcare professionals generate and access consist of text, numbers, signals, sounds, and images.

Public health data are data that are open to all users. Many public data sets are available on the Internet and can be easily downloaded for use.

Private health data are data not approved for public release because the data contain personally identifiable information that is considered confidential.

Local health data are basically divided into four basic types:

1. Patient demographics
2. Activity data which consists primarily of clinical data collected from various sources
3. Resource data (financial, workforce, material, and information from external sources)
4. Health service provider data [1]

11.2.1 Collection of Local Health Data

Collections of local health data are the processes by which data elements are accumulated.

Primary data are collected and recorded either manually or electronically. Manual data collection and recording has higher probabilities of errors due to the intervention of human factors. Electronic data collection may be achieved via bar code readers, optical mark readers, or point-and-click devices or directly from other electronic devices or databases. The electronic way of collection conforms to standards and ensures accuracy, consistency, and completeness of data.

National data sets define a standard set of information that is generated from care records, from any organization or system that captures the primary data. They are structured lists of individual data items, each with a clear label, definition, and set of permissible values, codes, and classifications. From this, secondary-use information is derived or compiled, which can then be used to monitor and improve services [2].

11.2.2 Warehousing of Local Health Data

Processes and systems are used to store and maintain data. A data warehouse is a repository of digital records in which data from multiple sources are brought together, integrated, unified, managed, and made accessible to multiple users [3]. Today local and national government organizations and agencies use healthcare data warehouses with data from different sources. All these local health data are kept independently by each organization and usually not shared or integrated. The quality of the data determines the quality of the database in which the data are stored. Unfortunately, the quality of data within an individual organization is essential but not sufficient. The growing costs of healthcare; the tremendous volumes of data; the fact that healthcare data have to be quality controlled, fresh, timely, and readily retrieved or used for any purpose; and their sensitivity and complexity increase the need to build healthcare data warehousing region-wide or nationwide. A local health data network provides statistical information that governments and community can use to promote discussion and make decisions on health and community services while at the same time being committed to ensuring that data privacy and confidentiality are not compromised.

11.2.3 Analysis of Local Health Data

Local health data are translated into information and knowledge utilized for an application. Health data are analyzed for many different reasons in order to support clinical, management, and policy decision making at various levels within the health system. Data analysis is achieved in a variety of ways, as the different types of data require different analytic methods appropriate to the form of the data and to the research question. There are numerous software packages that support all stages of health data analysis and facilitate presentation and dissemination of the results.

The resources of health data are categorized by type as follows:

A *primary data source* is an agency that collects data for its own purposes but is willing to provide those data to others for additional purposes. An example is a health department that collects vital statistics data from birth certificates for its own surveillance purposes and then makes those data available to others, such as researchers interested in assessing the relationship between low birth weights and environmental exposures [4].

A *secondary data source* is an agency that provides interpretation and analysis of primary data sources. An example is a business intelligence system at a regional health authority that collects data for hospitals and other health agencies, adopts innovative analytic approaches, and supports healthcare leaders in the decision-making process. Another example of secondary data source is the nationwide Diagnosis-Related Group (DRG) system, which accumulates primary health data and analyzes them, produces cost weights, and defines the hospitals' budgets [4].

11.2.4 Local Health Data Network

Local health data are produced by many different organizations at citywide, region-wide, and nationwide levels. A local health data network consists of the following:

1. Citywide and region-wide
   a. Healthcare professionals (electronic patient record: patient history, physical examination, and patient prescription)
   b. Hospitals (patient demographics, admissions and attendances, clinical data, laboratory results, health service provider data, financial data, workforce data, etc.)
   c. Diagnostic centers (laboratory results, images, and biosignals)
   d. Home care agencies (physical examination and biosignals)
   e. Primary healthcare agencies (prescriptions, patient history, physical examination, and laboratory results)
   f. Public health data (demographics, health promotion data, health prevention data, lifestyle factors, data on vaccine uptake, alcohol, smoking, and deaths)
   g. Local government organizations (mortality and fertility data)
2. Nationwide
   h. Government healthcare data and statistics organizations (ePrescription: hospital prescriptions, primary care prescriptions, hospital care admissions, and attendances; hospital care outcomes: deaths, discharges, readmissions, healthcare costs, and disease classification)
   i. Public health databases and registries (births, deaths, demographics, aging, alcohol, cancer, chronic diseases, smoking and tobacco, and life expectancy)
   j. Centers for disease control and prevention (epidemiological data for infectious disease)
   k. National organization for medicines (medicinal products, pharmacovigilance, recalls of medicines, and biocides)
   l. National transplant organization (national organ and tissue donor register and national register of organ recipient candidates)
   m. Institute of child health (newborn infants screening data for metabolic, genetic, and endocrine disorders)
   n. National statistics agency or authority (accidents, health prevention data, etc.)

There are the several aspects that are related to the application, collection, warehousing, and analysis of health data. The following presents all of these aspects:

- **Accuracy:** Ensure that the data are the correct values, valid, and attached to the correct patient record.
- **Accessibility:** Data items should be easily obtainable and legal to access with strong protections and controls built into the process.
- **Comprehensiveness:** All required data items are included. Ensure that the entire scope of the data is collected and document intentional limitations.
- **Consistency:** The value of the data should be reliable and the same across applications.
- **Current validity:** The data should be up to date.

- **Definitions:** Each data element should have a clear meaning and acceptable values.
- **Granularity:** The attributes and values of data should be defined at the correct level of detail.
- **Precision:** Data values should be just large enough to support the application or process.
- **Relevance:** The data are meaningful to the performance of the process or application for which they are collected.
- **Timeliness:** Timeliness is determined by how the data are being used and their context [5].

11.2.5 Challenges and Inefficiencies Associated with a Local Health Data Network

11.2.5.1 Data Complexity and Integration

Health data are more complex than other categories of data. The electronic patient record is a collection of many records of different types and health organizations manage millions of electronic patient records along with much other data about providers, medication, equipment, supplies, facilities, and diseases. All these healthcare data are coded and these codes are used inconsistently across the healthcare system and across time. The challenge of integration of health data concerns deeply those that have to load and integrate an unusually complex data model from numerous sources since identifying all data is often very complex, as different standards are used among sources, the healthcare records have long data life, and the identifiers of the people, organizations, facilities, and other entities are not necessary consistent across sources or over time. Another issue is temporal data and location data as time and location are semantic factors in healthcare and the analysis of time-related events or geographic data on a large scale is challenging.

11.2.5.2 Privacy, Security, and Patients' Consent

Privacy, security, and patients' consent are always challenges as they are the focus of healthcare policy makers and health informatics professionals and will be the focus of society for years to come. Another challenge is the arising issue of security and privacy in access and management of electronic health records that are often shared and integrated in healthcare cloud computing infrastructures. As health data and information exchange between healthcare providers and organizations increases, patient trust must be ensured and patients may more often be asked to make a "consent decision." This consent decision concerns the sharing and accessing of the patient's health information for treatment, payment, and healthcare operation purposes. Health providers and health IT implementers have to help patients make a consent decision meaningful.

11.3 Databases

11.3.1 Introduction

There are several database architectures currently in use. These databases are applied in the health industry as well as in many other fields such as business and commerce.

Therefore, here, a general introduction is given and the choice of the right architecture for storing health data relies on several criteria including the following:

- Where do the data and database management system (DBMS) reside?
- Where are the application programs executed (e.g., which central processing unit [CPU])? This may include the user interface.
- Where are the business rules enforced?

In this chapter there is a brief description of the main database architectures and a listing of some database system concepts on healthcare.

11.3.2 Database Architectures

11.3.2.1 Traditional Architectures

Centralized systems or centralized database systems are those that run on a single computer system and do not interact with other systems.

Client–server systems have their functionality split between a server system and multiple client systems.

11.3.2.2 Server System Architectures

Transaction servers systems, also known as query server systems, provide an interface to which clients can send requests to perform an action, in response to which they execute the action and send back results to the client.

A *data server* allows clients to interact with the servers by making requests to read or update data, in units such as files or pages.

In a *cloud-based server system architecture*, there are a number of servers owned by a third party that is neither the client nor the service provider.

11.3.3 Parallel Systems

Parallel database systems consist of multiple processors and multiple disks which are part of a fast interconnection network.

Parallel database architectures include the following:

- *Shared-memory architecture:* The processors and disks have access to a common memory, typically via a bus or through an interconnection network.
- *Shared-disk architecture:* All processors can access all disks directly via an interconnection network, but the processors have private memories. Shared-disk systems are also called clusters.
- *Shared-nothing architecture:* The processors share neither a common memory nor a common disk. Each node is independent and acts like a server for the disk it owns.
- *Hierarchical architecture:* This model is a hybrid of the preceding three architectures.

There are various types of parallelism including the following:

- *Input/output (I/O) parallelism:* The relations are partitioned among available disks so that they can be retrieved faster. Three commonly used partitioning techniques are round-robin partitioning, hash partitioning, and range partitioning.
- *Interquery parallelism:* In this case, different queries or transactions execute in parallel with one another. The primary use of this type of parallelism is to scale up a transaction-processing system to support a larger number of transactions per second.
- *Intraquery parallelism:* This type of parallelism refers to the execution of a single query in parallel on multiple processors and disks.

Interoperation parallelism: There are two forms of interoperation parallelism—pipelined parallelism and independent parallelism. In independent parallelism, different independent operations are executed in parallel, whereas in pipelined parallelism, processors send the results of one operation to another operation as those results are computed without waiting for the entire operation to finish. For example, the execution of a single query can be parallelized in two different ways: (1) processing of a query by parallelizing the execution of each individual operation, such as sort, select, project, and join, and (2) by executing in parallel the different operations in a query expression.

We can use intraoperation parallelism to execute relational operations, such as sorts and joins, in parallel. This type of parallelism is natural for relational operations, since they are set oriented.

Query optimization: In parallel databases, query optimization is significantly more complex than in sequential databases.

11.3.4 Distributed Systems

In distributed database architectures, the database is stored on several computers [6]. The computers in a distributed system do not have shared storage or main memory and communicate with one another through either high-speed dedicated networks or the Internet. The computers in this system may vary in size and function, ranging from workstations up to mainframes.

Replication and duplication are the two procedures which secure that the distributed databases stay up to date. Replication includes the usage of specific software which searches for potential changes in the distributive database. As soon as the changes have been recognized, then all the databases are made to look alike by the replication procedure. This procedure can be very complex and can take up a considerable time depending on the number of the databases.

The other process is duplication, as mentioned, which is less complex and copies the database which identifies as being the master (the authoritative database). This procedure is usually done at a set time and makes sure that each distributed site has the same data. Both procedures maintain the data up to date in all distributive places.

Significant considerations should be taken into account when dealing with distributed databases. First of all, it has to be ensured that the distribution is transparent, which means that all users should be capable of interacting with the system. Each action must retain integrity, meaning that all the data which are inserted into a database are correct and reliable.

When storing a reference k in a distributed database, we have to take into account replication and fragmentation. In replication, many identical replicas of the same reference k are kept by the system in different locations. In this way, data are more available and parallelism is enhanced when read demand is served. In fragmentation, the reference k is fragmented into different references ($k_1, k_2, k_3, \ldots, k_n$) but in a way that the initial reference can always be reconstructed from the fragments and then the fragments are thrown away to different sites.

Distributed databases have several advantages, such as enhanced credibility and availability, the easier expansion that they offer, and that data can be distributed and stored in different systems. They are secure and can be accessed over different networks and data can be updated from different tables which are located on different machines. Even if a system fails, the integrity of the distributed database is maintained. Also, a user does not know where the data are located physically and the database presents the data to the user as if it were located locally. Finally, the systems can be changed, added, or removed without influencing other systems and continuous operation is ensured even if some systems go off-line.

The disadvantages include reduction in performance since the data are accessed from a remote system and the difficulty of database optimization. Also, network traffic is increased in a distributed database and different data formats and DBMS products are used in different systems, which increase the complexity of the system. Finally, managing system catalog as well as distributed deadlock is a difficult and time-consuming task.

11.4 Database System Concepts in Healthcare

11.4.1 World Health Organization Classifications

There are some main data classifications of basic parameters of health. These classifications have been prepared by the World Health Organization (WHO) and approved by the organization's governing bodies for international use.

The main types of classifications include the following:

- International Classification of Diseases (ICD)
- International Classification of Functioning, Disability, and Health (ICF)
- International Classification of Health Interventions (ICHI)

11.4.2 General Online Health Databases

11.4.2.1 European Health for All Database

The European Health For All database (HFA-DB) provides a selection of core health statistics covering basic demographics, health status, health determinants and risk factors, and healthcare resources, utilization, and expenditure in the 53 countries in the WHO European Region [7]. It allows queries for country, intercountry, and regional analyses and displays the results in tables, graphs, or maps, which can be exported for further use.

11.4.2.2 The National Institutes of Health Intramural Database

The National Institutes of Health Intramural Database (NIDB) collects [8] and disseminates information about research being performed in the intramural programs of the institutes and centers of the National Institutes of Health of the United States.

11.4.2.3 Other European Online Health Databases

Apart from HFA-DB, there are many other European online databases, such as:

- Mortality indicator database (HFA-MDB): mortality indicators by 67 causes of death, age, and sex
- European detailed mortality database (DMDB)
- European hospital morbidity database (HMDB)
- European database on human and technical resources for health (HlthRes-DB)
- Centralized information system for infectious diseases (CISID)
- European inventory of national policies for the prevention of violence and injuries
- WHO European Database on Nutrition, Obesity, and Physical Activity (NOPA) [9]
- Tobacco control database
- European Information System on Alcohol and Health (EISAH)
- European Regional Information System on Resources for the Prevention and Treatment of Substance Use Disorders

11.5 Data Curation

Data curation refers to actively updating, cleaning, verifying, managing, collecting, and preserving data for our use as well as others. Data also have a special life cycle that they should go through, which should be taken into consideration by the researchers.

The activities associated with data curation include storing them while we are using them as well as when we are finished with them. The data should be stored on a secure server rather than on a flash drive and should also be described or documented [10]. This description is called metadata and includes components such as title of the research, creator, language, dates, and file formats. The provenance of this data must also be established by appropriate instruments and research methods to produce the data. The purpose of these activities is to add clarity and supportive information to our data so that these can be interpreted by others. If others are using the data, they should be able to identify these for the purpose of citation, for example. Finally, to ensure that the data survive, it is important to archive the related software and hardware in secure repositories. Some repositories can be general, while others are subject based [11].

Figure 11.1 shows the stages of the data life cycle [12]. Data can be considered as living things; they should not be stagnant. This cycle is not always exact as some developers might start at different places depending on what they are interested in or what the specifications of their jobs are but the important thing is that the data do go through evolving life cycles. A life cycle has eight components and the first step is to plan. This step involves description of the data that will be compiled and how the data will be managed and made accessible throughout their lifetime. Before we even start collecting the data, we have to decide what type of data we are collecting, how they are going to be described, the file type that will be used, and how they are going to be archived. Once the data management plan is completed, the next step is to collect data either by hand or with sensors or other instruments, express them into a digital form, and then assure their quality through checks and

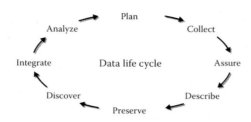

FIGURE 11.1
Stages of data life cycle.

inspections. The data quality is very important and should always be a priority. The collection and assurance of the data are followed by description, in which we always make sure that the proper metadata standards are used for the purpose and destination of data. The next step includes the preservation of the data, in which they are submitted to an appropriate long-term archive like a data center or repository. After we decide where we are going to send the data to be kept for long term, we have to discover or locate other data that could be relevant to the one in hand by using metadata and then integrate by combining data from different sources to form one homogeneous set of data that could be easily read and analyzed. The final step of course is the analysis through different tools and programs.

All the above regarding data management and curation are necessary for several reasons. The first one is preservation. Some research data are very valuable and unique and cannot be replaced or found if destroyed. Repositories and data centers are set up specifically to maintain and preserve such data. Open access is the second reason. Data sharing is vital and does not only help the researchers but other related fields as well. Another reason is contribution and comes as a result of the above, meaning that if we can preserve data so that we can look back at them, maintain them, and also share them, then we are all going to contribute to the field. Other reasons why we should share and archive our data include the verification of results, the security and reuse of data, and the increased citation and credits gained. It is known that papers that share and archive data publicly are cited 69% more than other papers that withhold data.

Specifically for digital curation, the latest approaches include the following [13–14]:

- *Conceptualization*: Examine websites, e-mails, publications, and different types of digital objects and develop storage choices.
- *Creation*: Assign all relevant metadata to digital objects.
- *Access and use*: Decide which material will be accessible publicly and which only by designated users.
- *Appraisal and selection*: Determine the relevance of digital objects, and follow potential guidelines or legal requirements.
- *Disposal*: Throw away digital objects which are not valuable to the institution.
- *Ingestion*: Transfer digital material to a predesignated storage repository or archive.
- *Preservation actions*: Use measures to guarantee long-term preservation of digital data.
- *Reappraisal*: Assess which digital objects are still relevant and return the ones which fail the test.
- *Storage*: Ensure safety of the data as described by prearranged guidelines.

- *Access and reuse*: Guarantee that material is still obtainable from assigned users.
- *Transformation*: Create different formats from the original digital material.

One of the main trends that have come into play nowadays concerning data curation is the use of both algorithms and humans together (algorithms + crowd; Figure 11.2). What we basically see in these types of deployments is that one can have a large collection of data sources that can be analyzed by data quality algorithms to find problems within that data. This algorithm can be complemented by using human computation or effective crowdsourcing, which will make some of the corrections that need to be made within those data. This ultimately results in clean data being produced at the end of the process. This is very much an emerging trend within the data curation space over the last couple of years and it has been used in a number of enterprises and also in some communities on the web.

The key things when we are thinking of working with humans and computers together in this mixed human–computer intelligence is how we coordinate a large group of workers to perform small tasks and be able to solve problems that computers or individual users cannot do individually. This is a very interesting area of research that is involving a number of different types of technologies and different types of trends coming together over the last decade. Related areas include collective intelligence, human computation, social computing, natural language processing, data mining, speech recognition, and computer vision.

The future of data curation lies in these technologies which are beginning to come together in these collaborative community-based data curation solutions. Very important also is to inform others about the importance of data sharing. It is very significant to help researchers and universities to understand how important it is to share data with others and how much more efficient research will be if we can have open access to each other's information. Also, we need to improve metadata awareness and training, making it mandatory to help others understand what metadata are and how they are used so that

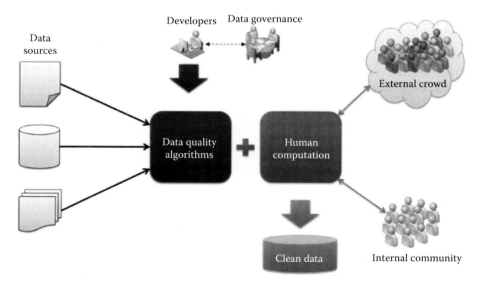

FIGURE 11.2
Algorithms and crowd.

they can be more equipped and better prepared for data curation [15]. Finally, we need to increase access to needed tools and provide incentives and motivations for others to get involved with data curation, particularly in the healthcare applications.

11.6 Interpretation of Health and Epidemiological Data—Biostatistics

The outcome of data analysis and the quality of data play an important role and have a lot of applications in public health practice. These include disease research, prevention assessment, population evaluation, program planning, potential future health problems, and hypothesis generation for study design. These data derive from local and national sources and include health inquiries, hospital and surveillance data, vital statistics, outbreak investigations, and general research. The most important thing in order to draw conclusions is the quality of our data which depends on accuracy and completeness.

Data interpretation refers to the process of comprehending and communicating the meaning of the data. We will focus on some basic concepts that are frequently used and are essential to interpret in order to evaluate public health data [16–17]. The first is prevalence, which is the number of cases of a disease at a specific time point or period in time. Prevalence is a percentage and is usually obtained from the surveys, and it represents a picture of the disease or health condition of interest.

Another measure of disease frequency is incidence, which is defined as the number of new cases of a condition during a defined time interval divided by the number of persons at risk of developing the condition over that time interval. The incidence of death in a specific population, rather than the incidence of disease, is a new, specific type of rate, called mortality rate, and is calculated by dividing the number of deaths in the population during a stated time period by the number of persons at risk of dying during that period. Mortality rates can be divided in cause-specific, age-specific, and crude mortality rates and can differ considerably by sex, ethnicity, race, and age. Case fatality is another type of measuring which is frequently used in health data and is estimated by dividing the number of deaths from a specific disease during a stated time period by the number of persons with the disease of interest and is an indication of the severity of the disease.

By measuring which person was affected, where the incident took place, and the time when it happened, we can thoroughly describe the distribution of diseases. Information about the individual carrying the disease should contain sex, age, religion, occupation, ethnic group, and marital status. The evaluation of the improvement in preventing communicable diseases is given from the comparison of the incidence rate of each disease at different times [18].

The application of statistics in public health and medicine in general is called medical biostatistics and helps us achieve a better understanding of the data [19]. Different statistical techniques are used to describe, analyze, or simplify data, which lead to a better understanding of the health of populations [20].

The use mean value is a very commonly used statistical measure and provides helpful information biostatics, and it corresponds to the average value of a set of data. It is obtained by adding several quantities together and dividing them by the number of quantities. The middle values of a set of numbers arranged in order from the smallest to the largest is called a median, resulting in half the values being above the median and half the values below it. Finally, the summary measure which occurs more frequently in a given list of numbers considering that at least one data point occurs more than once is called a mode.

Another important statistical tool is the confidence interval, which is a range of values used to indicate how certain we are that an estimated rate from a sample of data is actually the true rate in the population from which the sample is taken. The width of the confidence interval serves as a means to indicate how accurate the assessment is, meaning that a narrow confidence interval correlates with a more precise estimate, while a wide confidence interval indicates a less precise one. Another very helpful tool in biostatistics is the *p* value, which is a measure of the statistical significance of a difference between rates. In other words, we measure how probable it is that the results observed happened by chance. A very small *p* value means that the estimated differences in rates were very unlikely to have occurred by chance. When analyzing public health data and biostatistics in general, the threshold of *p* value is less than .05, which means that there is less than 5% chance that the differences happened by chance alone.

In biostatistics and epidemiological studies, is very important when analyzing and interpreting findings from a study to take into account how much of the observed association between an exposure and a result is due to the fact that it might have been influenced by mistakes in the design, transaction, or analysis. The following questions should be answered before we deduce that the observed association between exposure and result is a genuine cause and effect relationship [21]:

- Is the observed association the result of systematic faults in the way individuals were chosen (bias) or in the way information was acquired from them?
- Is the observed association the result of differences between the groups in the distribution of another variable that was not considered in the analyses?
- Is the observed association likely to have occurred by chance?
- Is the observed association likely to be causal?

Another aspect that is important in health data interpretation is the actual data presentation, which should aim to summarize and organize data as accurately as possible. Visual displays (as in Figure 11.3) are often more comprehensible, allowing the easier identification of disease frequencies, tendencies, and comparisons between groups as well as other potential connections in the data.

A table is a visual display of data arranged into rows and columns. Its main advantage is that it lets us display many patterns or differences depending on the data which are included in the table. Nearly any quantitative information can be included and presented in a table.

A line graph is a helpful data presentation tool which displays a long series of data and is also used for the comparison of different series of data in the same graph. The data are displayed in two dimensions, the x axis and the y axis. The independent variable, or x, is on the horizontal axis and the dependent variable, or y, on the vertical axis. Falls and rises in a line graph show how one variable is influenced by the other.

A bar graph is a graphical display of data and shows relationships between two or more variables by using bars of different heights. Data are presented either vertically or horizontally, allowing for the comparison of the displayed items, and each independent variable is discrete.

Finally, a pie chart is usually used to display how a section of something correlates to the whole. The fundamental design is a circle (in the shape of a pie) and the sectors

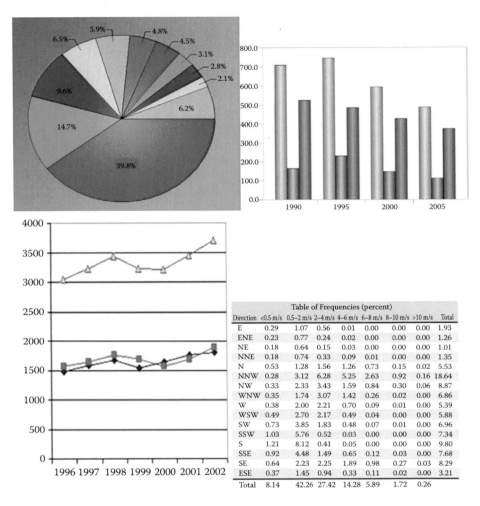

FIGURE 11.3
Visual displays: pie, bar, and line graphs and table.

(or slices of the pie) are the percentages of the different kinds of the variable and add up to 100% [22].

All the above contribute to an improvement in the interpretation of health data, which, in turn, helps professionals to comprehend more efficiently the community's health needs.

11.7 Global Health Data Management and Interpretation

Health data are growing in an exponential manner and nearly 85% are unstructured and clinically relevant. Health data exist in numerous forms, such as individual electronic medical records, medical correspondence, lab and physician notes, imaging and customer relationship management systems, and finance. Getting access and reshaping these precious data into clinical analytics is essential in order to improve healthcare worldwide. In this way, big data technology comes into play and provides invaluable help to healthcare organizations to seize all of the information regarding a patient in their effort to achieve a complete view of patient treatment and population health management. The successful use of big data helps to improve care and outcomes, to increase access to healthcare, and to build supportable healthcare systems.

The efficient processing of big quantities of data in sustainable time requires unique and outstanding technologies. These include [23] the following:

- **Crowdsourcing:** It comes from the words *crowd* and *outsourcing* and is the act of getting information or services by requesting contributions from a number of people mainly via the Internet.
- **Data fusion:** This is the process of integrating multiple data from different sources to generate more accurate representation, which can be more useful than the initial single-source data.
- **Genetic algorithms:** These are heuristic algorithms which emulate several of the processes noticed in natural selection.
- **Machine learning:** It is a form of artificial intelligence dealing with the construction and ability of computers to learn from data without being programmed. Computers train themselves and change behavior when exposed to new data.
- **Natural language processing:** It is the process referring to the ability of a computer to derive meaning from human or natural speech.
- **A/B testing:** It describes simple, accidental experiments with two variants A and B representing the treatment and the control, respectively.
- **Visualization:** It is the process of creating images or scientific data as diagrams to help in the interpretation of the data.
- **Time-series analysis:** It concerns the collection of data points obtained through repeated measurements at successive points in time.
- **Signal processing:** It is the technology which refers to different operations for the analysis, interpretation of information, and improvement of accuracy of analog and digital signals.
- **Simulation:** It describes the imitation of the behavior of a system or some situation over time.

All the above exceptional technologies contribute to the efficient processing not only of global health data but also of the big data which are constantly being gathered from many scientific areas and research projects. Data sets grow in size extremely rapidly, resulting in the fact that 90% of the world's data today were created in the last 2 years (Figure 11.4). The world's capability to store data has roughly doubled every 3 years since the 1980s [24] and as the cost of storage decreases the growth of information will continue to rise.

As discussed earlier, Global health data is very big and there is no standard and globally accepted methods to handle it. However, in principle, the requirements of such big data storage are the ability to handle large amounts of data and to provide the input/output operations per second (IOPS), which are required in order to deliver data to analytics tools. Supercompanies which produce vast amounts of data, like Apple, Facebook, and Google,

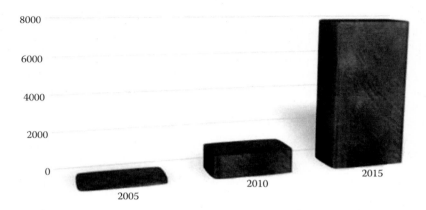

FIGURE 11.4
Rate of growth of data since 2005 to 2015.

run hyperscale computing environments. These include big amounts of commodity servers with direct-attached storage (DAS), meaning that they have storage which is connected
to one computer but is not accessible to other computers. Such environments use as analytics engines the likes of Cassandra [25], Hadoop [26], NoSQL [27], and flash storage medium
in addition to disk to reduce storage latency to a minimum. In this type of configuration,
there is no shared storage. These hyperscale computing environments are actually the
heart of the biggest web-based operations to date, but it is expected that even these sophisticated storage architectures will be surpassed in the near future.

Scale-out or clustered network-attached storage (NAS) [28] is an example of technology
that helps big data storage. This is file access shared storage which uses parallel file systems and can be scaled out to address increased computer requirements. These parallel
file systems are dispensed across multiple storage nodes which can manage billions of
files without the performance degradation that takes place with ordinary file systems as
they grow.

Object storage [29] is another storage format that is used for large numbers of files. They
contain data like files do and can scale to huge capacity. They overcome the problem of
becoming unwieldy when containing large numbers of files by giving a unique identifier to each file and also indexing the data and their location. This identifier provides the
end user with the ability to recover the object without having to know the actual physical
location of the data. This approach is very helpful in cloud computing environments for
automating data storage. So summarizing, low latency for analytics work and ability to
manage capacity are two key prerequisites for big data storage and can be achieved in
hyperscale environments like the big companies do or adopt object storage or NAS like
more conventional IT departments do.

However, inspirational to large global health data handling is the case of a European
Organization for Nuclear Research (CERN) project, which handles the one million giga-
bytes of data produced every single second from the Large Hadron Collider (LHC). This
large amount of information cannot be managed or analyzed by any single computer center alone. To share the load, CERN shares its data with more than 150 institutions in 36
countries around the world by using the Worldwide LHC Computing Grid (WLCG), thus
breaking down the work required to handle and store into manageable chunks [30]. The
advantages of a grid system are numerous and include the following:

- Copies of information can be maintained in different sites, providing access for all researchers involved in the project, regardless of their geographical location.
- Institutions and research centers in different time zones facilitate monitoring and expert support availability and reduce the chances of failure.
- Scientists can access resources and information from their computers, making their task easier and more effective.
- New ideas and techniques regarding computing and analysis are encouraged by independently managed research centers.
- New demands can easily be addressed by the system as more data are collected.
- The different centers can adopt innovative technologies which might offer improved usability or energy efficiency.

The WLCG is composed of four levels or tiers, which are called tier 0, tier 1, tier 2, and tier 3. These tiers analyze and store all LHC data between them. CERN, which is tier 0, sends data to 11 tier 1 locations, which are large computer centers all over the world, and they send data to more than 140 tier 2 locations, which are usually universities or institutes. Each tier 1 location shares data with a specific group of tier 2 locations. Through the tier system, big data can be stored, backed up, analyzed, and accessed more manageably. Grid computing is a very promising field for three purposes: first, it can reduce the cost of using a specific amount of computer resources; second, it can solve problems which require a vast amount of computing power; and finally, it proposes that computers should be harnessed and cooperated toward a common goal.

References

1. Mantas J. and Hasman A., *Textbook in Health Informatics—A Nursing Perspective*: IOS Press, p. 484, 2002.
2. Health and Social Information Centre, http://www.hscic.gov.uk/datasets. Accessed on March 11, 2014.
3. Ries D. and Winter R., Ed., Healthcare Data Warehousing in the Government. Winter Corporation, Cambridge, Massachusetts, 2012.
4. Comer K., Derr M., Kandris S., Ritchey M., Seyffarth C., and Thomaskutty C., Community Health Information Resource Guide, Volume 1: Data. The Polis Center, Indianapolis, Indiana, June 2011.
5. Statement on Quality Healthcare Data and Information, American Health Information Management Association, Reviewed and approved December 2007.
6. Özsu M. T. and Valduriez P., *Principles of Distributed Database Systems*, Third ed.: Springer, New York, 2011.
7. The WHO Family of International Classifications, http://apps.who.int/classifications/en/. Accessed on January 19, 2015.
8. NIH Intramural Database, http://intramural.nih.gov/index.tml. Accessed on August 30, 2014.
9. Databases, World Health Organization, Europe, http://www.euro.who.int/en/data-and-evidence/databases. Accessed on August 30, 2014.
10. Atkinson M., Guest Editorial: Research Data: It's What You Do With Them, *International Journal of Digital Curation*, Vol. 4, No. 1, pp. 1–5, 2009.
11. Gelernter J. and Lesk M., Use of Ontologies for Data Integration and Curation, *International Journal of Digital Curation*, Vol. 6, No. 1, pp. 70–78, 2011.

12. Constantopoulos P., Dallas C., Androutsopoulos I., Angelis S., Deligiannakis A., Gavrilis D., Kotidis Y., and Papatheodorou C., DCC&U: An Extended Digital Curation Lifecycle Model, *International Journal of Digital Curation*, Vol. 4, No. 1, pp. 34–45, 2009.

13. Yakel E., Digital Curation, *OCLC Systems & Services: International Digital Library Perspectives*, Vol. 23, No. 4, pp. 335–340, 2007. doi:10.1108/10650750710831466.

14. Higgins S., Digital Curation: The Emergence of a New Discipline, *International Journal of Digital Curation*, Vol. 6, No. 2, pp. 78–88, 2011.

15. Witt M., Carlson D., Brandt D., and Cragin M. H., Constructing Data Curation Profiles, *International Journal of Digital Curation*, Vol. 4, No. 3, pp. 93–103, 2009.

16. Silberschatz A., Korth H. F., and Sudarshan S., *Database System Concepts*, Sixth ed.: McGraw-Hill, 2010.

17. Abraham S., Kulkarni K., Madhu R., and Provan D., *The Hands-on Guide to Data Interpretation*: Wiley Blackwell, 2010.

18. Sullivan M. L., *Essentials of Biostatistics in Public Health*: Jones and Bartlett Publishers, 2011.

19. Wayne W. D. and Cross L. C., *Biostatistics: A Foundation for Analysis in the Health Sciences*: Wiley, 2013.

20. Rosner B., *Fundamentals of Biostatistics*: Cengage Learning Asia, 2010.

21. Friss H. R. and Sellers T., *Epidemiology for Public Health Practice*: Jones and Bartlett Publishers, 2013.

22. Evergreen D. H. S., *Presenting Data Effectively: Communicating Your Findings for Maximum Impact*: Sage Publications, 2013.

23. Manyika J., Chui M., Bughin J., Brown B., Dobbs R., Roxburgh C., and Byers A., Big Data: The Next Frontier for Innovation, Competition, and Productivity. McKinsey Global Institute, May 2011, http://www.mckinsey.com/insights/business_technology/big_data_the_next_frontier _for_innovation. Accessed on August 30, 2014.

24. Hilbert M. and López P., The World's Technological Capacity to Store Communicate and Compute Information. *Science*, Vol. 332, No. 6025, pp. 60–65, 2011.

25. Sharma S., *Cassandra Design Patterns*: Packt Publishing, 2014, https://books.google.com. Accessed on August 30, 2014.

26. Chaudhuri S., Dayal U., and Narasayya V., An Overview of Business Intelligence Technology. *Communications of the ACM*, Vol. 54, No. 8, pp. 88–98, August 2011.

27. Tweed R. and James G., A Universal NoSQL Engine, Using a Tried and Tested Technology, http://www.mgateway.com, 2010.

28. Noronha R., Ouyang X., and Panda D., Designing a High-Performance Clustered NAS: A Case Study with pNFS over RDMA on InfiniBand. *Lecture Notes in Computer Science* Vol. 5374, pp. 465–477, 2008.

29. Factor M., Meth K., Naor D., Rodeh O., and Satran J., Object Storage: The Future Building Block for Storage Systems. *Local to Global Interoperability—Challenges and Technologies*, pp. 119–123, 2005.

30. Brumfiel G., Down the Petabyte Highway. *Nature*, Vol. 469, pp. 282–283, January 2011.

12

Electronic Medical Records: Management and Implementation

Liping Liu

CONTENTS

12.1 Introduction

A medical record contains personal data, a summary of medical history, and documentation of medical events, including symptoms, diagnoses, treatments, and outcomes. Its main purpose is to provide essential facts for medical diagnoses. It also forms the first link in the information chain, producing the depersonalized aggregated data for knowledge discovery and statistical analysis.

A medical record is traditionally paper based, rendering it difficult to search for useful information and to reuse data for changing tasks. Often the paper chart is thick, tattered, disorganized, and illegible; progress notes, consultants' notes, nurses' notes, and radiology reports are all comingled in accession sequence. Practitioners claim to spend as much as 75% of their work time chasing specific data items on pieces of paper [1]. The records confuse rather than enlighten; they contain distorted, deleted, and misleading information [2] and provide a forbidding challenge to anyone who tries to understand what is happening to the patient [3]. At a time when the concept of "informed patient" is becoming the norm, the records can hardly be used to inform the patient with a coherent, consistent packet of information. Most importantly, it is nearly impossible to exchange data among healthcare providers, to accumulate data for knowledge discovery, and to connect medical records to the growing body of medical facts stored in medical expert systems.

To address these issues with paper-based medical records, there has long been a vision to store an entire medical record, or any part of it, electronically so that a paper-based chart is replaced by a digital record comprising a mix of written text, codes, images, and audio and video notes. This concept of medical records is now referred to as an electronic medical record (EMR) or other similar terms such as *computer-based patient records* or *electronic health records*.

While the concept of EMR is simple, there has been no consensus in the industry regarding exactly what EMR is with respect to scope of data coverage and distribution, system functionality, and interoperability [1].

With respect to the scope of data coverage, the most ambitious vision is to encompass all medical data from prenatal to postmortem information and cover all practitioners ever involved in a person's healthcare, independent of medical specialties. In contrast, a moderate vision is to include only "relevant" patient information, and the conservative vision requires only a simple change in current documentation habits from easy handwriting or dictation to computer input such as scanning, digitizing, and indexing.

With respect to data distribution and control, there exist five positions: (1) each patient being in charge of his or her health information, (2) one central organization holding all medical records of all patients, (3) multiple competing third parties holding data for disjunctive patient segments, (4) each enterprise holding medical records for its departmental healthcare providers and patients, and (5) each healthcare provider being responsible for its own records.

Positions 1 and 2 represent the most and the least centralization of data storage, respectively, while 3 to 5 represent positions in between.

With respect to interoperability, a revolutionary vision is to have complete interoperability between information systems in various locations, provider settings, and infrastructures. In contrast, a realistic approach is to create interoperability among the departmental systems within an enterprise.

With respect to system functionalities, there is also no consensus regarding which components or functions make up an EMR. The following are seven common components that a broad definition of EMR would typically include [4]: (1) clinical workstations for supporting nurse or staff order entry, physician order entry, clinical decision support, and results reporting; (2) clinical data repositories for storing EMR data including lab results, reimbursement codes (International Classification of Diseases [ICD] and Current Procedural Terminology [CPT] codes), clinical codes (Logical Observation Identifiers Names and Codes [LOINC], MEDCIN, Systematized Nomenclature of Medicine [SNOMED], etc.), audio memos and notes, and clinical images; (3) medical record document imaging systems for digitizing analog data such as dictations, X-rays, and ECG traces; (4) master patient index or enterprise directory for locating patient medical records; (5) integration/interface engines and communication networks for connecting the data repository to clinical workstations and departmental systems; (6) a data warehouse or secondary database of patient information for supporting retrospective analysis of outcomes, utilization, and clinical processes; and (7) online personal medical records accessible to the patient.

12.2 Detailed Functional and Data Requirements

To illustrate the concept of electronic medical records, in this section, a prototypical blueprint for implementing a conservative vision of EMR in a typical clinic setting is proposed. This section shows the typical data activities involved in managing medical records through use case models and sample data contents that make up medical records through class diagrams. Both models are expressed using the Unified Modeling Language, a standard language for modeling business requirements and system specifications. The reader may refer to a standard textbook for technical details on the language notations, syntaxes, and semantics.

12.2.1 Functional Requirements

In a typical clinic, the users of EMR include patients, doctors, nurses, and receptionists. To access EMR, they all need log-in credentials and retrieve the record as granted by the credentials. To ensure privacy, access control must be in place. To this end, a protocol such as the Bell–LaPadula model [5] may be followed. The protocol governs how a patient may grant and a doctor may request file access to certain portions of a record. Note that, in terms of user roles, a user is a generic type of users, and his/her associated use cases can be carried out by any users who play more specialized roles, including receptionists, nurses, doctors, or patients. By the same token, a physician plays a more specialized role than a nurse, who, in turn, plays more specialized roles than a receptionist, in the sense that whatever use cases that can be performed by receptionists can be performed by nurses, and whatever use cases that can be performed by nurses can be performed by physicians. The role map is shown in Figure 12.1 by the triangular arrows pointing from physician to nurse and from nurse to receptionist.

Receptionists perform basic data activities, including check-in, checkout, and managing patient profiles and appointments. Manage Patient Profile may optionally execute Create New Profile use case if the profile does not exist (Figure 12.2b). This use case can be often optionally activated when checking in patients (Figure 12.2a) or making appointments (Figure 12.2c).

As Figure 12.1 shows, nurses can perform whatever use cases that a receptionist can do. In addition, nurses are responsible for creating records on visits, managing test orders, recording prescriptions, and administering medications (see Figure 12.3).

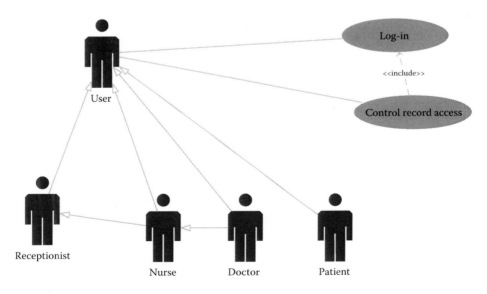

FIGURE 12.1
Control record access.

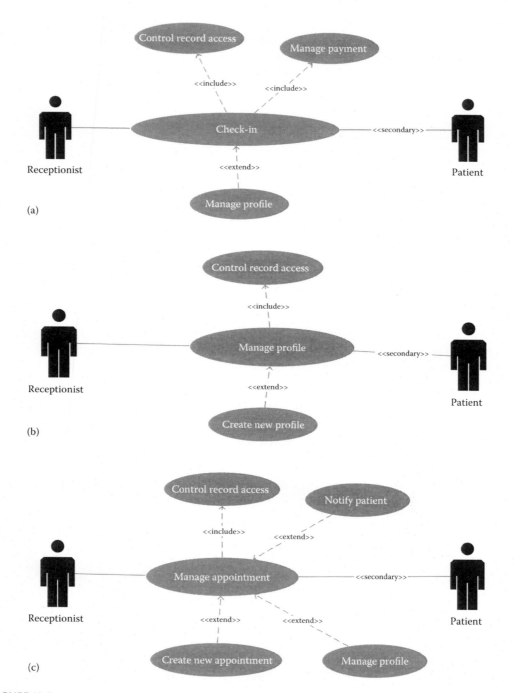

FIGURE 12.2
Use cases performed by receptionists: (a) check-in, (b) manage profiles and (c) manage appointments.

(Continued)

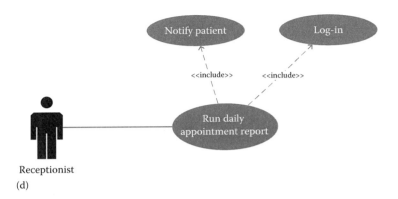

FIGURE 12.2 (CONTINUED)
Use cases performed by receptionists: (d) run daily appointment reports.

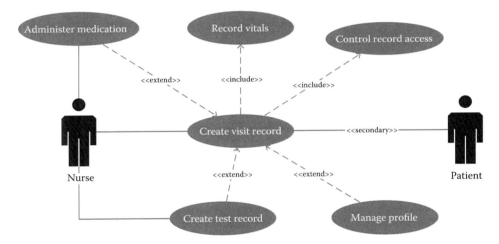

FIGURE 12.3
Create visit record and vitals.

A doctor shall be able to perform all of the tasks that a nurse can perform. A doctor may additionally enter diagnostic results (Figure 12.4a); create orders, including test orders and prescription of medications and treatments; and make referrals and request files (Figure 12.4b).

Finally, a patient accesses the system mostly for looking up his or her health record such as test results, requesting appointments or profile updates, and managing record access (Figure 12.5). Here we assume that patients cannot make appointments or update profiles directly as is the case in most clinics. Instead, they can request such, and it is up to a receptionist to grant or reject the request.

12.2.2 Data Requirements

At the heart of any EMR system are data sets or databases that organize, store, and secure the vast amount of medical records. A data model is the framework in which these records are organized. The most popular industrial database systems are relational, and the prominent data models are entity–relationship diagrams or relational models. However, to be

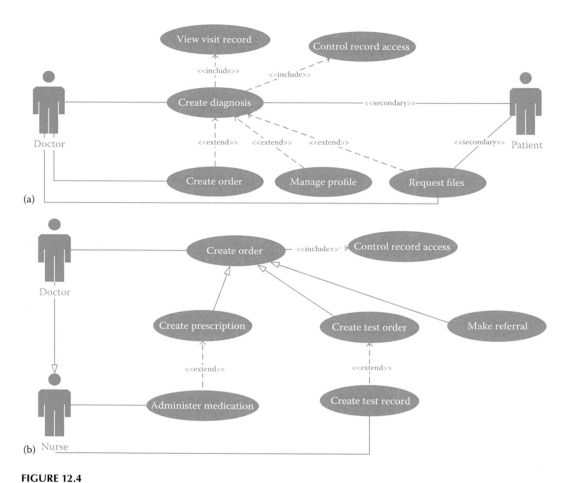

FIGURE 12.4

Use cases performed by physicians: (a) create diagnosis record and (b) create test orders, prescriptions, and referrals.

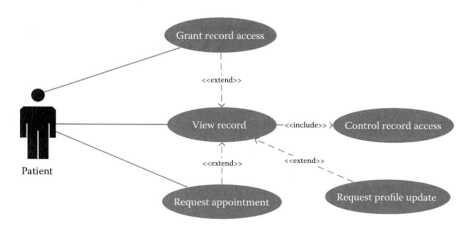

FIGURE 12.5

Use cases for patients.

forward looking and also consistent with modern systems development methodology, here, class diagrams as conceptual data models are employed for patient visits, physician orders, medication data, and capture patient histories.

Figure 12.6 shows a class diagram representing data on patient visits, including appointments, vitals, and diagnoses. Each visit may involve a group of medical professionals such as nurses and physicians, and the associative class PhysicianVisit represents their involvements and roles. Of course, a receptionist sets appointments, nurses take vitals, and physicians make diagnoses. Here orders are a concept unifying medicine prescriptions, test or treatment orders, and supply orders, as well as request for external files and records. Many

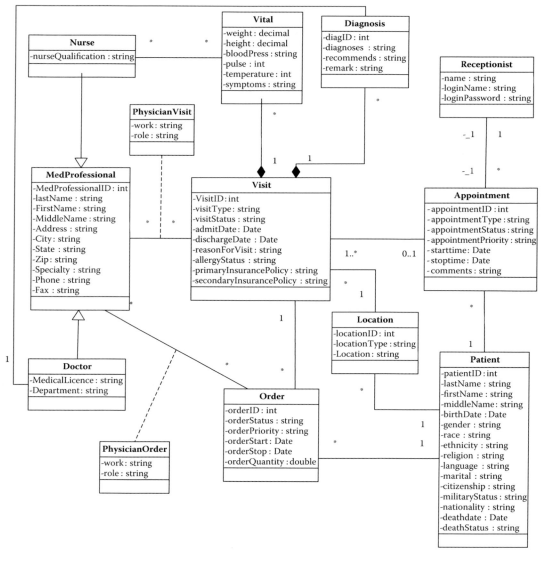

FIGURE 12.6
Class diagram for patient visits.

professionals may be involved in each; some create the order, some validate the order, and others may execute the order. The associative class PhysicianOrder helps record their works and roles.

There are four essential domains of data to be represented: medications, encounters or visits, orders, and patient history. Of course, there are some nonbusiness data to be stored. These include data on forms, users, activity work flows, system controls, and transaction logs. These data are dependent on implementation and ignored here. We also omit control- and log-related attributes for business data, such as date and user created or modified and values of old versions. A comprehensive data model representing all these data is beyond the scope of this chapter. An interested reader may refer to data models available in the public domain such as the ones for OpenMRS and PatientOS.

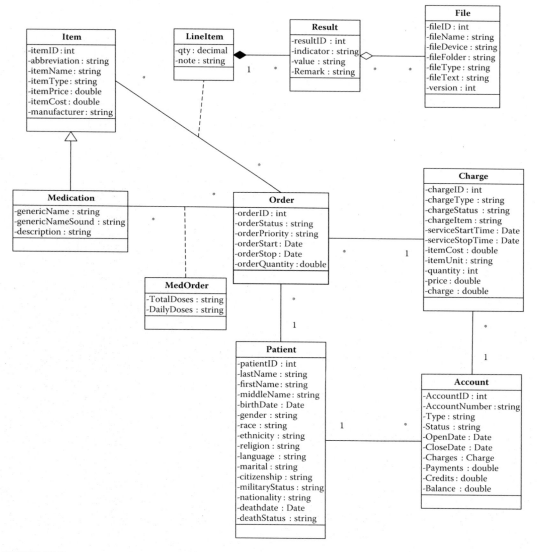

FIGURE 12.7
Class diagram for physician orders.

Figure 12.7 shows more details on orders, including prescription doses, test or treatment results, and charges to patient accounts. Here items to be ordered include test or treatment procedures, medical supplies, and medications. Each order may contain many items, and its quantity is recorded in LineItem associative class. In addition, for medications, the data on total dosage and daily dosage are kept in MedOrder. For each test or treatment procedure ordered, its results are stored in the Result table, which may have one or more files to be associated with. Of course, the cost of the orders will be charged to patient accounts, and it seems a common practice that each patient may have multiple accounts.

Medications are singled out and treated as a special kind of items for one reason: they constitute the most important domain of medical records. Figure 12.8 shows more details on medications, including medication effects and interactions between medicines, which

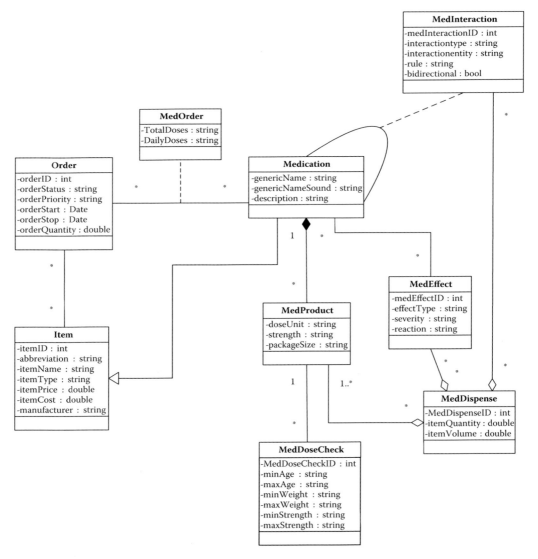

FIGURE 12.8
Class diagram for medication data.

may be observed during medicine dispenses or known as scientific facts. The knowledge of these effects and interactions will serve as warnings when physicians write prescriptions. Each medicine may have several generic drugs or MedProducts to be equivalent in gradients and effects. Each drug bears its own dosage check data, which also serve physicians and patients in filling prescriptions.

Among all the constituents of a medical record, a patient history is most complex to organize and difficult to define. When a patient has seen a doctor a few times, his or her past visit records will be a part of the patient history, useful for future diagnoses. However, these historical records are incoherent with the history data that may not be available from external healthcare providers. The history data are often captured during the patient's first visit through a lengthy form regarding what happened in the past prior to the first visit. The questions to be answered include prior diagnoses, treatments, medications, allergies, lifestyles, and genetic diseases. For these types of answers, some EMR systems try to restructure or classify them into different tables such as allergies, lifestyles, prior diagnoses, or genetic diseases. Often such attempts are problematic; the restructured data overlap with those captured into the tables in Figures 12.6 through 12.8, and yet have different levels of details and objectivity. Therefore, in this chapter, I propose organizing and capturing nonclinically observed history data in their original form of questions and answers. Figure 12.9 is a class diagram representing the structure of such data. A clinic may have various forms for patients to fill out. Each form has three types of questions: open-ended

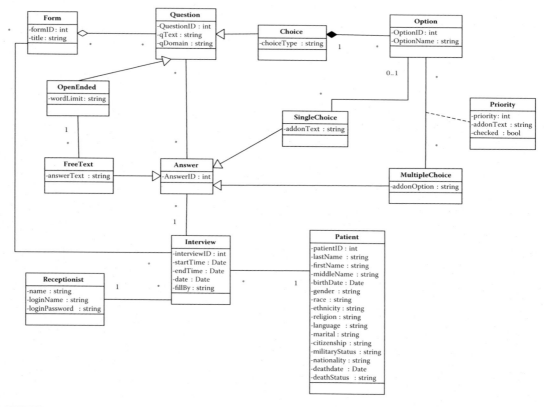

FIGURE 12.9
Class diagram for capturing patient histories.

questions for free-text answers, select-a-choice questions for single-choice answer with a possible free-text add-on field, and check-all-that-apply questions with the possibility that a patient may add other choices. Accordingly, answers will be classified into three types, with each being kept in a different table. For an open-ended question, a free-text answer is recorded. For a select-a-choice question, the answer will record which option a patient chooses and a possible free text qualifying the choice. For a check-all-that-apply question, the database will keep a record of whether a patient checked each option, an add-on text qualifying the choice, and a possible priority value if choices need to be ranked. With all the answers stored in their original form, no data item is lost because of restructuring, and queries may be created to list allergies, prior diagnosed diseases, and lifestyles by pooling data from all these separate tables and the tables shown in Figures 12.6 through 12.9.

Lifestyle is another aspect of a health record that requires a further study. It becomes complex due to the prevalence of mobile or distributed computing devices, which have profound impacts on what constitute medical records or the concept of EMR, and how to implement EMR. For example, should a patient's daily exercise and diet records be a part of EMR? If they should be, what about the patient's other activities, such as sleeping, working, driving, going to a movie, shopping, or even smiling? Mobile devices can collect these records with ease, and these records can serve as a snapshot of the patient's health or establish a long-term pattern of lifestyle that helps medical diagnoses and prescriptions.

12.3 Implementation Issues and Solutions

Compared to paper-based medical records, the EMR concept has clear advantages. It improves data availability; a medical record can be shared with multiple service providers and can be accessible from anywhere across a communication link by multiple simultaneous users, should a patient move or fall ill away from home. It improves data storage efficiency, data quality control, and flexibility of data abstraction and reporting. It allows records made by multiple providers in different locations and units to be linked, shared, and synchronized into a single record for each individual. The problem of record fragmentation can be resolved, patient care can be genuinely shared between providers, and there will be fewer redundant tests as results are shared by labs. It also enables direct links to knowledge-based tools and improves clinical decision support; some advanced systems can help doctors think by giving them digital prompts, such as warnings not to prescribe the wrong medication or suggestions about the latest treatments. For example, the current development of the Arden syntax and a library of medical decision logic modules will make it possible for any system to incorporate intelligent alerting flags to EMR users warning them of possible errors.

12.3.1 Implementation Issues

Despite its claimed benefits, the diffusion of EMR is still in the state of infancy, and only an estimated 20% of United States-based physicians used EMR a few years ago. According to industrial surveys [4,6] in 2002, only 30% of respondents indicated that they had installed some EMR functions in some or all of their departments. The diffusion rate varied, depending on the specialty of practice; about 42% of respondents working in an internal medicine setting reported using EMR, while only 8% in a pediatric practice indicated so.

With respect to installed EMR components, 25% of respondents indicated functions in nurse order entry, 11% in physician order entry, and 32% in results reporting. With respect to EMR data contents, 22% reported text and reimbursement codes; 11%, clinical codes (LOINC, MEDCIN, SNOMED, etc.); 10%, clinical images; and 3%, audio notes. Among those who stored SNOMED codes, 3% stored lab results and 1% stored radiology results. Ten percent of respondents reported having enterprise-wide master patient index and 23% reported having the index for a single site. Note that less than 3% of deployed systems provided web-based personal health records.

A few critical factors inhibit the diffusion of EMR [4]. The first is a lack of economic motivation and cost–benefit justification for EMR investments. The second is technical interoperability, which is concerned with the ability to communicate and share data with different EMR systems. The third is cognitive obstacles that hinder a change from a manual-driven environment to a computerized system. A few studies have empirically examined the impact of these cognitive factors on the acceptance of e-service–oriented EMR [7].

Economic motivation is concerned with costs and benefits. Why should healthcare providers spend substantial resources for EMR? In theory, EMR has many clear benefits. However, in practice, it is often not easy to translate these benefits into measurable indexes such as return on investment, reduction in medical errors, healthcare efficiency, and improved patient and employee satisfaction [4]. First, implementing EMR typically requires significant commitments to purchasing hardware and software, system administration and maintenance, end-user training and support, and sometimes staffing technical talents. The requirement can exceed what many small and midsize healthcare providers are able to afford. Second, providers often fail to reengineer care processes to reap full benefits of new technologies. Instead, they equate the EMR adoption to the additional time for typing notes. For example, some physicians estimate that adding 5 min of data entry to each patient means 20%–50% of their productivity loss.

Responding to the first aspect of economic motivation, the U.S. government recently passed the Health Information Technology for Economic and Clinical Health (HITECH) Act of 2009, which set aside $35 billion over 10 years to encourage healthcare providers to adopt and use EMR to a meaningful extent. The incentives include paying doctors up to $44,000 over 5 years by Medicare and up to $63,750 over 6 years by Medicaid. To date, more than 50% of eligible healthcare providers, mostly doctors, have claimed using EMRs, up from 17% in 2008, according to data released by the U.S. Department of Health and Human Services (HHS) in May 2013. About 86% of eligible hospitals reported to the federal department that they used EMRs to a meaningful degree, also a significant increase. HHS had already paid out $13.7 billion of incentives up to March 2013.

Technical interoperability is the greatest hurdle among all [4]. For a modest vision of EMR, interoperability means, for example, that all systems within an enterprise are interoperable; a patient's demographics are captured only once and every authorized practitioner should have full access to the patient's health information stored within an enterprise. For aggressive visions, it means interoperability independent of provider, medical specialty, geographic location, country system, legislation, etc. Technical interoperability has two vital requirements: (1) having the same semantics in codes, vocabulary, terminology, context, and other information representations and (2) communicating using the same syntactic and grammatical rules, i.e., communication protocols. Overall these two requirements dictate that each independent EMR system become a distributive unit of a whole and isolated EMR data islands become an integrated information repository. The reality is to the contrary, however. There are over 500 certified medical software companies in the United States selling over 1000 different software programs. Some are big, and some are niche players catering to subspecialties, but they have one thing in common; i.e., they are not interoperable. It is not uncommon that a dermatologist using the latest system from one company cannot send a record to the allergist in the practice down the hall who uses a competing product. It is also not uncommon that one clinic or department has to install, for performing different EMR functionalities, two or more of such systems that are informational silos. A physician usually has to carry over 10 passwords to log into different systems, e.g., a McKesson system for a radiology report, a General Electric (GE) system for an ECG, and yet another program for pathology report. Sometimes, even the systems produced by the same vendor do not talk to one another.

12.3.2 Technological Solutions

Cloud computing, a type of distributed computing over networks, is poised to be the platform for the delivery of most future IT services, including EMR. It originated from the idea of IT outsourcing in general and the two business models in particular: application service provision (ASP) and web service provision (WSP).

Since the watershed event of Kodak outsourcing in 1989 [8], IT outsourcing has become a strategic alternative against in-house IT deployment in IT-related decisions [9–10]. ASP is a business model of outsourcing applications over computer networks. The provider is a supplier of application services and assumes responsibility of buying, hosting, and maintaining a software application on its own facilities, publishing its user interfaces over networks, and providing its clients with shared access to the published user interfaces. The consumer, on the other hand, subscribes and receives the application services through the Internet or a dedicated network connection as an alternative to hosting the same application in house.

A provider may provide a mix of applications for managing medical records, including prescribing, charting, and office visit work flow, as well as coding, clinician scheduling, billing, and reporting. Some can even provide clinical alerts normally associated with expensive institution-based EMR systems, such as warnings of potential drug–drug interactions. The provider takes patient charts and medical records and keeps them on a centrally managed repository, which a healthcare organization can access from all over the world. With a patient's consent, physicians can review the patient's medication lists from all previous encounters and their prescription-filling habits. They can electronically retrieve prior health information, including clinical handwritten notes, lab reports, and photographs, and avoid redundant diagnostic tests.

The ASP model essentially allows organizations to hand over the responsibility of IT deployment or its execution to an outside vendor while still satisfying self-information needs. It reduces the complexity associated with the traditional make-or-buy model while allowing effective control of the deployment costs and risks [11]. On one hand, a provider can amortize expenditures over its entire client base, enabling it to improve quality of services, security, and risk-reduction measures that an individual clinic may find cost prohibitive. On the other hand, consumers or healthcare organizations do not incur the costs associated with traditional software implementation, including software license fees, hardware investment, and staffing and training of system administration personnel. They avoid nightly backup of patient data, monthly software updates, loss of data because of local hard drive or server failures, and the need to contract with a technical support group. By eliminating the need to manage hardware, software, information, and personnel, they can focus on their core businesses and free up resources for mission-critical applications. By eliminating the need to evaluate, purchase, deploy, and test hardware and software, applications can be up and running in a matter of weeks, instead of months or even years. According to surveys conducted by the ASP Industry Consortium and others, 60%–77% of respondents indicated that the extremely important factors for adopting the ASP model are reduced cost of application ownership, reduced risk of application deployment, improved ability to focus on strategic business objectives, and improved quality of data service.

The ASP model enables multiple healthcare providers to access the same set of patient data and, therefore, provides a solution to the interoperability problem. However, the solution works only if all the healthcare organizations subscribe to the same application. Unfortunately, this has never been the case. In fact, there are hundreds of competing providers such as Internet Logician, Hyper Charts, WebChart, and Practice Point Chart, who offer EMR application services. Note that most providers have struggled financially and never gained widespread acceptance for several reasons. First, with the advent of the Health Insurance Portability and Accountability Act (HIPAA), healthcare organizations have a pressing need for achieving compliance with government mandates, which require that all healthcare organizations adhere to a specific format for electronic transactions such as eligibility confirmation, treatment authorization, and referrals. However, no software vendor is big enough to cover every aspect of HIPAA compliance [12]. To be compliant with HIPAA, a healthcare organization may have to subscribe to multiple applications that are likely to be noninteroperable. Second, the expense of maintaining and updating host software has resulted in high overhead costs, diminishing the viability of the ASP as a business model [13]. For example, an ASP that maintains an EMR application must update data on drugs, tests, procedures, laws, and clinic equipment, as well as medical facts and discoveries constantly in order to provide relevant and timely data.

There have been many different approaches competing to be a platform for interoperability. These include Common Object Request Broker Architecture (CORBA), Distributed Component Object Model (DCOM), Remote Method Invocation (RMI), and Distributed System Object Model (DSOM) for communication protocols and Good European Health Record (GEHR), health level 7 (HL7) Clinical Document Architecture (CDA), openEHR, and the generic extensible markup language (XML)/ontology as a representation protocol. Unfortunately, these technologies do not resolve the interoperability problem. Healthcare is a typical many-to-many business. Sharing medical records is more than just connecting a hospital to a few branch clinics. Instead, each healthcare organization is an information node that sends and receives transactions to an array of internal and external information nodes. Using technologies like DCOM and CORBA, each party would have to incur significant expenses writing custom bridges to "hardwire" to the other nodes. For example, a dermatologist would need $10,000 for a bridge to an allergist next door. If one party changes its internal system, all other parties would have to respond. If a new party wants to join the network, all have to incur an enormous cost of entry to maintain the status of integration.

The interoperability problem has forced companies including Microsoft and International Business Machines (IBM) to coin a new open standard—web services—for distributed computing across applications, platforms, and devices. Unlike DCOM and CORBA, web services are loosely coupled software components that exchange XML-based data [14]. Web services have the following distinct characteristics [15]: First, each web service represents a business function or business service whose interface is exposed to the Internet and accessible by another program remotely [16]. It encapsulates a task; when an application passes a message (e.g., data or instructions) to it, the service processes the input and, if required, generates an output message back to the application. Second, all messages are written in XML-based text, which follows the simple object access protocol (SOAP) coding and formatting specifications, instead of cryptic binary strings. This enables web services to communicate with other applications that may be developed in different programming languages and reside on different platforms [13]. Third, web services are self-describing; each is accompanied by a description, written in the web services description language (WSDL), regarding what it does and how it can be used. Fourth, web services are discoverable; service consumers can search for and locate desired web services through the Universal Description, Discovery, and Integration (UDDI) registries.

Web services provide the healthcare industry with an ideal playground for sharing medical records. Each clinic, hospital, insurance company, or pharmacy can expose some or all functionalities of its internal legacy (open or proprietary) systems to the Internet as web services. These services can be as simple as making and scheduling appointments, validating credit cards, receiving lab results and physician orders, or submitting insurance claims. They can also be as complex as the functions carried out by an entire supply chain, customer relation management system, or eHealth applications. These services hide their internal complexities such as data types and business logics from their users, but expose their programming interfaces by using WSDL and their locations by using UDDI protocols. Since every service complies with one set of web services standards, there is no need for writing custom bridges in order to accommodate different computing platforms. Instead, clinics and hospitals can exchange patient data by directly invoking each other's data-exchange services. Clinics, hospitals, and pharmacies can leverage a common web service interface to asynchronously transmit claim data to an insurance company. If one participant were to change how a certain function is processed internally, as long as its programming interface does not change, the rest of the world can remain still. Thus, the cost of entry and exit will be greatly reduced.

WSP shares the same vision as the ASP model in terms of deploying software as Internet-based services and creating an economy of supplying and consuming the services. Of course, there are some differences [13]. First, an ASP usually offers large, complete applications with limited customization for individual clients. In contrast, web services can be smaller components performing specific functions. Second, web services are often developed and maintained by the same business unit, whereas most ASPs host software created and owned by others. These two differences suggest that WSP is a more efficient business model than ASP; it simplifies software maintenance and increases the flexibility of creating custom applications. Third, an ASP publishes user interfaces, which are meant for interactive use, whereas a web service publishes programming interfaces, whose use is programmable and complies with the WSDL. Thus, web services have better interoperability than application services.

The superiority of web services, however, does not necessarily drive out the market for application services. In fact, they are complementary business models and will coexist in the near future. On one hand, we expect that many ASPs would leverage web services to enhance the interoperability of their hosted applications. For example, they can employ web services provided by government agencies, research organizations, pharmaceutical manufacturers, and insurance companies for updating data on drugs, codes, procedures, etc., and, therefore, reduce their operating and maintenance overhead expenses. They can also assemble web services to create more comprehensive EMR applications that are fully HIPAA compliant. Some ASPs may even modify their technical infrastructures and business models to be more like WSPs.

On the other hand, we expect that application services will be still in demand. Although web services are easy to use for programmers, they are hard to consume for nontechnical users. Web services entail programming expertise to be understood and business knowledge to be assembled. It is analogous to assembling a computer from its parts; it is easy to do it for electronic engineers but challenging, if not impossible, for many others. Thus, we anticipate that the software industry will need both web services and application services, much like the computer industry that needs not only parts manufacturers, like Intel and 3Com, but also computer assemblers like Dell and Lenovo.

Besides the expertise requirement, the logistics of distributing services also concerns WSPs. Regardless of how easy it is to search for and locate a web service through a UDDI registry, after all, a consumer may have to contact the provider and negotiate a service contract. If there are thousands of web services to be subscribed, it is practically impossible for anyone to contact this many providers individually and renew contracts periodically. At the same time, WSPs are facing another dilemma. Since a web service is usually a small component, it may not be cost effective to spend money to market the service in a distinctive way. However, a lack of marketing effort will reduce consumer awareness about the service, which, in turn, will reduce the number of subscribers. Consequently, most providers cannot even afford the expense of maintaining their services and may have to exit, leading to a shrunken service market that diminishes the viability of web services as a business model.

How, then, do we resolve this logistic issue? One approach is to have a few large ASPs responsible for assembling and delivering suites of web services or packaged applications to consumers. This approach, if executed, will materialize our prediction that future application services and web services will coexist in the e-service market. The approach works but may be insufficient because these ASPs, like other consumers, still need to find and negotiate contracts for required services. What seems more desirable is to have a global market for efficient exchange of web services between service consumers and providers. This solution leads to the notion of service grids, which revolves around the idea of service creation and delivery through coordinated resource sharing and problem solving in dynamic, multi-institutional virtual organizations [17]. Peer-to-peer (P2P) computing was an early implementation of a service grid; it aggregates the unused computing power of individual personal computers into a computer power grid to create a virtual supercomputer [18]. With the advent of web services, now grid services and web services are rapidly converging to form a single set of standards and technologies as manifested in the Open Grid Services Architecture [19].

In addition to collaborative computing, grid services can be an effective market mechanism for distributing web services. A service grid may act as an intermediary between service providers and service consumers and break a typical many-to-many business (between providers and consumers) into simple one-to-many relationships. It buys web services from the providers and then sells the services to consumers. Then consuming web services becomes as easy as watching TV programs from a cable network or obtaining electricity from a power grid; requesting and delivering web services becomes as easy as plugging an appliance into the grid. In the meantime, those small WSPs do not have to incur prohibitive expenses to advertise and run their businesses. They can focus on their core business—developing and upgrading web services—and then plug the services into the grid to sell.

12.4 An Integrated e-Service Framework

Taking advantages of unique features offered by e-services, we proposed a new business model for implementing EMR as illustrated in Figure 12.10 [20]. As it shows, the model

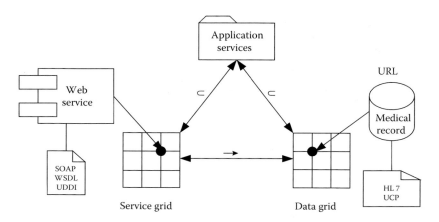

FIGURE 12.10
An integrated model of using e-services.

integrates application, web, and grid services into a unified architecture for managing and sharing EMRs. In particular, it proposes a separation between services and data and a mechanism for exchanging EMR services and sharing EMR data through services grids and data grids, as well as their assemblies—application services. The three components of the model are then triangulated through data flows (→) and physical compositions (⊂).

In the integrated framework, loosely coupled web services are the basic EMR functional elements. Individual web services participate in and form a process-oriented service grid, which acts as a global marketplace for healthcare providers and other individuals and organizations to exchange EMR services. Any clinic or hospital can plug its business processes, such as X-ray image readers and medical diagnosis experts, as web services into the service grid. These services will then be available to and consumed by other clinics and individual patients as well.

In the framework, personal information nodes are the basic EMR data elements. All information nodes join together to form a data-oriented P2P network, called *data grid*, which acts a distributed information repository of medical records. Each node within the grid has a universal resource locator (URL) address, which may be registered to a central registry. By providing a URL and appropriate authentication, any EMR service can access any information node.

A unique feature of our model is the separation between data and services. In particular, we propose that service providers own EMR services, whereas patients own and control EMR data. Note that the separation between services and data is logical. Physically, a patient may choose to host her data with an existing healthcare provider or government agent. She may colocate the data with an EMR service provider. She may also pool all record segments from an array of sources into her smartphone or chip, providing a unified node in the grid. These alternatives are particularly viable given that there have been a few decades of local, intermittent EMR initiatives that have warehoused a large volume of medical records in the computer. Of course, to integrate them into the data grid, it may be necessary to recode them or format their outputs by using standards such as HL7 CDA.

Despite the separation, services and data meet via two avenues. The first is data flows, which are labeled as → in Figure 12.10. In this channel, a web service retrieves existing data from and dispatches new or updated data to an information node. The second avenue is physical compositions, which are labeled as ⊂ in Figure 12.10. In this channel, small web services may be assembled into large applications and individual information nodes may be pooled together into large medical databases and warehouses. These applications and databases are centrally colocated and managed. Due to a lack of technical talents, many small and medium healthcare providers are likely to subscribe to these applications on a rental basis. Large providers will likely assemble and manage their custom applications and databases to sustain their competitive advantages. Of course, some applications may further join the service grid as web services. Similarly, medical databases may also join the data grid as information nodes. Therefore, the physical compositions shown in Figure 12.10 are all bidirectional.

12.4.1 Justifications

Neither web services nor grids are new to the medical informatics community. For example, the Cancer Biomedical Informatics Grid (caBIG) is an initiative that provides an informatics infrastructure for interdisciplinary collaborations across the field of cancer research (http://cabig.cancer.gov/). What is new here, however, is how we make these technologies work together by allowing each technology to capitalize on its strengths while compensating for the weaknesses of others. What is also different from all other EMR models is that our model emphasizes the reuse of existing services and records rather than reinventing the wheel and creating a brand new system from the ground up.

The business model we proposed is both economically and politically sound. The constraints to implementing EMR include patient privacy protection as dictated by HIPAA [12], costs of information capturing [4], and technical interoperability of data content. With respect to privacy protection, the P2P data grid gives a patient a full control over her own data; she can selectively release certain portions of her health records to a physician; she can also contribute historical data to a data warehouse or research archive while removing personal identifiers.

Economically, our model has its advantages by distributing the responsibility of implementing EMR to the participating stakeholders. On one hand, by releasing the responsibility of managing patient records, healthcare providers will be able to reduce costs associated with performing data-related activities and concentrate on what they do best. Assume that an average medical staff member spends 50%–75% of their time in retrieving and updating patient data. Off-loading these functions to patients means over 20%–50% of savings in healthcare costs. It allows thousands of small and medium healthcare providers to achieve HIPAA conformance without incurring prohibitive costs. In addition, the service grid allows technically capable healthcare providers, besides running their core business, to make additional revenues by contributing its unused computing power and in-demand technologies. It allows a more efficient distribution of computing resources, especially those with proprietary technologies, among the healthcare industry.

On the other hand, since it is patients who are responsible for their health, it is their best interest to ensure the availability and quality of their own records, invest in their own data, and take the ownership of the data as well as their own care as informed patients. Of course, patients may incur a fee to use clinical or public facilities to have their data digitized and updated. However, they are better off for improved healthcare due to improved data quality and availability of eHealth applications. By the way, a portion of cost savings to healthcare providers may be passed on to the patients to offset the costs. Furthermore, due to a widespread demand for record digitization and data entry as well as their storage and maintenance, there will be a market for a brand-new business sector specialized in these activities. Information capturing and use will gain efficiency due to the division of labor.

Our model also helps achieve large-scale distributed computing for medical research. With patients contributing their data resources, the data grid becomes a global repository of medical records that can be aggregated and researched. With healthcare providers contributing their business processes and computing power, the service grid offers a flexible infrastructure that is available on demand and as required for changing groups of people to interact and collaborate [18,21], and helps investigators identify research participants that meet precise and complex clinical trial criteria. It will also help them conduct large-scale clinical research across a broad range of diseases, complex diagnoses, and treatment categories. For example, seven million Americans suffer from chronic inflammation, such as arthritis, bursitis, and other joint diseases. In the future, physicians may be able to consult databases of historical data on millions of similar patients to determine the most effective course of treatment for an individual patient [22]. In fact, as one of early adopters of grid services, Mayo Clinic initiated a service grid project linking its own medical database with external public and private data sources. The grid could enable its entire medical staff, including 2400 physicians and scientists in more than 100 specialty areas, to quickly draw meaning from a wealth of medical data to support medical treatments [22]. The data include genome information from public and private databases and retrospective studies of millions of archived records collected from informed, consenting patients.

Note that the vision of self-controlled EMR is hardly new. Researchers and practitioners have long realized that sharing medical records does not have to be the responsibility of clinics and hospitals [4]. Indeed, back in the mid-1980s, there was a vision that a patient is given a card device like a smart card with a computer chip and that, being a connecting entity of all health information, the card is used when a person receives medical treatments. This vision failed because of the problems associated with device capacity and interoperability.

Now, with the availability of broadband connections and widespread adoption of personal computers, this vision needs a revisit; a data grid composed of personal information nodes becomes a viable option. Mobile phones and tablets are changing many aspects of our lives, including the delivery of healthcare services and the management of medical records. In a recent survey of 114 nations, the World Health Organization [23] found that many countries had established a varied range of mobile health initiatives, including the creation of health call centers to respond to patient inquiries, using short message service (SMS) for appointment reminders, using telemedicine, accessing patient records, measuring treatment compliance, raising health awareness, monitoring patients, and physician decision support. Using mobile devices to collect medical records is also on the horizon. Companies like Apple are developing applications such as Healthbook that allow patients not only to pool health data from their doctors into one place but also to track nutrition, fitness, and weight by monitoring the user's habits such as food intake, sleep cycle, and hydration levels.

12.4.2 Implementation

Reelaborating the web services architecture of Hagel and Brown [24], we proposed an implementation architecture consisting of three layers: resources, resource grids, and application services [20] (see Figure 12.11). The resources include both individual web services and personal information nodes. The resource layer consists of standards and protocols that establish a common language for these services and data. It essentially provides long-distance glue that binds the services and data together and allows information to be exchanged easily and securely between different services. Resource grids include

Application: Provides services to end users	Application services	
	Service grids	Data grids
	Web services	Data nodes

Presentation: Encodes and compresses data
Session: Creates and manages connections
Transport: Corrects errors with lost and missed packets
Network: Creates and dispatches packets across networks
Data link: Creates and delivers frames within one network
Physical: Transmits bits and bytes as signals

FIGURE 12.11
e-Service architecture.

both process-oriented service grids and data-oriented information grids. The resource grid layer consists of protocols and standards that dictate the development of these grids and provides shared utilities in support of the operation of the grids. Application services perform actual business functions, from medical diagnosis to physician order entry. The application service layer consists of standards and protocols on how these applications interface with resource grids and how they are assembled from EMR service and data elements.

As Figure 12.11 shows, the proposed architecture extends the Open Systems Interconnection (OSI) model by decomposing its application layer into three sublayers. Note that the application layer specifies communication at the message (rather than packet or signal) level and functions that provide computing services to end users (see, e.g., Goldman et al.'s book [25]). The protocols and standards in the three sublayers all focus on the communication of XML messages possibly along with non-XML attachments and, thus, fall into this layer.

The protocols at the resource layer are built on existing Internet application protocols. First, SOAP, also known as XML protocol, uses hypertext transfer protocol (HTTP) and simple mail transfer protocol (SMTP) for message transport. In particular, it uses HTTP to penetrate firewalls, which are usually configured to accept HTTP and SMTP service requests.* The XML Protocol Workgroup at the World Wide Web Consortium (W3C) is currently developing a few SOAP extensions, including attachments, routing/intermediaries, reliable messaging, security, transaction support, context/privacy, and quality of service. The extensions provide additional modules of functionality for developers to plug into SOAP when necessary [26]. Some of these extensions are in support of general purposes such as routing SOAP messages through intermediaries, protecting the security of e-services, and ensuring the delivery of SOAP messages. The others are of particular relevance to EMR. They include SOAP messages with attachments and the privacy of personal information nodes. The attachment extension will describe a standard way of attaching non-XML or binary files such as medical images and fax documents to a SOAP message.

---

* HTTP is the standard Internet protocol for transferring data between web servers and web browsers, while SMTP is for transferring electronic mail messages between mail servers and mail browsers.

The multipurpose Internet mail extensions (MIME) are likely to be the standard envelope format in this extension [26].

With respect to privacy, the Platform for Privacy Preferences Project (P3P) is currently under development by the W3C (see http://www.w3.org/p3p). Note that P3P aims to build intelligence in e-services so that their execution matches a user profile and fits the context that exists at the time of the execution. Of course, the use of context information raises the concern for privacy. However, currently there are no standards addressing this concern [26]. The need for implementing EMR brings the concern up to a higher level due to HIPAA. Thus, there is a need to establish a privacy control protocol (PCP) for protecting the privacy of information at each information node. The PCP would specify how medical records at each node might be accessed and secured. In particular, it specifies how patients assign system privileges and access permissions to users, groups, and roles and how to code access control files to be saved along with medical records at the outset shell of an EMR information node. It also specifies how an e-service may be programmed to authenticate against the control data to gain access to desired content. With these specifications, the PCP can then accommodate the separation of EMR data and services and allow any e-service to access any information node while observing its unique privacy rules.

Second, on top of SOAP and its extensions, WSDL specifies the abstract service interface, including messages and operations supported, and the concrete service description, including data format, network protocol, network address, and port of a specific installed web service. From the WSDL files, the application that intends to use a service can then identify which message formats and protocols are supported and forward data using a format and protocol appropriate to the service. Also, on top of SOAP, there is the need for a language that describes information nodes. With some modification, the HL7 CDA may serve the purpose. Currently HL7 defines how clinical documents, such as discharge summaries and patient records, will be coded (http://www.hl7.org). In this standard, patients can format their data in HL7 XML format so that all HL7-based web and application services understand the data content located at any information node in the data grid without custom instructions specific to the node. Note that, besides describing the representation of medical records, HL7 should incorporate PCP and describe metadata and access control at each information node. Eventually, HL7 allows any e-service within a service grid to connect to any information node within a data grid and retrieve from and dispatch data to the node. Therefore, although EMR services and data are separated in our business model, the implementation ensures a collaborative environment for implementing a revolutionary vision of EMR—there is complete interoperability between information systems in all locations, provider settings, and infrastructures—as well as aggregating computing resources for medical research.

Finally, on top of WSDL and HL7, the UDDI is used to find web services and information nodes. UDDI is a specification for registering potential network-accessible healthcare providers offering web services, the web services offered, and the technologies supported by the web services, including specific representation and control protocols, document types, and transaction sets, as described by WSDL. For example, it can register a service that accepts certain types of clinical patient information or certain types of control data. UDDI can be easily extended to register individual patients who hold information nodes.

Figure 12.12 shows the hierarchy of protocols at the resource layer. At the bottom of the hierarchy are HTTP and SMTP, which provides Internet connectivity for web services and information nodes. Then SOAP and its extensions build upon HTTP and SMTP and provide the specifications for message transport. On the third layer are the protocols governing how EMR services and information nodes are implemented, including WSDL and HL 7.

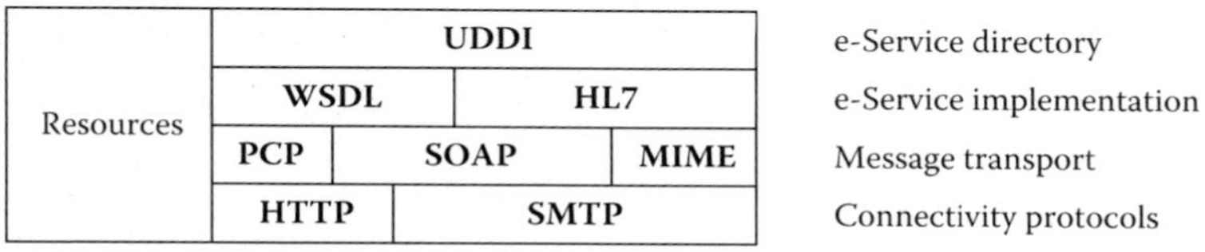

Resources	UDDI			e-Service directory
	WSDL		HL7	e-Service implementation
	PCP	SOAP	MIME	Message transport
	HTTP		SMTP	Connectivity protocols

FIGURE 12.12
The resource protocol layer.

On the top layer is the protocol for service directory. Note that this four-layer architecture builds upon W3C recommendations and is consistent with the framework currently in development by the newly formed Organization for the Advancement of Structured Information Standards.

Besides EMR service and data elements, the resource layer also consists of devices in support of the resource protocols. Such devices include, for example, utilities that convert existing medical records into HL7-compatible information nodes, programs for patients to code their medical records to conform to HL7, and programs that allow patients to assign access permissions in accordance with PCP. Of course, thanks to the use of XML, the resource protocols make EMR documents both machine readable and human readable so that they can be parsed and processed electronically as well as retrieved and used by human beings.

The resource grid layer is the middle layer of the architecture and builds on top of the resource layer; while the resource layer addresses how individual services and information nodes may behave and function, the resource grid layer specifies the function of a service or data grid as a whole in exchanging and distributing resources. It specifies how a provider may join a grid, offer grid service instances, and receive compensation from the grid. It specifies how a consumer may subscribe to the grid, consume its services and data, and make payment to the grid. It also addresses issues associated with quality of service and provides auditing and assessment of third-party performance. For example, the Open Grid Services Architecture [17] defines, in terms of WSDL interfaces, mechanisms required for creating and composing sophisticated distributed systems, including lifetime management, change management, and notification. It specifies uniform service semantics and standard mechanisms for creating, naming, and discovering transient grid service instances to provide location transparency, multiple protocol bindings for service instances, and integration with underlying native platform facilities [21]. In short, the main functions of the resource grid layer are to (1) help providers and consumers find and connect with each other and (2) create a trusted environment for offering or using services and data and for carrying out mission-critical business functions and transactions over the Internet [24]. Consequently, this layer glues together distributed e-services and information nodes together into service grids so that sharing and using EMR services and data becomes as easy as plugging appliances into a power grid.

Besides the protocols that dictate the operation of service and data grids, the resource grid layer also consists of a set of shared utilities—from security to performance assessment and to billing and payment—that implement the function of service and data grids. For example, security utilities provide authentication and authorization for one to use service grids, whereas assessment utilities ensure users of services to obtain agreed-upon levels of performance. The utilities may be further classified into three categories: resource management, service management, and transport management [24]. Resource management utilities provide directory and brokerage services and intelligent agents for updating resource registry. Service management utilities manage connections and usage, monitor service quality and performance, and ensure reliability and consistency. Transport management provides messaging services to service providers and consumers. The utilities include intelligent agents for contract negotiation and conflict resolution, middle ware that bridges resource elements, and orchestration utilities that help assemble application services from resources offered by different vendors.

The application service layer includes a diverse array of application services—from appointment scheduling to medical diagnosis—that automate particular business functions. Some applications may be proprietary to a particular company or group of companies, while others will be shared among all companies [24]. In some cases, companies may develop their own application services and then choose to sell them on a subscription basis to other enterprises, creating new and potentially lucrative sources of revenue.

The application service layer dictates how to assemble business applications based on resource grids. It consists of protocols and standards that specify how an application may interface with a resource grid. Analogous to an appliance, for example, a washer or dryer, which plugs into a power grid and a water grid, an application connects to a service grid and a data grid to be functional. To this end, it has to conform to some interface standards. Examples include encryption and privacy settings that each application service has to set and how an application may use shared utilities in a resource grid in order to determine the quantity and quality of consumed services and data. These standards do not yet exist and it is up to further research and development to define them. One possibility is to assign communication ports to certain utilities. For example, one may designate one port for a usage meter, another one for performance meter, and still another one for authentication. Finally, since the Internet is a public infrastructure, it is beyond the control of any individual organization or resource grid. To fit into this environment, an application service has to be tolerant to the execution speed of its component services and the characteristics of information flow such as flow speed and reliability [27]. Therefore, the application service layer may need to specify an appropriate communication mode, i.e., whether it is synchronous or asynchronous, and acceptable ranges of tolerance to information flow turbulence.

12.5 Conclusions

In this chapter, various concepts of EMRs are presented. To illustrate the core concept, a few use cases and class diagrams as a prototypical model of the functional and data requirements of EMR are proposed. These diagrams combine similar models scattered in books and open-source projects and may serve as references for actual implementation. Also, the issues regarding how to organize data on patient history and lifestyles are addressed.

Two problems with EMR implementation and management are discussed in depth: economic motivation and technical interoperability. Economic motivation is concerned with costs and benefits. As with most IT investments, it is often difficult to justify EMR in terms of measurable benefits [28]. The resource commitment to EMR is often beyond the affordability of many small to midsize healthcare providers. The technology interoperability entails both semantic standards for information representation and communication protocols for information exchange. Existing EMR systems do not provide such interoperability.

Bearing the two problems in mind then, how emerging e-services such as cloud computing might facilitate EMR implementation is discussed. Along this line, there are two important questions to be addressed. First, how do e-service models help solve the economic motivation and technical interoperability problems? Second, how will different e-service models be made to work together? As argued, ASP helps overcome the economic obstacle but does not offer the technical interoperability required for implementing a higher vision of EMR. On the other hand, web services provide a flexible architecture for interoperable computing over the Internet. However, there is a lack of expertise for small to midsize healthcare providers to use them and a lack of economic mechanism to distribute and exchange them. Therefore, these e-service models do not work at all if they do not work together.

To meet the challenge of putting the pieces together, a business model that ties both application services and web services into an integrated e-service framework for implementing EMR is proposed. Sharing EMR functions and data through distributed resource grids and centralized application services are detailed. The resource grids allow healthcare providers to share their business processes and individual patients to share their medical records. The resource grids play the role of a marketplace, resolving the logistic issue associated with selling and buying e-services. Application services build on top of resource grids and improve their interoperability by integrating with web service infrastructures. They provide readily usable solutions, enabling many small to midsize healthcare providers with no technical expertise to consume web services.

A separation between functions and data through two types of resource grids, process-oriented service grids and data-oriented information grids, is proposed. The separation is shown to be a viable alternative to EMR implementation. It corroborates with an old vision of self-controlled EMR and allows for full control of patient privacy. It achieves an optimal allocation of management responsibility among all the stakeholders and allows each participant to focus on what they do best. It advocates the division of labor and enhances the economy through the creation of a new service industry and the improved efficiency in EMR-related economic activities due to specialization.

To implement the business model, I extended the OSI model and proposed an e-service architecture consisting of standards, protocols, and devices at three layers: resources, resource grids, and application services. Resources include EMR functional and informational elements, i.e., e-services and personal data nodes. The resource layer consists of the protocols and standards at four sublayers for connectivity (HTTP and SMTP), message transport (SOAP and its extensions PCP and MIME), e-service implementation (WSDL and HL7), and e-service directory (UDDI). The resource grid layer consists of the standards and protocols on how service and data grids are operational as well as shared utilities in support of their operations. The aim is to provide a trusted infrastructure for people to connect with each other and offer and use e-services and data. The application services layer governs how eHealth applications interface with resource grids in order to take advantage of the infrastructure and a vast array of resources that resource grids offer.

e-Service architectures have already captured a lot of interest among researchers and developers. Among other issues, it is a well-recognized challenge that some important architectural layers are yet to be developed and finalized [29]. The proposed architecture has many advantages and makes contribution to understanding and applying e-services in healthcare. First, it aims to provide a solution to real healthcare issues involving e-services. It grows out of issues associated with EMR implementation and adoption. It is consistent with the IT strategy proposed by Hagel and Brown [24]. Thus, it is a technical solution that may work in business. Second, the architecture improves the web service architecture of Hagel and Brown [24]. It redefines the service grid layer and application service layer and allows for embedding the layers along with resource layer consistently into the OSI model. It also refines the protocol layer into a finer architecture of four sublayers, and shows how these sublayers support each other and support higher layers of the architecture. Third, our architecture bridges the gap in existing technologies and identifies several important but yet missing protocols to be developed or modified. For example, the vision of personal information nodes and the need for the privacy control lead to the need for the PCP that provides functionalities well beyond P3P currently under development. Similarly, many protocols and standards at the resource grid and application services layers are important but yet missing.

References

1. P. C. Waegemann, "An electronic health record for the real world," *Healthcare Informatics*, vol. 18, no. 5, pp. 55–60, May 2001.
2. J. F. Burnum, "The misinformation era: The fall of the medical record," *Annals of Internal Medicine*, vol. 110, pp. 482–484, 1989.
3. H. Bleich, "Weed and the problem-oriented medical record," *MD Computing*, vol. 10, pp. 70–71, 1993.
4. P. C. Waegemann, "Status Report 2002: Electronic Health Records," Medical Records Institute, Chicago, Illinois, Status Report 2002.
5. M. Bishop, *Computer Security: Art and Science*. Boston: Addison Wesley, 2003.
6. HIMSS (December 15, 2002), HIMSS/AstraZeneca Clinician Survey. Available: http://www.himss.org/content/files/surveyresults/Final_Final_Report.pdf.
7. Q. Ma and L. Liu, "The role of Internet self-efficacy in the acceptance of web-based electronic medical records." In *Contemporary Issues on End User Computing*, M. A. Mahmood (ed.), IGI Global, Hershey, PA, Chapter 3, pp. 54–76, 2007.
8. L. C. Applegate and R. Montealegre, *Eastman Kodak Company: Managing Information Systems through Strategic Alliances*, vol. 9-191-030. Boston: Harvard Business School Case, 1991.
9. M. C. Lacity and R. Hirschheim, *Information Systems Outsourcing: Myths, Metaphors, and Realities*. New York: Wiley, 1993.
10. L. Loh and N. Venkatraman, "Diffusion of information technology outsourcing: Influence sources and the Kodak effect," *Information Systems Research*, vol. 3, pp. 334–358, 1992.
11. D. T. Dewire, "Application service providers," *Information Systems Management*, vol. 17, pp. 14–19, 2000.
12. L. MacVittie, "Survivor's guide to 2003: Business applications," *Network Computing*, vol. 13, pp. 78–85, 2002.
13. H. M. Deitel, P. J. Deitel, B. DuWaldt, and L. K. Trees, *Web Services: A Technical Introduction*. Upper Saddle River, New Jersey: Prentice Hall, 2002.
14. M. Stal, "Web services: Beyond component-based computing," *Communications of the ACM*, vol. 45, pp. 71–76, 2002.
15. P. Flessner (August 15, 2001), "XML Web services: More than protocols and acronyms," *Software Development Times*. p. 31, 2001.
16. P. Fremantle, S. Weerawarana, and R. Khalaf, "Enterprise services," *Communications of the ACM*, vol. 45, pp. 77–82, 2002.
17. I. Foster, C. Kesselman, J. M. Nick, and S. Tuecke, "The Physiology of the Grid: An Open Grid Services Architecture for Distributed Systems Integration," The Globus Project, http://www.globus.org, Technical Report 2002.

18. D. P. Anderson, J. Cob, E. Korpela, M. Lebofsky, and D. Werthimer, "SETI@home: An experiment in public-resource computing," *Communication of the ACM,* vol. 45, pp. 56–66, 2002.

19. D. Gannon, K. Chiu, M. Govindaraju, and A. Slominski, "A Revised Analysis of the Open Grid Services Infrastructure," Indiana University, Bloomington, Indiana, Technical Report 2002.

20. L. Liu and D. Zhu, "An integrated e-service model for electronic medical records," *Information Systems and e-Business Management,* vol. 11, pp. 161–183, 2013.

21. I. Foster and C. Kesselman, *The Grid: Blueprint for a New Computing Infrastructure.* San Francisco: Morgan Kauffman, 1999.

22. A. Goyal (January 7, 2002). *The Future of Distributed Computing: Grid Services.* Available: http://e-serv.ebizq.net/dvt/goyal_1.html.

23. World Health Organization, "mHealth: New Horizons for Health through Mobile Technologies," *Global Observatory for eHealth Series,* vol. 3, 2011.

24. J. Hagel and J. S. Brown, "Your next IT strategy," *Harvard Business Review,* vol. 79, pp. 105–113, 2001.

25. J. E. Goldman, P. T. Rawles, and P. Rawles, *Applied Data Communications: A Business-Oriented Approach,* Third ed. Hoboken, New Jersey: John Wiley and Sons, 2000.

26. P. Cauldwell, R. Chawla, V. Chopra, C. Damschen, C. Dix, T. Hong et al., *Professional XML Web Services.* Birmingham, United Kingdom: Wrox Press, 2001.

27. R. Krovi, A. Chandra, and B. Rajagopalan, "Information flow parameters for managing organizational processes," *Communication of the ACM,* vol. 46, pp. 77–82, 2003.

28. K. Kumar, H. G. V. Dissel, and P. Bielli, "The Merchant of Prato—Revisited: Toward a third rationality of information systems," *MIS Quarterly,* vol. 22, pp. 199–226, 1998.

29. C. Ferris and J. Farrell, "What are web services?," *Communication of the ACM,* vol. 46, pp. 31–34, 2003.

13

Public Health Informatics in Australia and around the World

Kathleen Gray and Fernando Martin Sanchez

CONTENTS

13.1 Introduction

This chapter begins by introducing the essential role that information and communication have played in public health past and present. It goes on to map the emergence and strength of public health informatics as a specialized domain of knowledge and practice. It outlines the key data, information, and knowledge management concerns of public health informatics.

13.1.1 Information and Communication in Public Health

Public health is distinguished from other health-related human activities and knowledge domains by its focus on whole-of-population systematic efforts to prevent diseases, promote health, and prolong life [1].

Tools and methods for managing information and communication are central to the three main functions of public health: to assess and monitor the health of communities and populations at risk to identify health problems and priorities; to formulate public policies designed to solve identified local and national health problems and priorities; and to assure that all populations have access to appropriate and cost-effective care, including health promotion and disease prevention services.

In the modern era, common examples of monitoring functions include the inspection and enforcement of standards for air quality, water potability, and food handling. Motor vehicle safety standards, workplace health, and safety laws and regulations controlling the use of alcohol, tobacco, and other drugs are mainstream examples of public policy formulation. The access to care function is exemplified classically by programs of vaccination and infection control and by services for family planning and maternal and child health. Information and communication management challenges that typify such core functions include maintaining registries and records and disseminating participation and compliance messages.

Fundamental measures in public health go back to ancient times, for example, religious teachings about personal hygiene; schools of philosophy propounding diet and exercise regimes; and the construction of facilities for bathing and sanitation in early settlements. The rise of cities and of trade was accompanied by organized efforts to count births and deaths and to quarantine people with diseases such as leprosy, plague, and cholera, enhanced by scientific methods from the late 18th century. The 19th and 20th centuries extended public health concerns to replacing slum housing, making low-cost healthcare widely available, and extending the health and lifestyle education available to individuals and families. Arguably, the early treatises and social structures that enabled these measures to be propagated could be seen as the precursors of modern information and communication management strategies in public health [2–3].

Public health took a global turn as recognition of a world economy strengthened in the 20th century. International health agencies were established—the International Sanitary Bureau in 1902 and the World Health Organization (WHO) in 1948. Concerns emerged about fallout from nuclear weapons, the health status of indigenous communities, and human health risks of environmental degradation and climate change. Worldwide efforts were made against diseases such as smallpox, influenza, and HIV/AIDS. Information and communication also came to be used for widespread expression of criticisms of public health measures and ideologies during this period, for example, protest movements against fluoridation of water supplies, against childhood immunization, and against health programs seen to be disempowering disadvantaged groups. Figure 13.1 shows public health data flow and informatics solutions. From the 1980s, the term *population health* came to be

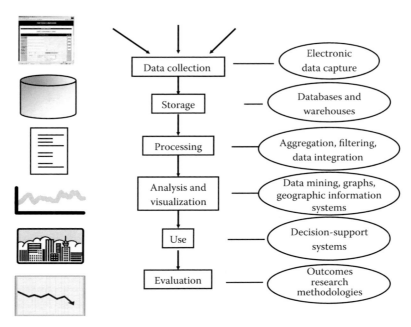

FIGURE 13.1
Data flow and informatics solutions in public health.

favored over *public health* in many quarters, because of perceptions that the expansion of the biomedical paradigm of health had overshadowed public health. Population health was seen as a stronger complement to the medical model of health, reenlarging the dwindling concept of public health to emphasize and integrate social, environmental, cultural, and physical determinants of health [4]. The rise of mass media, computers, and the Internet has had a powerful influence on public health information and communication since the latter part of the 20th century.

13.1.2 Evolution of Public Health Informatics

Public health informatics was defined first in 1995, in the United States, as the application of information science and technology to public health practice and research. It was said to reflect converging interests in health, healthcare reform, and the advent of the information age. It is hard to imagine that only two decades ago, its aim—to introduce electronic data and information systems, communications tools, and monitoring methods into public health—was still seen by many in the public health community as obscure, complex, impractical, expensive, overly specialized, and not obviously useful [5].

Public health informatics sits within the multidisciplinary and heterogeneous field called health, medical, or biomedical informatics, which encapsulates the application of information and computer science, knowledge management, and information and communication technology in research and practice in health [6–7]. Although public health has a long history, the recognition of a subspeciality of health informatics came relatively late, not until the 1990s [8]. Core competencies for public health informatics specialists were formulated only in 2009 [9].

Nevertheless, public health informatics themes can be seen throughout the emergence of the broader discipline of health and biomedical informatics. This history can be traced in North America to the 1879 creation of Index Medicus by medical librarians and to the

1928 establishment of the American Association of Medical Record Librarians (now the American Health Information Management Association). European records show that the terms *informatique de médecine* and *informatique médicale* were in use from the 1950s in France, and that health agencies established informatics departments in the 1960s in France, Holland, and Belgium [10].

Public health is both implicit and explicit in the vision of the International Medical Informatics Association (IMIA). IMIA, established in 1967 as a subcommittee of the International Federation for Information Processing, is now a nongovernment organization recognized by the World Health Organization as the peak body of informatics societies around the world, including North America, Europe, Latin America, and the Caribbean, Africa, the Middle East, and the Asia/Pacific region. IMIA's vision is "that there will be a world-wide systems approach for healthcare. Clinicians, researchers, patients and people in general will be supported by informatics tools, processes and behaviors that make it easy to do the right thing, in the right way, at the right time to improve health care for all. This systems approach will incorporate and integrate research, clinical care and public health" [11].

Although public health informatics is identified as an area of interest of health informatics societies around the world, it has developed its strongest identity in the United States [12]. The American Medical Informatics Association (AMIA) has over 500 members in its public health informatics working group. AMIA considers population health informatics to be a corollary of public health informatics; although its main focus is on informatics for classical functions such as biosurveillance, outbreak response, and electronic laboratory reporting, it notes the potential of the field to extend to aspects of agriculture, architecture, climate, and ecology [13]. The American Public Health Association too has a health informatics information technology interest group [14]. Since 2003, a coalition of United States city and county, state, and federal agencies has hosted an annual public health informatics conference, with over 1000 participants each year in recent years [15].

Public health informatics is attracting growing worldwide attention in the 21st century. The 2006 Davos World Economic Forum report highlighted the poor state of public health infrastructure as a major barrier to achieving global health goals, prompting further attention to public health informatics. Public health informatics was included as part of the Rockefeller Foundation's Making the eHealth Connection: Global Partnerships, Local Solutions 2008 Bellagio Centre conference series [16]. In 2011 the Health Metrics Network hosted by the World Health Organization announced two major public health informatics initiatives: Monitoring of Vital Events in support of the Millennium Development Goals, to strengthen countries' systems for recording every birth, death, and cause of death by utilizing advances in information technology; and the State of the World's Information Systems for Health, a concerted effort to track progress in this area [17].

In the 21st century the increasingly common use of the term *eHealth* is blurring some of the boundaries between public health informatics and other branches of health informatics. Figure 13.2 shows that eHealth not only encompasses collection, analysis, and exchange of health information and data on secure private networks but also particularly extends these activities using Internet and web-based communication methods, with major implications for public health practice, education, and research [18].

13.1.3 Key Concepts in Public Health Informatics

Public health informatics applies information science and technology to the management of the data, the information, and the knowledge associated with public health practice and research. Each aspect of management is outlined in turn.

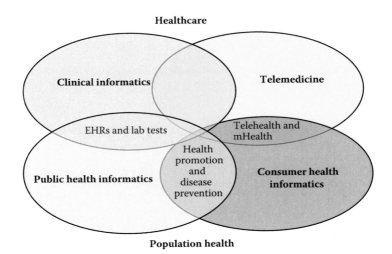

FIGURE 13.2
Public health informatics in the context of informatics, eHealth, and health disciplines.

13.1.3.1 Data Management in Public Health Informatics

Public health data management adds dimensions to the already complex data management challenges of clinical informatics. In public health informatics the emphasis is on generating, collecting, analyzing, and sharing data about populations, rather than individual patients. Like clinical informatics, it supports effective practice by optimizing access to representative, complete, and timely data from multiple sources; indeed, clinical and public health data management join forces in some areas of public health informatics, for example, chronic and communicable disease control, population health assessments, and healthcare policy [19]. Cognitive activities involving the large, heterogeneous, and complex bodies of public health data are beginning to benefit from sophisticated visual analytics tools [20]; however, there are some fundamental issues that need to be addressed first.

A major hurdle for public health data management is to improve data quality. It is still common to encounter inaccurate data, inconsistencies across data sources, and missing or incomplete data to support public health decision making. In cases of highly infectious diseases (for example, influenza), it is obvious that if complete data are not received quickly, then public health agencies are compromised in their ability to contain outbreaks. Some studies have shown that notifiable conditions (such as vaccine-preventable diseases like pertussis, commonly called whooping cough) were reported to public health authorities in fewer than half of the cases where they were diagnosed [21].

13.1.3.2 Information Management in Public Health Informatics

Public health information systems (PHISs), now considered to be in the third wave of their evolution, ideally function as more integrated knowledge systems than earlier information systems did. A modern PHIS integrates data, descriptive and analytical information, and evidence-based knowledge. It utilizes electronic data-exchange methods and standards. It is designed to meet the information needs of public health workers at different organizational levels, across organizations, and working in collaboration with various healthcare and government stakeholders. It is usable by public health professionals in a wide variety of office, community, and field practice settings [22]. Data warehouses, geospatial information services and simulations may be deployed in these systems [23], and real-time information from systems of this kind is now in sight [24].

Modern toxicology information systems offer good examples of these approaches to information management. Such systems have arisen from separate but intersecting information needs among clinicians who diagnose patients with symptoms of toxicity (commonly called poisoning) and of environmental health officers (EHOs) who assess exposure to toxic substances in a specific group of people. These systems must meet information needs that may not arise at the same time, in the same place, or in the same cultural or linguistic setting. They must filter information on a huge number of possible toxic substances and their physical properties as well as different pathophysiological effects on human health and prevention and treatment options [25].

13.1.3.3 Knowledge Management in Public Health Informatics

The aim of public health knowledge management is to build common frameworks that improve public health research and practice by enhancing information retrieval, data annotation and integration, semantic interoperability, and reasoning across public health jurisdictions. Examples of applications of public health knowledge management are public health legislation ontologies [26], public health reporting logic [27], and repositories of public health research evidence [28].

Overview knowledge of the types of information needs and categories of information systems is an essential part of ensuring continuity in public health operations. This level of knowledge is not only fundamental to routine service quality assurance but also has been shown to be particularly important when business as usual is disrupted, in major incidents or disaster scenarios, for instance. Types of information needs in such situations include synthesized information, centralized data access, coordination/incident command support, staff training/education, planning, plan testing/exercise support, interoperable external communication/alerting, internal communication/alerting, staff attendance/contact list management, resource tracking/capacity management, collaboration, remote work/portable data, and geographic data. Four categories of public health information systems address information needs in such situations—risk-assessment decision-support tools, networked and communications-enabled decision-support tools, dedicated training tools, and dedicated videoconferencing tools [29].

13.2 Public Health Informatics in Australia

This section reviews and illustrates the nature and scope of public health informatics in the context of public health and population health responsibilities at all three levels of government in Australia. This section concludes by summarizing the current status of public health informatics professional practice, peer-reviewed research, and specialist qualifications in Australia.

13.2.1 Australia's Public Health

It is important to double-check what is meant by *public health* in Australia because the term may be used in two ways. One way describes the population-level approach to health, which is the focus of this chapter. Another differentiates privately funded healthcare services from taxpayer-funded universal access to medical care, which is the cornerstone of Australia's healthcare system. This chapter avoids the latter interpretation in examining public health informatics.

Responsibility for Australia's public health like the rest of its healthcare system is shared among the federal (known as commonwealth) government and six state and two territory governments; the greater constitutional responsibilities and regulatory powers rest with the states and territories. Local governments, under the jurisdiction of states and territories, contribute substantially at the service delivery level; other resources are drawn from the primary healthcare system, from nongovernment and community organizations, and from universities and research institutes.

An example of public health information collaboration across these three levels of government is Communicable Diseases Intelligence reporting. This covers the Sentinel Practice Research Network, the Childhood Immunisation Register, the Gonococcal Surveillance Programme, the Meningococcal Surveillance Programme, the Creutzfeldt–Jakob Disease Registry, the Paediatric Surveillance Unit, HIV/AIDS surveillance, the Influenza Surveillance Scheme, the Notifiable Diseases Surveillance System, OzFoodNet, and the Sentinel Chicken Surveillance Programme, as well as including tuberculosis notifications, the Australian Mycobacterium Reference Laboratory Network, invasive pneumococcal disease surveillance, the National Arbovirus and Malaria Advisory Committee, and the Australian Rotavirus Surveillance Program [30].

At the national level, the commonwealth's Department of Health's Population Division works closely with the Office of Health Protection, the Mental Health and Drug Treatment Division, and the Office for Aboriginal and Torres Strait Islander Health and also works with other agencies such as the Health Regulatory Policy and Governance Division, the Therapeutic Goods Administration, the National Industrial Chemicals Notification and Assessment Scheme, and the Office of the Gene Technology Regulator. Their suite of population health initiatives spans disease prevention, screening, disease control, palliative and end-of-life care, regulatory policy, immunization, and public health [31]. Table 13.1 shows key commonwealth initiatives in 2013–2014 and gives examples of related information system needs [32].

13.2.2 Australian National Public Health Information Infrastructure

Interrelated national level responsibilities for Australia's public health information infrastructure are held by the Australian Population Health Development Principal Committee (APHDPC), the Population Health Information Development Group (PHIDG), and the Public Health Information Development Unit. The APHDPC, which operates under the Australian Health Ministers' Advisory Committee (AHMAC), works at a high level across disease groups and the primary and secondary health sectors, with a focus on building a framework across both the life course and the continuum of care [33].

The PHIDG, which provides advice to APHDPC on national public health information issues, has representatives from each government and from the Australian Bureau of Statistics and the Australian Institute of Health and Welfare. It has terms of reference which span the work of traditional public health informatics. Its functions are to develop the infrastructure for population health information; increase the scope, quality, and coverage of population health data collections; encourage the analysis of population health data to answer policy-relevant questions and to generate new knowledge; and enhance the transfer of knowledge to decision makers and the public. Its specific tasks are to support

TABLE 13.1

Australian Commonwealth Government Population Health Initiatives 2013–2014 with Examples of Related Information System Needs

Programs and Objectives	Needing Support of Information Systems
1. Prevention, Early Detection, and Service Improvement	
Reduce the incidence of chronic disease and promote healthier lifestyles	Best practice in chronic disease management in the primary and community care sector
Improve palliative care in Australia	Education, training, and quality improvement
Support early detection and prevention of cancer, especially prostate, bowel, breast, and cervical	Cancer-screening program registers
Manage and support chronic diseases, especially diabetes	Comprehensive patient-centered care models and coordinated access to multidisciplinary care services
2. Communicable Disease Control	
Reduce the incidence of blood-borne viruses and sexually transmissible infections	Education and prevention activities by nongovernment organizations
3. Drug Strategy	
Reduce the harmful effects of tobacco use	Management and evaluation of the plain packaging program
Reduce harm to individuals and communities from excessive alcohol consumption and use of illicit drugs	Enhanced access to materials on the National Drugs Campaign website
4. Regulatory Policy	
Develop food standards and food regulatory policy	Management and evaluation of a new food-labeling system
Ensure that therapeutic goods are safe, effective, and of high quality	Enhancement of the Australian Register of Therapeutic Goods reference database
Implement the Therapeutic Goods Administration Reform blueprint	Shared information and complaints systems
Establish the Australia New Zealand Therapeutic Products Agency	Joint Adverse Event Notifications Systems
Aid in the protection of the Australian people and the environment by assessing the risks of industrial chemicals and providing information to promote their safe use	Registration of identified introducers of industrial chemicals and provision of regular information updates, including optimal use of international information
Protect the health and safety of people and the environment by regulating dealings with genetically modified organisms (GMOs)	Records of GMOs and maps of field trial sites and public availability online
5. Immunization	
Strengthen immunization coverage	Dissemination to immunization providers of information for parents hesitant to immunize children and public release of immunization data that help to achieve targets
Improve the efficiency of the National Immunization Program	Centralized procurement of vaccines and contract management of procurement
6. Public Health	
Increase the evidence base for the development of targeted health programs	Analysis of the Australian Health Survey, as well as the Australian Longitudinal Studies on Male Health and on Women's Health
Improve child and youth health	Provision of up-to-date evidence-based guidance for health professionals to support pregnant women

(Continued)

TABLE 13.1 (CONTINUED)

Australian Commonwealth Government Population Health Initiatives 2013–2014 with Examples of Related Information System Needs

Programs and Objectives	Needing Support of Information Systems
Improve men's and women's health	Monitoring and management of progress informed by the National Male Health and National Women's Health policies
Promote healthy lifestyle choices, including healthy eating, healthy weight, and healthy workplaces	Development, management, and evaluation of health promotion campaigns

the development, collection, analysis, and interpretation of nationally consistent population health information for a range of monitoring, assessment, evaluation, and surveillance aims; provide expertise in the use of monitoring and surveillance techniques to measure the health and well-being of the Australian population; promote and support national consistency in key population health and related data collections; identify deficiencies and gaps in scope, coverage, access analysis and application of population health data collection; and monitor trends in the health and well-being of Aboriginal and Torres Strait Islander people [34]. The Public Health Information Development Unit assists PHIDG in the development of public health data, data systems, and indicators with emphasis on the development and publication of small-area statistics for monitoring inequality in health and well-being. It produces atlases, charts, data, and reports, including interactive maps and graphics [35].

Data set development is one example of a nationwide infrastructure initiative. The National Health Information and Performance Principal Committee (NHIPPC) has overall responsibility for advising AHMAC on information strategies and for facilitating collaboration between the Australian government and states and territories to implement them. NHIPPC's subcommittee, the National Health Information Standards and Statistics Committee (NHISSC), is responsible for overseeing the development of data standards for inclusion in the National Health Data Dictionary (NHDD); mandatory national minimum data sets for national implementation; and best-practice data set specifications for inclusion in the NHDD. The Australian Institute of Health and Welfare serves as the repository for national metadata standards for the health, community services, and housing assistance [36].

Data linkage is another example of a nationwide infrastructure initiative. The Population Health Research Network (PHRN), shared among Australian states and territories, supported by the commonwealth's National Collaborative Research Infrastructure Scheme, links health-related data from different collections to allow large-scale population health research to be conducted efficiently and effectively. With the ability to provide large sample groups, the PHRN facilitates a whole-of-population approach to health and health-related research. PHRN units themselves do not conduct health-related research based on these data sets; rather, they specialize in providing infrastructure for researchers who need to link data sets and exchange data collections securely and ethically [37].

13.2.3 Australian State and Territory Public Health Informatics Strategies

As well as participating in national networked approaches, the Australian state and territory governments exercise their responsibilities through diverse approaches to using public health information infrastructure and implementing informatics strategies. Table 13.2

TABLE 13.2

Australian State and Territory Public Health Informatics-Related Strategies: 2010–2014 Examples

Jurisdiction	Examples of Strategies
Australian Capital Territory	Improve availability of reliable information by improving governance and processes for accessing population health data sets. Design web presence that includes information on research priorities, funding opportunities, and processes for engaging in research activities. Develop process for capturing information (on priority research areas) and transferring knowledge to relevant stakeholders [38].
New South Wales (NSW)	Develop and maintain population health research infrastructure: • Streamline the electronic Chief Health Officer's Report (e-CHO), i.e., revise the user interface. • Invest in the SAS® Enterprise Business Intelligence server to streamline access to the Health Outcomes Information Statistical Toolkit (HOIST), i.e., revise the user interface for HOIST. • Invest in ArcGIS and Quick Locate products to allow analysis of data in relation to geographical position and geographical features. • Publish an outline of the major data sets held by the Department of Health (and links to data set websites) on the NSW Department of Health Research website, which will include data custodians; questions, data items, and data format; whether they can be linked to other data sets; how to request the data (and limitations on access); guidelines on the use of the data; existing and new initiatives that provide instruction on how to use data sets; and ethical considerations [39].
Northern Territory	Childhood immunization web portal: Expand the accessibility of critical health information including records of immunization and medications targeting 80% of all healthcare providers in the Northern Territory. Healthcare providers can use their web service to easily access the immunization register and obtain the current immunization status for a client [40].
Queensland	Ensure quality information for practice of and decision making (establishing and implementing sustainable systems) on • Reporting of state and local outcomes and health behavior information; • Collecting and reporting prevalence and trends relating to health attitudes, skills, and behaviors and major chronic diseases; • Regular analysis and reporting of burden of disease and injury; and • Systematic release of topic-specific health status fact sheets. Review current technology communication options, determine appropriate use, and implement a sustainable model for their use. Update Internet and intranet sites and population health resources to enhance access to timely, evidence-based information for staff, partners, and the public [41].
South Australia	The Digital Technology Health Lens project demonstrated that it is possible for health and well-being implications to be considered within the core business of a nonhealth government department. In particular, we felt that it encouraged the technological side of achieving greater broadband uptake to be balanced with consideration of the related social and equity issues, and for broadband to be seen as a broader social determinant of health [42].
Tasmania	Establishing accessible, meaningful, and timely population and social health information and information-sharing systems is fundamental to understanding how the social determinants influence the health and well-being outcomes of Tasmanians [43].

(Continued)

TABLE 13.2 (CONTINUED)

Australian State and Territory Public Health Informatics-Related Strategies: 2010–2014 Examples

Jurisdiction	Examples of Strategies
Victoria	Encourage community partnership and engagement opportunities provided by interactive Internet technology and digital media: • Develop a whole-of-government prevention web portal which captures the range of initiatives across government to improve health and well-being, and which provides evidence-based guidance and ideas for the roles of other sectors and institutions in prevention and partnership development. • Develop tools to enhance collaborative planning capacity and establish shared learning networks [44].
Western Australia	Environmental Health Tracking Project—Phase II: • The final report identifies issues with data access and use, securing partnerships, and information technology requirements. • The development of an environmental health tracking system will significantly advance the monitoring, review, and analysis of environmental health operations, enabling better communication of performance to stakeholders [45].

provides representative examples of public health informatics-related strategies from this level of government during the 5-year period of 2010–2014.

13.2.4 Australian Local Government Public Health Informatics Initiatives

The three cases here illustrate public health informatics concerns with data, information, and knowledge management in areas where local governments have responsibilities for service delivery—maternal and child health, disaster rapid response, and promotion of healthy lifestyles.

13.2.4.1 Systematizing Data for Child and Family Nursing

A public health service offered to all families with children up to 6 years of age focuses on prevention and early detection in child health, through health promotion and community support rather than treatment; an initial home visit follows nearly 100% of births and further home visiting services are available for families in need [46].

Each of Victoria's 79 local government areas provides this service either by directly employing maternal and child health nurses (MCHNs) or by contracting service provision to another agency, such as a community health center or hospital. MCHNs may be colocated with another health service or with an early-childhood education center or they may work alone.

A 2010 survey of 60% of Victoria's MCHNs identified the need to develop integrated, computerized approach to supporting the work of individual MCHNs and improving overall child and family health services. It found that, although MCHNs were capable users of information and communication technology, they often lacked the appropriate technology in their workplace to maintain a comprehensive set of electronic health records of important data such as child weights, type of feeding, and development issues; manage appointments and follow up new and existing clients; generate mail-outs to invite parents to new parent groups; access and disseminate consumer health information resources; monitor child and family health outcomes; and undertake their own continuing professional development.

The survey also found that the responsibility for keeping records of families with young children in a specified local government area made for an unwieldy record transfer process across jurisdictions when families move in or out. Additionally, in all cases, MCHNs were handwriting details into the parents' own copy of their child's health record. The personally controlled electronic health record (PCEHR) introduced by the commonwealth government in 2012 targeted parents with young children as a priority and could go some way to addressing the issues faced by MCHNs.

13.2.4.2 Immediate Information for Disaster Management

In local government areas in northern Queensland after natural disasters, it is the primary role of EHOs to determine the environmental health risks and priorities from a community perspective and to provide this information to decision makers in a timely manner.

Debriefing after major cyclones had pointed to the lack of evidence-based and objective tools for rapid community-level response to environmental health issues after these disasters and, thus, insufficient information being collected and collated to prioritize public health risks and interventions based on evidence.

So in 2011, the Cairns public health unit initiated a project to develop an integrated systematic approach to assessing environmental health issues and priorities after a disaster. Cairns worked with other northern Queensland local governments, indigenous and nonindigenous community agencies, the Red Cross, and the State Health Department [47].

The project developed two tools, both designed for use in electronic formats and also hard copies to ensure adaptability to unpredictable disaster situations. Both tools were based on review of international good practice and aimed to support EHOs to pass consistent information rapidly to an emergency management coordinator to use as a guide to decision making.

An environmental health rapid assessment tool for use by EHOs accessing affecting sites immediately postcyclone divides assessment into key themes of drinking water, sewerage, waste, asbestos, food safety, personal hygiene, vectors, and chemical hazards. The tool has a section for EHOs for prioritizing environmental health issues and priorities in line with the key themes.

An evacuation center checklist for EHOs to use before and after an evacuation center is opened has sections including general overview, power supply, sanitation, food and water, waste, vectors, health surveillance, and pets. EHOs use their expertise to complete the form; information such as the number of occupants is prepopulated by the evacuation center managers.

Communication between EHOs across northern Queensland has strengthened, a response guide was developed, and a rapid assessment tool and evacuation center checklists have been adopted. To complement these achievements, the development of a database indicating baseline public health information is required.

The process of developing these tools showed that the current understanding of the public health status of populations and infrastructure from an environmental health perspective was incomplete at the regional level. This pointed to the need to develop a database of more location-specific baseline information on, for instance, types of sewerage systems, sewage overflow points, water infrastructure including access to chemical suppliers, waste-disposal sites, and emergency food and water suppliers for buildings with asbestos-containing materials.

The project recommended that to mitigate public health risks after a disaster, local governments throughout the region should formally adopt the response system as part of their public health disaster management plans. The project noted that this would ensure that necessary training, maintenance, and improvement occurred before disaster struck again.

13.2.4.3 Knowledge Translation for Obesity Prevention

Under Victorian state law, local councils must have an evidence-based municipal public health plan; section five of the act states that decisions should be based on the best available evidence to ensure effective and efficient use of resources.

Local governments have assumed key responsibilities for healthy eating and physical activity in recent years, which they exercise through policy and regulation; moreover, obesity prevention is a secondary aim of many councils' work on food security, open space for physical activity, and public transport connections.

A research initiative sought to focus on obesity prevention as an illustration of how knowledge translation in this national public health priority area could make local government efforts more evidence based [48]. Council officers with responsibilities in this area might not have specific or up-to-date knowledge of effective approaches to obesity prevention. In large councils, staff turnover, or in small councils, staff isolation from professional peers, could further limit councils' capacity to apply the best available knowledge about obesity prevention to their operations.

The initiative, called Knowledge Translation for Local Government, was developed with public health staff in 79 councils in Victoria and implemented over a 2-year period. It aimed to increase public health workers' access to research evidence through provision of evidence summaries and additional individualized support such as tailored messages. It aimed to develop their skills in accessing research, assessing trustworthiness, and applying research evidence to their local context. It aimed to assist in developing an organizational culture in each council that supported evidence-informed public health decision making at the local government level.

13.2.5 The Discipline and Profession of Public Health Informatics in Australia

Despite Australia's demonstrated needs and initiatives, recognition of public health informatics as a specialized field of knowledge and practice is not as well developed in Australia as in some other parts of the world [49–52].

A review of Australian public health advances from 1901 to 2006, published in 2012, noted simply that health informatics was among a range of future challenges [53]. The Public Health Association of Australia (PHAA) 2014 annual conference echoed this theme, i.e., "New technologies also provide challenges and opportunities in public health, so how do we use these efficiently and effectively?" [54].

The PHAA some years ago noted law reformers' concerns about privacy of health information [55]; on another occasion it directed its members to public health informatics activities at the U.S. Centers for Disease Control and Prevention [56]. In 2011 a PHAA representative participated in an Aboriginal and Torres Strait Islander Health Informatics Roundtable [57]. Recently, the PHAA recognized eHealth as a public health issue, specifically making a submission to the commonwealth government [58] and publishing an opinion piece [59] about the government's introduction of the PCEHR.

Public health informatics is clearly named in the Certified Health Informatician Australasia competencies released in 2013, both among the epidemiology and basic health research skills core competencies and as a separate public health informatics specialized competency [60]. One university in Australia with a master of public health degree program has offered a core unit of study called public health informatics [61] and two other universities with such degrees offer broad health and biomedical informatics as elective subjects [62–63]. However, there are no offerings of university degrees or substantial continuing professional development in public health informatics in Australia.

The research literature in public health informatics in Australia is diffuse. Since the 1980s the *Australian and New Zealand Journal of Public Health* has published fewer than 20 papers which make any reference to informatics and only a handful in which this is a major focus [64]. Of the 2350 publications indexed in PubMed between 1990 and 2005 that were written by authors with Australian affiliations, only 17 clearly related to public health informatics [65]. However, a Google Scholar search shows that among nearly 3000 papers published worldwide 2005–2014 that mention public health informatics, mention of Australia co-occurs in nearly 20% of papers; so it seems that much research activity occurring in this field is unidentifiable as such.

13.3 Current International Perspectives on Public Health Informatics

Since 2006 the AMIA during its main conference, the AMIA Annual Symposium, has organized a year-in-review session mirroring those of other medical societies. In this session, an invited speaker provides a summary of notable new developments in the published literature during the year of the conference on different aspects of health informatics. In 2013, for the first time, representatives from AMIA working groups on public health informatics and global health informatics were invited to organize a year-in-review session on their work topics [66].

The selection of articles was based on independent searches of MEDLINE and input from working group members followed by full article review by multiple experts. In total, 65 articles were selected in the category of public health informatics and 82 articles in the category of global health informatics. The public health informatics articles were categorized under the following themes: meaningful use surveillance, immunization and other registries, methods for improving geospatial data, natural language processing, data quality for public health, continuous use, chronic disease prevention and management indicators, screenings, policy and infrastructure decision support and bidirectional communication, Internet usage, and consumer online behaviors and workforce, education, and training. The global health informatics articles were categorized under the following themes: telemedicine, mHealth, eHealth, education, evaluation, surveillance, and communication. The following selection of four case studies is drawn from the papers selected in this process, to represent high-quality up-to-date examples from international perspectives.

13.3.1 Biosurveillance Methods in England and Wales

Algorithms and statistical modeling have been successfully applied to the detection of outbreaks caused by infectious pathogens. Enki et al.'s paper [67] describes the analysis of 20 years of data related to 9 million isolates examined across 3303 distinct types of infectious agents in England and Wales. This information can be used to improve the design of systems for automated surveillance of multiple organisms.

13.3.2 Assessment of European Community Health Indicators

Kilpeläinen et al.'s paper [68] describes the first comprehensive assessment of the availability of general health data in Europe. Use of the European Community Health Indicators (ECHI) is not equally distributed across the 31 member states. The availability score ranges from 56% to 84%. Furthermore, many important ECHI indicators are not available in most European countries. Missing data usually pertained to health determinants, the provision and use of healthcare services, the quality of healthcare, injuries, and health promotion. The European Union would benefit from the development of a European health information system for public health monitoring.

13.3.3 Data Use Workshops in Tanzania

Decision makers' confidence in national health management information systems (HMISs) is very low in many developing countries. This makes health managers think that developing parallel data-collection systems is a potential solution. However, this generates a vicious circle, resulting in poor quality across all of these resources. Braa et al.'s paper [69] reported an increased use of the system after holding several workshops with health facilities staff in Zanzibar, in the United Republic of Tanzania. The participants were invited to review and critique the quality of the data. Input from workshops was used to improve the HMIS. The results of this intervention included improvements in data quality, analytical processing, and computer literacy. Braa et al.'s article emphasizes the value of user-centric design of public health informatics systems in low and medium income countries.

13.3.4 The Impact of Technology on Sub-Saharan Hospitals

Verbeke et al.'s article [70] explores how information and communication technology (ICT) tools can be applied to improve the efficiency and effectiveness of health services in five sub-Saharan countries. Thirty hospitals participated in the study, through business process analysis and semistructured interviews. Tools were implemented in 19 health facilities. Their impact was measured in terms of admission, patient identification, mortality, and morbidity. This research illustrates the secondary use of clinical data, enabling non-redundant reporting of health and care performance indicators. Patient identification, financial management, and structured reporting improved dramatically after implementation of ICT tools. Overall morbidity decreased, as well as the number of inpatient visits. The hospital workforce found the implementation of the system very positive, in terms of both quality of care and working conditions.

13.4 Directions for Public Health Informatics

Disruptive and pervasive changes brought about by new information and communication technologies and by environmental degradation and climate change are creating new opportunities and challenges for public health informatics, in Australia and around the world. This section outlines three directions that public health informatics is taking in the first part of the 21st century that would have been unthinkable when this knowledge domain was first recognized in the 1990s—influenced by social media, by ICT architectures of participation, and by exposome science. It concludes by summarizing a mature integrated public health informatics agenda formulated by the AMIA Public Health Informatics conference 10 years on from its inception.

13.4.1 Public Health 2.0

Some traditional modes of work in public health are being transformed by new practices that are variously described as consumer, personal, or participatory health informatics. There are two main drivers of these changes. Firstly, the digital revolution in civic participation in other sectors such as banking, insurance, leisure, and government has now reached healthcare. Secondly, affordable technologies such as wireless sensors, personal electronic health records, personal mobile devices, computer games, and social media now make it possible for members of the public to collect, manipulate, and share health data and health information with considerable autonomy [71]. In these circumstances public health informatics is sometimes rebadged as ePublic health or as public health 2.0 [72].

Reports of activities in this area are wide ranging, in health promotion [73], chronic disease prevention [74], improving access to medicines [75], and disaster management [76]. The Mo-Buzz system for preventing dengue offers an example of the integration of public health functions that can be achieved: Predictive surveillance uses a computer simulation to forewarn health authorities and the general public about impending disease outbreaks. Citizen participation occurs through the use of social media tools and smartphone technologies to report symptoms, mosquito bites, and breeding sites to health authorities. Health communication utilizes data gathered by the system from and about citizens to disseminate customized health awareness and disease prevention messages [77].

13.4.2 Bidirectional Communication

Traditionally, the monitoring function enabled by public health informatics has implied a unidirectional (bottom–up) data flow from general practitioners or public health laboratories toward central public health agencies. The objective of these communications is to report potential cases of risk to public health, such as adverse reaction to drugs (also called pharmacovigilance), infectious disease cases, or problems with toxic agents (for instance, food poisoning). However, in recent years, we are witnessing the development of ICT architectures that support bidirectional communication, allowing for communication of information regarding emerging health threats to be pushed from public health agencies to clinicians or healthcare provider organizations [78].

A successful example of bidirectional communication in public health informatics is when, in January 2011, the United States Centers for Disease Control and Prevention issued a drug recall for the antibiotic metronidazole. The central information system connecting 400 medical practices' EHRs was able to identify 62 patients who had been prescribed the drug during the previous year. A secure message was then sent to the providers' EHRs with instructions on how to use the system to identify patients for notification [79].

13.4.3 Exposome Informatics

In public health, there is a renewed fascination with the effect of the natural and built environment on human health, described as the exposome (in other words, every exposure to which an individual is subject from conception to death) [80]. The emergence of new technologies for measuring exposure, such as personal wearable monitors, high-throughput genomic tools, and source-to-dose modeling systems, has raised expectations about applying "exposure science" to refine public health interventions and improve outcomes [81]. Along with these expectations comes the need for new informatics methods and tools for supporting research into the causative elements associated with pathologies in order to improve risk profiling, thus contributing to advancing preventive medicine [82].

In the United States, a joint project between the Rollins School of Public Health at Emory University and the Georgia Institute of Technology was funded in 2013 by the National Institute of Environmental Health Sciences (NIEHS) of the National Institutes of Health to establish the Health and Exposome Research Center: Understanding Lifetime Exposures (HERCULES). This is the first exposome-based center created in the United States [83]. The Centers for Disease Control and Prevention outlined in 2013 the concept of exposome and exposomics and identified the three main priority areas for researching the occupational exposome [84].

The European Commission has also funded three large collaborative research networks in 2012 and 2013 to investigate different aspects of the exposome, and these rely heavily on the use of informatics and mobile technology. The Human Early-Life Exposome (HELIX) project coordinated by the Barcelona-based Centre for Research in Environmental Epidemiology aims to develop an early-life exposome [85]. The EXPOsOMICS project coordinated by Imperial College London is using smartphones and sensors to collect large-scale exposure data [86]. The Health and Environment-wide Associations Based on Large Population Surveys (HEALS) project has a very strong informatics component applied to the study of interactions between molecular and environmental factors with the purpose of predicting risk and associated health outcomes [87].

A case study of exposome informatics can be found in the Medical and Environmental Data Mash-up Infrastructure (MEDMI) partnership among researchers in climate, weather, environment, and human health and well-being for addressing human health concerns arising from environmental change. A central data and analysis resource, combining existing databases currently held in various locations and organizations and hosted on an Internet-based platform, will make possible the linking and analysis of complex meteorological, environmental, and epidemiological data. The platform will also support translation of research findings into epidemiologic, clinical, and commercial collaborative applications such as facilitating novel research into environmental exposures and health using integrated models; rapidly identifying hot spots (points in time and space with convergent increased environmental and human health risks) for targeted prevention, interventions, and research; providing healthcare practitioners, public health planners, and environmental managers with relevant information for improving services for locations and populations identified at risk; initiating and evaluating interventions to reduce the exposures and, thus, adverse health effects on individuals and whole populations; and making data widely and openly accessible to researchers, policy makers, and citizens [88].

13.4.4 Advancing the Agenda for Public Health Informatics

On the 10th anniversary of the 2001 meeting among the public health and health informatics communities that first defined a national agenda for public health informatics in the United States, the agenda was revisited and updated comprehensively. This process confirmed a five-track agenda for moving the field forward: technical frameworks; research and evaluation; ethics; education, training, and workforce development; and sustainability. It also identified three crosscutting needs: to enhance the communication and information sharing within the public health informatics community; to improve the consistency of public health informatics practice; and to nurture the leadership to drive public health informatics forward [89].

13.5 Conclusions

This chapter sets out key concepts in the discipline of public health informatics. It illustrates these concepts variously—through a multilevel account of public health informatics activities in Australia; with a selection of current international case studies; and by indicating emerging directions in the field. The chapter begins by introducing the essential role that information and communication have played in public health past and present. It ends by asserting that collaborations between public health and health informatics can continue to produce innovation and demonstrate value, so long as there is interest and investment in the health of humans at the population level.

References

1. World Health Organization. 2014. Trade, Foreign Policy, Diplomacy and Health: Public Health, http://www.who.int/trade/glossary/story076/en/. Accessed on March 1, 2014.
2. Science Museum. 2014. Brought to Life: Exploring the History of Medicine, http://www.sciencemuseum.org.uk/broughttolife/themes/publichealth.aspx. Accessed on March 1, 2014.
3. Mooney, G., History of Public Health, Baltimore, United States: Johns Hopkins University Bloomberg School of Public Health, 2005, http://ocw.jhsph.edu/courses/historypublichealth/PDFs/ReadingList.pdf. Accessed on March 1, 2014.
4. Kindig, D., and Stoddart, G., What is population health?, *American Journal of Public Health*, 93(3), pp. 380–383, 2003.
5. Friede, A., Blum, H. L., and McDonald, M., Public health informatics: How information-age technology can strengthen public health, *Annual Review of Public Health*, 16(1), pp. 239–252, 1995.
6. DeShazo, J., LaVallie, D., and Wolf, F., Publication trends in the medical informatics literature: 20 years of "Medical Informatics" in MeSH, *BMC Medical Informatics and Decision Making*, 9(1), p. 7, 2009.
7. Bernstam, E., Smith, J., and Johnson, T., What is biomedical informatics?, *Journal of Biomedical Informatics*, 43(1), pp. 104–110, 2010.
8. Yasnoff, W. A. et al., Public health informatics: Improving and transforming public health in the information age, *Journal of Public Health Management and Practice*, 6(6), pp. 67–75, 2000.
9. Karras, B. et al., *Competencies for Public Health Informaticians*, Atlanta, United States: University of Washington Center for Public Health Informatics and Centers for Disease Control and Prevention, 2009.
10. Collen, M., Origins of medical informatics, *Western Journal of Medicine, Medical Informatics* [Special Issue], 145, pp. 778–785, 1986.
11. International Medical Informatics Association. 2014. The IMIA Vision, http://www.imia-medinfo.org/new2/node/31. Accessed on March 1, 2014.

12. White, M., Public health informatics: An invitation to the field, *Bulletin of the American Society for Information Science & Technology*, 39(5), June/July 2013. http://www.asis.org/Bulletin/Jun -13/JunJul13_White.html. Accessed on March 1, 2014.

13. American Medical Informatics Association. 2014. Working Group: Public Health Informatics. http://www.amia.org/programs/working-groups/public-health-informatics/. Accessed on March 1, 2014.

14. American Public Health Association. 2014. Health Informatics Information Technology Group http://www.apha.org/membergroups/sections/aphasections/hiit/. Accessed on March 1, 2014.

15. Public Health Informatics Conference. 2014. ePublic Health: The Future Is Now. April 29–May 1, 2014, Atlanta, Georgia. http://phiconference.org/. Accessed on March 1, 2014.

16. Rockefeller Foundation, "Public health informatics," in *From Silos to Systems: An Overview of Ehealth's Transformative Power*, New York, United States: Rockefeller Foundation, pp. 14–19, 2010. http://www.rockefellerfoundation.org/uploads/files/7d1832b5-96a1-4a70-8089-288f0b3235e7 -silos-to.pdf.

17. Kieny, M.-P., Message from the Health Metrics Network Acting Executive Secretary. Geneva, Switzerland: World Health Organization Health Metrics Network, 2011. http://www.who.int /healthmetrics/news/message_acting_exec_sec_1mar2011/en/. Accessed on March 1, 2014.

18. Tan, J., Soto Mas, F. G., and Hsu, C. E., "E-public health information systems: E-technologies for public health: Preparedness and surveillance," in *E-Health Care Information Systems: An Introduction for Students and Professionals* (First ed.), Tan, J. K. H., Ed., San Francisco, California: Wiley, Josey-Bass Publishers, pp. 127–154, 2005.

19. Dixon, B. E., and Grannis, S. J., Why "What data are necessary for this project?" and other basic questions are important to address in public health informatics practice and research, *Online Journal of Public Health Informatics*, 3(3), ojphi.v3i3.3792, 2011.

20. Sedig, K., and Ola, O., The challenge of big data in public health: An opportunity for visual analytics, *Online Journal of Public Health Informatics*, 5(3), 2014.

21. Dixon, B. E. et al., Electronic health information quality challenges and interventions to improve public health surveillance data and practice, *Public Health Reports*, 128(6), pp. 546–563, 2013.

22. Reeder, B. et al., Reusable design: A proposed approach to public health informatics system design, *BMC Public Health*, 11(1), p. 116, 2011.

23. Buckeridge, D. L. et al., An infrastructure for real-time population health assessment and monitoring, *IBM Journal of Research and Development*, 56(5), pp. 2:1–2:11, 2012.

24. Jacquelinet, C. et al., "Public health decision support," in *Medical Informatics, e-Health*, Venot, A. et al., Eds., Paris, France: Springer, pp. 221–248, 2014.

25. Kilbourne, E. M., "Informatics in toxicology and environmental public health," in *Public Health Informatics and Information Systems*, Magnuson, J. A., and Fu, P. C., Eds., London, England: Springer, pp. 277–293, 2014.

26. Keeling, J., *Development of Systematic Knowledge Management for Public Health: A Public Health Law Ontology* (Doctoral dissertation, Columbia University), 2012.

27. Staes, C. J. et al., Evaluation of knowledge resources for public health reporting logic: Implications for knowledge authoring and management, *Online Journal of Public Health Informatics*, 3(3), 2011.

28. Dobbins, M. et al., A knowledge management tool for public health: health-evidence.ca, *BMC Public Health*, 10(1), p. 496, 2010.

29. Reeder, B., Turner, A., and Demiris, G., Use of technology to support information needs for continuity of operations planning in public health: A systematic review, *Online Journal of Public Health Informatics*, 2(1), 2010.

30. Australian Government Department of Health. 2012. Surveillance Systems Reported in CDI [Communicable Diseases Intelligence]. https://www.health.gov.au/internet/main/publishing .nsf/Content/cda-surveil-surv_sys.htm#nndss. Accessed on March 1, 2014.

31. Australian Government Department of Health. 2013. PHD [Population Health Division]. http://www.health.gov.au/internet/main/publishing.nsf/Content/health-pubhlth-index.htm. Accessed on March 1, 2014.

32. Australian Government Department of Health. 2013–2014 Health PBS attachments, Section 2— Department Outcomes and Planned Performance—2.1 Outcome and Performance Information— Outcome 1 Population Health. http://www.health.gov.au/internet/budget/publishing.nsf/Content /2013-2014_Health_PBS_sup1/$File/2013-14_DoHA_PBS_2.01_Outcome_1.pdf. Accessed on March 1, 2014.

33. Australian Government Directory. 2013. AHMAC Australian Population Health Development Principal Committee (APHDPC). http://www.directory.gov.au/directory?ea0_lf99_120.&orga nizationalUnit&28e1ceb7-3527-42e8-ba19-8c29bb791246. Accessed on March 1, 2014.

34. Australian Institute of Health and Welfare. 2013. Population Health Information Development Group (PHIDG). http://www.aihw.gov.au/phidg/. Accessed on March 1, 2014.

35. Public Health Information Development Unit. 2014. Home page, http://www.publichealth .gov.au. Accessed on March 1, 2014.

36. Australian Institute of Health and Welfare, *Creating nationally-consistent health information: Engaging with the national health information committees*, Cat. No. CSI 18, Canberra, Australia: Australian Institute of Health and Welfare, 2014.

37. Population Health Research Network. 2011. Home page, http://www.phrn.org.au. Accessed on March 1, 2014.

38. Australian Capital Territory. Population Health Research Strategy 2012–2015. http://www .health.act.gov.au%2Fc%2Fhealth%3Fa%3Dsendfile%26ft%3Dp%26fid%3D1393819614%26 sid%3D&ei=m88nU5aPE4PFkAWQ6oGgBQ&usg=AFQjCNEgi8qdwh6gVZwZa_O42s 6hjmIsRg&bvm=bv.62922401,d.dGI. Accessed on March 1, 2014.

39. New South Wales. Promoting the Generation and Effective Use of Population Health Research in NSW: A Strategy for NSW Health 2011–2015. http://www0.health.nsw.gov.au/pubs/2011 /pdf/pop_health_research_strat.pdf. Accessed on March 1, 2014.

40. Northern Territory. eHealthNT. http://www.ehealthnt.nt.gov.au/About_Us/index.aspx. Accessed on March 1, 2014.

41. Queensland. Strategic Directions for Quality Management 2009–2012. http://www.health.qld .gov.au/ph/documents/pdu/phstratdir_quality.pdf. Accessed on March 1, 2014.

42. Golder, W. et al., "Digital technology access and use as 21st century determinants of health: Impact of social and economic disadvantage," in Implementing Health in All Policies: Adelaide 2010, produced for the Adelaide 2010 Health in All Policies International Meeting cohosted by the Government of South Australia and the World Health Organization, Kickbusch, I., and Buckett, K., Eds., pp. 133–144, 2010.

43. Tasmania. Department of Health and Human Services Population Health. State of Public Health 2013. http://www.dhhs.tas.gov.au/__data/assets/pdf_file/0017/132263/State_of_Public _Health_2013_LR.pdf. Accessed on March 1, 2014.

44. Victoria. Victorian Public Health and Wellbeing Plan 2011–2015. http://docs.health.vic.gov.au /docs/doc/Victorian-Public-Health-and-Wellbeing-Plan-2011-2015. Accessed on March 1, 2014.

45. Western Australia. Environmental Health Directorate, Public Health Division, Department of Health WA. Year Book 1 July 2010–30 June 2011. http://www.public.health.wa.gov.au/cproot /4224/2/Year_Book_Nov_2011.pdf. Accessed on March 1, 2014.

46. Ridgway, L., Mitchell, C., and Sheean, F., Information and communication technology (ICT) use in child and family nursing: What do we know and where to now?, *Contemporary Nurse*, 40(1), pp. 118–129, 2011.

47. Ryan, B. et al., Environmental health disaster management: A new approach, *Australian Journal of Emergency Management*, 28(1), pp. 35–41, 2013.

48. Armstrong, R. et al., Knowledge translation strategies to improve the use of evidence in public health decision making in local government: Intervention design and implementation plan, *Implementation Science*, 8(1), p. 121, 2013.

49. Foldy, S., "National public health informatics, United States," in *Public Health Informatics and Information Systems*, Magnuson, J. A., and Fu, P. C., Eds., London, England: Springer, pp. 573–601, 2014.

50. Frisch, L. E. et al., "Public health informatics in Canada," in *Public Health Informatics and Information Systems*, Magnuson, J. A., and Fu, P. C., Eds., London, England: Springer, pp. 603–618, 2014.

51. Macfadyen, D., "Importance of informatics for public health and the work of the World Health Organization in Europe," in *Health in the New Communications Age*, Laires, M. F., Ladeira, M. J., and Christensen, J. P., Eds., Amsterdam, Netherlands: IOS Press, pp. 27–30, 1995.

52. Richards, J., Douglas, G., and Fraser, H. S., "Perspectives on global public health informatics," in *Public Health Informatics and Information Systems*, Magnuson, J. A., and Fu, P. C., Eds., London, England: Springer, pp. 619–644, 2014.

53. Gruszin, S., Hetzel, D., and Glover, J., *Advocacy and Action in Public Health: Lessons from Australia over the 20th Century*, Canberra: Australian National Preventive Health Agency, 2012.

54. Public Health Association of Australia. 2014. Annual Conference, http://www.phaa.net.au/documents/43rd_Annual_Conf_Call_For_Abstracts.pdf. Accessed on March 1, 2014.

55. Australian Law Reform Council, Health information and Internet privacy top youth privacy concerns, *In Touch: Newsletter of the Public Health Association Australia*, 24(4), p. 13, May 2007.

56. U.S. Department of Health and Human Services Centers for Disease Control and Prevention, 60 years of public health science at CDC, *In Touch: Newsletter of the Public Health Association Australia*, 24(1), p. 4, February 2007.

57. Public Health Association of Australia. Annual Report 2011–2012. http://www.phaa.net.au/documents/120827_Annual%20Report%202011-12-FINAL.pdf. Accessed on March 1, 2014.

58. Public Health Association of Australia. Submission on PCEHR—Draft Concept of Operations. June 2011, http://www.phaa.net.au/documents/110621_PHAA%20submission%20on%20PcEHR%20FINAL.pdf. Accessed on March 1, 2014.

59. McGowan, R., Australia's attempt at ehealth: Brave new world or safety first?, *Public Health Association Australia Primary Health Care Special Interest Group PHCSIG Update*, 2(1), p. 4, June 2012.

60. *Certified Health Informatician Australasia Health Informatics Competencies Framework*. Melbourne, Australia: Health Informatics Society of Australia, December 2013.

61. Edith Cowan University. 2014. University Handbook 2012, Course Information I62 Master of Public Health, http://handbook.ecu.edu.au/CourseStructure.asp?disyear=2012&CID=1431&USID=0&UCID=0&UID=0&Ver=1&HB=HB&SC=PG. Accessed on March 1, 2014.

62. Torrens University of Australia. 2014. Master of Public Health 2014, http://tua.edu.au/study/programs/public-health/master-of-public-health/#.Uyfad84WdsI. Accessed on March 1, 2014.

63. University of Melbourne. 2014. Handbook 244CW Master of Public Health 2014, https://handbook.unimelb.edu.au/view/2014/244CW. Accessed on March 1, 2014.

64. *Australian and New Zealand Journal of Public Health*. 2014. Home page, http://onlinelibrary.wiley.com/journal/10.1111/%28ISSN%291753-6405. Accessed on March 1, 2014.

65. Mendis, K., Health informatics research in Australia: Retrospective analysis using PubMed, *Informatics in Primary Care*, 15(1), pp. 17–23, 2007.

66. American Medical Informatics Association. Annual Symposium, Washington, D.C., November 16–20, 2013. http://www.amia.org/amia2013/panels. Accessed on March 1, 2014.

67. Enki, D. G. et al., Automated biosurveillance data from England and Wales, 1991–2011, *Emerging Infectious Diseases*, 19(1), pp. 35–42, 2013.

68. Kilpeläinen, K. et al., Health indicators in Europe: Availability and data needs, *European Journal of Public Health*, 22(5), pp. 716–721, 2012.

69. Braa, J., Heywood, A., and Sahay, S., Improving quality and use of data through data-use workshops: Zanzibar, United Republic of Tanzania, *Bulletin of the World Health Organization*, 90(5), pp. 379–384, May 1, 2012.

70. Verbeke, F., Karara, G., and Nyssen, M., Evaluating the impact of ICT-tools on health care delivery in sub-Saharan hospitals, *Studies in Health Technology and Informatics*, 192, pp. 520–523, 2013.

71. Martin Sanchez, F., Lopez Campos, G., and Gray, K., "Biomedical informatics methods for personalized medicine and participatory health," in *Methods in Biomedical Informatics: A Pragmatic Approach*, Sarkar, N., Ed., London, England: Academic Press, pp. 347–385, 2013.

72. Ossebaard, H. C., Gemert-Pijnen, L. V., and Seydel, E. R., "ePublic health: Fresh approaches to infection prevention and control," in *ThinkMind // eTELEMED 2011, The Third International Conference on eHealth, Telemedicine, and Social Medicine*, February 23–28, 2011, Gosier, Guadeloupe, France, Gemert-Pijnen, L. V., Ossebaard, H. C., and Hamalainen, P., Eds., pp. 27–36, 2011, http://www.thinkmind.org/index.php?view=instance&instance=eTELEMED+2011. Accessed on March 1, 2014.

73. Chou, W. Y. S. et al., Web 2.0 for health promotion: Reviewing the current evidence, *American Journal of Public Health*, 103(1), pp. e9–e18, 2013.

74. Ding, H., Varnfield, M., and Karunanithi, M., "Mobile applications towards prevention and management of chronic diseases," in *Web Technologies and Applications: 14th Asia-Pacific Web Conference, APWeb 2012*, April 11–13, 2012, Kunming, China, Cheng, Q. Z. et al., Eds., Berlin, Germany: Springer, pp. 788–791, 2012.

75. de Magalhães, J. L., Quoniam, L., and Boechat, N., Web 2.0 tools for network management and patent analysis for health public, *Revista de Gestão em Sistemas de Saúde*, 2(1), pp. 26–41, 2013.

76. Huang, C. M., Chan, E., and Hyder, A. A., Web 2.0 and Internet social networking: A new tool for disaster management?—Lessons from Taiwan, *BMC Medical Informatics and Decision Making*, 10(1), p. 57, 2010.

77. Lwin, M. O. et al., A 21st century approach to tackling dengue: Crowdsourced surveillance, predictive mapping and tailored communication, *Acta Tropica*, 130, pp. 100–107, 2014.

78. Dixon, B. E., Gamache, R. E., and Grannis, S. J. Towards public health decision support: A systematic review of bidirectional communication approaches, *Journal of the American Medical Informatics Association*, 20(3), pp. 577–583, 2013.

79. Buck, M. D. et al., The Hub Population Health System: Distributed ad hoc queries and alerts, *Journal of the American Medical Informatics Association*, 19(e1), pp. e46–e50, 2012.

80. Wild, C. P., The exposome: From concept to utility, *International Journal of Epidemiology*, 41(1), pp. 24–32, 2012.

81. Lioy, P. J., and Smith, K. R., A discussion of exposure science in the 21st century: A vision and a strategy, *Environmental Health Perspectives*, 121(4), pp. 405–409, 2013.

82. Martin Sanchez, F. et al., Exposome informatics: Considerations for the design of future biomedical research information systems, *Journal of the American Medical Informatics Association*, doi:10.1136/amiajnl-2013-001772, 2013.

83. HERCULES Center. http://humanexposomeproject.com/. Accessed on March 1, 2014.

84. U.S. Department of Health and Human Services Centers for Disease Control and Prevention, Exposome and Exposomics. 2014. http://www.cdc.gov/niosh/topics/exposome/. Accessed on March 1, 2014.

85. HELIX project. http://www.projecthelix.eu/. Accessed on March 1, 2014.

86. EXPOsOMICS. http://www.exposomicsproject.eu/. Accessed on March 1, 2014.

87. HEALS research project. http://www.heals-eu.eu/. Accessed on March 1, 2014.

88. Fleming, L. E. et al., Data mashups: Potential contribution to decision support on climate change and health, *International Journal of Environmental Research and Public Health*, 11(2), pp. 1725–1746, 2014.

89. Massoudi, B. L. et al., An informatics agenda for public health: Summarized recommendations from the 2011 AMIA PHI Conference, *Journal of the American Medical Informatics Association*, 19(5), pp. 688–695, 2012.

14

Ubiquitous Personal Health Records for Remote Regions

H. Lee Seldon, Jacey-Lynn Minoi, Mahmud Ahsan, and Ali Abdulwahab A. Al-Habsi

CONTENTS

14.1 Introduction

Healthcare in remote regions of the world, and in remote parts of many developed countries, is poorer than in urban areas. This is due to *shortages* of trained personnel, medical facilities, funds, data communications, and many other things, including health records. All these factors are not "ubiquitous," at least not in remote regions.

There have been and are numerous attempts to resolve this problem in developed countries. Almost all projects to improve health and healthcare in remote regions involve effective use of relevant health records.

This chapter is a discussion of personal health records (PHRs). Since there have been many publications and reviews of PHRs, this chapter concentrates specifically on the difficult task of making PHR systems that are capable of functioning everywhere, even in remote regions. So the aim is to examine current PHR systems and discover how to make ones which are ubiquitous for remote regions. Some of this is based on the experience of the authors in Southeast Asia, specifically in Malaysia.

14.1.1 Scope

We start by clarifying our definitions:

- Personal health record: A PHR is also called a personally controlled health record (PCHR), or a personal electronic health record (PEHR), among others. This is a record of health owned and maintained by an individual—the emphasis on ownership and maintenance is important. Strictly speaking, PHRs do not include records owned or maintained by healthcare workers—those are called electronic medical records (EMRs) or electronic health records (EHRs). The distinction is sometimes blurred, as in cases of EMRs or EHRs to which the patients have access—those may be called integrated PHRs or tethered PHRs [1]. Also, PHRs are certainly not restricted to "patients," individuals who at the time are ill; instead, PHRs are records of health maintained by mostly healthy, but sometimes ill, people. PHRs should contain the following:
  - General health data: Vital signs, ailments, injuries (even ones which do not require medical treatment), laboratory and imaging results, vaccinations, etc.
  - Monitored values of particular parameters, such as blood glucose, diet, exercise of diabetics (or healthy people).
  - Contextual information: The where and when or context of a health problem [2–3]. It includes alerts about epidemics or environmental factors which may contribute to a person's ailment.
- Ubiquitous: *Ubiquitous* means available everywhere (and at any time). For the PHR owner this means that the PHR is wherever the person is. For the rest of the world this means that the PHR is accessible, e.g., it may be on the web. Ubiquitous data collection may imply that PHR entries can be recorded at any time and wherever the PHR owner is, whereas ubiquitous data availability may imply that data can be retrieved whenever and wherever necessary.
- Remote region: Normally this means an area far away in terms of distance, but for PHRs remote also means not being well connected to the rest of the region or community by communications, i.e., no 24/7 broadband networks, no Internet, and

telephone system. Of course, geographically remote regions and communications-remote regions are often the same. Our examples come from Southeast Asia, the area which authors are familiar with.

The diagnostic value of PHR data bears particular importance. Diagnostic measurements which require great accuracy may be made by healthcare workers with approved and calibrated medical devices, but such devices are not generally available to or affordable by the general public, especially inhabitants of remote regions. However, when such measurements are made, the results may be recorded in a PHR. On the other hand, the history of a health problem is also extremely important for its diagnosis, as every medical student learns, and a good PHR can deliver a meaningful history.

The who, where, and how are ways to describe whether a system is a PHR and how ubiquitous it is:

- Who collects health data and where and when?
- Where, how, and by whom the data are stored?
- Where, how, and by whom the data are retrieved and viewed?

Given the speed of change of Internet technology nowadays, this chapter makes no attempt to list or describe all the ongoing PHR projects. Instead, it describes the considerations involved in developing a ubiquitous PHR for remote regions, and gives examples of systems which involve some or all of the necessary features.

14.2 Personal Health Record Data Collection and Storage: Ubiquitous or Not?

Because we are interested in PHRs for remote regions, we must investigate the steps involved in creating and using a PHR—recording data, storing data, and retrieving data—in the light of the applicability (constraints) in remote areas. Where and how data are collected and where and how they are stored determine if a record is strictly speaking a PHR, or if a PHR is ubiquitous for remote regions.

Health data are collected either by the individual or by healthcare workers who interview and evaluate the individual. The data are stored on paper and/or on a local electronic system and/or on a distant electronic system. Table 14.1 gives a few examples [4–5], including some which are "almost PHRs" or almost ubiquitous.

14.2.1 Constraints in Remote Regions: Healthcare Workers and the Web Are Not Ubiquitous

Even developed countries such as the United States and Australia acknowledge a shortage of qualified healthcare workers in remote regions. The situation in developing countries is not better. Therefore, PHRs which *require* the participation of healthcare workers struggle to be ubiquitous in such remote regions.

Internet coverage, as shown, e.g., by Wikipedia and the International Telecommunications Union (ITU), has gaps which correspond unsurprisingly with regions which are

TABLE 14.1

Examples of PHR Types

Example	Data Type; Collector	When?	Stored?	PHR?	Ubiquitous?
Paper booklets (Vietnam, Sarawak, Africa)	All types; healthcare workers	Episodic, when person visits clinic	Paper kept by person	"Owned," but not maintained by person, not continuous	Almost, but does not allow remote participation
International vaccination certificate	Vaccinations; healthcare workers	Episodic, when person visits clinic	Paper kept by person	Complete with regard to vaccinations	Almost, but does not allow remote participation
OpenMRS, Sana	What can be collected in remote areas; healthcare workers	Episodic, when healthcare worker visits person	Online database	Not maintained by person, not continuous (does not claim to be a PHR)	Ubiquitous on Internet, but not off-line
Ericsson Mobile Health (Sarawak, Malaysia)	BP, individual; telecenter managers, healthcare workers	Episodic, when healthcare worker and/or telecenter manager visits person	Online database	Not maintained by person	Ubiquitous on Internet, but not off-line
Zilant™ by Embedded Wireless (Sarawak, Malaysia)	BP, weight, oxygen level; individual, telecenter managers, healthcare workers	Episodic, when person visits the telecenter	Online database	Not maintained by person	Ubiquitous on Internet, but not off-line
Indivo, Microsoft HealthVault, WebMD, etc.	All text types; individual, health workers	Continuous	Online database	Yes	Ubiquitous on Internet, but not off-line
Mobile health apps	Many types; individual	Continuous	Mostly only on phone	Mostly not complete PHRs	Ubiquitous for the owner

poor and isolated geographically and in terms of communications [6–7]. In Malaysia, in 2011 Internet users were 82.2% urban versus 17.8% rural [8]. The broadband penetration rate in Sarawak, the largest state by area, but located on Northwest Borneo, was 53.5% in the fourth quarter of 2013, versus 111.7% (due to multiple subscriptions) in the capital region Kuala Lumpur. Therefore, PHRs which *require* Internet connections struggle to be ubiquitous in such remote regions.

One thing which nowadays is almost ubiquitous is mobile phones or cell phones, as they are called in some countries including Malaysia. The mobile network covers a much wider range than either the Internet or healthcare workers, although it is also not complete. Even a developed country like Australia does not have full coverage by mobile networks—the Australian government announced (2014) the Mobile Coverage Programme to try to extend coverage to more remote corners. In Malaysia, in 2012 Sarawak had 1.057 cell phones per inhabitant, versus 2.035 in the capital Kuala Lumpur [8]. Subscribers sent 17 billion SMS messages in the fourth quarter of 2013, or almost 400 per subscriber.

With these constraints in mind, we will describe various approaches to PHRs and their applicability in regions with shortages of healthcare workers and/or Internet coverage.

14.2.2 Personal Health Records on Paper: Not Quite Ubiquitous

The disadvantages of paper medical records filed in private clinics or hospital store rooms have been discussed many times, but those systems do not pretend to be PHRs. The paper records kept by patients in Vietnam, parts of Africa, or Sarawak (East Malaysia) are still not quite ubiquitous. Data entry is by healthcare workers during meetings with the patients, so it is episodic and not ubiquitous in time. The records are accessible wherever the patient is (or has them) but not by anyone remote from the patient, so they are not ubiquitous in location.

14.2.3 Web-Based Personal Health Records

Web-based personal health records (WWW-PHRs) can be roughly divided into two groups: (1) purely web-based for input and output and (2) web-based storage, but allowing a variety of inputs (and outputs). The first group is constrained by Internet coverage as mentioned above. The second group may overcome that constraint but may be constrained by other shortages.

Another classification for web-based PHRs is stand-alone versus tethered, whereby the latter means that the PHR is linked to an EMR or EHR system belonging to a specific healthcare organization [1].

Indivo (http://indivohealth.org/) was possibly the original web-based PCHR [9–10]. Figure 14.1 shows some of the history of the Indivo project, which now is based at the Boston Children's Hospital, Massachusetts, United States.

Indivo is an Internet-based PHR system. It includes, as do almost all PHRs, an authentication step which allows authorized users access to the system, but this step is set up to also allow applications to be authorized, and that, in turn, allows other systems to participate. Some of the systems to be mentioned derive directly or conceptually from the Indivo group.

Most of these systems allow data to be collected by the individuals (anytime) and by healthcare workers (during encounters with patients), which makes them ubiquitous in that respect. Some also allow data collection from registered devices. Being web based, for data retrieval they are ubiquitous as far as the web is concerned.

14.2.3.1 Purely Web-Based Personal Health Records Are Not Quite Ubiquitous

As mentioned above, this group extends as far as the web, so we will not consider it in detail. Examples include the following:

- Netherland's Nationaal ICT Instituut in de Zorg (Nictiz) (http://www.nictiz .nl/): Within the Netherlands, this is ubiquitous due to the complete penetration of the Internet throughout the country. However, the concept could not easily be extended to remote regions.
- MyOSCAR (http://www.myoscar.org/) and Dossia (http://www.dossia.org/): Both are derivatives of Indivo.

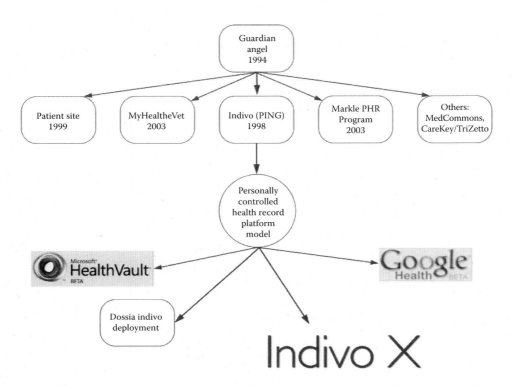

FIGURE 14.1
Indivo evolution. (From Indivo, Children's Hospital Informatics Program: Research, History of the Personally Controlled Health Record, retrieved on Aug. 14, 2014, from http://indivohealth.org/research, 2007.)

- WebMD (stand-alone PHR; http://www.webmd.com/): This is popular health-related site and a good web-based PHR system. The PHR was previously named Healthscape and was founded by Jim Clark and Pavan Nigam in 1996 in the United States. In 1999 WebMD acquired it and merged it with their service. People know WebMD for its public Internet site, where various kinds of health information like symptom checklist, drug information, and pharmacy information, are available. Also, their blog is very popular, as expert physicians regularly publish articles for specific health topics [11]. WebMD provides a basic web-based PHR system. People can register on their site and provide some basic information. There are visual tools so people can check their health condition and get some information, such as an overview, what to expect, risk factors, prevention, treatment, self-care, etc. WebMD also has a long list of health-related topics, drugs list, first aid, etc.

- Patient Ally (https://www.patientally.com/Main): This is an Internet-based free service for PHR. Users can store health data and communicate with their doctors, but the service is only for the United States. To communicate with patients the healthcare provider must be added by the patient as a provider; from there the patient and the provider can interact. The user can request appointments and medications refills. Patient Ally can be accessed only via a device that has Internet access.

- PatientSite (https://www.patientsite.org/) [12]: This appears to be another derivative of the Indivo family, based at Beth Israel Deaconess Hospital in Boston, United States, so it is a tethered PHR.

14.2.3.2 Web-Based Personal Health Records with Various Inputs and Outputs

The emphasis in this group is on more ubiquitous data-collection points rather than just on the web.

- OpenMRS (http://www.openmrs.org/) and Sana (http://www.sanamobile.org) are perhaps the prime and most successful examples of health data collection at remote sites using handheld devices. However, because the records are centrally managed and the patients themselves do not control the records, this system is not strictly a PHR system. This is described in more detail in Subsection 14.2.5.

- The Australian government started a nationwide PEHR in July 2012 (http://www .ehealth.gov.au/Internet/ehealth/publishing.nsf/content/home), after many less than successful attempts. According to the description, it includes phone-based access, although it is not clear what transport protocol is required.

- Microsoft HealthVault (stand-alone PHR; http://www.healthvault.com/), a web-based platform for storing and maintaining health records, started in October 2007. Individuals and healthcare providers can access the service via the website. There are various ways to contribute information. The individual can type by himself or upload documents and medical images. Also, an individual can directly store doctor fax records in HealthVault. Other services include collection of data directly from pharmacies, laboratories, clinics, hospitals, etc. Also, there are some compatible devices, like pedometer and blood pressure monitors, that can be used to add health data. However, many of these upload services are available only in the United States and the United Kingdom. Information retrieval is via the web and is governed by each account owner. For example, family members may share access to their records. It is also possible for other software packages to connect to the HealthVault platform via its application programming interface (API) and get results.

- Google Health (stand-alone PHR) started as a promising service, but was shut down at the end of 2011 (due apparently to insufficient popularity). People could enter their health data and images manually via the web, including via websites of partner organizations. Users did appreciate the Google Health interface [13].

- MobileHealth2U (MH2U) (tethered PHR; http://www.mobilehealth2u.com/) allows data collection in the associated hospital as well as by the individuals. Hospital patients can view their hospital records via the MH2U website. Everybody, including hospital patients and anybody who registers with the website, can upload data either manually or via the MH2U MediHome monitoring system. The latter is built around a small aggregator box which can be installed at home; it is easily portable physically, but it requires a wired TCP/IP connection to function. The aggregator supports up to five users. There are monitoring devices, such as a glucometer, BP cuff, handheld ECG recorder, and a fingertip pulse oximeter, which can record data anywhere and anytime (except the BP). The monitors then connect to the aggregator by universal serial bus (USB) cable or Bluetooth, and the aggregator retrieves data from the monitors and uploads them directly to the appropriate tables in the MH2U database. Therefore, data collection can be considered as ubiquitous. Records can be viewed via the website or via a smartphone application (which connects to the website), so record retrieval is ubiquitous only as far as the web. Similar packages are available from companies such as A&D (http://www .aandd.jp/) and Tunstall (http://www.tunstallhealthcare.com.au/).

- Ericsson Mobile Health (EMH) is another remote health platform for patients to monitor personal health (http://www.ericsson.com/hr/ict_solutions/e-health/emh/). This is undergoing trials in Serasot, Sarawak (East Malaysia). The health devices are based at the village telecenter, a government-run office with a broadband connection. The telecenter manager or local healthcare worker could use the devices to monitor patients at home or at the telecenter. The patients and healthcare providers (doctors who monitor the measurements) must be registered on the project website to access and manage the service. The EMH system uses an arm-cuff BP meter and allows the captured BP measurements to be automatically transmitted via a wireless 3G network to a central server; the patients and their medical doctors can view the data (health records) via the server's web interface. Patients can send e-mails to the doctors and vice versa. The EMH also comes with five other monitoring devices such as peak flow, pulse oximeter, weight scale, ECG recorder, and glucometer, but they were not used in this trial. Note that this system emphasizes the participation of healthcare workers and the patient–doctor relationship. The project encountered a perennial barrier to health monitoring in remote regions—lack of awareness about or interest in health and lifestyle. This is due to lack of relevant information (lifestyle campaigns do not reach remote regions), lack of literacy, lack of money, increasingly westernized diet, and simple inconvenience (monitoring health parameters cannot be done anywhere and anytime). It is interesting that very few health-monitoring projects mention these hindrances.

- Zilant system by Embedded Wireless (http://www.embeddedwireless.com/solutions /home-health/) was also tested at Serasot, Sarawak. This is a health self-management system that also comes with automated ambulatory devices such as blood pressure monitor, oximeter, glucometer, and weight scale. In this trial blood pressure, weight and oxygen level were taken. The system allows the measurements to be automatically transmitted via a 3G network to the cloud. Patients can log in to the system via a website or a mobile Android application and can view and update their own records, i.e., monitor their health parameters. A group of patients are assigned to a specific nurse, who can monitor their records. General practitioners (GPs) may view all patients' records. So this system is ubiquitous in terms of data collection (if there is a 3G network) and for retrieving data via the Internet over a 3G network.

- Palo Alto Medical Foundation Online (PAMFOnline) (http://www.pamf.org/) is part of the corresponding medical center. The portal allows a patient to view parts of his (or a family member's) PAMF EHR and to request an appointment or prescription renewal. Also, a patient can notify a physician about his health condition via the site. There is a smartphone application for iPhone (and Android phones at https://play.google.com/store/apps/details?id=epic.mychart.android&hl=en), which has much of the functionality of the web interface, under the name *MyChart*. MyChart, in terms of functions provided, is considered to be a subset of the online portal. The app can be used to view health information, contact healthcare providers, and manage appointments. Before being able to use it, the user must have an existing account in My Health Online [PAMF]. This app can be thought of as a mobile app version of the original website.

- EvisitMyDr.com (https://evisitmydr.com/) is an online service based in Minnesota, United States. It supports an online PHR with access for family members. It offers online encounters between patients and doctors, and these are billable in the United

States as an e-visit for insurance reimbursement (Health Insurance Portability and Accountability Act [HIPAA]). People can participate via all devices, which implies PCs, phones, tablets, etc.

- MphRx is short for My Personal Health Record Express (http://www.mphrx .com/). Their WWW-PHR is cloud based and accessible from a variety of computers such as PCs, phones, tablets, etc. The emphasis is on the ability to store all health-related data, including imaging scans, in the cloud and to have it very quickly accessible (through the Internet). The mobile app MphRx Mobile Connect (at https://play.google.com/store/) makes it possible to view and download data using wireless or mobile networks. This is done by connecting the app to the MphRx Secure Private Cloud.

Those systems with web plus smartphone applications may appear to allow wider access to their systems, but in many cases the phone applications require an Internet (Wi-Fi) connection (or advanced mobile network technology such as 3G/4G to access the Internet) in order to function, so in reality these are still purely web-based systems.

14.2.4 Personal Health Records on Stand-Alone Mobile Devices: Not Quite Ubiquitous

The PHR concept can be developed as a mobile application. A portable personal health record (PPHR) or a mobile personal health record (mPHR) could be a simple application that can run on a mobile phone or USB device where information can be stored [14].

This includes a plethora of mobile apps for both Apple iOS– and Google Android–based smartphones. Health-related apps rank among the popular ones for smartphones of any kind. This type of PHR usually contains very basic demographic data and has limited features and limited facility to upload image-based information. Information is saved only in the particular mobile phone's application [14] and cannot be easily transferred to another device or database. Thus, they satisfy the ubiquitous criteria for presence wherever the individual owner/user is but may fall short if remote access is required.

Such mobile apps are mostly targeted at self-monitoring, either of chronic conditions, such as diabetes or hypertension, or of diet or exercise.

14.2.4.1 Diabetes or Hypertension Management

A search for *diabetes* on the Google Play online database yields hundreds of results. Most of them have common features like manually recording blood glucose levels; fewer include tracking food intake, exercise (calories), etc. Very few interact with Bluetooth-enabled glucometers such as those used in MobileHealth2U. Few or none claim to be a complete PHR. Diabetes-management applications can be categorized into two main classes: one which is limited to manual user input and one that can be connected to glucometers. OnTrack Diabetes (at https://play.google.com/store/) is considered to be one of the more comprehensive diabetes-management apps; in fact it can be seen almost as a general PHR since it maintains a record of many things besides blood glucose—like food intake, weight, medications, blood pressure, and pulse. This application has the ability to generate graphs and reports that can be useful for physicians. It also has reminders that can keep the diabetic patient on track with their condition. iBGStar® Diabetes Manager (at https://itunes.apple.com /us/app/) is one of the few applications that capture readings from a Bluetooth-enabled glucometer. The application can also record related information like carbohydrate intake and exercises. Like many health-management tools, IBGStar offers analytical analyses through charts, logbooks, and statistics. The chart tool shows a graphical representation of sugar level over time. The logbook shows readings organized by relationship to meals intake. The statistics can show averages over specific periods of time.

The number of mobile apps for hypertension is much smaller than that for diabetes. Also, the type of data collected tends to be restricted to blood pressure and a few related measurements. Hypertension-management applications can be classified based on how the blood pressure readings are acquired. The simplest way to get the readings is by the user entering the readings manually. The second way is by acting as data collector for a recording device that sends the readings via Bluetooth. Finally, some applications claim to be able to measure the blood pressure; the accuracy of these readings is very questionable, however. A typical app allows for manual insertion of data as well as collecting data from another recording device. The "Blood Pressure Watch" application (at https://play.google .com/store/), for example, allows for manual insertion of data only. The user must provide systolic and diastolic blood pressure readings, the pulse rate, date and time, weight, and the posture of the body in which the reading was taken. The app offers some analytical tools that can track the readings over different periods of time (e.g., weekly or monthly). All the analyses are represented graphically. DynaPulse® Blood Pressure (at https://play .google.com/store/) is an application that makes it possible for Android devices to collect readings from another device (i.e., DynaPulse Recorder) via Bluetooth. A brief Bluetooth setup must be done first before using the app with the device; the app also requires the user to pay attention to their arm movements during the recording. Most of these stand-alone applications do not have the ability to send data to other entities, but some of them allow the user to export the readings locally in the phone as an XML file, which can be used later on by the user.

14.2.4.2 Diet and Exercise

Calorie-counting apps are also legion, but the great majority are just that and not PHRs. In contrast, diet-management apps are mostly diet plans and recommendations. Diet apps can be specific depending on the objective of following a diet plan. Some diet apps are directed toward diabetic patients. For example, the app Fooducate (at https://play.google .com/store/) has a comprehensive database that allows the user to scan the product bar code and grade the product to indicate how much sugar it has or how healthy it is to consume that product. This clearly addresses part of a PHR.

Mobile apps for exercise tend to be related to gaining fitness or muscle mass or losing weight. There are some extended fitness apps like Argus (at https://itunes.apple.com/us /app/). Argus can monitor activities such as walking, running, and cycling. The app also takes into account food intake and calories burnt during activities. The app keeps track of personal workouts and exercises, hydration, and even sleeping patterns—some of the data which could appear in a PHR.

14.2.4.3 Personal Health Records

Although many apps appear in such a search, few of them are comprehensive PHRs. Some of those are specialized or are related to specific healthcare organizations. One, motion-PHR Health Record Mngr (at https://play.google.com/store/), claims to be a ubiquitous

PHR—a comprehensive health record on Android-based smartphones, with connectivity to Microsoft HealthVault; however, this application appears to have lost support, with no development since 2010.

14.2.5 Personal Health Records on Connected Devices: Maybe Ubiquitous

Connected devices can upload health records to a central online server. Whether such systems are ubiquitous depends on where the data are stored, and that depends partly on who collects the data (Table 14.1). Healthcare workers mostly work for organizations which have web-based health records, as the following examples show; they rarely store PHRs on mobile devices. In addition, the devices used by healthcare workers generally stay with them, so are not available to their individual patients at any time.

14.2.5.1 OpenMRS, OpenRosa, JavaRosa, and Sana

The OpenMRS project (http://www.openmrs.org/) is an outstanding example of this approach to ubiquitous health records. Its purpose is to provide healthcare support in developing countries, and one of its outstanding features is the facility for remote data collection using handheld devices. Although at first it was developed to provide tracking and treatment of HIV in Africa, later it became general-purpose software that supports many kinds of medical treatments.

Its most remarkable feature is the emphasis on mobile data collection in remote regions. This is done with the OpenRosa, JavaRosa, Open Data Kit (e.g., http://opendatakit.org/ or https://bitbucket.org/javarosa/javarosa/overview), or Sana (http://sana.mit.edu/) platform. It facilitates data collection easily via mobile devices, using the World Wide Web Consortium (W3C) XForms standard. With the open platform based on Java 2 Micro Edition (J2ME™), a healthcare worker encounters a patient and determines the type of problem, then loads the appropriate XForm and fills in the information, then sends (or queues to send) the completed form to the OpenMRS system via an Internet (Wi-Fi) connection [15]. The records are available to authorized users via a web interface. The XForms can accommodate clinical notes, allergies, laboratory and imaging results, social economic condition, etc., depending on the health problem being addressed.

Other important features of OpenMRS include the following: the software follows some medical standards like HL7, integrated exchange format (IXF), and LOINC; it also supports some open-source tools, e.g., openXdata and HTMLForm entry; and it has a modular structure, so programmers can easily add new functions without modifying the core code [16]. Although it is ubiquitous in its data collection, the OpenMRS system does not allow the individuals (patients) to own and manage their records, so strictly speaking it is not a PHR.

Sana is an extension of the OpenMRS remote data collection to Android-based smartphones (http://sana.mit.edu/). Data are collected as for the J2ME platforms: a health worker interviews a patient and determines the purpose or type of the encounter, downloads the pertinent form, and records the patient's answers—including text, images, and audio and video streams. The health worker does this for a number of patients, and then uploads the information to OpenMRS [17]. Medical experts can review the information and return a diagnosis to the health worker via the Sana application. Similar to JavaRosa, Sana sends data to the central server; it does not have local storage, and the patient does not have access to or a copy of the data.

14.2.5.2 EPI Life and EPI Mini

EPI Life from Singapore (http://www.epimhealth.com/) is a customized mobile phone with applications that is offered by EPI Mobile Health Solutions Pte Ltd. There are various useful health-monitoring applications installed in the phone. HeartSuite is an application designed for EPI Life to keep track of heart conditions, including the ability to measure a patient's ECG, as well as blood pressure, glucose level, and cholesterol. Either the patient can save the ECG report in the phone or he can send it to the EPI Life online database, where a doctor can evaluate it. EPI Mini is a separate handheld ECG recorder which connects to an Apple iPhone or an Android-based phone via Bluetooth to upload ECG traces. The phone application can forward the data to the EPI base (in Singapore).

14.3 A Ubiquitous Personal Health Record for Remote Regions Must Involve Individuals and Include Phone-Based Records

The lack of ubiquitous availability of the web and of healthcare workers implies that the only way to achieve a ubiquitous PHR for remote regions, such as that shown in Figure 14.2 [18–19], is to utilize what is available, i.e., the people themselves and the existing communication network (which in most places is at least a GSM network). This happens to be quite compatible with many government initiatives toward patient participation and self-monitoring.

There are, however, several problems in creating such a system. These will be discussed in light of an example.

14.3.1 Examples: Portable Personal Health Records

Figure 14.3 [20] shows a system that has been under development since about 2005, first at Monash University in Melbourne, from 2007 at Swinburne University of Technology Sarawak campus (Kuching, Malaysia), and from 2010 at the Multimedia University in Malacca, Malaysia. This addresses the ubiquitous problem by using the ubiquitous mobile phones or smartphones, being maintained by the individual, and communicating with an online server via the available transport protocol (including GSM, i.e., SMS), and storing the PHR in both the phone and the online database.

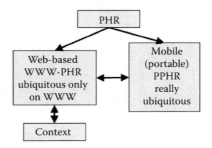

FIGURE 14.2
Hybrid PHR system model. (Adapted from Seldon, H. L., *Electron. J. Health Inform.*, 8, e1, 2014; and Ahsan, M., Integrating Contextual Information into a Personal Health Record System. MSc thesis, Faculty of Information Science and Technology, Multimedia University, Melaka, Malaysia, 2014.)

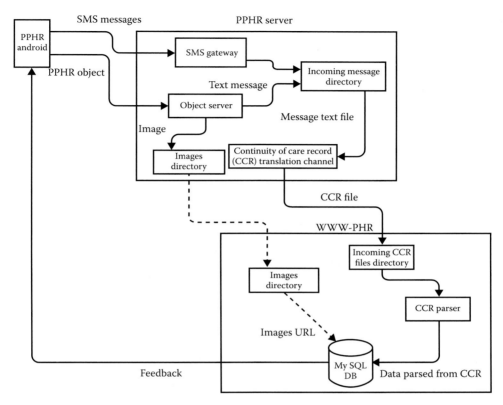

FIGURE 14.3
PPHR communications system. (From Al-Habsi, A.A.A., *Communications between Portable and Web-Based Personal Health Record Systems.* MSc thesis, Faculty of Information Science and Technology, Multimedia University, Melaka, Malaysia, 2014.)

14.3.1.1 *Portable Personal Health Records: Mobile Records*

Health record entry is made—by individuals—with a mobile phone or smartphone. This PPHR is available in a J2ME version for all old, Java-enabled cell phones (which until recently were ubiquitous throughout SE Asia and much of the world), and in an Android version for such smartphones (which are becoming ubiquitous in Southeast Asia). The records are stored as text on the phone; the smartphone version also allows inclusion of photographs.

One problem for individuals to maintain a PHR is that it must be in a language which they understand, and that excludes much professional medical terminology and for many inhabitants of remote regions it excludes English. But the records must still be meaningful to healthcare workers, so the terminology must be sufficiently professional. In this PPHR the user interface is in Malay, English or (Malaysian), Chinese. The organization and vocabulary are based on the International Classification for Primary Care, version 2e (ICPC-2e) [21]. This is a list of less than 800 terms which cover most requirements of primary care; it has been translated into the local languages and further simplified for use by nonmedical rural inhabitants. The original vocabulary includes links to the 10th revision of the International Statistical Classification of Diseases and Related Health Problems

(ICD-10) terms, which have also been extended to cover many terms in the PPHR and provide more professional terminology.

The early J2ME version of this PPHR has been described by Wee [22] and Seldon et al. [19]. A few screenshots are shown in Figure 14.4.

14.3.1.2 *Portable Personal Health Records: Web-Based Records*

This split-PHR system requires a web-based PHR to back up the phone-based records. Such systems already exist, as described above, so it is not necessary to create more. The PPHR system described here connects to the MH2U system described above, which was adapted to allow Continuity of Care Record (CCR) (American Society for Testing and Materials [ASTM] 2369-05) standard input [23].

14.3.1.3 *Portable Personal Health Records: Ubiquitous Communications*

Data communications in remote regions are a big problem. As seen in Figure 14.4, the phone-based record can be transferred to the WWW-PHR via SMS, which is the basic,

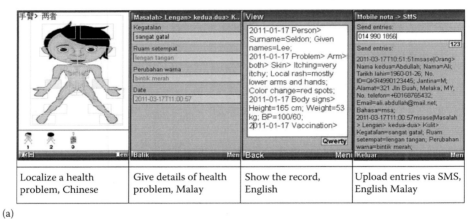

FIGURE 14.4
Samples of screen displays from our PPHR. (a) Java 2 Micro Edition (J2ME) version (taken from the NetBeans 6.9 screen emulator) and (b) Android version.

almost ubiquitous text-transfer mode available on all mobile networks, or via the Internet protocol transmission control protocol (TCP)/Internet protocol (IP) (Wi-Fi). The latter uses basic socket programming, so it is very fast, and also allows transmission of images, but, of course, can only be utilized when and where there is Internet access. For this reason, the restriction of PPHR entries to text and small images allows transmission to finish within a second, so it works where the Internet access is only intermittent. The choice of the communication medium is done by scanning the available connections (e.g., Wi-Fi, general packet radio service [GPRS], enhanced data rates for GSM evolution [EDGE], and GSM), but the priority will be given to Wi-Fi if it is available.

Matching the PPHR data format and vocabulary to that of the web-based PHR is a problem. To solve this, the PPHR data is converted to the international CCR standard format [23] to be ubiquitously available to health record databases [24]. Thereby, the available ICD-10 and other codes are attached to the corresponding PPHR entries to allow more professional comprehension, especially of entries in a local language.

14.4 Contextual Information: The Icing on the Personal Health Records

Information from outside the person, which might contribute to the person's health or condition, is contextual information [2]. The context is always ubiquitous—all health or health problems are found in a context. This relates to environmental factors, public health warnings, etc., and is important for medical diagnosis but is often omitted. For example, if a person is suffering from a new problem and visits the doctor, the doctor will ask when and where it started, or what the patient had eaten—these questions enable the doctor to possibly make a link with known environmental factors such as epidemic or food poisoning warnings, and environmental pollution. Table 14.2 gives examples of the relationship between symptoms and environmental context [25].

In order to be ubiquitous, a PHR must include this information. But just as medical personnel seldom record such information, so also normal people, including those in remote regions, seldom note it or know about it. So the PHR system must try to retrieve the information without burdening the user too much.

There are three types of context which may relate to health problems; they are location, time, and history.

Examples of historically relevant contextual information include vaginal cancer, linked to use of diethylstilbestrol by victims' mothers during pregnancy [26], and mesothelioma, linked to asbestos exposure many years prior to the appearance of the disease. This type of information is extremely difficult to find in health records.

TABLE 14.2

Relationship between Context and Health Symptoms

Environmental Dimensions	Environmental Hazards, Exposures	Possible Reactions (Symptoms and Signs)
Air	Allergens, pollution	Itching, sneezing, drowsiness, nasal stuffiness, watery eyes, etc.
Water	Arsenic, pollution	Nausea, vomiting, diarrhea, abdominal pain, decreased production of red and white blood cells

Source: Lau, T.C., *A Web-Based Environmental Health Information Source for Malaysian Context*, MSc Thesis, Swinburne University of Technology, Kuching, Malaysia, 2013.

Examples where locations have been linked to disease include the following:

- Tanneries in Bangladesh [27].
- Gold mines in Papua New Guinea [28].
- Nuclear reactors or waste-processing plants like Sellafield in England [29].
- Asian poultry farms or markets [30].
- Polluted air, e.g., particulate matter, sulfur dioxide, nitrogen dioxide, hydrogen sulfide, and photochemical smog [31].

Examples of time-dependent contextual information include epidemics and weather, among others:

- Severe acute respiratory syndrome (SARS) in Southeast Asia in 2002–2003; bird flu in Southeast Asia in 2010.
- Cholera in Haiti in 2012; polio in Syria in 2013.
- Winter or heat waves in Europe resulting in increased mortality as well as respiratory or food-borne diseases [32].
- Time-dependent air pollution has been demonstrated by the "haze" which starts with massive fires in Indonesia and blankets Singapore and large parts of Malaysia, most recently in June 2013 [33].

14.4.1 Contextual Information on the Web

Knowledge of possible environmental factors, especially time- and location-dependent ones, must usually come from sources outside the individual, mainly from the mass media. The media publish reports of pollution, epidemics, food poisoning, etc., although the thoroughness of the reports may vary [34–35]. Nowadays much of the reporting is either done on the web or duplicated on the web or other Internet media. For example, the monitoring website myehms.com derives its input from Really Simple Syndication (RSS) feeds from Malaysian newspapers and such.

14.4.2 Retrieval of Contextual Information

There are several possible ways to retrieve relevant contextual information from the web. They all involve similar steps: (1) identify reliable sources of information; (2) query those sources about information related to an individual's symptom at a place and time; (3) in the query results, identify that information which is most specific and relevant to the symptom, place, and time; and (4) store the contextual information so that it can be linked to the original health problem. The first step must still involve manual selection by a human being; the others have been automated using a search engine API such as Google Custom Search API [2,18,36].

14.4.3 Integration into a Ubiquitous Personal Health Record System

The contextual information is stored online, separately from a WWW-PHR (Figure 14.2), as one context may apply to many PHR owners. (Thus, the contextual information can become epidemiological information.) There is a many-to-many relationship between contexts and WWW-PHR entries containing health problems. The PHR user whose symptom triggered a search for contextual information is notified either via e-mail or SMS if the search returned a result.

14.5 Conclusions

The aim of this chapter has been to try to demonstrate that making PHRs really ubiquitous in remote regions requires extra effort. The regional shortages of healthcare personnel and of Internet connectivity mean that PHRs must be in the hands of the local inhabitants, on devices which are always available to them (i.e., mobile phones), and that PHRs must be able to communicate with the online world through simple mobile networks. PHRs should also include important contextual information for diagnosis and epidemiology.

References

1. Emani, S., Yamin, C.K., & Bates, D.W. (2012). Patient perceptions of a personal health record: A test of the diffusion of innovation model. *Journal Medical Internet Research*, 14(6), p. e150.
2. Ahsan, M., Seldon, H.L., & Shohel, M. (2012). Personal Health Records: Retrieving Contextual Information with Google Custom Search. In *Global Telehealth 2012*, Smith, A.C. et al. (Eds.), IOS Press. doi:10.3233/978-1-61499-152-6-10, pp. 10–18.
3. Lau, T.C., Seldon, H.L., & Lau, B.T. (2012). A web-based system for environmental health information in a Malaysian context. In *Proceedings of Third International Conference on Recent Trends in Information Processing and Computing (IPC). Springer Lecture Notes in Computer Science*. Berlin Heidelberg: Springer Verlag.
4. Trân, V.A., Seldon, H.L., Chu, H.D., & Nguyên K.P. (2006). Electronic healthcare communications in Vietnam in 2004. *International Journal Medical Informatics*, 75(10–11), pp. 764–770.
5. Seldon, H.L. (2014). Personal health records in Southeast Asia Part 1—A way to computerize healthcare? *electronic Journal of Health Informatics*, 8(1), p. e1. Retrieved on Aug. 14, 2014, from http://www.ejhi.net/.
6. Global Internet usage. (2014). In *Wikipedia*. Retrieved on Aug. 15, 2014, from http://en.wikipedia.org/wiki/Global_Internet_usage.
7. International Telecommunications Union. (2014). Retrieved on Aug. 15, 2014, from http://www.itu.int/.
8. Malaysian Communications and Multimedia Commission. (2014). *Communications and Multimedia Pocketbook of Statistics* Q4 2013. ISSN 2180-4656.
9. Mandl, K.D., Simons, W.W., Crawford, W.C., & Abbett, J.M. (2007). Indivo: A personally controlled health record for health information exchange and communication. *BMC Medical Informatics and Decision Making*, 7(1), p. 25.
10. Indivo. (2007). Children's Hospital Informatics Program: Research, History of the Personally Controlled Health Record. Retrieved on Aug. 14, 2014, from http://indivohealth.org/research.
11. WWAY News Channel 3. (Sept. 7, 2007). More turn to Internet for medical advice. Retrieved on May 9, 2012, from http://www.wwaytv3.com/more_turn_to_Internet_for_medical_advice/09/2007.
12. Halamka, J.D., Mandl, K.D., & Tang, P.C. (2008). Early experiences with personal health records. *Journal of the American Medical Informatics Association*, 15(1), pp. 1–7.

13. Sunyaev, A., Chornyi, D., Mauro, C., & Kremar, H. (2010). Evaluation framework for personal health records: Microsoft HealthVault vs. Google Health. In *2010 43rd Hawaii International Conference System Sciences (HICSS)*, pp. 1–10.

14. Santos, J., Pedrosa, T., Costa, C.M., & Oliveira, J.L. (2010). Modelling a Portable Personal Health Record. In *Third International Conference on Health Informatics*, INSTICC—Institute for Systems and Technologies of Information, Control and Communication, pp. 465–468.

15. Dimagi, Inc. (2012). Open Source Mobile Data Collection. Retrieved on Aug. 14, 2014, from http://www.dimagi.com/javarosa/.

16. Seebregts, C.J., Mamlin, B.W., Biondich, P.G., Fraser, H.S., Wolfe, B.A., Jazayeri, D. et al. (2009). The OpenMRS implementers network. *International Journal of Medical Informatics*, 78(11), pp. 711–720.

17. OpenMRS. (2011). Open Source Health IT for the Planet. Retrieved on May 9, 2012, from http://openmrs.org/.

18. Ahsan, M. (2014). *Integrating Contextual Information into a Personal Health Record System*. MSc thesis, Faculty of Information Science and Technology, Multimedia University, Melaka, Malaysia.

19. Seldon, H.L., Moghaddasi, H., Seo, W.J., & Wee, J.N.S. (2014). Personal health records in Southeast Asia Part 2—A digital portable health record. *electronic Journal of Health Informatics*, 8(1), p. e2. Retrieved on Aug. 14, 2014, from http://www.ejhi.net/.

20. Al-Habsi, A.A.A. (2014). *Communications between Portable and Web-Based Personal Health Record Systems*. MSc thesis, Faculty of Information Science and Technology, Multimedia University, Melaka, Malaysia.

21. Verbeke, M., Schrans, D., Deroose, S., & De Maeseneer, J. (2006). The International Classification of Primary Care (ICPC-2): An essential tool in the EPR of the GP. *Studies in Health Technology and Informatics*, 124, pp. 809–814.

22. Wee, J.N.S. (2012). *Providing Personal Health Records in Malaysia—A Portable Prototype*. MSc thesis, Swinburne University of Technology Sarawak, Kuching, Malaysia.

23. ASTM. (2006). Standard Specification for Continuity of Care Record. Retrieved on Aug. 14, 2014, from http://www.astm.org/Standards/E2369.htm.

24. Al-Habsi, A.A.A, & Seldon, H.L. (2013). A communication module for mobile personal health record. In *2013 IEEE 7th International Power Engineering and Optimization Conference (PEOCO)*, pp. 727–731.

25. Lau, T.C. (2013). *A Web-based Environmental Health Information Source for Malaysian Context*. MSc Thesis, Swinburne University of Technology, Kuching, Malaysia.

26. National Cancer Institute. (2011). Diethylstilbestrol (DES) and Cancer. Retrieved on Jan. 28, 2013, from http://www.cancer.gov/cancertopics/factsheet/Risk/DES.

27. Karim, M., Manshoven, S., Islam, M., Gascon, J.A., Ibarra, M., Diels, L. et al. (2012). Assessment of an urban contaminated site from tannery industries in Dhaka City, Bangladesh. *Journal of Hazardous, Toxic, and Radioactive Waste*, 17(1), pp. 52–61.

28. Jell-Bahlsen, S., & Jell, G. (2012). The trans-national gold curse of Papua New Guinea. *Dialectical Anthropology*, 36(3–4), pp. 317–341.

29. Rahman, R., Plater, A.J., Nolan, P.J., Mauz, B., & Appleby, P.G. (2013). Potential health risks from radioactive contamination of saltmarshes in NW England. *Journal of Environmental Radioactivity*, 119, pp. 55–62.

30. Wong, L.P., & Sam, I.-C. (2011). Knowledge and attitudes in regard to pandemic influenza A (H1N1) in a multiethnic community of Malaysia. *International Journal of Behavioral Medicine*, 18 (2), pp. 112–121.

31. Chen, Z., Huang, X., & Wang, Q. (2009). The effect of air pollution on human health in China: A macro evaluation. In *3rd International Conference on Bioinformatics and Biomedical Engineering, 2009, ICBBE 2009*, pp. 1–4.

32. Islam, M., Chaussalet, T., Ozkan, N., & Demir, E. (2010). An approach to exploring the effect of weather variations on chronic disease incidence rate and potential changes in future health systems. In *2010 IEEE 23rd International Symposium Computer-Based Medical Systems (CBMS)*, pp. 190–196.

33. *The Star Online*. (2013a). Malaysia, Indonesia and Singapore to discuss haze mitigation. Retrieved on Dec. 18, 2013, from http://www.thestar.com.my/News/Nation/2013/06/30/Malaysia-Indonesia-and-Singapore-to-discuss-haze-mitigation.aspx.

34. *The New Straits Times Press*. (Nov 4, 2013). Malaysia warns on dengue as deaths spike. Retrieved on Aug. 14, 2014, from http://www.nst.com.my/latest/malaysia-warns-on-dengue-as-deaths-spike-1.

35. *The Star Online*. (2013b). Klang Valley air quality improves further. Retrieved on Dec. 18, 2013, from http://www.thestar.com.my/News/Nation/2012/06/20/Klang-Valley-air-quality-improves-further.aspx.

36. Google Inc. (2014). Google Search API. Retrieved on Mar. 7, 2014, from https://developers.google.com/custom-search/docs/api.

15

Education and Training for Supporting General Practitioners in the Use of Clinical Telehealth: A Needs Analysis

Sisira Edirippulige, Nigel R. Armfield, Liam Caffery, and Anthony C. Smith

CONTENTS

15.1 Introduction

Increase in chronic disease, the rising cost of care, and fragmentation of care delivery have put significant pressures on general practice [1]. The use of information and communication technologies (ICTs) has been recognized as having a role in addressing at least some of these challenges. It has been stated that ICT can potentially be used to "redesign primary care" to maximize efficiency of service delivery and patient outcomes [2].

Telehealth, the use of ICT to deliver healthcare at a distance, can improve both accesses to primary care and patient outcomes [3]. Recognizing the benefits, in 2011 the Australian government added item numbers for video-based consultations to the Medicare Benefits Schedule (MBS).

Several studies have shown that general practitioners (GPs) are interested in using telehealth in their practices [4–5]. Evidence also suggests that clinicians support the use of telehealth in the primary care sector [6–8]. Despite this support, the use of telehealth in the primary care setting is so far limited both in Australia and globally [9–11].

Research suggests that a lack of telehealth-related education and training is an important barrier to the adoption of telehealth [6,12–13].

This study aims to understand Australian GPs' current levels of telehealth use and their knowledge and understanding of telehealth and to identify their telehealth-related educational and training needs.

15.2 Methods

An online survey was conducted using SurveyMonkey (http://www.surveymonkey.com) one year after the introduction of Medical Benefit Schedule (MBS) item numbers for video-based consultations. MBS is part of the wider Medicare Benefits Scheme managed by the Department of Health and administered by Department of Human Services in Australia. MBS Online contains the latest MBS information and is updated as changes to the MBS occur.

15.2.1 Participants

General practitioners in all Australian states and territories were invited to participate in the survey. The invitation was provided by e-mail via the Medicare Local organizations. The survey opened in May 2012 and a reminder e-mail was sent to all participants in June 2012.

TABLE 15.1

Survey Questions Identifying Characteristics of Participant's Practice

Question	Valid Responses
1. Which state(s)/territory(ies) do you practice in?	Queensland; New South Wales; Australian Capital Territory; Victoria; South Australia; Tasmania; Northern Territory; Western Australia
2. What best describes the setting of your practice?	Rural; regional; metropolitan
3. How would you describe the size of your practice?	Small; medium; large
4. Staff working in your practice include:	GP; GP with specialist training; nurse; nurse practitioner; allied health professional; indigenous health worker; administrative support

TABLE 15.2

Current and Planned Use of Telehealth

Question	Valid Responses
1. Has the addition of MBS item numbers for online consultations motivated you to consider practicing by telehealth?	Yes; no
2. Do you or any of your practice colleagues currently use, or plan to use in the near future, telehealth for specialist consultations?	Definitely; maybe; definitely not; unsure
3. Do you currently, or do you plan to, use telehealth for consultations with patients in their homes?	Definitely; maybe; definitely not; unsure
4. What proportion of work do you estimate could potentially be done by telehealth?	Nil; <5%; 5%–10%; 11%–20%; >20%; unsure

15.2.2 Survey Questions

The survey collected characteristics of the participants' practices (Table 15.1); current and planned use of telehealth within the participants practice (Table 15.2); and telehealth-related education and training needs (Table 15.3).

15.2.3 Analysis

Data was downloaded from SurveyMonkey and response frequencies were tabulated. Frequencies of education and training interest areas were charted.

TABLE 15.3

Telehealth Education and Training Needs

Question	Valid Responses
1. How would you describe your understanding of telehealth?	None; minimal; moderate; considerable
2. Do you believe that education and training relating to clinical telehealth would be useful?	Yes; no
3. Have you ever had any *formal* education or training in relation to clinical telehealth (if so, please describe)?	Yes; no; free text
4. Have you ever had any *informal* education or training in relation to clinical telehealth (if so, please describe)?	Yes; no; free text
5. Given the appropriate opportunity, would you like to receive education and training (or further education and training) relating to clinical telehealth?	Yes; no
Questions 6 to 11 Were Presented to Participants Who Answered Yes to Question 5	
6. What form of education and training would be of most interest to you?	University qualification (e.g., a graduate certificate/diploma/master's level); vocational short course
7. What would be your preferred way of receiving education and training?	Face to face; online; a combination of the two; free text for further comments
8. Would you be interested in attending a practical component to gain hands-on skills and experience in clinical telehealth?	Yes; no
9. Would CPD points for clinical telehealth–related education and training be of interest to you?	Yes; no
10. Which of the following areas would you be interested in receiving education and training on?	History; applications of clinical telehealth (case studies); technical options; establishing and running a clinical telehealth service; telehealth and the MBS; clinical guidelines for telehealth; evidence base for clinical telehealth; privacy and security issues; medicolegal issues; research and evaluation; free text for further comments
11. Would you be willing to pay for clinical telehealth education and training?	Yes, no
Question 12 Was Presented to Participants Who Answered No to Question 5	
12. Please comment on why education and training for clinical telehealth is not of interest.	Free text

Note:　CPD: continuing professional development.

15.2.4 Ethics

The study was approved by the University of Queensland Human Research Ethics Committee (reference 2011001430).

15.3 Results

15.3.1 Characteristics of the Responding Practices

A total of 62 completed surveys were received. One responder did not provide details of their location or size of practice. The largest number of responses were received from Victoria ($n = 21$; 34.4%) and the least from Western Australia ($n = 3$; 4.9%). Most responses ($n = 51$; 83%) were from practices in rural and regional areas. Most respondents ($n = 30$; 49.2%) described their practice as medium or small ($n = 21$; 34.4%). No responses were received from practices in the Australian Capital Territory or the Northern Territory. The characteristics of all responding practices are summarized in Table 15.4.

Almost a third of the respondents ($n = 18$; 29%) had GPs with specialist training on staff. Most practices ($n = 54$; 87.1%) employed nursing staff, with $n = 16$ (25.8%) employing a nurse practitioner. Allied health professionals ($n = 23$; 37.1%) and administrative support ($n = 41$; 66%) were often were often employed. Only one practice from rural New South Wales reported employing an indigenous health worker.

A small number of practices reported employing other health staff, including a pharmacy assistant (one small practice, rural Tasmania); medical specialists and a diabetes educator (one large practice, metropolitan Victoria); a pharmacist (one large practice, metropolitan Queensland); and a dentist, psychologist, drug and alcohol team, and public health team at one large practice in rural New South Wales.

15.3.2 Current and Planned Use of Telehealth

The average response rate to questions related to current and planned use of telehealth was 90%. For the majority of respondents ($n = 47$; 80%), who expressed an opinion, the introduction of item numbers for video-based telehealth consultations was a motivation to consider practicing by telehealth (Table 15.5). Responses related to current and planned usage of telehealth for specialist consultation are detailed in Table 15.6 and to current and planned usage of patient–provider telehealth are detailed in Table 15.7. Estimates of GP workload conducive to telehealth are detailed in Table 15.8.

15.3.3 Education and Training

The average response rate to questions in this category was 79.1%. Respondents' understanding of telehealth is detailed in Table 15.9.

Most respondents had not received any kind of formal ($n = 49$; 96%) or informal ($n = 40$; 78.4%) education or training in telehealth but believed that such education would be useful ($n = 15$; 83%) and they would like to receive such education ($n = 47$; 92.2%). The following sources of informal educational and training experiences were described by some participants: "reading Royal Australian College of General Practitioners (RACGP) guidelines to conducting telehealth, sharing experiences with colleagues;" "conference, emailing with

TABLE 15.4

Survey Responses by Location and Size of Practice

			Size of Practice					
	Responses		Small		Medium		Large	
Practice Location	*n*	%	*n*	%	*n*	%	*n*	%
Queensland								
Rural	2	3.3	2	3.3	–	–	–	–
Regional	6	9.8	–	–	6	9.8	–	–
Metropolitan	4	6.6	1	1.6	1	1.6	2	3.3
Total	12	19.7	3	4.9	7	11.5	2	3.3
New South Wales								
Rural	10	16.4	4	6.6	5	8.2	1	1.6
Regional	4	6.6	1	1.6	2	3.3	1	1.6
Metropolitan	–	–	–	–	–	–	–	–
Total	14	23	5	8.2	7	11.5	2	3.3
Victoria								
Rural	7	11.5	5	8.2	2	3.3	–	–
Regional	9	14.8	1	1.6	7	11.5	1	1.6
Metropolitan	5	8.2	2	3.3	2	3.3	1	1.6
Total	21	34.4	8	13.1	11	18.1	2	3.3
Western Australia								
Rural	1	–	–	–	1	–	–	–
Regional	1	–	1	–	–	–	–	–
Metropolitan	1	–	–	–	1	–	–	–
Total	3	4.9	1	1.6	2	3.3	–	–
Tasmania								
Rural	3	4.9	2	–	1	–	–	–
Regional	3	4.9	1	–	1	–	1	–
Metropolitan	–	–	–	–	–	–	–	–
Total	6	9.8	3	4.9	2	3.3	1	1.6
South Australia								
Rural	1	1.6	–	–	1	1.6	–	–
Regional	1	1.6	1	1.6	–	–	–	–
Metropolitan	3	4.9	–	–	–	–	3	4.9
Total	5	8.2	1	1.6	1	1.6	3	4.9
Total	**61**	**100**	**21**	**34.4**	**30**	**49.2**	**10**	**16.4**

TABLE 15.5

MBS Item Numbers as a Motivator

	Yes		No	
Question	*n*	%	*n*	%
Has the addition of MBS item numbers for online consultations motivated you to consider practicing by telehealth? (response *n* = 56; 90.3%)	47	84	9	16.1

TABLE 15.6

Current and Planned Uses of Telehealth in the Practice

Question	Definitely		Maybe		Definitely Not		Unsure	
	n	%	*n*	%	*n*	%	*n*	%
Do you or any of your practice colleagues currently use, or plan to use in the near future, telehealth for specialist consultations? (response *n* = 56; 90.3%)	29	51.7	16	28.6	2	3.6	9	16.1

TABLE 15.7

Current and Planned Uses of Telehealth at Patients' Homes

Question	Definitely		Maybe		Definitely Not		Unsure	
	n	%	*n*	%	*n*	%	*n*	%
Do you currently, or do you plan to, use telehealth for consultations with patients at their homes? (response *n* = 55; 88.7%)	11	20	23	41.8	9	16.4	12	21.8

TABLE 15.8

Estimates of Telehealth Work

Question	Nil		<5%		5%–10%		11%–20%		>20%		Unsure	
	n	%	*n*	%	*n*	%	*n*	%	*n*	%	*n*	%
What proportion of work do you estimate could potentially be done by telehealth? (response *n* = 57; 91.9%)	2	3.5	18	31.6	26	45.6	3	5.3	3	5.3	5	8.8

TABLE 15.9

Understanding of Telehealth

Question	None		Minimal		Moderate		Considerable	
	n	%	*n*	%	*n*	%	*n*	%
How would you describe your understanding of telehealth? (response *n* = 52; 83.9%)	11	21.2	24	46.2	14	26.9	3	5.8

patients, interacting with colleagues;" "discussion with others who had done it;" "information pamphlets distributed by Medicare Australia;" "previous recreational personal experience with Skype;" and "occasional videoconferencing use."

Education and training that earned continuing professional development (CPD) points was attractive to most respondents (*n* = 39; 83%), as was the potential for attending a practical component (*n* = 38; 74.5%) (Table 15.10). Most respondents (*n* = 42; 93.3%) preferred a vocational offering to a university-based academic course (Table 15.11).

There was no strong preference for delivery modality (e.g., face to face, online, or hybrid approach) (Table 15.12). Three respondents provided free-text preferences on how they would like to receive education and training: "short continuing professional development kind of courses with some hands-on training would be good (preferably evening or

TABLE 15.10

Education and Training in Telehealth

Question	Yes		No	
	n	%	*n*	%
Do you believe that education and training relating to clinical telehealth would be useful? (response *n* = 51; 82.3%)	43	84.3	8	15.7
Have you ever had any formal education of training in relation to clinical telehealth (if so, please describe)? (response *n* = 51; 82.3%)	2	3.9	49	96.1
Have you ever had any informal education of training in relation to clinical telehealth (if so, please describe)? (response *n* = 51; 82.3%)	11	21.6	40	78.4
Given the appropriate opportunity, would you like to receive education and training (or further education and training) relating to clinical telehealth? (response *n* = 51; 82.3%)	47	92.2	4	7.8
Would you be interested in attending a practical component to gain hands-on skills and experience in clinical telehealth? (response *n* = 51; 82.3%)	38	74.5	13	25.5
Would CPD points for clinical telehealth related education and training be of interest to you? (response *n* = 47; 72.6%)	39	83	8	17

TABLE 15.11

Forms of Telehealth Education

Question	University		Vocational	
	n	%	*n*	%
What form of education and training would be of most interest to you? (response *n* = 45; 72.6%)	3	6.7	42	93.3

TABLE 15.12

Modes of Delivery of Telehealth Education

Question	Face to Face		Online		Combination	
	n	%	*n*	%	*n*	%
What would be your preferred way of receiving education and training? (response *n* = 47; 75.8%)	12	25.5	18	38.3	17	36.2

TABLE 15.13

Willingness to Pay for Education and Training

Question	Yes		No	
	n	%	*n*	%
Would you be willing to pay for clinical telehealth education and training? (response *n* = 48; 77.4%)	29	60.4	19	39.6

weekend sessions)"; "a two hour intensive"; and "some short face to face sessions on technical aspects would be useful."

Twenty-nine (60%) respondents indicated that they would be willing to pay for education and training (Table 15.13). Figure 15.1 shows the respondents level of interest in education and training by topic area.

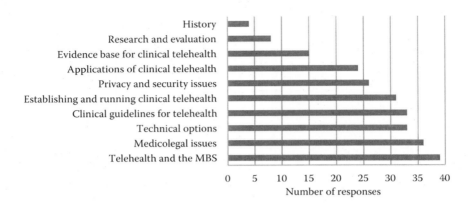

FIGURE 15.1
Respondents areas of interest for education and training in clinical telehealth (response $n = 47; 75.8\%$).

15.4 Discussions

It would appear that the introduction of the MBS payment has encouraged the use of telehealth for GP to specialist consultation. Over 70% of respondents provided a favorable response to idea of practicing telehealth and less than 4% of respondents indicated strongly that they would not practice telehealth.

While video-based consultations directly between a GP and their patients are currently not reimbursed, nearly 62% of respondents provided favorable responses to practicing provider–patient telehealth. Around 16% of respondents indicated that they would not practice provider–patient telehealth compared with only 4% of respondents unwilling to practice provider–specialist telehealth.

Other studies have also found a link between reimbursement and willingness to practice. One such study concluded that GPs did not feel they were able to conduct patient–provider video consultations because such consultations are not subsidized by government funding [14]. The reluctance is perhaps due to government funding being associated with regulatory acceptance of a practice. The introduction of reimbursement for patient–provider consultation may present a further opportunity to increase the uptake of telehealth.

A study examining the level of telehealth use by Victorian GPs in Australia found that GPs had limited experience using telehealth and were concerned about medicolegal aspects and the reimbursement issues relating to practicing telehealth [6]. These issues were also confirmed in our study, featuring highly in our respondents' areas of interest for telehealth training.

15.4.1 Telehealth Training

Our results show the main areas of interest are the practical how-to aspects of setting up and running a telehealth service and the regulatory aspects of telepractice (Table 15.14). We have weighted the responses by summing the number of respondents who were interested in each area of training interest divided by the product of the number of discrete areas of interest and the total number of respondents ($n = 47$).

TABLE 15.14

Weighting of Training Categories

Training Category	Areas of Interest	Weighting
Practical "how to"	Telehealth and the MBS, technical issues, establishing and running clinical telehealth, applications of clinical telehealth	67%
Regulatory aspects	Medicolegal issues, clinical guidelines, privacy and security issues	67%
Evidence and research	Evidence base for clinical telehealth, research and evaluation	16%
History	History	9%

This was a small study with low response rates; hence, care must be taken in generalizing the results. Despite the low response rate, there were a number of strong emerging themes. Firstly, there is definite willingness to practice telehealth but there may be limited scope to do so with a large majority of GPs believing between 1% and 10% of their workload is conducive to telehealth. The inability to examine the patient is the often cited reason for GPs not conducting a video consultation [14]. Secondly, the results support the notion that telepractice requires a specific skill set which is not part of current training for GPs [15]. Thirdly, most GPs appear very willing to undertake training in clinical telehealth.

15.5 Conclusions

This study indicates that GPs are willing to both practice telehealth and undertake education and training in telehealth. There is, however, a lack of availability of formal training programs. GPs are most interested in receiving training in the practical aspects of establishing and running a telehealth service—for example, reimbursement and technical issues—and the regulatory aspects of telehealth—for example, medicolegal issues.

References

1. Harris MF and Zwar NA, Care of patients with chronic disease: The challenge for general practice. *Medical Journal of Australia*, 187(2):104–107, 2007.
2. Schoen C, Osborn R, Doty MM et al., A survey of primary care physicians in eleven countries, 2009: Perspectives on care, costs, and experiences. *Health Affairs*, 28(6):w1171–w1183, 2009.
3. Cottrell E, Chambers R, and O'Connell P, Using simple telehealth in primary care to reduce blood pressure: A service evaluation. *BMJ Open*, 2(6), 2012.
4. Goodyear-Smith F, Wearn A, Everts H et al., Pandora's electronic box: GPs reflect upon email communication with their patients. *Informatics in Primary Care*, 13(3):195–202, 2005.
5. Neville RG, Reed C, Boswell B et al., Early experience of the use of short message service (SMS) technology in routine clinical care. *Informatics in Primary Care*, 16(3):203–211, 2008.
6. Hanna L and Fairhurst K, Using information and communication technologies to consult with patients in Victorian primary care: The views of general practitioners. *Australian Journal of Primary Health*, 19(2):166–170, 2013.

7. Kim J-E, Song Y-M, Park J-H et al., Attitude of Korean primary care family physicians towards telehealth. *Korean Journal of Family Medicine*, 32(6):341–351, 2011.

8. Richards H, King G, Reid M et al., Remote working: Survey of attitudes to eHealth of doctors and nurses in rural general practices in the United Kingdom. *Family Practice*, 22(1):2–7, 2005.

9. Schattner P, Pleteshner C, Bhend H et al., Guidelines for computer security in general practice. *Informatics in Primary Care*, 15(2):73–82, 2007.

10. Hendy J, Fulop N, Reeves BC et al., Implementing the NHS information technology programme: Qualitative study of progress in acute trusts. *BMJ: British Medical Journal*, 334(7608):1360, 2007.

11. Davis K, Doty MM, Shea K et al., Health information technology and physician perceptions of quality of care and satisfaction. *Health Policy*, 90(2):239–246, 2009.

12. Hanna L, May C, and Fairhurst K, Non-face-to-face consultations and communications in primary care: The role and perspective of general practice managers in Scotland. *Informatics in Primary Care*, 19(1):17–24, 2011.

13. McConnochie KM, Potential of telemedicine in pediatric primary care. *Pediatrics in Review*, 27(9):e58–e65, 2006.

14. Jiwa M and Meng X, Video consultation use by Australian general practitioners: Video vignette study. *Journal of Medical Internet Research*, 15(6):e117, 2013.

15. Picot J, Meeting the need for educational standards in the practice of telemedicine and telehealth. *Journal of Telemedicine and Telecare*, 6 Suppl 2:S59–S62, 2000.

Further Reading

1. Lamb GS and Shea K, Nursing education in telehealth. *Journal of Telemedicine and Telecare*, 12(2):55–56, 2006.

2. Edirippulige S, Smith AC, Armfield NR, Bensink M, and Wootton R, Student perceptions of a hands-on practicum to supplement an online eHealth course. *Journal of Medical Internet Research*, 14(6), 2012.

3. Edirippulige S, Smith AC, Beattie H, Davies E, and Wootton R, Evaluation of nursing students' knowledge, understanding and readiness to practice e-health. *Journal of Telemedicine and Telecare*, 13 Suppl 3:37–39, 2007.

4. Edirippulige S, Armfield NR, and Smith AC, A qualitative study of the careers and professional practices of graduates from an e-health postgraduate programme. *Journal of Telemedicine and Telecare*, 18(8):455–459, 2012.

Section IV

Business Opportunities, Management and Services, and Web Applications

16

Delivering eHealthcare: Opportunities and Challenges

**Deborah A. Helman, Eric J. Addeo, N. Iwan Santoso,
David W. Walters, and Guy T. Helman**

CONTENTS

16.1 Introduction

eHealth refers to the delivery of healthcare services remotely, e.g., via the Internet. These services are delivered through smartphones, medical devices, apps, and cloud computing. The growth of eHealth has implications for a range of industries and stakeholders. In this chapter we explore eHealth as a solution to the problems posed generally by rising healthcare costs and specifically by aging populations. Focusing on implementation, we argue that synergies can be derived from a convergence of technology, patient-and-doctor relationships, and regulatory and business drivers, and explore how the eHealthcare value-chain network, as a business model, can be configured to deliver both feasible and viable eHealthcare services to the end user.

The value chain network model enables us to address the practical problem solving required for effective implementation of eHealthcare. We link the value drivers of performance management, time management, and cost management to value delivery through the value proposition. Thus, ICT eHealthcare applications make the whole delivery process efficient and effective and provide greater value capture through competitive advantage.

A survey of the literature indicates a number of challenges; the first relates to nomenclature and the second to the vast scope of the field. We have adopted the term *eHealthcare* or *eHealth* and use this to describe the delivery of healthcare services via the Internet—there are numerous other terms in use, such as *telehealth, telemedicine, mobile health, digital health, connected care, EHRs,* and *PHRs* (personal health records). This is confusing, but temporary, as the *e* becomes an unnecessary appendage. The second point relates to the explosion of the field of literature encompassing everything related to combinations of medicine, technology, and business. eHealthcare extends to many stakeholders, disciplines, methodologies, and issues. And as a consequence, the field of literature has grown like topsy-turvy—it is highly fragmented and specialized with entirely separate terminology belonging to the disparate disciplines. To that end, in this chapter we attempt to address multiple stakeholders' interests and focus on the very practical problem of the implementation of eHealthcare. This chapter begins by examining the context of the evolution of healthcare and is followed by a description of the value chain network model that we argue should underpin any business model that is developed in the pursuit of the effective delivery of healthcare. We examine some specific value drivers of eHealthcare that result in adoption and the challenges resulting in nonadoption, provide case studies, discuss the technology implications, and conclude with the directions that we anticipate being of consequence in the future.

16.2 Context: The Evolution of eHealth

"We are already designer humans living longer, better lives than any humans before. Technology allows us to avoid the lottery of nature—we should choose our destiny" [1]. Savulescu's [1] observation relating to designer babies might seem obvious to some and morally challenging to others; however, what it does not take into consideration is the rather prosaic problem of integrating certain kinds of technology into medicine and making this politically palatable and a viable value proposition from a business perspective.

One perspective on the evolution of eHealth is to view it from the perspective of the three revolutions that have impacted us from midway through the 20th century and onward.

Phenomena such as eHealth have arisen as a result of the interactions between structural, expressive, and information technology revolutions, producing changes in inter- and intraorganizational power and exchange relationships and changes in conceptions of time and space and the nature of purposeful activity. As a consequence, the patient-consumer desires more control in negotiating relationships with healthcare providers and demands access to information, products, and services anywhere, anytime, and with security. And, of course, healthcare providers, the individual physician, medical groups and hospitals, insurance companies, pharmaceuticals, and governments are to varying extents subject to these forces as well and are responding to the challenge of the empowered patient-consumer. The value of the three revolutions' perspective is to help us think through the big changes we have experienced and how these changes have impacted consumers and organizations—it helps us reflect back on how change occurred and anticipate the patterns of future change that will shape healthcare and the business models required to deliver this care. A patient-centric approach has emerged in the eHealth domain [2]. Additionally, Wilson [2] observed the need to develop business models based on experiences in the financial services sector, in the absence of appropriate models in healthcare.

The most significant change arising from the interaction of the three revolutions that we have experienced is the advent and growth of the Internet. As a consequence, consumers now inhabit a variety of electronic landscapes [3]; an adapted model of the e-landscape is shown in Figure 16.1. This enables us to visualize the synergistic intersections of consumer demand, business, and technology and the opportunities that arise from these interactions. If we focus on the consumer as patient, the patient-consumer, we can anticipate the nature of demand as consumers navigate through their everyday lives in the various

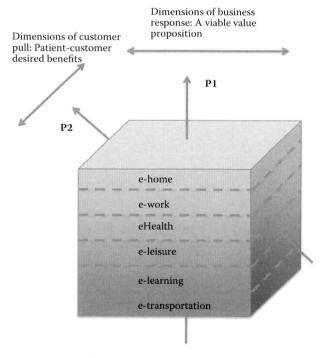

FIGURE 16.1
The dimensions of customer pull and business responses in the e-landscapes.

e-landscapes they traverse. The smartphone has emerged as the ubiquitous platform through which multiple tasks can be performed and integrated into the pattern of everyday life. The pathway taken by the patient-consumer at P1 suggests a patient-consumer who is fully integrated in terms of devices and networks. The patient could be suffering from diabetes but is able to seamlessly connect from mobile platforms as well as wired connections to healthcare providers, pharmacies, and home from work or the car to exchange health-related data, adjust diet and medications, and fill new prescriptions. By contrast, the patient at P2 is less well connected and, therefore, less integrated and must rely on other means of communication and will experience less efficiency and higher costs in meeting the demands of his medical condition. A business model must be developed with the patient-consumer needs as the central organizing concern, to which the organizations' capabilities and resources are matched in the e-landscapes. This is particularly critical in relation to the opportunities for collaboration that will be required to effectively deliver these complex product–service combinations that are required.

16.2.1 The Multidimensional Landscape of Healthcare Delivery: Associated Driving Forces

Focusing on the eHealth layer in the e-landscape, we can consider the multiple stakeholders involved in delivering healthcare. The model in Figure 16.2a indicates a response to the rising demand for healthcare and rising costs, and we can see opportunities for alternative online solutions to emerge (examples of which are teleradiology companies vRad and NightHawk). Among the stakeholders we encounter both enabling forces or drivers and resistances. The various forces operating in concert or against each other will reconfigure the landscape, and over time, a more sustainable state could emerge.

The landscape of healthcare delivery and associated driving forces can be characterized in terms of (a) quality, (b) cost, and (c) the balance of *online–on-site* healthcare delivery. However, the relative weight of each of these factors is dependent upon the rate of adoption and implementation of enabling technologies into the healthcare delivery process. The driving forces considered in this model include (a) cost reduction, (b) quality improvement, (c) physician resistance/readiness, (d) patient resistance/readiness, (e) enabling technologies, and (f) business opportunities. Regulation, law, and politics (e.g., malpractice law and tort reform) may act as constraints and indirect driving forces. The combined forces can be defined as a resultant (or composite) vector $\mathbf{F}_t$ as follows:

$$\mathbf{F}_t = \mathbf{W}_1 \cdot \mathbf{F}_{\text{cost}} + \mathbf{W}_2 \cdot \mathbf{F}_{\text{quality}} + \mathbf{W}_3 \cdot \mathbf{F}_{\text{Physicians}} + \mathbf{W}_4 \cdot \mathbf{F}_{\text{Patients}} + \mathbf{W}_5 \cdot \mathbf{F}_{\text{Technologies}} + \mathbf{W}_6 \cdot \mathbf{F}_{\text{Opportunities}} \quad (16.1)$$

16.2.2 Feasible Models

While the possibilities of feasible evolution paths are essentially unlimited, we will evaluate some likely scenarios. There are many ways to represent the relationship between the variables we have selected to visualize this as the relationship between costs, quality, and technology adoption into eHealth in general and telemedicine in particular. Of course, in our analysis only one variable is changed, while the rest of the variables are assumed to be constant:

1. The first general model, Figure 16.2a, shows a feasible landscape where all forces exist in a rather balanced fashion. $\mathbf{F}_t$ will serve as the prime mover in the evolution of the current healthcare system (within the boundaries of the current constraints)

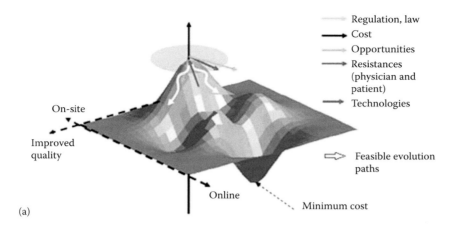

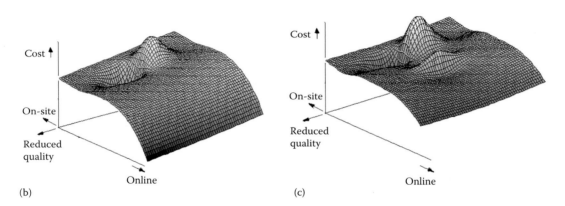

FIGURE 16.2
The landscape of healthcare delivery: Cost vs. driving forces and resistances; (a) when all forces are relatively balanced; (b) when all forces are synergistically supportive; and (c) when all resistances are more dominant than the need to lower healthcare costs.

toward a new equilibrium state. Many new equilibrium states, as shown in Figure 16.2, may emerge as solutions. Each offers different benefits, challenges, and opportunities to healthcare providers.

2. The second model, Figure 16.2b, shows a landscape where all factors are supporting the move toward telemedicine. There is minimal resistance from physicians and patients; governments and related regulations are supporting the implementation of eHealth/telemedicine and enough opportunities keep funding the evolution. With the proper implementation and structure, we expect eHealth/telemedicine will reduce costs significantly until it reaches its new equilibrium. Naturally, we will expect the quality will be slightly reduced until better diagnostic technologies become available and/or cheaper.

3. The third model, Figure 16.2c, shows a landscape where resistances are more dominant than the need to lower healthcare costs. Immediate benefits/savings of moving toward eHealth cannot be immediately seen; only with significant adoption of

eHealth/telemedicine does cost reduction (as well as other benefits) becomes discernible. Regulations are not favorable for the implementation, and adoption of the new technologies into eHealth/telemedicine is slow. Other measures may provide more immediate savings and manifest as a dip on the left of the peak.

In general, the landscape is less accepting of the advancement of eHealth/telemedicine. The system is resisting the changes. Only a significant increase in F_t will drive the system over the hurdles before it can provide any cost–benefit advantages.

16.2.3 Physicians' and Patients' Resistance and Readiness

The most complex driving forces in the evolution of healthcare delivery are physicians' and patients' resistance and readiness to accept a "different" way of practicing medicine. On one side the physician, as a patient advocate, is eager to implement various aspects of eHealth; however, at the same time, they are reluctant to change beliefs that online services may depersonalize the patient–physician relationship. This coupled with the absence of information usually obtained during clinical evaluation deprives the physician of some important decision-making inputs used in the diagnostic process. There are concerns that if a patient chooses to be consulted by the quickest available physician (online), this will undermine the importance of patient history and will reduce the quality of care.

In the United States, barriers to implementation also exist in relation to payment and regulations. Existing health insurers offer different medical coverages (provider list, pharmaceutical, imaging, laboratory, other procedures, and fee schedule); coordinated care requires full-time monitoring, which, in turn, increases the costs significantly. Many regulations also contribute to healthcare costs, such as the stringent controls of insurers, HIPAA, loosely defined payment schedules, and legal responsibility for practicing telemedicine. On the other side of the ledger, enormous potential for cost savings exist:

- Eighteen percent of all children/adolescents have special healthcare needs; these needs represent 80% of all healthcare expenditures for this group.
- Five percent of the U.S. population are chronically ill and this group accounts for 50% of healthcare costs.
- Using eHealth/telemedicine, patients can actively participate and manage their own health needs (commensurate with their cognitive ability), which, in turn, will increase cooperation and compliance; thus, fewer medical complications arise and less hospitalization or direct medical care is required.
- Populations will increasingly become more technology savvy and more familiar with electronic communication/interaction.
- The administrative costs of healthcare will be reduced.
- The medical provider's overhead costs will be reduced.

Implementation of eHealth and telemedicine is only possible with a solid EHR infrastructure. While logically, these new technologies can only improve and reduce the cost of healthcare delivery, the implementation outcome may differ between countries. The implementation of eHealth in the United States has been difficult and expensive without any clear immediate outcome [4]. Some determining factors are the initial cost of implementing the technology, technology updates and glitches which add to the total implementation cost, nonstandard insurance claim systems, different payment schedules between insurance carriers, and the economically prohibitive cost of high-security technology. Many of the government measures for improving quality and reducing cost, such as the "Meaningful Use" regulation, cause short-term cost increase to the providers. Meaningful Use is one of many government regulations introduced in recent years to improve patient care. To receive an EHR incentive payment, providers have to show that they are meaningfully using their EHRs by meeting thresholds for a number of objectives determined by Centers for Medicare and Medicaid Services (CMS) [5]. Satisfying more than 30 objectives of Meaningful Use reduces efficiency, which, in turn, increases cost.

Maintaining a secure EHR with connection to various ever-changing platforms has proven to be quite expensive and complicated; the healthcare provider is fully responsible for any occurrence of a security breach at the provider's office. A current survey [4] estimates that 54% of the EHR users are not satisfied with their system interconnectivity. Furthermore, there is a dismal effect on doctor–patient relationships, with diminished productivity, efficiency, and revenue generation while contending with the high cost of EHR adoption, estimated at $5000–$50,000 per physician.

Lack of standardization between American states adds constraints to servicing patients outside the state where the physician practices since a medical license is issued by each state. Each state has different rules about the practice of telemedicine, malpractice insurance coverage, and prescription limitations. Payment for telemedicine services is also not currently well defined. All of these add significant resistance to the adoption of eHealth practices.

While some states have laws that cap specific types of compensatory damage awards in medical malpractice cases, many loopholes remain. Other states still do not have any means to control the frivolous medical malpractice lawsuit. This legal climate forces the physician to practice "defensive medicine," which generates a large number of unnecessary medical procedures, which are not in the patient's best interest. Together with the legal cost, defensive medicine healthcare costs the United States significantly.

We have examined the evolution of eHealth with reference to change on a revolutionary scale, but before we move on to our consideration of business models, we should consider further the minutiae of everyday life. The patient-consumer armed with devices and connected to numerous networks has the opportunity to shape their personal destiny with technology. Three driving forces underpin the growth of eHealth [3]: *digitalization*, driven forward by Moore's law, provides cheaper, smaller, faster devices and digital content; and *expanding networks*, following Metcalf's law, where network value grows with network size. These also include *individual interactivity*, fueled by the attraction of novelty, growing expertism and user fascination, the desire to manipulate time and space, and the desire to upset the traditional balance of power governing ownership and access to information. Focusing attention on the domain of individual activity or, rather, interactivity, we find that the activities of everyday life for the aging patient have been summarized in a model of social performance levels, developed pre-Internet [6]. Over time and with illness, the patient's ability to maintain these daily activities is undermined. The task of healthcare providers is to extend support to maintain the rings and prevent hospitalization for the long term; it is immediately apparent that the use of the devices and networks immediately alters the ability of the patient or caregiver to impact the outcome here. In developing a viable business model in eHealthcare, we need to begin with an understanding of the importance of routine and the activities of daily life, as well as the medical conditions and metrics in current use related to disease, fitness, and wellness.

"Avoiding the avoidable," in terms of illnesses and managing microevents, has been identified by the Oregon Medical Laboratories as an approach to lifestyle management where the individual enabled with technology is able to defer the onset of major illness and so reduces the costs of healthcare over a lifetime by between 60% and 80%. These cost savings can be partially realized through the use of devices and applications. In marketing, as in medicine, the focus is currently on personalization and utilizing vast databases to target markets of one. Topol has addressed the issue of digitizing humans and the capabilities of science to address the medical needs of the individual: the science of individuality [7].

If we survey the marketplace, it is apparent that healthcare is expanding well beyond the traditional delivery mechanisms and institutions. This is in line with the idea of the "diagnosis difference" tracked by Pew Research. Pew Research has been exploring the evidence associated with the notion that health outcomes improve with diagnosis and self-measurement. Following a diagnosis, individuals actively seek out information on their condition on the Internet [8].

Tracking health data is not a new phenomenon. Englishman Ian Clements has recorded 400-plus measures per day since 1974. The technology-driven Quantified Self movement founded by Gary Wolf and Kevin Kelly, from the *Wired* magazine, began in 2007. The movement promotes self-monitoring by using technology to capture data related to every-day life measures of inputs (food), states (mood), and performance (physical activity). Data are captured with wearable technology that typically relies on smartphones and sensors. Devices track physical activity, calorie intake, sleep, posture associated with wellness, and track activity at work to increase productivity and are also used to track an infant's prog-ress: the quantified baby (see quantifiedself.com). The goal is to capture data for analy-sis to improve performance levels. Activity-tracking devices currently in the marketplace include Fitbit, Nike+ FuelBand, Jawbone's UP, and the Misfit Wearables device Shine. The Quantified Self movement provides advice regarding fitness and health tracking and builds an online community among devotees interested in the following:

- Chemical body load counts
- Personal genome sequencing
- Life logging
- Metabolic monitoring
- Self-experimentation
- Behavior monitoring
- Location tracking
- Sleep tracking
- Mood and emotion tracking
- Medical self-diagnostics

16.3 Delivering eHealthcare: Practical Applications

A range of business models for delivering eHealthcare are emerging. The complexity of these models varies with the focus of the business activity and the scale of the operation. At one end of the spectrum, there are eHealthcare-related start-ups producing medical devices and apps, many of which are implementing a lean start-up model [9] and at the other end of the spectrum entire healthcare systems that are integrating eHealthcare into operations. And in between an extensive range of pilot studies and experiments, general and specialized services are delivering eHealth remotely.

16.3.1 Children's National Medical Center: Specialized Services

eHealthcare is an innovative technology that can and should be pursued to bridge the gap between expert opinion and practitioners with little experience in rare, complex neurologic disorders. Pilot studies in medically underserved areas have shown improvement in clinical outcomes and this could be the basis for addressing the regulatory issues that have been a barrier to implementation in the United States [10]. Recent use of telemedicine in neurodevelopmental disorders in the United Arab Emirates in conjunction with a number of site visits provided clinical and multidisciplinary expertise in complex cases, as well as diagnostic workup, management recommendations, and procedural interventions [11]. This was also an invaluable method of providing educational conferences for providers to develop local expertise to support patient care.

The study presented by Pearl et al. [11] demonstrates the reliability and clinical effectiveness of multidisciplinary care incorporating telemedicine in neurodevelopmental disabilities. The program was funded by the United Arab Emirates government and was implemented through a two-step process involving site visits for initial assessment of healthcare needs and curriculum development for healthcare workers, followed by weekly videoconferencing sessions. Forty-eight diagnoses were found in 48 patients (some patients had overlapping primary diagnoses) covering a spectrum of neurogenetic, neuromuscular, and epileptic syndromes. Specific management recommendations and training for the implementation of these strategies was provided by videoconferencing. Additionally, recommendations in diagnosis and therapy through the multidisciplinary team assembled allowed the complex care paradigm to be addressed.

A number of eHealth tools are designed to be instrumental in preventive medicine and lifestyle management. Its use in neurologic conditions, where many of the underlying genetic bases is unclear or unknown, provides an impetus for improving many technologies associated with lifestyle management. Complex care of these patients must find ways to involve multidisciplinary tools and improve the consultation tools for patients in medically underserved areas.

16.3.2 Kaiser Permanente: A Healthcare System

Kaiser Permanente (KP) as an integrated managed healthcare consortium, operating in the United States, provides an example of a healthcare system that has moved toward full adoption of eHealthcare. KP is among the country's largest not-for-profit health providers, with 9.1 million members, of which 7 million members reside in California. KP has 38 hospitals in nine states and the District of Columbia. Physicians number 16,942; nurses, 48,701; and employees, 174,259. Kaiser's revenue has increased steadily over the past 5 years, from $40.3 billion in 2008 to $50.6 billion in 2012 [12].

Between 2004 and 2010, Kaiser implemented a nationwide comprehensive EHR system referred to as KP HealthConnect. The patient portal is integrated with KP HealthConnect and empowers patients to manage their own healthcare, such as the ability to securely e-mail doctors. For KP, the evolution of the integrated healthcare model was realized through the adoption of EHR. As a consequence of the investment of $4 billion in building an EHR system, patient data are made available at all Kaiser hospitals and offices and the data collected creates a massive database for driving quality of care initiatives that includes preventive care programs, metrics concerning practitioner performance, and productivity and patient satisfaction. The system is designed to provide decision support and alerts to practitioners and patients, e.g., screening procedures and follow-up appointments. As well as guidelines and information for aiding diagnosis, data are captured to track side effects of medications, to the extent that the problem with the arthritis medication Vioxx was identified by the Kaiser system [12].

The Kaiser approach to the EHR and the broader health information technology (HIT) system provides significant cost savings and convenience; it enables at least a third of all patients who have adopted the technology access to their medical data and the ability to communicate with doctors via secure e-mail, and as a consequence office visits have decreased by a quarter. Patients have direct access to laboratory results. It is estimated that 100,000 patients connect with the system on a daily basis. A major flaw in this system is that it does not communicate well with non-Kaiser systems; it is a closed-loop system. However, efforts are underway to link more efficiently with external institutions [7]. There are scant data available on the effectiveness of patient-to-doctor e-mail communications as a realized benefit of integrated healthcare systems. However, a 2010 Kaiser study of diabetes and hypertension patients indicated improvements for patients using this communication tool. The integrated Kaiser system overcomes a key barrier to adoption by doctors of eHealthcare that relates to reimbursement for these forms of communications [13].

16.3.3 Misfit Wearables: Start-Ups

Misfit Wearables produces wearable technology; the company is based in San Francisco and has manufacturing facilities in Korea and Vietnam. The following is based on a presentation given by one of the company's founders [14]: The company is a product of evolution through three start-ups, which has critically shaped the organization. As a consequence, the company philosophy and business model is underpinned by three key ideas: the ability to adapt or pivot, the ability to focus on users and products, and the culture. FireSpout, the initial start-up business failed, but the team learned how to productize and the importance of the lean approach to start-ups. The second company, AgaMatrix, successfully developed blood glucose sticks and a device that could be used with the iPhone. The team learned how to prototype, develop business plans, and attract finance through a demonstration of skills and integrity. As the company evolved, the team learned how to solve product-related problems and identified where the profit potential was, and this led to a series of adaption/pivots, or changes in direction. The team recognized the need to outsource production, to get help with developing the consumer brand, and to find partners and effectively became an integrator. AgaMatrix attracted huge success as it was launched, largely because it was able to benefit from the huge resources of its partners (Apple, Sanofi, and Walgreens). The team learned the value of having an obsessive focus on user experience and essentially redesigned the product category (glucose monitors) by developing a deep understanding of the product in use; the product needs to be easy to use. The company began with focus groups as a research method but discovered that this did not provide meaningful insights—the team discovered the importance of being the user and developed an anthropological approach that gathered data on the product in use. By distributing cameras among diabetics, the team discovered aspects of the user experience that needed to be addressed to create ease of use. One manifestation of this was to make zippers on the monitor cases easy to open. The team learned to gather the data and then to respond with rapid prototyping.

The third company, Misfit Wearables, has benefited from the foregoing learning experiences, and in the process of developing wearable technology, the company identified two important criteria: ease of use and motivation. The company discovered that consumers are more likely to adopt a product if those criteria are satisfied. In terms of ease of use of wearable technology, the team learned that the product must be either beautiful or invisible; products that do not meet these criteria are not purchased or used. In addition, the company holds a new product to the "turn-around test"—would a customer go home for it if they forgot it? The company has developed an activity-monitoring device called Shine, a small disk made out of aluminum that tracks activity and is used in conjunction with the iPhone. The device can be worn anywhere and anytime—even when swimming or at a formal event. The team determined consumer interest in the product by posting the device on a crowdfunding site, Indiegogo, and immediately sold 8000 devices that were not even in production at that point in time.

16.4 The Value Chain Network Business Model

Since the marketing concept supposedly replacing production-driven management business models emerged in the 1960s, the customer has been seen as at the core of business processes and customer satisfaction the key objective. The creation of value has then taken on strong customer-centric perspectives together with awareness that other groups should be considered. The notion of stakeholder management suggests that the interests of all participants within a transaction process are considered. The view of the business model as a networked sequence of value-producing interactions between purchasers and vendors and the interests of investors (shareholders), local communities, industry associations, and government regulators (e.g., an international trade bloc) has also emerged. The value chain network (as it is known) offers a number of advantages to business organizations; if all of its features are used, it can create increased value for end-user customers. These value benefits may be categorized as increases in performance, time, and cost; these are available because networked businesses can specialize in products and services in which they have specific expertise. Many organizations operating within value chain structures are quite small in terms of number of employees, but they are known internationally because their specialist products or services offer unique characteristics. A full account of the evolution of the value chain concept is given by Walters and Rainbird [15].

The value chain network business model differs from the traditional approach to business decision making in a number of ways. The first is the emphasis on the end-user customer as a focal point and the assumption that they are purchasing results rather than products or services; the approach facilitates a closer, more detailed attention to specific customer expectations. Access to partners' expertise reduces the investment required to meet customer specificity; this is particularly evident in capital goods markets, for example, aerospace, where the large manufacturers Boeing and Airbus contract with their suppliers to design, as well as manufacture, complete systems that are incorporated into their products. Often innovative approaches emerge; recently the aeroengine manufacturers introduced the concept of procuring "output" rather than hardware (the engines), arguing that this offers a considerable investment reduction, freeing up capital for use elsewhere in the customers' businesses. The recent, and ongoing, developments in ICT have facilitated performance, cost, and time-management benefits. *Performance management* is enhanced by digital control systems that monitor and manage quality and quantity on an ongoing basis, anywhere. Manufacturing management is now possible using ICT systems on a global 24-hour basis; control centers manage satellite production facilities close to markets and have built-in facilities for responding to configuration differences in the local markets. *Cost management* is becoming increasingly possible with the use of robotics. Many United States and European businesses are now able to repatriate outsourced manufacturing from offshore locations, thereby reducing quality control and inventory costs and increasing the overall control of manufacturing and distribution. The application of ICT to business operations management offers benefits where *time management* is essential, often critical. A Qantas A380 experienced an engine fire soon after departure from Singapore. The remote diagnostics system notified the airline management and Rolls-Royce Engines, the manufacturer, at the same time as it indicated the problem to the aircrew; this ensured that all possible required expertise was brought into focus and effect a safe solution.

Possibly the most significant change that has occurred has been that of being able to consider and evaluate alternative solutions to strategic and operational problems. Knowledge, technology, and process-management applications have changed not only management thinking and applications but so too the ways in which businesses manage relationships between internal and external organizations. A "lexicon" of types of relationships has evolved and in doing so has broadened the interactions between these entities. Concepts such as *cocreativity* (suppliers working with customers' design and development staff), *coproductivity* (sharing operations tasks with suppliers and customers), and *coopetition* (using competitors' manufacturing, distribution, and selling capabilities and capacities) have become common management parlance and practice.

Before exploring the capabilities of the value chain model in the healthcare context, we should discuss the importance of establishing a workable view of value.

16.4.1 Value, Value Drivers, and Value Propositions

Anderson and Narus [16] suggested that the successful suppliers in business markets are successful because they have developed *customer value models,* which are data-driven representations of the worth, in monetary terms, of what the supplier is doing, or could do, for their customers. Customer value models are based on assessments of the costs and benefits of a given market offering in a particular customer application. *Value* is defined as follows [16]:

> Value in business markets is the worth in monetary terms of the technical, economic, service, and social benefits a customer company receives in exchange for the price it pays for a market offering.

Value is expressed in monetary terms. Benefits are net benefits; any costs incurred by the customer in obtaining the desired benefits except for the purchase price are included. Value is what the customer gets in exchange for the price it pays. Anderson and Narus add an important perspective concerning a market offer [16]. A market offer has two "…elemental characteristics: its value and its price. Thus raising or lowering the price of a market offering does not change the value such an offering provides to a customer." And, finally, value creation takes place within a competitive environment; if no competitive alternative exists, the customer always has the option of making the product rather than buying it. This serves to remind us that value as a concept has two perspectives—it is an important transaction consideration suggesting that purchasers have product–service preferences, which if ignored may result in a competitor satisfying their needs; it is also critical to the vendor who also has expectations concerning the longevity of his business and the profitability and cash flow required from transactions to ensure long-term survival.

The Anderson and Narus model [16] identifies *benefits* sought and acquired by the customer and acknowledges the fact that customers are confronted with *acquisition costs*. The customer will reach a purchase decision based upon *net benefits* that are the optimal gap (for them); between the benefits, they receive *less* the costs involved in their acquisition. Clearly, this varies by customer, or perhaps market segment, and the role of service support becomes an important consideration. While some customers may rely heavily on service support such as installation, staff training, and maintenance, others may prefer to be able to purchase only the product because they have an adequate internal service infrastructure. This is often referred to as "bundling" and "unbundling"—both product and support services are made available as a product–service package or simply as a product-only option. At first this may appear irrelevant in healthcare but a second pass at this will suggest there to be no difference—healthcare is a transaction between user and supplier (and possibly a number of intermediaries)—and as a transaction it shares the product–service characteristics of other, nonrelated product–service categories: the difference is one of emphasis.

The notion of a customer value model has been addressed by a number of authors. Heskett et al. proposed a *customer value equation* (Equation 16.2), which, in addition to customer benefits and acquisition costs, also includes process quality and price [17]. The model is described by

$$\text{Value} = \frac{\begin{array}{l}\text{Results or benefits} \\ \text{produced (\textit{Value-in-Use})} \quad + \quad \text{Process quality} \\ \text{for the customer}\end{array}}{\begin{array}{l}\text{Price to the} \\ \text{customer} \quad + \quad \begin{array}{l}\text{Costs of acquiring} \\ \text{the product-service benefits}\end{array}\end{array}} \qquad (16.2)$$

In Heskett et al.'s model, *value produced for the customer* is based upon results, not products or services, that produce results—the actual value benefits delivered and the costs of acquiring the benefits. Identifying **value in use** is a helpful approach; the notion that an end user should consider all aspects of a product–service purchase, not simply the price to be paid, enables both vendors and purchasers to identify all of the elements of the procurement–installation–operation–maintenance–replacement continuum. The process encourages both parties to look for trade-off situations such as high acquisition costs with low operating and maintenance costs, together with relevant supplier services packages. This approach introduces the possibility of integrated activities in which the supplier–customer relationship expands from a one-to-one relationship into a fragmented, but economically viable, value-delivery system. A number of companies use the value-in-use concept to arrive at a **value proposition**. Another benefit of value-in-use offers to vendors is the possibility of introducing alternative value production and delivery processes—an alternative business model.

Before accepting Heskett et al.'s model as a basis to build upon, some other considerations are required. We have seen a number of important considerations take on increasing importance in the business environment. Healthcare is a business; this is more obvious in some countries than it is in others—those without a social welfare system that includes a measure of state provided health services—and within which an important partner/supplier is the health insurance provider. Both sectors are cognizant of the impact of their role on the socioeconomic fabric of the environment within which they operate. Accordingly we should modify Heskett et al.'s model [17] to include this increasingly important consideration; see the following socially responsible value equation:

Value =

$$\text{Value} = \begin{array}{ccc} \begin{array}{c}\text{Results or "benefits"} \\ \text{produced (\textit{Value-in-Use})} \\ \text{for the customer}\end{array} & + \quad \text{Process quality} \quad + & \begin{array}{c}\text{Social and economic} \\ \text{benefits received}\end{array} \\ & \textit{less} & \\ \begin{array}{c}\text{Price to the} \\ \text{customer}\end{array} & + \quad \begin{array}{c}\text{Costs of acquiring} \\ \text{the product-service} \\ \text{benefits}\end{array} \quad + & \begin{array}{c}\text{Opportunity costs} \\ \text{of alternative social and} \\ \text{economic benefits forgone}\end{array} \end{array}$$

(16.3)

The model now includes consideration of the social and economic benefits received as a "positive" consideration; here we might consider increased economic productivity from a labor force that is provided with accessible and affordable healthcare. The acquisition costs are increased by opportunity costs, created by the loss of contribution from alternative projects, which could not be funded and that may have made significant, but alternative, contributions. Figure 16.1 reflects the more recent view of the business environment. In a business context, value now considers stakeholder satisfaction, which is a broader consideration than simply customer satisfaction. Stakeholder satisfaction ensures meeting not only customers' expectations but also those of employees, suppliers, shareholders, and the investment market influencers, the community, and the government.

16.4.2 Identifying Value Drivers

Understanding the importance of **value** to customers and other stakeholders helps strengthen relationships between, and among, customers, suppliers, shareholders, and investors and an organization, as these **value-based relationships** are the link an organization needs if it is to develop a strong competitive position. To do so it clearly needs to identify the **value drivers** (and **value builders**) that are important to the end-user customer to structure a value-delivery system that reflects these *and* those of the other value-chain participants. In applying Slywotzky and Morrison's **customer-centric approach** to the value network/value chain, it is apparent that the things that are so important to customers are the customers' *value drivers* and the important value drivers are those adding *significant value* to customers [18].

Phelps approaches value from a corporate performance perspective [19]. He argues that it is insufficient to simply measure outputs to know if we are creating value; the value drivers of present and future value must also be measured. Measuring output indicates success (or perhaps lack of it), whereas understanding (and measuring) what it is that drives value provides management with an indication of the success of resource allocation. Phelps also argues that it is important to distinguish between factors that drive **current value** (value drivers—suggesting cost reduction as an example) and those responsible for **creating future value** (value builders—such as brand development and research, design, and development). He makes the point that overlap may occur; value drivers may well contribute to building both current and future values. He considers a generic perspective of value drivers to be *performance management, cost management,* and *time management*: from these it is possible to construct a detailed and focused **value driver profile**.

Vendors as well as their customers have value drivers. For vendors they are based upon the longevity of the business and the profitability and cash flow required from transactions to ensure long-term survival; clearly these can be mapped onto the generic value driver model. Here it is important to make a significant point: vendor and customer value drivers are connected. The vendor must identify the target customer value drivers; not to do so jeopardizes any success of generating the necessary cash flow and profitability required for long-term success, employees and suppliers could not be paid, nor could interest and dividends be paid to the investor owners. A moment's thought will identify the relationship of these factors to the generic value drivers. Figure 16.3 may help explain this relationship in the healthcare sector.

Figure 16.3 illustrates the process of identifying *critical customer healthcare value drivers* and their transformation into an implementable *value proposition response*. The process would require a review and analysis of all available options to ensure that the *performance, cost*, and *time expectations criteria* are achievable at an acceptable cost. Friedman describes a number of global ICT partnerships in which X-ray and bone scan data are evaluated overnight by offshore services that are thousands of kilometers distant [20]. The value proposition response identifies and details the suppliers' understanding of customer expectations of its target market; furthermore, it serves to identify partners' input requirements if the value proposition is to be delivered successfully. Scott reminds us that creating value creates cost and, therefore, is an essential component of the value-creation process, ensuring that the value proposition is economically viable [21]. An evaluation of the value-delivery costs, e.g., critical healthcare supplier cost drivers, is an essential task, one that should be completed during the development of the value proposition response. Clearly the **value proposition** is a critical component of the overall **value-delivery process** and requires further discussion.

16.4.3 The Value Proposition

Satisfying customer expectations requires an understanding of their product–service purchasing and applications activities and a *proposition* that spells out a response to these. Webster contends that positioning and the development of the value proposition must be based on an assessment of the product offering and of the firm's distinctive competencies *relative to competitors* [22]. Hence, the **value proposition** should make clear its *relative competitive positioning*. In doing so, it should communicate to the target customer the distinctive competence portfolio of the value chain network participants, demonstrating that it extends its collective skills and resources beyond the current dimensions of competitive necessity into creating competitive advantage that, in turn, offers customers an opportunity to do likewise.

A vital component of the value proposition is the role it performs in identifying the *roles and tasks of partner organizations*. Here the issue is one of communication to both partners and customers. If the roles and tasks are made explicit, this serves to create credibility for the value chain within the eyes of the customers and increases their confidence in dealing with the organization. Four aspects should be considered: cooperation, commitment, coproduction, and coopetition. *Cooperation* implies an agreement within the value chain structure that an ongoing *commitment* to improving product–service offer and value production and delivery processes exists, and that this is prosecuted for the benefits of both customers and partners. *Coproduction* seeks to identify where the production process is most effectively conducted. The IKEA example used by Normann and Ramirez is widely published and is an excellent example of dispersed production across distributed assets to

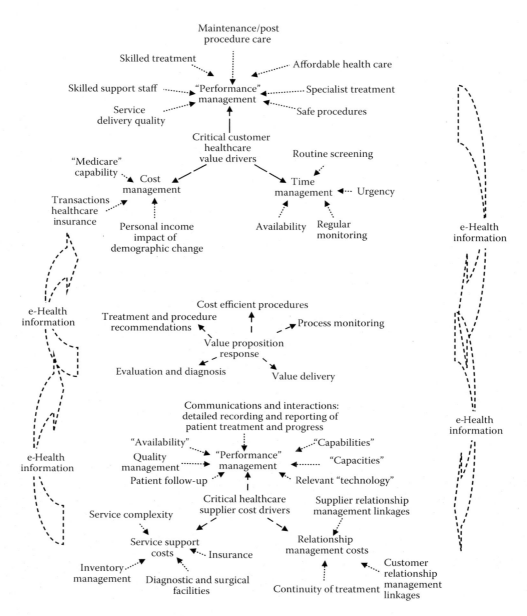

FIGURE 16.3
Managing healthcare interactions using eHealth information technology.

the benefit of all stakeholders. *Coopetition* occurs when competitors work with each other, sharing capability and capacity resources [23].

Anderson et al. suggest four considerations when creating a value proposition [24]:

1. Understand the customers' businesses: Invest in time and effort to understand the customers' businesses and identify their specific procurement needs and product–service applications. The authors provide an example of a resin manufacturer that

enrolled managers in courses where they learned how painting contractors estimate jobs. Focus groups and field research studied product–service performance
on *customer critical criteria*. It asked customers to identify performance trade-offs
they were willing to make and their willingness to pay for product–service characteristics that enhanced performance. It also participated in industry associations between manufacturers and distributors.

2. Document value-delivery success: AkzoNobel (a chemical manufacturer) conducted a 2-week pilot study on a production reactor at a prospective customer's
   facility in order to study the performance of its high-purity metal organics product relative to the next best alternative in producing compound semiconductor
   wafers. The study proved that AkzoNobel's product was as good as, or better,
   than the rivals' *and* that it significantly lowered energy and maintenance costs.

3. Substantiate value claims: Performance and economy claims require to be proven.
   Rockwell Automation calculated the cost savings from reduced power usage that
   would result from the use of their pump solution compared with that of a competitor.

4. Make the customer value proposition (and your company's supplier implications)
   a corporate business skill: Quaker Chemical conducts a value proposition training
   program annually for chemical program managers. The managers review case
   studies from industries the company serves and participate in simulated customer interviews to gather information needed to devise proposals. Incentives are
   awarded to the managers who produce the most viable value propositions.

In relation to healthcare, the notion of the value proposition is concerned with satisfying
patient expectations, making explicit the roles of partners, and establishing the credibility
of the value chain entity (the connected healthcare system).

16.4.4 Managing in the Value Chain Network

John Del Vecchio, a financial/educational raconteur writing for Fool.com, suggests that a
value chain is "a string of companies working together to satisfy market demands" [25].
The value chain typically consists of one or a few primary value (product or service) suppliers and many other suppliers that add value to the primary value that is ultimately
presented to the end-user purchaser.

Microsoft and its Windows operating systems, the nucleus of the personal computer
desktop for which much business software is developed, is often cited as a prime example
of a company and product that drives a value chain. The businesses who buy personal
computer software may spend far more on the add-on software than on the essential operating system that is the de facto standard for running the software. To the extent that companies standardize on Windows, Microsoft is said to control a value chain. This particular
value chain was reported in a McKinsey study to be worth $383 billion in 1998. Although
Microsoft's share of the value chain was reported to be only 4% of the total, that was still
$15.3 billion. A company that develops a product or service that engenders a value chain
by providing a platform for other companies is considered more likely to increase its market share than a company that tries to provide the entire value chain on its own. Figure
16.4a shows value chain activities and processes. Figure 16.4b demonstrates how, during
its development stages, alternatives are identified and evaluated to ensure that not only
cost-effective alternatives are identified and evaluated but that they are organizationally

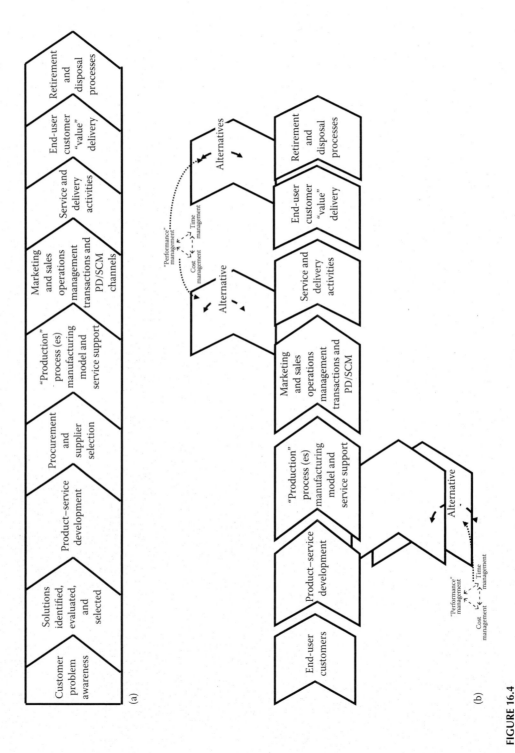

FIGURE 16.4

The value chain business model: (a) value chain activities and processes and (b) a "working model" in which alternatives are evaluated to ensure that customer value drivers can be met.

compatible and *fit* the culture and management style of the partners that are creating the value delivery system.

16.4.5 Value and the Value Chain

Slywotzky and Morrison introduced the term *customer-centric thinking* [18]. They consider the *traditional value chain*, which begins with the company's core competencies and its assets and then moves to consider other inputs and materials, to a product offering through marketing channels, and then finally to the end user. In customer-centric thinking, the *modern value chain* reverses the approach. The customer becomes the first link and everything follows. This approach changes the traditional chain such that it takes on a customer-driven perspective. Slywotzky and Morrison suggest that

> In the old economic order, the focus was on the immediate customer. Today business no longer has the luxury of thinking about just the immediate customer. To find and keep customers our perspective has to be radically expanded. In a value migration world, our vision must include two, three, or even four customers along the value chain. So, for example, a component supplier must understand the economic motivations of the manufacturer who buys the components, the distributor who takes the manufacturers products to sell and the end-user consumer.

As organizations become components of value chain networks (and as such operate to create value for the network as well as to create value to meet their own stakeholder needs), there is an incumbent responsibility to consider interorganizational network system added value as a means of increasing productivity within the network as well as for the end-user customer.

Within the context of the value chain, value drivers assume a sixfold significance. One is clearly that of the role of the process of adding *relevant value* for customers and its ability to differentiate the value offer such that it creates competitive advantage for both the customer and the supplier organization. The second is that like their customers, suppliers also have value drivers, and creating value creates costs for supplier organizations, thereby raising questions on the impact on the value *and* cost drivers of the supply/vendor organization. The third questions the impact on the value *and* cost drivers of the supply/vendor organization network. A fourth is that supplier organizations also have value drivers. The fifth consideration is that value drivers create costs for both customers *and* suppliers. The sixth consideration follows logically; accordingly, an analysis of the impact of enhancing a customer value driver should be accompanied with an evaluation of its impact on the supplier's own value drivers *and on* the impact of the suppliers cost of focusing on the customer's value driver(s). Clearly these two issues are linked as a supplier's value drivers include financial performance, and it follows that unless the marginal revenue generated by enhancing a value driver exceeds its marginal cost, there would be little point in pursuing the proposal.

The overwhelming benefit offered by the value chain approach is its ability to offer management the facility to consider alternative ways and means of creating value. Is healthcare in any way different? Clearly the answer is no. Patients are customers. They have value needs, which are met by specific resources and services. Healthcare has marketing channels and has specialist partnership infrastructures that deliver value (or treatment services). Healthcare managers should be mindful of the fact that they should, and can, identify alternatives to meet patient needs and take into account the alternative combinations of performance, cost, and time requirements and select the relevant suppliers to affect an appropriate value-delivery system.

16.4.6 Applying the e-Value Chain to Healthcare

Earlier we mentioned the importance of *service quality* in the value-delivery process that was identified by Heskett et al. [17]. The authors suggest a strategic value vision model that offers a useful approach for any organization seeking to pursue the eHealth value-chain network concept. Essentially, their model is based upon identifying a target market (segment or niche) and using the customer expectations to create a value proposition (the authors see this as a value strategy), identifying the operations processes and organization structure to commission for its delivery, and designing the delivery system. Figure 16.5a uses the model to consider how it could be used to create an eHealth-driven value chain.

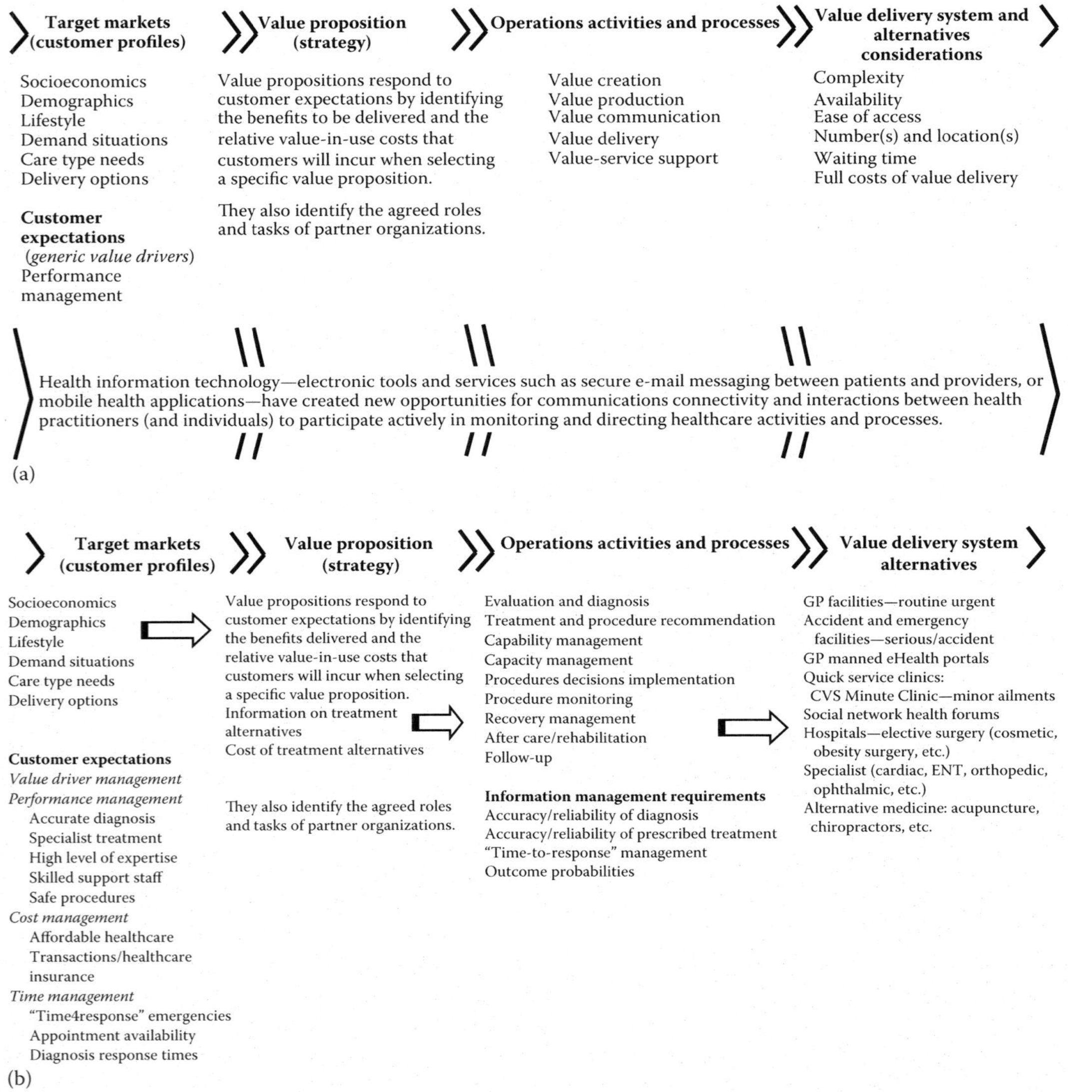

Target markets (customer profiles)	Value proposition (strategy)	Operations activities and processes	Value delivery system and alternatives considerations
Socioeconomics Demographics Lifestyle Demand situations Care type needs Delivery options	Value propositions respond to customer expectations by identifying the benefits to be delivered and the relative value-in-use costs that customers will incur when selecting a specific value proposition.	Value creation Value production Value communication Value delivery Value-service support	Complexity Availability Ease of access Number(s) and location(s) Waiting time Full costs of value delivery
Customer expectations (*generic value drivers*) Performance management	They also identify the agreed roles and tasks of partner organizations.		

Health information technology—electronic tools and services such as secure e-mail messaging between patients and providers, or mobile health applications—have created new opportunities for communications connectivity and interactions between health practitioners (and individuals) to participate actively in monitoring and directing healthcare activities and processes.

(a)

Target markets (customer profiles)	Value proposition (strategy)	Operations activities and processes	Value delivery system alternatives
Socioeconomics Demographics Lifestyle Demand situations Care type needs Delivery options	Value propositions respond to customer expectations by identifying the benefits delivered and the relative value-in-use costs that customers will incur when selecting a specific value proposition. Information on treatment alternatives Cost of treatment alternatives	Evaluation and diagnosis Treatment and procedure recommendation Capability management Capacity management Procedures decisions implementation Procedure monitoring Recovery management After care/rehabilitation Follow-up	GP facilities—routine urgent Accident and emergency facilities—serious/accident GP manned eHealth portals Quick service clinics: CVS Minute Clinic—minor ailments Social network health forums Hospitals—elective surgery (cosmetic, obesity surgery, etc.) Specialist (cardiac, ENT, orthopedic, ophthalmic, etc.) Alternative medicine: acupuncture, chiropractors, etc.
Customer expectations *Value driver management* *Performance management* Accurate diagnosis Specialist treatment High level of expertise Skilled support staff Safe procedures *Cost management* Affordable healthcare Transactions/healthcare insurance *Time management* "Time4response" emergencies Appointment availability Diagnosis response times	They also identify the agreed roles and tasks of partner organizations.	**Information management requirements** Accuracy/reliability of diagnosis Accuracy/reliability of prescribed treatment "Time-to-response" management Outcome probabilities	

(b)

FIGURE 16.5
Providing communications and interactions facilities in healthcare by (a) using eHealth information technology and (b) using a value chain approach to healthcare value delivery.

Figure 16.5a identifies the important topics requiring management consideration and decisions by using similar headings but adding emphasis to the need to evaluate options for managing the performance, cost, and time value criteria. More details are added to Figure 16.5b. Figure 16.5a identifies the eHealth tools that are available, while Figure 16.5b adds suggestions for topics that should be considered during the construction of the model.

Finally we offer a hypothetical eHealth value chain model in Figure 16.6. While it is hypothetical, it serves to demonstrate how an eHealth ICT model can coordinate and integrate healthcare delivery by offering a template for future work.

16.5 Value Drivers and Value-Led Productivity: A Network Perspective

Our discussion so far has dealt with the broad context of value drivers and has not dealt with their application to network performance planning: in any network structure the partners are a series of suppliers and customers, and at each transaction point both expect to meet the performance requirements of their value drivers.

16.5.1 Value Drivers as Network Performance Drivers

Essentially, value drivers are *performance drivers for each network partner* and transactions are negotiated around the notion of optimizing the performance of the outcome of the transaction such that each partner has the result that meets the satisfaction of internal stakeholders (employees and shareholders) and external shareholders (end-user customers). Without this outcome, the longevity of the network structure is in doubt; either it will fail due to the lack of overall economic viability or it will undergo structural (or repositioning activities) in order to find a structure that meets the returns expectations of all partners.

The *value drivers* in any business depend on the specific setting, competition, and market structure(s). Their time perspective is clearly short term given they are factors that "drive present value." A focus on adjustments to the value drivers in response to changing customer expectations results in short-term improvements in performance. Value drivers can be considered to be *performance drivers*, reflecting the productive use of the "components" of an organization's *resources capability profile*. An organization's capability resource profile reflects how well it can respond to customer (and market) expectations. For example, in defense organizations, "capability" is used to reflect the ability to meet a range of specified situations, such as the obvious peacekeeping roles that are typically associated with military organizations and also in other nonmilitary roles, such as humanitarian activities involving disaster relief. In healthcare they may reflect broad expectations of quality of life and longevity expectations.

16.5.2 Value Drivers as Productivity Components

Given the relationships of generic value/performance drivers (i.e., performance, cost, and time management) throughout networks, their role in influencing productivity is an essential feature of the success of the network. Productivity has long been considered to be a measure of the performance achievement of selected inputs. Typically, governments measure productivity internationally by comparing relative labor costs across industries or by

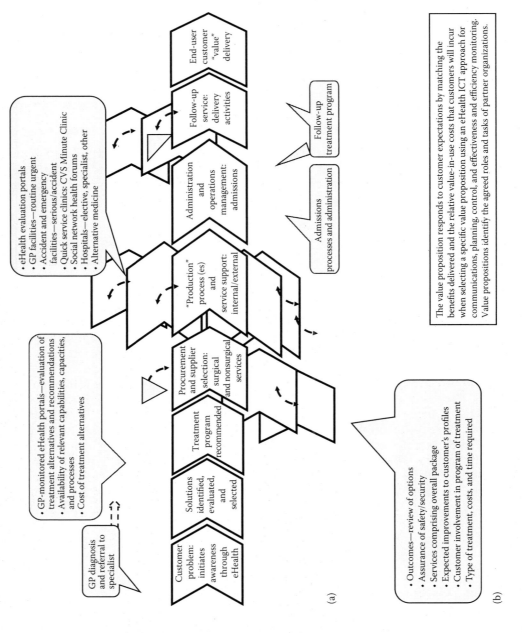

FIGURE 16.6

An eHealthcare-monitored value chain.

measuring the time taken to produce specific items (e.g., hours taken to produce an automobile). While relative productivity comparisons are interesting, it could be argued they are not particularly helpful. For example, if labor rates per hour in an emerging economy are significantly lower than those in advanced economies, it is hardly surprising that significant cost differences will emerge. Furthermore, there may well be considerable differences in the quality and features of the end product as well as in the methods of production: while this suggests the involvement of research design and development inputs, productivity is often *process led*. Stephen Mills, chair of the Australian Business Foundation, gave a timely reminder of the somewhat confused perspectives that exist concerning what productivity actually is. Mills argues that Australian industry requires "a greater ability to absorb and to apply the knowledge we already have" [26]. He suggests that productivity does not rely solely upon research and development (R&D) expenditures; rather, it is about finding *solutions to problems* (authors' italics) by using innovative approaches. Mills's contribution suggests a much needed focus on process innovation in an attempt to increase productivity; and that "rather than recognizing innovation as knowledge sharing and collaboration across the value chain ... most commentators have fallen back into traditional postures on either side of the old divide separating industry and research." There are a number of interesting examples supporting Mills's assertion:

> In 1976, Dr. Govindappa Venkataswamy retired from performing eye surgery at the Government Medical College in Madurai, Tamil Nadu, a state in India's south. He decided to devote his remaining years to eliminating needless blindness among India's poor and founded the Aravind Eye Hospital. The world's largest provider of eye care has found success by directly adapting the management practices of a "mass producer", a fast-food brand, one that is not often associated with good health: McDonald's. Dr. Venkataswamy, however, has produced a sustainable business model because of the other major influence on his thinking: McDonald's and Sri Aurobindo. These are an unlikely pair. But Aravind can practice compassion successfully because it is run like a McDonald's, with assembly-line efficiency, strict quality norms, brand recognition, standardization, consistency, ruthless cost control, and, above all, volume. Each surgeon works on two operating tables, a set of nurses prepares the patient while the surgery is conducted, and by the time the surgery is done with one patient, the other patient is ready to be operated upon. This ensures that the surgeons' time is effectively used by eliminating the bottlenecks and he or she can focus on high value addition. The same philosophy is followed in each and every department—of reducing the low value added activities by the most expensive resources. In 1992, in order to enhance the treatment of cataracts by using an implanted lens in the patient's eye, Aravind set up Aurolab, which now makes lenses (for $2 apiece), sutures and medicines. Aurolab is now a major global supplier of intraocular lenses and has driven down the price of lenses made by other manufacturers as well [27].

Moreover:

> Heart surgery has received similar treatment. Devi Shetty's Narayana Hrudyalaya Hospital business model *lowers the cost of healthcare to India*. The approach has transformed health care in India through a simple premise that works in other industries: economies of scale. By driving huge volumes, even of procedures as sophisticated, delicate, and dangerous as heart surgery, Dr. Shetty has managed to drive down the cost of healthcare in India. Devi Shetty's team accounts for 15% of all heart surgeries in India. Original equipment makers do not sell directly to hospitals unless they obtain volume orders. Medicines and equipment account for 40% of revenue outflows. The Narayana

Hrudyalaya hospital has attractive volume throughput: It handles 30 heart surgeries and at least 1000 walk-in patients a day and is able to convince its suppliers to supply at a low cost. Shetty continually searches for ways to reduce costs; his microbiology department makes hand-wash and disinfectants in-house, bringing down the monthly cost of supply considerably. Shetty does something else to cut costs. Every ICU patient has dedicated nurses watching over him, 24 h a day. They work 8 h shifts, standing in front of the patient. Shetty doesn't provide chairs: "The moment you provide a chair, the efficiency of the nurse goes down by at least 30%." He encourages attrition among them: "As they grow older, they don't contribute as much towards patient care, but their salary keeps going up." To keep salary costs low, he hires people with basic college education and trains them for jobs like reading radiology charts [28].

These examples both address the generic performance/value drivers; clearly there is a strong emphasis on defined cost management and performance. The attention to input costs management demonstrates a necessary approach with performance management influenced by both qualitative (internationally expected results) and quantitative criteria (patient throughput numbers), the latter being an essential link to the economies of scale that influences suppliers. It also suggests that productivity measurement should be based upon critical value driver criteria. Time management can be critical in responding to healthcare demand, for example, the delivery of transplant items, and cost may be a secondary consideration.

16.5.3 Assessing the Productivity and Competitive Advantage of the Value Proposition

To be an effective management tool, productivity management should be linked to a data-information decision-making model. Figure 16.7 proposes an eHealth data-driven model that identifies customer/patient value expectations (value proposition), the availability of resource requirements and their deployment, and the relationship between an acceptable (viable) value proposition and organizational profitability, productivity, and competitive advantage.

Step 1 in Figure 16.7 identifies a process in which a possible value proposition is explored through end-user interactions. Typically there is a supplier/end-user iterative process during which expectations become realistic and economically possible. Figure 16.3 suggests some possible value proposition characteristics. Step 2 assesses capabilities and resource requirements together with their costs. Again there will be some iteration as a "fit" between expectations and the costs of meeting them are explored. This can be a protracted activity as alternative sources of supply are sought from external suppliers. Part of the analysis involves a comparison between the costs of meeting the value proposition characteristics and alternative supplier arrangements and the revenues likely to be realized that will contribute toward the productivity of alternative resource deployment/value delivery options.

Step 3 in Figure 16.7 extends the analysis to evaluate the economic viability of the alternative resource formats. Of particular importance is the extent to which the value proposition alternatives offer productivity and profitability returns and, importantly, competitive advantage. This is particularly important as a strong competitive advantage will suggest that the value proposition will offer marketing sustainability when compared with the value propositions of the competitors. Step 4 offers an opportunity to validate the value proposition by exploring the relationship between resources productivity, profitability, and competitive advantage in a limited test market activity. This is essential, as it offers management the opportunity to make changes prior to implementing the value proposition in step 5.

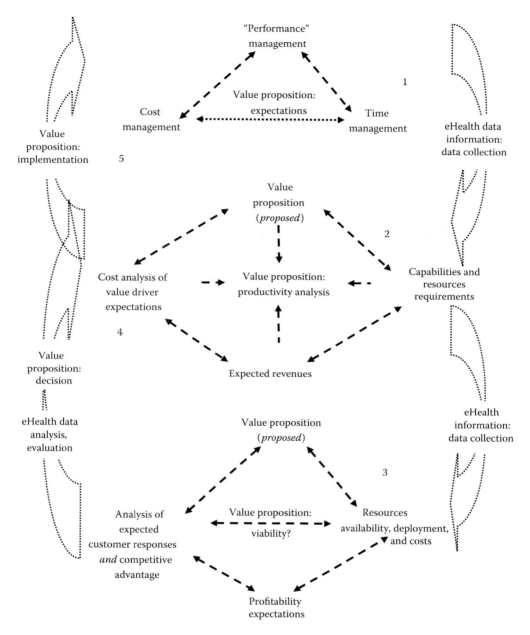

FIGURE 16.7
Assessing the productivity of the value proposition.

16.6 Technology Perspective

Synergistic advances in all areas of technology from sophisticated signal-processing systems and low-power sensor technologies to the development of broadband wired and wireless communication systems, the existence of increasing end user–friendly appliances and the emergence of networked cloud computing and storage services suggests a dramatic shift in provisioning healthcare in the 21st century. Indeed, advances in microelectronics, computing, and signal processing and the increasing availability of broadband end-to-end networking suggest that sensors and wireless body area networks can be used to biometrically monitor the state of the human body (and mind) in a more timely and efficient manner. Indeed, advances in sensing and communication technologies along with advances in software engineering make it possible to synthesize new solutions for wearable healthcare systems (mobile health or mHealth appliances), thus enabling the development of healthcare-centered smart homes that can sense the relative state of health of persons living in the home.

The application of these sophisticated and networked biometric sensor systems suggests that aging Internet citizens and those with preexisting health conditions can remain in their own home while healthcare providers can remotely monitor and advise them on how to improve their well-being from a distance. These composite capabilities promise delivery of quality healthcare—from anywhere and at anytime. The following subsections include discussions on key eHealth enablers including cloud computing, smart healthcare personal assistants, and mechanisms for security and privacy.

16.6.1 Cloud Computing

A key eHealth enabler is the rapidly evolving landscape of networked cloud computing infrastructures that will provide storage, processing, and the communications fabric needed to intelligently interconnect patients with medical practitioners, caregivers, and family members. Cloud computing is a model that enables users to have access to a shared pool of configurable computing resources (network, software, databases, and hardware) at anytime and from anywhere. From the service point of view, this model includes three service models (Figure 16.8): software as a service (SaaS), platform as a service (PaaS), and infrastructure as a service (IaaS). The personality of cloud computing architectures comprises five essential characteristics, including on-demand self-service, heterogeneous network access, resource pooling including sharing of databases, modularity, and extensibility, that ensure orderly changes with modular expansion capabilities and quality-of-service guarantees. Cloud computing architectures naturally support four deployment models, including private, public, and community models and hybrid variations of these models. Increasingly, organizations are moving to the cloud in order to leverage off the salient advantages offered by networked cloud computing architectures.

Healthcare organizations will likely not be the exception to this trend. Cloud computing has the potential to bring tremendous benefits to healthcare organizations, such as the promise of high-level patient care, cost reductions, accessibility to the medical data, and support to medical research. However, cloud computing brings many unique challenges that need to be carefully addressed. The spectrum of these challenges varies from technical and organizational to medical, legal, and public policy issues. A detailed discussion of these challenges is beyond the scope of this chapter.

The architecture given in Figure 16.8 enables end users to have access to a shared pool of configurable computing resources (networks, software, and hardware), anytime and from anywhere. It is expected that healthcare organizations and service providers could and likely will assimilate this approach on a broad scale since cloud computing holds the promise of achieving a very high level of patient care while facilitating cost reductions if cloud services are shared amongst geographically dispersed healthcare organizations and private physician office buildings.

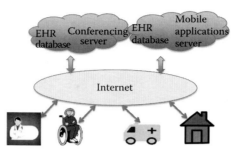

FIGURE 16.8
Cloud computing information networked architecture.

It is expected that the communications, storage, and application-aware capabilities of a powerful cloud computing information-networked infrastructure will likely be multiplied many times to enable a new and ubiquitous regime of medical care, unconstrained by geographical borders, to smoothly deliver quality care to medical patients living in Brooklyn, New York, or patients residing in rural areas of Sri Lanka. These systems will likely leverage multimedia databases and patient record systems with integral capabilities for real-time multimedia and multipoint conferencing systems. With these systems, healthcare providers will be able to collaborate remotely, remotely monitor, and advise their patients on how to improve their well-being, and in general have the capability to provide quality healthcare at a distance.

The target capabilities of eHealth cloud computing systems include the following:

- Service provisioning that integrates capabilities for medical information and patient EHR coupled with hospital information and inventory management systems
- Being easy to use and having a spectrum of capabilities that hospitals, private physicians, and medical labs will find compelling to use on a regular basis
- An infrastructure that meets national, global, and end-user standards for privacy and security
- A platform that is compatible with standardized EHR systems and standardized imaging systems (e.g., EHR and digital imaging and communication in medicine [DICOM])

Cloud computing brings many unique challenges that need to be addressed. These challenges vary from technical and organizational to legal and medical. Related to technical challenges are the following:

1. *Service/data availability*: Users of eHealth services, such as doctors and patients, demand high availability of the services and data. In other words, medical data and services must be available to users whenever they are required.
2. *Service/Data reliability*: The information systems should be error-free—eHealth services should secure the validity of the data regardless of any software, hardware, or network failures.
3. *Flexibility*: The cloud that eHealth services use should be flexible enough to change its services and infrastructures or add new ones, according to the different requirements of healthcare providers. These requirements focus on the quality of the services, operations, auditing, and management.

4. *Interoperability*: This characteristic refers to the ability of the system to provide integrated services and data integration among different cloud service providers from both internal and external clouds, coupled with the ability for easy migration to different systems.

5. *System security*: Security is the most critical issue for eHealth services due to the sensitive information that it copes with, such as medical records. It deals with access control, audit control, authentication, authorization, transmission security, and storage security issues.

6. *Privacy*: Privacy is of paramount importance to all end users. It is essential that eHealth cloud systems provide strict access control to data (authorization) and use security standards to provide privacy and security of patient data as discussed in subsequent paragraphs.

Protecting privacy-sensitive data in the cloud is seen as a controversial topic, especially when discussing information management within the eHealth sector. Protecting the data from malicious unauthorized access, misuse, or public availability are key factors concerning privacy. Moreover, inconsistencies and loss of data are seen as huge risks that can result in severe consequences. A challenge that still needs to be overcome is the interoperability between service platforms, considering that sharing privacy-sensitive data on a larger scale requires integration of service platforms like, e.g., Microsoft HealthVault.

Legal issues can play a vital role in the adoption of cloud computing in the health sector. Health data are private information, confidential between patients and their doctors. Privacy laws may differ from country to country but the main goal is to protect the rights of an individual. One of the main challenges is that only authorized medical personnel have access to patient data. Assuring the cloud provider's compliance with the privacy and security requirements imposed on healthcare organizations remains a significant challenge. eHealth service delivery is by its very nature highly dependent on a new information-networked infrastructure. Transformations are necessary from the eHealth viewpoint to make the right decisions at an organizational level. Benefits unfold as healthcare adopts cloud computing; as a result, challenges arise on an enterprise scale. Hence, organizational change is of great importance to eHealth services.

As mentioned earlier, cloud computing offers the promise of having data accessible anywhere and at all times. Information technology is ubiquitous in healthcare. However, these resources are necessarily housed outside of the hospital's own data center. On the organizational level, major decisions have to be made concerning investments up-front or recurring costs for hosted cloud services. Cloud computing offers a pay-as-you-use model, which allows providers to potentially achieve better resource utilization and enables users to avoid the costs of resource overprovisioning through dynamic scaling. It is true that from the technology side, the implementation of cloud healthcare has clear cost advantages. However, from the healthcare organization side, these advantages sometimes may not outweigh the cost of implementing healthcare solutions. Economic challenges and trade-offs associated with adoption of cloud computing need to be understood and quantified in the larger scheme of medical care. Using IT systems to improve work flows is nothing new. However, it is essential that the organization focus special attention on the actual core process of eHealth that is being influenced by the "improved work flows."

16.6.2 Smart Healthcare Personal Assistants

An emerging solution to alleviate the burden of frequent visits to healthcare providers is to empower patients with eHealth self-management technology, of the type suggested in Figure 16.9. There is increasing consensus among healthcare professionals and patients alike that many chronic disorders can be better managed in the home environment than in an outpatient clinic or hospital. Self-management may involve taking self-measurements and behavioral change encouragement, such as engaging in regular exercise, weight control, and dietary change. Mobile technology and intelligent medical applications may play an important role in achieving this.

Networked smart healthcare assistants are seen as promising technology, where *smart* refers to the ability of drawing clinical conclusions about the patient's health status without, or with limited, human intervention. The use of smart self-management healthcare appliances hold the promise of achieving sustainable, high-quality, and cost-effective care of a spectrum of medical problems ranging from the need for transient episodes involving physical rehabilitation to care of long-term chronic diseases. The general architecture consists of devices that collect information directly from the user, or directly from EHRs that contain user data. The essence of this architecture is that the user obtains appropriate feedback from the mobile device through the interpretation of available data. Feedback is provided by an embedded decision-support system. The mobile computing device is a networked mobile computing platform that supports collecting information about symptoms and environmental conditions directly from the patient and the patient's surroundings, interpreting the data and providing feedback about the patient's health status. In addition, the device communicates the patient's data to a web center, and/or a personal caregiver. A smartphone or tablet can act as the user device.

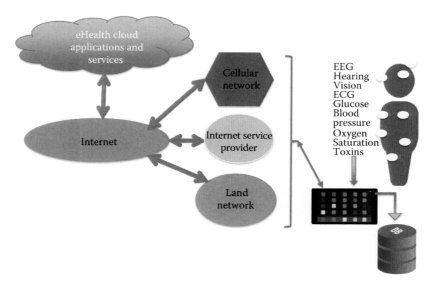

FIGURE 16.9
Example of a networked smart appliance that monitors ECG, pulse rate, blood oxygen, saturation, etc.

16.6.3 Security and Privacy in eHealth

Security threats may be classified as confidentiality attacks, attacks to integrity, and threats to availability. Confidentiality attacks occurs when electronic access to patient records, access to health-related databases, access to health-related measurements such as blood pressure, and even remote diagnosis and treatment are granted without authenticating the end user. Examples of confidentiality attacks include eavesdropping, location tracking, and activity tracking. An example of a confidentiality attack is *location tracking*—a person's current or past location could be tracked. This is particularly problematic in mobile health systems because messages about a person's location are regularly transmitted.

Attacks with which an unauthorized user may modify or intercept medical data cause threats to the integrity of the data. This type of attack includes modification of medical data or generation of false alarms. Data modification can obviously cause serious problems for the patient. Threats to eHealth also include making a resource unavailable for access. A specific threat against availability is denial of service (DOS). DOS makes eHealth service unusable by jamming, or by overloading, the system.

Main security solutions proposed to address security requirements of patient-monitoring systems reside in the areas of encryption and secure communication. Access-control techniques need to be developed to address the privacy issues in eHealth applications. In particular, procedures for authentication (user identification) and authorization (access right checking) are essential for acceptable and sustainable eHealth applications. The authentication function ensures that users are indeed who they claim to be. Therefore, an unauthorized party should not be able to receive or access a patient's medical data. Various techniques being considered include a combination of different authentication atomics. These atomics include biometric information (fingerprint, retina scan, typing pattern, etc.) and information the user knows (mother's maiden name), user possessions (e.g., smart card identification systems), and may also include current location and/or the end-user terminal identity (media access control [MAC] hardware address) and time of access.

16.7 Conclusions and Future Challenges

eHealth can become a proactive process for avoiding the avoidable with the promise of support from emerging wireless technologies and sensor networks and the emergence of secure and high-performance end-to-end communications infrastructures. Moreover, it is highly likely that eHealth technologies can potentially come to the rescue of aging world citizens with the promise of remote biometric monitoring and delivery of healthcare that is independent of geographical boundaries. However, solutions that are modulated and influenced by the various market forces discussed in this chapter will emerge.

A spectrum of challenges remains regarding adoption and broad deployment of eHealth applications and services. The diversity of standards and the associated interoperability issues are a significant set of challenges. Indeed, the integration of standardized electronic patient health records from all levels of medical facilities, pharmacies, and private physician offices remain a significant barrier to broad scale deployment. Further research is needed to assess the willingness and potential acceptance of eHealth technologies by medical practitioners and patients. Security mechanisms that are perceived as trustworthy need to be adopted. Private, reliable, and secure provisioning of eHealth applications is essential for broad application in the marketplace.

Thinking ahead, two key groups, patients and politicians, can impact the adoption of eHealthcare. The patients need to be educated as to the benefits eHealth can deliver—many are suspicious and want to retain face-to-face involvement with physicians; they are also concerned over data security. As populations age, there is a need to introduce efficiencies in healthcare. Monitoring is one way of doing this; sensors and monitors are increasingly cost efficient. This brings us to the politicians; budget deficits are increasing and there appears to be determination by most governments for immediate returns to "surpluses"; while laudable, this is impractical. eHealth offers cost-efficient alternatives; perhaps the benefits are not fully understood by the largest target market (aging patients) or the legislators.

What emerges from the forgoing is a sense that effective business models for eHealthcare are evolving; however, adoption of eHealthcare by users, ranging from individual patients to entire healthcare systems is impacted by a complex range of factors—and solving these problems will lead to improvements in the quality of care and the development of both viable and feasible business models. In the future, creative approaches to project funding, effective delivery, recognition of benefits through education, and a full understanding of the process of developing value chain networks with partners to deliver product–service combinations to users will drive adoption.

Abbreviations

CMS	Centers for Medicare and Medicaid Services
DOS	denial of service
EHR	electronic health record
HIPAA	Health Insurance Portability and Accountability Act
HIT	Health Information Technology
IaaS	infrastructure as a service
ICT	information and communication technology
KP	Kaiser Permanente
PaaS	platform as a service
PHR	personal health record
SaaS	software as a service

References

1. Savulescu, J., Should we be allowed to choose the sex of babies? *Sunday Herald*, February 1, 2004, accessed on March 10, 2014.
2. Wilson, E.V., ed., *Patient-Centered E-Health*, Hershey, PA: Medical Information Science Reference, 2009.
3. Helman, D.A., de Chernatony, L., and Addeo, E.J., An evolving e-Landscape: Synergistic intersections of technology and brand management, *Thought Leaders in Brand Management Conference Proceedings*. Università della Svizzera italiana, Lugano, Switzerland, April 18–20, 2010.
4. Kane, L., and Chesanow, N., Survey data source, *Medscape EHR Report*, 21, 202, 2012.

5. Centers for Medicare and Medicaid Services, http://www.cms.gov/Regulations-and-Guidance/Legislation/EHRIncentivePrograms/Meaningful_Use.html, accessed on March 1, 2014.
6. Williams, E.I., A model to describe social performance levels in elderly people, *Journal of the Royal College of General Practitioners*, 36, 290, pp. 422–423, September 1986.
7. Topol, E., *The Creative Destruction of Medicine*, New York: Basic Books, 2012.
8. Pew Research, http://www.Pew.org, accessed on February 1, 2014.
9. Blank, S., Why the lean start-up changes everything, *Harvard Business Review*, 91, 5, pp. 63–72, May 2013.
10. Rogove, H.J., McArthur, D., Demaerschalk, B.M., and Vespa, P.M., Barriers to telemedicine: Survey of current users in acute care units, *Telemedicine Journal and e-Health*, 18, pp. 48–53, 2012.
11. Pearl, P.L., Sable, C., Evans, S., Knight, J., Cunningham, P., Lotrecchiano, G.R., Gropman, A. et al., International telemedicine consultations for neurodevelopmental disabilities, *Telemedicine Journal and e-Health*, 20, 6, pp. 1–4, June 2014.
12. Kaiser Permanente, http://www.kp.org, accessed on March 1, 2014.
13. Zhou, Y., Improved quality at Kaiser Permanente through email between physicians and patients, *Health Affairs*, 29, pp. 1370–1375, July 2010.
14. Vu, S., Keynote Presentation, George Washington University Business Plan Competition, 2013, http://www.youtube.com/watch?v=XO5buwQQ-Rw, accessed on March 1, 2014.
15. Walters, D.W., and Rainbird, M., *Strategic Operations Management: A Value Chain Approach*, London: Palgrave, 2013.
16. Anderson, J.C., and Narus, J.A., Business marketing: Understand what customers value, *Harvard Business Review*, 76(6), pp. 53–65, November/December 1998.
17. Heskett, J.L., Sasser Jr., W.E., and Schlesinger, L.A., *The Service Profit Chain*, New York: The Free Press, 1997.
18. Slywotzky, A.J., and Morrison, D.J., *The Profit Zone*, Wiley: New York, 1997.
19. Phelps, B., *Smart Business Metrics*, Harlow: FT/Prentice Hall (Pearson), 2004.
20. Friedman, T.L., *The World Is Flat: The Globalized World in the Twenty First Century*, London: Penguin Books, 2006.
21. Scott, M., *Value Drivers*, Chichester: Wiley, 1998.
22. Webster, F.E., *Market Driven Management*, New York: Wiley and Sons, 1994.
23. Normann, R., and Ramirez, R., *Designing Interactive Strategy: From Value Chain to Value Constellation*, New York: Wiley, 1994.
24. Anderson, J.C., Narus, J.A., and van Rossum, W., Customer value propositions in business markets, *Harvard Business Review*, 84(3), pp. 90–99, March 2006.
25. Del Vecchio, J., http://www.fool.com/research, 2013, accessed on March 1, 2014.
26. Mills, S., Labs aren't the only ones with ideas, *The Australian Financial Review*; Walters, D., Competition, collaboration, and creating value chain, in Modelling Value, Selected papers on the *1st International Conference on Value Change Management*, Eds. Jodlbauer, H., Olhager, J. and, Schonberger, R. J., Physica-Verlag, Springer, Berlin Heidelberg, 2012.
27. Rosenberg, T., A hospital network with a vision, http://youthvoices.net/discussion/health-care-and-my-opinion, accessed on March 1, 2014.
28. Mahajan-Bansal, N., Spin-offs from the Devi Shetty Model, http://india.forbes.com/article/magazine-extra/spinoffs-from-the-devi-shetty-model/1922/1, accessed on March 1, 2014.

17

Mobile Healthcare User Interface Design Application Strategies

Ann L. Fruhling, Sharmila Raman, and Scott McGrath

CONTENTS

17.1 Introduction

Mobile device usage is growing at an exponential pace mainly due to rapid advancement in mobile technology and the availability of a vast group of features and new capabilities. In 2014, 497 million mobile devices were activated, bringing the global total up to 7.4 billion; with this upward trend, the number of mobile devices will exceed the entire human population in 2015 and by 2019 there will be nearly 1.5 mobile devices per capita. Not only are the numbers of devices in use growing but so, too, is mobile data usage. Global data consumption grew 69% from 1.5 exabytes perminth at the end of 2013 to 2.5 exabytes (EB) per month at the end of 2014. It is projected that data usage will surpass 24.3 EB by 2019. To help put this into perspective, the entire connected Internet in 2000 transmitted 1 EB of information. In 2013, approximately 18 EB was transmitted to mobile devices alone (Cisco 2014).

The devices themselves are also becoming more advanced and powerful. The ability to send SMS messages (or text messages) is a feature found on all phones sold today. In addition, cameras and embedded sensors on these devices also allow for robust data collection. Modern smartphones are equipped with proximity sensors, accelerometers, and GPS; some higher-end models are also equipped with gyroscopes, barometers, and heart-rate sensors. Technologies like Bluetooth low energy (Bluetooth Special Interest Group Inc. 2014) and near-field communication (NFC) (NearFieldCommunication.org 2014) allow these smartphones to extend their reach (up to 5–10 m) by connecting to wearable devices. These wearable devices can be specialized to monitor parameters such as blood glucose and oxygenation. When connected in such a fashion, the mobile phone can be the nexus of a body area network (BAN) that is able to collect data for trend analysis and alerts when levels deviate outside of normal values (Patrick et al. 2008).

In the current fast-paced business environment, the adoption of innovative mobile technologies has a crucial role in an organization's business strategy and delivery of significant results. To gain competitive advantage, many organizations are expanding their services into the wireless space to enable their workforce to gain access to vital information anytime, anywhere. Organizations in various sectors, including healthcare, retail, and banking, are looking for ways to harness the power of mobile technologies and streamline business processes with future mobile applications (apps). This is expected to improve productivity and lower operational costs (Pervagus Inc. 2002).

The wireless trend for medical information systems (ISs) and ITs has also moved beyond just mobile phones. For example, ambulances in Sweden have been equipped with mobile computing, wireless networks, and GPS. While in transit, patient vitals and the ambulance's GPS coordinates can be sent to the hospital, providing real-time status updates. Finland has set up a network where patient records, health consulting, and prescriptions can all be accessed by authorized individuals wirelessly on connected devices (Wu et al. 2007). While these advances are a net benefit for healthcare, there still exist hurdles that need to be considered, including how to best design and implement the mobile user applications' interfaces.

In North America alone the healthcare IT market is anticipated to grow at a compound annual growth rate of 7.4% and reach $31.3 billion in 2017 from its previous level of $21.9 billion in 2012 (North American Healthcare IT Market by Application Delivery Mode & Component—Forecasts to 2017, 2013). A main driving factor of the healthcare information market is the changing government regulations and initiatives on minimizing healthcare costs (Chong et al. 2004). The U.S. government has provided many incentives that were aimed to have a significant impact on the development and deployment of various healthcare information systems, including mobile applications. One key incentive was the American Recovery and Reinvestment Act (ARRA) of 2009. In addition, the Health Information Technology for Economic and Clinical Health (HITECH) Act was created to fund and provide support to research and development activities related to the creation of a paperless national health information network through the adoption of electronic EHRs (Peterson 2006).

The healthcare industry is benefiting from increasing innovation in the mobile and tablet platforms. More and more healthcare organizations are realizing the potential of implementing mobile extensions to various information technology systems they have in place, from patient appointment request systems to distribution of medical test results to patients to physician-to-physician consultations. Mobile technology is being used in innovative ways and offering a wide spectrum of projects that can help healthcare organizations enhance their services and connect healthcare providers while achieving business goals. Although, the ARRA has boosted the advancement of information technology in the healthcare sector, many of the applications have not yet reached their potential and there is still more work to be done.

The continuing innovation in the bandwidth and the speed of Internet and intranet networks is another driver of the expected increase in demand for healthcare-related mobile apps. Further, using cloud technology within the boundaries of HIPAA regulations (U.S. Department of Health and Human Services 2014) will only further widen the data availability that mobile apps can provide. Moreover, the standardization of data-exchange protocols between systems (e.g., IP), as well as the improvement in structuring of medical data with initiatives like HL7 (Health Level Seven International 2014) and the implementation of various data security laws, is making it possible for medical information systems in different locations operating on diverse platforms to understand one another and work together. These applications are not only useful for exchanging data but also for providing connectivity to various healthcare experts.

Even with the recent improvements in mobile technology, compared to developers of desktop information systems, mobile applications developers often need to use different strategies on how to create the best user experience within the constraints of smaller displays, limited user input capacity, and reduced network bandwidth and data access. An area that can have a significant positive impact on the user experience is the design, usability, and work flow of the mobile app user interface.

One of the most important success factors in the adoption of any information system is designing an effective and efficient user interface. The user interface is one of the main components of any interactive computing system. It is the part of the information system through which the end user interacts with the system either physically, perceptually, or conceptually (Nielsen 1994) and it often determines most users' perceptions of the system. User interface design is vital to successful information system adoption. Information technology projects invest a significant amount of human resources in the design and construction of the user interfaces; in some cases 80%–90% of the code of a system involves elements of the user interface. If a user interface is poorly designed, the user's effectiveness and efficiency will decrease, defeating productivity gains envisioned for the system. Designing and implementing mobile healthcare application user interfaces is an evolving area and calls for continual study.

Usability is a means to making user interfaces easy to learn and easy to use and includes the consistency and ease with which the user can manipulate and navigate the app, the clarity of interaction, ease of reading, and the arrangement of information as well as the speed and layout (Nielsen 1994). Research has asserted for a long time that the study of usability factors, such as organization, presentation, and interactivity, is key to the successful design and implementation of user interfaces (Shneiderman 1987; Jeffries et al. 1991; Nielsen 2012). Prior research overwhelmingly suggests that usability is associated with many positive outcomes, such as a reduction in the number of errors, enhanced accuracy, a more positive attitude on the part of the user toward the target system, and increased usage of the system by the user (Lecerof and Paternò 1998; Nielsen 2012). Understanding how usability impacts the user is instrumental to the design, development, and deployment of the mobile app. Even though, in general, the mobile apps' functionality is a subset of their desktop counterparts, the user interface design of the desktop counterpart cannot directly be replicated onto a mobile device. This is because the physical and operating system environments are quite different. Thus, the mobile app user interface design needs to be modified to adapt to the constraints of the mobile environment. In this chapter we provide strategies and design principles that will help streamline the process of designing and developing mobile apps with the healthcare provider user in mind. This information will be especially useful for practitioners to use as a reference as they improve and build new user interfaces for mHealthcare applications.

The structure of the chapter is as follows: We begin by presenting key design and development strategies for healthcare mobile user interfaces and applications. Next, we demonstrate how many of these strategies can be implemented using an example of a real-world public health emergency response (PHER) mobile app for microbiologists and healthcare providers as the foundation of our research. The PHER mobile app is currently employed as a supplement delivery and communication option for state public health microbiology laboratory experts. It serves state public health laboratories in three states that assist 60-plus rural hospital and clinical public health laboratories (Fruhling 2006).

17.2 Mobile Healthcare App User Interface Design Strategies

Expanding information system application delivery from the desktop to mobile platforms is not a trivial undertaking. As one would expect, there are significant differences between how a user interface is designed for a large desktop monitor and how a small mobile display screen is designed. For example, the small displays on mobile devices have a limited number of input and output facilities, reduced processor power, and less memory and bandwidth (Dilmaghani and Rao 2008). The wide range of displays and sizes offered by today's mobile device manufacturers has provided consumers a plethora of choices; however, the lack of standardization does make human–computer interaction design slightly more difficult. It is possible to purchase mobile devices with screens that may be as small as 144 × 168 pixels (the wearable Pebble watch), some phones have screens running at 1920 × 1080 pixels (Sony Z Ultra), and there are tablets that can support up to 2048 × 1536 pixels (Apple iPad Air). Designers need to ensure a uniform experience across all devices for their applications. When adapting visualizations and medical images from desktop-based systems to these smaller screens, these limitations require new approaches. Some techniques used to circumvent the limited real estate a smaller screen provides are changes in navigation techniques (remapping scroll and zoom functions), visual references to off-screen areas, and use of predictive algorithms that highlight potential areas of interest (faces, for example, or visual patterns) (Chittaro 2006).

In most cases, following general user interface design guidelines is a good first step; however, in our research we have found several key strategies that are especially important for healthcare mobile app design, development, and implementation. We present them in the following.

17.2.1 Focusing on Essential Functions in the Mobile Environment

Designing a good mobile app involves many aspects. It is especially important to identify the essential capabilities and features expected by the target users. In the case of expanding a desktop-based information system that already exists, the IT analysts should begin by itemizing all the capabilities of the desktop information system and organizing them into functional categories. Next, the current user group needs to be examined to help the IT analyst understand who the target users will be for the mobile app. This is necessary because the user community for the desktop system may not be the same as the mobile users. Once identified, the target users should assist in prioritizing which capabilities are essential for them to do their tasks when using the mobile application. Most likely there will be cases when not all of the counterpart desktop system's capabilities are needed on the mobile app. In fact, it is often not advantageous to provide all things to all users in the case of a specialized healthcare mobile app. Prioritizing capabilities by impact (how critical is it to the healthcare service the user is providing) and frequency (how often is the capability used on a routine basis) is an important consideration. Next, the system needs to be designed and programmed to meet the identified capability requirements. If the features that are essential are provided in an efficient and effective manner, the users will generally be satisfied. Accurately identifying the essential capabilities for the mobile application is a key to successful implementation.

17.2.2 Ease of Use

As with all information systems' design intentions, the goal is to design a user interface that is easy to use. Ease of use has been identified as one of the most important factors that determine the level of adoption of the application by the users (Davis 1989; Venkatesh et al. 2003). User interface usability testing on how easy the mobile app is to use is a must. Usability testing is a methodical approach to determining whether an information system meets usability criteria for specific types of users carrying out specific tasks through interactions with the system's user interface. User interfaces are most commonly critiqued using the following four techniques: heuristic evaluation, usability testing, software guidelines, and cognitive walkthroughs (Jeffries et al. 1991).

17.2.3 Intuitive Interaction

Intelligent interfaces that provide decision-support functionality are one way to make a user interface more intuitive. It is especially important in mobile healthcare applications that the user can quickly be able use the system with minimal or no training. It is often the case that the target users will have limited time for training. Therefore, target users will expect that the mobile app is as easy to use as a retail or banking app such that they can download the app with a few clicks and that it is fully functional. Additionally, when errors occur, the user needs to have a clean way to exit or have a clear message on how to correct the problem. Should a capability or update take more than a few seconds to process, the app should include a feedback mechanism letting the user know that the task is in progress.

To further reduce the introduction of potential errors, the user interface design should avoid long text entries and instead provide options that allow the user to select prepopulated input values. Moreover, mobile apps that interact with the user in the format of a friendly conversation are also perceived as more intuitive (Microsoft 2014).

17.2.4 Consistency within a Family of Applications

The mobile app should mimic its desktop counterpart in look and feel, naming conventions, color schemes, and fonts. To increase the productivity of the target user from the very beginning, the mobile app should present content, navigate, and respond in a similar fashion as the desktop counterpart. Following the practice of naming icons and labels in such a way that they explicitly specify the exact operation that they will perform is extremely important for mobile apps. This will reduce the time for the user to become proficient in using the app.

If a specific task can be completed in two or more ways, this might lead to unnecessary confusion. Therefore, it is better to provide only one way to do a task (Microsoft 2014).

17.2.5 Matching Routine Work Flow

Effective user interfaces allow users to perform a task in a way that matches their routine/ daily work flow. One way to determine if the user interface is attuned to the users' work flow is by incorporating a design that correlates with the users' mental models. A mental model is "a conceptual representation within a person's mind that is used to help the person understand the world and to help the person interact with the world." The term *mental model* was presented by Johnson-Laird (Johnson-Laird et al. 1998) and it refers to an internalized, mental representation of a device or idea. (Johnson-Laird credits Craik [Craik 1967] with the initial idea of mental representations.)

When applied to information systems, mental models require a bridge of trust to be constructed. The user of a proposed information system needs to know if the information system is trustworthy. In order to conclude that the information system is credible, the user needs to know who backs the claim that it can be trusted and understand what exactly it is trusted to do (Yee 2004). Security and trust are elements that need to be firmly established in the information system if software developers expect users to adopt it into their daily routine.

For new mobile applications, user interface designers need to consider the most common user interfaces that are routinely used on a daily basis (e.g., checkout process for an online purchase, site search capabilities, and system status) that users are familiar with and adopt design strategies that translate to a healthcare mobile application environment, for example, by taking advantage of the touch screen interfaces and utilizing the shortcuts built into the mobile operating system (for swipes, scrolls, and enlarging). It may be possible to design the application so that common tasks can be mapped out to a touch interface. Deleting an entry could be mapped to swiping left on the item, or a refresh of the displayed information may be requested with a downward drag with a finger. The display layout should also be arranged to account for the orientation of the device. One layout should be presented if the device is being held horizontally or vertically.

User interfaces designed that tap into users' mental models is a strategy that generally increases intuitive processes and reduces application errors (Johnson et al. 2005). Moreover, efforts made to attune a user interface to the users' mental models often reduces the time needed for training. Together these benefits help to improve user productivity and reduce costs (Preece et al. 1994).

The user interaction with the mobile user interface should follow common navigation practices, for example, implementing forward and backward arrows that take as little display space as possible. Navigation patterns should be consistent throughout the system, thus making the system more intuitive.

17.2.6 Limiting Menu/Layer Display Structure

Designing a single-layered user interface is desirable as this reduces the need to scroll or drill through layers before selecting an option. Moreover, navigating multilevel menus can be stressful and unproductive. This decreases the ease of use and increases the cognitive overload for the user (Kallio and Kaikkonen 2005). The multilayer structure of the mobile user interface design makes it difficult for the user to use the various functionalities that are present on the device as they are not always visible. In a multilayered structure the user has to know where the desired function is and how to access it (Chong et al. 2004). Users generally do not prefer to browse through long lists of complex menu features. Searching lists can be time consuming and requires additional effort to look for menu features that are not readily available or visible.

17.2.7 Minimalist Aesthetics

The design of the mobile app should follow a minimalist aesthetic approach by reducing nondata "ink" that does not convey information (such as graphics and "eye candy") (Tufte 1986; Few 2006). As a second measure, colors should be used conservatively (Few 2006). Most of the interface should employ neutral colors, such as white, gray, or black. Colors should be used to convey information, statuses, or alerts, such as using colored markers to indicate information status and highlighting urgent/emergency data in red. The size of the font and font type are also important considerations. Thorough user testing to determine the optimum font size with the selected display should be conducted.

To make the app simple to use, the features need to be as explicit as possible. In addition, providing automatic data entry options when possible is advised. One way to implement this is by providing features to remember options that are frequently used or the user's password (Microsoft 2014).

17.2.7.1 Log-In/Log-Out Guidelines

Many users prefer the log-in and password process be modified for sensitive and/or small keyboard data entry. Users often have problems knowing if the password code was entered correctly. To remedy this situation, password characters should be made visible for a few seconds, so that the user can verify that the correct letter, number, or symbol was entered, and then be masked with asterisks. Moreover, for applications that require a secure log-in, a visible log-out option needs to be implemented. In the case of healthcare mobile apps, this is especially relevant as healthcare providers need to a way to make sure that improper access to patient information does not happen. Further, a system time-out policy should be determined and implemented. When there is no activity after a period of time, the system should automatically log out the user and close the app.

Another log-in design strategy that is useful and efficient for the user is to implement a mobile app icon for quick access to the app on the device. Lastly, policy makers should consider allowing the log-in credentials to be the same for the desktop and the mobile application. This reduces user cognitive load by eliminating the need for the user to remember another login identification (ID) and password.

17.2.8 Leveraging Agile Development Practices

Following a system development methodology using iterative prototypes is an efficient and effective way of evaluating various user interface designs with users. Agile development practices help ensure that the mobile app capabilities are meeting the anticipated user expectations (Fruhling 2006; Fruhling and Vreede 2006; Fruhling and Tarrell 2007).

Agile software development methods, like Scrum (Schwaber and Beedle 2001), and extreme programming (Beck 2000), are intended to further streamline the development process and bring significant improvements, such as timely delivery of required functionality of information systems. Agile system development claims several anticipated benefits. Agile methods can improve the design process. Iterative releases being used by customers, even those having had little design input, can serve as ongoing usability tests. Agile development also allows automated usage tracking and testing tools to be brought into play sooner and in a more relevant context. Lower-risk release cycles can encourage design experiments and reduce, if not eliminate, writing lengthy specification documents. The fast release pace gives an ongoing sense of accomplishment. Since there are closer team interactions, shared goals, and less solitary time invested in elaborate design, individuals are less defensive about their designs (Armitage 2004). Agile system development styles also purport to embrace situations involving changing or unclear requirements while also promoting user and developer interaction (Beck et al. 2001). Agile methods seek to produce finished, working, reliable code in an iterative, incremental fashion in response.

Agile methods exist to mitigate product development risk. They are more empirical than other information system development methods, essentially using trial and error to reduce the risk of building the wrong thing. Since users are often likely to alter their requirements once they see and test the system, successful projects involve user feedback on a regular and frequent basis. Developers accept the expense of having more routine code rework and having to maintain all development code close to release quality to achieve the gains that come from the user-centered approach, where the user plays a central role during the design process (Armitage 2004).

17.3 Using Mobile Device Simulators for Testing

Testing is an important aspect of the entire mobile development process. It is important to test the app in different environments and using different devices to check compatibility.

One efficient way to test during the development phase is to use software that simulates the mobile device environment, including the physical display. This helps target users to evaluate the display screen layouts quickly and provide immediate feedback to the system developer. The developer can then easily make changes to the simulation software and show the user the new layout. After making the design satisfactory, the more time-consuming effort to write the executable code for the mobile device platform can be done *once* instead of several times. Using simulation software often speeds up the development process.

17.3.1 Aiming for Quick Response Time

The user's acceptable system response time expectation should be carefully discussed and understood. In the case where the system response is taking longer than what is acceptable, the app should provide feedback on the status to the users. Feedback is usually provided if the wait time exceeds more than a few seconds. Should the user decide that the response time is taking too long, the app should provide the ability for the user to cancel the task (Microsoft 2014).

An important consideration for designing a robust mobile user interface design is to employ code-optimization techniques wherever possible. By optimizing software program executables, it is possible to create programs that have a fast execution speed and short response times as well as to minimize the amount of code. Code optimizing is a vital strategy considering the smaller screens found on mobile devices and their limited memory and processing power.

17.3.2 Physical Device Selection Considerations

There are several aspects that need to be considered when selecting the physical mobile device. Mobile devices come in a variety of shapes and sizes, with a multitude of different features. It is possible to purchase devices with hardware keyboard or software-emulated keyboards on screen. The form factor of the device may be designed to be comfortably held in a single hand or with two hands. For healthcare providers in particular, the size of the device is very important. The device must be able to be easily and securely carried by the provider and stowed in a provider's coat pocket. The goal is to have the maximum display size and as lightweight a device as possible.

17.4 Example of Applying Healthcare Mobile Development Strategies

In this section we present a public healthcare information technology project that implemented a new mobile application which was based on an existing desktop/browser information system implementation. We begin by providing background about the emergency response information system for public health; next, we discuss why adding a mobile delivery option was needed and the anticipated benefits; and then we present how we employed various user interface design and development strategies.

17.4.1 Public Health Mobile Application Background

This case study focuses on a PHER information system called Secure Telecommunications Application Terminal Package (STATPack™). STATPack enables microbiologists and pathologists to securely send microscopic and macroscopic digital images of suspicious or unknown organisms electronically to state public health laboratory experts for further diagnostic consultation. The system uses a specialized telehealth network to securely transmit deidentified patient microbiology data and images. The overarching purpose of STATPack was to reduce diagnostic time and eliminate the inherent risks of having the sample delivered by hand through courier to the state public health laboratory (Fruhling 2006, 2009). The STATPack system is shown in Figure 17.1.

The STATPack consultation process involves capturing macroscopic and/or microscopic digital images of culture samples, or possibly original clinical specimens, and sending them electronically to experts at state public health laboratories for consultation. STATPack

FIGURE 17.1
STATPack system. (Courtesy of the STATPACK Group [www.statpack.org], Peter Kiewit Institute, Nebraska, Omaha.)

enables microbiology laboratories around the state to send pictures of suspicious organisms to the state public health laboratory, instead of the samples themselves, thus lessening the risk of spreading potentially deadly bioterrorism agents or infectious diseases and likewise decreasing the feedback time of the consultation process. After viewing the images via STATPack, the state public health laboratory may decide if the sample needs to be sent to the laboratory as a precautionary measure for Centers for Disease Control and Prevention (CDC) confirmatory testing and subsequent reporting. The STATPack system includes an alert process that is bidirectional and has various levels of priorities (drills, emergency, urgent, and routine). STATPack has many more capabilities. For more information, please see www.statpack.org.

Many hospital and private reference laboratories use what are called the rule-out-and-refer (ROAR) guidelines as defined and made available by the CDC. The laboratories that use these procedures are referred to as sentinel laboratories in the CDC's Laboratory Response Network (LRN) (Centers for Disease Control and Prevention 2014). The guidelines are utilized to rule out the presence of biological agents with the use of just a few rapid method biochemical reactions. If the organism cannot be ruled out, the sentinel laboratory's role is not to identify the organism but rather is required to refer to the state's public health laboratory, where confirmation can be achieved by advanced biochemical tests, gene amplification, and other molecular methods.

Experts are trained to recognize the colony morphology and other trigger factors, which assist in recognizing the hazard. Many sentinel hospital laboratories in rural settings are not equipped with such training. Nor do they have medical staff specialized in microbiology. More often, the laboratorians are generalists who cover all sectors of the laboratory. The STATPack system can be essential in affording expertise and consultation in these circumstances.

What was especially unique and challenging about this project is that microscopic and macroscopic images must also be projected at a level where the clarity and quality of the image is at a level that microbiology experts can confidently provide consultation to the laboratorians. The user must also be able to pan across and magnify the image at various intervals.

Expanding the public health emergency response information system to a mobile platform was requested so that the new smartphone technology could be leveraged. Public health microbiologists who were traveling to rural or remote locations could provide timely assistance and not have to wait until she/he has Internet access to a desktop computer. STATPack mobile was requested by the public healthcare provider community to expedite consultation among healthcare providers in pathology, microbiology, and epidemiology at the state public health laboratories and frontline microbiologists and laboratorians in hospitals and clinics using smartphones.

17.4.2 Mobile Solution

One of the key anticipated benefits of the proposed mobile app was that it would help experts in the field respond even more quickly in the case of a public health threat or emergency. The current desktop system is a browser-based system and requires public health experts to have a computer and Internet access to respond. However, this is not always convenient or possible due to the microbiology experts' other duties that frequently take them outside of the office and have them visit rural hospitals. The proposed STATPack mobile app would provide the state public health training coordinator the capability to immediately retrieve the microscopic and/or macroscopic images with the help of a mobile device, in this case, a smartphone. The state coordinator is able to respond within seconds upon receiving the alert, thus saving valuable time, reducing anxiety, and minimizing concern of those involved. In summary, the proposed mobile app would increase the state public health training coordinators' mobility and allow other public health experts respond to the needs of laboratorians across the state within a short span of time. This increased timeliness would be especially beneficial during an emergency or public health crisis.

The corresponding desktop STATPack functionality that the developers wanted to provide via a mobile device was currently available in a desktop delivery mode. Following the same overarching architecture seemed to make the most sense instead of starting from scratch (e.g., building a new mobile app). The developers first needed to determine what functionality should be included and then analyze how to design the user interface so that enough information was presented (in a timely fashion) and that the images to be viewed were of a quality that was needed to confidently assess and consult.

STATPack mobile was developed based on the requirements gathered from the target users about the key features that they expected from a mobile-based public health emergency response system. The IT analysts carefully analyzed the target user work flow as part of this process. It is important to note that STATPack mobile was not intended to replace the PHER STATPack desktop system, but rather enhance the overall PHER STATPack information system.

STATPack mobile was designed to be a thin-client application and when appropriate the existing code was used. By using the existing server code as much as possible, the development time was minimized and process consistency between the two systems was preserved. By extending the server code to the mobile platform, the development team was able to optimize their time and avoid introduction of new bugs or errors that would have required additional resources and time to fix. The functionality for the STATPack mobile system was prioritized based on the previous information from the requirements' study for the desktop STATPack PHER system. STATPack mobile included the functions that were most likely to be used during an emergency consultation. Additional effort was taken to ensure that the selected functionality completely covered all essential aspects of the PHER desktop system.

The seven key functions identified were as follows:

1. *System log-in*: Single log-in credentials are used for both the desktop system and the mobile system. This reduced the need to create a new user log-in for the mobile system. This also ensured that a single set of user credentials would be maintained in the database.
2. *Main menu*: List of all the available features.
3. *Message history*: Provides a list of the messages sent by users to the state health laboratory experts.
4. *STATPack message details*: Provides the message details, such as the microscopic or macroscopic images, related laboratory data, and level of urgency.
5. *Image details*: Displays the image, which can be zoomed, panned, and annotated.
6. *Sending message*: Provides an option to the public health experts to respond to the messages sent by the various users.
7. *Message sent status*: Provides the status of the message sent.

To help the development process go smoothly, data-flow diagrams and scenarios were developed based on these seven functions. Scenarios for each of these functions were used to develop user interface design by using "wireframe" technology. The scenarios were also the used for benchmarking usability testing.

The STATPack mobile app was an implemented module. Each module included a specific set of functionalities and the corresponding user interface. Each module was integrated with the next module that was implemented. The integrated modules were tested and the user approved every iteration throughout the process. Regular STATPack development tools included GNU Compiler Collection (GCC) editor, Subversion, and GTK. Programming languages included C++, hypertext markup language (HTML), PHP, JavaScript, and MySQL. STATPack mobile's tools included BlackBerry® Java® Development Environment (JDE) 4.5.0 and BlackBerry and smartphone simulators.

17.4.3 Mobile User Interface and System Guidelines

In this subsection we present the guidelines with short descriptions and the strategies on how we implemented the guidelines.

Guideline	STATPack Mobile Implementation Comments
Minimize scrolling If screens are too long (five-plus pages), this can become a source of frustration for the user. Scrolling through these many pages is not efficient and can overwhelm users when they are looking for something in particular.	The content is organized in a manner such that it occupies the central part of the screen on which the user concentrates. The screens are also designed to minimize scrolling. Most of the screens require no scrolling, and if scrolling was needed, no more than two scrolls per screen were allowed.

Guideline	STATPack Mobile Implementation Comments
Ease of use The system should be easy to learn, intuitive, helpful, and encourage repeat usage.	The STATPack mobile application design gave high priority to consistent navigation. The users were familiar with several other applications that they use on a daily basis. We considered this and have designed the navigation to be similar. Our design included a button that assists the user with navigation on the right-hand corner of the screen. This was consistent with respect to all of the screens, which helped the user navigate easily. This enabled users to subconsciously click buttons as they are used to this form of navigation.
Menu simplification Menu options should be designed with simplicity in mind. Designing overly complex menus or functionalities should be avoided (Jung 2005).	The screen layouts were designed with simplicity in mind. Complex submenus were avoided and only one or two capabilities were presented on a screen. Data were organized in a tabular manner.
Clear error prompts Error messages are useful only if they can be properly interpreted and help resolve issues. The aim should be to generate clear, concise error messages that inform the user on how to correct the problem.	The STATPack mobile app provided meaningful error messages. For example, when the user enters an incorrect username or password or invalid data, the field in error is highlighted and a short error message is displayed. This helps users to easily correct their mistakes.
Screen optimization In order to help accomplish the scrolling minimization, the information being displayed should be maximized. Overplaying this can lead to an overly crowded screen that can be hard to read. A proper balance needs to be struck between compressing the information on a smaller screen and maintaining a high degree of functionality.	We conducted a user review session to determine the important data that needed to be displayed on STATPack mobile screens. The numbers of screens were decided according to the criteria of one function per screen. Feedback from the users indicated that we have the right balance. See Figures 17.2 and 17.3.
Uniform navigation All methods of navigating the system should be standardized and intuitive.	For STATPack mobile we standardized the navigation by consistently positioning a button on the top of the every screen.
Clearly defined options The buttons provided in the application need to be named in such a way that explicitly specifies the exact operation that they would perform.	The button-naming conventions are the same as those for the desktop application. In-depth usability analysis was done to ensure the best naming selected for the desktop system (Fruhling 2006). However, in some cases, the text in STATPack Mobile is abbreviated. See Figures 17.4 and 17.5.
Home page directory Provide a list of all the features in an easily accessible manner.	STATPack mobile consists of a new home screen that helps the user access the feature of their choice. They can either view only the newest messages received or view the message history.
Shared security ID and password Limit the number of new user IDs and passwords. This helps ease the user cognitive load and also decrease user ID and password administration.	The users can use the same log-in credentials they use for the STATPack desktop application.
Password entry Follow industry standard security measures for data protection and encryption.	STATPack mobile is implemented in such a manner that facilitates the user to view the keyed-in characters for the password for a few seconds, after which they change to the asterisk format.
Feedback There should always be a method to let the users know the system status (e.g., feedback).	STATPack mobile provides feedback when a delay results during the process of sending a message. The user is kept informed about the current status of the message by using a progress bar.

Guideline	STATPack Mobile Implementation Comments
Uniformity Adapting from the desktop information system to the mobile app should be seamless. The mobile application should be instantly familiar to users. The two systems should look and behave in a similar manner.	The STATPack mobile's color schemes and naming conventions are consistent with those of the desktop information system. One change we did was remove images in the logo so that the app would load faster and the display screen would have less clutter.

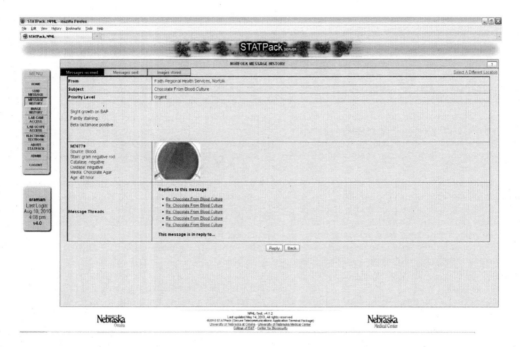

FIGURE 17.2
Desktop image screen showing the layout of the image detail on the desktop platform.

FIGURE 17.3
Mobile image detail screen. The message details and the image if any are displayed.

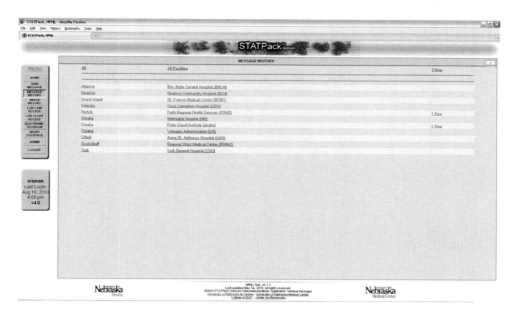

FIGURE 17.4
Desktop facilities screen showing list of healthcare facilities; the corresponding list of new messages received is displayed.

FIGURE 17.5
Mobile facilities screen. STATPack mobile provides an additional screen, which displays only the new messages. The user can opt to view just the new messages or can view the complete list of messages by clicking on the Message History button.

The following discusses the guidelines that we implemented, though we needed to make modifications to the guideline for our environment.

Guideline	Implementation Comments	Modifications
Text entry should be limited Methods should be followed to reduce the instances where the user needs to enter long text data (e.g., a long universal resource locator [URL]). A list of commonly entered text messages will be available.	A STATPack mobile executable can be downloaded onto a mobile device. This allows the user to just click on the STATPack logo to start the application, thus eliminating the need for typing in a lengthy URL and reducing the possibility of introducing typos and frustration.	The STATPack message field can also be lengthy. It is difficult to avoid this, as this is an important aspect of the application. As much as possible, the fields are prepopulated. However, this guideline could not be fully implemented.
Home page link Provide access to the main screen throughout the application.	In the transition from a desktop application to a mobile version, there was a need to establish a landing page.	Currently, there is no main (home) screen because of our goal to have a single-layer structure. With user feedback, the message screen became the default "home" page.
Device testing Test the mobile application on a wide variety of devices to ensure uniform user experience.	The STATPack mobile app was customized to a variety of BlackBerry devices (8800, 8820, 8830, 8900, 9500, 8330, and 8350) and Android devices. Specifically selecting the devices provided a better look and feel to users who use these devices.	In our project we customized the mobile app for BlackBerry devices because at that time BlackBerry devices were restricted by the state government approvals. The targeted user base does not currently have a wide variety of devices. The design still provides the possibility of using the application on other devices because of its thin-client architecture.
Simple navigation Menu items or task options are clearly stated and are easily accessible to the user.	To simplify the navigation, the user is provided with two viewing options: (1) view only the new messages that have been sent and (2) view a complete history of all messages.	Task options are limited to provide a minimal interface that is intuitive and aligned with the user's cognitive and work flow.

17.4.4 Mobile Thin Client

The mobile thin client depends on the server when it comes to tasks that involve the CPU and tasks that are data intensive. The client is browser based and sends user requests to the server and displays the responses on the screen. By adopting this architecture, mobile thin clients do tend to be less powerful and are limited in the number of features they can provide. However, the advantage of running a mobile thin client is a lower cost of ownership, increased security, and a simpler architecture to maintain (Tang et al. 2011).

17.4.4.1 STATPack Mobile Implementation

The web, application, and database servers operate behind a state public health laboratory firewalls and all communication to the clients is over the standard hypertext transfer protocol secure (https) + secure sockets layer (SSL) protocols. Dedicated servers are used to deploy the various components.

STATPack mobile was implemented module by module in an incremental fashion. Each module consisted of a prioritized set of functionality based on user feedback. Each newly developed module was integrated with previous one in an incremental manner. The integrated modules were tested with the help of BlackBerry simulators.

Smartphone simulators proved to be beneficial, especially during testing. The thin-client application can be used in a variety of devices and buying all those devices to test is an expensive affair. Hence, using simulators of various devices to test the application served as an effective way of testing before the users tested it on an actual device. After this was implemented, additional user feedback was gathered, basically focusing specifically on the presentation and ease of use of the graphical user interface. Downsizing of the screens was necessary to fit in all the features and content of a specific feature. Wireframes were created and user feedback was gathered to figure out the most important content necessary to complete the specific task associated with the functionality. The image size, clarity, and quality were carefully evaluated. The screen with just sufficient content to carry out the task effectively was provided to the users.

17.5 Conclusion

In this chapter we present numerous strategies and guidelines that specifically focus on eHealth mobile application design, development, and implementation. The guidelines selected emphasized ways to design the user interface to create the best user experience. The mobile app strategies are aimed to ensure that the new mobile app meets the users' expectations and the implementation and deployment of the mobile app goes smoothly. To help the practitioner better understand how to operationalize the guidelines and strategies, we presented an illustration of a real-world example of a public health emergency response mobile app. The future is bright for the healthcare sector to benefit from information technology innovations, especially in the mobile environment.

Acknowledgments

We greatly appreciate the comments and suggestions from our user community that helped us design and implement a cutting-edge mobile app when mobile apps were just beginning to be introduced. We are indebted to Greg Hoff and Bettina Lechner for their ongoing technical support and to the public health experts at the Nebraska State Public Health Laboratory. They are Anthony Sambol and Karen Stiles. Their ongoing enthusiasm and support of the STATPack system is one of the reasons this project was so successful.

References

Armitage, J., 2004, Are agile methods good for design? *Interactions* 11 (1):14–23.

Beck, K., 2000, *Extreme Programming Explained: Embrace Change*. Addison-Wesley Professional.

Beck, K., M. Beedle, A. V. Bennekum et al., 2001, Manifesto for agile software development.

Bluetooth Special Interest Group Inc., n.d., Available from http://www.bluetooth.com/Pages/low-energy-tech-info.aspx [cited July 22, 2014].

Centers for Disease Control and Prevention, n.d., *Laboratory Network for Biological Terrorism*. Available from http://emergency.cdc.gov/lrn/biological.asp [cited July 22, 2014].

Chittaro, L., 2006, Visualizing information on mobile devices. *Computer* 39 (3):40–45.

Chong, P. H. J., P. L. So, P. Shum, X. J. Li, and D. Goyal, 2004, Design and implementation of user interface for mobile devices. *IEEE Transactions on Consumer Electronics* 50 (4):1156–1161.

Cisco, 2014, *Cisco Visual Networking Index: Global Mobile Data Traffic Forecast Update, 2014–2019 White Paper*. Available from http://www.cisco.com/c/en/us/solutions/collateral/service-provider/visual-networking-index-vni/white_paper_c11-520862.html [cited July 22, 2015].

Craik, K. J. W., 1967, *The Nature of Explanation*. CUP Archive.

Davis, F. D., 1989, Perceived usefulness, perceived ease of use, and user acceptance of information technology. *MIS Quarterly* 319–340.

Dilmaghani, R. B., and R. R. Rao, 2008, A wireless mesh infrastructure deployment with application for emergency scenarios. Paper read at Fifth International ISCRAM Conference.

Few, S., 2006, *Information Dashboard Design: The Effective Visual Communication of Data*. O'Reilly Media Inc.

Fruhling, A., 2006, Examining the critical requirements, design approaches and evaluation methods for a public health emergency response system. *Communications of the Association for Information Systems* 18 (20).

Fruhling, A. L., 2009, STATPack—An emergency response system for microbiology laboratory diagnostics and consultation. In *Information Systems for Emergency Management*, edited by Van de Walle, B., M. Turoff, and S. R. Hiltz. Routledge.

Fruhling, A. L., and A. E. Tarrell, 2007, *Best Practices for Implementing Agile Methods: A Guide for DOD Software Developers*. IBM Center for the Business of Government.

Fruhling, A., and G. J. De Vreede, 2006, Field experiences with eXtreme programming: Developing an emergency response system. *Journal of Management Information Systems* 22 (4):39–68.

Health Level Seven International, n.d., *Introduction to HL7 standards*. Available from http://www.hl7.org/implement/standards/ [cited July 22, 2014].

Jeffries, R., J. R. Miller, C. Wharton, and K. Uyeda, 1991, User interface evaluation in the real world: A comparison of four techniques. Conference on Human Factors in Computing Systems. Reaching Through Technology, CHI '91, pp. 119–124, 1991.

Johnson, C. M., T. R. Johnson, and J. Zhang, 2005, A user-centered framework for redesigning health care interfaces. *Journal of Biomedical Informatics* 38 (1):75–87.

Johnson-Laird, P. N., V. Girotto, and P. Legrenzi, 1998, Mental models: A gentle guide for outsiders. *Sistemi Intelligenti* 9 (68):33.

Jung, J., 2005, The research of mobile user interface design components from the standpoint of universal design for learning. Paper read at IEEE International Workshop on Wireless and Mobile Technologies in Education, 2005: WMTE 2005.

Kallio, T., and A. Kaikkonen, 2005, Usability testing of mobile applications: A comparison between laboratory and field testing. *Journal of Usability Studies* 1 (4–16):23–28.

Lecerof, A., and F. Paternò, 1998, Automatic support for usability evaluation. *IEEE Transactions on Software Engineering* 24 (10):863–888.

Microsoft, n.d., *How to Design a Great User Experience*. Available from http://msdn.microsoft.com/en-us/library/windows/desktop/dn742462(v=vs.85).aspx [cited July 22, 2014].

NearFieldCommunication.org, n.d., Available from http://www.nearfieldcommunication.org/ [cited July 22, 2014].

Nielsen, J., 1994, Usability inspection methods. Paper read at Conference Companion on Human Factors in Computing Systems.

Nielsen, J., 2012, January 4, *Usability 101: Introduction to Usability.* Available from http://www.nngroup.com/articles/usability-101-introduction-to-usability/ [cited July 22, 2014].

North American Healthcare IT Market by Application Delivery Mode and Component—Forecasts to 2017, 2013. Available from http://www.researchandmarkets.com/research/ksbfll/north_american [cited July 22, 2014].

Patrick, K., W. G. Griswold, F. Raab, and S. S. Intille, 2008, Health and the mobile phone. *American Journal of Preventive Medicine* 35 (2):177.

Pervagus Inc., 2002, *Exploring Mobile Applications Technology Landscape.* Available from http://www.idii.com/wp/pmtExploringMobile.pdf [cited July 22, 2014].

Peterson, C., 2006, Be safe, be prepared: Emergency system for advance registration of volunteer health professionals in disaster response. *OJIN: The Online Journal of Issues in Nursing* 11 (3).

Preece, J., Y. Rogers, H. Sharp, D. Benyon, S. Holland, and T. Carey, 1994, *Human–Computer Interaction.* Addison-Wesley Longman Ltd.

Schwaber, K., and M. Beedle, 2001, *Agile Software Development with Scrum.* Prentice Hall PTR.

Shneiderman, B., 1987, User interface design for the Hyperties electronic encyclopedia (panel session). Paper read at Proceedings of the ACM conference on Hypertext.

Tang, W., J. H. Lee, B. Song, M. Islam, S. Na, and E. N. Huh, 2011, Multi-platform mobile thin client architecture in cloud environment. *Procedia Environmental Sciences* 11:499–504.

Tufte, E. R., 1986, *The Visual Display of Quantitative Information.* Graphics Press.

U.S. Department of Health and Human Services, n.d., *Health Information Privacy.* Available from http://www.hhs.gov/ocr/privacy/ [cited July 22, 2014].

Venkatesh, V., M. G. Morris, G. B. Davis, and F. D. Davis, 2003, User acceptance of information technology: Toward a unified view. *MIS Quarterly* 425–478.

Wu, J. H., S. C. Wang, and L. M. Lin, 2007, Mobile computing acceptance factors in the healthcare industry: A structural equation model. *International Journal of Medical Informatics* 76 (1):66–77.

Yee, K. P., 2004, Aligning security and usability. *IEEE Security & Privacy* 2 (5):48–55.

18

Epidemic Tracking and Disease Monitoring in Rural Areas: A Case Study in Pakistan

Hammad Qureshi, Arshad Ali, Shamila Keyani, and Atif Mumtaz

CONTENTS

18.1 Introduction

Universal healthcare remains a distant dream for most of the developing world. The great increase in the population seen in the last 3 decades and a general lack of resources remain as one of the main causes. Every year the situation gets worse, with the current infrastructure coming under more and more pressure, with very few new resources becoming available. The rising costs of healthcare and an aging population are making provisioning of quality healthcare more expensive even, in the developed world. A great challenge faced by most governments and social sector organizations in the current century is the timely provisioning of quality healthcare. Application of community medicine and focus on controlling disease spread and epidemic prevention are seen as a means of mitigating the problem and achieving quality and sustainable public health. However, the model is difficult to implement in countries with a large population. But the rapid advance of modern ICT has made it possible today to remotely monitor diseases and also track epidemics. In this chapter, we present a case study that has been successfully applied in Pakistan. In this study, a computing system which is used to track epidemics and monitor disease outbreaks was developed. The system helps to manage better the healthcare of about half a million people living in a remote rural region.

Over the last few decades there has been considerable improvement in healthcare, but it is still far from satisfactory, requiring many improvements. As stated earlier, an effective and universal healthcare system is still unreachable, even in the greater part of the developed world. A general lack of available funds in much of the developing and underdeveloped world leads to a situation where states tend to fail to fulfill the most basic needs of the public [1]. Moreover, a rapid growth in population has also adversely affected healthcare [2], even in the developed countries. The condition of hospitals in the urban areas tends to be very poor while the situation gets even worse in rural areas, where even the most basic facilities may not be available. In Pakistan we tried to address the problem by implementing a disease-tracking and incidence-reporting system using the existing ICT infrastructure. Disease incidence tracking is important since it can help in controlling an epidemic in its initial stages by using preventive measures. This, in turn, leads to prevention, which can be instrumental in keeping the healthcare costs to manageable levels. The system developed is referred to as the Jaroka Tele-Healthcare System (JTHS), which uses mobile communication systems and computing facilities, along with Google Maps, to geographically track the incidence of diseases to provide real-time information on disease incidences.

The majority of the population in Pakistan live in rural areas (about 65% of the total population) which do not have access to any kind of healthcare. Moreover, the situation in hospitals in adjoining semiurban and urban areas is not much better due to lack of staff and equipment. In our work, we tried to address these problems by using modern information and communication technology, since in recent years, the country has experienced a rapid rise in the ICT infrastructure. There has been a substantial increase in Internet and mobile phone usage all over the country. The JTHS has been designed such that it can make use of this existing infrastructure to provide quality healthcare [3]. However, in the process the JTHS has quickly evolved into a full-fledged epidemic-tracking system.

Pakistan as a country is well suited for this study due to at least two reasons: (1) the fact that there is a vast network of government-sponsored health workers already working in the county, with most of the rural population in Pakistan depending upon these community health workers for their healthcare needs; and (2) the reach of telecommunication networks has expanded exponentially over the last decade, leading to 90% penetration nationwide. Therefore, mobile telephony is available in far reaches of the country. Therefore, building a system that uses mobile and Internet technology is possible. Moreover, the availability of a network of community health workers has provided an opportunity to reach easily to vast populations in remote or underserviced areas.

An interesting program was developed in Pakistan in 1994; it was called the Community and Lady Health Workers (CLHWs) program. Although CLHWs had been set up 20 years ago to broaden the reach of health services to rural areas, the structure has not allowed them to communicate with doctors and healthcare experts. They could not communicate effectively, even in the cities, just to report serious cases. Traditionally, CLHWs members go from door to door to collect patient data and attempt to provide preliminary medical treatment in rural areas and remote villages. Clearly, the collected data have the potential to be used for developing a modern epidemic-tracking and disease-surveillance system. Moreover, recently the CLHWs are also provided with mobile phone connectivity by the Ministry of Health of Pakistan as a part of the Mobilink GSM Development Fund Program (2008) [4]. However, the mobile phone connectivity has not successfully been used to this date for the provision of healthcare. The existing network of CLHWs provides an opportunity to implement an ICT-based program for improving a wider and efficient healthcare access.

It is clear that an effective utilization of the existing ICT infrastructure has the potential to bring improvements in the healthcare sector. In this regard, our telehealthcare system, JTHS, is a contribution toward development of an epidemic-tracking and disease-surveillance system that can be used to improve the health of communities in urban areas as well as rural areas of the country. Available telehealthcare mostly uses Internet-based systems [5], but nowadays the trend has shifted toward the use of mobile telemedicine for surveillance and monitoring purposes [6].

The JTHS has been developed by broadly using mobile phone platforms to provide healthcare services, and it has been deployed in the rural areas of Mardan, a district in Pakistan. The SMS-based system provides the services for registering patients, reporting symptoms, and acquiring prescriptions while the patient is still in the field. The incoming data are entered by the CLHWs from remote locations and used to track the spread of diseases. Google Maps has also been used to show the geographical location of patients and associated diseases. The system can also be used to acquire diagnosis for patients, which is reported in detail by Keyani et al. [3].

18.2 Jaroka Tele-Healthcare System: A System for Disease Surveillance

It is common to use geographical disease representation on maps to record disease incidence and disease occurrence rates. To detect an emerging geographical cluster on the map due to sudden occurrence of diseases is of great importance for public health [7]. This can be used to discover rising epidemics, track spreading diseases, and unravel previously unknown diseases that can pose an emerging threat. To this date, no systems for disease incidence tracking existed in Pakistan either in public or in the private sectors which can collect, process, analyze, and present the information in a visual form [8]. There are various conventional ways to visualize data, such as tables, histograms, pie charts, and bar graphs [9]. But these tend to be slow techniques and are not instantaneous. Moreover, in these techniques managing large data is difficult due to the recent exponential increase in the amount of available data. Hence, there is a need for effective methods that quickly and easily optimize the way of handling instantaneous data.

Recently, the geographic information system (GIS) has emerged as one of the revolutionary techniques for geographically locating diseases and studying their spread. It has been used for surveillance and monitoring of almost all types of applications and it is used frequently for demonstrating exposures [10–11]. In our work, we have employed geographical mapping techniques to record and track diseases as they occur. However, before we explain how the mapping is done, it is important to introduce the JTHS in detail.

18.2.1 How Does the Jaroka Tele-Healthcare System Work?

Figure 18.1 depicts the overall working mechanism of the JTHS. Online web servers are connected to the workstations at the rural Mardan medical facility in Zahidabad through the Internet. The Internet connectivity was provided by Cybernet as a corporate social responsibility initiative. Round-the-clock electricity is maintained using generators as the area suffers from power outages for almost 12 h on a daily basis. The systems in Zahidabad are connected to the servers at the National University of Sciences and Technology (NUST), where all the patient records are stored and managed. The system is also coupled with a mobile-based SMS system that allows the workers in the field to record patient demographics, register symptoms, and in some cases also acquire diagnosis from medical professionals in the cities and in the Zahidabad medical facilities. In some of the severe cases, we are able to engage physicians from the Association of Pakistani Physicians of North America. who review symptoms of the patients that are available on our system and provide diagnosis and advice.

The system is designed using open-source technologies, namely, Linux, OpenEMR, and Kannel. The server is deployed on Linux as this operating system (OS) is highly secure and stable and the majority of the servers in the world are hosted on Linux. OpenEMR is a renowned software system used by hospitals and medical professionals worldwide for electronic health record storage and exchange and medical practice management. This system is free to use and has many useful features, which include patient demographics, scheduling, electronic medical records, computer-assisted prescription management, clinical decision

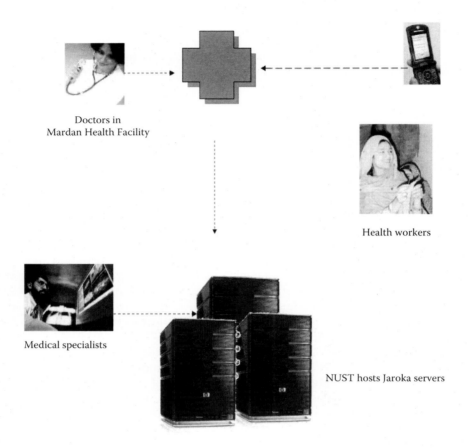

FIGURE 18.1
Overview of how the JTHS works.

rules, patient portal, and multilanguage support. Kannel is a powerful open-source wireless application protocol (WAP) and SMS gateway. It is used by thousands of organization world-wide for implementing SMS, WAP push service, and mobile Internet connectivity. Kannel coupled with an short message service center (SMSC) device is used for exchanging messages and communicating with the people in the field. Figure 18.2 shows a block diagram of how the various components of JTHS interact.

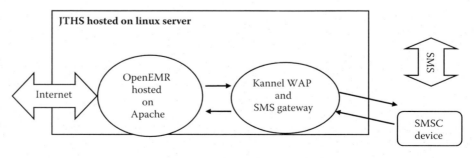

FIGURE 18.2
Major components of JTHS working in tandem.

The disease data for patients in a geographical area are recorded using the system described above. Subsequently these data are utilized for depicting the disease incidence per location where the data were collected or the address of a patient if it is known. The data used in this chapter have been obtained from the JTHS database for the Zahidabad clinic in Mardan. This database stores the data of patients who visit the rural Mardan hospital facility and also patients who get registered over mobile phones through the health workers. This facility has been in operation since 2008 and the patient data available for analysis are vast; therefore, a subset of the data is used for analysis here.

18.2.2 Mapping

The simplest way to characterize the spread of diseases is in the form of a rate, i.e., to compute crude rates by dividing the number of people affected by the total population [12–13]. Although crude rates are good for indicating areas where the frequency of occurrence of a disease is high [4,14], they have a major drawback for not accounting for the differences in the population densities in different areas. Often referred to as the small-number problem, spurious variation in rates can result from small denominator data (i.e., population) [15]. Extremely high disease rates can be noted in rural areas with small populations where a very small segment of the population may be affected by a particular disease. Furthermore, great fluctuations are seen in disease rates with the addition of one or two extra cases in low-population areas. To overcome this problem, we display the actual figures rather than the ratio of people affected in the total population.

To examine the patterns spatially, we employ local cluster analysis of disease outbreak for rural areas of Mardan. Such local statistics are useful for analyzing the spatial variation of clusters. Figure 18.3 depicts the incidence of a disease by using Google Maps. The

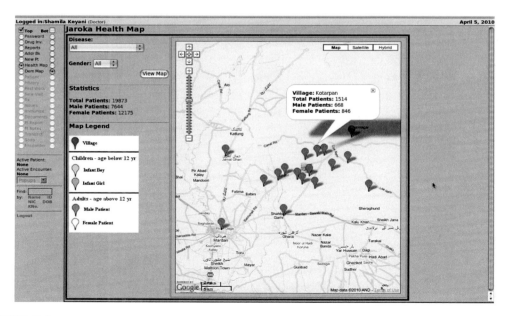

FIGURE 18.3
Number of patients for a certain disease in a defined geographical area, such as a village.

patients are depicted at the respective geographical location and this map is referred to as the Jaroka health map. Such mapping may be used to track disease incidence and to subsequently develop plans for mitigating the associated health problems.

Figure 18.4 shows incidence of a particular disease in a certain geographical area. It is interesting to see that the disease incidence among men, women, and children is treated differently and also displayed using different colors. The data are subsequently clustered and shown as balloons, as shown in Figure 18.3. Such a geographical mapping of disease incidence may be used by health professionals and governments to study disease spread and also analyze and predict whether an epidemic is in the making. To test for the presence of clusters of diseases, we employ the aggregation function for gathering all the patients falling into a single disease category and locating their villages from the data to identify the position of the clusters. The coordinates of the locations are stored in our database and the grouped data are displayed at the location in the form of spatial clusters. The Google Maps application programming interface (API) is used to display the results.

Clustering allows us to identify areas of high disease incidence. To detect if there is a significant association between the environmental exposure and clusters of diseases, the magnitude of the association must be large and the supposed exposure should be pervasive in the community [16]. The areas with high rates contain an increased number of patients with particular diseases. The results can be viewed in the form of disease category, gender, and age group. We use different color markers for distinguishing between patients, as shown in Figure 18.4. The JTHS is the first of its kind in Pakistan that is capable of depicting disease incidence geographically.

The patients were mapped, using the JTHS data, to visually show the geographic distribution of diseases. We displayed the data as per the regional scale, showing the villages

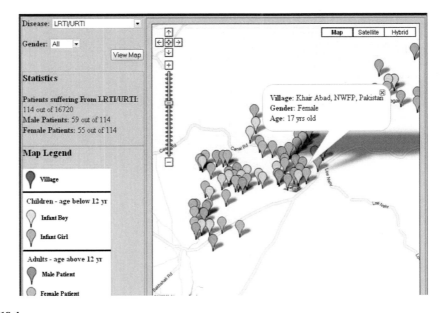

FIGURE 18.4
Spatial pattern with the disease category selected (e.g., lower respiratory tract infection [LRTI]/upper respiratory tract infection [URTI]).

TABLE 18.1

Patient Statistics in Rural Areas of Mardan for 2009–2010

Village Name	Male Patients	Female Patients	Total
Mardan	327	473	803
Bakhshali	356	666	1022
Gujrat	1009	1840	2849
Bhai Khan	28	58	86
Khair Abad	96	172	268
Katta Khat	1040	1367	2407
Kotarpan	639	797	1436
Surkh Dheri	5	12	17
Cham Dheri	18	35	33
Chargulli	411	745	1156
Narshak	3	2	5
Rustam	707	1103	1819

and the surrounding rural areas of Mardan. Comparing the aggregated (summing up the number of males and females suffering from a particular disease) regional level map to the maps with crude rates serves two analytical purposes. First, the ratios between the crude and the aggregated regional maps can be compared to report with confidence that both show a similar pattern for a particular disease. Second, each area depicts its own disease aggregate and compares it with the crude rates, which helps in identifying which areas contain more patients with a particular disease. Figure 18.3 shows the regional-level map, which not only provides a stable disease map for reference purposes but also presents disease estimates that can be used by local managers and public health officials.

The regional-scale statistics are shown in the form of information windows that are associated with each village. The information shown consists of the total patients affected and categorizes them into gender-based distributions as well. The figures are acquired for the rural areas near Mardan, shown in Table 18.1. These statistics show the number of patients from a particular village who have visited our Mardan facility for a checkup or have been registered by the CLHWs over the mobile phone. The identification of clusters of elevated diseases may be achieved through local spatial statistics rather than simple visual interpretation because size and potential for spurious rate variation can create an illusion of clusters that may not be statistically significant [17].

18.3 Disease Trends

To better interpret the data from JTHS we have classified the diseases into 46 disease categories, as given in Table 18.2. These categories were provided by the medical doctors employed at the rural Mardan healthcare facilities. In order to tag a patient to a disease category, the disease category for the patient is stored along with the subjective, objective, assessment, and plan (SOAP) record.

The classification of diseases is used to generate disease statistics on a monthly basis. These statistics reflect the overall health situation of the area and which disease incidence is on the rise and which is declining. Figures 18.5 through 18.7 show the disease trends for

TABLE 18.2

Generalized Disease Categories

Serial No.	Disease Category
1	LRTI/URTI
2	Pulmonary tuberculosis
3	Bronchial asthma
4	Chronic obstructive pulmonary disease (COPD)
5	Occupational lung diseases
6	Gastroenteritis
7	Worm infestation
8	Dyspepsia/ulcer
9	Enteric fever
10	Acute viral hepatitis
11	Chronic liver diseases
12	Skin infections
13	Scabies
14	Eczemas
15	Urinary tract infection
16	Conjunctivitis
17	Cataract
18	Dental diseases
19	Menstrual irregularities
20	Vaginal candidiasis
21	Antenatal
22	Postnatal
23	Chronic suppurative otitis media (CSOM)/serous otitis media (SOM)
24	Deviated nasal septum (DNS)/polyps
25	Tonsillitis
26	Sinusitis
27	Psychiatric diseases
28	Diabetes mellitus
29	Thyroid diseases
30	Cushing's disease
31	Malnutrition
32	Delayed milestones
33	Congenital defects
34	Hypertension
35	Ischemic heart disease
36	Acute appendicitis
37	Nephrolithiasis
38	Cholelithiasis
39	Backache, body ache
40	Hernias
41	Incision and drainage
42	Traumas
43	Benign prostatic hyperplasia
44	HIV, AIDS
45	Polio
46	Valvular heart disease

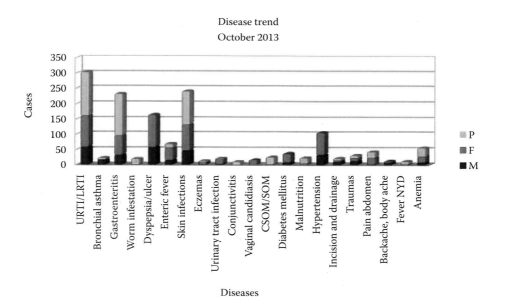

FIGURE 18.5

Disease trend for the month of October 2013 showing disease incidences in children (P), women (F), and men (M). NYD stands for "not yet diagnosed."

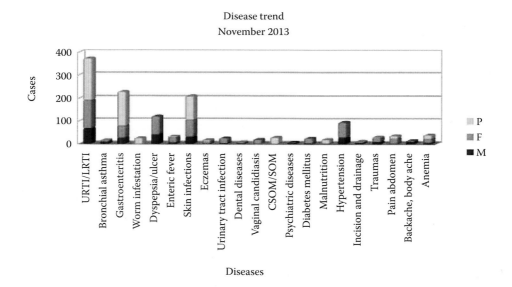

FIGURE 18.6

Disease trend for the month of November 2013 showing disease incidences in children (P), women (F), and men (M).

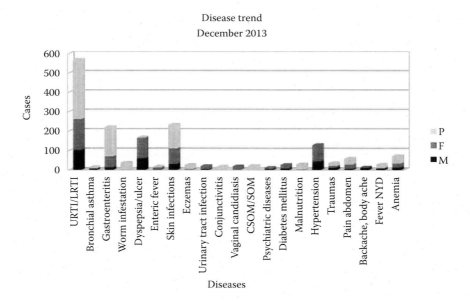

FIGURE 18.7

Disease trend for the month of December 2013 showing disease incidences in children (P), women (F), and men (M).

3 months toward the end of 2013. Statistics just as these are published on a monthly basis on our website http://www.tele-healthcare.org. These statistics can then be used to plan interventions and also apply community medicine in order to improve the overall health of the community. This, in turn, brings the cost down and the overall health of the community improves.

Figures 18.5 through 18.7 show that URTI/LRTI is on the rise among children, men, and women. On the other hand, gastroenteritis, ulcers, and skin infections are also clearly prevalent among the population. These diseases can be controlled by making interventions in key areas such as by improving the quality of drinking water. Skin infections can also be catered for by working on improving the hygiene of the people through education and provisioning of hygiene kits. Hence, most of the diseases on the rise may be controlled or completely removed using simple precautionary measures. This would, in turn, lead to lowering of the high patient volume at hospitals and tertiary care centers and overall improvement of public health at a more reasonable cost.

18.4 Conclusions

Patients' disease data are fraught with inconsistencies across databases. Different practice management systems have used different design structures to store the related data. Due to the lack of a standardized disease classification scheme, disease outbreaks are not counted in the same way for any two databases [18]. In fact, disease occurrence varies geographically. Pakistan's Health Management Information System generates yearly reports for disease epidemics for all provinces of Pakistan. Mostly the epidemics are caused by

TABLE 18.3

Report on Epidemic Diseases for October–December 2006

Province/Region	No. of Reports	Reporting Rate	Suspected Cholera	Suspected Meningococcal Meningitis	Probable Poliomylelitis	Probable Measles
Punjab	7709	86%	64	27	13	14
Sindh	2633	87%	45	14	19	31
NWFP	2796	90%	60	0	0	146
Balochistan	2590	88%	135	17	1	115
Federal	21	44%	0	0	0	0
AJK	1314	88%	0	2	0	1
NAs	658	83%	3	26	0	14
Grand total	17,721	81%	307	86	33	321

Source: Health Management Information System, Health Division, Islamabad, Pakistan.
Note: AJK: Azad Jammu and Kashmir; NWFP: North-West Frontier Province.

sanitation problems and lack of good water resources in rural areas. Table 18.3 presents the epidemic disease rates for the year 2006. The data in Table 18.3 are the latest data available to us. It represents information over wide geographical areas in Pakistan only. On the other hand, JTHS allows for real-time and instantaneous monitoring of the diseases in a geographical area it targets: the cities, towns, or villages.

The tracking of disease spread incidence and epidemics has become very important in the present age, in which epidemics spread quickly and healthcare provisions are short. Rising populations and the rising cost of healthcare coupled with a general lack of healthcare infrastructure in the developing world have led to a virtual breakdown in public health. Hence, as shown in this case study, information and communication technology can be used to help improve local healthcare in many countries, and the sustained use of such technologies may someday lead to a dream of universal effective healthcare.

Acknowledgments

The authors would like to acknowledge the funding provided by the Higher Education Commission, Pakistan, and the United States Agency for International Development (USAID) for supporting this work.

References

1. M. Akram and F. J. Khan, Health Care Services and Government Spending in Pakistan, Pakistan Institute of Development Economics Working Paper No. 32, Pakistan Institute of Development Economics, Islamabad, February 2007.
2. Population Association of Pakistan, Population Statistics, Available at: http://www.pap.org .pk/statistics/population.htm#fig1.3. Accessed on May 3, 2011.

3. S. Keyani, A. Mumtaz, H. Mushtaq, and A. Hussain, Affordable and accessible tele-healthcare to rural areas of Pakistan through web and mobile based technologies, in *Proceedings of 6th IEEE International Symposium on High-Capacity Optical Networks and Enabling Technologies (HONET)*, pp. 110–114, 2009.

4. Rural Communications for Health Programme Overview, Lady Health Worker and GSMA Development fund, Available at: http://www.gsmworld.com/documents/lady_health_worker _pakistan.pdf. Accessed on January 10, 2008.

5. B. Tulu and S. Chatterjee, Internet-based telemedicine: An empirical investigation of objective and subjective video quality, *Decision Support Systems*, 45 (4), pp. 681–696, 2008.

6. U. Varshney, Pervasive healthcare and wireless health monitoring, *Springer Science and Business Media*, (12), pp. 113–127, 2007.

7. G. Jamtvedt, J. M. Young, D. T. Kristoffersen, M. A. Thomson O'Brien, and A. D. Oxman, Audit and feedback: Effects on professional practice and health care outcomes, *Cochrane Database of Systematic Reviews*, 6, June 13, 2012.

8. D. A. Griffith, *Spatial Autocorrelation and Spatial Filtering: Gaining Understanding through Theory and Scientific Visualization*, Berlin: Springer-Verlag, 2003.

9. V. Friedman, *Data Visualization: Modern Approaches*, Available at: http://www.smashingmaga zine.com/2007/08/02/data-visualization-modern-approaches/. Accessed on January 25, 2015.

10. G. A. AvRuskin, G. M. Jacquez, J. R. Meliker, M. J. Slotnick, A. M. Kaufmann, and J. O. Nriagu, Visualization and exploratory analysis of epidemiologic data using a novel space time information system, *International Journal of Health Geographics*, 3 (26), 2004.

11. K. Elgethun, R. A. Fenske, M. G. Yost, and G. J. Palcisko, Time–location analysis for exposure assessment studies of children using a novel Global Positioning System instrument, *Environmental Health Perspectives*, 111 (1), pp. 115–122, January 2003.

12. L. Anselin, N. Lozano, and J. Koschinsky, Rate transformations and smoothing, *Geographical Analysis*, 38 (1), pp. 5–22, Spatial Analysis Laboratory, Department of Geography, University of Illinois at Urbana–Champaign, 2006.

13. P. A. Buescher, Age-adjusted death rates, *Statistical Primer*, 13 (1), pp. 3–7, North Carolina Department of Health and Human Services, 1998.

14. K. A. Borden and S. L. Cutter, Spatial patterns of natural hazards mortality in the United States, *International Journal of Health Geographics*, 7 (64), 2008.

15. R. Haining, *Spatial Data Analysis: Theory and Practice*, Cambridge: Cambridge University Press, 2003.

16. P. B. English, An introductory guide to disease mapping, *American Journal of Epidemiology* 154 (9), pp. 881–882, 2001.

17. K. Osnes, Iterative random aggregation of small units using regional measures of spatial autocorrelation for cluster localization, *Statistics in Medicine*, 18 (6), pp. 707–725, 1999.

18. W. W. Carder and N. K. Kwak, An analysis of standardized disease classification system for Health Care Planning, *Journal of Medical Systems*, 1 (2), pp. 151–160, 1997.

19

mHealth and Web Applications

Javier Pindter-Medina

CONTENTS

19.1 Introduction

An essential part of human evolution has been the development of scientific knowledge in almost any aspect of human life, with the main intention of improving the quality of life and preventing diseases, among others. Today, advances in the field of healthcare have brought a significant increase in life expectancy and all of that is a result of years of research and development. In the 13th century, the average life span at birth was only about 35 years [1], whereas now estimations suggest that the life expectancy of the global population will be 73.16 years by 2020. While living longer is desirable; the constantly growing and aging population that currently lives in the world (about 7.7 billion humans in July 2014, est. [2]) brings consequences that impact directly on World economy. Every year, governments designate higher resources to attend the demand of more complex health services. Furthermore, the existence of multiple armed conflicts related to cultural differences, religious intolerance, and political issues, among others, has made it extremely difficult to cover the right to health of every human being [3], mainly in countries or regions underdeveloped in contrast with developed ones.

International organizations like the World Health Organization (WHO) and other entities work day-by-day promoting the use of technology with the goal that one day the access to quality health services will be available and ensured to all human beings on the globe. The use of telecommunication infrastructure along with health technologies has meant a breakthrough in the aim to reach this objective; the use of telemedicine systems has been available for nearly 30 years. Initial applications were based on the use of fixed phone lines for patients to communicate with doctors; as soon as technology allowed it, the support of videoconferencing was added [4]. Nowadays the possibilities and options for remote medical assistance have widened, from a simple phone call to remote surgery [5]. Therefore, telemedicine as a concept has been limited and it has been included in a new revolutionary concept called **eHealth**. The term eHealth was used for the first time in November 1999 at the Seventh International Congress on Telemedicine and Telecare, in London, United Kingdom [6]. In general, eHealth covers all telemedicine and telecare aspects with the addition of new standards for handling medical records and any other element of the information technologies used for medical purposes. In short, eHealth is a group of systems and infrastructure for providing medical services at a distance which allow constant monitoring of the patients from the shelter of their homes and workplaces or even when they are on the move.

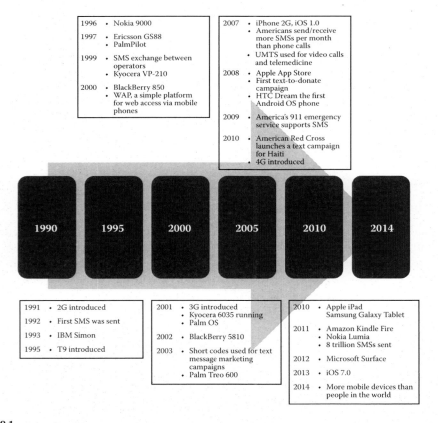

FIGURE 19.1
A time line of how text messaging and cellular networks evolved during the last 25 years.

Along with the advances in the field of medicine and health, the telecommunications technology has been growing exponentially during the last 25 years (see Figure 19.1). Moreover, due to its relatively low cost, according to the International Telecommunications Union (ITU), active wireless lines exceeded the number of people in the world in 2014, representing over 7.1 billion subscribers. With those facts, the scientific community started using mobile and wireless technologies [7] to support the achievement of health objectives. In this case and for the first time, these technologies bring the potential to transform the approach of how those health service are delivered across the globe, even for developing regions. These strategies ended with the creation of a new term known as **mobile health,** or **mHealth**. The growth of mHealth is directly related with the coverage of mobile cellular networks and with its capability of being integrated into existing eHealth services. The GSM Association reports that commercial wireless signals cover over 85% of the world's population, extending far beyond the reach of the electrical grid [8]. For the first time WHO's Global Observatory for eHealth (GOe) has sought to determine the status of mHealth in its member states; its 2009 global survey contained a section specifically devoted to mHealth. Completed by 114 member states, the survey documented for analysis four aspects of mHealth: adoption of initiatives, types of initiatives, status of evaluation, and barriers to implementation. Fourteen categories of mHealth services were surveyed: health call centers, emergency toll-free telephone services, management of emergencies and disasters, mobile telemedicine, appointment reminders, community mobilization and health promotion, treatment compliance, mobile patient records, information access, patient monitoring, health surveys and data collection, surveillance, health awareness raising, and decision-support systems [9].

19.2 mHealth Today

In contrast with eHealth, an mHealth system is intended to provide medical services anytime and anywhere a wireless connection is available. The mHealth system could be as simple as a network based on mobile phones' built-in basic GSM support for text messaging, or as complex systems as needed that combine smartphones with portable biomedical instruments (PBIs) with advanced telecommunication systems where PBIs based on low-power mixed-signal microcontroller and wireless devices will allow end users or medical staff with basic training to install, measure, and/or record specific biological signals at remote locations, even though a patient is on the move. Those mHealth services are available to patients depending on demanded services, locations, and the availability of telecommunications systems. In a complex context where health services are becoming more expensive, sophisticated, and personalized due to population aging, complex diseases, patients' mobility, remote population distribution, and diverse ways of life, mHealth systems represent an intelligent and affordable solution for public and private organizations.

Today mHealth systems are a hot research topics for universities, the industry, and standard organizations where PBIs, telecommunication systems, data-processing systems, data privacy, and online health and on-the-go telemedicine services, databases, and access to databases are under constant development. Inasmuch as any wireless technology can be used to achieve mHealth, it is logical to isolate the concept in terms of services and technological approaches. Of course, mHealth is not limited only to humans; the animal kingdom is also benefitted by this technology. At least today, dogs, cats, and even horses already have mHealth devices; e.g., the AliveCor Veterinary Heart Monitor, which works with iPhone 4 or 5S devices, bringing the healthcare of animals to the next level [10].

19.2.1 mHealth Based on Text Messaging

One of the most common services for mHealth used in both developed and developing countries might be the SMS since this solution provides a relatively cheap and easy way of giving people a first approach to real and quality health services; more importantly, this technology does not require a high-end smartphone or the last-generation Long-Term Evolution (LTE) network to work. Furthermore, today 100% of mobile phones, smart or not, are now SMS enabled, and in major cases all carriers offer to their postpaid users' unlimited texting [11]. Much better, today many companies like Konekt and Particle offer contract-less global SIM cards that bring endless possibilities for the development of health solutions.

mHealth systems based on SMS technology could work as a broadcasting service in which any subscriber receives the same SMS message with general information about a subject of common interest. In contrast, more advanced systems could treat each subscriber as a unique patient in order to create a health record and return to the subscriber, via SMS messages (Figure 19.2), specific and relevant information related to his/her condition.

Each country has its own needs regarding the approach to healthcare services to its citizens. For instance, some of the most popular text-based services in the United States are those provided by Voxiva (considered as one of the pioneers using SMS focused on mHealth) [11]. About 4 years ago, Voxiva launched Text4baby with the purpose of reaching pregnant women and new mothers to educate and accompany them during all the maternity process. Due the success of Text4baby, four more services are available from this company:

- Text4kids: Used to help parents raise their children as happy and healthy as possible
- Text4health: A coaching service for helping people to understand the importance of physical activity in order to have healthier lives

FIGURE 19.2
A simple diagram showing an SMS mHealth service.

- Text2quit: A coaching service for helping people who want to quit their smoking habits
- Care4life: Used to educate people about diabetes and helps them to manage their illness; this service is available in English and Spanish languages

The private initiatives sometimes promote the governments to take care of its citizens and start their own public healthcare services. The U.S. Department of Health and Human Services started a couple years ago the Text4health projects with the intention of providing text-based health applications. Later, on December 6, 2012, the Federal Communications Commission of the United States announced that the service named text-to-911 will be available by June 30 for almost all of the United States [12].

Developed countries are not the only ones that offer SMS-based services; there are some international organizations that offer free solutions for emerging countries. In those cases, the services provided were classified into four modes of intervention: one-way communication, two-way communication, with or without incentives. and with educative games. Most of the applications developed are focused to help people with HIV, but some also are oriented toward the following:

- Sexual and reproductive health
- Malaria
- Diarrhea
- Others

Some of the best nonprofit or free applications for mHealth based on SMS services according to mHealthAfrica.com are shown in Table 19.1.

There are some concerns about using SMS as a platform for mHealth services. While it is a cheap and affordable technology, it is also limited to 160 characters (140 characters in some regions and for Tweets), so the text-based applications should take that restriction into consideration. Moreover, security is also a great issue because SMS messages cannot be encrypted in the fly without the use of a third-party application that should be cross platform in order to ensure the maximum compatibility between users and the health service provider, which may cause a significant increase in costs and development times.

Despite the stated concerns, SMS could be the best way to achieve one of WHO's millennium objectives, which is to bring universal access to reproductive health in 2015. And that is particularly important because the segment of the population in most need of mHealth solutions is typically those who can least afford it. Perhaps in the near future the SMS will be replaced by modern services like WhatsApp or similar services.

19.2.2 mHealth and Smartphones

The immersion of smartphones into the market, along with faster mobile networks, meant a major step forward for mHealth applications. Every day more advanced technologies are available and accessible for more people around the globe; these increase the potential for further healthcare delivery. Moreover, far from ubiquitous, the spread of smartphones and the Internet-of-Things open up the doors for more sophisticated mHealth projects, such as technology for self-diagnosis, real-time monitoring and telemedicine with support for videoconference, web browsing, and GPS/IPS tracking of patients with Alzheimer's disease, among others. A good example of how mHealth should interact with smartphone

TABLE 19.1

List of SMS-Based mHealth Applications for Attending Some Needs of Developing Countries

Application	Details	Users	Website
Ushahidi	Health facilities' locations, epidemic mapping	Schwab Foundation, United Nations (UN) Haiti, *Washington Post*	http://ushahidi.com/
FrontlineSMS	Health workers' networking, health-related m-learning	Georgetown University's Institute for Reproductive Health (IRH), Plan International	http://www.frontlinesms.com/
OpenMRS	Electronic medical record, health demographics, HIV treatment	Google, Rockefeller Foundation	http://openmrs.org/
RapidSMS	Health workers' communications, automated treatment alerts, nutritional surveillance	ChildCount, UNICEF	http://www.rapidsms.org/
Sana	Remote diagnosis, protocol management	Wellbody Alliance, Boston Children's Hospital	http://sana.mit.edu/
Text to Change	Health training, point-of-care reporting	UNICEF, Airtel, Health Child, African Medical and Research Foundation (AMREF)	http://www.texttochange.org/

platforms and take advantage of their technology occurred in 2009 when Apple showed at its Worldwide Developers Conference the potential of its iPhone 3G for being a suitable base for mHealth applications. Apple demonstrated on stage that different apparatuses, in that case a blood pressure monitor along with a glucose meter, could interact with the iPhone 3G via its dock connector or Bluetooth [12]. That event paved the path for the development of a whole new universe of mHealth-related wearable devices. Many of those devices are now commercially available and their manufacturers are seeking for approval from the most important regulatory organizations, like the FDA in the United States.

19.2.3 Five Years of History of Smartphone-Based mHealth Devices

After showing the capabilities of smartphones in the field of mHealth, it seemed like a good decision to take a brief look back at the history of smartphone-based mHealth devices over the course, starting from 2009 through late 2014 [13].

19.2.3.1 2009

In early 2009, iTMP Technology Inc. launched the SMHEART LINK, which acts like a wireless bridge between an iPhone and digital heart monitors from different vendors. This device is oriented toward fitness. Later, on March 17, 2009, at Apple's Worldwide Developers Conference, an iPhone 3G was used as a platform for mHealth applications. On April 8, CellScope, first developed in 2007, won a Vodafone Americas Foundation award. CellScope was focused on the idea of helping physicians detect ocular diseases, tuberculosis, blood diseases, and parasitic worms [14]. The current commercial version of the CellScope is the CellScope Oto, which converts an iPhone into a digital otoscope. Entra Health Systems, on July 16, 2009, launched the MyGlucoHealth Clinical Point-of-Care System, which creates a Bluetooth interface between the glucose meter and a smartphone. The first application was created for the iPhone, but in 2010 an app was launched for Nokia. On September, a French company named Withings announced the first Wi-Fi–enabled body scale with a built-in web browser including support for tablets and smartphones.

19.2.3.2 2010

AgaMatrix, a U.S. company, claimed that its WaveSense JAZZ meter could be the first medical device capable of being directly connected to Apple's iOS platform, including iPad, iPhone, and iPod touch. Today, this company, along with its partner Sanofi-Aventis, has the first FDA-cleared glucose meter (iBGStar) that is fully compatible with Apple's iOS platform. During the fall, BodyMedia announced its armband with apps made for Android and iOS platforms. The device is Bluetooth based and it is focused on three aspects: calories burned, calories consumed, and sleep quantity and quality. Finally, on December 30, Dr. David E. Albert presented the first ECG monitor based on the iPhone. Today, the device is selling via AliveCor's website and supports some Apple devices and Samsung and HTC Android-based devices [15].

19.2.3.3 2011

iHealth Lab and Withings announced their respective blood pressure monitor for the iOS platform. Later, on February, the FDA granted clearance to Mobisante MobiUS, a smartphone-based ultrasound system, which used Microsoft Windows Mobile [13]. During the second half of the year, Proteus Biomedical publicized that it received a patent for an ingestible biomedical sensor for monitoring medication adherence, Raisin, which is capable of sending information via Bluetooth by using a special receptor attached on the patient's skin to an iPhone.

19.2.3.4 2012

At the beginning of this year, Nike launched the Nike+ FuelBand, which interacts with iOS-based devices in order to monitor calories burned and steps taken. Note that every year, more and more mHealth devices are launched with specific orientation toward fitness. Later, a company called Vignet announced the first iOS app that met the Continua Health Alliance interoperability standards, which allows the platform to interact with multiple medical devices, such as weight scales, pedometers, and blood pressure monitors. At the end of the first quarter of the year, a German company named Medisana introduced the ThermoDock®, an infrared thermometer compatible with the iOS platform. Asthmapolis, on July 11, received a clearance from the FDA for its asthma and child seat presence and orientation detection (CPOD) sensor, which uses an app to track the time and location of each medication discharge coming from the patient's inhaler. This year, EyeNatra continued developing a smart app that today is known as Netra^G to help people with any ophthalmologic knowledge to perform an eye test using a smartphone; some of the tests are for near sightedness, farsightedness, and astigmatism. In the last month of this year, Masimo released the iSpO2, an iOS-enabled pulse oximeter. Finally, SHL Telemedicine received FDA approval for its smartheart, which was the first personal 12-lead ECG that is Bluetooth enabled for sharing data via a user's smartphone.

19.2.3.5 2013

This is the year in which buccal health started aiming at mHealth. Beam Technologies started selling a Bluetooth-based toothbrush in companion with an Android app for monitoring brushing habits. Moreover, a young Mexican company named PindNET R&D, along with the buccal health professional Isaac Kably, DDS, started with the development of an mHealth buccal monitor.

The iExaminer System was developed by Welch Allyn; it is a device that allows doctors to use their iPhones' camera to take pictures of the interior surface of the eye. Later this year, Zinc Software started developing a device that clips to a user's earlobe; with the use of an iOS-based app, this device will be able to provide biofeedback in order to reduce blood pressure and stress. In mid-April a Lebanese startup, Instabeat, built a heart-rate tracker for swimmers that uses optical sensors. At the American Telemedicine Association was announced the first Bluetooth 4.0 pulse oximeter; on May 8 the product was under review by the FDA. In the same month, Dario, a smartphone-based diabetes management system, was under development by the Israeli company LabStyle Innovations. This smartphone does not have a Conformité Européenne (CE) mark or an FDA approval. Later, on May 10, researchers at the University of Illinois showed a prototype of an iPhone-based spectrometer which can detect viruses, bacteria, toxins, proteins, and allergens in food. By June 3, Polar launched its first running stride sensor built-in Bluetooth Smart technology and it claimed that the device can record data without a GPS.

Lumo BodyTech created the Lumo Back, which monitors the user's posture and sends a feedback to an Android or iPhone device in order to track the evolution of the user over time. Later, Neumitra introduced a smart watch called neuma, which helps to measure and manage the autonomic nervous system, it also sends the information to a smartphone.

In the past 5 years the use of smartphones as mHealth platforms caused an exponential growth in the development of medical devices. Nonetheless, the use of smartphones carries with it issues about costs, development time, and compatibility, as it is noticed that many of the solutions are device dependent; most of them require an iOS-based device, which people from low-income regions cannot afford. Moreover, similar to the text-based mHealth solutions, the security and privacy of the data collected will remain a concern.

19.2.4 mHealth and Other Technologies

Beyond mobile phones, wireless-enabled laptops and specialized health-related software applications are currently being developed, tested, and marketed for use in the mHealth field. Many of these technologies, while having some application in developing nations, are developing primarily in developed countries. However, with broad advocacy campaigns for free and open-source software (FOSS), applications are beginning to become adapted and available for all countries. Some other mHealth technologies include the following:

- Patient-monitoring devices
- Mobile telemedicine/telecare devices enabling video support
- Music players for m-learning
- Microcomputers
- Data-collection software

The use of technologies based on cross platforms and complete operating systems for mobile device hardware helps to ensure confidentiality and integrity and builds trust between patients. This may foster greater adoption of mHealth technologies and services, by exploiting lower-cost multipurpose mobile devices such as tablets and smartphones and, even better, those platforms born from the Open Hardware initiative. For instance, in August 2013 the web store cooking-hacks.com launched its e-Health Sensor Platform V2.0 for Arduino and Raspberry Pi and Intel Galileo. This platform supports 10 different sensors: pulse, oxygen in blood, airflow, body temperature, ECG, glucometer, galvanic skin response (GSR-sweating), blood pressure (sphygmomanometer), and EMG. The platform offers up to six connectivity options: Wi-Fi, 3G, GPRS, Bluetooth, ZigBee, and 802.15.4. When used with 3G or Wi-Fi, it is possible to attach a camera for enabling teleconsultation. Undoubtedly, the existence of such platforms allows any individual, such as a hobbyist, a student, or an independent professional, to develop any kind of technology that can impact on a big scale.

Another approach concerning the development under open platforms is the availability of operating systems that should be agile and evolve to effectively balance and deliver the desired level of service to an application and an end user while managing display real estate and power consumption without forgetting security posture. With advances in capabilities such as integrating voice, video, and web 2.0 collaboration tools into mobile devices, significant benefits can be achieved in the delivery of healthcare services to as many people as possible. Moreover, the immersion of new sensor technologies, which are not commonly used by health applications, such as high-definition (HD) video and audio capabilities, accelerometers, GPS/IPS, ambient light detectors, barometers, and even inertial measurement units (IMUs), can break the paradigm and enhance the methods of describing and studying cases close to the patient or consumer of the healthcare service. This could include autodiagnosis of heart diseases without the need for a teleconsultation, education, treatment, and on-the-move monitoring. According to the Vodafone Group Foundation, on February 13, 2008, a partnership for emergency communications was created between the group and the United Nations Foundation. Such a partnership will increase the effectiveness of the information and communications technology response to major emergencies and disasters around the world [15].

19.2.5 Emerging Trends and Areas of Interest in mHealth

Emerging trends and areas of interests in mHealth are listed in the following:

- Emergency response systems
- Human resources coordination, management, and supervision
- Mobile synchronous and asynchronous telemedicine diagnostic and decision support to remote clinicians
- Pharmaceutical supply-chain integrity and patient safety systems
- Clinical care and remote patient monitoring, even if the patient is on the move
- Health extension services
- Health service monitoring and reporting
- Health-related m-learning for the general public
- Training professionals in mHealth technologies for working in difficult areas

- Health promotion and community mobilization
- Support of long-term conditions
- GPS tracking and remote monitoring of Alzheimer's disease patients
- Enhancing the peer-to-peer experiences related to telecare
- The use of visible light communication, iBeacon, and other new technologies for IPS tracking

19.2.6 Health Informatics: The European Committee for Standardization ISO/IEEE 11703 Standards

Since there are lots of platforms for bringing mHealth and eHealth services, some international regulatory organizations released a group of standards for health informatics which determines the rules about how all this universe of devices, infrastructure, communications, and others should interact with each other in order to obtain a sustainable and maintainable environment. The European Committee for Standardization (CEN) ISO/IEEE 11073 standards for health informatics are a group of standards that give the directives on enabling communication between all participants in the healthcare universe; these standards can be divided into the following sections [16]:

- Core standard
  - Nomenclature: At this point, all nomenclature codes are defined with the idea of identifying objects and attributes at the programming level.
  - Domain information model: Here all definitions for "vital" data transmission are set. This is the platform for the service model of a standardized communication. Furthermore, at this stage all data are segmented into packages:
    - Medical package: This stores medical vital signals, e.g., ECG data. The storage can be performed in different structures.
    - Alert package: This includes additional information about alert parameters of a particular medical package.
    - System package: This package stores in blocks the information of all the components of a medical device, i.e., battery, clock, and memory.
    - Control package: Objects of the remote control of a medical device are defined.
    - Extended services package: This element is used to store gusts of data from scan devices, including scan interval, scan lists, and scan period.
    - Communication package: This contains the basic communication profile.
    - Archival package: This is used to store a patient's data, online and off-line; in other words, it is an EHR.
    - Patient package: Only the demographics data of a patient are available in this confidentially kept package. More information can be retrieved from the patient's archival package.
  - Base standard: In this stage, the information receives the transmission model, the format, and the syntax. This standard also describes layers 5 to 7 of the Open Systems Interconnection (OSI) model.
- Agent/manager principle: Here, all the roles are defined for each member. This allows setting of the parameters of communications, so that each member can

interact and understand the other participants. This scenery is represented by agents and managers. This principle has the following stages:

- Agent application process: This module is the bridge between proprietary protocols and the CEN ISO/IEEE standards.

- Medical data information base (MDIB): Managed medical objects (MMOs) are stored in a tree structure defined by the domain information model.

- Association service control element: This module is a supervisory control and data acquisition (SCADA) element related to the ISO/International Electrotechnical Commission (IEC) 15953 and 15954 standards. No MMOs are transmitted in this stage.

- Common medical device information service element (CMDISE): The services for MMOs' exchange between agents and manager are given here. This implements the functions of persistent storage, like create, update, and delete.

- Presentation layer: This layer contains the encoding information of the attributes of an MMO.

- Session layer: It controls connection parameters.

- Communication model: This gives all the parameters for establishing the communication sequence. A finite-state machine is used to regulate the synchronization between agents and managers in different conditions. An initialization occurs according to the MDIB parameters of both the agent and the manager. Later, a process called data exchange through services occurs based on the CMDISE and a GET service is set up, and here is when the agent retrieves a list of attributes identifiers. Finally, the process of data exchange through scanner objects uses the create service of the CMDISE and the manager requests the agent to create a scanner object; this action creates a channel in which all updates occur automatically.

One example of the implementation of the CEN ISO/IEEE 11073 standards is the BluetoothHealth class for Android 4.0 systems, which implements a proxy object for controlling Bluetooth service via interpersonal communication (IPC) for those medical devices' built-in Bluetooth health device profile (HDP) [17]. The HDP is also a traditional Bluetooth profile but it is designed to facilitate transmission and reception of medical device data. The APIs of this layer not only interact with the lower-level multichannel adaptation protocol (MCAP) layer, but also perform service discovery protocol (SDP) behavior to connect to remote HDP devices.

19.3 Wireless Technologies Used in mHealth

There are several technologies that can be used by mHealth devices to achieve data transmission (Figure 19.3). However, in the majority of cases the technology depends directly on the media used for that purpose; e.g., most of the smartphones are Bluetooth enabled, and a few others have Wi-Fi hot spot support. Although Bluetooth is very common in smartphones and tablets, it might not be the best option for mHealth devices, which require

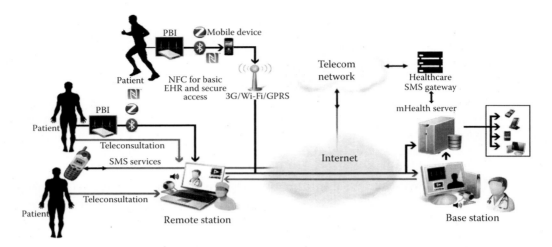

FIGURE 19.3
A hybrid scenario which combines multiple wireless technologies for mHealth applications.

long record periods due to power consumption limitations unless the device uses the new Bluetooth 4.0 low-energy technology.

Another suitable technology could be ZigBee, since its inception was created for low-power applications and for building sensor networks, which results as an ideal solution for mHealth applications. However, the problem is that the TPH-One by TazTag is one of the only smartphone in the market that is Bluetooth 4.0, near-field communication (NFC), Wi-Fi, and ZigBee enabled. This presents many problems during the development of mHealth devices and applications. The suggested solution is to use a bridge between ZigBee and Bluetooth in order to use a smartphone as a transmission device.

Next, a brief comparison between Bluetooth and ZigBee is given [18]:

- **Origin:** Bluetooth was launched in the year 1994 by Ericsson, while ZigBee was conceived around the year 1998.

- **IEEE standards:** Bluetooth is defined under the 802.15.1 standard, while ZigBee is 802.15.4 standard.

- **Management alliances:** ZigBee is managed by the ZigBee Alliance, which tests and certifies ZigBee-based devices. Bluetooth is managed for standards and devices by the Bluetooth Special Interest Group (SIG).

- **Protocol updates:** Bluetooth has been updated to versions 2.0, 2.1, 3.0, and 4.0, while ZigBee received updates in 2004, 2006, and 2007. Now ZigBee has two types: ZigBee and ZigBee PRO, which differ in terms of addressing algorithms.

- **Frequency:** Bluetooth works in frequencies below 2.4 GHz. ZigBee works under the unlicensed 2.4 GHz band and has regional operation in the 915 MHz (Americas) and 868 MHz (Europe) bands.

- **Channel bandwidth:** Bluetooth consumes bandwidths of up to 1 MHz, while ZigBee-based networks consume bandwidths starting at 0.3 up to 2 MHz, depending upon the frequency at which networks are communicating.

- **Geographic scale:** Both Bluetooth and ZigBee have been designed to communicate within a personal area network, up to 10 m, in most cases. However, they can reach a range of 100 m along the line of sight.

- **Maximum number of cell nodes:** In Bluetooth, up to 8 cell nodes can be connected to each other, while ZigBee can handle about 65,000 cell nodes.

- **Self-healing networks:** ZigBee-based systems are inherently known to keep up with self-healing network technology, while Bluetooth does not have this technology defined in its description.

- **Network topologies:** Bluetooth connects in point-to-point master–slave method, with one master and up to seven slave piconet networks. Bluetooth can link two or more piconets (scatter networks). On the other hand, ZigBee supports star, mesh, and other generic topologies. Different ZigBee network topologies can be connected to each other. There are three types of nodes in ZigBee topologies: ZigBee coordinator, ZigBee router, and ZigBee end point.

- **Data-transfer rates:** Bluetooth has a data transfer rate of up to 1 Mbps, while ZigBee data-transfer rates are up to 250 Kbps.

- **Energy consumption:** For any network, power consumption depends on the type of data being exchanged, distance between transmitter and receiver, desired power to be retained by signal and other factors. Bluetooth is a protocol known for exchanging almost all types of data, such as text and multimedia. In contrast, the ZigBee protocol is precisely for operational instructions and not much variety of data is known to be exchanged using it. Usually, ZigBee devices are 2.5–3 times more efficient than those working with Bluetooth.

- **Security:** ZigBee has built-in 128-bit Advanced Encryption Standard (AES) encryption and includes additional security features in its applications layer. Bluetooth supports 64-bit and 128-bit encryptions.

After showing a comparison between Bluetooth and ZigBee and due to the fact that nowadays Wi-Fi adapters are more common, even in embedded systems, talking about this technology, therefore, seems to be pertinent as it can represent a suitable option for enhancing the scope of mHealth. Wi-Fi is a trademark of the Wi-Fi Alliance that may be used with certified products that belong to a class of wireless local area network (WLAN) devices based on the IEEE 802.11 standards. Wi-Fi allows local area networks (LANs) to be deployed without wires for client devices, typically reducing the costs of network deployment and expansion. As mentioned before, wireless network adapters are now built into most smartphones and tablets. The price of chipsets for Wi-Fi continues to drop, making it an economical networking option included in even more devices. An advantage of Wi-Fi is that unlike mobile phones, any standard Wi-Fi device will work anywhere in the world. Nevertheless, similar to Bluetooth, this technology requires more power to operate and it depends on a router to get access to the Internet. On the plus side, this technology can be the best solution to bringing a better infrastructure for performing teleconsultation in the mHealth field. Of course, the organizations need to work harder with governments and private initiative to grow the coverage of Wi-Fi Internet access to remote communities. The first approach to reach that goal might be Google's project Loon, which is in the early stage and was conceived with the idea of bringing Internet access to rural or isolated communities through a high-altitude balloons network [19]. Finally, NFC is starting to take a place in the mHealth arena by providing the framework and infrastructure for optimizing health records and bringing an appropriate channel to exchange data securely. The time and the experience will dictate the position of this technology and the interoperability among the current technologies used in mHealth. Furthermore, with the introduction of new technologies, such as the Apple's iBeacon (based on BLE), interoperability should become more reliable.

19.4 Web Applications

Web applications are application programs that can be accessed over the Internet through a standard web browser. Web applications enable information-processing functions to be initiated remotely on the server. A typical web application consists of a client layer, an application layer (on a server), and perhaps a database layer (on a server). The development of web applications oriented to bring mHealth services represents a very important matter in terms of costs, development and deployment time, and quality of service. One of the main advantages of deciding to build web applications instead of smart apps is the ubiquity of their nature. In major cases a web application is cross platform and cross browser, which represents a better solution to approaching mHealth services to more people, considering that in the market exist tablets, laptops, mobile phones, and smartphones with Android, iOS, BlackBerry OS, Symbian OS, Linux-based OS, and Windows Phone, among others. Web apps can run on any mobile web browser and can be developed using JavaScript, HTML5 (fifth revision of the hypertext markup language), and CSS3 (cascading style sheets level 3) so that device-specific customization can often be achieved easily. Making smart decisions about how to create an mHealth web application is crucial to guarantee security, performance, and success. A good approach to architectural abstraction is the one suggested by Open mHealth, which defines the architecture as modular software units:

> Open mHealth is non-profit startup building open software architecture to break down the barriers in mobile health to integration among mHealth solutions and unlock the potential for mHealth. Through a shared set of open APIs, both open and proprietary software modules, applications and data can be 'mixed and matched', and more meaningful insights derived through reusable data processing and visualization modules [20].

- Data-storage units (DSUs): DSUs are any web-accessible applications which store user's data. These data should be divided into payloads. A payload is an abstraction of sets of definitions; each definition has a unique ID and numeric version. The use of DSUs allowed creating new registers with more or less data without losing the original data associated with a specific user (previous version of a payload). An API is required to implement a DSU. The preferred schema to represent complex data structures is the use of JavaScript Object Notation (JSON), XML in some cases, because it is natively supported by JavaScript. In addition there exist many open-source and robust libraries for different programming languages.

- Data-processing units (DPUs): A DPU is the unit of work to be done on JSON data. Of course, this unit requires a comprehensive, stateless, and well-documented API. A DPU should be represented as a library or framework that includes at least hypertext transfer protocol (HTTP) or hypertext transfer protocol secure (HTTPS) service for data exchange.

- Data-visualization units (DVUs): After a returned call from a DSU or a process done by a DPU, a DVU will perform the visualization in a web browser. The DVU in major cases should be written using JavaScript and HTML5 in order to assure interoperability and compatibility across web browsers and devices. A set of DVUs can be used as a backbone of applications for interactions among clinicians, physicians, health professionals, and patients. An API for a DVU is desirable in order to support multiple Open mHealth systems.

Note that the previous architecture is a first approach trying to unify mHealth systems. However, this architecture can be compared to the well-known pattern defined as model–view–controller (MVC), shown in Figure 19.4, used for dividing the internal representation data from how these data are presented or shown to the end user:

- Model: contains the core of functionality of the application and is controller–view independent
- View: presents the data to the end user; can access the model, but cannot change the model's state
- Controller: updates the state of the model and executes the pertinent action

This kind of architecture represents a good way to create and maintain mHealth applications. It can be easily used to incorporate different technologies for data storage and the possibility of allowing multiple programming languages for the same application.

XML is a markup language developed by the World Wide Web Consortium used for representing data structures. Due to its simplicity and usability over the Internet, XML is widely used for the implementation of web services. This technology should be used to create or represent health records dynamically. On the other hand, beyond the representation of patient's data, it is important to keep each record in the proper manner. There are several ways to store data, but it is important to have a protocol or standard that dictates

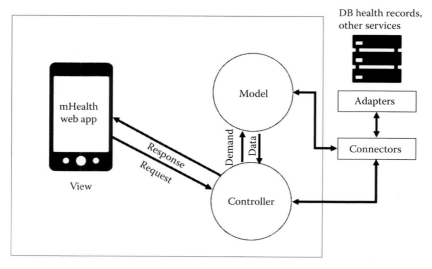

FIGURE 19.4
A sample of mHealth web application using the MVC design pattern.

how the patient's data should be stored and presented. Today different structures exist that dictate the proper manner to store and present patient data with the finality of ensuring its integrity, privacy, security, and interoperability. An EHR or EMR is a set of electronic health information about an individual patient or population which is theoretically capable of being shared across different healthcare platforms and different connection technologies. EHRs may include a range of data, including demographics, medical history, medication and allergies, immunization status, laboratory test results, radiology images, vital signs, personal statistics like age and weight, and billing information [21]. The idea is to create a standard to represent accurate and updated data for the patient at all times. It reduces the chances of data replication as there is only one modifiable file, which means the file is constantly up to date. Obviously, healthcare infrastructure represented only by EHRs entails the creation of a world or at least, national centralized server, which represents a tremendous challenge about security and privacy. The following is a list of some of open-source EHRs and health platforms:

- ClearHealth is a PHP cross platform that covers five basic categories of a healthcare system: scheduling, patient registration, EHRs, prescription writing, and billing.
- CottageMed is a cross platform EHR system based on FileMaker and is released under the GNU General Public License (GPL).
- Epi Info is Windows-based statistical software for epidemiology and was developed by the Centers for Disease Control and Prevention (CDC). This software has been in the market for over 20 years. There is a web version called OpenEpi.
- FreeMED is a cross platform EHR system based on Linux, Apache, MySQL, and PHP (LAMP). It is released under the GPL and hosted by SourceForge.
- GaiaEHR is a free and cross platform EHR system oriented toward web applications and based on LAMP. It is programmed in PHP at server side and uses Ext JS at client side.
- GNUmed is a WxPython application which uses PostgreSQL as a database engine.
- GNU Health is a centralized and scalable system oriented toward hospital infrastructure, including EHR support, a hospital information system, and a health information system. Jamaica is the first country that adopted GNU Health in the entire health system of the nation [22].
- Hospital OS is a research and development project implemented in Thailand. Today, this system provides services to at least 5 million patients [23].
- Mirth is a piece of software that enables HL7 of international healthcare informatics for interoperability standards. This software is cross platform and supports multiple protocols for exchanging healthcare messages, such as TCP/minimum lower layer protocol (MLLP), simple mail transfer protocol (SMTP), HTTP, and simple object access protocol (SOAP), among others [24].
- OpenEHR is an open-standard specifications for health informatics that is used to describe the process of the management, storage, retrieval, and exchange of EHRs from different sources [25].
- OpenEMR is a PHP-based EHR system. This system also supports prescription writing and medical billing.
- OpenMRS is a Java-based enterprise-level EHR framework, very extensible and scalable.

- OSCAR McMaster is an EHR software developed to work with Canadian health-care providers.

- THIRRA is a web-based EHR system released under Mozilla Public License (MPL) and is considered as a pioneering project for eHealth and telemedicine which also incorporates the biosurveillance mode for being used by the public health departments with the purpose of monitoring and investigating infectious diseases outbreaks in specific regions. Another feature is that it can be used off-line since data can be stored in the mobile device. THIRRA was programmed in PHP, using the CodeIgniter framework, jQuery user interface (UI) for presentation, and PostgreSQL as the database engine. It incorporates support for LOINC and ICD-10 [26].

- The Zambia Electronic Perinatal Record System (ZEPRS) is similar to THIRRA. This EHR system is web based and now is currently used in public obstetric clinics in Lusaka, Zambia. In contrast to THIRRA, this was developed in Java. It incorporates asynchronous JavaScript and XML (AJAX) technology for the data exchange and presentation and MySQL as the database engine. Another important feature is that ZEPRS components support a high-speed peer-to-peer (P2P) voice and data wireless network [27].

- SmartCare is a C# Windows-based EHR which incorporates the uses of smart cards, taking into consideration that the main target of this system is developing countries with limited or poor connectivity infrastructure. This system is used by the governments of Zambia, Ethiopia, and South Africa [28].

- Spatiotemporal Epidemiological Modeler (STEM) is a Java-written framework under the Eclipse Public License (EPL) that is used for creating spatial and temporal models of infectious diseases.

In addition to those open-source EHR platforms, there are Linux-based distributions focused on healthcare administration: BioLinux, Debian Med, and Ubuntu-Med.

In conclusion, the use of web applications will depend strictly on the technical and functional requirements, the budget, resources constraints, and time. For instance, some of the advantages of smart apps are the ability to leverage elements of their native OS, so in most cases they have a more responsive and attractive interface and can be run off-line [29]. Moreover, smart apps can easily take advantage of a smartphone's hardware. Obviously, as was mentioned before, smart apps are platform dependent so for each target smartphone, smart apps require a complete redesign. Note that some scenarios have led to the creation of hybrid apps that provide cross platform solutions with the required abstraction via use of the same native APIs so that developers can write apps by using JavaScript and HTML5 [30]. However, hybrid apps may not provide adequate performance compared to native apps due to the overhead introduced as a result of the abstraction. Probably in the future more sophisticated solutions will emerge to overcome this problem.

19.5 mHealth Challenges and Ethics

As exciting, disruptive, and transformative as mHealth is to the entire healthcare ecosystem, no one should forget or underestimate the challenges that still need to be solved.

There are open discussions about how through mHealth healthcare was modernized. Moreover, there are open questions which in the near future should be answered, such as how to determine quality of medical apps, the role of specialized organizations in regulating mobile apps, how these enabling technologies will play a crucial role in the healthcare industry, and in cases of patients with chronic diseases, how these services will remain affordable for them. Furthermore, the integrity, privacy, and security of the data will remain an active concern.

The mHealth technology has enabled the capability of gathering in real time astonishing amounts of data that are housed in massive databases, sometimes in the mobile device. But getting meaningful information and conveying it to the average end user is a big challenge that needs to be solved. Sometimes mHealth apps are downloaded by millions of people, but are those apps reaching the correct population groups? For instance, if a person uses a smartphone-based ECG with its respective apps that can also include some diet tips, how does the patient know if the app is suitable for his needs? Finally, a very important question to ask is, how do we know mHealth technologies work? Randomized clinical trials typically take several years to answer the question of whether an intervention is having the desired effect. That is an eternity in today's world, especially with how quickly technologies become irrelevant. It is the job of the authorities, researchers, private initiatives, and also the end user at different levels to perform the necessary tasks in order to ensure the limits of each mHealth technology and clearly determine to which segment of the population each device is oriented. All those questions and challenges are derived from ethical issues that in general terms are covered in the following list:

- Financial and commercial issues
  - Who pays?
  - Who is the beneficiary?
  - How long will the mHealth service remain free?
- Safety and privacy
  - Which entity or entities are responsible for guaranteeing the safety and privacy of all records?
  - For a centralized healthcare server, how is the integrity of all records guaranteed?
- Technical issues
  - With regard to the accuracy and reliability of measurements, how can we determine at which point there is a human mistake?
  - What should be done during a connection or power failure, hardware/software malfunctions, or other situations under real-time monitoring, primarily with patients on critical conditions?
- Quality of knowledge (know-how)
  - Do patients or health professionals have the skill to use the technology in safe conditions?
  - Does the technology offer sufficient information in order to make judgments?
- Quality of service
  - How do we determine the frontier between telecare and face-to-face care?

- Responsibility
  - Who takes the responsibility in case of a failure: the patient, the health professional, or both?
  - The more participants there are, the less transparency.
- Social, religious, and political issues
  - Equity, justice, and trust must be held between patients and health professionals.

The last issue to explore might be how to face the ethical issues related to the future of mHealth. The best way to do this is by taking scientific bases and facts when future scenarios are discussed.

19.6 Conclusions

Smartphone-based medical devices will represent the main driver for mHealth [31]. Moreover, mHealth applications will be tailored specifically for smartphones or tablets and will be native rather than web-based applications, according to a report from Research2Guidance. It is very likely that, as indicated in many publications, the mHealth market will be a mass market with a potential to reach billions of smartphone and tablet users, even in developing countries. By that time, 50% of these users will have downloaded mHealth applications. By the end of 2017, the total mHealth market revenue will have grown by 61% compound annual growth rate (CAGR), reaching the multibillion dollar level. The main sources of revenue will not come from application download revenue itself, but from mHealth services and hardware sales, as applications will serve as platforms to sell other health services and hardware. "This growth projection is based on the assumption that private buyers will continue to be the primary spenders in the next five years, but that the integration of mHealth applications into traditional health care systems will become more and more common during that time," the report noted. The market for mHealth applications is developing in three ways. Currently, mHealth vendors have managed to exit the initial trial phase and have entered the commercialization phase of the market. This phase can be characterized by a massive increase of offered solutions, the creation of new business models, and the concentration of private, health-interested people, patients, and corporations as major target groups. The study includes a forecast of mHealth and total app user base until 2017, an analysis of the user number by mHealth service type, and an mHealth sensor shipment projection until 2017. The report also notes that the general sophistication of today's mHealth applications is low to medium, and many of the mHealth-categorized applications provide a limited benefit for patients and health professionals. But it is not possible to forget the challenges, mentioned previously in this chapter, which represent the major barriers for the mHealth market to enter the next market phase of integration. In this phase, mHealth applications will become an integrated part of doctors' treatment plans and health insurers will become the main payer, especially for the more advanced mHealth solutions, the report said. The report also includes a forecast for mHealth market revenue until 2017, a breakdown of future revenues by revenue type, a forecast of mHealth app downloads by business model, both free and paid, and an analysis of the average downloads per mHealth user per year until 2017. At the end of the day it should be a balance between all technologies, smart apps, web-based applications, etc. A good example about what is coming further is the Scanadu Scope, which is a futuristic device that aims to become the first handled diagnostic device similar to the *Star Trek* tricorder [32]. The final objective of this device is to create a new way to diagnose individuals with the integration of novel noncontact and noninvasive technology incorporated in a smartphone. Finally, this chapter presents and discusses mHealth technology for advanced medical services. The mHealth systems have considered the integration of different technologies as advanced wireless communication, wireless devices, data processing, telecommunication, and portable biomedical instruments. All of those are a potential solution for bringing health services to the entire population. Moreover, different scenarios were discussed in order to show how mHealth is responding to the constantly increasing demand for advanced medical services, covering multiple scenarios like the ones that are required to be online, off-line, and on the move. For sure, it can be said that this is just the beginning of mHealth.

References

1. BBC News, 1998, "A millennium of health improvement." *BBC News—Health.* Accessed January 13, 2014. http://news.bbc.co.uk/2/hi/health/241864.stm.
2. Central Intelligence Agency, 2014, "World," *The World Factbook.* Accessed June 22, 2015. https://www.cia.gov/library/publications/the-world-factbook/geos/xx.html.
3. World Health Organization, 2013, "The right to health." *Media.* Accessed January 13, 2014. http://www.who.int/media/factsheets/fs323/en/.
4. Pidgeon, T., 2004, *Satellite Services and Telemedicine in Rural Alaska.* Accessed January 14, 2014. http://agency4.org/t/telemedicine-cart-solutions-s630/.
5. Ramos, A. C., 2003, "Cirugía Robótica." *Asociación Mexicana de Cirugía General, A. C.* 25(4), pp. 314–320, http://www.medigraphic.com/pdfs/cirgen/cg-2003/cg034f.pdf.
6. Della Mea, V., 2001, "What is e-Health (2): The death of telemedicine?" *Journal of Medical Internet Research.* 3(2), p. e22, http://www.jmir.org/2001/2/e22/.
7. Frejill, P., D. Chambers, and C. Rotariu, 2007, "Using ZigBee to Integrate Medical Devices." *Proceedings of the 29th Annual Conference of the IEEE EMBS*, Lyon, France. 2007, pp. 6718–6721. http://www.ncbi.nlm.nih.gov/pubmed/18003568.
8. International Telecommunication Union, 2008, "Implementing e-Health in developing countries." *Guidance and Principles.* Accessed January 16, 2014. http://www.itu.int/ITU-D/cyb/app/docs/e-Health_prefinal_15092008.PDF.
9. World Health Organization, 2009, "2009 Survey." *Global Observatory for eHealth.* Accessed January 16, 2014. http://www.who.int/goe/survey/2009/2009survey/en/.
10. AliveCor, n.d., "Using the veterinary heart monitor." Accessed March 4, 2014. http://www.alivecorvet.com/#subworks.
11. Elliot, G., December 18, 2013, "The case for using SMS in mHealth." *mHealthNews—The Voice of Mobile Healthcare.* Accessed January 16, 2014. http://www.mhealthnews.com/blog/case-using-sms-mhealth.
12. Federal Communications Commission, 2015, "What you need to know about Text-to-911" Guides. Accessed June 20, 2015. https://www.fcc.gov/text-to-911.
13. Pai, A., J. Comstock, and B. Dolan, 2013, "Timeline: Smartphone-enabled health devices." *mobihealthnews.com.* Accessed January 16, 2014. http://mobihealthnews.com/22674/timeline-smartphone-enabled-health-devices/.
14. Vodafone Americas Foundation, 2009, "Winner 2009: CellScope." Accessed January 16, 2014. http://vodafone-us.com/wireless-innovation-project/past-competitions/2009/2009-winners/cellscope/.

15. AliveCor, "AliveCor Heart Monitor." Accessed January 16, 2014. http://www.alivecor.com /how-to-set-up.
16. OpenHealth, n.d., "Our ISO/IEEE 11073-20601 standard implementation." Accessed March 6, 2014. http://openhealth.libresoft.es/node/45.
17. Bluetooth, n.d., "Health device profile (HDP)." *Developer Portal.* Accessed March 6, 2014. https:// developer.bluetooth.org/TechnologyOverview/Pages/HDP.aspx.
18. Houda, L., A. Hossam, and C. de Santis, 2007, *Wi-Fi™, Bluetooth™, ZigBee™ and WiMax™.* The Netherlands: Springer.
19. Google, n.d., "Balloon-powered Internet for everyone." *Project Loon.* Accessed February 16, 2014. http://www.google.com/loon/.
20. Open mHealth, 2013, "About." Accessed February 18, 2014. https://openmhealth.org/about /history/.
21. HealthIT.gov, n.d., "What is an electronic health record (EHR)?" Accessed January 16, 2014. http://www.healthit.gov/providers-professionals/faqs/what-electronic-health-record-ehr.
22. Falcón, L., n.d., "GNU health/Introduction." *GNU Health.org.* Accessed March 20, 2014. http:// en.wikibooks.org/wiki/GNU_Health/Introduction.
23. Hospital-OS.com, 2011, "What is Hospital OS?" *FAQ.* Accessed March 20, 2014. http://www .hospital-os.com/en/faq.php#W_HospitalOS.
24. Mirth, 2014, "Mirth connect." *Powering Healthcare Transformation.* Accessed March 20, 2014. http://www.mirthcorp.com/products/mirth-connect.
25. openEHR, 2014, "What is openEHR?" Accessed March 20, 2014. http://www.openehr.org/.
26. Tan, B. T. J., 2009, "THIRRA Electronic Health Records System." Accessed March 20, 2014. http://sourceforge.net/projects/thirra/.
27. GMC, 2008, "Improving perinatal care." Accessed March 20, 2014. http://www.ictedge.org /node/169.
28. SmartCare, 2010, "SmartCare—Zambia." Accessed March 20, 2014. http://www.smartcare.org .zm/.
29. Pindter-Medina, J. et al., 2009, "Proposal for an m-Health system." *Electronics, Robotics and Automotive Mechanics Conference.* CERMA, pp. 55–59.
30. Selvarajah, K. et al., 2013, "Native apps versus web apps: Which is best for healthcare applications?" *Human-Computer Interaction.* 2(8005), pp. 189–196.
31. Eddy, N., 2013, "Mobile health market driven by smartphone, tablet adoption." *eWeek.* Accessed March 24, 2014. http://www.eweek.com/mobile/mobile-health-market-driven-by-smartphone -tablet-adoption.html/.
32. Comstock, J., 2012, "Scanadu unveils smartphone-enabled home diagnostics." *mobileHealth-News.* Accessed March 24, 2014. http://mobihealthnews.com/19288/scanadu-unveils-smart phone-enabled-home-diagnostics/.

20

Investigation and Assessment of Effectiveness of Knowledge Brokering on Web 2.0 in Health Sector in Quebec, Canada

Moktar Lamari and Saliha Ziam

CONTENTS

20.1 Summary

Knowledge brokering is defined as an activity involving intermediaries (individuals, organizations, and networks) that act as "connectors," or liaison agents, between the producers and users of new knowledge (Dobbins et al., 2009). This chapter provides an overview of knowledge brokering in the digital age in the context of the health sector in Canada. Despite the proliferation of theoretical and inductive research that has examined knowledge brokering, there are still some questions need to be answered, inspired by the works and conclusions of Ward et al. (2009a) and Dobbins et al. (2009): what technologies are used by knowledge brokers, what interactions are initiated with the stakeholders involved, and what about the effectiveness of knowledge brokering? Drawing from a robust literature review and based on a survey of a representative sample of knowledge brokers using web 2.0 social networks, our investigation attempts to measure the challenges and the determinants of knowledge brokering, with the aim of helping healthcare policy makers and researchers and allow them to better understand the process behind knowledge brokering. Our empirical finding characterizes knowledge brokering activities (e.g., behaviors, interactions, brokerage instruments used, and target audience), identifies the individual attributes of knowledge brokers (age, gender, experience, training, and preferences), specifies the types of knowledge shared, and assesses the perceived effectiveness of knowledge brokering on the performance of healthcare organizations, as well their leeway in public health interventions and decision making, in Quebec, Canada. Filling a gap in knowledge brokering research, our findings illustrate the nature of knowledge brokering in healthcare organizations and demonstrate that its effectiveness is proportionate to (i) the quality of knowledge exchanged, (ii) the intensity of interactions initiated with the stakeholders, and (iii) the degree of connectivity brokers have in web 2.0 social networks.

20.2 Introduction

Knowledge brokering takes advantage of web technologies to disseminate health sector information digitally and create and maintain websites, blogs, Facebook pages, etc. The use of Twitter, YouTube, and wikis is frequent. The trend appears to be growing rapidly, reaching more and more subscribers and potential users. In Canada, particularly in the health sector, knowledge brokering is sparking increasing interest in innovative manner for transferring and exchanging new knowledge (Dobbins et al., 2009; Harrington et al., 2008; Landry et al., 2003; Lomas, 2007; Pentland et al., 2011; Ziam, 2010) and is evolving radically by leveraging the technologies from web 2.0. These new technologies offer innovative options for tracking new knowledge throughout the world and imparting it instantaneously while interacting with potential users of such knowledge and networking in a user-friendly and seamless manner. Knowledge brokers value these technologies as a way to diversify their instruments of dissemination, as well as bolster interactions with their partners in order to intensify the use of new healthcare knowledge in decision making.

This chapter presents an empirical and analytical overview of the instruments of dissemination used by brokers, the social interactions initiated as part of their brokering activities, and the impact of these instruments and interactions, such as on the performance of organizations and public policies of the health sector in Canada (Dobbins et al., 2009; Lomas, 2007; Ward et al., 2009b,c). This chapter is based on the results of a survey of a representative sample of knowledge brokers operating within the health sector in Canada and using digital technologies. We might point out that it is in the wake of the studies carried out by Canadian researchers dealing with knowledge transfer (Dobbins et al., 2009; Landry et al., 2003; Lavis, 2006; Lomas, 2007) that the structures of knowledge brokering using the Internet have grown so swiftly, building on the development of platforms and applications first from web 1.0 and then web 2.0 (O'Reilly, 2005). Indeed, knowledge brokering has been greatly enhanced by applications and platforms from the web (websites, Facebook, blogs, Twitter, e-newsletters, wikis, YouTube, LinkedIn, podcasts, chat, Really Simple Syndication [RSS], mashups, social bookmarks, P2P, etc.) (Chiang et al., 2009). In light of this circle of influence, new vocations and professions dedicated to knowledge brokering (transfer, exchange, and brokerage) have emerged. Thus, the knowledge broker business has begun to take shape and is becoming more professional, leading brokers to be increasingly recognized as *knowledge brokers* (Dobbins et al., 2009; Robeson et al., 2008). Lomas (2007) revealed the issues and challenges of the professionalization of knowledge brokering and noted the complexity of knowledge brokering activities and the interactions that it implies. It concerns the networking of actors and separate communities that do not spontaneously interface (due to differences in cultures, languages, interests, rationales, etc.) and that belong to different environments: academic, governmental, and media communities of practice (frontline community service actors, professionals, nonprofit organizations, associations, and so on.). Knowledge brokering and web 2.0 technologies are powerful levers that bring these environments together and spur on their networking.

In concrete terms, this chapter comes on the heels of requests recently made by Ward et al. (2009b; 2010) and Dobbins et al. (2009) that call for scientific researchers to provide empirical support for the practice of knowledge brokers and assess its impact on the performance of the healthcare sector. Our chapter is divided into three parts. First, we conduct a general theoretical perspective, highlighting the concepts underlying knowledge brokering theory and principles; then we introduce the methodology adopted to investigate the effectiveness of knowledge brokering in Canada and present the models, data, hypothesis, etc.; and, lastly, state and discuss the findings and results. These findings and their implications are summarized in Section 20.5.

20.3 General Approach to Knowledge Brokerage, Theory, and Definitions

There are many studies that emphasize the contribution that brokers and brokering knowledge have made to innovations and organizational performances (Hargadon, 1998; Ward et al., 2009a). Many of those studies focus mainly on how brokers acquire and disseminate knowledge information that is likely to instigate solutions or initiate promising innovative practices (Dobbins et al., 2009; Lomas, 2007). Hence, the knowledge brokers are actively involved in conveying the knowledge and better matching the stakeholders involved in the supply and demand of knowledge (Lomas, 2007; Martinez and Campbell, 2007; Ward et al., 2009c). Murray et al., (2011) describe knowledge brokers as intermediaries: "They are individuals who provide a specialised interface between the internal system and external knowledge sources. They can also span boundaries within the organisation. They monitor the environment and translate external information into a form understandable by the organisation. The gatekeeping function may have structured centralised capacity or may be diffused across many individuals" (Murray et al., 2011).

In the same vein, the structures of knowledge brokering are defined as "a set of resources (human, material and technology) dedicated to the collection, analysis, management and dissemination of information, centering on improving knowledge and foreseeing strategic issues for a given group" (Khénissi and Gharbi, 2010). The emergence of these structures is in line with the changing patterns of production and dissemination of new knowledge. Gibbons et al. (1994) have analyzed the changing patterns in the production of science, distinguishing between two modes: (i) a mode called *traditional* (mode 1), often monodisciplinary, academic, hierarchized, and focused on providing knowledge ("science push"), and (ii) a mode termed *contemporary* (mode 2), characterized by interdisciplinarity, interactivity, networking, and the need for the enhancement of new knowledge. As can be seen in

Table 20.1, knowledge brokering is based primarily on the dynamics of mode 2 (Armstrong et al., 2006; Dobbins et al., 2009; Pentland et al., 2011).

It is in this transition of the modes of scientific production that knowledge brokering has emerged as a key concept. Admittedly, this concept has theoretical ramifications that are rooted in three major corpus of theory.

The first corpus is knowledge management (KM), which regards knowledge as being a productive resource in its own right, which must be managed to offset the asymmetry of the knowledge market and ensure better coordination of the various specializations of individual knowledge (Grant, 1997; Liebowittz, 2005; Lamari, 2010). For Grant (1997), knowledge management allows, on one hand, mobilizing of tacit knowledge and, on the other, mitigating the risk of retention of explicit knowledge. In this corpus, the broker assumes the role of a manager capable of mobilizing, managing, and disseminating knowledge (tacit, codified, formal, informal, old, new, and so on).

The second corpus of theory relates to the *social networks of knowledge* (Landry et al., 2003; Lomas, 2007; Rogers, 2003; Ward et al., 2009b, 2011). In this corpus, the knowledge broker appears as a liaison officer and connector with sufficient interpersonal skills and credibility to network and mobilize stakeholders interested in the production and use of knowledge (Clark and Kelly, 2005). This takes place at the interface between separate realms opposing different communities, each with its institutional culture and its own language when it comes to the enhancement of knowledge. Therefore, the skills required of knowledge brokers continue to diversify to include scientific skills, initiative, independence, and involvement in social networks (Lomas, 2007; Kramer and Cole, 2003).

The third corpus relates to *skills development* ("capacity building") through knowledge (Rogers, 2003). Here, the knowledge broker acts as a human development officer, acting to "educate," distribute, and make knowledge accessible for decision making (Morley, 2006).

On another level, let us recall that knowledge brokering differentiates between the notion of knowledge and that of information (Blumentritt and Johnston, 1999; Cohen and Levinthal, 1990). Information often refers to data that convey an informational message issued by a transmitting source intended for a receptor source. Such information may consist of statistical evidence, observed facts, or factual and specific events. As shown in Figure 20.1, factual data supply information, and information, in turn, supplies knowledge. Data and information occupy the first two steps of the continuum of knowledge production (Miller and Morris, 1999; Kaipa, 2000; Scharmer, 2001). However, before morphing into

TABLE 20.1

Modes of Production and Dissemination of Scientific Knowledge

Mode 1	Mode 2
Research issues conceived and solved in terms of academic interests	Problem solving and collaborative research with regard to applications and user needs
Disciplinary and uni-institutional activities	Multidisciplinary and multi-institutional activities
Homogeneous realm, introverted and focused on excellence in the supply of knowledge	Heterogeneous realm, extroverted and focused on the application of science and knowledge brokering
Hierarchical approach (top–down) and based on the interest of research organizations	Nonhierarchical, interactive approach, involving often divergent interests
Academic control of the supply of science	Social control aligned along collective and participatory governance of science and innovation

Source: Gibbons, M. et al., *The New Production of Knowledge: The Dynamics of Science and Research in Contemporary Societies*. Sage, London, 1994. With permission.

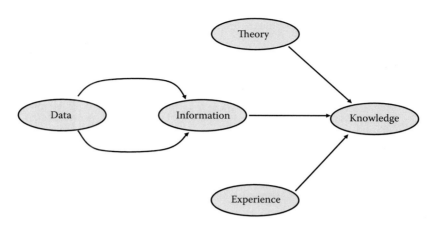

FIGURE 20.1
The continuum of knowledge creation: data, information, theory, and experiences.

knowledge, information (data, facts, statistics, etc.) should be inoculated with an additive mixing conceptualization and referencing originating from the experiential background, practices, and knowledge already acquired by the actors involved. In other words, knowledge is developed from the combination of available information about the experiential backgrounds of individuals (including values, norms, and references) and the capacity for abstraction, which incorporates conceptual and methodological constructs (Nonaka, 1994). Clearly, knowledge requires a level of articulation and abstraction that cannot be easily demystified by laypeople whose capacity to absorb knowledge is relatively limited to begin with (Cohen and Levinthal, 1990).

Seen from this viewpoint, it is obvious that knowledge brokering encompasses various conceptual constructs which deserve operational and empirical examination. Despite the proliferation of theoretical and inductive research that has examined knowledge brokering, Ward et al. (2009a) still lament about "the lack of evidence about how knowledge brokering works, the contextual factors that influence it and its effectiveness."

This chapter seeks to advance knowledge as applied to health sector by answering the following questions posed by Ward et al. (2009a) and by Dobbins et al. (2009): (i) What are the instruments of dissemination mobilized by knowledge brokers? (ii) What are the interactions initiated by knowledge brokers in their networking activity? (iii) What are the perceived impacts of brokering knowledge on the performance of public health interventions? Let us look at the methodological guidelines included in this framework.

20.4 Public Health-Related Survey, the Data, and Data Analysis

This section is based on data obtained from a mail survey consisting of 25 questions, most of which are open-ended questions and measured quantitatively (Likert scales, binary or continuous variables). With an average completion time of 20 min, the questionnaire was tested before its use to confirm comprehension, measurement consistency, and responsiveness issues. The study was validated from an ethical viewpoint and received an ethics compliance certificate from the ethics committee of the university with which the researchers are affiliated.

In order to determine the group of knowledge brokers to survey, we developed a comprehensive inventory of all knowledge brokering structures using the web and operating in conjunction with the health sector in the province of Quebec, Canada. This inventory was the result of extensive searches conducted on the Internet and on the websites of organizations related to health and social services, as well as in collaboration with the brokers identified through this search. The inventory obtained was submitted for improvement and validation to two independent experts with extensive experience in knowledge transfer in the field of healthcare in Quebec. It was improved through certain additions and modifications, following proposals from the two experts. In all, 45 structures of knowledge brokering using web technologies were retained. Brokers operating in these structures were asked to complete our questionnaire on a voluntary basis and return it by mail (in a prepaid envelope sent to them). A reminder notice was also sent to ensure a higher response rate. Of the 45 questionnaires sent, 31 were completed and returned, representing a response rate of 69%, which is considered to be an expected response rate for a mail survey. Data collection took place in the winter of 2011. Different statistical analyses were conducted: descriptive analyses, exploratory factorial analyses, and interpretive analyses (linear regression) using the SPSS software. Descriptive analyses allowed us to characterize brokering practices and interactions. Factorial analyses helped to group several variables around latent federating variables (factors), especially when the time came to deal with the various perceived impacts of knowledge brokering. To identify the determinants of the impacts on knowledge brokering, a set of analytical assumptions was developed. Based on the literature review on the conditions for successful knowledge brokering (Dobbins et al., 2009; Lomas, 2007; Ward et al., 2009a), our research tested three empirical hypotheses regarding (i) *the quality of knowledge via brokers*, (ii) *the networking interactions with potential users* of this new knowledge, and (iii) *the connectivity of knowledge brokers* in web 2.0 social networks. Considering the number of observations made (31), it was difficult to introduce a greater number of explanatory variables without compromising the validity of the explanatory model chosen to identify the determinants of brokering impacts (statistical degree of freedom of the linear regression used). The hypotheses are as follows:

H1: The more convincing the data conveyed by knowledge brokering, the stronger the impact of knowledge brokering on (i) the improvement of organizational performance (H1a), and (ii) the creation of a new leeway for changes in public health policies (H1b).

H2: The more knowledge brokering is involved in networks and interactions linking knowledge brokers to potential users of new knowledge, the stronger the impact of knowledge brokering on (i) the improvement of organizational performance (H2a) and (ii) the creation of new leeway for changes in public health policies (H2b).

H3: The more connected a knowledge broker is to web 2.0 social networks, the greater the impact of these knowledge brokers on (i) the improvement of organizational performance (H3a), and (ii) the creation of a new leeway aimed at changing public health policies (H3b).

Three independent variables were calibrated to empirically test these hypotheses in the context of an explanatory model supported by the equation formulated as follows. The first variable relates to the quality (QUAL) of knowledge as carriers of added value. It is binary and weighs the innovative nature of the data and potential contributions to the advancement of knowledge. The second variable relates to the relationship of interaction (INTER) between knowledge brokers and the managers and professionals concerned by the potential use of new knowledge being exchanged (Pentland et al., 2011). The third variable gauges the connectivity to social networks (SOCN) web 2.0 (Facebook, Twitter, blogs, YouTube, websites, etc.). The importance of such determinants in the process of knowledge brokering is described qualitatively (Gallezot and Le Deuff, 2009), without being demonstrated by statistics that measured, to a significant degree, their influence on the impacts of the knowledge brokering process. The dependent variables are related to the impacts of knowledge brokering on organizational performance and the perceived impacts on the margin of maneuver available to organizations. These two dependent variables are the product of a factorial analysis which is described further in this chapter. Two equations were conceived to test these hypotheses:

$$I_{performances} = \beta_0 + \beta_1 QUAL + \beta_2 INTER + \beta_3 SOCN + \mu, \qquad (20.1)$$

$$I_{decision\ making} = \beta_0 + \beta_1 QUAL + \beta_2 INTER + \beta_3 SOCN + \mu. \qquad (20.2)$$

β_i ($i = 0, 1, 2, 3$) are the regression coefficients and μ is an error term.

After having established this theoretical framework and methodology, it is now time to answer the questions posed.

20.4.1 Survey Results, Findings, and Interpretations

Let us begin with a description of the group of knowledge brokers surveyed. The survey data allow us to characterize the attributes of professionals and organizations involved in knowledge brokering. Organizations that have set up structures of knowledge brokering in connection with the health sector are very diverse: nearly two in five organizations (39%) are public institutions that operate directly in the health sector; 16% are government agencies; 13% are university institutions; and almost one-third (32%) are made up of nonprofit organizations and private or hybrid institutions, all operating in connection with healthcare and health determinants. Such diversity suggests that the development of healthcare knowledge brokering is of interest to several communities and fields: academia, government circles, and practice settings (community, private, nonprofit organizations, etc.). In the organizations taking part in the survey, professionals are assigned to knowledge brokering for most of their duties. Some organizations will set aside 60 h a week dedicated to knowledge brokering, assigning more than one full-time professional to the task.

The survey also reveals that digital knowledge brokering activity is still a fledging activity in organizations, with the average duration under 5 years. Three out of five brokers are women (60%) and the average age of brokers is 48 years. Knowledge brokers are generally highly qualified professionals. Four out of five brokers (80%) hold a postgraduate degree (master's or doctorate or PhD) in two main disciplines: information sciences (35% of cases) and health sciences (29% of cases). The remaining brokers have rather varied training. In addition, work experience in knowledge brokering is close to a 5-year average. This suggests that brokers performed other activities prior to working in knowledge brokering. They reported having chosen knowledge brokering practice to stimulate their careers and for "prestige," allowing better ties contacts with the knowledge "producers" and the decision makers. Overall, our data suggest that the knowledge brokering structures have influence over 70,000 subscribers and users involved in knowledge brokering in the health field; the average is 2000 subscribers per knowledge brokering structure working in the healthcare sector of Quebec. The median is around 700.

20.4.1.1 Instruments of Health-Related Knowledge Dissemination

The data also show that the structures of knowledge brokering use—to varying degrees— instruments of dissemination and exchange of new knowledge. Table 20.2 provides an inventory of the various media used according to their popularity with knowledge brokers.

Publishing a regular newsletter using electronic communications (e.g., e-mail LISTSERV) to a subscriber list is the prevailing practice for one of two brokerage structures (52%). The contents of these bulletins may address a specific health issue (for example, obesity, prevention, diabetes, immunization, safety, and nutrition) or several themes deemed to be relevant jointly and this is by making an inventory of the latest publications of interest during a given cycle (quarter, month, or week). Such newsletters often convey the results of an electronic watch of the latest publications or events that bring new knowledge (conferences, conventions, etc.). The use of websites is also widespread practice (45% of cases). In these websites, dealers have a wide variety of topics ranging from information to data storage and publications considered to be relevant. The use of blogs comes in third place, at 38%.

Other related web 2.0 platforms (Facebook, Twitter, YouTube, etc.) have been adopted by one broker in four (25%). Then there are the info letters, newsletters, or e-letters that are sent by e-mail to a list of loyal subscribers of knowledge brokering. Some knowledge brokers produce press releases, selecting the latest information publicized, to be able to briefly summarize news of interest to their clients.

The frequency of distribution and update of these materials was also analyzed. In fact, one out of four brokers (25%) say that they disseminate new knowledge on a daily basis. Nearly 16% of brokers do so weekly; 22%, on a monthly basis; 16%, on a quarterly basis; and 20%, irregularly. Furthermore, over 70% of brokers have on their websites a search engine to view, by key words, all the documents produced by the knowledge brokers. About 84% of brokers surveyed said they follow up with users to review their documents and to learn of the satisfaction of their subscribers about the knowledge thus conveyed using periodic surveys. But to better understand these practices, our survey identified the potential users and partners targeted by knowledge brokers.

20.4.1.2 Beneficiaries of Knowledge Brokerage

Table 20.3 shows the agencies and actors who benefit from knowledge brokering. Community organizations active in the healthcare sector are shown to be the chief beneficiaries

TABLE 20.2

Instruments of Dissemination and Exchange of New Knowledge

Instruments of Dissemination and Exchange of New Knowledge	% of Brokers Who Use Them
Digital watch newsletters and periodicals (LISTSERV)	51.6
Websites	45.2
Blogs	38.7
Other web 2.0: Facebook, Twitter, YouTube, etc.	25.0
Newsletters and e-letters	22.6
Current events, news, press reviews	19.4

TABLE 20.3

Clientele Targeted by Knowledge Brokers

Position	Knowledge Brokers' Clientele	Proportion Estimated by Brokers
1	Public health network agencies	36%
2	Provincial/federal health departments	33%
3	Community/private health organizations	17%
4	Universities, research chairs, and units	13%
5	Citizens	11%

(36%). They are likely less equipped (in budget and skills) than other agencies to have recourse to brokerage services that produce information bulletins automatically aimed at their organization. Then come the various departments of both levels of government (federal and provincial), accounting for one-third. Firms and private for-profit organizations come in third place (17%), ahead of the academic research community (13%) and citizens, who are probably the least concerned by knowledge brokers, coming in last (11%). We can conclude from this that individuals make up the lowest proportion of beneficiaries targeted by knowledge brokers since the knowledge brokering facilities consulted are mandated by their organizations to help with decision making and organizational decisions requiring new knowledge. Given this context, ordinary citizens are not necessarily considered a privileged clientele, even though more and more citizen groups demand better access to information and knowledge that have a bearing on their health and well-being. With that said, the beneficiaries of the knowledge conveyed directly by knowledge brokers do not comprise all the partners dealing with knowledge brokers. Knowledge brokers carry out several interactions with a more diverse network of partners.

20.4.1.3 Networking and Interactions of Knowledge Brokers

The data reveal the diversity and intensity of the networking activities of knowledge brokers. Table 20.4 describes the possibilities for networking of knowledge brokers and partners operating in the healthcare sector. Clearly, almost three brokers out of five (58%) maintain frequent interactions with researchers connected to a university milieu. In addition, nearly all knowledge brokers (93%) claim to personally know academic researchers

TABLE 20.4

Interactions Initiated by Knowledge Brokers

Rank	Partners with Whom Brokers Have Ongoing Interactions	% of Brokers Having Forged Direct Links
1	Researchers, professors, university professional staff	58
2	Health agency staff	35.5
3	Staff at the department of health and social services	26
4	Staff at private firms and groups of interest	13
5	Staff at community health organizations	10
6	Staff at other government departments (federal/provincial)	6.5
7	Citizens	3.2

working in their fields of interest. The close link thus forged with university researchers is notable, especially because it allows brokers to seek new knowledge from the source, directly and quickly, often several months before its dissemination in scientific publications. However, the personal relationships between brokers and potential users of new knowledge appear to be not as strong when it comes to staff at the department of health and workers at health and social service agencies. Our data indicate that only one broker in four (26%) confirms maintaining direct and ongoing interaction with administrative staff at the department of health and social services, and one in three brokers does the same with health and social service agencies. This result should be interpreted with care, as knowledge brokers rely on their instruments of dissemination and the newsletters and documents sent out regularly to their subscribers; they do not feel it necessary to maintain additional personal ties with the professionals of organizations that provide health services. The network of knowledge brokers extends, to a lesser degree, to healthcare professionals working in the private sector (firms and interest groups), in the community sector, and, to a lesser degree, with citizens, in proportions ranging, respectively, from 13%, 10%, to 3%. Here too, connections of knowledge brokers appear to be underdeveloped with citizens, particularly because of the nature of mandates for knowledge brokers, wanting to serve instead the decision-making and administration levels of the healthcare sector. However, the lessons drawn from our research tend to distinguish between the knowledge brokering clientele and all partners with knowledge brokers who are continuously interacting. Table 20.4 provides an informative mapping of the networking links forged by brokers as part of knowledge brokering.

20.4.1.4 Perceived Impacts of New Knowledge on Decision-Making Process

When asked about the impact of new knowledge on the decision-making process of users of this knowledge, brokers ranked the perceived impacts in two categories. The first category of impacts deals with organizational performance. The second category deals with opening leeway and options for decision makers and concerned organizations. As seen in Table 20.5, factorial analysis using varimax rotation with Kaiser normalization was performed to classify and group the items measuring the perceived impact of knowledge brokering. Both eigenvalues (greater than 1.0) and factor loadings produced two separate factors with 0.5 and above factor loadings for all items selected. Kaiser–Meyer–Olkin (KMO) tests and Bartlett's test of sphericity have been used to confirm that the data are appropriate for factor analysis. Throughout the analysis, the KMO test always turned out highly adequate and Barlett's test, significant ($p < .00$).

The descriptive analysis revealed that knowledge brokers (i) are almost unanimous in saying that the beneficiary clientele considers knowledge brokering credible (90% of cases); (ii) are, in a proportion of three out of four (75%), inclined to believe that knowledge brokering in relation to health policy is varied and useful; (iii) in a proportion of two out of three (66%), believe that knowledge brokering improves the effectiveness of organizations and public policies; and (iv) in a proportion of one out of three (33%), believe that knowledge brokering improves the efficiency of public policies. It is not surprising that efficiency comes last, since it is difficult to measure in terms of costs and tangible, visible benefits for knowledge brokers over the short term. These items make up the first factor (F1), which gauges the performance of knowledge brokering, and account for 26% of the variance in the perceptions of impact of knowledge brokering. A second set of items belong to the second factor (F2), which measures the impact of knowledge brokering in decision-making latitude. The second factor explains 23% of the variance in perceived impacts and purports

TABLE 20.5

Factorial Analysis of Perceived Impacts of Knowledge Brokering

	Impact of Brokering (% of Positive Responses)	Factor Loadings
F1: performance • Variance explained 26.6% • Eigenvalue = 2.1	• Users deem knowledge brokering to be credible in health sector (90.4%)	0.639
	• Users deem knowledge brokering to be useful for the health sector decision making (75%)	0.648
	• Knowledge brokering improves effectiveness in health sector (65%)	0.738
	• Knowledge brokering improves efficiency in health sector (36%)	0.62
	• Knowledge broking is relevant and adapted to user needs in health sector (68%)	0.585
F2: margin of maneuver (leeway) • Variance explained 23.3% • Eigenvalue = 1.6	• Knowledge brokering promotes innovation in the public health policies (77.4%)	0.701
	• Knowledge brokering provides new options for public health policies (42%)	0.66
	• Knowledge brokering promotes a change in the design of health public policies (10%)	0.735

Note: Total variance explained 49.9%; KMO = 0.939.

that knowledge brokers believe that knowledge brokering (i) promotes innovation in public interventions related to health (in 77% of cases), (ii) allows the identification of new options for public decisions (in 42% of cases), and (iii) participates in the change of design in public policies (in 10% of cases). This last result is somewhat disappointing, since it insinuates that knowledge brokering does not influence change in the design of public health policies. It suggests that such uses of new knowledge are akin to a rather symbolic utilization, therefore becoming a matter of change in the design of public health policies (Weiss, 1979).

20.4.1.5 Determinants of the Perceived Impacts

Delving further into the analyses of the impacts of knowledge brokering, it was considered useful to identify the determinants involved. The impacts explained are those described previously through factorial analysis (Table 20.5), which brought together items measuring the impact of knowledge brokering into two factors. F1 measures the performance generated by knowledge brokering in government health interventions and includes four items measuring, on the Likert scale (ranging from 1, strongly disagree, to 5, totally agree), the perceptions of the impacts of knowledge brokering on the credibility of knowledge brokering (item 1), the usefulness of knowledge brokering (item 2), the effectiveness of public interventions (item 3), and the efficiency of public interventions (item 4). The index created by adding unweighted scores of items considered has excellent internal validity (Cronbach's alpha of 0.757) and varies from 0 to 25, with an average of 15 and a standard deviation of 2.3. F2 measures the impact of brokering knowledge on the creation of new latitude for decision making in public interventions. An index was created by adding the scores of items related to three distinct items and measured by Likert scales ranging from 1, strongly disagree, to 5, totally agree. As mentioned earlier, item 1 measures the impact of knowledge brokering on public intervention, item 2 measures the impact of brokering on opening up new options for public intervention, and item 3 measures the impact of knowledge brokering on a change in the design of public policies. The analysis of internal consistency of the index gives a very satisfactory Cronbach's alpha (0.730) for an index ranging from 0 to 15, and with an average $M = 10$ and a standard deviation SD = 2.6. Table 20.6 presents the results of the statistical model identifying the determinants of the impacts of knowledge brokering.

This explanatory model of a linear regression type (ordinary least squares) gives consistent results, supported by sufficiently high adjusted R^2: 0.68 and 0.43, respectively, along with significant F statistics. The data in the second and third columns of Table 20.6 illustrate that the perceived impacts of knowledge brokering on the performance of public health interventions are associated positively and statistically significantly with (i) the quality of the knowledge conveyed by knowledge brokers, (ii) the presence of collaborative links between knowledge brokers and the users concerned, and (iii) the involvement of knowledge brokers in web 2.0 social networks dealing with health-related knowledge. Clearly, these findings tend to confirm the three hypotheses H1a, H2a, and H3a. In other words, our results suggest that the impacts of knowledge brokering on the performance of health interventions are stronger and more significant when the knowledge conveyed is of high quality (from a scientific viewpoint), when knowledge brokers maintain direct consultations with the potential users of such new knowledge, and when knowledge brokers are active in web 2.0 social networks related to the mobilization of new knowledge in the health policy field.

Table 20.6 also presents the regression findings of the second regression model explaining the impacts of knowledge brokering on creating leeway in public policy decision making. The findings obtained suggest that only two variables have a positive and statistically significant influence, the variable related to consultation and the variable related to the

TABLE 20.6

Determinants of Perceived Impacts

Variable	Model 1: Impact on Performance (Effectiveness, Efficiency, etc.)		Model 2: Impact on Decision-Making Leeway	
	B Nonstandardized (Standard Error)	p	B Nonstandardized (Standard Error)	p
Constant	13.333*** (0.744)	.00	9.5 *** (0.49)	.00
QUAL: I disseminate evidence that advances scientific knowledge (yes = 1; if not = 0)	3.348*** (0.96)	.01	0.75 (0.78)	.361
INTER: I consult potential users on useful knowledge they need (yes = 1; if not = 0)	2.591** (1.054)	.045	2.25** (1.05)	.05
SOCN: I am active on web 2.0 social networks (yes = 1; if not = 0)	0.545* (0.21)	.059	1.750* (1.052)	.097
R^2 adjusted	68%		43%	
F statistical	8.29		4.24	
Statistical significance	0.01***		0.03**	
Number of observations (N)	29		31	

*$p < .1$; **$p < .05$; ***$p < .01$.

involvement of knowledge brokers in web 2.0 social networks. Paradoxically, the variable measuring the quality of knowledge conveyed by brokers does not appear to be significantly associated with the creation of decision-making leeway. This finding can be interpreted by considering that, a priori, new knowledge is not meant to be always accompanied by opportunities opening up room for maneuvering and new options for policymakers. This finding reminds us of the theory of Cohen et al. (1972), described in the famous metaphor "a garbage can model of organizational choice," which emphasizes the complexity of the choice of useful knowledge and the apparent erratic irrationality of many public decisions unwilling to take into account convincing evidence from scientific research. Obviously, this type of result needs further empirical examination to better qualify the complexity by connecting the availability of new knowledge and decision making with different contexts marked by contingencies and uncertainties surrounding the decision-making process. The second explanatory model and the findings tend to confirm two of the three hypotheses, namely, H2b and H3b, with hypothesis H2a being rejected.

20.5 Conclusion

As its title indicates, this chapter examines knowledge brokering in the era of web 2.0 in the context of the healthcare sector in Canada. The data used are derived from a survey of knowledge brokers making great use of platforms found on the web 2.0. Our analyses are advancing knowledge by responding to current issues (Dobbins et al., 2009; Lomas, 2007; Ward et al., 2009b,c) relating to knowledge brokering and attempting to describe the instruments of dissemination used by knowledge brokers, grasp the interactions initiated by knowledge brokers, and explain the perceived impacts associated with knowledge brokering, mainly with respect to policies and organizations operating in the health sector. The findings obtained provide valuable evidence and suggest courses of action for improving knowledge brokering and enhancing its positive impact on the performance of public policies.

Our findings corroborate the importance of knowledge brokering conveyed by Internet platforms. Knowledge brokering takes advantage of web technologies to disseminate digital watch newsletters, create and maintain websites, blogs, Facebook pages, etc. The use of Twitter, YouTube, and wikis is frequent. The trend appears to be growing rapidly, reaching more and more subscribers and potential users. Knowledge brokers are using these instruments in a complementary and nonexclusive fashion. Moreover, thanks to these technologies, knowledge brokers act to disseminate knowledge on a frequent basis, as more than 40% of brokers transmit, at least once a week, newsletters on or updates to their flow of knowledge transmitted. The clientele of knowledge brokers is largely different from the partners and stakeholders in the network of knowledge brokers. The interactivity with decision makers keen on acquiring new knowledge proves to be a winning condition in the knowledge brokering process. Our research also explored the effectiveness of knowledge brokering impacts on the performance (effectiveness, efficiency, innovation, etc.) of public health decisions. The results suggest that the impacts of knowledge brokering are more important when the knowledge conveyed is of acknowledged quality, when knowledge brokers maintain ongoing interactions with decision makers, and when brokers are heavily involved in social networks from the web. The level of commitment of knowledge brokers to social networks appears to be a condition for successful brokerage. However, our research demonstrates the complexity and diversity of the tasks entrusted to brokers. It also stresses the importance of training and support to strengthen brokers' skills as well as infrastructures to convey information in the digital media that is useful for enhanced decision making and public health policy.

Training is required to bolster the skills of brokers who tap into new digital technologies (creation, maintenance, readability, update, etc.), which are increasingly the main instrument of dissemination and exchange of new knowledge. Investments are needed to subsidize the acquisition of these new technologies that are renewed very quickly, condemned to obsolescence both in regards to the equipment and skills required and in regards to the equipment and technologies with web 2.0 platform connectivity.

According to Cohen and Levinthal (1990), the effectiveness of knowledge brokers apparently does not depend solely on their own capacity for knowledge management but also on that of their recipients or users (absorptive capacity), hence the value of the second part of this report that is relevant to users of digital watch reports.

Moreover, further research ought to be conducted to examine the extrapolation of knowledge brokering practices observed in the health sector to other sectors, especially to other organizations acting in an intermediary position between research and development, those who invent new knowledge and new technologies and those who translate it into promising innovations of wealth and well-being for the entire community.

References

Armstrong, R., Waters, E., Roberts, H., Oliver, S. and Popay, J. (2006). The role and theoretical evolution of knowledge translation and exchange in public health. *Journal of Public Health*, 28(4), pp. 384–389.

Blumentritt, R. and Johnston, R. (1999). Towards a strategy for knowledge management. *Technology Analysis and Strategic Management*, 11(3), pp. 287–300.

Chiang, I.P., Huang, C.Y. and Huang, C.W. (2009). Characterizing Web users degree of Web 2.0-ness, *Journal of the American Society for Information, Science and Technology*, 60(7), pp. 1349–1357.

Clark, G. and Kelly, L. (2005). *New Directions for Knowledge Transfer and Knowledge Brokerage in Scotland*. Scotland: Scottish Executive Social Research.

Cohen, M.D., March, J.G. and Olsen, J.P. (1972). A garbage can model of organizational choice. *Administrative Science Quarterly*, 17(1), pp. 1–25.

Cohen, W. and Levinthal, D.A. (1990). Absorptive capacity: A new perspective on learning and innovation. *Administrative Science Quarterly*, 35(1), pp. 128–152.

Dobbins, M., Robeson, P., Ciliska, D., Hanna, S., Cameron, R., O'Mara, L., DeCorby, K. and Mercer, S. (2009). A description of a knowledge broker role implemented as part of a randomized controlled trial evaluating three knowledge translation strategies. *Implementation Science*, 4, pp. 1–9.

Gallezot, G. and Le Deuff, O. (2009). Chercheurs 2.0. *Les Cahiers du Numérique*, 5(2), pp. 15–32.

Gibbons, M., Limoges, C., Nowotny, H., Schwartzman, S., Scott, P. and Trow, M. (1994). *The New Production of Knowledge: The Dynamics of Science and Research in Contemporary Societies*. London: Sage.

Grant, R.M. (1997). The knowledge-based view of the firm: Implications for management practice. *Long Range Planning*, 30(3), pp. 450–454.

Hargadon, A.B. (1998). Firms as knowledge brokers: Lessons in pursuing continuous innovation. *California Management Review*, 40(3), p. 209.

Harrington, A., Beverly, L., Barron, G., Pazderka, H., Bergerman, L. and Clelland, S. (2008). *Knowledge Translation: A Synopsis of the Literature*. Edmonton: Alberta Mental Health Research Partnership Program.

Kaipa, P. (2000). Knowledge architecture for the twenty-first century. *Behaviour and Information Technology*, 19(3), pp. 153–161.

Khénissi, M.G. and Gharbi, J.-E. (2010). La veille stratégique, bilan de la culture numérique, la veille du 2.0. *Les Cahiers du Numérique*, 6(1), pp. 135–156.

Kramer, D.M. and Cole, D.C. (2003). Sustained intensive engagement to promote health and safety knowledge transfer to and utilization by workplaces. *Science Communication*, 25(1), pp. 56–82.

Lamari, M. (2010). Le transfert intergénérationnel des connaissances tacites: Les concepts utilisés et les évidences empiriques démontrées. *Télescope*, 16(1), pp. 39–65.

Landry, R., Lamari, M. and Amara, N. (2003). The extent and determinants of the utilization of university research in government agencies. *Public Administration Review*, 63(2), pp. 192–205.

Lavis, J.N. (2006). Research, public policy making, and knowledge translation processes: Canadian efforts to build bridges. *Journal of Continuing Education in the Health Professions*, 26(1), pp. 37–45.

Liebowitz, J. (2005). Linking social network analysis with the analytical hierarchy process for knowledge mapping in organizations. *Journal of Knowledge Management*, 9(1), pp. 76–86.

Lomas, J. (2007). The in-between world of knowledge brokering. *British Medical Journal*, pp. 129–132.

Martinez, N.R. and Campbell, D. (2007). Using knowledge brokering to promote evidence-based policy-making, *Bulletin of the World Health Organization*, 85(5), A–B.

Miller, W.L. and Morris, L. (1999). *4th generation R&D: Managing Knowledge, Technology and Innovation*. New York: John Wiley and Sons.

Morley, M. (2006). *Knowledge for Regional NRM: Connecting Researchers & Practitioners*. Canberra: Land and Water Australia.

Murray, K., Roux, D.J., Nel, J.L., Driver, A. and Freimund, W. (2011). Absorptive capacity as a guiding concept for public sector management and conservation of freshwater ecosystems. *Environmental Management*, 47(5), pp. 917–925.

Nonaka, I. (1994). A dynamic theory of organizational knowledge creation. *Organization*, 5(1), pp. 14–37.

O'Reilly. T. (2005). *What is Web 2.0*, O'Reilly Media, Inc. [Online: http://oreilly.com/web2/archive/what-is-web-20.html, last consulted 2012].

Pentland, D., Forsyth, K., Maciver, D., Walsh, M., Murray, R., Irvine, L. and Sikora, S. (2011). Key characteristic of knowledge transfer and exchange in healthcare: Integrative literature review. *Journal of Advanced Nursing*, 67(7), pp. 1408–1425.

Robeson, P., Dobbins, M. and DeCorby, K. (2008). Life as a knowledge broker. *Journal of the Canadian Health Libraries Association*, 29, pp. 79–82.

Rogers, E.M. (2003). *Diffusion of Innovations*. New York: Free Press.

Scharmer, C.O. (2001). Self-transcending knowledge: Sensing and organizing around emerging opportunities. *Journal of Knowledge Management*, 5(2), pp. 137–151.

Ward, V., House, A. and Hamer, S. (2009a). Developing a framework for transferring knowledge into action: A thematic analysis of the literature. *Journal of Health Services Research & Policy*, 14(3), pp. 156–164.

Ward, V., House, A. and Hamer, S. (2009b). Knowledge brokering: Exploring the process of transferring knowledge into action. *BMC Health Services Research*, 9, pp. 1–6.

Ward, V., House, A. and Hamer, S. (2009c). Knowledge brokering: The missing link in the evidence to action chain. *Evidence and Policy*, 5(3), pp. 267–279.

Ward, V., House, A. and Hamer, S. (2011). Exploring knowledge exchange: A useful framework for practice and policy. *Social Science & Medicine*, pp. 1–8.

Ward, V., Smith, S., Foy, R., House, A. and Hamer, S. (2010). Planning for knowledge translation: A researcher's guide. *Evidence and Policy*, 6(4), pp. 527–541.

Weiss, C. (1979). The many meanings of research utilization. *Administration Review*, 39(5), pp. 426–431.

Ziam, S. (2010). Knowledge brokers and how to communicate knowledge in 2010. *Allergy, Asthma & Clinical Immunology*, 6(4), pp. A3–A4.

Section V

Examples of Integrating Technologies: Virtual Systems, Image Processing, Biokinematics, Measurements, and VLSI

21

Virtual Doctor Systems for Medical Practices

Hamido Fujita and Enrique Herrera-Viedma

CONTENTS

21.1 Introduction

Medical service systems have become a burden to many governments due to a shortage in medical practitioners, a rapidly aging population, a shortage of medical resources in regional communities, and overburdened public health insurance systems. Particularly, the shortage of healthcare workers has brought a big burden to both developing and developed nations. For example, Japan had 2.2 doctors and 9.5 nurses per 1000 people in 2009 according to the Organisation for Economic Cooperation and Development (OECD) [1]. However, in India the proportions were 0.7 doctors and 0.9 nurses per 1000 people. The health worker shortage has reached a crisis level in 57 countries [2]. The demand for medical care is much higher in urban areas. The burden among the developed countries is the healthcare, while among developing countries it is the economic imbalance.

The above-mentioned issues require a need to adapt technology and intelligent systems to provide a thinking-support system and the necessary tools that help to provide solutions for the intelligent healthcare services and providers.

One of the solutions can be the Virtual Doctor System (VDS), a project supported by the Ministry of Internal Affairs and Communications of Japan. It aims to provide tools that could utilize the experiences of medical doctors (MDs) and ongoing practices in medical service [3–6]. It will help in the optimization of doctor time and in the reduction of medical expenses. The VDS has already been used and tested in diagnosis in simple medical cases. VDS systems have been exhibited in public demonstrations and experimented by medical experts. It could provide good practice in helping medical practitioners facilitate outpatients based on the concept of screening through the VDS before they meet with the practitioner [7]. The VDS could facilitate outpatients and perform decision making (DM) based on their symptoms and weight aggregation functions representing the physical and mental data articulated by the patients.

Recently, in another active project, we have expanded the VDS into medical diagnosis and screening in car driving situations to help in understanding the causes of accidents by the elderly in predefined predicted scenarios. These applications are computed based on aggregation operators that are used to compute the weights of the physical characteristics of the symptoms and the mental conditions of the drivers by mapping these symptoms onto the drivers' performances.

In this chapter, we outline the VDS applied in healthcare practices for medical diagnosis together with medical doctors. We discuss the state-of-art of related work on VDS innovations; moreover, on the subjective intelligence in understanding a patient's mental situation. The subjective criteria of patients helps in decision making by the practitioners. The related aggregation of other effecting factors useful in decision making are also presented.

Healthcare policy planners [8] consider the shortage problems [9] and the quality of medical doctors' practices that impact on the healthcare economy and the quality of life of individuals [10–11]. However, a feasible medical information system, such as the VDS, can provide a suitable information for a patient based on collective observation in relation to symptoms of physical and mental disorders. In most cases, the analyses are done using medical diagnosis systems that relate to physical symptoms. However, the subjectivities of patients play a big role in having medical doctors and experts filter through the possibilities of causes based on subjective analysis and observations of the patients. This involves some aspects of the uncertainty issues that are part of the analysis of medical diagnosis by the VDS. These uncertainty-related solutions addressed by VDS could provide optimized solutions when the medical situation is subjectively projected on a patient. For example, Fujita et al. [4] have provided a case study showing the importance of a well-functioning relationship between the physician and the patient, which can provide an effective way to improve the ability to show empathy with the patient's concerns based on mental analysis projected on the identified physical symptoms. The subjective evaluations of medical diagnosis are unique to the patient characteristics and profile [5,12]. Medical practitioners' style and communication with a patient [7] play a major role in establishing a good positive interaction [6] that could lead to a successful and satisfactory outcome. There are many different types of contextual interactions between patients and doctors that specify the relationship in terms of medical analysis and provide knowledge about symptoms, thus clarifying uncertainty issues between the patients and the practitioners. Both actors (patients and practitioners) are decision makers, engaged in a collaborative manner in providing specific knowledge on particular health scenarios. In most cases, these health scenarios help in decision making in relation to health issues. These human nature health scenarios are of two types: verbal and nonverbal. However, so far, the analyses of verbal and nonverbal communications in medical diagnosis have received little attention from researchers. Nonverbal communications are body language, physical appearances, gestures, and facial expressions. In addition, the doctor's appearance plays a part in the establishment of a positive engagement between the patient and the doctor. We have examined these verbal and nonverbal communications by mimicking the doctor's face through graphical software called "avatar" that resembles the face of a real doctor, the facial expressions, and the sounds generated by the doctor. One aspect behind this implementation was to provide a trust between the avatar and the patient [13].

Medical diagnosis is an important practice and is related to optimized evaluations having many variables. Diagnosis is complex due to imprecise criteria and fuzzy alternatives involved in decision making [14]. The process involves complex mental exercises and a state space search of medical knowledge, which could become complicated as the variables involved are complex and outpatients' symptoms are fuzzy with related alternatives. In addition, patients are not precise in expressing their conditions in a predictable mathematical manner, as they use ambiguous terms and languages to express their health aligned with different perspectives (contexts). Due to different backgrounds and experiences, practitioners would have to interpret patient symptoms in different manners, resulting in possibly arbitrary priorities and alternatives [15]. This is due to subjective uncertainties in providing appropriate interpretation of a patient's health condition descriptions.

This conundrum compounds when a particular pathological process presents ambiguous symptoms that can be similar to those of other conditions, as in the case of fever, or in situations when expert medical practitioners are in short supply and pressured for quick decisions [9]. In order to improve the possibility of early and accurate diagnosis of sickness, there is a need for the application of computational intelligence or other decision-support systems (DSSs) in the diagnosis process, to improve practitioner performance, reduce costs, and improve outcomes for the patients.

A number of expert technology-oriented systems have been tried to mitigate these challenges as well as to address the acquisition, representation, and utilization of the knowledge in medical diagnosis. However, the problem of managing the imprecise knowledge still exists. Clinical decision-support systems utilize a patient's data and some inference procedures to generate case-specific advice and suggestions to the practitioner [11].

The VDS [4,16–18] shown in Figures 21.1 and 21.2 has contributed in solving some of the issues discussed above. VDS is designed to resemble and act to interact with the patient-user just like a real human doctor to establish narrative diagnosis. The interoperability is represented by utilizing medical diagnosis cases from doctors, and it is presented in machine-executable fashion based on outpatient interactions with avatar.

The avatar or VDS works as a preliminary diagnosis to classify the patients conditions, based on the criticality, emergence-related parameters, and diagnosis scenarios' outcomes for decision making in healthcare services.

In relation to our studies on VDS implementation, so far, we have not seen similar implementation. However, related to this study, another technology has recently emerged [19]. It is the virtual-people factory, which was created with a web-based authoring tool for educating people in interpersonal skills. The virtual-people factory has some similarity in providing natural or common-sense knowledge of collaboration, but it does not provide intelligent methods to support interpersonal interactions. There are recent tools related to virtual patient concepts [20] created as education tools to develop the basic communication skills of medical students with patients; however, the doctor's medical knowledge and subjective experience, as in the case of avatar, are not considered.

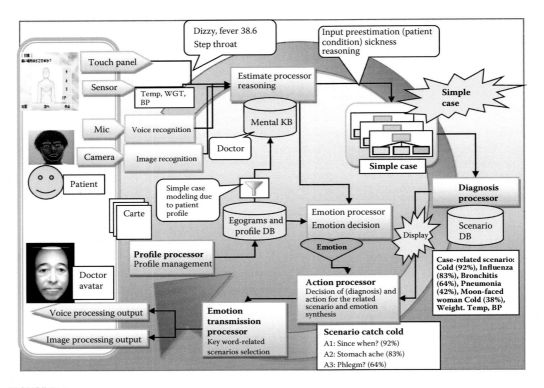

FIGURE 21.1
VDS outline.

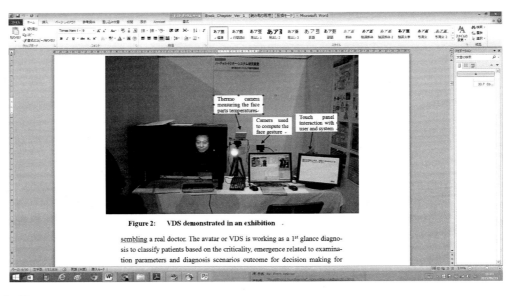

FIGURE 21.2
Components of VDS, demonstrated in an exhibition.

21.2 Outline of Virtual Doctor System

The VDS is an innovation reflecting the state of the art in system interaction that resembles the user's mental and psychological behavior through a face. We called the concept mental cloning and it is reported in other publications of Fujita et al. [3–5] and Kurematsu et al. [18,21]. Mental cloning is used to provide a means of representing the human mental aspect in an interactive style through avatar. The animated face images created in real time on avatar resemble the emotional behavior of that person articulated in the same manner that humans interact with a certain environment (specific cultural and personality preferences) with predefined certain invocations. This is represented by using the person's ego state [4,21] as the initial state and transactional analysis [4] and on ego-based interactions.

Figure 21.1 represents the implemented system outline. Figure 21.2 shows the system as exhibited in an experimental demonstration in a public booth in Japan. The input data collected from the sensors—blood pressure, temperature–thermoscan grids of the face parts, and others—are illustrated in Figure 21.3. A touch panel is provided for the patient to answer some questions. The camera is used for estimating facial patterns to predict the emotional state of the patient and estimate stress accordingly. All these data collected by the system modules are for the purpose of estimating the patient's physical and mental state. These two types of classified symptoms are used for reasoning and decision making, as explained later in this chapter.

The microphone is used for measuring the pitch and power of the patient's voice to estimate the breathing sounds and throat soaring patterns [18]. The importance of monitoring breathing sounds is reported in the following subsection, explaining how we can use sound for symptom extractions.

21.2.1 Health Symptom Estimation from Breathing Sound

The human voice consists of sounds made using the lungs, the vocal cords, the throat, and the lips. The human voice reflects unhealthy features of persons when one or more organs are damaged. Cracking voice, breathy voice, and cough are well-known unhealthy indicators that can be observed by human voices [22]. When a person senses unhealthy features from another person's voice, he thinks that that person has some respiratory problems [23]. That is, everyone can predict somebody's health condition from his voice. Doctors and nurses are trained to identify unhealthy features in human voice that can be related to a

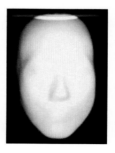

FIGURE 21.3
The avatar screen creation.

particular sickness. In this study, we classify human voices on doctors' experiences. We call human voices evaluated by doctors as the sample voices. In order to classify human voices, we convert a human voice to a frequency series to compare the similarities with the frequency series of the sampled voices. After that, we rank the human voice in a descending order by the degree of similarities. The correlation coefficient between a voice and a sample voice is used as similarity in computation.

We convert every sample voice to the frequency series using fast Fourier transform (FFT) with 500 Hz as the initial frame rate [18] to work out the correlation coefficient between all possible combinations of the frequency series. Grubbs' test (maximum normed residual test) indicates the states of correlation coefficients between a particular person's voice class and others as outlier. Therefore, we use frame rates to convert human voices to the frequency series by filtering and repeating the process while increasing 500 Hz windows until the frame rate is greater than 20 kHz.

Afterward, we collect the correlation coefficients among the frequency series with the filtered frame rates, and then make a frequency distribution of the entire process. After frequency distribution is made, we set the class interval of the frequency distribution as 0.1 thus showing the feature in each human voice class.

Next, we convert a target human voice to the frequency series by using FFT with repeated frame rates. In our study, a target human voice is the human voice that we try to classify. We make the frequency distribution with 0.1 class interval from the correlation coefficient between all combinations of the frequency series of the target voice and other sample voices resulting in the target human voice and the sample voice with the closest frequency distribution. Finally, we show the possibility of human voice classes ranked in descending order by similarity.

In this chapter an experiment on lung sounds is given as an illustration. Although doctors or nurses can hear lung sounds by using a stethoscope, it is not easy for the VDS to do the same in a similar practical manner. Therefore, we use training lung sounds pre-installed on a compact disc (CD) often for students in nursing schools [24]. There are more than 25 different types of lung sounds, including normal sounds. In experiments, we selected "coarse crackle," "fine crackle," "wheeze," and "rhonchus" because these are well-known unhealthy lung sounds, according to the advice of our doctors' committee.

We divided each lung sound by cycles and identified five sample data from each lung sounds for testing others.

Table 21.1 shows the number of filtering frame rates picked up from sample lung sounds, and they were converted in the form of frequency series for each rate.

The frequency distributions of the correlation efficient among the frequency series of the sample lung sounds resulted in the correlation coefficients among the frequency distribution of the test lung sounds and the frequency distribution of the sample lung sounds. Table 21.2 shows the experimental results.

TABLE 21.1

Filtering Frame Rates of Lung Sounds

	Type				
	Coarse Crackle	Fine Crackle	Wheeze	Rhonchus	Normal
Number of frames	33,000	9000	16,800	27,500	11,500

TABLE 21.2

Results of the Experiment

Class of Target Lung Sounds	Class of Sample Lung Sounds				
	Normal	Coarse Crackle	Fine Crackle	Wheeze	Rhonchus
Normal	82.2 (1)	71.4	74.5	57.6	18.0
Normal	89.4 (1)	55.1	70.7	65.5	72.0
Normal	74.8 (2)	66.3	75.0	63.3	53.0
Normal	74.9 (2)	66.1	81.3	44.0	53.2
Coarse crackle	68.1	71.4 (2)	79.3	44.1	53.6
Coarse crackle	83.1	68.6 (4)	78.6	63.1	74.2
Coarse crackle	78.9	55.1 (4)	78.6	43.2	69.2
Coarse crackle	85.9	67.6 (3)	70.4	50.5	62.7
Fine crackle	78.6	71.4	86.3 (1)	74.5	70.2
Fine crackle	72.7	66.1	84.9 (1)	59.5	32.2
Fine crackle	82.1	66.3	80.9 (2)	75.4	54.8
Fine crackle	75.5	71.4	90.0 (1)	51.4	52.8
Wheeze	65.4	41.4	76.9	51.0 (4)	56.1
Wheeze	68.9	63.8	81.3	41.5 (5)	46.3
Wheeze	72.4	55.1	84.9	56.9 (4)	68.4
Wheeze	50.9	71.4	94.1	68.4 (4)	95.4
Rhonchus	84.8	66.7	74.9	63.9	83.2 (2)
Rhonchus	74.3	66.7	70.0	64.6	82.8 (1)
Rhonchus	68.4	66.1	81.1	67.2	11.3 (5)
Rhonchus	70.6	66.3	83.5	67.9	63.3 (5)

Note: Ranking is indicated by the parenthesizes.

In Table 21.2, the name of one sound is shown in the row heading and the other is shown in the column heading, and each value in the parentheses shows the ranking. The average rank of human voice classes is 2.7. The average rank of fine crackle and normal lung sounds are under 1.5, while the average ranking of others is over 3.0. The reason why this happens is that most lung sounds, such as fine crackle, are similar to the normal lung sounds. The quality and quantity of sample data influence the accuracy of this approach. Although the quality and quantity of sample data in these experiments are not perfect, still, the experimental results show that it is possible to classify these lung sounds, especially fine crackle, based on the frequency series. Machine learning techniques [22,25] could enhance this approach by analyzing lung sounds with medical expert knowledge, to improve the quality, in the same manner as explained in Section 21.5.

21.2.2 Avatar Screen Generation

The VDS was first designed to facilitate and classify the outpatients in a hospital. During the development of the VDS, we have set a team of collaborators in various hospitals. We formed a committee of medical doctors who accepted collaboration, thus allowing us to utilize their experiences in our system. The VDS reads the physical data from different sensors and extracts mental analysis from facial expressions, etc. The avatar intending to replace a real doctor as a part of the VDS would be able to establish natural harmony between the patient and the avatar. The fourth (right hand side) panel in Figure 21.3 shows a real doctor whose face is copied (masked) by a special material that has been reproduced to create the screen, such as the one shown in the third panel. The system projected on the face is for creating a real doctor animated avatar, as shown in the first and second panels.

The system, as shown in Figure 21.2, creates a virtual three-dimensional face model through which a patient can interact with a virtual version of a medical doctor. It should be noted that the avatar projects the image of an actual doctor that the patient knows. The idea is to bring trust between the avatar and the patient. Trust is an important part of the diagnosis quality. The black box shown in Figure 21.2 is the system that creates an avatar reflecting the three-dimensional generated images on the mask screen with the actual image of a doctor. In addition, the VDS speaks with a voice mimicking the practitioner's *natural* accent and emotions based on the patient's extracted mental state articulated from the patient's profile and facial analysis. Data are measured and sent online to the VDS together with the mental status of the user as well as the estimated ego, all selected from the databases from universal templates [5].

21.2.3 Virtual Doctor System Interaction Based on Universal Templates

To construct databases for face-based mental state analysis, a survey has been carried out to collect data from the Japanese population with different genders and different ages. Each person entered responses to the questionnaires, which include Tokyo University Egogram (TEG)–based questions, as shown in Figure 21.4. The user, after entering his/her profile information (gender and age range), would answer the TEG questions with yes, yes/no, or no. The scoring answers as specified by TEG would be 2, 1, and 0, respectively [26]. The collected answers would represent their ego state as a number; an egogram is shown in Figure 21.5. The highest value would represent the best estimation of self-state of that person as one of the any state shown in Figure 21.5. These egogram and related values are stored and indexed in a database. We have considered six categories of age class; child (6–12 years), junior young (12–18 years), senior young (18–26 years), adult (26–45 years), senior adult (45–60 years), and old (61 years and older). These categories are set according to the accepted educational system and working structure in Japan. The first category is related to the elementary school class; the second category is related to intermediate school and high school. The third category is related to the university class and young inexperienced workers. The fourth category is related to middle-class workers of different slices. The fifth class is related to experienced workers and the advanced middle class. The last class is the class of elderly and retired people. Such classifications reflect the relative social characteristics of people as general classes reflecting each class personality and its relative characteristics of a specific specialization of the class. The user when visiting VDS would sit on a chair (Figure 21.6) where there is a touch panel beside him/her. Then enters information by clicking on a check box, a gender selection box, and an age class box. The system uses a generative program of universal templates to compute the user's personality.

We use the universal templates to select the patient's face match templates from the database by using three search keys, extracted from the facial image of the patient, namely, X, Y, and Z, as shown in Figure 21.7. The universal templates have been constructed based on a database of 330 Japanese faces (Figure 21.8). The best-fitting templates are selected from the database based on matching templates that are indexed and ranked, as the X, Y,

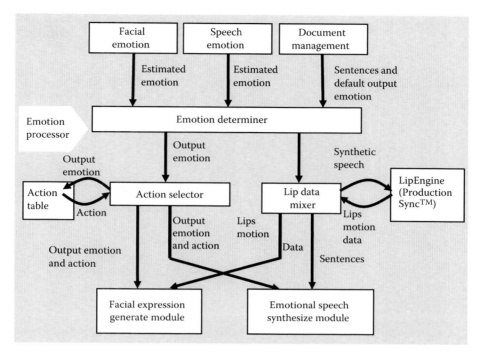

FIGURE 21.4
Block diagram of the emotion processor.

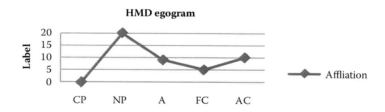

FIGURE 21.5
Tokyo University egogram example. CP: critical parent; NP: nurturing parent; A: adult; FC: free child; AC: adapted child.

and Z parameters shown in Figure 21.7, by matching the three classification keys. The face templates indexed on these three keys are used for emotional reasoning by calculating the correlation distance of emotional points; for more details on the universal templates, please refer to Fujita et al.'s paper [5].

The selected templates matching the patient will be used to reason on the mental state of the patient as combination computation values in six Ekman universal emotions: "sadness," "happiness," "disgust," "surprise," anger," and "fear." By using the OKI human sensing division [27], we could also extract the emotional feature from the movements of the human face parts.

There are feature points that detect the movement of the face parts as shown in Figure 21.9. This detection is done using the tools and library of OKI FSE Face Recognition Middleware

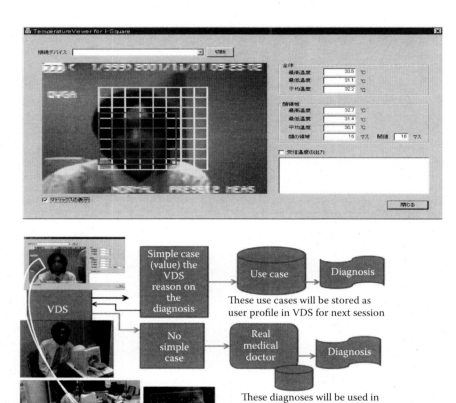

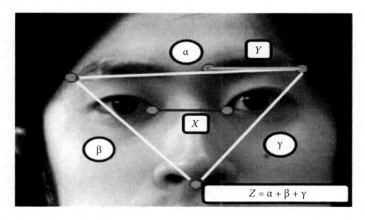

FIGURE 21.6
Physical ontology and VDS-related simple case.

$$Z = \alpha + \beta + \gamma$$

FIGURE 21.7
Parameter setting of a universal template.

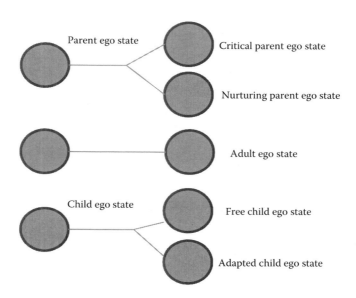

FIGURE 21.8
The five ego states.

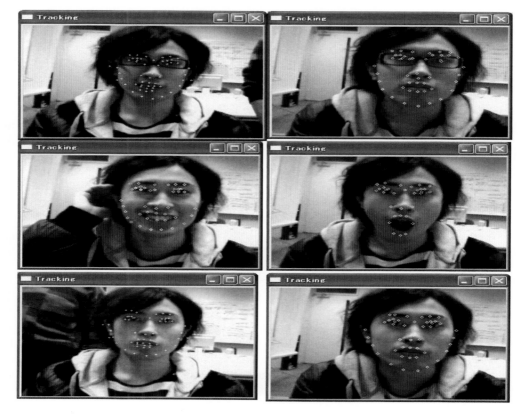

FIGURE 21.9
Facial analysis using OKI FSE, SDK.

for Embedded Systems V4 [28] and software development kit (SDK) [27]. For the eyes' centers, there is 1 feature point for each eye, there are 5 feature points for eyelids of each eye, and there are 3 points for each eyebrow. Also, there are 8 feature points for the mouth and 3 feature points for localizing the nose. In addition, there are 12 feature points specifying the contour of the face from the ears down. Using these feature points, we can track the emotional feature of the human user based on the value difference between the tracked point and the labeling dictionary in the database specifying the best emotional estimation and the selection of the emotional state among the six Ekman universal emotional states [3,5]. The emotional processor as shown in Figure 21.4 computes the received input data from the following three modules:

1. Facial emotion estimation module: Estimates emotion from facial expression.
2. Speech emotion estimation module: Estimates emotion in speech. This and the previous module examine the estimated emotion by using the emotion processor.
3. Document management module: This module provides sentences (resembling the interaction scenarios) and emotion to the processor. The processor makes output emotion in accordance to the estimated emotion provided by the emotion estimation modules. Therefore, we call this as the default output emotion. The documentations are organized according to the scenario that we need to establish so that the VDS can react based on the initial ego and patient emotional state. The emotion processor consists of an emotion determiner, an action selector, and a lip data mixer.

21.2.4 Transactional Analysis

Transactional analysis (TA) is a theory of personality and systematic psychotherapy for assisting personal growth and personal change [29]. TA proposes that we each have an *inner parent*, an *inner child*, and an *inner adult* personality part that collectively determine our feelings, beliefs, and behavior [30]. Berne [31] noted that analyses of transactions between ego states are the fundamental activity of a transactional analyst. He focused on ego states and transactions because they are eminently observable. Ego states and their representation as three stacked circles are the icons of transactional analysis, as shown in Figure 21.8. "The egogram is a visual symbol that represents the total personality of any human being by separating into its various aspects thus clearly showing which parts are "weak" and which are "strong." The egogram is like a psychological fingerprint; each person has a unique profile which can be seen and measured" [32]. It is pointed out that an egogram may have high scores when a person is in a stressful situation [32]. The Tokyo Medical University provided measures for egogram based on an ego state checklist (as in the example shown in Figure 21.10) on questionnaires used in diagnosis of psychosomatic illnesses [33].

It is stated by Murakami et al. [34] that persons with psychosomatic disorders are almost nurturing parent (NP) and adapted child (AC)–dominating ego state reflecting self-inhibitory lifestyle (Figure 21.8). These are measured from the egogram shown in Figure 21.5.

Because the VDS compiles a real practitioner, the ego state of the VDS would be the same ego state of the real actual doctor. This would initiate transactional interaction between the avatar based on the ego of the medical doctor and the ego of the template selected from the database of universal templates. This interaction is an interpersonal competency for establishing and maintaining effective communication.

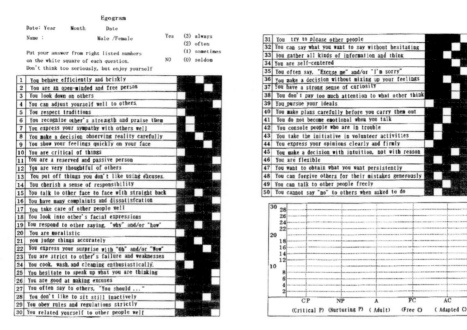

FIGURE 21.10

Tokyo Medical University Egogram example. (From Sachiko S. S., "How transactional analysis can be used in terminal care," *International Congress Series*, Elsevier 1287, pp. 179–184, 2006.)

21.2.5 Patient Interaction with Virtual Doctor System Avatar

The avatar interacts with the user (patient) based on the mental cloning of the subject person (medical practitioner), and according to it we have the human user (patient) to interact with the system based on TA. TA was first used by American Airlines to improve communication between staff and clients and then to be adopted by British Airlines [35]. It was later used in healthcare [36–37] studies on nurse faculty members' ego states as effective communication with patients to understand which of the nurses' ego states were most often applied in communication with patients [30,38]. Based on the analysis of the egograms of patients [34], we can establish effective communication narrative scenarios by having the avatar select an effective ego state that fits with the psychosomatic parameter of patients to establish subjective communication [36], like establishing critical parent (CP) (patient) → adapted child AC (VDS), or adult (A) → A. From an empirical analysis of local hospitals, we could extract the most frequently used interaction based on TA, as shown in Figure 21.11, which shows a diagram of states transactions between the five states of the avatar, for effective communications in health practices in two hospitals where the VDS is employed. The communication skills presented by the avatar are stimulated by the personality adaptations presented by Wilson [39]. The functional interaction between the patient and the avatar (mimicking the ego state of the doctor) is stated on the mechanism of ego analysis, as reflected in Figure 21.11. The action scenario between the patient and the VDS is based on the styles and wordings that are used by practitioner. If the patient's user emotion state is neutral, this reflects that the user is in adult state, in the terminology of TA (Figure 21.12). Rule 2 states that if the patient is in happy state, then the user is in nurturing parent state.

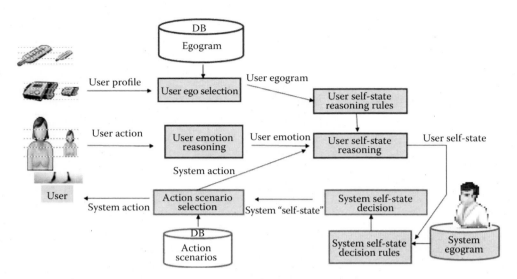

FIGURE 21.11
Outline of the system in relation to the egogram.

Based on rule 3, if both the virtual medical doctor (VMD) system emotion state and the estimated patient emotion state are the same, then the patient's user emotion state is NP. Based on rule 4, if the estimated patient's user emotion state is surprise, then the user state is free child (FC). In rule 5, if the estimated patient's user emotion state is anger or sadness or disgust (negative response), then the patient's user emotion state is CP. The priority sequence of these rules is rule 1 > rule 2 > rule 3 > rule 4 > rule 5. These rules were implemented in the system based on the doctor's experiences in the hospital. However, these rules could be changed based on the doctor's experience, ego, and style of interaction with the patient. The following was extracted empirically from local hospitals:

Self-State	Condition
VDS_NP → VDS_AC	Patient_CP > 4
VDS_AC → VDS_NP	Patient_CP ≤ 4
VDS_AC loop	Patient_CP > 4
VDS_NP → VDS_FC	Patient_NP > 7 or Patient_FC > 7
VDS_FC → VDS_NP	Patient_NP ≤ 7 or Patient_FC ≤ 7
VDS_FC loop	Patient_NP > 7 or Patient_FC > 7
VDS_NP → VDS_A	Patient_A > 5
VDS_A → VDS_NP	Patient_A ≤ 5
VDS_A loop	10 > Patient_A > 5
VDS_A → VDS_CP	Patient_A > 9
(Transition finish)	
※ VDS_CP → VDS_A and VDS_CP loop not exist	

VDS_NP loop Patient_CP ≤ 4 or Patient_NP ≤ 7 or Patient_FC ≤ 7 or Patient_A ≤ 5. Figure 21.12 shows the state transition of these rules and the relation of ego-based interaction between the patient and the doctor.

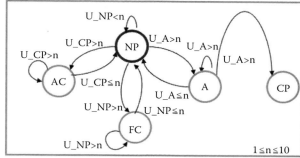

State transition of Kenji self-state decision

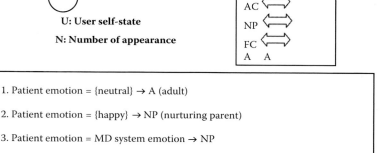

1. Patient emotion = {neutral} → A (adult)

2. Patient emotion = {happy} → NP (nurturing parent)

3. Patient emotion = MD system emotion → NP

    (nurturing parent)

4. Patient emotion = {surprise} → FC (emotional)

FIGURE 21.12
Patient/doctor self-state reasoning rules.

21.3 Outline of Virtual Doctor System Diagnosis

The medical scenarios used in medical diagnosis are defined based on formalization and customization in general guidelines according to the doctor's experience and related specialization on medical practices. This system is to help the doctor by filtering the outpatient (when they come to the hospital). The virtual doctor sees patients by interacting with them and performs decision making for category-based analysis of simple cases and complex cases. The simple medical case in our context is defined as cases that the doctor can diagnose, as a state that the patient can recover from by getting some rest, or taking simple medical supplements, or cases resulting from stress, heavy work, tiredness, etc. The simple-case treatment relies on the patient to get some rest and come back in a few days if recovery is not achieved or a correlated physical sign has reemerged or has not been relieved. We use the system to help doctors to classify medical cases of outpatients based on criticality issues. Criticality issues are estimations of the outpatient's sickness state. This is based on his/her mental and physical reasoning that is reasoned collectively by the VDS. There are two concepts that VDS uses for reasoning; *physical ontology* (PhO), and *mental ontology*

(MeO). We define the concept of pain as part of mental ontology. The two aligned ontologies produce a vector of the list of parameters for information retrial in medical knowledge, represented as the diagnosis processor, Figure 21.1. The action processor is decision making that uses fuzzy reasoning, in which the membership of index terms is defined to do decision making and accordingly direct the action processor selects the best scenarios that would read out the diagnosis fitting, in the same manner as a doctor does. In other works [40], we provided several videos that demonstrate how the VDS interacts with the patient and how the system's camera is eyeing the patient's face to extract values to reason the patient's state, as shown in Figure 21.2. The case studies examined are based on analysis provided by six medical doctors working with us on the project. Figure 21.13 shows the doctors in a session to provide advice and support in integrating different parts of the system to make it available to support their medical service for diagnosis of simple cases. The medical doctors' committee examines the scenarios created by the research team and the VDS. The committee expresses keen interest in the integration of mental criteria along with the physical symptoms' and they agree this practice as an important potential medical services.

However, most of the computer support diagnoses use physical symptoms based on case-based reasoning. These cases are objective oriented but not subjective to the patient's situation.

At an early stage of our experiment, the reasoning of medical cases we implemented was based on using a Bayesian network computation model [7,16].

As shown in Figure 21.14 the reasoning network is visualized using the tool Netica [41]. This figure shows a sample of the reasoning among different sickness probabilities the system is aggregating on observed symptoms and answers from patient through the touch panel. The results diagnosis is shown in Figure 21.14, which also shows the aggregated results for reasoning (see also Figure 21.15). This first experiment was carried for diagnosis of simple cases for actual medical cases [16]. The nature of observed medical diagnosis made us to be aware of the medical symptoms in the form of a fuzzy representation related to medical knowledge. These are represented by linguistic values (as will be shown in Section 21.4). Decision making in the context of the VDS is further studied in the context of decision-support system employing subjective criteria, such as mental issues, in multicriteria decision making.

FIGURE 21.13
A group of medical doctors reviewing the VDS.

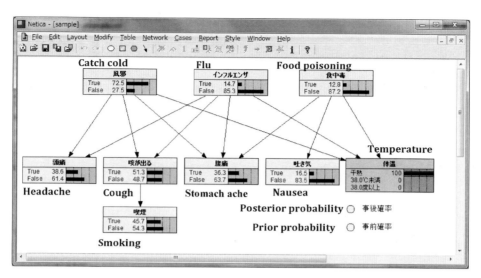

FIGURE 21.14
Applying Netica in an MD-provided simple case.

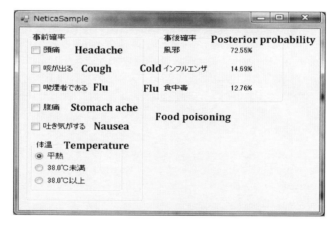

FIGURE 21.15
Results of simple cases of Figure 21.14.

21.4 Review of Literature on Decision Support for Medical Diagnosis

The literature of related to medical DSSs is divided as subjective and objective criteria. In the literature in this section we show that subjective reasoning must be aggregated with the objective criteria.

21.4.1 Decision-Support Systems

DSS technologies are mostly based on objective analysis of sickness. The analysis outcome is ranked on criteria that are weighted by the experts' opinions in relation to their experiences and on the medical data. The ranking is based on statistical analysis and experiences of medical experts. However, the medical experts may have imprecise knowledge on how to weigh these alternatives or criteria. Decision makers use ordinal and linguistics values to express preferences and these values but they can be computed as independent values operators having mutually exclusive values. Various forms of aggregation functions and operators were developed [42], as an extension to the additive aggregation operators [43]. These operators do not provide subjective relation to criteria nor to decision makers who define the weights of these criteria. This is because criteria relationships are assumed to be independent for analysis and computation purposes. However, these approaches do not take into account the subjective information in which criteria values could be adjusted. In addition, some researchers tend to provide theoretical solution without looking directly to the uniqueness of these criteria. There are called mental models that reflect the setting of each criterion in an independent fashion in relation to a set of subjective models that provide collective reflection to those criteria. Such reflection is used to have multiplicative weight on the objective criteria for decision making to achieve better selection and ranking. Power average–based aggregation functions are defined by many authors with multiplicative preference relations [44]. In practice this is not realistic as the achieved aggregated preferences do not reflect the subjective issues of possible changes that decision makers may impose in regard to criteria dynamics. This becomes more complex if a moderator is involved. The aim of a moderator is to moderate the decision makers' preferences. A support function based on arithmetic mean [45] can assist on regulating consensus by the use of power average [46] method; however, this does not take into account the subjective relations of these preferences when the contents of criteria have some relation to decision makers during the implementation of aggregation methods [12,47]. These systems provide decision makers a set of object related solution.

21.4.2 Subjective Intelligence

In medical services, patient preferences and personality traits are essential to identify accurate medical scenarios [48]. Also, efficient communication skills between practitioners and patients are essential for appropriate and efficient diagnosis. This is evident from the survey carried by Ommen et al. [49]. An effective practitioner and patient communication provides trust and better outcomes in quality and service [50]. Educating health personnel on effective practices in having them better perceived by the patients is one of the major characteristics in evaluations and quality services [51].

Looking to only objective analysis collected from medical knowledge and type of diseases would not be complete without projecting it subjectively on the patient's situation. Therefore, understanding the patient's personality and it's impact his/her health can provide parametric analysis that doctors can consider for a better judgment on the patient's health condition and its possible causes of deterioration.

Idiosyncratic preferences and profiles [52] become an essential part in providing better understanding in value-based decision making and choice-based decision making among a set of elements projected on objective criteria related to sickness. As stated earlier, we studied a concept, called mental cloning [3], which can capture subjectively people's mental states based on estimating their ego states by using the Tokyo egogram, and relating with the interaction between avatar and patient through transactional analysis [5] using facial movements [17,53] and voice analysis [18,21].

21.5 Reasoning Framework

In this section, we consider medical diagnosis based on Doctor A. So collecting the user mental state aims to adapt to changes that a user would have.

Described in a paper by Fujita et al. [54], the voices' and emotional features are examined to reflect the patient's state of health, like sore throat or pain or other emotions. This is done for the system to participate in the dialogue between the patient and the doctor with an emotional voice synchronized with the patient's mental situation. We have built the system to resemble a medical doctor interacting with the patient based on the framework as explained above for diagnosis on patient at clinics in Japan. MD style of reasoning is examined [12] and presented in the VDS system based on simple cases of medical practice data exhibiting a fuzzy nature.

21.6 Fuzzy-Based Reasoning

In this section, we discuss aspects of using fuzzy techniques for realizing multiple criteria in decision making for medical diagnosis. The ontologies PhO and MeO are extended to have fuzzy index terms governed by fuzzy logic that is used to do decision making based on the medical diagnosis.

The conceptual reasoning framework is based on representing two types of ontologies, reflecting patient's physical status as seen by the VDS (Figure 21.16). We have primarily used a combination of fuzzy techniques to do decision making on *simple-case* diagnosis. The target of the system is identifying the patient case as *simple*, with *weight* (*Simple_?*), where "?" is *high, medium, low,* or *not-simple*. This is done by ranking the symptoms based on attributes that have fuzzy membership in the ontology (Figure 21.17).

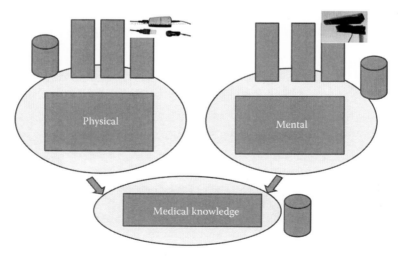

FIGURE 21.16
Outline of the classified criteria.

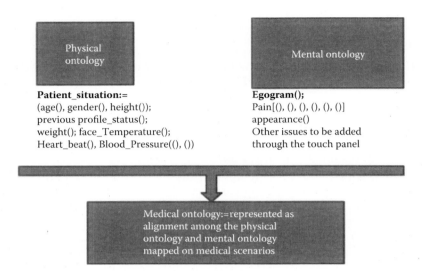

FIGURE 21.17
Physical criteria and mental criteria.

We have defined two types of ontological views: PhO and MeO representing causal relation articulated from the physical view analysis and the mental view analysis. In a previous work [16], we had used a probabilistic model to reference and infer doctor diagnosis
[55]. The probabilistic model makes use of alignment of the two defined ontologies and
performs calculation to compute values that are used to make decisions related to special
cases using the Bayesian network. The Bayesian network [54] (as belief network) made
use of these two ontologies. Logical and probabilistic approaches were also applied to the
diagnosis for decision making (DM) [55–57].

However, due to uncertainty of human cognition, it is not easy to provide suitable numerical values for the criteria needed to make appropriate DM. We need to provide fuzzy
linguistic attributes for decision making doctors or experts. Medical knowledge represents symptoms of patients in fuzzy manner, like high temperature, young or very young,
and so on. To do that among alternatives [22–23,47,58], we use a fuzzy ordinal linguistic
approach [22–23] to represent the assessments of symptoms of patients and the relative
importance weights associated with the symptoms. The goal is to combine all symptom
attributes in each ontology and then to fuse both ontologies to provide a final diagnosis.

21.6.1 Fuzzy Linguistic Approach to Representing User Assessments

The fuzzy linguistic approach is an approximate technique which represents qualitative
aspects as linguistic values by means of linguistic variables, that is, variables whose values are
not numbers but words or sentences as in the case of a natural languages [24]. Its application is
beneficial because it introduces a more flexible framework, which allows us a representation
of the information in a more direct and adequate way when we are unable to express it with
numerical precision. In such a way, the burden of quantifying a qualitative concept is removed.

The choice of the linguistic term set with its semantics is the first goal to satisfy the
requirements of any linguistic approach in solving a problem. It consists of establishing
the linguistic variable or linguistic expression domain with a view to providing the linguistic performance values.

Definition 1

A linguistic variable used in here is characterized by a quintuple $(L, H(L), U, G, M)$, in which L is the name of the variable; $H(L)$ (or simply H) denotes the term set of L, i.e., the set of names of linguistic values of L, with each value being a fuzzy variable denoted generically by X and ranging across a universe of discourse U, which is associated with the base variable u; G is a syntactic rule (which usually takes the form of a grammar) for generating the names of values of L; and M is a semantic rule for associating its meaning with each value of L, $M(X)$, which is a fuzzy subset of U.

From a practical point of view, we can find two possibilities of choosing the appropriate linguistic descriptors of the linguistic term set and their semantics [22–23]:

1. *Traditional fuzzy linguistic approach*: The first possibility defines the linguistic term set by means of a context-free grammar, and the semantics of linguistic terms is represented by fuzzy numbers described by membership functions based on parameters and a semantic rule [24,59].
2. *Ordinal fuzzy linguistic approach*: The second possibility defines the linguistic term set by means of an ordered structure of linguistic terms, and the semantics of linguistic terms is derived from their own ordered structure which may be either symmetrically distributed on the [0, 1] interval [60–61] or not [62].

In both possibilities, in order to establish the linguistic descriptors of a linguistic variable, an important aspect to analyze is the granularity of uncertainty, i.e., the cardinality of the linguistic term set used to express the information. The cardinality of the term set must be small enough so as not to impose useless precision on the users, and it must be rich enough in order to allow a discrimination of the assessments in a limited number of degrees. Typical values of cardinality used in the linguistic models are odd ones, such as 7 or 9, with an upper limit of granularity of 11 or no more than 13, where the midterm represents an assessment of "approximately 0.5," and with the rest of the terms being placed symmetrically around it [63–64]. In the first possibility, the granularity of uncertainty is not easily under control, and we can achieve inadequate cardinalities (very high). However, in the second one, we can control this aspect and supply users with a few but meaningful linguistic descriptors.

In this chapter, we assume the second method in order to reduce the complexity of defining a grammar and a semantic rule and to control and supervise the granularity of uncertainty. We consider a finite and totally ordered label set $S = \{s_i\}$, where $i \in \{0, \ldots, T\}$ in the usual sense, i.e., $s_i \geq s_j$ if $i \geq j$, and with odd cardinality (7 or 9 labels). The midterm represents an assessment of approximately 0.5, and the rest of the terms being placed symmetrically around it, and assuming that each label for the pair (s_i, s_{T-i}) is equally informative. The semantics of the labels is given by fuzzy numbers defined on the [0, 1] interval, which are described by linear trapezoidal membership functions represented by the quadruple (a_i, b_i, c_i, d_i) (the first two parameters indicate the interval in which the membership value is 1.0; the third and fourth parameters indicate the left and right widths of the distribution). For example, we can use the following set of seven labels to provide the user evaluations on the variable temperature:

$S = \{$M = Maximum = (1, 1, 0.16, 0), VH = Very_High = (0.84, 0.84, 0.18, 0.16), H = High = (0.66, 0.66, 0.16, 0.18), M = Medium = (0.5, 0.5, 0.16, 0.16), L = Low = (0.34, 0.34, 0.18, 0.16), VL = Very_Low = (0.16, 0.16, 0.16, 0.18), M = Minimum = (0, 0, 0, 0.16)$\}$.

In any linguistic approach, it is necessary to define some operators that can be applied to linguistic information. An advantage of the ordinal fuzzy linguistic approach is the simplicity of its computational model for computing with words. It is based on a symbolic computation [60,64–66]. This technique acts by direct computation on labels by taking into account the order of such linguistic assessments in the ordered structure of linguistic terms.

This symbolic tool seems natural when using the fuzzy linguistic approach, because the linguistic assessments are simply approximations, which are handled when it is impossible or unnecessary to obtain more values that are accurate. Usually, the ordinal fuzzy linguistic model for computing with words is defined by establishing (i) a negation operator, (ii) comparison operators based on the ordered structure of linguistic terms, and (iii) adequate aggregation operators of linguistic information. In most ordinal fuzzy linguistic approaches, the negation operator is defined from the semantics associated to the linguistic terms as

$$\text{Neg}(s_i) = s_j | j = T - i;$$

and there are defined two comparison operators of linguistic terms:

- Maximization operator, $\text{MAX}(s_i, s_j) = s_i$ if $s_i \geq s_j$
- Minimization operator, $\text{MIN}(s_i, s_j) = s_i$ if $s_i \leq s_j$

An interesting class of linguistic aggregation operators is based on the ordered weighted averaging (OWA) operators [67]. Some examples of useful linguistic OWA operators to combine linguistic information are the linguistic ordered weighted averaging (LOWA) [60] and the linguistic weighted averaging (LWA) operators [68].

The LOWA operator is an operator of fuzzy computing with words which is used to aggregate nonweighted linguistic information [60] (i.e., linguistic information values with equal importance). ∎

Definition 2

Let $A = \{a_1, \ldots, a_m\}$ be a set of labels to be aggregated, then the LOWA operator Φ is defined as

$$\Phi(a_1, \ldots, a_m) = W \cdot B^T = C^m\{w_k, b_k, k = 1, \ldots, m\} = w_1 \odot b_1 \oplus (1 - w_1) \odot C^{m-1}\{\beta_h, b_h, h = 2, \ldots, m\},$$

where $W = [w_1, \ldots, w_m]$ is a weighting vector, such that $w_i \in [0, 1]$ and $\Sigma_i w_i = 1$; $\beta_h = w_h/(\Sigma_{k=2}^{m} w_k)$, with $h = 2, \ldots, m$; and $B = \{b_1, \ldots, b_m\}$ is a vector associated with A, such that $B = \sigma(A) = \{a_{\sigma(1)}, \ldots, a_{\sigma(m)}\}$, where $a_{\sigma(j)} \leq a_{\sigma(i)} \forall i \leq j$, with σ being a permutation over the set of labels A. C^m is the convex combination operator of m labels and if $m = 2$, then it is defined as

$$C^2\{w_i, b_i, i = 1, 2\} = w_1 \odot s_j \oplus (1 - w_1) \odot s_i = sk,$$

such that

$$k = \min\{T, i + \text{round}[w_1 \cdot (j - i)]\}, \quad s_j, s_i \in S, \quad j \geq i,$$

"round" being the usual round operation, and $b_1 = s_j$, $b_2 = s_i$. If $w_j = 1$ and $w_i = 0$ with $i \neq j \, \forall \, i$, then $C^m\{w_i, b_i, i = 1, \ldots, m\} = b_j$.

The LOWA operator is an OR–AND operator [60]; therefore, it can carry out a soft computing in the modeling of MAX and MIN linguistic operators. In order to classify OWA operators in regard to their localization between AND and OR, Yager [67] introduced a measure of orness, associated with vector **W**, which is defined as

$$\text{orness}(W) = \frac{1}{m-1} \sum_{k=1}^{m} (m-k)w_k,$$

An important question of the LOWA operator is the determination of **W**. A possible solution consists of representing the concept of *fuzzy majority* by means of the weights of **W**, using a nondecreasing proportional *fuzzy linguistic quantifier* [69] Q in its computation:

$$w_i = Q(i/m) - Q[(i-1)/m], \quad i = 1, \ldots, m,$$

being the membership function of Q with definition parameters (a, b), with $a, b, r \in [0, 1]$:

$$Q(r) = \begin{cases} 0 & \text{if } r < a, \\ \dfrac{r-a}{b-a} & \text{if } a \leq r \leq b, \\ 1 & \text{if } r > b. \end{cases}$$

Some examples of a nondecreasing proportional fuzzy linguistic quantifier are "most" (0.3, 0.8), "at least half" (0, 0.5) and "as many as possible" (0.5, 1). When a fuzzy linguistic quantifier Q is used to compute the weights of LOWA operator Φ, it is symbolized by Φ_Q.

The LWA operator [68] is another important operator of fuzzy computing with words, which is based on the LOWA operator. It is defined to aggregate weighted linguistic information, i.e., linguistic information values with not equal importance. ∎

Definition 3

The aggregation of a set of weighted linguistic opinions, $\{(c_1, a_1), \ldots, (c_m, a_m)\}$, where $c_i, a_i \in S$, according to the LWA operator Π is defined as

$$\Pi[(c_1, a_1), \ldots, (c_m, a_m)] = \Phi(h(c_1, a_1), \ldots, h(c_m, a_m)),$$

where a_i represents the weighted opinion, c_i is the importance degree of a_i, and h is the transformation function defined depending on the weighting vector W assumed for the LOWA operator Φ, such that $h = \text{MIN}(c_i, a_i)$ if $\text{orness}(W) \geq 0.5$, and $h = \text{MAX}(\text{Neg}(c_i), a_i)$ if $\text{orness}(W) < 0.5$. ∎

21.6.2 Medical Reasoning in a Fuzzy Linguistic Context

This section outlines issues related to how to do linguistic fuzzification for the attributes that are used in reasoning. These attributes are represented using the aforementioned ordinal fuzzy linguistic approach.

Next, the medical cases are analyzed according to the a set of n symptoms $M = \{MA_1, MA_2, MA_3, ..., MA_n\}$, where $MA_i \in S$. In such a way, we build a database of medical cases. The set of symptoms are classified into two main sets. One is related to physical set, and the other is related to mental set. All these symptoms are represented as ordered weighted attribute represented by linguistic assessments; i.e., for each symptom MA_i we have a weight $w(MA_i) \in S$. Fuzzy rules are based on classifying these attributes values and order them based on related aggregation functions. The distances' measurement is related to the aggregation function used in either ontology. This is based on the similarity coefficient between those attributes related to decision making correlated with medical knowledge. The order weight average resembles the symptom that has the major relation (central tendency attribute) to sickness main property and has other criteria that converge on it in an aggregating manner. This representation is what is called *consensus attributes aggregations* (of objective attributes and the subjective attributes). In our theory, the objective attributes are those related to physical attributes contributed to define the physical properties of a sickness. The subjective attributes are those related to specify the mental characteristic of patient (user) in relation to the ego state and other mental characteristics that contributed to that sickness.

Then suppose that we have a set of symptoms of a user i characterized by a vector of linguistic values $M_i = \{MA_1, MA_2, MA_3, ..., MA_n\}$ representing his/her physical state and a set of weights indicating the importance of each symptoms to identify a possible medical cases, $W(M) = \{W(MA_1), W(MA_2), W(MA_3), ..., W(MA_n)\}$. Then, to compute the similarity with a real medical case j stored in the knowledge base and characterized by a characteristic linguistic assessments previously provided by medical experts, $K_j = \{KA_1, KA_2, KA_3, ..., KA_n\}$, we could apply the linguistic aggregation operator LWA (Π) in order to obtain a physical similarity coefficient ($\text{psc}_{ij} \in S$) with the stored physical state as

$$\text{psc}_{ij} = \Pi[(W(MA_1), a_1), ..., (W(MA_n), a_n)] = \Phi_Q(h(W(MA_1), a_1), ..., h(W(MA_n), a_n)),$$

with $a_h = \text{Label}(|I(MA_1) - I(KA_1)|)$, with I as a numerical index function that returns the index associated to a linguistic term and Label as that linguistic function with a reverse behavior.

On the other hand, we could compute a mental similarity coefficient ($\text{msc}_{ij} \in S$) for the case of stored mental state. These above two situations each are reflected and represented on ontology [70] reflecting the medical ontology and then using both systems identifies the possible diagnostic by comparing both coefficients with those established in the knowledge base in order to obtain a diagnostic degree ($d_i \in S$).

The medical case is defined in conceptual view and generalized from the specialization due to the type definition of medical preference cases according to the doctor's experiences represented on medical ontology. The variables outcome would infer the medical cases and invoke certain scenarios. All criteria are computed based on mental and physical observation in the model. Criteria are computed and collected by initiating scenarios with the outpatient to collect these causes' fuzzy linguistic values to be used for decision making. With the doctors' support, we have established several cases resembling different types of configuration of variables treated as linguistic membership functions. These configurations can resemble different types of simple medical cases. We can create more complicated configuration in the same manner; however, this would make the linguistic membership functions more complicated.

21.6.3 Medical Reasoning in a Multigranular Fuzzy Linguistic Context

In many situations, we find that the assessments need to be expressed in different linguistic term sets. For example, in group decision-making (GDM) problems with multiple experts, it is common that experts have different degrees of knowledge about the alternatives, and so they may prefer to use different linguistic term sets with different levels of distinction to express their preferences. We call this fuzzy linguistic context as multigranular fuzzy linguistic context. In such cases, we say that the GDM is defined in a multigranular fuzzy linguistic context [71–73].

This is the case in the VDS, where we have different kinds of information that would be better represented in a multigranular fuzzy linguistic context, i.e., by using a set of linguistic term sets to represent the different assessments: $S_j = \{s_i^j\}$, $i \in \{0, ..., T_j\}$, where $T_j + 1$ is the cardinality of the linguistic term set S_j. Then we could assume that each symptom is assessed on a different linguistic term set, i.e., $M = \{MA_1, MA_2, MA_3, ..., MA_n\}$, where $MA_j \in S^j$. Similarly, we could assume that the set of importance weights associated with symptoms could also be assessed on different linguistic term sets, i.e., $W(M) = \{W(MA_1), W(MA_2), W(MA_3), ..., W(MA_n)\}$, where $W(MA_k) \in S^k$.

In order to operate with multigranular linguistic information, the first step is to unify it in a unique linguistic expression domain [71]. All linguistic assessments are conducted (using a transformation function) in a single domain called basic linguistic term set (BLTS) and denoted as S_T [71]. The BLTS should be able to maintain the uncertainty degrees associated with each individual domain S^j. Hence, in a general multigranular linguistic context, to select S_T we proceed as follows [71]:

1. If there is only one linguistic term set from the set of different domains with maximum granularity, then we choose that one as the BLTS, S_T.
2. If there are two or more linguistic term sets with maximum granularity, then the selection of S_T will depend on the semantics associated with them:
   a. If all the linguistic term sets have the same semantics (with different labels), then any one of them can be selected as S_T.
   b. If two or more linguistic term sets have different semantics, then S_T is defined as a linguistic term set with a number of terms greater than the number of terms a person is able to discriminate, which is normally 7 or 9.

Once the BLTS has been selected, the multilinguistic information is unified in the BLTS by means of a multigranular transformation function.

Definition 4

If $S = \{l_1, l_2, l_3, ..., l_q\}$ and $S_T = \{s_1, s_2, s_3, ..., s_g\}$ are two linguistic term sets, with $g \geq p$, then a multigranular transformation function, τ_{SST} is defined as

$$\tau_{SST}: S \to F(S_T),$$

such that $\tau_{SST}(l_i) = \{(s_h; \alpha_h) | h = 0, \ldots, g\}; \forall\, l_i \in S$, with

$$\alpha_h = \max_y \min\{\mu_{li}(y); \mu_{sh}(y)\},$$

where $F(S_T)$ is the set of fuzzy sets defined on S_T and $\mu_{li}(y)$ and $\mu_{sh}(y)$ are the membership functions of the fuzzy sets associated with the linguistic terms l_i and s_h, respectively.   ■

Example 1

Let $S = \{l_0, l_1, l_2, l_3, l_4\}$ and $S_T = \{s_0, s_1, s_2, s_3, s_4, s_5, s_6\}$ be two term sets, with 5 and 7 labels, respectively, and with the following semantics associated:

$l_0 = (0; 0; 0; 0.25)$	$s_0 = (0; 0; 0; 0.16)$
$l_1 = (0.25; 0.25; 0; 0.5)$	$s_1 = (0.16; 0.16; 0; 0.34)$
$l_2 = (0.5; 0.5; 0.25; 0.75)$	$s_2 = (0.34; 0.34; 0.16; 0.5)$
$l_3 = (0.75; 0.75; 0.5; 1)$	$s_3 = (0.5; 0.5; 0.34; 0.66)$
$l_4 = (1; 1; 0.75; 1)$	$s_4 = (0.66; 0.66; 0.5; 0.84)$
	$s_5 = (0.86; 0.84; 0.66; 1)$
	$s_6 = (1; 1; 0.84; 1)$

The fuzzy set obtained after applying τ_{SST} for l_1 is

$$\tau_{SST}(l_1) = \{(s_0; 0.39); (s_1; 0.85); (s_2; 0.85); (s_3; 0.39); (s_4; 0); (s_5; 0); (s_6; 0)\}.$$

Then suppose that we have a set of symptoms of a user i characterized by a vector of linguistic values, $M_i = \{MA_1, MA_2, MA_3, \ldots, MA_n\}$, with $MA_j \in S^j$ representing his/her physical state, and a set of weights indicating the importance of each symptoms to identify a possible medical cases, $W(M) = \{W(MA_1), W(MA_2), W(MA_3), \ldots, W(MA_n)\}$, with $MA_k \in S^k$. To make a diagnostic decision we have to compute the similarity with a real medical case p stored in the knowledge base and characterized by a characteristic linguistic assessments previously provided by medical experts, $K_p = \{KA_1, KA_2, KA_3, \ldots, KA_n\}$, with $KA_h \in S^h$. In such a case, we choose the BLTS S_T from all linguistic term sets used to provide the linguistic assessments, i.e., $\{S^j, S^k, S^h; j, k, h = 1, \ldots, n\}$, and then we transform all assessments in S_T by means of linguistic transformation functions τ_{SST}^j, τ_{SST}^k, and τ_{SST}^h. For example, each linguistic assessment of a user i, $MA_j \in S^j$ could be transformed in a fuzzy set MA_j^* defined on $S_T = \{s_1, s_2, s_3, \ldots, s_g\}$ as

$$MA_j^* = s_{ST}^j(MA_j) = \left\{ (s_t, \alpha_t^j) \middle| t = 0, \ldots, g \right\},$$

such that

$$\alpha_t^j = \max_y \min\{\mu_{MAj}(y); \mu_{st}(y)\}, \quad \alpha_t^j \in [0, 1].$$

By simplifying, we symbolize each MA_j^* by means of its membership values, i.e., $\left(\alpha_0^j, \alpha_1^j, \alpha_2^j, \ldots, \alpha_g^j \right)$.

In order to compute the similarity between the symptoms of users and those stored in the medical knowledge database, we define a procedure to compute distances between the linguistic assessments $\left(\alpha_0^j,\ \alpha_1^j,\ \alpha_2^j, ...,\ \alpha_g^j\right)$ and $\left(\alpha_0^h,\ \alpha_1^h,\ \alpha_2^h, ...,\ \alpha_g^h\right)$ for all j and h:

1. Compute the *central value* (cv) of fuzzy sets $\left(\alpha_0^j,\ \alpha_1^j,\ \alpha_2^j, ...,\alpha_g^j\right)$ and $\left(\alpha_0^h,\ \alpha_1^h,\ \alpha_2^h, ...,\alpha_g^h\right)$ [74]:

$$\mathrm{cv}\left(MA_j^*\right) = \left(\frac{\sum_{v=0}^g v\alpha_v^j}{\sum_{v=0}^g \alpha_v^j}\right), \qquad \mathrm{cv}\left(KA_h^*\right) = \left(\frac{\sum_{v=0}^g v\alpha_v^h}{\sum_{v=0}^g \alpha_v^h}\right),$$

where $\mathrm{cv}(MA_j^*)$ and $\mathrm{cv}(KA_h^*)$ represent the centers of gravity of the information contained in the fuzzy sets MA_j^* and KA_h^*, respectively.
2. Compute the similarity $s(MA_j^*,\ KA_h^*)$ as

$$\mathrm{s}(MA_j^*,\ KA_h^*) = 1 - \left[\frac{\mathrm{cv}(MA_j^*) - \mathrm{cv}(KA_h^*)}{g}\right].$$

The closer $s(MA_j^*,\ KA_h^*)$ to 1, the more similar MA_j^* and KA_h^* are; while the closer $s(MA_j^*,\ KA_h^*)$ to 0, the more distant MA_j^* and KA_h^* are.

Then, to compute the similarity between the symptoms of a user i with a real medical case p stored in the knowledge base, we could apply the linguistic aggregation operator LWA (Π) in order to obtain a physical similarity coefficient ($\mathrm{psc}_{ip} \in S_T$) with the stored physical state as

$\mathrm{psc}_{ip} = \Pi[(\mathrm{Label}(g\mathrm{cv}(W(MA_1)^*)), \mathrm{Label}(g\mathrm{s}(MA_1^*, KA_1^*)), ..., (\mathrm{Label}(g\mathrm{cv}(W(MA_n)^*)), \mathrm{Label}(g\mathrm{s}(MA_n^*, KA_n^*))].$

Similarly, it is computed the mental similarity coefficient ($\mathrm{msc}_{ip} \in S_T$) for the case of stored mental state.

21.7 Conclusions

This chapter references a research project conducted by the authors on VDS, or Virtual Doctor System. It presents the framework for reasoning based on medical cases analyzed on physical view reasoning (objective-based criteria) and mental view reasoning (subjective-based criteria). The avatar has provided a platform for interaction with the patient by mimicking a doctor in terms of narrative-generated sound and facial dynamics. The mental model reasoning on a patient could provide specific means to estimate the emotional state of the patient that has subjective projection on the physical criteria. The decision making for simple cases was tested based on a Bayesian network using knowledge bases provided by doctors in regional hospitals. The results obtained show an accuracy of 80% in diagnosis of simple medical cases. The results were encouraging to expand the system to more complex medical cases. The reasoning mechanism is expanded to have criteria be represented by fuzzy linguistic values that could provide better mechanism than the Bayesian network due to uncertainty issues, and have better analysis and accuracy in representing medical symptoms.

The linguistic values of physical-type attributes (objective) are represented using the aggregation functions of fuzzy logic attributes. The mental attributes (subjective) are classified to those related to ordered weighted attributes based on assigned linguistics values (for mental state). Collectively evaluating these values into collective alignment to reason on medical knowledge based on similarity distance computation provides suitable reasoning structure mimicking a doctor's hierarchical reasoning for better services.

For a future challenge, this chapter can provide relevance to the research objective in the Horizon 2020 Work Programme 2014–2015 in the area of health, demographic change, and well-being [75]. The objective is to provide innovative technology that can participate in providing eHealth as a feasible service in cloud-based knowledge acquisition from blog data.

References

1. http://dx.doi.org/10.1787/health_glance-2011-en.
2. http://www.who.int/workforcealliance/forum/2011/SecondGlobalForumHRH_Presspack .pdf.
3. Fujita, H., J. Hakura, and M. Kurematsu, "Intelligent human interface based on mental cloning-based software," *International Journal on Knowledge-Based Systems*, 22 (3), pp. 216–234, 2009.
4. Fujita, H., J. Hakura, and M. Kurematsu, "Virtual Medical Doctor Interaction Based on Transactional Analysis," *Frontiers in Artificial Intelligence and Application Series*, Volume 199, *New Trends in Software Methodologies, Tools and Techniques (SoMeT_09)*, IOS Press, ISBN: 978-1-60750-049-0, pp. 503–517, 2009.
5. Fujita, H., J. Hakura, M. Kurematsu, S. Chida, and Y. Arakawa, "Empirical Based Techniques for Human Cognitive Interaction Analysis: Universal Template Design," *Proceedings of the 7th New Trends in Software Methodologies, Tools and Techniques (Proceedings of SoMeT_08)*, pp. 257–277. IOS Press, ISBN: 978-1-158603-916-5, pp. 339–360, 2008.
6. Fujita, H., J. Hakura, and M. Kurematsu, "Virtual Doctor System (VDS): Framework on Reasoning Issues," *Frontiers in Artificial Intelligence and Applications*, Vol. 217, *Proceedings of the 9th New Trends in Software Methodologies, Tools and Techniques (Proceedings of SoMeT_10)*, IOS Press, ISBN: 978-1-60750-628-7, pp. 481–489, 2010.
7. Fujita, H., M. Kurematsu, and J. Hakura, "Virtual Doctor System (VDS): Aspects on Reasoning Issues," *New Trends in Software Methodologies, Tools and Techniques (SoMeT_11)*, IOS Press, Saint Petersburg, Russia, pp. 293–304, Sept. 2011.
8. "Coping with the doctor shortage," *The Japan Times*, http://www.japantimes.co.jp/text/ed20071001a1 .htm, accessed on August 14, 2014.
9. Lennon, B., "Medical Workforce Expansion in Australia—Commitment and Capacity," Ninth International Medical Workforce Collaborative Conference; Nov. 16–19, 2005; Melbourne, http://www.health.nsw.gov.au/amwac/amwac/9conf.html, accessed on August 14, 2014.
10. Ishikawa, T., H. Ohba, Y. Yokooka, K. Nakamura, and K. Ogasawara, "Forecasting the absolute and relative shortage of physicians in Japan using a system dynamics model approach," *Human Resources for Health*, 11(1), Art. no. 41, 2013.
11. "Japan: Universal health care at 50 years," *The Lancet*, 30 Aug. 2011, http://www.thelancet .com/japan, accessed on August 14, 2014.

12. Steinhausen, S., O. Ommen, S. Thüm, E. Neugebauer, and H. Pfaff, "Physician empathy and subjective evaluation of medical treatment outcome in trauma surgery patients," *Patient Education and Counseling*, 95(1), pp. 53–60, April 2014.

13. Cooper, L., D. Roter, R. Johnson, D. Ford, and D. Steinwachs, "Patient-centered communication: Ratings of care, and concordance," *Annals of Internal Medicine Journal*, 139, pp. 907–915, 2003.

14. Boegl, K., K. P. Adlassnig, Y. Hayashi, T. E. Rothenfluh, and H. Leitich, "Knowledge acquisition in the fuzzy knowledge representation framework of a medical consultation system," *Artificial Intelligence in Medicine*, 30(1), pp. 1–26, 2004.

15. Barberá-Tomás, D., and D. Consoli, "Whatever works: Uncertainty and technological hybrids in medical innovation," *Technological Forecasting and Social Change*, 79(5), pp. 932–948, 2012.

16. Fujita, H., J. R. Imre, J. Fodor, M. Kurematsu, and J. Hakura, "Fuzzy Reasoning for Medical Diagnosis-Based Aggregation on Different Ontologies," *IEEE 7th International Symposium on Applied Computational Intelligence and Informatics*, pp. 137–146, 2012.

17. Hakura, J., M. Kurematsu, and H. Fujita, "Facial Expression Invariants for Estimating Mental States of Person," *Frontiers in Artificial Intelligence and Application Series*, Volume 199, *New Trends in Software Methodologies, Tools and Techniques (SoMeT_09)*, IOS Press, ISBN: 978-1-60750-049-0, pp. 518–530, 2009.

18. Kurematsu, M., M. Ohashi, O. Kinoshita, J. Hakura, and H. Fujita, "A Study of How to Implement a Listener Estimate Emotion in Speech," *Frontiers in Artificial Intelligence and Application Series*, Volume 199, *New Trends in Software Methodologies, Tools and Techniques (SoMeT_09)*, IOS Press, ISBN: 978-1-60750-049-0, pp. 531–540, 2009.

19. Rossen, B., and B. Lok, "A crowd sourcing method to develop virtual human conversational agents," *International Journal of Human-Computer Studies*, 70(4), pp. 301–319, 2012.

20. Deladisma, A. D., M. Cohen, A. Stevens, P. Wagner, B. Lok, T. Bernard, C. Oxendine, L. Schumacher, K. Johnsen, R. Dickerson, A. Raij, R. Wells, M. Duerson, J. G. Harper, and D. S. Lind, "Do medical students respond empathetically to a virtual patient?," *The American Journal of Surgery*, 193(6), pp. 756–760, 2007.

21. Kurematsu, K., H. Chiba, J. Hakura, and H. Fujita, "A Framework of Emotional Speech Synthesize Using a Chord and a Scale," *New Trends in Software Methodologies, Tools and Techniques (SoMeT_10)*, Volume 217, IOS Press, Sept. 2010. ISBN: 978-1-60750-628-7, pp. 500–508.

22. Herrera, F. and E. Herrera-Viedma, "Linguistic decision analysis: Steps for solving decision problems under linguistic information," *Fuzzy Sets and Systems*, 115, pp. 67–82, 2000.

23. Herrera, F., S. Alonso, F. Chiclana, and E. Herrera-Viedma, "Computing with words in decision making: Foundations," *Trends and Prospects, Fuzzy Optimization and Decision Making*, 8(4), pp. 337–364, 2009.

24. Zadeh, L. A., "The concept of a linguistic variable and its applications to approximate reasoning," Part I. *Information Sciences*, 8, pp. 199–249; Part II. *Information Sciences*, 8, pp. 301–357; Part III. *Information Sciences*, 9, pp. 43–80, 1975.

25. Wang, Y., X. Xiaoyan, H. Zhao, and Z. Hua, "Semi-supervised learning based on nearest neighbor rule and cut edges," *Knowledge-Based Systems*, 23(6), pp. 547–554, 2010.

26. Kuboki, T. et al., "Multidimensional assessment of mental state in occupational health care— Combined application of three questionnaires, Tokyo University Egogram (TEG), time structuring scale (TSS), and profile of mood states (POMS)," *Environmental Research*, 61, pp. 285–298, Academic Press, 1993.

27. Hakura, J., M. Kurematsu, and H. Fujita, "An Automatic Facial Expression Recognition Method Using Situational Information," *The 7th New Trends in Software Methodologies, Tools and Techniques (SoMeT_08)*, IOS Press, pp. 290–306, 2008.

28. OKI FSE Ver. 4, a Face Recognition Middleware for Embedded System, http://www.oki.com /en/press/2008/07/z08019e.html. Accessed on August 14, 2014.

29. Stewart, I., and V. Joines, *TA Today: A New Introduction to Transactional Analysis*, Lifespace Publishing, 1987.

30. Keçeci, A., and G. Taşocak, "Nurse faculty members' ego states: Transactional analysis approach," *Nurse Education Today Journal*, 29(7), pp. 746–752, 2009.
31. Berne, E., "Transactional Analysis in Psychotherapy," *A Systematic Individual and Social Psychiatry*, Souvenir Press (Originally published by Grove Press, 1961), 1961.
32. Dusay, J. M., *Egogram*, Harper and Row Publishers, 1977.
33. Sachiko, S. S., "How transactional analysis can be used in terminal care," *International Congress Series*, 1287, pp. 179–184, 2006.
34. Murakami, M., T. Matsuno, K. Koike, S. Ebana, K. Hanaoka, and T. Katsura, "Transactional analysis and health promotion," *International Congress Series*, 1287, pp. 164–167, 2006.
35. Turner, C. M., Interpersonal Skills Paper No. 5: Transactional Analysis. Coombe Lodge, 1978.
36. Booth, L., "Observations and reflections of communication in health care—Could Transactional Analysis be used as an effective approach?," *Radiography*, 13(2), pp. 135–141, 2007.
37. Shirai, S., "How transactional analysis can be used in terminal care," *International Congress Series*, 1287, pp. 179–184, 2006.
38. Rowe, J., "Self-awareness: Improving nurse-client interactions," *Nursing Standard*, 14(8), pp. 37–40, 1999.
39. Wilson, S., "Using transactional analysis as a psychotherapist: How to use TA in clinical situations," *International Congress Series* 1287, pp. 173–178, 2006.
40. http://www.somet.soft.iwate-pu.ac.jp/system_news/News_2009.MPG, accessed on August 14, 2014; http://www.somet.soft.iwate-pu.ac.jp/system_news/System_flow_operation.wmv, accessed on August 14, 2014; http://www.somet.soft.iwate-pu.ac.jp/system_news/VDS_Sample2.mpg, accessed on August 14, 2014; http://www.somet.soft.iwate-pu.ac.jp/system_news/VDS_sample.wmv, accessed on August 14, 2014.
41. Netica software package: NORSYS Software Corporation, http://www.norsys.com/, accessed on February 11, 2015.
42. Fujita, H., and E. Herrera-Viedma, Special Issue: Intelligent Decision Making Support Tools, *Knowledge-Based Systems*, 58, pp. 1–2, 2014.
43. Fujita, H., "Fuzzy reasoning for medical diagnosis based on subjective attributes and objective attributes alignment," *Proceedings of the 2013 Joint IFSA World Congress and NAFIPS Annual Meeting, IFSA/NAFIPS*, Art. no. 6608528, pp. 950–955, 2013.
44. Yager, R. R., and N. Alajlan, "A generalized framework for mean aggregation: Toward the modeling of cognitive aspects," *Information Fusion*, 17 (1), pp. 65–73, 2014.
45. Yager, R. R., and D. P. Filev, "Induced ordered weighted averaging operators," *IEEE Transactions on System, Man, and Cybernetics*, 29, pp. 141–150, 1999.
46. Yager, R. R., "The power average operator," *IEEE Transactions on Systems, Man and Cybernetics*, 31, pp. 724–731, 2001.
47. Xu, Z., "Approaches for multiple attribute group decision making based on intuitionistic fuzzy power aggregation operators," *Knowledge Based Systems*, 24, pp. 749–760, 2011.
48. Wang, H. H., S. Z. Wu, and Y. Y. Liu, "Association between social support and health outcomes: A meta-analysis," *Kaohsiung Journal of Medical Science*, 19, pp. 345–351, 2003.
49. Ommen, O., S. Thuem, H. Pfaff, and C. Janssen, "The relationship between social support, shared decision-making and patient's trust in doctors: A cross-sectional survey of 2197 inpatients using the Cologne Patient Questionnaire," *International Journal of Public Health*, 56, pp. 319–327, 2011.
50. Stewart, M. A., "Effective physician–patient communication and health outcomes: A review," *Canadian Medical Association Journal*, 152, pp. 1423–1433, 1995.
51. Adams, R., K. Price, G. Tucker, A. M. Nguyen, and D. Wilson, "The doctor and the patient—How is a clinical encounter perceived?," *Patient Education Counseling*, 86, pp. 127–133, 2012.
52. Sugawara, K., and H. Fujita, "Intelligent decision support for business workflow adaptation due to subjective interruption," *Acta Polytechnica Hungarica*, 10(8), pp. 5–26, 2013.
53. Hakura, J., M. Kurematsu, and H. Fujita, "An Exploration toward Emotion Estimation from Facial Expressions for Systems with Quasi-Personality," *International Journal of Circuits, Systems and Signal Processing*, 1(2), pp. 137–144, 2008.

54. Fujita, H., J. Hakura, and M. Kurematsu, "Multiviews Ontologies Alignment for Medical based Reasoning," *11th IEEE International Symposium on Computational Intelligence and Informatics (CINTI 2010)*, pp. 15–23, 2010.

55. Pearl, J., "Probabilistic Reasoning in Intelligent Systems," *Networks of Plausible Inference*. Morgan Kaufmann, 1988.

56. Ding, Z., "BayseOWL: Probabilistic Framework for Semantic Web", PhD thesis, University of Maryland, Baltimore, 2005.

57. Savage, J. L., "Elicitation of personal probabilities and expectations," *Journal of American Statistical Association*, 66, pp. 783–801, 1971.

58. Xu, Z. S., and R. R. Yager, "Intuitionistic fuzzy Bonferroni means," *IEEE Transactions on Systems, Man and Cybernetics*, 41, pp. 568–578, 2011.

59. Bordogna, G. and G. Pasi, "A fuzzy linguistic approach generalising Boolean information retrieval: A model and its evaluation," *Journal of the American Society for Information Science*, 44, pp. 70–82, 1993.

60. Herrera, F., E. Herrera-Viedma, and J. L. Verdegay, "Direct approach processes in group decision making using linguistic OWA operators," *Fuzzy Sets and Systems*, 79, pp. 175–190, 1996.

61. Yager, R. R., "An approach to ordinal decision making," *International Journal of Approximate Reasoning*, 12, pp. 237–261, 1995.

62. Herrera-Viedma, E., and A. G. López-Herrera, "A model of information retrieval system with unbalanced fuzzy linguistic information," *International Journal of Intelligent Systems*, 22(11), pp. 1197–1214, 2007.

63. Bonissone, P. P., and K. S. Decker, "Selecting Uncertainty Calculi and Granularity: An Experiment in Trading-Off Precision and Complexity," *Uncertainty in Artificial Intelligence*, North-Holland, pp. 217–247, 1986.

64. Herrera, F., E. Herrera-Viedma, and J. L. Verdegay, "A sequential selection process in group decision making with a linguistic assessment approach," *Information Science*, 85, pp. 223–239, 1995.

65. Alonso, S., I. J. Pérez, F. J. Cabrerizo, and E. Herrera-Viedma, "A linguistic consensus Model for Web 2.0 communities," *Applied Soft Computing*, 13(1), pp. 149–157, 2013.

66. Herrera-Viedma, E., "Modeling the retrieval process for an information retrieval system using an ordinal fuzzy linguistic approach," *Journal of the America Society for Information Science and Technology*, 52(6), pp. 460–475, 2001.

67. Yager, R. R., "On ordered weighted averaging aggregation operators in multicriteria decision making," *IEEE Transactions on Systems, Man, and Cybernetics*, 18, pp. 183–190, 1988.

68. Herrera, F., and E. Herrera-Viedma, "Aggregation operators for linguistic weighted information," *IEEE Transactions on Systems, Man and Cybernetics*, Part. A: Systems and Humans, 27, pp. 646–656, 1997.

69. Zadeh, L. A., "A computational approach to fuzzy quantifiers in natural languages," *Computers and Mathematics with Applications*, 9, pp. 149–184, 1983.

70. Kalfoglou, Y., and M. Scholermmer, "Ontology mapping: The state of the art," *The Knowledge Engineering Review*, 18(1), pp. 1-31, 2003.

71. Herrera, F., E. Herrera-Viedma, and L. Martínez, "A fusion approach for managing multi-granularity linguistic term sets in decision making," *Fuzzy Sets and Systems*, 114, pp. 43–58, 2000.

72. Chang, S. L., R. C. Wang, and S. Y. Wang, "Applying a direct multi-granularity linguistic and strategy-oriented aggregation approach on the assessment of supply performance," *European Journal of Operational Research*, 177(2), pp. 1013–1025, 2007.

73. Chen, Z., and D. Ben-Arieh, On the fusion of multi-granularity linguistic label sets in group decision making," *Computers and Industrial Engineering*, 51(3), 2006.

74. Klir, G. J., and B. Yuan, "Fuzzy sets and fuzzy logic: Theory and applications," Prentice-Hall PTR, 1995.

75. http://ec.europa.eu/research/horizon2020/pdf/work-programmes/health_draft_work_pro gramme.pdf.

22

Synthetic Biometrics in Biomedical Systems

Kenneth Lai, Steven Samoil, Svetlana N. Yanushkevich, and Adrian Stoica

CONTENTS

22.1 Introduction

Modeling and simulation of biometric data are emerging technologies for educational and training purposes (public safety and healthcare, security access, and forensic systems). To model data for the study of patient's biometrics, as well as for training purposes, simulated, or synthetic, biometric data can be used.

This chapter reviews an example of the application of synthetic biometrics for training users in healthcare systems and a number of examples are given in Section 22.8, such as: the facial nerve disorder modeling, the avatar systems, and rehabilitation applications. The modeling and simulation of biometric data is used for efficient support of medical personnel training in dealing with the patient data processing under conditions of uncertainty. Another example is remote monitoring of patients' biometrics (physiological and behavioral patterns) in hospitals or care units. Such modeling requires developing specific training methodologies and techniques, including virtual environments.

Computer vision and *computer graphics* provide methodology for the above area. Computer vision tools perform an automated analysis of images and attempt to generate a computer model from a sequence of such images.

Computer graphics deal with, for instance, a three-dimensional model of a face and attempt to render or project this model onto a two-dimensional surface to create an image of the face.

Computer graphics is a very advanced and well-developed field, driven forward by the gaming, movies, advertising, and computer-aided design and other related industries.

Biometric technologies attempt to generate computer models of humans' appearance or behavior with the purpose of personal identification using computer vision techniques.

On the other side of the spectrum, *biometric synthesis* aims at rendering biometric phenomena from their corresponding computer models. Its methodology lies in computer graphic techniques. For example, a synthetic face can be generated from a model that includes mesh and texture data, and can be altered in order to simulate facial expressions.

Training systems that use simulation of biometric data have been developed in many domains such as the flight simulations [1], as well as biomedical areas such as surgical simulations.

A relatively new area of biometric technologies extends to designing systems that acquire and process biometric data, or biometrics. Biometrics include physiological and behavioral characteristics of humans in the form of visual, audio, or other information that enables identifying or verifying individuals (primarily for security-relevant applications), or using the data for healthcare purposes (to monitor biomedical parameters of patients).

Simulators of biometric data are emerging technologies for educational and training purposes (immigration control, banking services, police, justice, and healthcare). They emphasize on the decision-making skills in nonstandard and extreme situations.

Synthetic biometrics are understood as generated biometric data that are meaningful for biometric systems. These synthetic data represent corrupted or distorted data. For example, facial images acquired by video cameras can be corrupted due to their appearance variation, as well as environmental conditions such as the lighting and camera resolutions.

Another example is in biomedical facilities for remote monitoring of patient biometrics (physiological and behavioral patterns) in hospitals or care units. Simulation of various monitored data, as well as extreme scenarios, is aimed at developing the particular decision-making skills of the system personnel. Such modeling requires developing specific training methodologies and techniques, including virtual environments.

22.2 Biometric Data and Systems

Biometric data, or biometrics, include visual, infrared (IR), and acoustic data for identification of both physical appearance (including aging and intentional changes, or camouflage), physiological characteristics (e.g., temperature and blood flow rate), and behavioral features (e.g., voice) of humans. Other examples of biometrics are gaits, keystroke patterns, signatures, fingerprints, ears, iris, and retina.

Biometrics provide many sources of information that can be classified with respect to various criteria; in particular, spectral criteria (visible, IR, and acoustic), behavioral and inherent criteria, contact and noncontact, correlation factors between various biometrics (multibiometric), computing indirect characteristics, and parameters such as drug and alcohol consumption, blood pressure, breathing function, and temperature.

Various signal-processing and pattern-recognition techniques are used for analysis of these data for the purposes of human identification or verification. Similar techniques can be deployed in behavioral screening, as well as biomedical applications, except that the purpose is not identification of an individual but analysis of the individual's biometrics for security-related issues and medical prediagnostics or diagnostic purposes.

Besides visual spectrum light, the human face and body emit IR light in both the mid-IR (3–5 μm) and the far-IR range (8–12 μm). Most light is emitted in the longer-wavelength, lower-energy far-IR band. Below the mid-IR range, at about 12 μm, the human skin reflects light. This is the basis for skin detection by the thermogram [2]; skin detection precedes temperature evaluation. IR images provide information for recognizing certain skin diseases [3,4], emotions [5], numerous psychological features, and gender [6].

In Ref. [3], heat-conduction formulas at the skin surface are derived. For example, a high-temperature spot in the IR image of the face, caused by severe acute respiratory syndrome (SARS), are important to detect early for further immediate isolation of those infected to prevent outbreaks of such an epidemic. The thermodynamic relation between the blood flow rate at the skin level, blood temperature at the body core, and the skin temperature is used to convert IR intensity to temperature and, therefore, to the blood flow rate. In addition, the breathing rate can be estimated during the dialogue of the patient or user with the personnel [7].

A behavioral biometric, such as gait, is used to determine the gender of a walker [6]. However, a physical condition can affect gait, for example, pregnancy, affliction of the legs or feet, and drunkenness.

22.3 Synthetic Biometrics

In a modeling system, the real biometrics can be partially or fully substituted with simulated data. This simulated, or synthesized, data are called *synthetic biometrics* [8,9].

There are several reasons for using synthetic data:

- Modeling human biometrics and behavior is a keystone for developing simulators, or virtual environments, used in biomedical applications, including avatars for training medical personnel or talking to patients in an automatically monitored healthcare facility; it can also be used in physiotherapy environment simulation.

- Real data, used for training medical personnel, are not always available, or it is laborious or impossible to collect a statistically meaningful amount of data for testing the biometric device or system.

- It may not be possible to collect the data representing the extreme, or boundary, case in biomedical situations. An example of extreme biomedical data is a high-temperature spot in the IR image of the face that is caused by SARS.

- There are privacy concerns, especially in biomedical and healthcare applications, about using a customer's or patient's data for training purposes.

The issue of protecting privacy in biometric systems has inspired the direction of research referred to as *cancelable biometrics* [10]. Cancelable biometrics are aimed at enhancing the security and privacy of biometric authentication through the generation of "altered" biometric data, that is, synthetic biometrics. Instead of using a true object, for example, a finger or the face, the fingerprint or face image is intentionally distorted in a repeatable manner, and this new print or image is used.

22.4 Synthetic Face

Face-recognition systems detect patterns, shapes, and shadows on the face. The reverse process, face reconstruction, is a classic problem of not only criminology but also reconstructive medicine or surgery.

22.4.1 Analysis by Synthesis in Face Recognition

Many biometric systems become confused when identifying the same person smiling, aged, with various accessories (moustache, glasses), and/or poorly lit (Figure 22.1). Facial recognition tools can be improved by training with a set of synthetic facial expressions and appearance/environment variations generated from real facial images.

Synthetic biometrics have been used for a while in the *analysis-by-synthesis* approach, in which synthesis of biometric data can verify the perceptual equivalence between original and synthetic biometric data. For example, facial synthesis can be formulated as deriving a realistic facial image from a symbolic facial expression model.

22.4.2 Three-Dimensional Facial Images

A face model is a composition of the various submodels (eyes, nose, etc.) The face model consists of the following facial submodels: *eye* (shape, open or closed, blinking, iris size, movement, etc.), *eyebrow* (texture, shape, and dynamics), *mouth* (shape, lip dynamics, and tooth and tongue position), *nose* (shape and nostril dynamics), and *ear* (shape) [8].

A 3D face model includes two constituents: a face shape model (represented by a 3D geometric mesh) and a skin texture model (generated from 2D images).

The main advantage of 3D face modeling is that the effect of variations in illumination, surface reflection, and shading from directional light can be significantly decreased. For

FIGURE 22.1

Modeling of a face. *Left to right:* face model derived from a photo, aging, drunk, poorly lit face, and with facial accessories.

example, a 3D model can provide controlled variations in appearance while the pose or illumination is changed. Moreover, the estimations of facial expressions can be made more accurately in 3D models compared with 2D models.

A face shape is modeled by a polygonal mesh, while the skin is represented by texture map images (Figure 22.2; Ref. [11]). Any individual face shape can be generated from the generic face model by specifying 3D displacements for each vertex. Synthetic face images are rendered by mapping the texture image on the mesh model.

This approach can also be used to assist modeling faces in the visible and infrared bands (Figure 22.3) [12]. The texture maps represent the hemoglobin and melanin contents (in the visible spectrum) and the temperature distribution of the facial skin. These maps are the output of the face analysis and modeling module. This information is used for evaluating the physical and psychoemotional state of a person.

The most recent approach to 3D modeling of faces takes advantage of advances in sensor technologies: 3D modeling of faces can be done using a variety of different image types, including infrared, near-infrared, depth, and red–green–blue (RGB) images. Using the Kinect™ v2 prototype, infrared, depth, and RGB images can be obtained simultaneously. Examples of the three types of images are included in Figure 22.4 (disclaimer: the Kinect v2 Development Kit is preliminary software and/or hardware and APIs are preliminary and subject to change).

RGB images are the normal color images that can be taken with almost any camera. Near-infrared images are taken using the light spectrum that is almost at the infrared

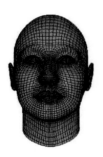
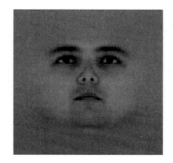
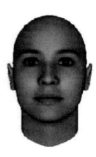

FIGURE 22.2
Left to right: generic 3D polygonal mesh, skin texture, and the resulting 3D rendered model.

Visible band		Infrared band	
Original image	3D face model	Original image	3D face model

FIGURE 22.3
Face images in the visible and infrared bands and their 3D models.

FIGURE 22.4
Sample images of the different possible Kinect outputs. *Left to right:* RGB, depth, and near-IR images.

point. This allows the camera to "see in the dark," allowing for usage of motion tracking and other features in a dark environment. The depth image is generated by creating a three-dimensional map of the room and then showing the distance to the object as different shades of gray.

Using these images and modeling software such as the FaceGen Modeller software, a 3D model of the photo subject can be generated. The results of this are shown in Figure 22.5.

22.4.3 RGB-D Technologies

Since 2010, a team at Microsoft Research Cambridge (http://research.microsoft.com/en-us /labs/cambridge/default.aspx) has used the Kinect for Windows hardware (http://www .microsoft.com/en-us/kinectforwindows/) in the project, called Touchless Interaction in Medical Imaging (http://research.microsoft.com/en-us/projects/touchlessinteraction medical/) [13].

The project is aiming at implementation in a virtual medical imaging environment, in order to give surgeons the ability to obtain a virtual peek inside the human body. Such modeling and rendering would enable a surgeon, who cannot use a mouse or a keyboard because it is unsterile and poses a risk of infection, to use simple hand gestures to change, move, or zoom in on MRIs, CT scans, and other medical images. In initial development, the Kinect for the Windows-based system has thrilled surgeons who have seen it and who believe that it could help make surgery faster and more accurate. It is expected that

FIGURE 22.5
Models created using the different possible Kinect outputs as inputs to FaceGen. *Left to right:* RGB, depth, and near-IR images.

the Kinect system will deliver better outcomes to patients when fully field-tested and approved.

22.4.4 Two-Dimensional Facial Gesture Tracking

Another application of tracking in images was created by the developers at Nouse. The main product offered by Nouse is an application that can be run on a computer with a webcam to be used to control the computer. This is accomplished by tracking the user's nose and head motions to control the mouse of the computer [14].

22.4.5 Modeling the Aging Face

A survey of age synthesis and estimation techniques was conducted in 2010 by Fu et al. [15]. One of the methods of age synthesis discussed by the authors was a paper written by Lanitis et al. (2002) on creating automatic ways of simulating aging faces. The method used by Lanitis et al. (2002) was to have a set of training images for various different ages to create a model representing the ages (Figure 22.6). To age an image of a face, the simulation technique requires that the age of the image be known. Using the known age, the image can be modified to represent the new age by using the model created from the training images [16]. Previous researches into age estimation have also been conducted by Fu and Huang [17] and also by Guo et al. [18].

22.4.6 Face Reconstruction from DNA

In a 2014 paper by Claes et al. [19] the possibility of generating a 3D model of a face using deoxyribonucleic acid (DNA) was explored. For the experiments, the authors created 3D depth images of many participants and removed noise from the images. Using the clean depth images of the face, the authors applied principle component analysis (PCA) and extracted a set of features that, when combined, described the majority of the faces. To discover the relationship between face features and DNA, the authors then turned to a tool called bootstrapped response-based imputation modeling (BRIM), which is used to model relationships. Using a variety of inputs, including gender and specific features of a person's genome, the authors were able to discover statistically relevant relationships between face features and DNA. Figure 22.7 shows the process used by Claes et al. (2014) to prepare the 3D models before using PCA. The authors started with a raw depth image and then ended with a usable mask of the face [19].

FIGURE 22.6
Aging modeling (neutral facial expression) using the FaceGen application. *Left to right:* age 30, age 40, age 50, and age 60.

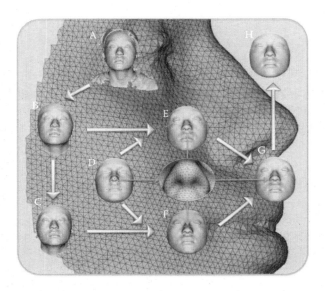

FIGURE 22.7
A: original depth data; B: data of the face only; C: reflected image of B; D: info of the important regions of the face; E: regions mapped back to the face; F: same as E except that it is the mirror image; G: final result that can be used for PCA. (Reproduced from Claes, P. et al., *PLOS Genet.*, 10, e1004224, 2014 [http://www.ncbi.nlm.nih.gov /pmc/articles/PMC3961191/] under the Creative Commons Attribution License. With permission.)

22.4.7 Behavioral Facial Synthesis: Expressions

Visual band images along with thermal (infrared) images can be used in this task [5,20,21]. Facial expressions are formed by about 50 facial muscles [22] and are controlled by dozens of parameters in the model (Figure 22.8). The facial expression can be identified once the facial action units are recognized. This task involves facial feature extraction (eyes, eyebrow, nose, lips, and chin lines), measuring geometric distances between the extracted points/lines and then recognition of facial action units based on these measurements.

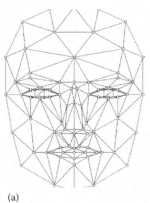

Action unit	Muscular basis
Inner brow raiser	Frontalis, pars medialis
Outer brow raiser	Frontalis, pars lateralis
Upper lid raiser	Levator palpebrae, superioris
Cheek raiser	Orbicularis oculi, pars palebralis
Lip corner puller	Zygomatic major
Cheek puffer	Caninus
Chin raiser	Mentalis
Lip stretcher	Risorius
Lip funneler	Orbicularis oris
Lip tightener	Orbicularis oris
Mouth stretcher	Pterygoid, digastric
Lip suck	Orbicularis oris
Nostril dilator	Nasalis, pars alaris
Slit	Orbicularis oculi

(a)   (b)

FIGURE 22.8
(a) 3D facial mesh model; (b) fragments of corresponding facial action units.

Psychologists distinguish two kinds of short-term facial expressions: *controlled* and *non-controlled* facial expressions [21].

Controlled expressions can be fixed in a facial model by generating control parameters, for example, a type of smile. Noncontrolled facial expressions are very dynamic and are characterized by short-time durations. Visual pattern analysis and classification can be carried out in 100 ms and involves a minimum of 10 synaptic stages from the retina to the temporal lobe (see, for example, Ref. [23]).

22.4.8 Animation as Behavioral Facial Synthesis

An example of a direct biometric problem is identifying speech given a video fragment without recorded voice. The inverse problem is mimicry synthesis (animation) given a text to be spoken (synthetic narrator) [24–26]. Behavioral biometric information can also be used in evaluation of truth of answers to questions, or evaluation of the honesty of a person in the process of speaking [22]. It is also possible to generate an animated digital avatar using synthetic biometric features (for example, face expressions and voice) for use in interacting with people. One such application was developed by Nunamaker et al. [27] as an automated kiosk for interacting with people. They developed an animated synthetic face for a user of the kiosk to interact with as a way of providing automated help. Such a device would allow automation in places that a person is currently required [27].

Research on behavior and animation related to creating robots that interact more naturally with humans can also be conducted. The goal of the work was to create a robot software application that could keep memories and act in a more social manner, such that the robot could recognize the faces of humans [28]. Facial recognition was used so that the robot was able to recognize people that interacted with it. In the 2010 paper by Mavridis et al., the robot was used to interact with people and use the images it found on Facebook to recognize who the person was. As of 2010, results of identifying a person correctly had a success rate close to 50% [28].

22.5 Synthetic Fingerprints

Cappelli et al. [29–30] have developed a commercially available synthetic fingerprint generator called SFinGe. In SFinGe, various models of fingerprint topologies are used: shape, directional map, density map, and skin deformation models. To add realism to the image, erosion, dilation, rendering, translation, and rotation operators are used. A similar approach to the continuous growing of an initial random set of features by using Gabor filters with polar transforms has been reported by Yanushkevich et al. [8]. This method itself can be used for designing fingerprint benchmarks with rather complex structural features.

Figure 22.9 shows examples of acceptable and unacceptable synthesized fingerprints, since not all compositions produce useful macroprimitives, even if the topological constraints and design rules are applied.

Kücken and Newell [31] have developed a method for synthetic fingerprint generation based on natural fingerprint formation and modeling based on state-of-the-art dermatoglyphics, a discipline that studies epidermal ridges on fingerprints, palms, and soles. The methods of fingerprint synthesis can be applied to synthetic palm print generation. Note that the generation of synthetic hand topologies is a trivial problem.

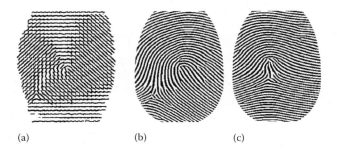

(a)   (b)   (c)

FIGURE 22.9
(a) Invalid and (b, c) valid topological compositions of fingerprint primitives.

22.6 Synthetic Iris and Retina Images

Synthetic iris patterns have been used by ocularists in manufacturing glass eyes and contact lenses for a while. The ocularist's approach to iris synthesis is based on the composition of painted primitives and utilizes layered semitransparent textures built from topological and optic models [32]. These methods are used today to create contact lenses with fake iris patterns printed onto them.

Iris-recognition systems scan the surface of the iris to compare pattern [33]. Retina-recognition systems scan the surface of the retina and compare nerve patterns, blood vessels, and similar features. To the best of our knowledge, automated methods of *iris* and *retina image reconstruction*, or *synthesis*, have not been developed yet except for an approach based on generation of iris layer patterns [8].

A synthetic image can be created by combining segments of real images from a database. Various operators can be applied to deform or warp the original iris image: translation, rotation, rendering, etc. Various models of the iris, retina, and eye used for improving recognition can be found in Refs. [34–36]. In Ref. [10], a cancelable iris image design is proposed for the problem as follows: The iris image is intentionally distorted to yield a new version. For example, a simple permutation procedure is used for generating a synthetic iris.

An alternative approach is based on the synthesis of patterns in the iris layers followed by superposition of the layers and the pupil (black center) [8].

An example of generating of a synthetic collaret topology modeled by a randomly generated curve is shown in Figure 22.10a. Figure 22.10b illustrates three different patterns obtained by this method. Other layer patterns can be generated based on wavelet, Fourier, polar, and distance transforms, as well as Voronoi diagrams [8].

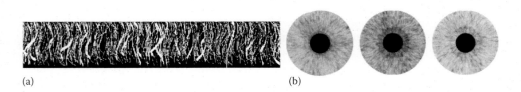

(a)   (b)

FIGURE 22.10
Synthetic collaret topology modeled by a randomly generated curve. (a) Spectral representation of a synthetic iris; (b) three different synthetic iris patterns.

22.7 Synthetic Signatures

Imitation of human handwriting is a typical inverse problem of graphology. The simplest method of generating synthetic signatures is based on geometrical models. Spline methods and Bézier curves are used for curve approximation, given some control points. Manipulations of the control points give variations of a single curve in these methods.

An algorithm for signature generation based on deformation has been introduced by Oliveira et al. [24]. Hollerbach [37] has introduced the theoretical basis of handwriting generation based on an oscillatory motion model. In the Hollerbach model, handwriting is controlled by two independent oscillatory motions superimposed on a constant linear drift along the line of writing. There have been many papers on the extension and improvement of the Hollerbach model.

A model based on combining the shape and the physical models in synthetic handwriting generation was developed by Wang et al. [38]. The so-called *delta–lognormal model* was developed by Plamondon and Guerfali [39]. This model can produce smooth connections between characters and can also ensure that the deformed characters are consistent with the models. Choi et al. [40] proposed that character shapes could be generated by Bayesian networks. By collecting handwriting examples from a writer, a system learns the writer's writing style. An example of a combined model based on geometric and kinematic characteristics (in-class scenario) is illustrated in Figure 22.11.

22.7.1 Voice Synthesis

Voice represents biometrics in the acoustic domain. The pitch, loudness, and timbre of an individual's voice are the main relevant parameters. Voice carries emotional information by *intonation*.

The *voice tension* technology utilizes varying degrees of vibration in the voice to detect emotional response in dialogue. *Synthetic speech* has evolved considerably since the 1960s, including improving the audio quality and the naturalness of speech, developing techniques for emotional "coloring," and combining it with other technologies, for example, facial expressions and lip movement [41].

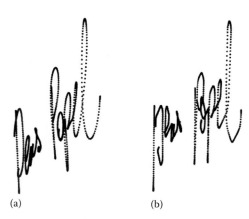

(a)   (b)

FIGURE 22.11
In-class scenario: (a) the original signature; (b) the synthetic one. (Courtesy of Prof. D. Popel, Baker University, Baldwin City, Kansas.)

22.8 Examples of the Usage of Synthetic Biometrics

In this section, we consider several applications of synthetic biometrics.

22.8.1 Example: Facial Nerve Disorder Modeling

In an undergraduate project done by Lo et al. [42], 3D models were used as a way to model facial nerve disorders such as Bell's palsy. This was done by creating a 3D model and then editing the mesh. By changing certain points of the mesh, the 3D model could be changed so that it was very similar to the face of a real person with the condition. In the project originally done by Lo et al. [42], the models were created using the FaceGen Modeler software development kit (SDK). Using the SDK allowed access to the underlying polygons and their vertices. Shifting specific vertices allowed the creation of a 3D model that closely represented different facial nerve disorders. The ability to move parts of the face allows quite a bit of freedom in the ability to simulate many different situations [42]. An example of the model and the underlying mesh for right Bell's palsy can be seen in Figure 22.12.

22.8.2 Decision-Making Support Systems

A registration procedure is common in the process of person identification, such as in immigration control, in patient prescreening in hospitals, and in other places where secure physical admission is practiced. Such systems can be enhanced using intelligent support using additional biometric data measured at a distance, such as a patient's temperature, facial expression, voice pitch, gait, and other behavioral patterns. The primary sources of information for such data acquisition include infrared and visual band images of the screened person, acoustics, and odor. The screening and monitoring system can analyze these data and provide the analysis results to the medical personnel, in order to support them in decision making regarding patient's admission and care procedure.

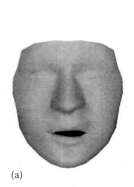
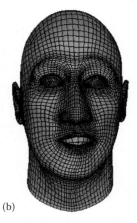

(a)   (b)

FIGURE 22.12
Facial modeling of a person with right Bell's palsy and a surprised expression by using data from Lo et al. [42]. (a) 3D model (generated using FaceGen); (b) the underlying mesh.

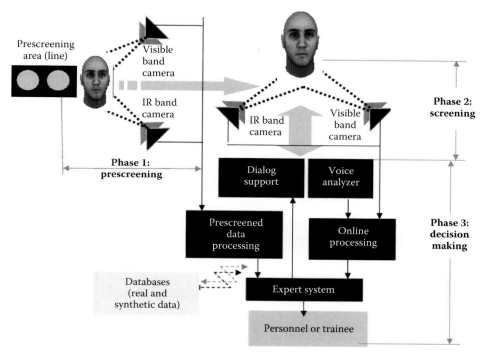

FIGURE 22.13

A biometric-based screening system: data on individuals is sensed during prescreening (before admission point); these data from the visible and IR bands are then analyzed to extract preliminary information (phase 1) and help the personnel to make a decision (phase 2).

The idea of modeling of biometric data for decision-making support has been explored, in particular, at the Biometric Technologies Laboratory at the University of Calgary (http://enel.btlab.ucalgary.ca). The prototype system (Figure 22.13) consists of the following modules: *cameras* in visible and infrared bands placed at two levels of observation, *processors* of preliminary information and online data, *expert system* for supporting conversation based on preliminarily obtained information, and a *personal file*-generating module. Two-level surveillance is used in the system: surveillance of the line, or admission area, and next to the personnel desk. A facial model of a tested person is captured and can be manipulated to mimic the changes in visual and infrared bands caused by physiological changes during the questioning period. This can be compared against generic models of change created based on statistical data. This approach can also be used to train personnel involved in questioning procedures. The decision-making–support system utilizes synthetic models for rendering the biometric data to support decision making.

22.8.3 Databases of Synthetic Biometric Information

The collection of large databases of biometric data, such as faces, is troublesome for many researchers due to the protection of personal information. Imitation of biometric data allows the creation of databases with tailored biometric data without expensive studies involving human subjects [43]. An example of a tool used to create databases for fingerprints is SFinGe developed at the University of Bologna (http://bias.csr.unibo.it /research/biolab/sfinge.html). The generated databases were included in the Fingerprint Verification Competition 2004 (FVC2004) and perform just as well as real fingerprints [44].

There are also several other databases available for use with synthetic biometrics. One such example is the Chinese University of Hong Kong (CUHK) Face Sketch Database (CUFS). This database contains both photos of a person and their sketches drawn by an artist. The example of using this database is creating synthetic sketches [45].

Synthetic databases have also been created for MRI images of the brain. The BrainWeb: Simulated Brain Database is a database created for use in studying MRI images of the brain. A use case for this database presented by the creators is for the evaluation of different image-analysis techniques [45].

22.8.4 Medical Personnel Training

Simulators of biometric data are emerging technologies for educational and training purposes (healthcare, immigration control, banking service, police, justice, etc.). A simulator is understood as a system for modeling specific conditions for taking in processing biometric data. An example of such a system is a simulator for training customs officers (supported by a signature imitator, face imitator, and fingerprint imitator) [43].

In Ref. [46], the biometric-based access control system is used as a training system (with minimal extension of tools) without changing of the place of deployment. The goal of training is to develop the user's decision-making skills based on biometric information collected during screening (noninvasive biometrics, such as video or thermal video monitoring) or during the facility check-in procedure (may be invasive, such as taking fingerprints or iris scan). The training system directly benefits users of biometric-based systems covering a broad spectrum of social activities, including surveillance and control, border control, hospitals, important public events, and banking.

This system must be capable of simulation, in particular, simulation of extreme scenarios aimed at developing the particular skills of the personnel. Such modeling of extreme situations requires developing specific training methodologies and techniques, including virtual environments.

In the development of the training system, most of the features of the physical access system itself are inherited. In Figure 22.14, the topological and architectural properties of both of these systems are shown. However, the training system is different from the actual system in the following features: (a) the biometrics of prescreened and screened individuals are simulated by the system, (b) the appearance of biometric data can be modeled by the user by using modeling tools, and (c) the local and global databases are simulated.

In the training system, the following biometric data are used: (a) file data from databases which are well structured and standard; (b) appearance, or visual, data acquired mainly in real time; (c) other appearance information, such as IR images, which requires specific skills for understanding and analysis; and (d) various forms of behavioral biometrics (facial expressions, voice, signature, etc.) that are acquired in real time (during prescreening or check-in procedure).

Researchers at the University of California, Irvine [47], developed a Kinect–Wheelchair Interface Controlled (KWIC) Trainer that helps children with disabilities to learn how to

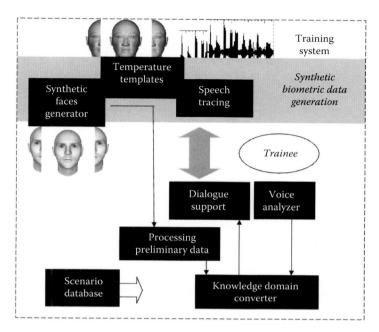

FIGURE 22.14
The training system architecture.

use a powered wheelchair. This training device allows a caregiver to use natural gesture commands to assist a child being trained by using Kinect's gesture control (Figure 22.15).

22.8.5 Avatar Systems

Virtual environments refer to interactive systems that provide the user with a sense of direct physical presence and direct experience. These systems include visual and feedback techniques and allow natural interaction by manipulation of computer-generated objects [48]. An example of such systems is a conversational avatar-based kiosk for automated interviewing [27]. Such systems interview individuals and detect changes in arousal, behavior, and cognitive effort by using psychophysiological information systems. The avatar-based kiosks use heterogeneous sensors to detect both physiological and behavioral biometrics during interactions, and they affect their environment by influencing human behavior by using gender and demeanor, messages, and recommendations.

22.8.6 Rehabilitation Applications

A study was conducted by Lange et al. in 2011 into the development of a game for use in balance rehabilitation using a Kinect sensor. While statistical results were not provided in their paper, overall impressions of the patients were provided. In general, it appears that the users of their developed system much preferred using it over more traditional rehabilitation techniques (such as going to the gym) [49].

FIGURE 22.15
A wheelchair connected to a laptop and Kinect for gesture control. Both the laptop and Kinect would be attached to the wheelchair so that the user is able to control where the wheelchair goes by using gestures only.

22.9 Conclusions

Biometric-based systems is a unique area that combines various methodologies, concepts, and techniques from many areas of the natural and social sciences, in particular image processing and pattern recognition, virtual environment design and synthetic biometric data generation, distributed and avatar system design, human–machine interaction, integrated knowledge-intensive system design, communication, and psychology. In many applications, analysis of biometric data (in particular, analysis of biomedical data) is combined with the inverse problem, that is, synthesis of biometric data.

Data generated by various models are used as databases (for example, databases of synthetic fingerprints) of synthetic biometrics for the testing of biometric hardware and software. The other application is biometric-based decision-making support systems for security, banking, and forensic applications. Generators of synthetic biometric information (for example, an aged or surgically changed face) are a vital component of such systems. Yet another recently emerged application is the creation of simulators for training highly qualified personnel in biometric-based physical access systems such as airport gates and hospital registration.

In applications such as training users of such system, artificial, or synthetic, biometrics are used in order to create the training environment which simulates regular, as well as boundary, or critical, situations.

References

1. B. Knerr, R. Breaux, S. Goldberg, and R. Thrurman, "National Defense," *Handbook of Virtual Environments: Design, Implementation, and Applications,* Taylor & Francis, pp. 857–872, 2002.
2. C. K. Eveland, D. A. Socolinsky, and L. B. Wolff, "Tracking Human Faces in Infrared Video," *Image and Vision Computing,* Elsevier, pp. 579–590, 2003.
3. I. Fujimasa, T. Chinzei, and I. Saito, "Converting Far Infrared Image Information to other Physiological Data," *IEEE Engineering in Medicine and Biology Magazine,* Vol. 10, No. 3, pp. 71–76, 2000.

4. N. Tsumura, N. Ojima, K. Sato, M. Shiraishi, H. Shimizu, H. Nabeshima, S. Akazaki, K. Hori, and Y. Miyake, "Image-Based Skin Color and Texture Analysis/Synthesis by Extracting Hemoglobin and Melanin Information in the Skin," *ACM Transactions on Graphics (TOG)*, Vol. 22, No. 3, pp. 770–779, 2003.

5. Y. Sugimoto, Y. Yoshitomi, and S. Tomita, "A Method for Detecting Transitions of Emotional States Using a Thermal Facial Image Based on a Synthesis of Facial Expressions," *Robotics and Autonomous Systems*, Vol. 31, No. 3, pp. 147–160, 2000.

6. J. P. Foster, M. S. Nixon, and A. Prügel-Bennett, "Automatic Gait Recognition using Area-Based Metrics Pattern," *Recognition Letters*, Vol. 24, pp. 2489–2497, 2003.

7. I. Pavlidis and J. Levine, "Thermal Image Analysis for Polygraph Testing," *Engineering in Medicine and Biology Magazine, IEEE*, Vol. 21, No. 6, pp. 56–64, 2002.

8. S. N. Yanushkevich, A. Stoica, V. P. Shermko, and D. V. Popel, *Biometric Inverse Problems*, CRC Press, 2005.

9. S. N. Yanushkevich, A. Stoica, and V. Shermko, "Synthetic Biometrics," *IEEE Computational Intelligence Magazine*, Vol. 2, No. 2, pp. 60–69, 2007.

10. R. Bolle, J. Connell, S. Pankanti, N. Ratha, and A. Senior, *Guide to Biometrics*, Springer, 2004.

11. "FaceGen" [Online]. Available: http://www.facegen.com/, accessed on February 11, 2015.

12. O. R. Boulanov, M. L. Gavrilova, A. Poursaberi, M. Spicher, V. P. Shmerko, S. N. Yanushkevich, and P. S. P. Wang, "Biometric-Based Intelligent Agent Systems," *IADIS International Conference Intelligent Systems and Agents*, pp. 162–164, 2011.

13. A. Taylor and H. Mentis, "Imaging the Body: Embodied Vision in Minimally Invasive Surgery," *Proceedings of the SIGCHI Conference on Human Factors in Computing Systems*, pp. 1479–488, 2013.

14. "Nouse" [Online]. Available: http://www.nouse.ca/en/, accessed on February 11, 2015.

15. Y. Fu, G. Guo, and T. S. Huang, "Age Synthesis and Estimation via Faces: A Survey," *IEEE Transactions on Pattern Analysis and Machine Intelligence*, Vol. 32, No. 11, pp. 1955–1976, 2010.

16. A. Lanitis, C. J. Taylor, and T. F. Cootes, "Toward Automatic Simulation of Aging Effects on Face Images," *IEEE Transactions on Pattern Analysis and Machine Intelligence*, Vol. 24, No. 4, pp. 442–455, 2002.

17. Y. Fu and T. Huang, "Human Age Estimation with Regression on Discriminative Aging Manifold," *IEEE Transaction on Multimedia*, Vol. 10, pp. 578–584, 2008.

18. G. Guo, Y. Fu, C. Dyer, and T. Huang, "Image-Based Human Age Estimation by Manifold learning and Locally Adjusted Robust Regression," *IEEE Transactions on Image Processing*, Vol. 17, No. 7, pp. 1178–1188, 2008.

19. P. Claes, D. K. Liberton, K. Daniels, K. M. Rosana, E. E. Quillen, L. N. Pearson, B. McEvoy, M. Bauchet, A. A. Zaidi, W. Yao, H. Tang, G. S. Barsh, D. M. Absher, D. A. Puts, J. Rocha, S. Beleza, R. W. Pereira, G. Baynam, P. Suetens, D. Vandermeulen, J. K. Wagner, J. S. Boster, and M. D. Shriver, "Modeling 3D Facial Shape from DNA," *PLOS Genetics*, Vol. 10, No. 3, 2014. http://www.ncbi.nlm.nih.gov/pmc/articles/PMC3961191/, accessed on September 25, 2014.

20. N. Oliver, A. P. Pentland, and F. Berard, "Lafter: A Real-Time Face Lips Tracker with Facial Expression Recognition," *Pattern Recognition*, Vol. 33, No. 8, pp. 1369–1382, 2000.

21. M. Pantic and L. J. M. Rothkrantz, "Automatic Analysis of Facial Expressions: The State of the Art," *IEEE Transactions on Pattern Analysis and Machine Intelligence*, Vol. 22, pp. 1424–1445, 2000.

22. P. Ekman and E. L. Rosenberg, eds., *What the Face Reveals: Basic and Applied Studies of Spontaneous Expression Using the Facial Action Coding System*, Oxford University Press, 1997.

23. E. T. Rolls, "Brain Mechanisms for Invariant Visual Recognition and Learning," *Behavioural Processes*, Vol. 1, No. 1, pp. 113–138, 1994.

24. C. Oliveira, C. Kaestner, F. Bortolozzi, and R. Sabourin, "Generation of Signatures by Deformation," in *Advances in Document Image Analysis*, Springer, pp. 283–298, 1997.

25. R. W. Sproat, *Multilingual Text-to-Speech Synthesis*, Kluwer Academic Publishers, 1997.

26. E. Yamamoto, S. Nakamura, and K. Shikano, "Lip Movement Synthesis from Speech Based on Hidden Markov Models," *Speech Communication*, Vol. 26, No. 1, pp. 105–115, 1998.

27. J. F. Nunamaker Jr, D. C. Derrick, A. C. Elkins, J. K. Burgoon, and M. W. Patton, "Embodied Conversational Agent-Based Kiosk for Automated Interviewing," *Journal of Management Information Systems*, Vol. 28, No. 1, pp. 17–48, 2011.

28. N. Mavridis, M. Petychakis, A. Tsamakos, P. Toulis, S. Emami, W. Kazmi, C. Datta, C. BenAbdelkader, and A. Tanoto, "FaceBots: Steps towards Enhanced Long-Term Human-Robot Interaction by Utilizing and Publishing Online Social Information," in *Paladyn*, Vol. 1, Springer, pp. 169–178, 2010.

29. R. Cappelli, "Synthetic Fingerprint Generation," in *Handbook of Fingerprint Recognition*, D. Maltoni, D. Maio, A. K. Jain, and S. Prabhakar, Springer, pp. 203–232, 2003.

30. R. Cappelli, "SFinGe: Synthetic Fingerprint Generator," in *Proceedings of the International Workshop on Modeling*, Calgary, Canada, 2004.

31. M. Kücken and A. C. Newell, "A Model for Fingerprint Formation," *Europhysics Letters*, Vol. 68, No. 1, pp. 141–146, 2004.

32. A. Lefohn, B. Budge, P. Shirley, R. Caruso, and E. Reinhard, "An Ocularist's Approach to Human Iris Synthesis," *IEEE Computer Graphics and Applications Magazine*, Vol. 23, No. 6, pp. 70–75, 2003.

33. J. Cui, Y. Wang, J. Huang, T. Tan, Z. Sun, and L. Ma, "An Iris Image Synthesis Method Based on PCA and Super-Resolution," *Proceedings of 17th International Conference on Pattern Recognition*, Vol. 4, pp. 471–474, 2004.

34. A. Can, C. V. Steward, B. Roysam, and H. L. Tanenbaum, "A Feature-Based, Robust, Hierachical Algorithm for Registering Pairs of Images of the Curved Human Retina," *IEEE Transactions on Analysis and Machine Intelligence*, Vol. 24, No. 3, pp. 347–364, 2002.

35. T. Moriyama, J. Xiao, T. Kanade, and J. F. Cohn, "Meticulously Detailed Eye Model and Its Application to Analysis of Facial Imaging," *IEEE International Conference on Systems, Man, and Cybernetics*, pp. 629–634, 2004.

36. A. Azizi and H. Pourreza, "A New Method for Iris Recognition Based on Contourlet Transform and Non Linear Approximation Coefficients," in *Emerging Intelligent Computing Technology and Applications*, Springer, pp. 307–316, 2009.

37. J. M. Hollerbach, "An Oscillation Theory of Handwriting," *Biological Cybernetics*, Vol. 39, No. 2, pp. 139–156, 1981.

38. J. Wang, C. Wu, Y.-Q. Xu, H.-Y. Shum, and L. Ji, "Learning-Based Cursive Handwriting Synthesis," in *Proceedings Eighth International Workshop on Frontiers in Handwriting Recognition, 2002*, 2002.

39. R. Plamondon and W. Guerfali, "The Generation of Handwriting with Delta-Lognormal Synergies," *Biological Cybernetics*, Vol. 78, No. 2, pp. 119–132, 1998.

40. H. Choi, S. J. Cho, and J. H. Jin Kim, "Generation of Handwritten Characters with Bayesian Network Based," in *Proceedings of the Seventh International Conference on Document Analysis and Recognition, 2003*, Edinburgh, Scotland, 2003.

41. Y. Du and X. Lin, "Realistic Mouth Synthesis Based on Shape," *Pattern Recognition Letters*, Vol. 23, No. 14, pp. 1875–1885, 2002.

42. B. Lo, H. Lee, M. Ing, and D. Chen, "Modelling of Facial Nerve Disorders," Biometric Technologies Laboratory, University of Calgary, Calgary, 2006.

43. S. N. Yanushkevich, A. Stoica, S. N. Srihari, V. P. Shmerko, and M. L. Gavrilova, "Simulation of Biometric Information: The New Generation of Biometric Systems," in *Proceedings of the International Workshop Modeling and Simulation in Biometric Technology*, 2004.

44. "The Fingerprint Verification Competition" [Online]. Available: http://bias.csr.unibo.it /fvc2004/databases.asp, accessed on September 25, 2014.

45. X. Wang and X. Tang, "Face Photo-sketch Synthesis and Recognition," *IEEE Transactions on Pattern Analysis and Machine Intelligence*, Vol. 31, No. 11, pp. 1955–1967, 2009.

46. S. N. Yanushkevich, A. Stoica, and V. Shmerko, "Fundamentals of Biometric-Based Training System Design," in *Image Pattern Recognition: Synthesis and Analysis in Biometrics*, S. N. Yanushkevich, M. L. Gavrilova, P. S. P. Wang, and S. N. Srihari, eds., World Scientific, 2007.

47. D. K. Zondervan and D. J. Reinkensmeyer, "Kinect-Wheelchair Interface Controlled (KWIC) Robotic Trainer for Powered Mobility, Demo Competition Abstract," in *34th International Conference of the IEEE Engineering in Medicine and Biology Society*, San Diego, 2012.
48. G. C. Burdea and P. Coiffet, *Virtual Reality Technology*, John Wiley & Sons, 2003.
49. B. Lange, C. Chang, E. Suma, B. Newman, A. S. Rizzo, and M. Bolas, "Development and Evaluation of Low Cost Game-Based Balance Rehabilitation Tool Using Microsoft Kinect Sensor," *Engineering in Medicine and Biology Society*, pp. 1831–1834, 2011.

23

Performance Analysis of Transform-Based Medical Image-Compression Methods for Telemedicine

Sujitha Juliet and Elijah Blessing Rajsingh

CONTENTS

23.1 Introduction to Telemedicine

With the increasing population, affording primary healthcare for the human community, especially through advanced medical facilities, is one of the major challenges today. Telemedicine is a good solution to link patients in remote areas and medical professionals all over the world. The efficient transfer of medical information from one location to another is made possible through telemedicine. It generally refers to the delivery of clinical care across distances. Telemedicine systems are based on the following:

- Store-and-forward telemedicine
- Two-way interactive telemedicine
- Remote monitoring

23.1.1 Store-and-Forward Telemedicine

Store-and-forward telemedicine involves acquiring medical images and biosignals like CT and MRI images, ECG, and blood pressure and then transmitting the information to the medical expert at a convenient time for evaluation off-line. This system is generally used in nonemergency situations [1–2]. Advances in telemedicine, viz., teleradiology, teledermatology, and telepathology, are examples of store-and-forward telemedicine. Properly organized medical data, preferably in electronic form, is a component of asynchronous telemedicine.

23.1.2 Two-Way Interactive Telemedicine

Two-way interactive telemedicine provides real-time interactions between patients and medical experts. This system requires videoconferencing devices and network infrastructures at both sides to provide excellent face-to-face interaction and efficient transfer of medical information [3].

23.1.3 Remote Monitoring

Remote monitoring facilitates medical professionals in observing patients' physiological signals remotely through technological devices. This system is mainly used to take care of elderly and disabled patients with chronic diseases [4]. It offers remarkable health benefits similar to those of traditional face-to-face encounters with patients and provides good satisfaction to elderly patients. Generally, remote monitoring systems monitor physiological signals regularly and alert the medical professionals during emergencies. The monitoring system should have few batteries, reliability, good power conservation, and privacy.

23.2 Challenges in Telemedicine

While recent advances in information and communication technology provide excellent means to access medical data, the transmission of medical information by using telemedicine remains challenging due to the following reasons:

- **Security:** Telemedicine involves exchange of medical information that passes through the network. Since the medical information is transmitted through wireless channels, it needs to be kept private. Therefore, security is a critical challenge to be addressed [5–6].

- **Patient monitoring:** By providing improved clinical care through telemedicine, patient monitoring has become one of the major requirements, particularly for aged and disabled patients who need individual care at home. Real-time monitoring of patients using multipurpose instruments to monitor physiological signals is essential in telemedicine. The monitoring system helps the medical professional to keep track of patients with critical illness [7].

- **Network accessibility:** The telemedicine network infrastructure often faces challenges in the delivery of the required quality of service, such as bandwidth requirements and minimum delay due to network congestion. Since a large volume of medical information needs to be transferred through the telemedicine network with the shortest delay possible, network accessibility is also an important challenge to be addressed in telemedicine [8].

- **Expert systems:** An automated decision-support system that incorporates a knowledge-based system and an artificial intelligence technique is also considered to be of great significance in telemedicine. This expert system helps the patients and physician for effective treatment [9].

- **Medical image compression:** Medical images acquired from imaging modalities occupy a significant portion of a patient's health record. These medical images need to be conveniently transmitted across the telemedicine network for diagnosis. Such medical images are extremely data intensive, causing a high storage cost and a heavy increase in network traffic during transmission. In this regard, the development of efficient compression methods that result in less storage requirement and better consumption of bandwidth is one of the major challenges in telemedicine [10].

All the above-mentioned challenges are important to be addressed in telemedicine. However, the transmission of a huge amount of medical data with comparatively low bandwidth is the critical requirement in telemedicine. Hence, the compression of medical images is the major challenge to be addressed in telemedicine for efficient transmission of medical data [11–12]. This chapter accordingly focuses on the development and analysis of efficient compression methods for medical images.

23.3 Challenges of Image Compression in Telemedicine

Recent years have seen great development in the field of medical imaging. Almost 10 terabytes (TB) of medical images needs to be stored in medical image archives per annum, and this burdens the current storage [13]. While the cost of storage devices is expected to decrease over time, it remains expensive with the increasing demand for storage [14].

In addition, network transmission bandwidth between healthcare institutions would have to be expanded to permit medical practitioners to access the images and reports. This will also increase the cost [15]. Hence, it has become extremely important to lessen the demand for storage and bandwidth by utilizing image-compression methods for effective storage and transmission of medical data [5].

23.4 Overview of Transform-Based Image-Compression Methods

Image-transform coding comprises three main components, viz., transformation, quantizer, and symbol encoder. A linear transform is applied to map the pixels onto a set of transform coefficients. These coefficients are then quantized and entropy encoded. The general block diagram of transform-based image coding is shown in Figure 23.1.

In transform coding, the transformation process converts a large set of highly correlated pixels into a smaller number of decorrelated transform coefficients, such that energy is concentrated into as few coefficients as possible. The transform should have decorrelation, linearity, and orthogonality properties for efficient compression.

The decorrelated images are then compressed using a symbol encoder. Prior to symbol coding, the decorrelated images are usually quantized to improve the compression performance. The quantization stage selectively removes or more coarsely approximates the coefficients that have the least information. In lossless coding methods and in medical image compression, no quantization is applied to the decorrelated data.

23.5 Quality Control in Telemedicine

At present, medical imaging modalities are significantly utilized in clinical investigations. These modalities provide greater means of observing physiological signals and anatomical cross sections of a patient's body. Moreover, they may lessen the radiation dosages and examination of trauma. Since MRI and CT images provide detailed visuals and greater contrast between different tissues of the body, these anatomical images are widely used in telemedicine. These medical images need to be conveniently transmitted across the network for diagnosis. The huge amounts of medical data cause a high storage cost and heavy increase in network traffic during transmission.

With great development in the field of medical imaging, compression of medical images plays a vital role [16]. The increase in network traffic may slow down the usage of telemedicine if medical images are not sufficiently compressed. Furthermore, preserving medical information for many years may become unaffordable for the majority of hospitals due to the storage requirements [17]. Compression methods reduce the file size and transmission time, thus improving overall clinical care.

Compression could be lossless if it achieves exact reconstruction of the original image and lossy when it includes quantization. In general, lossless compression is preferred in medical applications in order to maintain the data integrity and to expedite accurate diagnosis. However, lossless compression provides relatively low compression ratios when compared to lossy methods. Therefore, certain applications such as telemedicine suffer from this limitation [18].

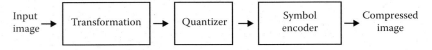

FIGURE 23.1
General block diagram of transform-based image compression.

Lossy compression of medical images is also preferable provided it preserves diagnostically critical information [19]. The Canadian Association of Radiologists (CAR) and the American College of Radiology (ACR) have recognized lossy medical image compression under monitored conditions. Lossy compression is also acknowledged by the DICOM standard. In practice, near-lossless compression is widely used for observing medical images [20].

23.6 Transform-Based Medical Image Compression

This section deals with three transform-based image-compression methods for the compression of MRI and CT images.

23.6.1 Ripplet Transform-Based Medical Image Compression

The ripplet transform is a higher-dimensional generalization of wavelet transform. It represents two-dimensional signals or images at different scales and different directions [21]. It is optimal for representing objects with C^2 singularities (twice–continuously differentiable curves). The introduction of support c and degree d yields anisotropy capability of exhibiting singularities along arbitrarily shaped curves.

Ripplet functions decay very fast in the spatial domain and have compact support in the frequency domain. Therefore, ripplet functions are properly localized in both spatial and frequency domains. Since the ripplet transform successively approximates images from coarse to fine resolutions, it provides a hierarchical representation of images. The general scaling and support result in anisotropy of ripplet functions that guarantees acquisition singularities along arbitrary curves. Ripplet coefficients decay faster than the coefficients of other transforms. Therefore, the ripplet transform has higher energy-concentration ability.

Traditional transforms like the wavelet transform provide efficient representations for smooth images but suffer from curve discontinuities. Therefore, it is desirable to have a compression method that provides efficient representation of edges by exhibiting singularities along arbitrarily shaped curves, which may lead to improvements in compression performance. The compression procedure using the ripplet transform is formulated as follows [22]:

Step 1: Input the medical image $f(x, y)$ of size 256×256.

Step 2: Decompose the input image into a set of frequency sub-bands:

$$f(x, y) \mapsto (P_0 f(x, y), \Delta_1 f(x, y), \Delta_2 f(x, y),\ldots) \tag{23.1}$$

where $P_0 f(x, y)$ is the lowest-frequency component and $\{\Delta_1 f(x, y), \Delta_2 f(x, y),\ldots\} \in \Delta_s f(x, y)$ represent high-frequency components.

The decomposed wavelet bands j are partially reconstructed into ripplet sub-bands s as $j \in \{2s, 2s, + 1\}$.

Step 3: Dissect the high-frequency band into small partitions by defining a grid of dyadic square:

$$Q_{(s,k_1,k_2)} = \left[\frac{k_1}{2^s}, \frac{k_1+1}{2^s} \right] \times \left[\frac{k_2}{2^s}, \frac{k_2+1}{2^s} \right] \in Q_s \tag{23.2}$$

where Q_s defines dyadic squares of the grid.

Multiplying the high-frequency band $\Delta_s f(x, y)$ with a windowing function w_Q produces a smooth dissection of the function into squares of sides $2^{-s} \times 2^{-s}$:

$$h_Q = w_Q \cdot \Delta_s f(x, y) \tag{23.3}$$

Step 4: Renormalize each resulting dyadic square by centering each dyadic square to the unit square $[0,1] \times [0,1]$. For each dyadic square Q, the operator T_Q is defined as

$$[T_Q f(x, y)] (x_1, x_2) = 2^s f(2^s x_1 - k_1, 2^s x_2 - k_2) \tag{23.4}$$

Each dyadic square is renormalized as

$$g_Q = T_Q^{-1} h_Q \tag{23.5}$$

Step 5: Analyze each square in the ripplet domain:

$$R_{(Q,a\vec{b}\theta)} = \left\langle g_Q, \rho_{a\vec{b}\theta} \right\rangle = \int g_Q(\vec{x}) \overline{\rho_{a\vec{b}\theta}(\vec{x})} d\vec{x} \tag{23.6}$$

where $R_{(Q,a\vec{b}\theta)}$ are ripplet coefficients and $\rho_{a\vec{b}\theta}$ is the ripplet function, which is generated as

$$\rho_{a\vec{b}\theta}(\vec{x}) = \rho_{a\vec{0}0}\left(R_\theta(\vec{x} - \vec{b}) \right) \tag{23.7}$$

where

$$R_\theta = \begin{bmatrix} \cos\theta & \sin\theta \\ -\sin\theta & \cos\theta \end{bmatrix}$$

is the rotation matrix that rotates θ radians; $\vec{x}$ and $\vec{b}$ are 2D vectors; and $\rho_{a\vec{0}0}(\cdot)$ is the mother ripplet function in the frequency domain.

The discrete ripplet transform is given as

$$R_{j,\vec{k},l} = \sum_{x=0}^{M-1} \sum_{y=0}^{N-1} g_Q(x, y) \overline{\rho_{j,\vec{k},l}(x, y)} \tag{23.8}$$

where $R_{j,\vec{k},l}$ are the ripplet coefficients and $\rho_{j,\vec{k},l}(x,y)$ are the ripplets or ripplet functions in discrete domain; $(\bar{\ })$ denotes the conjugate operator.

Step 6: Encode the resulting coefficients by using set partitioning in hierarchical trees (SPIHT) encoder.

23.6.2 Bandelet Transform-Based Medical Image Compression

Developing an efficient image-compression method often requires building sparse representation where a signal is absolutely approximated with a lesser number of coefficients. Such a representation takes advantage of geometric regularity of edges in images. Even though, recent developments in image analysis community such as curvelet transform [23] and contourlet transform [24] are able to capture the geometric regularity, these methods are unable to construct orthogonal bases of regular functions which are greatly required for image compression. Similarly, image representation in separable orthonormal bases such as wavelets do not possess the benefits of geometric regularity. Integration of geometric regularity of image structures in image representation is one of the major challenges to enhance the performance of image compression.

The bandelet transform [25] utilizes the benefits of geometric regularity of image structure for efficient sparse image representation. It decomposes the input image along multiscale vectors elongated in the focus of geometric flow. In the bandelet transform, the geometry of images is described with the geometric flow of vectors rather than being described through edges. The geometric flow vectors represent the directions in which the image gray level has regular variations. The bandelet construction is orthogonal and the commensurate basis functions are regular. The compression algorithm using bandelet transform is formulated as follows [26]:

Step 1: Decompose the input medical image *f(x, y)* through the two-dimensional wavelet transform to obtain wavelet decomposition of sub-bands:

$$
\begin{aligned}
a_j(x,y) &= \left\langle f(x,y), \varphi_j(x)\varphi_j(y) \right\rangle, \\
d_j^H(x,y) &= \left\langle f(x,y), \psi_j(x)\varphi_j(y) \right\rangle, \\
d_j^V(x,y) &= \left\langle f(x,y), \varphi_j(x)\psi_j(y) \right\rangle, \\
d_j^D(x,y) &= \left\langle f(x,y), \psi_j(x)\psi_j(y) \right\rangle.
\end{aligned}
\tag{23.9}
$$

Step 2: Obtain dyadic squares S of constant size 2^k by recursively segmenting the wavelet-transformed image.

Step 3: For each square region S_i, construct geometric flow by determining the image sample values along the flow lines.

The geometric flow in S_i is a two-dimensional vector field $\vec{v}(x,y)$ defined over the image sampling grid. If the geometric flow is parallel in the vertical direction, then $\vec{v}(x,y) = \vec{v}(x)$ and if the geometric flow is parallel in the horizontal direction, then $\vec{v}(x,y) = \vec{v}(y)$.

Step 4: Warp the discrete wavelets along the flow lines with an operator W and compute the warped wavelet coefficients. For the vertical parallel flow, the warped wavelet coefficients are obtained as

$$\left\{ \begin{array}{l} \varphi_j^{m_1}(x)\psi_j^{m_2}(y - c(x)), \\ \psi_j^{m_1}(x)\varphi_j^{m_2}(y - c(x)), \\ \psi_j^{m_1}(x)\psi_j^{m_2}(y - c(x)) \end{array} \right\}_{j, m_1, m_2} \tag{23.10}$$

and the warped wavelet coefficients for the horizontal parallel flow are given as

$$\left\{ \begin{array}{l} \varphi_j^{m_1}(x - c(y))\psi_j^{m_2}(y), \\ \psi_j^{m_1}(x - c(y))\varphi_j^{m_2}(y), \\ \psi_j^{m_1}(x - c(y))\psi_j^{m_2}(y) \end{array} \right\}_{j, m_1, m_2}. \tag{23.11}$$

Step 5: Bandeletize the warped wavelet basis by replacing $\left\{\varphi_j^{m_1}(x)\psi_j^{m_2}(y)\right\}$ with $\left\{\psi_l^{m_1}(x)\psi_j^{m_2}(y)\right\}$ through 1D discrete wavelet transform.

Step 6: Encode the resultant bandelet coefficients by using the SPIHT encoder.

23.6.3 Radon Transform-Based Medical Image Compression

The Radon transform is an integral transform composed of integrals of two-dimensional functions over straight lines and further in the case of three dimensions, it is integrated over planes. It describes an image as a group of projections along several directions. It plays a vital role in medical image processing that comes under the heading of tomography. Tomography is the problem of reconstructing the interior of an object by passing the X-ray beam and recording the resulting intensity over a range of directions. The tomographic image is a cross-sectional image of an object. The Radon transform finds multiple line integrals of absorption along parallel paths and reiterates the procedure for various angles of incidence around a radial path. It defines the projection of image intensity around a circular line adapted to a definite angle. It is an effective method for analyzing signals between the spatial domain and its projection domain.

Applying the Radon transform to an image $f(x, y)$ for a given set of angles can be used to compute the projection of the image along the given angles. It maps the Cartesian rectangular coordinates (x, y) to a distance and the angles (s, θ) termed as *polar coordinates*. The Radon transform of a given image $f(x, y)$ can be represented as $g(s, \theta)$, which denotes the line integral along the angle of incidence θ with regard to the x axis and a distance s from the origin. The Radon transform operation is depicted in Figure 23.2.

In the Radon transform, the spatial domain (x, y) is mapped with the projection domain (s, θ). Each coordinate in the projection domain becomes a straight line in the spatial

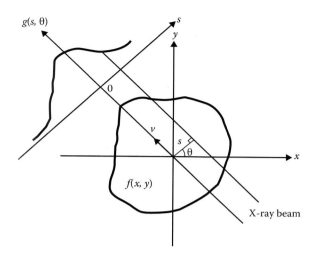

FIGURE 23.2
Radon transform in tomography.

domain and contrarily each coordinate in the spatial domain corresponds to a sine curve in the projection domain. The Radon transform of a function $f(x, y)$ can be expressed by $g(s, \theta)$, given as follows:

$$g(s,\theta) \overset{\Delta}{=} \int_{-\infty}^{\infty} \int_{-\infty}^{\infty} f(x,y)\delta(x\cos\theta + y\sin\theta - s)\,dxdy, \tag{23.12}$$

where δ represents the Dirac function, s denotes the perpendicular distance of the line from the origin, and θ represents the angle between the x axis and the normal from the origin to the line. Every coordinate (s, θ) in the Radon domain becomes the sum of intensity values along a certain straight line in the image domain. The dimension-reducing property allows the conversion of the two-dimensional signal $f(x, y)$ to a set of simple one-dimensional signals independently on each of the projections.

For image compression, the input medical image $f(x, y)$ is first applied to the Radon transform which maps the Cartesian rectangular coordinates (x, y) to polar coordinates (s, θ). Applying the Radon transform for a set of angles computes the projection of the image along the given angles. It maps the line integrals along the angle of incidence into a related collection of projections. The Radon projections are computed at various orientations and directional features of the input image are captured. When Radon coefficients are computed, the pixel intensities are also considered. Therefore, variations in pixel intensities are preserved and improved. The compression procedure is given as follows:

Step 1: Input the medical image $f(x, y)$ of size 256×256, 8 bits per pixel (bpp).

Step 2: Compute the projection of distance s and vector v in the given plane.

Step 3: Apply the Radon transform to map the spatial domain coordinates (x, y) to the projection domain coordinates $g(s, \theta)$:

$$g(s,\theta) = \int_{-\infty}^{\infty} \int_{-\infty}^{\infty} f(x,y)\delta(x\cos\theta + y\sin\theta - s)\,dx\,dy. \tag{23.13}$$

Step 4: Encode the resulting Radon coefficients by using the SPIHT encoder.

23.7 Set Partitioning in Hierarchical Trees Encoder

This section deals with the encoder used for encoding the transformed coefficients generated in the previous section. All the transformed coefficients (ripplet, bandelet, and Radon) are encoded using the SPIHT encoder [27], which further improves the compression performance. The transformed coefficients are further coded using the SPIHT algorithm which encodes coefficients by spatial orientation tree. This algorithm defines the spatial relationships between transformed coefficients at various decomposition levels. It organizes the coefficients into significant and insignificant portions based on the significance test as follows:

$$s_n(B) = \begin{cases} 1 & \text{if } \max_{(x,y)\in B}\left(\left|B_{x,y}\right|\right) \geq 2^n \\ 0 & \text{otherwise} \end{cases} \tag{23.14}$$

where $s_n(B)$ is the significance of a set of coordinates B, and $B_{x,y}$ represents the resultant transformed coefficients at coordinates (x, y).

In the spatial orientation tree, each coefficient is progressively encoded from the most significant bits to the least significant bits. The coding starts from highest-magnitude coefficients at the lowest pyramids. The SPIHT encoding utilizes three sets of lists based on the significance of coefficients: (i) the list of insignificant pixels (LIP), (ii) the list of insignificant sets (LIS), and (iii) the list of significant pixels (LSP). LIP and LSP contain nodes with single pixels and LIS consists of nodes with descendants. This algorithm utilizes two stages of operation: sorting and refinement stages. The maximum number of bits needed to represent the highest coefficients in the spatial orientation tree is obtained and expressed as $n_{\max}$, where

$$n_{\max} = \left\lfloor \log_2\left[\max_{x,y}\left(\left|B_{x,y}\right|\right)\right]\right\rfloor \tag{23.15}$$

During the sorting stage, the pixels in LIP are tested for $n_{\max}$ and the significance result of $s_n(B)$ is transferred to the output. Significant pixels in LIP are transferred to LSP. Each set of pixels in LIS is also evaluated based on a significance test. If the set is observed as significant, then it is removed from the list and portioned into subsets. If the subsets with a single coefficient are observed to be significant, they will be transferred to LSP or else they will be included in LIP. LSP now consists of pixel coordinates that are inspected in the refinement stage which produces the nth most significant bit of $s_n(B)$. The value of n is decreased by 1 and the sorting and refinement procedures are repeated until all the transformed coefficients are evaluated completely.

23.8 Results and Discussions

The performances of proposed transform-based compression methods are analyzed on a set of gray intensity medical images collected from medical imaging modalities such as MRI and CT and from different body districts. These images are represented in different planes, viz., sagittal, coronal, and axial.

Two different types of image sets are used in this work. The first set consists of MRI images which include the sagittal section of T1-weighted image (T1WI) MRI lumbar spine, sagittal section of MRI knee, axial section of T2-weighted image (T2WI) MRI brain, axial section of half-Fourier acquisition single-shot turbo spin-echo imaging (HASTE) T2WI MRI abdomen, axial section of MRI brain, sagittal section of T1WI MRI brain, coronal section of T1WI MRI chest, coronal section of T1WI MRI leg, axial section of T2WI MRI brain, and sagittal section of MRI cerebral.

The second set consists of CT images which include axial sections of the CT abdomen, CT scan of chest (left side) pulmonary arteries, CT scan of brain cranioverterbral (CV) junction, axial section of CT abdomen with oral contrast, and axial section of CT skull. Both the image sets are collected from KG Hospital, Coimbatore, Tamil Nadu, India. The test images are of the size 256 × 256, 8 bpp, with their file sizes varying from 20 to 656 KB. Figure 23.3 shows the sample of medical images used for evaluation.

The quality of the compressed images were assessed in terms of peak signal-to-noise ratio (PSNR) (dB), structural similarity index measure (SSIM), compression ratio, and computational time (s). The efficiency of the proposed methods are evaluated on comparison with discrete cosine transform (DCT) [28], Haar wavelet transform [29], contourlet transform [30], curvelet transform [31], and JPEG compression [32]. In order to implement and analyze the proposed and existing compression methods, the image processing toolbox of MATLAB® software is used and all the algorithms are executed on a general personal computer with a 2.67 GHz Intel® i5 processor and 4 GB random-access memory (RAM).

The following subsections present thorough experimental investigations on the overall behavior of the proposed transform-based compression methods. Subsection 23.8.1 analyzes the quality of the compressed images in terms of PSNR and Subsection 23.8.2 evaluates the image quality in terms of SSIM. Subsection 23.8.3 tabulates and analyzes the compression ratios and bit rates and Subsection 23.8.4 analyzes the computational times utilized by the proposed compression methods for the compression of medical images. Finally, Subsection 23.8.5 presents subjective evaluations on the diagnostic quality of compressed images.

23.8.1 Analysis of Image Quality Based on Peak Signal-to-Noise Ratio

The major design objective of the image-compression method is to obtain the best visual quality of images with the minimum number of bits. PSNR is one of the ample parameters

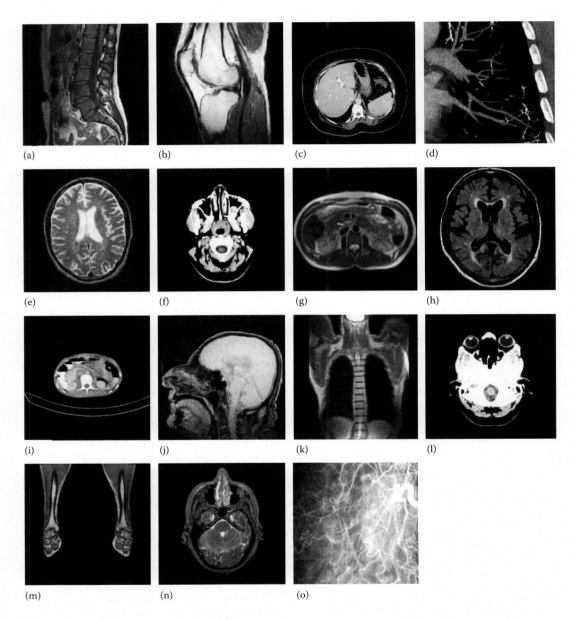

FIGURE 23.3
Examples of test images used for evaluation: (a) sagittal section of T1WI MRI lumbar spine, (b) sagittal section of MRI knee, (c) axial section of CT abdomen, (d) CT scan of chest (left side) pulmonary arteries, (e) axial section of T2WI MRI brain, (f) CT scan of brain CV junction, (g) axial section of HASTE T2WI MRI abdomen, (h) axial section of MRI brain, (i) axial section of CT abdomen with oral contrast, (j) sagittal section of T1WI MRI brain, (k) coronal section of T1WI MRI chest, (l) axial section of CT skull, (m) coronal section of T1WI MRI leg, (n) axial section of T2WI MRI brain, and (o) sagittal section of MRI cerebral.

for evaluating the quality of compression. It is a pixel-wise error metric extensively implemented to judge image quality based on mean squared error (MSE) measure. Higher PSNR value leads to an improved compression performance. It is defined as

$$PSNR = 10 * \log_{10}\left[\frac{(255)^2}{MSE}\right] \tag{23.16}$$

The MSE in Equation 23.16 represents the mean squared error of the image and is defined as

$$MSE = \frac{1}{M*N} \times \left\{\sum_{x=0}^{M-1}\sum_{y=0}^{N-1}[f(x,y)-F(x,y)]^2\right\} \tag{23.17}$$

where $M*N$ represents the size of the image, $f(x, y)$ represents the original image and $F(x, y)$ denotes the compressed image. Bit rate (bpp) is defined as the ratio of the size of the compressed image in bits to the total number of pixels.

Table 23.1 shows comparisons of the performances of the ripplet transform-based method and existing methods in terms of PSNR gain. In the ripplet transform-based method, the ripplet transform searches over (c,d) pairs for the highest PSNR. The anisotropy property of the ripplet transform provides the effectiveness to capture 2D singularities along curves in images and yields efficient representation of edges. For the axial section of T2WI MRI brain, on average, the ripplet transform-based method brings a gain of 9.72 dB with respect to the DCT method, 2.85 dB with respect to the Haar wavelet method, 4.64 dB with respect to the contourlet method, 5.2 dB with respect to the curvelet method, and 7.27 dB with respect to the JPEG compression method.

The contourlet transform utilizes a Laplacian pyramid to capture point discontinuities. In addition, it uses directional filter banks to relate point discontinuities into linear structures. However, the non-subsampled process in the contourlet transform causes

TABLE 23.1

Average PSNR Gains of Ripplet Transform-Based Method Compared with Existing Methods

	Average PSNR Gain (dB)					
	Axial Section of T2WI MRI Brain	Sagittal Section of MRI Cerebral	Coronal Section of T1WI MRI Chest	Axial Section of CT Abdomen with Oral Contrast	CT Scan of Brain CV Junction	CT Scan of Chest (Left Side) Pulmonary Arteries
Over DCT-SPIHT	9.72	8.71	8.19	8.42	10.41	9.66
Over Haar-SPIHT	2.85	3.15	1.95	1.54	5.84	4.22
Over contourlet-SPIHT	4.64	1.84	3.37	2.74	4.05	5.25
Over curvelet-SPIHT	5.2	2.67	3.71	4.15	3.64	4.67
Over JPEG	7.27	8.33	6.87	6.68	6.6	7.75

frequency aliasing, which leads to a relatively low PSNR. DCT and JPEG suffer from blocking artifacts caused by discontinuities. Even though the curvelet transform provides effective representation of edge discontinuities in images, it is not favorable for sparse approximation of features beyond C^2 singularities. Also in the Haar wavelet transform, since the Haar window is only two elements wide, all high-frequency changes are not reflected in the high-frequency coefficient spectrum. Therefore, these methods achieve relatively low PSNR. In the ripplet transform-based method, the nonnegative windowing function and the sub-band filtering procedures yield exact reconstruction. Hence, the ripplet transform-based method achieves relatively high PSNR compared to the existing methods.

Table 23.2 shows the average PSNR gains of the bandelet transform-based compression method compared to existing methods. This method provides efficient sparse representation by capturing the geometrics in images. For the sagittal section of T1WI MRI brain image, on average, the bandelet-based method brings a gain of 10.96 dB with respect to the DCT method, 4.66 dB with respect to the Haar wavelet method, 4.97 dB with respect to the contourlet method, 5.86 dB with respect to the curvelet method, and 6.24 dB with respect to the JPEG compression method.

Even though the curvelet transform and the contourlet transform are capable of capturing the geometric regularity of image structures, they are unable to build orthogonal bases of regular functions. This impediment leads to relatively low PSNR. Since the correlation across block edges is not removed in block-based DCT and JPEG compression, it results in blocking artifacts. Since the Haar wavelet is inefficient in representing geometric features with line and surface singularities, it does not enjoy the benefit of geometric regularity of image structures. Owing to these setbacks, the reference methods achieve relatively low PSNR. The bandelet functions are regular and, hence, offer no blocking artifacts. Therefore, the bandelet-based method achieves high PSNR compared to existing methods.

Table 23.3 shows the average PSNR gains of the Radon transform-based compression method compared to the existing methods. Since the exact invertibility of the Radon transform prevents the appearance of artifacts, it results in high PSNR. For the axial section of CT abdomen with oral contrast, on average, the Radon-based method brings a gain of 10.33 dB with respect to the DCT method, 3.45 dB with respect to the Haar wavelet method, 4.65 dB with respect to the contourlet method, 6.06 dB with respect to the curvelet method, and 8.59 dB with respect to the JPEG compression method.

TABLE 23.2

Average PSNR Gains of Bandelet Transform-Based Method Compared with Existing Methods

	Average PSNR Gain (dB)					
	Sagittal Section of T1WI MRI Brain	Coronal Section of T1WI MRI Leg	Sagittal Section of T1WI MRI Lumbar Spine	Axial Section of CT Skull	CT Scan of Brain CV Junction	Axial Section of CT Abdomen
Over DCT-SPIHT	10.96	8.57	7.48	7.96	9.3	7.91
Over Haar-SPIHT	4.66	2.99	2.85	2.27	4.94	1.52
Over contourlet-SPIHT	4.97	4.94	2.98	1.62	3.46	2.27
Over curvelet-SPIHT	5.86	4.99	3.74	3.6	4.41	3.86
Over JPEG	6.24	6.9	6.32	6.62	7.69	6.58

TABLE 23.3

Average PSNR Gains of the Radon Transform-Based Method Compared with Existing Methods

	Average PSNR Gain (dB)					
	Coronal Section of T1WI MRI Leg	Axial Section of T2WI MRI Brain	Sagittal Section of MRI Cerebral	Axial Section of CT Abdomen with Oral Contrast	Axial Section of CT Skull and Orbits	CT Scan of Chest (Left Side) Pulmonary Arteries
Over DCT-SPIHT	8.67	7.71	7.38	10.33	8.79	9.77
Over Haar-SPIHT	3.09	1.42	1.83	3.45	3.1	4.32
Over contourlet-SPIHT	5.04	1.72	0.52	4.65	2.45	5.36
Over curvelet-SPIHT	5.09	2.61	1.34	6.06	4.43	4.77
Over JPEG	7	5.89	7	8.59	7.45	7.86

Since the contourlet transform possesses fewer directional features, this leads to artifacts in compression. In the Haar wavelet transform, the lack of translation invariance and poor energy compaction yield relatively low PSNR. The DCT and JPEG compression divide the original image into 8×8 sub-blocks and the coefficients in each block are quantized individually. The quantization results in pseudoborders between the sub-blocks. In addition, curvelet transforms are not effective for image curve structures outside C^2 singularities. Owing to these setbacks, the PSNR achieved by the reference methods are relatively low. Since the Radon transform-based compression method preserves the deviations in pixel intensities, it achieves relatively high PSNR compared to existing methods.

23.8.2 Analysis of Image Quality Based on Structural Similarity Index Measure

The SSIM is an objective image quality metric used for the measurement of similarity between two images based on the characteristics of the human visual system. It measures the structural similarity rather than error visibility between two images. SSIM is defined as

$$\text{SSIM}(x,y) = \frac{(2\mu_x\mu_y + C_1)(2\sigma_{xy} + C_2)}{\left(\mu_x^2 + \mu_y^2 + C_1\right)\left(\sigma_x^2 + \sigma_y^2 + C_2\right)} \tag{23.18}$$

where x and y represent spatial patches (windows); μ_x and μ_y are the mean intensity values of x and y, respectively; σ_x^2 and σ_y^2 denote the standard deviations of x and y, respectively; and C_1 and C_2 are constants. In Equation 23.18, SSIM (x, y) is equal to unity if and only if $x = y$. The mean intensity and standard deviations are calculated as follows:

$$\mu_x = \frac{1}{M \cdot N} \sum_{i=0}^{M-1} \sum_{j=0}^{N-1} [x(i,j)] \tag{23.19}$$

$$\mu_y = \frac{1}{M \cdot N} \sum_{i=0}^{M-1} \sum_{j=0}^{N-1} [y(i,j)] \tag{23.20}$$

$$\sigma_x = \left\{ \frac{1}{M \cdot N - 1} \sum_{i=0}^{M-1} \sum_{j=0}^{N-1} [x(i,j) - \mu_x]^2 \right\}^{1/2} \tag{23.21}$$

$$\sigma_y = \left\{ \frac{1}{M \cdot N - 1} \sum_{i=0}^{M-1} \sum_{j=0}^{N-1} [y(i,j) - \mu_y]^2 \right\}^{1/2} \tag{23.22}$$

$$\sigma_{xy} = \left(\frac{1}{M \cdot N - 1} \sum_{i=0}^{M-1} \sum_{j=0}^{N-1} \left\{ [x(i,j) - \mu_x][y(i,j) - \mu_y] \right\} \right)^{1/2} \tag{23.23}$$

In Equation 23.18, C_1 and C_2 are added to avoid instability when $\mu_x^2 + \mu_y^2$ or $\sigma_x^2 + \sigma_y^2$ are very close to zero. These constants can be adjusted by $C_1 = (K_1 L)^2$ and $C_2 = (K_2 L)^2$, where K_1 and K_2 are small constants and L denotes the dynamic range of pixel values. In this work, K_1 is set to 0.01 and K_2 is set to 0.03. Since 8-bit gray-level images are used in this work, $L = 255$. The image qualities of the proposed transform-based methods are analyzed in terms of SSIM for various bit rates and Figure 23.4a through c shows the SSIMs achieved for test images by using different methods.

Figure 23.4 shows that the proposed transform-based methods outperform other methods by yielding higher SSIM value (close to 1) for all test images. It shows that the compressed image is close to the original in terms of structural shapes. For example, for the CT scan of chest (left side) pulmonary arteries, on average, the ripplet transform-based method outperforms DCT by 40.51%, the Haar wavelet by 11.66%, the contourlet by 14.27%, the curvelet by 9.89%, and JPEG by 18.96%. For the sagittal section of T1WI MRI Lumbar spine, on average, the proposed bandelet-based method outperforms DCT by 31.76%, the Haar wavelet by 13.25%, the contourlet by 12.58%, the curvelet by 9.29%, and JPEG by 16.34%. The Radon transform for the CT scan of brain CV junction outperforms, on average, DCT by 24.09%, the Haar wavelet by 11.47%, the contourlet by 13.76%, the curvelet by 7.76%, and JPEG by 14.7% in terms of SSIM.

In medical images, the edges are usually curved instead of straight. Haar wavelets are efficient in representing point singularities. However, they are unable to provide effective representation of higher-dimensional singularities due to their poor orientation selectivity. The contourlet transform is capable of providing a flexible number of directions and able to capture the inherent geometrical structure of images. But the combination of directional filter with the Laplacian pyramid results in pseudo–Gibbs phenomenon around the singularities, leading to a relatively low similarity index. Similarly, JPEG and DCT also suffer from block-shaped artifacts near edge regions in images. Therefore, these compression methods provide relatively low SSIM. The curvelet transform utilizes the benefits of geometric regularity of image structures by decomposing the image in a fixed basis. However, it is effective only for piecewise C^α functions with $\alpha = 2$ and is not optimal for functions with $\alpha > 2$ or for bounded variation functions. Therefore, if the edges of an image

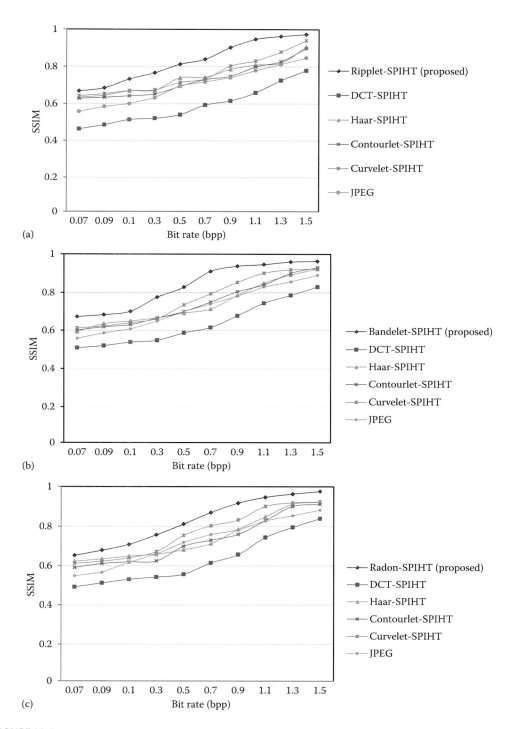

FIGURE 23.4
SSIM values for test images at various bit rates using different methods: (a) CT scan of chest (left side) pulmonary arteries (ripplet transform), (b) sagittal section of T1WI MRI lumbar spine (bandelet transform), and (c) CT scan of brain CU junction (Radon transform).

are along irregular curves, then the curvelet approximations are not accurate. This results in relatively low similarity index.

The ripplet functions orient at various directions and exhibit scaling with arbitrary support and degree. Therefore, ripplets localize the singularities more accurately and the value of SSIM closely reaches 1. The bandelet-based compression method with adaptive techniques provides greater performance for images with textures beyond C^2 singularities. Since it provides multiscale representation of the image geometry and represents sharp image transitions such as edges efficiently, it achieves a relatively high similarity index when compared to existing methods. Moreover, the tunable redundancy of the Radon transform distributes the data over a predefined projection set and exploits both the periodicity within each projection and the similarities between projections, which leads to a higher SSIM value than that of other methods.

23.8.3 Analysis of Compression Ratio

The compression ratio enumerates the reduction in image size yielded by the compression method. It is the ratio of the uncompressed file size to the compressed file size. It is calculated as the number of bits occupied by each pixel before compression divided by the number of bits occupied per pixel after compression. For a grayscale image, the compression ratio = (1/bpp) * 8.

A comparative analysis on the performances of the proposed transform-based methods and reference methods is made in terms of the compression ratio. Table 23.4 compares the compression ratios and bit rates among a set of medical images. From Table 23.4, it is shown that the ripplet transform provides better performance for the compression of both MRI and CT images. Since the ripplet transform successively approximates images from coarse to fine resolutions and is highly directional to represent the orientations of singularities, the proposed method using the ripplet transform performs well in the compression of medical images such as axial section of MRI brain, coronal section of T1WI MRI chest, axial section of CT skull, and axial section of CT abdomen.

It has also been inferred from Table 23.4 that the dimension-reducing property of the Radon transform results in a smaller number of significant coefficients, thus yielding a higher compression ratio. This method performs well on the compression of CT images like axial section of CT abdomen with oral contrast, CT scan of brain CV junction, and CT scan of chest (left side) pulmonary arteries. It performs well on the sagittal section of T1WI MRI lumbar spine also. In addition, it has been observed that the bandelet-based method provides better performance for the compression of MRI images like the coronal section of T1WI MRI leg, sagittal section of T1WI MRI lumbar spine, axial section of MRI brain, coronal section of T1WI MRI chest, and sagittal section of MRI cerebral. Since the wavelet decomposition of geometric regularized input image results in a smaller number of significant coefficients, the bandelet transform-based method yields a higher compression ratio compared to the existing methods.

23.8.4 Analysis of Computational Time

The computational time depends on the complexity of the compression algorithm, the efficiency of the implementation of the algorithm, and the speed of the processor hardware. All the algorithms are executed on a general personal computer with 2.67 GHz Intel i5 processor and 4 GB RAM. Table 23.5 compares the computational times utilized by the processor to compress the medical images in using different methods.

TABLE 23.4

Comparative Analysis of Proposed Transform-Based Compression Methods (Ripplet, Bandelet, and Radon) with Existing Methods in Terms of Compression Ratio and Bit Rate

Medical Image (256 × 256:8)	DCT-SPIHT	Haar-SPIHT	Contourlet-SPIHT	Curvelet-SPIHT	JPEG	Ripplet-SPIHT	Bandelet-SPIHT	Radon-SPIHT
				Compression Ratio (Bit Rate [bppl])				
Sagittal section of T1WI MRI lumbar spine	8.71:1 (0.92)	9.62:1 (0.83)	9.96:1 (0.80)	9.53:1 (0.84)	8.21:1 (0.97)	11.49:1 (0.70)	10.37:1 (0.77)	12.55:1 (0.64)
Axial section of MRI brain	10.36:1 (0.77)	10.57:1 (0.76)	11.05:1 (0.72)	10.89:1 (0.73)	9.82:1 (0.81)	12.72:1 (0.63)	11.38:1 (0.70)	12.68:1 (0.63)
Coronal section of T1WI MRI chest	9.83:1 (0.81)	9.41:1 (0.85)	9.74:1 (0.82)	10.31:1 (0.78)	9.27:1 (0.86)	12.64:1 (0.63)	10.86:1 (0.74)	10.21:1 (0.78)
Sagittal section of MRI cerebral	9.7:1 (0.82)	10.22:1 (0.78)	9.73:1 (0.82)	9.24:1 (0.87)	9.48:1 (0.84)	10.41:1 (0.77)	10.45:1 (0.77)	11.13:1 (0.72)
Coronal section of T1WI MRI leg	12.07:1 (0.66)	10.29:1 (0.78)	12.22:1 (0.65)	11.14:1 (0.72)	10.85:1 (0.74)	12.18:1 (0.66)	12.22:1 (0.65)	12.15:1 (0.66)
Axial section of CT abdomen with oral contrast	8.22:1 (0.97)	9.73:1 (0.82)	9.92:1 (0.81)	9.41:1 (0.85)	8.86:1 (0.90)	10.25:1 (0.78)	9.89:1 (0.81)	10.98:1 (0.73)
CT scan of brain CV junction	9.37:1 (0.85)	9.62:1 (0.83)	10.21:1 (0.78)	11.93:1 (0.67)	8.92:1 (0.90)	11.88:1 (0.67)	11.08:1 (0.72)	12.49:1 (0.64)
Axial section of CT abdomen	9.59:1 (0.83)	10.21:1 (0.78)	10.33:1 (0.77)	9.78:1 (0.82)	9.47:1 (0.84)	12.16:1 (0.66)	10.14:1 (0.79)	11.83:1 (0.68)
Axial section of CT skull	9.22:1 (0.87)	9.37:1 (0.85)	9.84:1 (0.81)	9.61:1 (0.83)	8.23:1 (0.97)	11.49:1 (0.70)	9.79:1 (0.82)	11.16:1 (0.72)
CT scan of chest (left side) pulmonary arteries	8.51:1 (0.94)	8.67:1 (0.92)	10.28:1 (0.78)	10.49:1 (0.76)	8.74:1 (0.92)	10.06:1 (0.80)	10.46:1 (0.76)	12.35:1 (0.65)

Results from Table 23.5 show that the average computational time spent on this processor to compress the given medical image by using the ripplet transform-based method is 0.62 s. Radon transform utilizes 0.37 s and bandelet transform takes 1.46 s to compress the images; whereas in the case of DCT, it is 0.49 s, and for Haar wavelet, it is 0.46 s. Contourlet takes 0.54 s, curvelet utilizes 0.65 s, and JPEG takes 0.79 s to compress the images.

Since the contourlet transform is implemented with pyramidal bandpass decomposition and multiresolution directional filtering, it is directly constructed in the discrete domain. Therefore, in the contourlet transform, there is no need for transformation from the continuous time–space domain. Therefore, the computational time is relatively low. In the Haar wavelet transform, the Haar basis functions have variable lengths and there is no demand for the input image to be decomposed into nonoverlapping two-dimensional blocks. Therefore, the Haar transform utilizes relatively low computational time. Since JPEG and DCT-based compression methods consider a stream of 8 × 8 blocks of image samples and perform compression by concentrating most of the signal in the lower spatial frequencies, they require relatively less time to compress the given image. Ripplet and curvelet transforms are highly redundant since blocks must be overlapped together to avoid the boundary effect. Therefore, they utilize relatively high computational time.

Since an exhaustive search algorithm is required for image segmentation and searching of the best directions on each scale, the computational time of the bandelet-based method is relatively high compared with other compression methods. Since the Radon transform uses fast computational algorithm which computes pair-by-pair differences between all projections, the computational time is relatively very low when compared to existing methods.

TABLE 23.5

Computational Times Utilized by Different Compression Methods

Medical Image	Computational Time (s)							
	DCT-SPIHT	Haar-SPIHT	Contourlet-SPIHT	Curvelet-SPIHT	JPEG	Ripplet-SPIHT	Bandelet-SPIHT	Radon-SPIHT
Sagittal section of T1WI MRI lumbar spine	0.42	0.39	0.51	0.67	0.83	0.62	1.41	0.33
Coronal section of T1WI MRI leg	0.53	0.47	0.49	0.64	0.72	0.58	1.48	0.39
Axial section of MRI brain	0.56	0.55	0.58	0.58	0.76	0.62	1.27	0.41
Axial section of CT abdomen with oral contrast	0.45	0.39	0.69	0.61	0.89	0.55	1.55	0.36
Axial section of CT skull	0.53	0.42	0.47	0.69	0.75	0.69	1.38	0.33
CT scan of chest (left side) pulmonary arteries	0.44	0.51	0.55	0.72	0.79	0.67	1.64	0.42

23.8.5 Analysis of Subjective Assessment

The main objective of the image-compression method is to reproduce images with no visible distortions. In order to analyze the visual quality of compressed medical images, an extensive subjective evaluation was carried out on the following parameters, viz., visual quality of the image, representation of edges in images, image sharpness, image contrast, ability to view the difference between bone and soft tissue, ability to view the difference between air and water, and image sharpness of the soft-tissue window. The appearance of artifacts, vignetting (darkening near the corners), and visible distortion of information signal (aberration that causes straight lines to curve) were also analyzed.

The subjective assessment was carried out by a cardiac radiologist from KG Hospital, Coimbatore, India. The radiologist was requested to provide his opinion on the perception quality of the images by using a linear rating scale. The rating scale was categorized into four equal regions described with the adjectives "excellent" (extremely or desirable high quality), "good" (high quality), "average" (acceptable quality), and "poor" (poor quality). The appearance of artifact level was classified into "none," "mild," "moderate," and "severe."

The result of subjective assessment indicates that the proposed transform-based methods preserve the diagnostic quality of medical images efficiently and provide excellent representation of edges in images. Opinions recorded at the end of the evaluation revealed that most of the images were ranked as excellent and not a single image was rated with poor quality. The compression methods preserved the diagnostic quality of medical images efficiently and the differences between air and water and between bone and soft tissue were clearly visible.

23.9 Conclusions

In this chapter, three transform-based image-compression methods for medical images have been presented and analyzed. The main focus of the ripplet transform-based method is to provide high-quality compressed images by representing images at different scales and directions and to achieve a high compression ratio. The compression method using the ripplet transform provides anisotropy capability to represent singularities along arbitrarily shaped curves, and combines with the SPIHT encoder to improve the compression performance. Experimental results demonstrate that besides providing high PSNR and high directionality, the ripplet-based method outperforms DCT by 20.61%, the Haar wavelet by 17.98%, the contourlet by 11.61%, the curvelet by 12.66%, and JPEG by 25.51% in terms of compression ratio. Since the ripplet transform successively approximates images from coarse to fine resolution and is highly directional to capture the orientations of singularities, the compression method using the ripplet transform performs well on the compression of both MRI and CT images.

The main objectives of the bandelet-based method are to provide high-quality compressed images by decomposing the input image over the basis of bandelets adapted to image geometry, and to achieve a high compression ratio. The bandelet functions are regular and, hence, offer no blocking artifacts. The wavelet decomposition of geometric regularized data results in a smaller number of significant coefficients, thus yielding high compression performance. Experimental results demonstrated that besides achieving high PSNR, the bandelet-based method outperforms DCT by 11.53%, the Haar wavelet by 9.11%, the contourlet by 3.2%, the curvelet by 4.24%, and JPEG by 16.1% in terms of compression ratio.

The main focus of the Radon-based method is to utilize the Radon transform, which represents the images as a set of projections. The exact invertibility of the Radon transform prevents the appearance of artifacts, thus yielding good visual quality. Experimental results on a set of MRI and CT images demonstrated that the Radon-based method provides high PSNR and less computational time and outperforms the DCT coder by 22.96%, the Haar wavelet by 20.29%, the contourlet by 13.78%, the curvelet by 14.85%, and JPEG by 27.95% in terms of compression ratio.

23.10 Future Scope

This chapter envisions several ideas for which this work can be extended in order to improve the performance of transform-based compression methods. These compression methods can be extended for transmission of medical data over wireless networks susceptible to error with various bandwidth capacities. With the great development in telemedicine technology, mobile devices have been recently incorporated into picture archiving and communication systems in order to allow instant diagnosis by a medical expert at any time and in any place.

Most of the work on medical image compression has been focused on visualization of medical images with high compression ratio, whereas little work has been done on designing error-protection techniques to access and transmit such compressed data over wireless networks vulnerable to error. Since medical images usually depict one or more clinically relevant regions of interest over a clinically irrelevant background, the region of interest and the background may be separately coded at different qualities and then transmitted using different error-protection levels. Data compression with priority scheduling that minimizes the transmission delay for crucial and vital physiological signals can also be investigated in the future.

Acknowledgments

We would like to thank Dr. Thiruvijayan, cardiac radiologist at KG Hospital, and R. Jerald, manager of the radiology department of KG Hospital, Coimbatore, India, for providing the medical image data sets and performing subjective evaluation.

References

1. Pandian P. S., Safeer K. P., Shakunthala D. T. I., and Gopal P., "Store and forward applications in telemedicine for wireless IP based networks," *Journal of Networks* 2(6), pp. 58–65, 2007.
2. Sudhamony S., Nandakumar K., Binu P. J., and Niwas S. I., "Telemedicine and telehealth services for cancer care delivery in India," *IET Communications* 2(2), pp. 231–236, 2008.

3. Bolle S. R., Larsen F., Hagen O., and Gilbert M., "Video conferencing versus telephone calls for team work across hospitals: A qualitative study on simulated emergencies," *BMC Emergency Medicine* 9, pp. 9–22, 2009.

4. Apiletti D., Baralis E., Bruno G., and Cerquitelli T., "Real time analysis of physiological data to support medical applications," *IEEE Transactions on Information Technology in Biomedicine* 13(3), pp. 313–320, 2009.

5. Huang Y. M., Hseieh M. Y., Hung S. H., and Park J. H., "Pervasive, secure access to a hierarchical sensor based healthcare monitoring architecture in wireless heterogeneous networks," *IEEE Journal on Selected Areas in Communications* 27(4), pp. 400–411, 2009.

6. Boukerche A. and Ren Y., "A secure mobile healthcare system using trust-based multicast scheme," *IEEE Journal on Selected Areas in Communications* 27(4), pp. 387–399, 2009.

7. Villalba E., Salvi D., Ottaviano M., Peinado I., Arredondo M. T., and Akay A., "Wearable and mobile system to manage remotely heart failure," *IEEE Transactions on Information Technology in Biomedicine* 13(6), pp. 990–996, 2009.

8. Feng S., Liang Z. L., and Zhao D., "Providing telemedicine services in an infrastructure-based cognitive radio network," *IEEE Wireless Communications* 17(1), pp. 96–103, February 2010.

9. Shazia K. and Imran S. B., "Clinical decision support system based virtual telemedicine," Third International Conference on Intelligent Human–Machine Systems and Cybernetics, pp. 16–21, 2011.

10. Bairagi V. K. and Sapkal A. M., "ROI-based DICOM image compression for telemedicine," *Indian Academy of Sciences* 38(1), pp. 123–131, 2013.

11. Scholl I., Aach T., Deserno T. M., and Kuhlen T., "Challenges of medical image processing," *Computer Science—Research and Development* 26, pp. 5–13, 2011.

12. Fong B., Fong A. C. M., and Li C. K., *Telemedicine Technologies: Information Technologies for Medicine and Telehealth*, Wiley, 2011.

13. Choong M. K., Logeswaran R., and Bister M., "Cost-effective handling of digital medical images in the telemedicine environment," *International Journal of Medical Informatics* 76(9), pp. 646–654, 2007.

14. Kesavamurthy T. and Thiyagarajan K., "Lossless volumetric colour medical image compression using block based encoding," *International Journal of Medical Engineering and Informatics* 4(3), pp. 244–252, 2012.

15. Babel M., Pasteau F., Strauss C., Pelcat M., Bedat L., Blestel M., and Deforges O., "Preserving data integrity of encoded medical images: The LAR compression framework," *Advances in Reasoning-Based Image Processing Intelligent Systems*, Springer Berlin Heidelberg, pp. 91–125, 2012.

16. Sanchez V., Abugharbieh R., and Nasiopoulos P., "Symmetry-based scalable lossless compression of 3D medical image data," *IEEE Transactions on Medical Imaging* 28(7), pp. 1062–1072, 2009.

17. Bernabe G., Garcia J. M., and Gonzalez J., "A lossy 3D wavelet transform for high-quality compression of medical video," *Journal of Systems and Software* 82(3), pp. 526–534, 2009.

18. Sahba N., Tavakoli V., Behnam H., Ahmadian A., Ahmadinejad A., and Alirezaie J., "An optimized two-stage method for ultrasound breast image compression," *4th Kuala Lumpur International Conference on Biomedical Engineering* 21, pp. 515–518, 2008.

19. Seeram E., "Irreversible compression in digital radiology—A literature review," *Radiology* 12(1), pp. 45–49, 2006.

20. Clunie D., "DICOM support for compression: More than JPEG," 5th Medical Imaging Informatics and Teleradiology Conference, MIIT 2009.

21. Xu J., Yang L., and Wu D. O., "Ripplet—A new transform for image processing," *Journal of Visual Communication and Image Representation* 21(7), pp. 627–639, 2010.

22. Juliet S., Rajsingh E. B., and Ezra K., "A novel medical image compression using ripplet transform," *Journal of Real Time Image Processing*, July 2013.

23. Starck J. L., Candes E. J., and Donoho D. L., "Curvelets, multiresolution representation, and scaling laws," *IEEE Transactions on Image Processing* 11, pp. 670–684, 2000.

24. Do M. N. and Vetterli M., "The contourlet transform: An efficient directional multiresolution image representation," *IEEE Transactions on Image Processing* 14(12), pp. 2091–2106, 2005.

25. Penneca E. L. and Mallat S., "Sparse geometric image representation with bandelets," *IEEE Transactions on Image Processing* 14(4), pp. 423–438, 2005.

26. Juliet S., Rajsingh E. B., and Ezra K., "A novel image compression method for medical images using geometric regularity of image structure," *Signal, Image and Video Processing* 9(7), pp. 1691–1703, 2015.

27. Said A. and Pearlman W., "An image multiresolution representation for lossless and lossy compression," *IEEE Transactions on Image Processing* 5(9), pp. 1303–1310, 1996.

28. Chen Y. Y., "Medical image compression using DCT-based subband decomposition and modified SPIHT data organization," *International Journal of Medical Informatics* 76(10), pp. 717–725, 2007.

29. Minasyan S., Astola J., and Guevorkian D., "An image compression scheme based on parametric Haar-like transform," *IEEE International Symposium on Circuits and Systems* 3, pp. 2088–2091, 2005.

30. Eslami R. and Radha H., "Translation invariant contourlet transform and its application to image denoising," *IEEE Transactions on Image Processing* 15(11), pp. 3362–3374, 2006.

31. Iqbal M., Javed M. Y., and Qayyum U., "Curvelet-based image compression with SPIHT," International Conference on Convergence Information Technology, pp. 961–965, 2007.

32. Wallace G. K., "The JPEG still picture compression standard," *Communication of the ACM* 34, pp. 31–44, 1991.

24

Tracking the Position and Orientation of Ultrasound Probe for Image-Guided Surgical Procedures

Basem F. Yousef

CONTENTS

24.1 Introduction

Prostate cancer is the second major cause of death among men due to cancer in North America. According to the Canadian Cancer Society, about 20,000 men are diagnosed with prostate cancer annually in Canada and about 4000 die from it [1]. The various methods for the treatment of prostate cancer [2] include surgical intervention, external radiotherapy, high–dose rate brachytherapy, cryotherapy (freezing therapy), RF (heat) therapy, and low–dose rate brachytherapy. The low-dose rate brachytherapy has proven to be effective and successful and, thus, has some advantages over surgical intervention because it does not cause impotence or incontinence that is associated with surgical intervention due to damage of tissues in close proximity with the prostate. In addition, it avoids the deleterious effects of external radiation on surrounding healthy tissues (e.g., seminal vesicles, essential nerve bundles, bladder, rectum, and the urethra).

In a two-stage manually performed procedure, radioactive seeds (e.g., iodine-125 or palladium-103) are implanted permanently into the prostate to irradiate the cancerous cells over a period of time. In the first stage, volumetric assessment of the prostate gland is performed using a transrectal ultrasound (US) probe that is inserted into the patient's rectum. Then, a dose plan for radiation therapy is developed during this stage. Based on the dose plan, the intraoperative seed implantation is performed during the second stage. The procedure is performed with the patient in the lithotomy position and the ultrasound probe is placed in the rectum to monitor the manual seed placement. The success of the procedure depends heavily on the accuracy of seed implantations. Moreover, the consistency and efficiency of this treatment procedure are highly dependent on the clinicians' skills. However, there are many drawbacks and limitations associated with the current procedure due to the nature of the manually performed procedure and the tools used. Furthermore, it is often the case that the position of the prostate during the planning stage differs from its position during seed implantation, thus resulting in seed misplacement.

Several research investigators believe that the disadvantages associated with the current prostate brachytherapy procedure can be eliminated or reduced, and the clinical outcome of the procedure can be significantly improved by developing robotic systems that can autonomously or semiautonomously perform the seed-delivery stage, taking advantage of the capability of robots to perform high-precision and fatigue-free movements with accuracy that is beyond that of a surgeon. Performing image-guided robot-assisted prostate brachytherapy allows the two stages of the current procedure to be combined into one session since robot-assisted brachytherapy can provide the tools needed for dynamic (i.e., online) dose planning.

Ng et al. [4] used a modular robotic platform for seed implantation for prostate cancer. Phee et al. [5] developed a robotic system for accurate and consistent insertion of a percutaneous biopsy needle into the prostate. Wei et al. [6] used an industrial robot to test the performance of a robot-assisted 3D transrectal ultrasound-guided prostate brachytherapy. Bassan et al. [7] describe a microrobot that operates under ultrasound image guidance to perform needle insertion in prostate brachytherapy. At the Urology Robotics Laboratory at the Johns Hopkins University, several robotic systems have been developed for use in prostate brachytherapy and other procedures. For instance, the percutaneous access to the kidney (PAKY) needle driver [8] and the remote-center-of-motion (RCM) robot [9] have been developed. While the PAKY consists of a passive arm and a novel needle-insertion mechanism which can accurately introduce a needle into the kidney, the RCM allows for precise needle insertion under radiological guidance. Other urological robots developed in the laboratory are described by Stoianovici [10].

Since a majority of prostate brachytherapy treatments, manually performed or robot assisted, are carried out under US image guidance, the US probe, which is one of the primary tools needed for the procedure, is mounted on a holder, also called a stabilizer, that is used to manipulate, position, and lock the probe in place. Different types of US probe stabilizers and precision steppers have been proposed such as the stabilizer assembly mechanisms presented by Ellard and Knudsen [11], and a US probe stepper apparatus designed by Ellard [12]. Whitmore et al. also designed another form of US stepping device [13].

Moreover, real-time three-dimensional imaging apparatus and techniques have been developed in an attempt to enable performing online dynamic seed location planning during an implantation procedure. As described by Tong et al. [14], Burdette and Komandina [15], and Nelson and Pretorius [16], a 3D US image can be constructed from a series of 2D US images.

These advances in US imaging and robotic systems can be utilized to perform improved image-guided robot-assisted prostate brachytherapy, which may lead to more accurate and convenient ways of delivering the radioactive seeds, thus translating into improved clinical outcomes, reduced side effects, and reduced radiation exposure time.

However, automating image-guided therapy and registering a medical image to the patient requires knowledge of the locations of both the medical image source (e.g., ultrasound) and the robot end effector with respect to a global coordinate system that is known relative to the patient. To achieve this, it is essential to unify the coordinate systems of the robot end-effector frame and the image source (e.g., US probe) frame in order for the robot to know its target point inside the patient through the US image. For example, referring to the schematic diagram in Figure 24.1, the robot can find its target point in the goal frame {G} provided that (1) the relationship between the tool frame {T} and the station frame {S} is known (note that this can be easily obtained from the kinematic equations of the robot); (2) the goal frame {G} is known with respect to the US probe frame {P} (this can be provided by the ultrasound imaging software); and (3) frame {P} must be known with respect to frame {S}. Satisfying (3) entails the use of a US probe holder that can acquire the position and orientation of the probe.

Unfortunately, neither the available stabilizers nor the stepper apparatus can acquire this information, and, although some available steppers enable precise US probe motion and rotation [12–13], the probe's location is still unknown relative to a reference coordinate system. In addition, these devices are not sturdy enough to firmly lock in place the ultrasound probe. Therefore, such devices are unsuitable for use in robot-assisted surgeries. Furthermore, the bulky structures of the available stabilizers increase the potential for robot–robot collisions and occupy a large space in the limited work space of the brachytherapy procedures. These cumbersome and heavy devices take considerable time and effort to set up and they limit the clinician's access to the patient.

This chapter presents a novel design of a compact-sized precalibrated ultrasound probe holder that can be used to easily manipulate, position, and lock in place the probe as well as to provide its location and orientation relative to a chosen reference coordinate frame on the operating table. This enables a robot in image-guided robot-assisted prostate brachytherapy to find its target point inside the prostate. It should be noted that, although the device is discussed with reference to seed implantation for prostate brachytherapy, it is by no means limited to this procedure. Such a device has applicability in other image-guided surgical procedures such as robot-assisted cryosurgery, thermotherapy, and laser surgery. Moreover, a wide range of minimally invasive surgeries that use radiological instruments or probes, such as treating benign prostatic hyperplasia (BPH) by laser ablation or microware therapy, can also benefit from this device.

In addition, we present a practical concept to accurately track an object's position and orientation in 3D space. Our approach overcomes the disadvantages associated with the most popular tracking systems used in image-guided surgery, e.g., optical and electromagnetic trackers. Our tracker can be used in environments with metallic

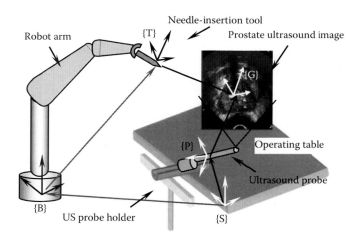

FIGURE 24.1
Schematic diagram of a robotic system for image-guided robot-assisted brachytherapy.

components, where electromagnetic trackers lose accuracy due to the distortion of the electromagnetic field. Furthermore, unlike optical tracking systems, the tracker does not require uninterrupted line of sight between the object and the tracker's sensor. Furthermore, it can be used to track the position and orientation of mechanisms of complex designs where direct installation of sensors on the joints is either difficult because of one or more ball joints or infeasible due to the complexity of the design or due to a large number of joints.

The following sections describe in detail the different components and features of the design.

24.2 Mechanism Description

The proposed US probe holder comprises two mechanisms (Figure 24.2): a sturdy stabilizer and a position and orientation tracker. Although coupled as one integrated mechanism, each mechanism can be used independently. The stabilizer joints are loosened to freely manipulate and insert the US probe into the patient's rectum, where the loose-jointed tracker follows the stabilizer and records the probe's final position and orientation through its encoders, regardless of the link arrangement and pose of the stabilizer.

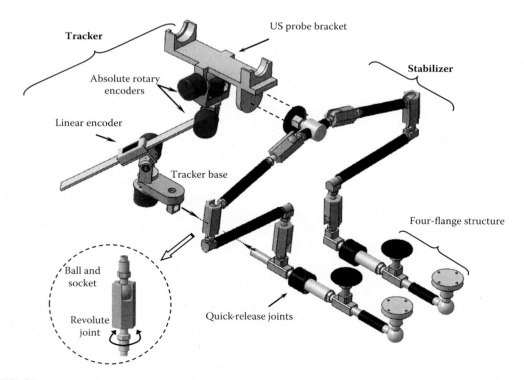

FIGURE 24.2
Computer-aided design (CAD) model of the proposed rectal ultrasound probe holder (stabilizer-tracker) mechanism.

24.2.1 Stabilizer Design

Figure 24.2 shows a CAD model of the stabilizer that consists of five links connected by six ball joints, which can be manually locked by tightening a lube knob for each of them. Each ball joint consists of a ball and socket at one end and a revolute joint at the other end, as shown in Figure 24.2. One can notice that the stabilizer is symmetric with three ball joints on each side. Although one side is sufficient to carry and lock the US probe in position, the use of two sides strengthens the structure and makes it sturdier, which is essential for robotic applications.

The stabilizer is attached to the operating table through a four-flange structure, which is permanently bolted to the operating table. Since it is desirable to be able to remove the mechanism for sterilization or to free the operating table for other procedures and to allow for accommodating different surgical tools/equipment necessary for other procedures, the stabilizer connects to the flange structure through two quick-release joints to enable attaching it to and detaching it from the operating table in a very short time, typically 1–2 min by one person. This feature is advantageous compared to other stabilizers/holders that need more than one person to set up and takes considerably longer setup time.

The US probe mounts on a specially customized bracket, which is attached to the stabilizer structure. When the lube knobs on the ball joints are loose, the clinician can easily and smoothly manipulate the position and orientation of the probe, and then can tighten it to lock the probe at the desired position. This architecture allows free motion in six degrees of freedom, thus enabling insertion of the probe into the patient's rectum freely without applying any force that may result in tearing the inner wall of the rectum or causing trauma to the patient.

24.2.2 Tracker Design

The tracker consists of lower-joint and upper-joint assemblies as shown in Figure 24.3. While the lower-joint assembly is attached to the tracker base, which is rigidly tightened to the stabilizer (Figure 24.2), the upper-joint assembly is attached to the US probe bracket.

It should be noted that the tracker base position and orientation are fixed with respect to the holder flange structure and, thus, to the operating table. The two joint assemblies are connected by a linear position sensor (encoder) that measures the distance between them.

The lower-joint assembly (Figure 24.4), which allows free rotations about two axes (C and D), is equipped with two absolute rotary encoders E1 and E2. While E1 can read the rotation angle θ_1 of part 1 with respect to part 2, E2 can read the rotation angle θ_2 of part 2 with respect to the tracker base that is fixed with respect to the table. Note that one end of the linear encoder stick is attached to part 1 through the lower stick adapter.

The upper-joint assembly (Figure 24.5), which allows free rotations about three axes (H, I, and K), is equipped with three absolute rotary encoders E3, E4, and E5. E3 reads the rotation angle θ_3 of part 3 with respect to the US bracket. In addition, while E4 reads the rotation angle θ_4 of part 4 with respect to part 3, E5 reads the rotation angle θ_5 of part 5 with respect to part 4. Since part 5 connects with the other end of the linear encoder stick, one can notice that the upper-joint assembly provides the roll–pitch–yaw orientations of the US probe bracket with respect to the linear encoder stick. However, the linear distance L_3, from the tip of the stick to the lower joint, is provided by the linear encoder.

The information acquired by the five rotary encoders and the linear encoder can be utilized to calculate the position and orientation of the US probe with respect to an arbitrary origin frame on the operating table as follows.

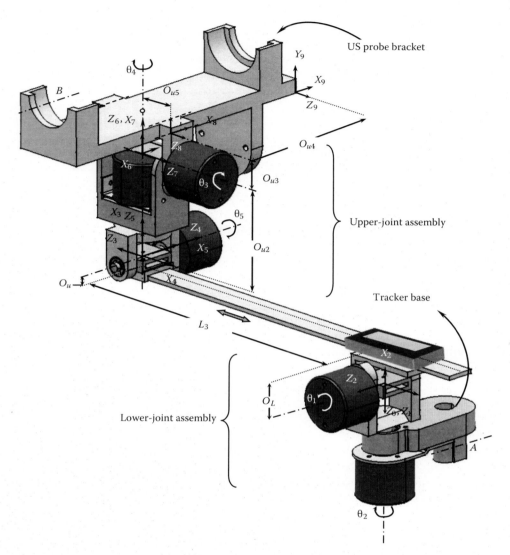

FIGURE 24.3
Tracker assembly and frame assignment.

Coordinate frames are attached to each encoder to read the rotational angles of the joints, and intermediate frames are used to extract the offset distances of the joint assemblies and part thicknesses, e.g., frames 6, 8, and 9 (Figure 24.3). The frame assignment shown in Figure 24.3 is used to generate the Denavit–Hartenberg (D-H) parameters given in Table 24.1 that are used to derive the forward kinematics equations [17] that describe the position and orientation of any frame affixed at an arbitrary point on the holder bracket with respect to the origin frame on the operating table. For example, the frame assignment in Figure 24.3 can be used to obtain the position and orientation of frame {9} with respect to frame {0}, which is known relative to a frame {origin} on the operating table.

The stabilizer gains its high dexterity from its ball joints, but, unfortunately, since it is very difficult to install position sensors on ball joints, this makes tracking the end effector

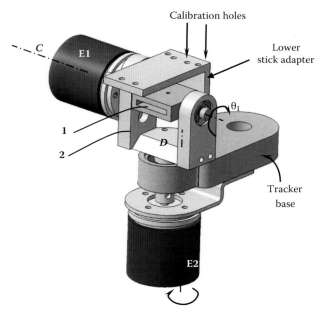

FIGURE 24.4
Tracker's lower-joint assembly.

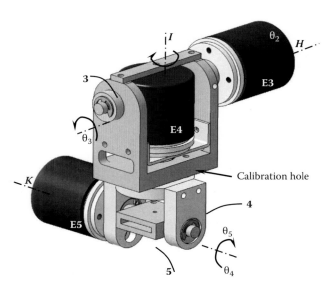

FIGURE 24.5
Tracker's upper-joint assembly.

TABLE 24.1

Denavit–Hartenberg Parameters Used for Derivation of Forward Kinematic Equations for the Tracker

i	α_{i-1}	a_{i-1}	d_i	θ_i
1	0	0	0	θ_2
2	$-\pi$	0	0	$\pi + \theta_1$
3	$-\pi$	$O_L + O_u$	L_3	0
4	$-\pi$	0	0	$\pi + \theta_5$
5	π	0	0	$\pi - \theta_4$
6	0	0	O_{u2}	0
7	π	0	0	$\pi - \theta_3$
8	0	O_{u3}	O_{u5}	$-\pi$
9	0	O_{u4}	0	0

of any mechanism that adopts ball joints extremely difficult, thus making these mechanisms useless for the applications where knowledge of the end-effector position and/or orientation is needed. The tracker presented in this chapter can be used not only with our stabilizer but also with any mechanism/stabilizer of complex design provided that precautions are taken to avoid collisions between the two mechanisms.

24.3 Tracker Calibration

Since the tracker base frame is fixed with respect to the operating table, the tracker sensors are calibrated with respect to the base frame. The flange structure should be bolted to the operating table such that the tracker base upper and side surfaces are parallel to those of the operating table. "Calibration" denotes mastering the mechanism to zero encoder values. In other words, first-time calibration of the tracker requires knowledge of the encoder readings that correspond to the zero angles according to the D-H parameters in Table 24.1; these readings can be obtained by the following:

1. Locking the tracker firmly in the pose depicted in Figure 24.3, i.e.,
   a. The linear encoder stick and the US probe bracket must be parallel to the tracker base.
   b. Axis B and the θ_1 or θ_5 axes must be parallel to axis A (Figure 24.3).
   c. The θ_4 axis must be perpendicular to the US probe bracket.
2. Obtaining the readings of the encoder that correspond to this tracker pose and using them as the encoder's zero values.

In order to achieve the first requirement, two precision *through-all* holes (ϕ 4 mm) are machined in each of the lower- and upper-joint assemblies with each of the holes passing through all of the components of the joint assembly such that when precision calibration shafts with the same diameter are tapped into the holes, they accurately lock the tracker in the desired pose, allowing the encoder readings to be obtained such that they correspond to the zero position as per the D-H parameters (Table 24.1). After the zero-position encoder readings are obtained, the calibration shafts are removed permanently.

24.4 Materials and Dimensions

To reduce the weight of the holder assembly, the tracker brackets and the holder links are made of aluminum, which resists corrosion as well. Furthermore, the ball joints are made of brass, which is known for its high corrosion resistance.

Since "homing" a six-DoF passive-jointed mechanism is very difficult and requires special software and hardware arrangement, we used absolute encoders at all joints, thus allowing the system to be started without additional precalibration steps after the first calibration procedure as described previously. The rotary sensors are 12-bit rotary encoders and the encoder at the prismatic joint is a linear digital encoder of 0.01 mm accuracy. This encoder is selected with a special output port that enables acquiring the reading to the computer through a special cable and data-processing unit to convert the encoder reading to any data format readable by the computer for further computations or processing.

The dimensions of the links are chosen so that the maximum width of the holder does not exceed the width of most standard operating tables, which is 60 mm. This frees the sides of the operating table and allows the accommodation of other equipment that is necessary for the procedure such as the patient's leg supports (Figure 24.6), which are needed for the prostate brachytherapy procedure. It also allows the use of a specially designed robot for the procedure [18–19] which extends from the side of the operating table.

To verify the machining tolerances and accuracy, the dimensions and offset distances of all tracker parts were cross-checked using a 0.01 mm accurate digital caliper. The mechanism's work envelope corresponds to 250 mm along the width of the table, 200 mm above the table, and 230 mm along the length of the table.

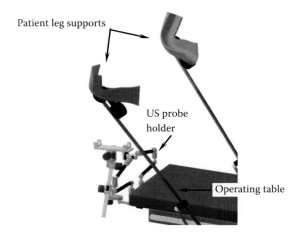

FIGURE 24.6
Probe holder design and dimensions allow the accommodation of other mechanisms necessary for prostate brachytherapy, such as leg supports and other robot arms that may extend from the sides.

24.5 Validation and Results

In order to quantify the displacement and angular accuracies of the integrated stabilizer-tracker assembly, the US probe bracket was moved and locked at arbitrary positions and at each position, the actual location and orientation of the bracket were compared against the corresponding theoretical location and orientation.

Referring to Figure 24.7a, which shows a picture of the actual constructed prototype of the stabilizer-tracker system, in order to obtain the actual location of any point on the bracket, the holder joints are loosened to enable dragging the bracket smoothly and then locking the holder at an arbitrary pose i. Using a three-axis stage which has an accuracy of 2 μm, the coordinates (x_{ij}, y_{ij}, z_{ij}) of three selected points, typically corners, on the bracket are measured relative to a chosen origin on the operating table (Figure 24.7a and b), where $i = 1, 2, \ldots, n$ denote pose points and $j = 1, 2, 3$ are arbitrary points/

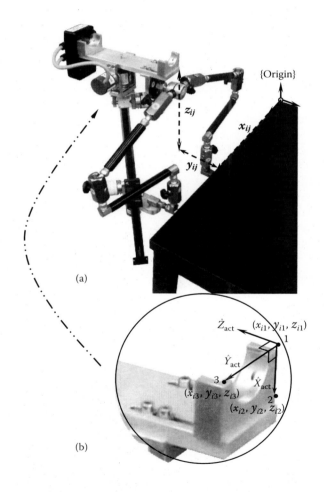

FIGURE 24.7

(a) The actual constructed prototype of the ultrasound probe holder positioned at a target point i and the {origin} reference frame is attached to the table. (b) For each target point i, the coordinates of three arbitrary points were determined and used to define a frame {act} that describes the actual orientation of the end effector.

corners on the bracket. The three points are chosen so that the two edges/lines connecting between them form a right angle. This is because these points are used to define a coordinate frame {act} that is affixed to the bracket and is used to define the bracket's actual orientation.

On the other hand, the corresponding theoretical values can be obtained as follows: After reading the joint angles from the encoders, and using the link frame assignment shown in Figure 24.3 and the D-H parameters (Table 24.1), it is easy to derive both the forward kinematics equations [17] that provide the theoretical locations of the selected points on the bracket based on the encoder angles, and the rotation matrix which can be used to describe the orientation of a frame {theo} that corresponds to the coordinate frame {act}.

When the US probe bracket is at pose i, the displacement error E_i is calculated as

$$E_i = \sqrt{\left(\overline{\Delta x_i}\right)^2 + \left(\overline{\Delta y_i}\right)^2 + \left(\overline{\Delta z_i}\right)^2}, \tag{24.1}$$

where $\left(\overline{\Delta x_i}\right)^2 = \dfrac{1}{3}\displaystyle\sum_{j=1}^{3}(x_{ij} - x_{ij\,\text{theo}})$ is the mean of the error in the x direction for the three selected points on the bracket at pose i; x_{ij} is the x coordinate of point j on the bracket at pose i obtained using the three-axis stage; $x_{ij\,\text{theo}}$ is the x coordinate of point j on the bracket at pose i obtained using the forward kinematics equations; and $\left(\overline{\Delta y_i}\right)^2$ and $\left(\overline{\Delta z_i}\right)^2$ are calculated similarly.

The mean error $\overline{E}_t$ for the US probe holder tested at $n = 8$ locations/poses was calculated using

$$\overline{E}_t = \frac{1}{n}\sum_{i=1}^{n} E_i. \tag{24.2}$$

The data acquired as described above were used to quantify the angular error as follows:

1. Frame {act} is attached to the bracket with the origin at point 1, where the x axis $\hat{X}_{\text{act}}$ points in the direction from point 1 to 2 and the y axis $\hat{Y}_{\text{act}}$ points from point 1 to 3. Since points 1, 2, and 3 are selected so that $\hat{X}_{\text{act}}$ and $\hat{Y}_{\text{act}}$ are at right angles, the z-axis direction $\hat{Z}_{\text{act}}$ can be obtained from the cross product of the vectors defining $\hat{X}_{\text{act}}$ and $\hat{Y}_{\text{act}}$ (Figure 24.7b), i.e.,

$$\hat{Z}_{\text{act}} = \hat{X}_{\text{act}} \times \hat{Y}_{\text{act}}. \tag{24.3}$$

Similarly, an imaginary frame {theo} is formed using the theoretical locations of points 1, 2, and 3, which are obtained using the forward kinematics equations [17]. Thus, the axes $\hat{X}_{\text{theo}}$, $\hat{Y}_{\text{theo}}$, and $\hat{Z}_{\text{theo}}$ of the frame {theo} are defined in the same way {act} is defined. It should be noted that frames {act} and {theo} define the actual and the theoretical orientations, respectively, of the US probe bracket at an arbitrary pose i.

2. The 3×3 rotation matrices $_{act}^{o}R$ and $_{theo}^{o}R$ that describe the orientations of frames {act} and {theo}, respectively, with respect to the reference frame {origin} can be obtained by

$$_{act}^{o}R = \left[\hat{X}_{act} \quad \hat{Y}_{act} \quad \hat{Z}_{act} \right], \tag{24.4}$$

$$_{theo}^{o}R = \left[\hat{X}_{theo} \quad \hat{Y}_{theo} \quad \hat{Z}_{theo} \right]. \tag{24.5}$$

Accordingly, the rotation matrix $_{act}^{theo}R$ that describes the orientation of frame {act} with respect to frame {theo} is given by

$$_{act}^{theo}R = {}_{theo}^{o}R^{-1} {}_{act}^{o}R. \tag{24.6}$$

3. There are many methods [17] that can be used to quantify the angular error of frame {act} with respect to frame {theo}. We used the "X-Y-Z fixed angles" method, sometimes called the "roll–pitch–yaw angles" method, in which frame {theo} is fixed and frame {act} is rotated about $\hat{X}_{theo}$ by an angle γ, then about $\hat{Y}_{theo}$ by an angle β, and finally about $\hat{Z}_{theo}$ by an angle α. In order to obtain the angles α, β, and γ, let the elements of the rotation matrix $_{act}^{theo}R$ obtained in Equation 24.6 be

$$_{act}^{theo}R = \begin{bmatrix} r_{11} & r_{12} & r_{13} \\ r_{21} & r_{22} & r_{23} \\ r_{31} & r_{32} & r_{33} \end{bmatrix}; \tag{24.7}$$

then the angles can be calculated by [17]

$$\beta = \operatorname{atan} 2\left(-r_{31}, \sqrt{r_{11}^2 + r_{21}^2}\right), \tag{24.8}$$

$$\alpha = \operatorname{atan} 2 \left(r_{21}/\cos(\beta), r_{11}/\cos(\beta)\right), \tag{24.9}$$

$$\gamma = \operatorname{atan} 2 \left(r_{32}/\cos(\beta), r_{33}/\cos(\beta)\right). \tag{24.10}$$

The US probe holder was tested at eight poses of different positions and orientations. Table 24.2 shows that the tracker can perform with high accuracy given by a mean displacement error $E_t = 0.66$ mm and mean angular errors $\bar{\alpha} = 0.24°$, $\bar{\beta} = 0.38°$, and $\bar{\gamma} = 0.19°$.

24.6 Conclusions

An integrated stabilizer-tracker mechanism has been designed to perform as a holder that can be used to carry, manipulate, lock in place, and accurately acquire the position and

TABLE 24.2

Mean Displacement Error and Angular Errors

i	$\overline{\Delta x_i}$	$\overline{\Delta y_i}$	$\overline{\Delta z_i}$	E_i (mm)	α°	β°	γ°
1	0.14	0.15	0.13	0.24	0.31	0.62	0.28
2	0.43	0.24	0.28	0.57	0.22	0.05	0.21
3	0.63	0.21	0.26	0.71	0.13	0.55	0.25
4	0.61	0.35	0.28	0.76	0.52	0.07	0.03
5	0.10	0.80	0.48	0.94	0.37	0.55	0.06
6	0.30	0.48	0.29	0.64	0.06	0.59	0.23
7	0.24	0.51	0.05	0.57	0.22	0.20	0.23
8	0.12	0.71	0.43	0.84	0.07	0.41	0.20
Mean				0.66	0.24	0.38	0.19

orientation of an ultrasound probe for use in prostate brachytherapy procedures. A reliable validation technique using forward kinematics was used to evaluate the performance of the holder. The improved sturdiness demonstrated by the compact-sized stabilizer and the high accuracy of the tracking mechanism makes the integrated holder mechanism well suited for use in image-guided robot-assisted prostate brachytherapy [20]. It is anticipated that this will lead to improvement in accuracy and clinical outcomes for the procedure. The novel tracker can also be used to acquire the positions and orientations of other passive mechanisms of complex designs.

Acknowledgments

This research was supported by the Natural Sciences and Engineering Research Council (NSERC) of Canada under the Collaborative Health Research Project Grant No. 262583-2003.

References

1. Canadian Cancer Society, 2006, "Prostate Cancer Stats," http://www.cancer.ca/ccs/internet/standard/0,3182,3172_14471_371299_langIden,00.html, accessed on August 26, 2014.
2. Townsend, M., Beauchamp, R., Evers, B., and Matox, K., *Sabiston Text Book of Surgery*, 16th edition, W.B. Saunders Company, Philadelphia, pp. 1673–1681, 2001.
3. Whitemore, III, W., Barzell, W., and Wilson, R., "Omni-Directional Precision Instrument Platform," U.S. Patent No. 5961527, 1999.
4. Ng, W. S., Chung, V. R., Vasan, S., and Lim, P., "Robotic Radiation Seed Implantation for Prostatic Cancer," *Proceedings of the 18th Annual International Conference of the IEEE Engineering in Medicine and Biology Society*, **1**, Amsterdam, pp. 231–233, 1996.
5. Phee, L., Xiao, D., Yuen, J., Chan, C. F., Ho, H., Thng, C. H., Cheng, C., and Ng, W. S., "Ultrasound Guided Robotic System for Transperineal Biopsy of the Prostate," *Proceedings of IEEE International Conference on Robotics and Automation*, Barcelona, pp. 1315–1320, 2005.
6. Wei, Z., Wan, G., Gardi, L., Mills, G., Downey, D., and Fenster, A., "Robot-Assisted 3D-TRUS Guided Prostate Brachytherapy: System Integration and Validation," *J. Med. Phys.*, **31** (3), pp. 539–548, 2004.

7. Bassan, H., Patel, R. V., and Moallem, M., "A Novel Manipulator for 3D Ultrasound Guided Percutaneous Needle Insertion," *IEEE International Conference on Robotics and Automation*, Rome, pp. 617–622, 2007.

8. Cadeddu, J., Stoianovici, D., Chen, R. N., Moore, R. G., and Kavoussi, L. R., "Stereotactic Mechanical Percutaneous Renal Access," *J. Endourol.*, **12** (2), pp. 121–126, 1998.

9. Stoianovici, D., Cadeddu, J. A., Whitcomb, L. L., Taylor, R. H., and Kavoussi, L. R., "A Robotic System for Precise Percutaneous Needle Insertion," *Thirteenth Annual Meeting of the Society for Urology and Engineering*, San Diego, 1998.

10. Stoianovici, D., "URobotics—Urology Robotics at Johns Hopkins," *Comput. Aided Surg.*, **6** (6), pp. 360–369, 2001.

11. Ellard, T., and Knudsen, S., "Stabilizer Assembly for Stepper Apparatus and Ultrasound Probe," U.S. Patent No. 6179262, 2001.

12. Ellard, T., "Stepper Apparatus for Use in the Imaging/Treatment of Internal Organs Using an Ultrasound Probe," U.S. Patent No. 5871448, 1999.

13. Whitemore, III, W., Barzell W., and Wilson, R., "Ultrasound Probe Support and Stepping Device," U.S. Patent No. 5931786, 1999.

14. Tong, S., Downey, D., Cardinal H., and Fenster, A., "A Three-Dimensional Ultrasound Prostate Imaging System," *Ultrasound Med. Biol.*, **22** (6), pp. 735–746, 1996.

15. Burdette, E., and Komandina, B., "Radiation Therapy and Real Time Imaging of a Patient Treatment Region," U.S. Patent No. 6512942, 2003.

16. Nelson, T., and Pretorius, D., "Three-Dimensional Ultrasound Imaging," *Ultrasound Med. Biol.*, **24**, pp. 1243–70, 1998.

17. Craig, J. J., *Introduction to Robotics: Mechanics and Control*, Third edition, Pearson Education, Upper Saddle River, 2005.

18. Yousef, B., Patel, R. V., and Moallem, M., "Macro-Robot Manipulator for Medical Applications," *IEEE International Conference on Systems, Man and Cybernetics*, Taipei, pp. 530–535, 2006.

19. Yousef, B., *Design of a Robot Manipulator and an Ultrasound Probe Holder for Medical Applications* [PhD Thesis]. The University of Western Ontario, London, Ontario, Canada, 2007.

20. Yousef, B., Patel, R. V., and Moallem, M., "An Ultrasound Probe Holder for Image-Guided Robot-Assisted Prostate Brachytherapy," *IEEE International Conference on Robotics and Automation*, Rome, pp. 232–237, 2007.

25

Biokinematics for Mobility: Theory, Sensors, and Wireless Measurements

Atila Yilmaz and Tuna Orhanli

CONTENTS

25.1 Introduction

Wireless and remote monitoring of human movements find many applications and is an active research area including in medical robotics; remote activity pattern tracking of the young as well as the elderly; and remote monitoring of the movements of patients with limited mobility or diagnoses of back and neck injuries, for patients suffering from identified problems such as mental illness, coronary-related sicknesses, circadian rhythms, sleep apnea, autism, and arterial diseases [1]. As far as kinematic analyses are concerned, there are many approaches which employ wearable and mobile devices capable of providing quantitative results with an acceptable degree of accuracy. Such approaches are highly dependable on wireless technologies together with appropriate sensors. Wireless sensor networks (WSNs) play an active part in these systems for collecting and analyzing data unobtrusively in more natural environments.

Human movement is a result of complex and highly coordinated mechanical interaction between bones, muscles, ligaments, and joints using nerve synchronization signals. It has always been a field of interest to understand mobility disorders resulting from obesity, stroke, chronic pulmonary disease, multiple sclerosis, and Parkinson's disease [1], life quality parameters like gait analysis and assessments for osteoarthritis patients [2], etc.

In this chapter, current studies in the frame of human movement that use modern technology-based tools and engineering approaches are discussed and the results of these studies are reported supported by some examples. The importance of kinematics for wireless application will be introduced first and some of the underlying mathematical principles on biokinematics will be highlighted in detail. A review of measuring and analysis methods will shortly be introduced together with examples of measurement-related applications.

25.1.1 General Review

Problems and efforts in biokinematics research are combined within a frame termed *kinematics* in order to describe and interpret the common underlying principles of *motion* to refer to moving parts (either body extremities or ridged mechanics attached to body extremities). In this respect, the terms and tools defined reflect the way of developing only the geometric displacement of motions specifically observed in medical science and clinical applications related to anatomy and (muscle) physiology. Kinematic variables considered mostly cover linear and angular displacements, velocities, and accelerations.

Kinematics-based knowledge is used widely in medicine and biology. Since the knowledge of accurate kinematics of natural human joints is essential for understanding the functionality and underlying dynamics for many clinical applications, accurate measurements of the kinematics of skeletal motion is essential. For example, kinematic analysis and measurements of human joints by using CT images that are processed or compared with gold-standard Roentgen Stereophotogrammetric Analysis (RSA) is one of most active research areas [2–5].

Considering gaits, for example, measurement and analysis techniques aim to produce reliable information on flexion/extension, abduction/adduction, and external knee rotation angles in a comparative manner in the form of less computational and more accurate presentation of data.

Common tools developed for these studies provide an analysis perspective by considering the independent generalized coordinates for description of rotation and translation motions and transformations utilizing Cardan angles [6]. Such examples of concept of biokinematics can be extended easily to many other applications, including bioinspired swimming robots like snakes, jellyfish, and other fish models, and dynamics of bio-inspired swimming systems using different kinematic metrics [7–8]. Each kinematic analysis begins with tracking and/or recording activities reliably and accurately. Kinematic measurements commonly apply two different methods: the *direct measurements,* and the *image-based measurements* [9].

Some measurements require all streams of kinematic data from direct and image-based systems in order to get instant and quality measurements for, say, evaluating a patient's gait dysfunction, classifying the severity of the patient's disability, predicting future status, determining the need for adaptive or supportive equipment, and assessing the effect of interventions. Among many others, the GaitMat II (GM) is an example of commercially available options for special gait measures in clinical studies [10–11]. However, there still is a need for a standard measurement system for simple events like calculating ankle or knee angles. Additional information about capturing the positions of the interested part of a skeletal system is found in a series of valuable reviews referring to stereophotogrammetric methods performed using markers or body sensors attached to the skin [12–15]. Supporting measurements such as EMG, oxygen consumption, and foot pressure are also measured in order to assess gait attributes.

Some studies present the design and implementation of a system that monitors the physical activities of a human user wearing a portable device equipped with inertial sensors. The common basis for this type of system is inertial sensor data for an automatic classification of human activities. Such dedicated portable devices are commonly used for both recording sensory information and remote application platform with a wireless communication interface. There are various physical performance metrics like activity type, number of steps, activity period, and the amount of energy expended during the action [16–19]. Those studies are particularly important in the context of home care for elderly people or monitoring teen time-use for parental concerns, as well as in activities related to obesity analysis. More specialized activity tracking and kinematic derivations are performed through motion-capture systems for motion analysis of various patient groups, prosthesis and orthosis designs, and sports and exercises to address leisure and clinical issues [20].

Robotics and wearable technologies are increasingly playing an important role in neuro-rehabilitation of people with nervous system injuries, i.e., stroke, traumatic brain injuries, or spinal cord injuries. It is obvious that an inability to perform even basic functional tasks with their limbs poses many difficulties for an autonomous daily life. There are different therapeutic measures for improvement and recovery of functional and motor capabilities. Robotic rehabilitation, part of which considers using an exoskeleton, does support the human joints. They are expected to be wearable comfortably and comply with safety regulations. Recent studies presented a high degree of freedom (high-DoF) model of the human arm kinematics as well as development, test, and optimization stages of the kinematic structure of a human arm interfacing an exoskeleton. Similar devices can interact together or separately with motions in upper and lower extremities [21–27]. Physical human–robot interaction at the joint level has problematic controlling and mechanical design issues which arise predominantly from kinematic complexity and variability of the muscle–skeleton system [21,28]. Self-alignment mechanism for wearable robots which affects position and orientation of the axes of any human articulation requires special attention on compensation for undesired effects of soft tissue (i.e., skin and muscle) deformations and inter- and intrasubject variabilities [21,28–30].

Kinematic models are also used in designs of smart hand and hand prostheses in order to track independently the motion in each degree of freedom. For example, fingers as a part of serial kinematic chain are composed of revolute joints. Different motions are considered, such as abduction/adduction and flexion/extension, together with pronation/supination of the wrist and fingers for the better manipulation and grasping motion of the hand [31–33]. Most of actual motion-tracking models represent the joint as a spherical joint, that is, an arbitrary rotation matrix located at a point. Even implanted parts like knee prostheses take advantage of sensor technology and kinematics measurements just to monitor the function of the knee in daily routines [34].

25.1.2 Basic Definitions

Kinematics apply classical mechanical methods that describes the motion of bodies. Kinematic variables or measurements specify the motion without consideration of the causes [35]. Many studies from biomechanics to astrophysics use kinematic variables for mainly determining the geometry of motion [36].

The discipline of biomechanics, a branch of kinematics, can be used in study of movement, and at a more general level, *biomechanics* can be defined as "the science involving the study of biological systems from a mechanical perspective" [37]. The two terms *kinesiology* and *biomechanics* might be also confusing in some sense. Biomechanics is related to studies of mechanical processes of human movement as a branch of the larger field of kinesiology.

Kinematics can simply be divided into two categories: linear motion and angular motion. Linear kinematics is the study of motion in terms of displacement, velocity, and acceleration in a linear manner, whereas angular kinematics is the study of rotational motion. Studying of a runner can be given as an example of linear kinematics and a swing in gymnastics in the Olympics is the example of angular kinematics [38].

In a kinematic model, different link and joint structures are used to describe the movement of mechanical frame. All the joints are represented by a z_i axis. Generally, the axis of a revolute joint (θ_i) or a prismatic joint (x_i) is denoted by z_i if the joint connects the links i and $i + 1$ as shown in Figure 25.1 [39]. If we have a revolute (rotary motion) joint, we rotate about θ_i, and if we have a prismatic joint, we translate along z_i by x_i. Thus, the joint variables (θ_i and x_i) represent the relative displacements between concatenated links. Any kinematic representation of a system concerned should cover a set of values for the joint variables in a vector denoted by q that may take the values of either θ or x. The number of joints determines the number (n) called degrees of freedom and gives the minimum number of parameters to configure the system with joint variables in a vector:

$$q_i = [q_1, \ldots, q_n]^T. \tag{25.1}$$

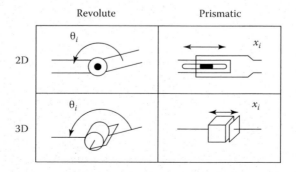

FIGURE 25.1

Types of joints and their symbolic representations. (Spong, M.W., Hutchinson, S., and Vidyasagar, M.: *Robot Modeling and Control*. 2005. Copyright Wiley-VCH Verlag GmbH & Co. KGaA. Reproduced with permission.)

The revolute joint descriptive parameter is angle of rotation and the prismatic joint descriptive parameter is displacement, where each parameter is denoted by q_i, defined as

$$q_i = \begin{cases} \theta_i & \text{if joint } i \text{ is revolute,} \\ x_i & \text{if joint } i \text{ is prismatic.} \end{cases} \tag{25.2}$$

Rotational and translational motions defined between two rigid bodies (femur and tibia bones) and necessary three-dimensional joint coordinate system can be seen be in Figure 25.2.

The coordinate system shown in Figure 25.2 is suitable for knee-joint motions and other knee-joint models.

25.1.3 Anatomical Reference System

Understanding and interpreting the kinematics of human movement requires specialized terminology that precisely identifies the body positions and directions. For this purpose, anatomical reference axes have been identified and standardized. The anatomical reference position is an erect standing position with the feet slightly separated and the arms hanging relaxed at the sides, with the palms of the hands facing forward [40–41]. The three imaginary cardinal planes bisect the mass of the body in three dimensions, depicted in Figure 25.3 [42]. A plane is a two-dimensional surface with an orientation defined by the spatial coordinates of three discrete points not all contained on the same line. It may be thought of as an imaginary flat surface. The sagittal plane divides the body vertically

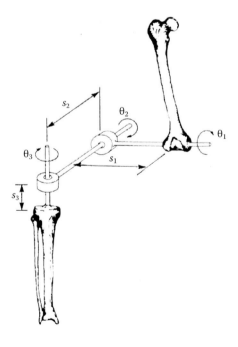

FIGURE 25.2
Joint angles are defined by rotations by occurring about the three joints coordinate axes. (From Grood, E.S., and Suntay, W.J., *J. Biomech. Eng.*, 105, pp. 136–144, 1983.)

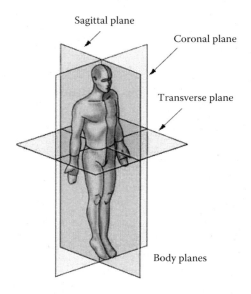

FIGURE 25.3
Sagittal plane, coronal plane, and transverse plane. (From Harmony Medical, *Joint Anatomy and Basic Biomechanics*, n.d.; Winter, D.A., *Biomechanics and Motor Control of Human Movement*, 4th ed., John Wiley & Sons, Hoboken, New Jersey, 2009.)

into left and right halves, with each half containing the same mass. The lateral plane, also referred to as the coronal plane, splits the body vertically into front and back halves with equal mass. The transverse plane separates the body into top and bottom halves of equal mass. For an individual standing in anatomical reference position, the three cardinal planes all intersect at a single point known as the body's center of mass or center of gravity.

25.2 Types of Kinematics

The complexity of the human body in standard kinematic analysis studies demands a simplified human skeleton model that is represented by links and connected joints [43]. The DoF defines the complexity of this model in which one DoF refers to one rotational axis. The human body can be represented by a hierarchical top–down skeleton model as presented in Figure 25.4a and the kinematic tree is given in Figure 25.4b [44].

25.2.1 Forward Kinematics

Forward kinematics is a relatively easier task in which motion parameters, like position and orientation of moving parts, can be determined in a specified time course. The motion of the end effector in robots (or the arm or the hand in biokinematics) is determined by indirect manipulations of transformations in root adjoints (or rotations at shoulder and elbow joints). By this method, it is possible to determine the cumulative effect of the entire set of joint variables defined in general complicated nonlinear functions. Once the orientation matrix of each joint is calculated, the final position of the end effector can be

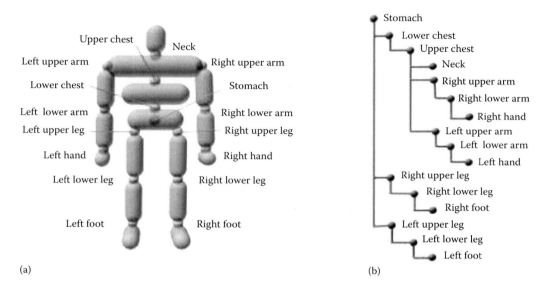

(a) (b)

FIGURE 25.4
(a) Human skeleton model and (b) kinematic tree. (From Filkorn, R., and Kocan, M., Simulation of human body kinematics, http://www.cescg.org/CESCG-2000/RFilkorn/.)

determined by using each transformation matrix in defined hierarchies [45]. For example, in the case of leg motion, given the rotation angles of hip and knee joints, the position of the foot can be calculated by using concatenated matrices with position and rotation information of the hip and the knee.

In order to work on the articulated structure, a matrix representation should be defined. Sims and Zeltzer proposed an alternative method, called axis–position (AP) representation [46]. In this model, the position of the joint, the orientation of the joint, and the pointers to the nodes are defined to represent the structure.

In order to explain the basics, arm architecture is given as an example in Figure 25.5 to calculate the end-effector position for the associated joint parameters. In Figure 25.5, the coordinate variables are defined by the pairs (x_0, y_0), (x_1, y_1), and (x_2, y_2) for each joint,

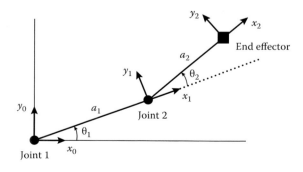

FIGURE 25.5
Articulated arm model for implementing forward-kinematics equations.

whereas θ_1 and θ_2 represent the joint angles between the joint position and the previous x axis in order to describe the positions of joints in the coordinate frames. In addition, a_1 and a_2 are the lengths of the links, respectively.

Coordinate frames, presented in Figure 25.5, define the possible moves of the links that can be modeled by a set of mathematical descriptions as orientation calculations and a rotation matrix. The end-effector position (x, y) for the coordinate frame $(0_2, x_2, y_2, z_2)$ is found by the equation

$$x = x_2 = a_1 \cos \theta_1 + a_2 \cos(\theta_1 + \theta_2),$$
$$y = y_2 = a_1 \sin \theta_1 + a_2 \sin(\theta_1 + \theta_2). \tag{25.3}$$

Considering $(0_0, x_0, y_0, z_0)$ as a base frame, the orientation of the end-effector frame relative to the base frame can be found as

$$\begin{bmatrix} x_2 \cdot x_0 & y_2 \cdot x_0 \\ x_2 \cdot y_0 & y_2 \cdot y_0 \end{bmatrix} = \begin{bmatrix} \cos(\theta_1 + \theta_2) & -\sin(\theta_1 + \theta_2) \\ \sin(\theta_1 + \theta_2) & \cos(\theta_1 + \theta_2) \end{bmatrix}. \tag{25.4}$$

The coordinate frames used more in "homogenous transformation" are important in kinematic analysis. A coordinate frame is defined for every different link rigidly. For the link i, the coordinate frame $(0_i, x_i, y_i, z_i)$ is employed. The Denavit–Hartenberg convention allows us to apply easy transformations for each joint under coordinate frames. Homogenous coordinates and homogenous transformations are used in order to simplify the kinematic relations among coordinate frames [39].

25.2.2 Inverse Kinematics

As mentioned previously, the main task is to determine the values of joint variables based on predefined position and the orientation of the end effector. It sometimes is referred to as goal-directed process and is known to be more difficult to solve due to inherent nonlinearity and the possibility of having more than one solution. In a similar structure, the end-effector coordinates x and y can be determined with respect to given joint angles θ_1 and θ_2 as seen in Figure 25.6. The angles θ_2 and θ_1 can be calculated using basic trigonometry as

$$\theta_2 = \tan^{-1} \frac{\pm\sqrt{1 - D^2}}{D}, \tag{25.5}$$

$$\theta_1 = \tan^{-1}\left(\frac{y}{x}\right) - \tan^{-1}\left(\frac{\alpha_2 \sin \theta_2}{\alpha_1 + \alpha_2 \cos \theta_2}\right). \tag{25.6}$$

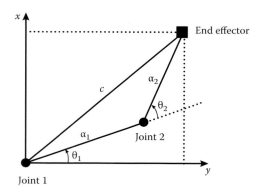

FIGURE 25.6
Articulated arm model for implementing inverse-kinematics equations.

25.2.3 Joint Velocity Kinematics

The velocity relationships are generally given by the Jacobian matrix. It plays an important role in almost every aspect of robotic studies, from planning of trajectories to the derivation of the dynamic equations of motion. It is also required for tracking a contour at constant velocity. There are different approaches to deriving the Jacobian of manipulators, for detailed information referring to the text on cross-product form is strongly recommended [47]. Clearly, it is possible to represent velocities for x and y by differentiating Equation 25.3, as

$$\dot{x} = -\alpha_1 \sin\theta_1 \cdot \dot{\theta}_1 - \alpha_2 \sin(\theta_1 + \theta_2)(\dot{\theta}_1 + \dot{\theta}_2), \tag{25.7}$$

$$\dot{y} = \alpha_1 \cos\theta_1 \cdot \dot{\theta}_1 + \alpha_2 \cos(\theta_1 + \theta_2)(\dot{\theta}_1 + \dot{\theta}_2). \tag{25.8}$$

These equations can be reformed by using vector notation:

$$x = \begin{bmatrix} x \\ y \end{bmatrix}, \quad \boldsymbol{\theta} = \begin{bmatrix} \theta_1 \\ \theta_2 \end{bmatrix},$$

$$\dot{x} = \begin{bmatrix} -\alpha_1 \sin\theta_1 - \alpha_2 \sin(\theta_1 + \theta_2) & -\alpha_2 \sin(\theta_1 + \theta_2) \\ \alpha_1 \cos\theta_1 + \alpha_2 \cos(\theta_1 + \theta_2) & \alpha_2 \cos(\theta_1 + \theta_2) \end{bmatrix} \dot{\boldsymbol{\theta}},$$

or

$$\dot{x} = J \cdot \dot{\boldsymbol{\theta}}, \tag{25.9}$$

where the matrix J is called the Jacobian matrix of the manipulator, which is a quite fundamental concept for this theory.

End-effector velocities give the user a linear relationship between end-effector velocity and joint velocity. Thus, the joint velocities can be found by using the relation

$$\dot{\boldsymbol{\theta}} = J^{-1}\dot{x}, \qquad (25.10)$$

where J^{-1} is the inverse Jacobian,

$$J^{-1} = \frac{1}{\alpha_1 \alpha_2 \sin\theta_2} \begin{bmatrix} \alpha_2 \cos(\theta_1 + \theta_2) & \alpha_2 \sin(\theta_1 + \theta_2) \\ -\alpha_1 \cos(\theta_1) - \alpha_2 \cos(\theta_1 + \theta_2) & -\alpha_1 \sin(\theta_1) - \alpha_2 \sin(\theta_1 + \theta_2) \end{bmatrix}. \qquad (25.11)$$

In order to simplify all operations that can be drawn for each coordinate frame, homogeneous transformation matrices can be used to combine those to represent all quantities from different coordinate frames [39].

25.3 Measurements of Human Motion Kinematics

Monitoring physical activity with determining of associated kinematic parameters is important since, as mentioned before, many researches in this area directly describe links between physical activities and overall health conditions. Dedicated systems and solutions may vary from inexpensive devices, such as pedometers counting only the number of steps during selected periods, to more complex choices ones pact with accelerometers and gyroscopes positioned on selected points on the body. Many modalities utilizing images are also available to assess the movements. As a result, some devices include observation and physical science technologies (foot switches, gait mats, force plates, and optical motion analysis) and others evaluate even simple diaries and questionnaires, but all serve the same task of kinematic motion measurements.

Kinematic motion measurements aim monitor information streams concerning time courses of measured variables used also for other purposes like kinetic or myographic data expression. In Figure 25.7, we have a general view of the present classification status of measurement systems in this field. Resistive, inertial, electromagnetic, and image-based measurement systems are very active research areas to track and verify kinematic derivations. Motion-measurement systems in general can be categorized into two major classes: direct measurements and image-based measurements that will be explained next.

25.3.1 Image-Based Measurement Techniques

Image-based measurement systems employ optical tracking to follow the displacement of markers placed at particular anatomical sites on the body links and segments. This technique is based on photogrammetry and it is essential for reconstructing the position of the object in three-dimensional spatial frames. With respect to marker types, the image-based systems can be divided into two categories: the *passive* systems and the *active* systems. Passive systems consist of markers that reflect light back to the sensor, whereas active systems use markers that contain the source of light for better sensory resolution [48].

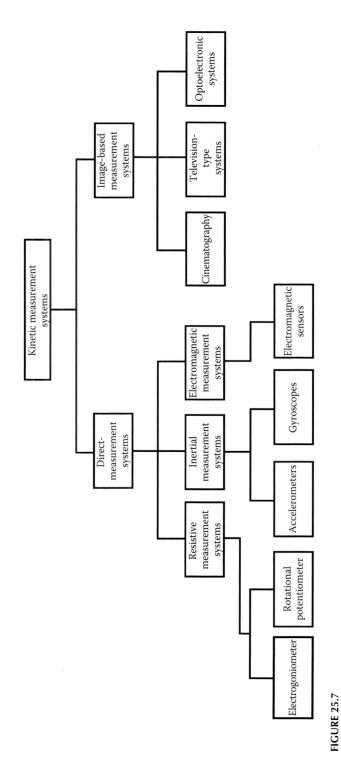

FIGURE 25.7
Main kinematic measurement systems.

While recording the kinematic parameters of an associated segment or joint, an instantaneous position vector for the link or instantaneous rotation matrix for the joint can be determined simultaneously or off-line analysis can be conducted in order to examine the movement comprehensively [49]. Through this process, systematic image-processing routines and digital filters enhance the signal quality for better estimation of the relative location of the marker in reference coordinate space. In Figure 25.8a, three-dimensional reflective markers are attached to the knee joint to determine various kinematical parameters of the shank and the knee joint [50]. In Figure 25.8b, local coordinates of the segment links of the right leg are defined [51].

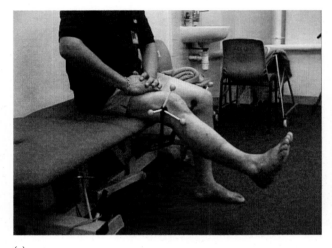

(a)

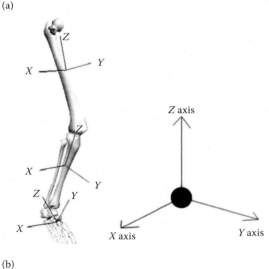

(b)

FIGURE 25.8
(a) Three-dimensional markers attached to knee joint (From Me, D.C.W.H.M., Kee, T.C.K, Cheung, B.H.C., Sai, M.L.B., and Yun, F.L.W., available at http://www.clinicalgaitanalysis.com/faq/reliability/. *Gait Lab, Princess Margaret Hosp. (PMH), Hong Kong*, n.d., Accessed on July 10, 2014.) and (b) three-dimensional marker geometry (From Jones, P.L. et al., *J. Med. Biol. Eng.*, pp. 184–188, 2009.).

In Figure 25.8b, the Z axis is determined by the unit vector directed from the distal to the proximal segment end, the Y axis is determined by the unit vector that is perpendicular to both the frontal plane and the Z axis, and finally the X axis is determined by the application of the right-hand rule (orientated in the medial–lateral direction).

In recent studies reflecting the progress in the fields of body area network (BAN) and the wireless body area network (WBAN) applications, effective solutions have been offered for the complications arising from using wires. Reconstructing the orientation of the object in three dimensions is the other positive merit of the imaging techniques as the video-acquisition procedure and video-processing routines for detection of the position of a link, or a joint, or a segment in three dimensions differ and improve those from using only the single plane, e.g., sagittal plane for gait analysis.

Kinematic variables through three-dimensional reflective markers and associated displacement trajectories can be obtained based on simultaneously recorded two or more camera frames. Detailed analysis about principles and possible limitations about video-based motion measurement systems have been reported comprehensively in the literature. Trade-offs between higher frame resolutions and memory allocation are the motion artifacts of captured fast objects in frames, and deinterlacing studies in scanning problems which are the topics worth special considerations [52]. Video-based motion measurement systems, also called stereometric systems, are divided into two categories by Cappozzo: the photography method and the optoelectronic stereophotogrammetry method [12–14]. In a broad sense, however, the image-based kinematic parameter–measurement systems can be divided into three categories [9]: cinematography, television, and optoelectronic. Passive and active image-based motion measurement systems are available commercially. The Ariel, HiRes, Motus, ProReflex, ElitePlus, and Vicon 370 systems are some examples of systems with passive markers, the Impulse X2 System, and the PTI Visualeyez motion-capture systems use active marker technology.

25.3.1.1 Cinematography

The motion of the associated link or joint is captured with a movie film camera in order to obtain kinematic displacement variables since the film cameras are commonly available in different technical specifications. Recordings of high-speed events are captured electronically using either a charge-coupled device (CCD) or a complementary metal–oxide–semiconductor (CMOS) active-pixel sensor. The recording is typically over 1000 frames/s [53]. One needs to take aperture and shutter speed alignments into consideration to get perfect exposures. The configuration with slow shutter speed and large aperture is suitable in situations where there is minimal light [54].

25.3.1.2 Television-Type Systems

Television-type systems use cameras or camcorders that have lower resolution and fixed frame rate. The accuracy level tends to be low due to deformations that occur on the shape and location of the markers that the camera is not able to deal with [55]. The scanning procedure of the television camera starts from the top of the frame and moves to the bottom. This causes a time delay between the top and the bottom lines of the image, which can result in calculation and orientation errors.

Recent studies showed that the choice of the CCD as main component in the system improves the acquired frame quality and gives higher-accuracy solutions to users. Infrared cameras deserve special credit among other systems because they do not require visible light and are not affected by reflection problems.

After markers are placed appropriately, the stream of video frames is processed to capture the central points of the markers, as shown in Figure 25.9.

Analysis of almost all walking processes can be projected into the sagittal plane, despite the fact that some research topics require more sensitive gait analysis defined on two planes including transversal motion. In most cases, expensive image-based solutions can generate clean frames with obvious markers due to dedicated camera systems and powerful devices. However, it is also possible to co-operate simple camera systems and associated software developed for experimental studies as a less expensive alternative. An ordinary example of a multistage frame process for this type of attempt is given in Figure 25.10 for knee prosthesis design [56].

As an example, with the backup of appropriate software and the application of trigonometric relations, relative knee angles for 200 samples of gait measurement for 8 s are demonstrated in Figure 25.11.

Several commercial products are available for such interest but the Vicon human analysis system based on infrared camera is the most popular in academic studies and leads the market.

25.3.1.3 Optoelectronic Measurements

In this system, the accuracy of detecting the orientation of the marker is increased with a set of infrared markers and special cameras. Sequentially flashing lights are detected more easily than other marker modalities. The system with this motion-capture facility tracks and assesses kinematics and dynamic motion in real time so that it introduces exceptional spatial and temporal accuracy features. The Optotrak Certus® motion-capture system

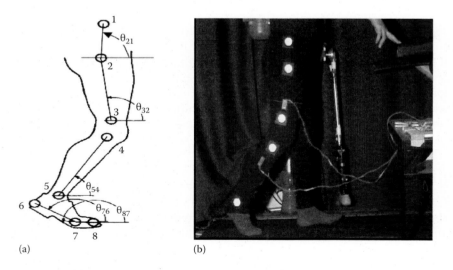

(a)   (b)

FIGURE 25.9
(a) Winter's proposed marker formation (Winter, D.: *Biomechanics and Motor Control of Human Movement*, vol. 2. 1990. Copyright Wiley-VCH Verlag GmbH & Co. KGaA. Reproduced with permission.) and (b) marker locations in the experiment.

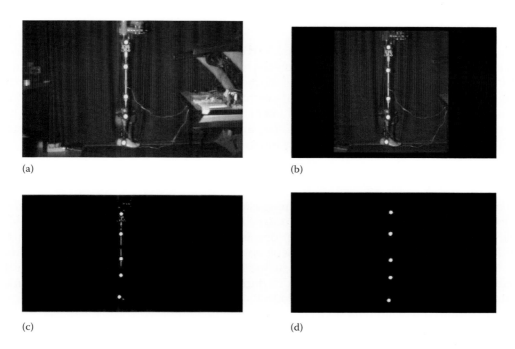

FIGURE 25.10
Frame processing stages: (a) original frame, (b) black-and-white conversion, (c) background elimination, and (d) marker detection.

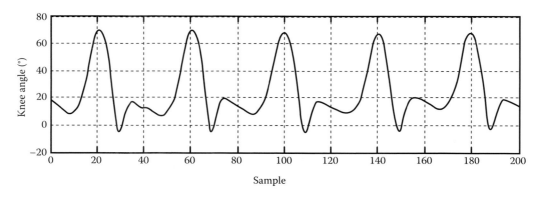

FIGURE 25.11
Knee angles derived from image-based motion measurement system.

claims maximum flexibility for motion-capture applications as a commercial product, and is a standard among research-oriented users. Some commercial systems are able to track up to 500 tiny markers scattered on the body to give a feasible solution to the analyst to derive kinematic movement parameters efficiently. However, despite being more accurate and reliable than cinematographic systems, they are costly for ordinary kinematic parameter estimation.

25.3.2 Direct-Measurement Systems

Mounting sensors on body parts in a task-driven manner ensures recording of data continuously for periods of days, weeks, and even months [1,57]. Moreover, lightweight, small, and low-power sensors are preferred for unencumbered working conditions. Inertial sensor-based systems are typically lightweight and portable, which facilitate more freedom in movements of the subject and do not confine data collection to a laboratory environment. Due to their small sizes, low cost, and suitability to portability, these sensors became an attractive option for wearable motion-analysis studies [58].

25.3.2.1 Resistive Measurement Systems

A basic resistive measurement method uses electrogoniometry, which has been standard in many applications for deriving the linear displacements since the beginning of 1970s [49]. The sensor output also carries the information about the angle of an anatomical point on the monitored body part. At the present, analog and digital electrogoniometers, which are based on variable resistors, for measuring linear displacement with an output in the form of a continuous voltage waveform or a digitized bit sequence reflecting the resolution of the sensor that can commercially be found.

As an example, a custom monoaxial electrogoniometer built with a 10 kΩ linear potentiometer was placed laterally to the knee to acquire the knee-joint angle to study the movements of the spinal cord patients and healthy volunteers. This attachment is depicted in Figure 25.12 [59].

Alteration of the segment positions generates a linear variation on the sensory output. Difficulty about positioning the goniometer onto the body parts requires a repeated

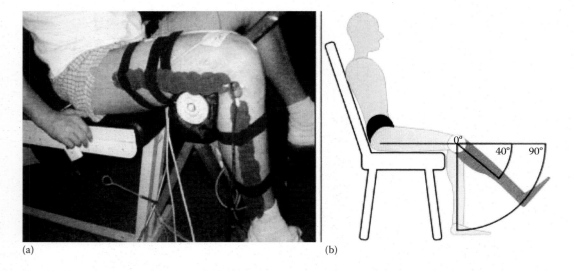

(a)   (b)

FIGURE 25.12
(a) A monoaxial electrogoniometer attached to knee joint and (b) knee-joint angle positions. (From Krueger, E. et al., *Rev. Bras. Eng. Biomédica*, 29, pp. 144–152, 2013.)

calibration procedure for every new conducted experiment. The power consumption on operation mode is the other main shortfall of the resistive measurements [60].

Body motion processing via wearable wireless sensor networks, especially with electrogoniometer studies, nowadays draws special attention. Most of the current prosthetic researches use dedicated ambulatory low-cost resistive measurement system together with vision-based measurements for validations and verifications [61].

25.3.2.2 Inertial Sensors

The development of the current technologies in aerospace, industrial, and robotic engineering is leading to more promising studies in the field-of-motion analysis. Accelerometers and gyroscopes are the two basic inertial sensors that play important role in bridging the gap between large laboratory facilities and applied clinical systems. Recent developments in integrated microelectromechanical systems (iMEMSs) have excited the research community for reducing the cost and size of accelerometers [47].

Gyroscopes measure angular velocity through using a vibrating element and the Coriolis effect. Murata's ENC-03J gyroscope and Analog Devices' ADXRS150 are two examples used in developing shoe-integrated wireless gait analysis [62].

Accelerometers that are used for kinematic studies can be divided into three categories: piezoelectric, piezoresistive, and differential capacitors, all of which implement the same basic principle of the spring–mass system. Piezoelectric accelerometers consist of an element which has a piezoelectric effect to generate an electrical output which is proportional to the applied acceleration according to earth's gravity. Piezoresistive accelerometers are based on the piezoresistive effect, which depends on the changing resistivity of a semiconductor due to applied acceleration forces giving voltage proportional to acceleration. Differentiable capacitor accelerometers are dependent on their capacitance, which varies with respect to the applied acceleration. The capacitance of parallel-plates varies through vertical motion of one plate with respect to the other and it is processed as an electrical output [1]. Uniaxial accelerometers can record the acceleration in a single direction but there are also biaxial and triaxial options to provide acceleration information in two or three dimensions.

25.3.2.2.1 Functions and Considerations

Gyroscopes are devices that measure rotational motions and provide absolute angular velocity with reference to the active axes while accelerometers provide acceleration according to earth's gravity with respect to the x, y, or z axis. Hence, their positions and placements on the human body become prominent in biokinematic measurements [63].

Applied acceleration according to gravity along a sensitive axis is the measure of rate and intensity of motion in up to three planes, namely; anterior–posterior, mediolateral, and vertical. Accelerometers, which can measure the corresponding x, y, or z axis as mentioned, have aspects superior to those of actometers or pedometers since they respond to both the frequency and the intensity of the movement. This is the reason why the accelerometers are used as tiltmeters along their sensitive axis [64]. The measuring tilt on the x or y axis leads to determining sagittal kinematic variables of human motion such as knee angle of a gait or bend angle of a moving arm.

Unencumbered structure, ability to record data continuously for long periods of time, and low costs make iMEMS-based accelerometers and gyroscopes popular especially in portable devices. Relatively low current demand is also a decisive parameter that favors these accelerometers. By using additive coupling capacitors, it is possible to cover a broad frequency content (0.6–5.0 Hz) expected for standard gait patterns [65].

There is a wide range of inertial sensor-based commercial products. Xsens MVN Biomech is a full-body and half-body measurement system based on inertial sensors (accelerometers and gyroscopes). InvenSense's MPU-9150, which is a nine-axis device, includes a gyroscope, an accelerometer, and a compass. This MotionTracking™ device is a complete set of inertial sensors that is used for inertia-based biokinematic studies. InterSense's InertiaCube BT™ provides real-time orientation data via a standard Bluetooth interface to a computer. With an integrated rechargeable battery and secure mounting straps, InertiaCube BT is the ideal wireless sensor for human-movement analysis. Recently announced products like Analog Devices' ADXL345 BCCZ 14-terminal Land Grid Array (14LGA), which is capable of measuring ±16 g, and InvenSense's ITG-3200 triple-axis gyroscope provide digital, low-cost, and yet powerful alternatives for inertial sensors for WBAN systems.

Patient activity monitoring is one of the most interesting research fields and it is fair to say that there are also assertive solution packages incorporating mainly accelerometers in the market. These monitoring devices can be attached to predefined spots on the body like thighs or waist depending on the application.

Some of the current accelerometer-based commercial technologies can be listed as follows: the RT3 triaxial research tracker kit, the activPAL™ professional activity monitor, ActiGraph's GT1M, Cyma's StepWatch3, Dynastream's AMP 331, Össur's PAM™ (Prosthetic Activity Monitor), and IDEEA® (Intelligent Device for Energy Expenditure and Activity).

25.3.2.2.2 Case Study: Kinematic Measurements by Inertial Sensors

In a case study, accelerometers were attached to the leg and the thigh to study relative knee angle alteration during walking. Similarly, accelerometers are attached to the wrist for monitoring the positioning of arm or hand prostheses. The system using triaxial accelerometers on the thigh and the shank, shown as A_1 and A_2 in Figure 25.13b, measures the difference between linear acceleration and gravity in each axis. The accelerations depicted as $S_{X_{1,2}}$ and $S_{Z_{1,2}}$ with orthogonal axes of sensors $u_{X_{1,2}}$ and $u_{Z_{1,2}}$ are selected to describe gait on the sagittal plane. The angles of limbs can be estimated by using accelerometers, so-called inclinometers, measuring the angle between sensitive axes and gravity. Nevertheless, soft tissue like flesh or fat causes alignment problems for axes of sensors $u_{X_{1,2}}$ with limb axes given by the M_2M_3 and M_4M_5 lines in Figure 25.13b. As a result of this mounting hindrance, an offset occurs between imaging and direct measurements.

By using the relevant equations similar to the ones explained above, the kinematic variables such as thigh angle, shank angle, or knee angle can be calculated from the integral of the angular velocity obtained from gyroscope. A combined accelerometer-and-gyroscope pair is used for decreasing noise level that especially occurred because of high spikes according to vibration during motion. The lower noise feature of the gyroscope makes it useful for kinematic analysis studies.

While the subject is walking at a speed of 3 km/h on treadmill, associated knee angle is derived from both the accelerometer and the gyroscope–accelerometer pair shown in Figure 25.14.

In a different application [66], it was necessary to study whole-body movements in which sensors were located close to the center of mass of the body.

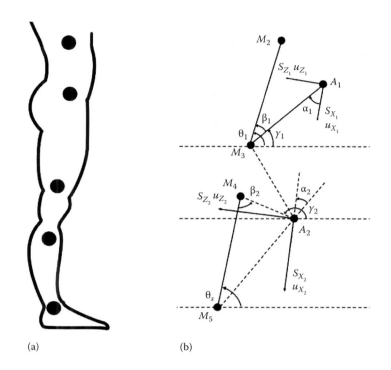

(a) (b)

FIGURE 25.13
(a) Marker orientation and (b) limb angles and placement of sensors.

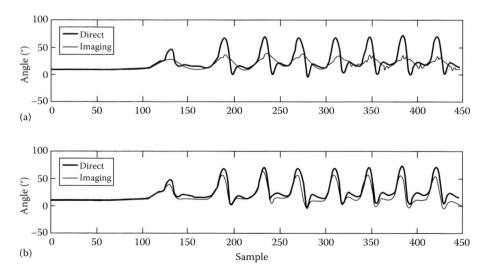

FIGURE 25.14
Relative knee angles: (a) accelerometer data and (b) gyroscope and accelerometer data.

25.3.2.3 Electromagnetic Systems

Three-dimensional orientation of a body part or a rigid body can also be determined with an electromagnetic motion measurement system. The basis of the system depends on magnetic field technology to interpret the interaction between the magnetic fields of three sets of orthogonal coils. Appropriate alignment and attachment of the source and sensor to anatomical structures lead to the determination of the accurate orientation of body parts [54]. These systems are accurate and do not have the line of sight restriction unlike optical tracking systems.

Electromagnetic tracking systems (ETSs) calculate and determine relative position and orientation by the relative magnetic flux generated by the three orthogonal coils on each transmitter and each receiver. Range and orientation determination is performed by measuring the relative intensity of the voltage or the current of the three coils [43]. Markers may be susceptible to magnetic and electrical interference from surroundings like rebar or coils that disturb the associated magnetic fields.

As an advantage over optical tracking systems, ETSs are significantly smaller and they allow applying whole-body kinematic analysis easily [43]. If the problems mentioned are addressed safely, the accuracy and reliability of position and orientation results provided by ETSs are comparable with those of image-based systems. The Flock of Birds® is a six-DoF tracker used for tracking one to four sensors simultaneously and Polhemus Fastrak six-DoF electromagnetic motion-tracking systems are examples of electromagnetic systems [67].

25.4 Wireless Measurement Systems for Biokinematics

Nowadays, WSNs and WBANs are used more often due to developments in wireless network technology and miniaturized electrical sensors and actuators. In these networks, sensing activities can be recorded from the skin surface, clothing attachments, or implants under the skin of human body. The wide variety of sensors and wireless nature of the networks leads to numerous new practical and innovative applications to improve the efficiency and quality of kinematic measurements.

A WBAN allows continuous monitoring of a set of physiological parameters, some of which are useful for kinematic analysis. Heartbeat, body temperature, electrocardiogram, and blood pressure levels are some examples of inner-body parameters, whereas the kinetic and kinematic parameters of the limbs or motion of the whole body can be considered as the outer physical parameters. Using a WBAN extends the border of experimental platform or monitored area and avoids cumbersome wirings and connections on the platform as well as avoiding the high cost of deployment and maintenance [68]. In order to realize communication between sensors and recording units, WSNs can also be used. However, the typical properties of a WBAN and current protocols designed for WSNs might cause problems in supporting a WBAN as there are some basic differences between a wireless sensor network and a wireless body area network. The WBAN is a dedicated biomedical system with special requirements concerning reliability and power demands.

25.4.1 Background on Wireless Measurement Systems

Custom developed wireless sensors are often unobtrusive and can capture activity in the best possible conditions. Systems for ambulatory monitoring of user activity may also configure off-the-shelf wireless sensor platforms and custom sensor modules. These platforms can be utilized for user demands and integrated into a wireless system. In the most general form, the structure of a wireless kinematic data transfer unit is given in Figure 25.15. In this representation, kinematic sensors collect, digitize, and prepare data for filtering or transmitting. After the transmission is completed, the receiving procedure begins with the associated wireless data transfer protocols. Transferring data to a healthcare station or a research center finalizes the process as a last step. From a practical point of view, three major wireless data protocols are listed below:

- Wi-Fi: exchanging data or connecting by 2.4 GHz ultrahigh-frequency(UHF) and 5 GHz superhigh-frequency (SHF) radio waves
- ZigBee: having high-level communication protocols for personal area network (PAN); based on an IEEE 802.15 standard
- Bluetooth: exchanging data over short distances (using short-wavelength UHF radio waves in the industrial, scientific, and medical [ISM] band from 2.4 to 2.485 GHz) from fixed and mobile devices and building PANs

The results drawn from the analysis can be used as performance criteria for human-movement assessments or detecting functional movement disorders in pathological conditions. In general, inertial sensors are used to obtain biokinematic variables such as 3D acceleration, 3D magnetization (earth's magnetic field), and 3D angular speed. Linear or angular displacement variables can also be determined with goniometers or various other displacement sensors; however, these can also be derived by applying trigonometric and mathematical equations to acceleration or velocity data, as explained above. Associated

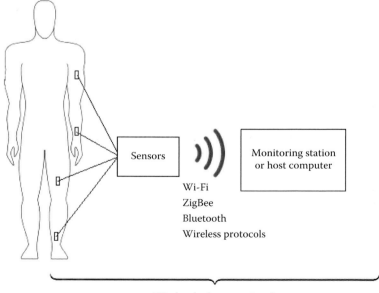

FIGURE 25.15
WBAN for biokinematic measurements.

digitized sensor data are transmitted via some wireless data-transfer protocols and they are converted to meaningful biokinematic human-motion parameters at the monitoring stations.

25.4.2 Applications Related to Wireless Kinematic Measurements

Wireless portable and wearable devices play an important role in a successful rehabilitation phase as they offer uninterrupted real-life conditions. For instance, wireless quantitative gait analysis to describe a person's gait provides the measurements of a set of parameters, namely, kinematic, kinetic, and temporal–spatial parameters, like walking distance, speed, cadence, and stride length. An example of such study has been reported for gait analysis developed for an objective assessment of gait quality [69]. The system consists of triaxial accelerometer-based sensor units on commercially available patellar tendon straps and integrated USB ANT transmitter and receiver. The algorithms provide real-time limping detection between steps and are used for searching critical features of a gait such as toe-off, heel-strike, and toe-strike and other events in stance and swing phases. These are all required to evaluate the raw gait data for further classification as pathological patterns in the concerned gait.

A system using wireless sensors manufactured by Shimmer and connected via means of WBANs presents an alternative to monitoring patients going through long periods of rehabilitation. The Shimmer sensor is a low-power nine-DoF wireless sensor equipped with built-in accelerometers, magnetometers, and gyroscopes.

A wireless body area network is composed of small sensors attached or implanted in the body which establishes a communication link with the data-acquisition unit. This attribute allows the user to track required data over a long period. If biokinematic studies are concerned, there are different sensor possibilities for understanding and developing the geometry of the associated motion. Interconnection between the user and the environment is handled by a computer, a smartphone, or a PDA [68].

A wireless data network which uses the Bluetooth network protocol has some special properties, which are pointed out in the literature studies. When the associated parameters are sensed from the body or skin, the microprocessor encodes and transfers the digitized data to the Bluetooth module via an internal serial link for the radio transmission to the receiver system [70].

There are different study fields in which wireless body area networks take part in transmitting or receiving kinematic variables. Some studies present wireless networks for event detection and building location-awareness systems. They are particularly useful when a kind of emergency situation occurs when real-time kinematic parameters are crucial. For example, a fall-detection system, which consists of a wireless body network, monitors human activities based on predefined event descriptions and detects events like a fall in living environment under network surveillance. The wireless network continuously transmits such activities to an in-house network based on a certain type of wireless network protocol. Additionally, event-detection algorithms using kinematic variables are still a serious research area and still require attention for better assist or alert performance for a whole monitoring facility [71].

In order to extract the complete human-movement kinematics, some sensor combinations are assembled onto a human body for determining the orientation of the associated body segment. In a recent movement kinematic study, a nine-DoF inertial sensor unit composed of an accelerometer, a gyroscope, and a magnetometer with a wireless data-transfer unit was mounted onto wrists and ankles; this sensor unit can be seen in Figure 25.16 [72].

FIGURE 25.16
Wrist and ankle sensors and kinematic measurement setup with a wireless sensor network.

The ideal body locations for motion sensor nodes are important as the interpretation of the measured data requires consistent sensor positioning. The network has its own difficulties as well since the body in motion (e.g., walking, running, and twisting) may trigger some unexpected problems like channel fading and shadowing effects [73]. For example, it is reported that arm motions to the front and side of the body can have a small impact on the received power. However, if the arms block the line of sight between two antenna elements, then the movement becomes more significant as the attenuation can go up to 20 dB. For some cases in which the personal device communication operates at 868 MHz, the loss rate reaches more than 50% in various arm movements [74].

Wireless kinematic measurements can give an idea to a clinician about the measurement system's reliability. The reliability of 3D kinematic parameter estimation-based gait analyses or the accuracy of the measured variables of the system can be measured with wireless sensor networks mounted onto the user. These experiments are conducted for detecting differences between consecutive trials with different examiners or different velocity and terrain conditions. Wireless sensor networks present flexible solutions to clinicians for high-range gait tests. In general, accelerometers are used as sensors for networks because of their unencumbered, lightweight, and portable structure. In addition, they are cost effective and are able to capture data from many gait cycles. Discrete-time filtering applications and different calibration procedures can give better analyzed results to clinicians [70].

25.5 Biodriven Hands, Prostheses, and Exoskeletal Ortheses

Another way to describe biokinematics from a different perspective is to consider motion-driven physiological signals, e.g., neural and muscular. The neural functions of the body and the mechanical interface are of long-lasting interest in terms of research and development of intelligent systems able to replace some lost limbs.

We begin by describing the mechanisms of how muscles work as a part of a prosthesis in the context of a robotic artificial hand. The same issue can be extended to other prostheses that have been brought up by many different researchers in the field. The nervous system has specific sensory systems, one of which is the feeling of touch, which is a perception that naturally results from activation of neural receptors on the skin. Receptors detect stimulus, converting into information and directing to the central nervous system by afferent neurons.

The specific prosthetic devices used for the upper body are called bionic arms. Usually these prosthetic devices are strongly influenced by research and development in robotics. Some leading studies from the Shadow Robot Company, located in London, United Kingdom, resulted in developing an advanced dexterous hand. The Shadow Hand has touching and position sensors in order to assess required kinematics and predefined movements. This structure is actuated by air muscles in 24 degrees of freedom. It is reported that the Shadow Hand uses state-of-the-art sensors such as position sensors based on Hall effect with a resolution of 0.2° and tactile sensors (quantum tunneling composite [QTC], a new type of conductive material developed by Peratech) [75–76].

One of the biggest problems that any type of arm or hand prosthesis carries is how to provide a feedback on what that arm is performing due to lack of information for space position, let alone considering the grip or hold kinetic problems. In 2005, the Defense Advanced Research Projects Agency (DARPA) made a program to carry out a very important research for a prosthetic arm. The first phase had been initiated for a neural interface in 2007; the second phase was planned for a neurally controlled arm in 2009 clinical trials in the following 4 years of research. They thus intend to develop with a sizable research budget a fully functional upper limb that responds to direct neural control [77]. Dean Kamen's DEKA Research and Development Corporation introduced the arm called the "Luke Arm" [78], which can be seen in Figure 25.17, with 18 degrees of freedom [78–79]. DEKA has a two distinct ways to operate the prosthesis. The first one is indirect control through a hidden unit in the soles of shoes. The second but more interesting and challenging way is the one with a so-called thought-controlled approach.

FIGURE 25.17
Luke Arm by DEKA (http://www.dekaresearch.com/deka_arm.shtml). (Courtesy of DEKA Research and Development Corporation, Manchester, New Hampshire.)

The second approach was realizable only when the innovative surgery leading to targeted muscle reinnervation (TMR) was developed by Dr. Todd Kuiken [80–81]. In its own right, TMR is a specific method by which the targeted muscle of an amputated patient is denervated and then reinnervated with residual nerves of the amputated limb. By this way, the patient is expected to think about moving his arm, and the signals are collected to search some nerve activities, i.e., myopotentials, in the chest muscles for further contraction. In final stage, the research setup has been designed to pick up those contractions through myoelectric sensors to generate command signals for actuators of the prosthesis. The shoulder socket hides some tiny balloons for providing some comfort and feedback sensing proportional to grip positions. Another feedback-sensing solution that has been reported was vibrating tactile sensors on fingertips. This summarized project prototype was successful and prepared to be in market after having a series of clinical trials [82]. After almost 8 years of research and trial period, the robotic arm for amputees, the so-called Luke Arm, has been approved for commercialization by the U.S. FDA [83].

Another approach has been proposed by a team lead by the research center called the Applied Physics Laboratory (APL) of Johns Hopkins University. Under DARPA support, the recent bionic arm Proto 2, which evolved from Proto 1, has been built with 25 DoFs with exceptional mechanical agility. This research promoted four levels of overcoming nerve signal processing according to amputation severity degree and developed two implantable sensors. One of the sensors is called injectable myoelectric sensor (IMES), developed to collect nerve signals for individual control of fingers. IMESs are implanted in the muscle bundles of the remaining limb just to detect much cleaner signals for second- and third-level amputation on the basis of TMR. In the same prosthesis research program at the last or fourth level for more severe amputations in which bodies have no part for interacting with an artificial arm, an implantable needle lookalike device with 5 mm square grid of 100 electrodes has been developed by the University of Utah and called the Utah Slant Electrode Array (USEA) to use the body's natural neural signals. According to their report, these sensors are placed in the brain's motor cortex on the top of the skull in order to harness the brain signals. Dedicated signal-processing methods translate and classify collected signals based on known intentions that are defined to control the mechanics of arm movements in real time [84].

There are also commercially successful bionic hands which use myoelectric principles, such as Touch Bionics' i-limb and the bebionic3. Of course, these are more affordable than the sophisticated upper-limb prosthesis developed for DARPA. They all interface with myoelectric sensors. As a usual approach, myoelectric signals from remaining muscles are recorded by specialized electrodes then interpreted and transmitted to motors in the artificial hand. For above-knee prosthesis, myoelectric signals are also among the other measured signals like the relative knee angular position and speed from the prosthetic hip and stump on the amputated side; force sensors are installed within the prosthesis. They all help to estimate the present phase of the gait and the intention of the next phase (intention recognition) and to develop control strategies for prosthetic damper [85].

Today, wearable and nonwearable rehabilitation devices exist and they both have their own advantages and disadvantages concerning movement therapy. Disadvantages originate from ignoring the biological mechanical properties of attached human limbs. In principle, wearable robotic orthoses support the entire limbs of the patient and imitate the kinematics of the impaired extremity. In robot rehabilitation studies, exoskeletal rehabilitation devices are designed for impairment of muscular weakness or loss of muscle geometry. These rehabilitation robots resolve the human factor problems by being independent of the judgements of the therapists. Robots appear to have been better suited for

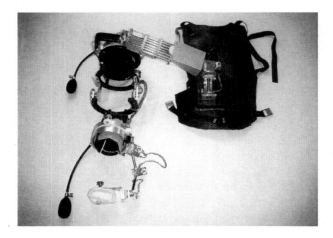

FIGURE 25.18
An ergonomic exoskeleton prototype interfacing the right arm. (From Schiele, A., and van der Helm, F.C.T., *IEEE Trans. Neural Syst. Rehabil. Eng.*, 14, pp. 456–469, 2006.)

performing consecutive tasks than humans [21] since they perform specified tasks with sensing the body part's current position and reaching relatively easily the desired position. In general, the sensing step is also implemented by inexpensive resistive measurement systems, which can be seen in Figure 25.18, in a rehabilitation upper-arm device. Position sensors are aligned with each axis to track performed movements. The hand is inserted in the glove provided by DAHO Hand Orthesis system of Biedermann Motech, Germany.

As a final remark, we can see that, any ergonomic design requires a good model of the kinematics and dimensions of the interacting human limb covering its Cartesian space motion. Then, the model can be used to define realistic ergonomic movement trajectories in 3D Cartesian space to understand and assist human movements. A set of simulations can estimate the ability of a robotic human–machine interface and permit modifying its kinematic structures [21] and the wireless technologies offer solutions to achieve such goals.

25.6 Conclusion

The field of biokinematics is a melting pot where motion-analysis methods used in robotics that blend with clinical requirements of movement specialists. This chapter has emphasized prosthetists and orthotists under the support of emerging technology. Biokinematics is a broad frame in which almost all efforts toward understanding, describing, measuring, and monitoring the defined variables of motion as well as developing devices for a wide variety of purposes in biology and medicine should be taken into consideration. We should underline that any kinematic analysis may use only a small fraction of available theory developed in this frame. However, complicated projects or a complete analysis like an assessment of energy consumption during the amputee's gait may require all possible usable kinematic instruments. For a complete picture of mathematical formulation and movement evaluation, the analysis is expected to be supported by a kinetic study including moments and force considerations along with mass and segment knowledge drawn by anthropometric studies.

Acknowledgments

The authors would like to thank Associate Professor Karl Shoemaker for his careful review and comments given in the writing-up stage of this chapter.

References

1. Godfrey, A., Conway, R., Meagher, D., and ÓLaighin, G., Direct measurement of human movement by accelerometry. *Med. Eng. Phys.*, 30, pp. 1364–1386, 2008.
2. Brandes, M., Schomaker, R., Möllenhoff, G., and Rosenbaum, D., Quantity versus quality of gait and quality of life in patients with osteoarthritis. *Gait Posture*, 28, pp. 74–79, 2008.
3. Muhit, A.A., Pickering, M.R., Ward, T., Scarvell, J.M., and Smith, P.N., A comparison of the 3D kinematic measurements obtained by single-plane 2D-3D image registration and RSA. *Conf. Proc. IEEE Eng. Med. Biol. Soc.*, 2010, pp. 6288–6291, 2010.
4. Fregly, B.J., Rahman, H.A., and Banks, S.A., Theoretical accuracy of model-based shape matching for measuring natural knee kinematics with single-plane fluoroscopy. *J. Biomech. Eng.*, 127, pp. 692–699, 2005.
5. Scarvell, J.M., Pickering, M.R., and Smith, P.N., New registration algorithm for determining 3D knee kinematics using CT and single-plane fluoroscopy with improved out-of-plane translation accuracy. *J. Orthop. Res.*, 28, pp. 334–340, 2010.
6. Grood, E.S., and Suntay, W.J., A joint coordinate system for the clinical description of three-dimensional motions: Application to the knee. *J. Biomech. Eng.*, 105, pp. 136–144, 1983.
7. Burton, L.J., The dynamics and kinematics of bio-in swimming systems 2013. http://dspace.mit.edu/handle/1721.1/81692, Accessed on January 23, 2015.
8. Zhou, C., Tan, M., Cao, Z., Wang, S., Creighton, D., and Nahavandi, S., Kinematic modeling of a bio-inspired robotic fish. *2008 IEEE Int. Conf. Robot. Autom.*, pp. 695–699, 2008.
9. Winter, D., *Biomechanics and Motor Control of Human Movement*, vol. 2, Wiley, 1990.
10. Barker, S., Craik, R., Freedman, W., Herrmann, N., and Hillstrom, H., Accuracy, reliability, and validity of a spatiotemporal gait analysis system. *Med. Eng. Phys.*, 28, pp. 460–467, 2006.
11. O'Sullivan, S., and Schmitz, T., *Physical Rehabilitation*, Fifth edition, F.A. Davis Company, 2006.
12. Cappozzo, A., Della Croce, U., Leardini, A., and Chiari, L., Human movement analysis using stereophotogrammetry—Part 1: Theoretical background. *Gait Posture*, 21, pp. 186–196, 2005.
13. Chiari, L., Della Croce, U., and Leardini, A., Cappozzo, A., Human movement analysis using stereophotogrammetry—Part 2: Instrumental errors. *Gait Posture*, 21, pp. 197–211, 2005.
14. Leardini, A., Chiari, L., Della Croce, U., and Cappozzo, A., Human movement analysis using stereophotogrammetry—Part 3: Soft tissue artifact assessment and compensation. *Gait Posture*, 21, pp. 212–225, 2005.
15. Della Croce, U., Leardini, A., Chiari, L., and Cappozzo, A., Human movement analysis using stereophotogrammetry—Part 4: Assessment of anatomical landmark misplacement and its effects on joint kinematics. *Gait Posture*, 21, pp. 226–237, 2005.

16. Thiruvengada, H., Srinivasan, S., and Gacic, A., Design and implementation of an automated human activity monitoring application for wearable devices. *IEEE Int. Conf. Syst. Man Cybern.*, IEEE, 2008, pp. 2252–2258, 2008.

17. Lester, J., Choudhury, T., and Borriello, G., A practical approach to recognizing physical activities, *Pervasive Comput.*, pp. 1–16, 2006.

18. Maurer, U., Smailagic, A., Siewiorek, D.P., and Deisher, M., Activity recognition and monitoring using multiple sensors on different body positions. *Int. Work. Wearable Implant. Body Sens. Networks*, IEEE, pp. 113–116, 2006.

19. Bouten, C.V., Koekkoek, K.T., Verduin, M., Kodde, R., and Janssen, J.D., A triaxial accelerometer and portable data processing unit for the assessment of daily physical activity. *IEEE Trans. Biomed. Eng.*, 44, pp. 136–147, 1997.

20. Lu, T.W., and Chang, C.F., Biomechanics of human movement and its clinical applications. *Kaohsiung J. Med. Sci.*, 28, pp. S13–S25, 2012.

21. Schiele, A., and van der Helm, F.C.T., Kinematic design to improve ergonomics in human machine interaction. *IEEE Trans. Neural Syst. Rehabil. Eng.*, 14, pp. 456–469, 2006.

22. Riener, R., Nef, T., and Colombo, G., Robot-aided neurorehabilitation of the upper extremities. *Med. Biol. Eng. Comput.*, 43, pp. 2–10, 2005.

23. Hidler, J.M., and Wall, A.E., Alterations in muscle activation patterns during robotic-assisted walking. *Clin. Biomech. (Bristol, Avon)*, 20, 184–93, 2005.

24. He, J., Koeneman, E.J., Schultz, R.S., Huang, H. et al., Design of a robotic upper extremity repetitive therapy device, *9th Int. Conf. Rehabil. Robot: ICORR 2005*, IEEE, pp. 95–98, 2005.

25. Sanchez, R., Reinkensmeyer, D., Shah, P., Liu, J. et al., Monitoring functional arm movement for home-based therapy after stroke. *Conf. Proc. IEEE Eng. Med. Biol. Soc.*, 7, pp. 4787–4790, 2004.

26. Wu, G., Siegler, S., Allard, P., Kirtley, C. et al., ISB recommendation on definitions of joint coordinate system of various joints for the reporting of human joint motion—Part I: Ankle, hip, and spine. *J. Biomech.*, 35, pp. 543–548, 2002.

27. Wu, G., van der Helm, F.C.T., (DirkJan) Veeger, H.E.J., Makhsous, M. et al., ISB recommendation on definitions of joint coordinate systems of various joints for the reporting of human joint motion—Part II: Shoulder, elbow, wrist and hand. *J. Biomech.*, 38, pp. 981–992, 2005.

28. Becker, J., Thakor, N., and Gruben, K., A study of human hand tendon kinematics with applications to robot hand design, *Proc. 1986 IEEE Int. Conf. Robot. Autom.*, vol. 3, IEEE, pp. 1540–1545, 1986.

29. Neckel, N., Wisman, W., and Hidler, J., Limb alignment and kinematics inside a Lokomat robotic orthosis. *Conf. Proc. IEEE Eng. Med. Biol. Soc.*, 1, pp. 2698–2701, 2006.

30. Cempini, M., De Rossi, S.M.M., Lenzi, T., Vitiello, N., and Carrozza, M.C., Self-alignment mechanisms for assistive wearable robots: A kinetostatic compatibility method. *IEEE Trans. Robot.*, 29, pp. 236–250, 2013.

31. Duraisamy, K., Isebor, O., Perez, A., Schoen, M.P., and Naidu, D.S., Kinematic synthesis for smart hand prosthesis, *First IEEE/RAS-EMBS Int. Conf. Biomed. Robot. Biomechatronics, 2006: BioRob 2006*, IEEE, pp. 1135–1140, 2006.

32. Zengin, E., Biokinematic Analysis of Human Arm. İzmir Institute of Technology, 2006.

33. Shen, H., Yang, T., and Ma, L., Synthesis and structure analysis of kinematic structures of 6-dof parallel robotic mechanisms. *Mech. Mach. Theory*, 40, pp. 1164–1180, 2005.

34. Arami, A., Simoncini, M., Atasoy, O., Ali, S. et al., Instrumented knee prosthesis for force and kinematics measurements. *IEEE Trans. Autom. Sci. Eng.*, 10, pp. 615–624, 2013.

35. Beggs, J.S., *Kinematics*, CRC Press, 1983.

36. Hibbeler, R.C., *Engineering Mechanics: Dynamics*, Prentice Hall, 2010.

37. Nelson, R.C., Learning dextrous manipulation skills using the evolution strategy, *Proc. Biomech. Symp.*, pp. 4–13, 1980.

38. Schleihauf, R., *Biomechanics of Human Movement*, AuthorHouse, 2004.

39. Spong, M.W., Hutchinson, S., and Vidyasagar, M., *Robot Modeling and Control*, Wiley, 2005.

40. Pennock, G.R., and Clark, K.J., An anatomy-based coordinate system for the description of the kinematic displacements in the human knee. *J. Biomech.*, 23, pp. 1209–1218, 1990.

41. Hall, S.J., *Basic Biomechanics*, McGraw-Hill Higher Education, 2012.

42. Winter, D.A., *Biomechanics and Motor Control of Human Movement*, 4th ed., John Wiley & Sons, Hoboken, New Jersey, 2009.

43. Filkorn, R., and Kocan, M., Simulation of human body kinematics. http://www.cescg.org /CESCG-2000/RFilkorn/, Accessed on June 23, 2015.

44. Ballan, L., and Cortelazzo, G.M., Marker-less motion capture of skinned models in a four camera set-up using optical flow and silhouettes, *Proc. 3DPVT '08—Fourth Int. Symp. 3D Data Process. Vis. Transm.*, 2008.

45. Memişoğlu, A., Human Motion Control Using Inverse Kinematics. Bilkent University, 2003.

46. Sims, K., and Zeltzer, D., A figure editor and gait controller for task level animation. *SIGGRAPH Course Notes #4 Synth. Actors Impact Artificial Intell. Robot. Animat.*, pp. 164–181, 1988.

47. Culhane, K.M., O'Connor, M., Lyons, D., and Lyons, G.M., Accelerometers in rehabilitation medicine for older adults. *Age Ageing*, 34, pp. 556–560, 2005.

48. Richards, J.G., The measurement of human motion: A comparison of commercially available systems. *Hum. Mov. Sci.*, 18, pp. 589–602, 1999.

49. Medved, V., *Measurement of Human Locomotion* [Hardcover], First edition, CRC Press, 2000.

50. Me, D.C.W.H.M., Kee, T.C.K, Cheung, B.H.C., Sai, M.L.B., and Yun, F.L.W., http://www.clinical gaitanalysis.com/faq/reliability/. *Gait Lab, Princess Margaret Hosp. (PMH), Hong Kong*, n.d., Accessed on July 10, 2014.

51. Jones, P.L., Kerwin, D.G., Irwin, G., and Nokes, L.D.M., Three dimensional analysis of knee biomechanics when landing on natural turf and football turf. *J. Med. Biol. Eng.*, pp. 184–188, 2009.

52. Orhanli, T., Finite State Control of Semiactive Knee Joint with Pneumatic Damper and a Test Platform Design for Testing Knee Prosthesis. Hacettepe University, 2014.

53. Balch, K., *High Frame Rate Electronic Imaging*, 1999. http://www.motionvideoproducts.com /MVP%20papers/HSV%20White%20Paper.pdf, Accessed on January 23, 2015.

54. Digital Photography School, *Shutter Speed*, http://digital-photography-school.com/shutter-speed, Accessed on January 23, 2015.

55. Akdoğan, K, E., Design and Comparison of Electronic Above Knee Prostheses Employing Pneumatic and MR Cylinders by Motion Measurement Systems. Hacettepe University, 2011.

56. Yilmaz, A., and Orhanli, T., Gait motion simulator for kinematic tests of above knee prostheses, *IET Measurement Sci. Technol.*, 9, pp. 250–258, 2014.

57. Foerster, F., Smeja, M., and Fahrenberg, J., Detection of posture and motion by accelerometry: A validation study in ambulatory monitoring. *Comput. Human Behav.*, 15, pp. 571–583, 1999.

58. Cuesta-Vargas, A.I., Galán-Mercant, A., and Williams, J.M., The use of inertial sensors system for human motion analysis. *Phys. Ther. Rev.*, 15, pp. 462–473, 2010.

59. Krueger, E., Scheeren, E.M., Nogueira-Neto, G.N., Neves, E.B. et al., Relationship between peak and mean amplitudes of the stimulating output voltage for functional control of the knee by spinal cord patients and healthy volunteers. *Rev. Bras. Eng. Biomédica*, 29, pp. 144–152, 2013.

60. Williamson, R., and Andrews, B.J., Detecting absolute human knee angle and angular velocity using accelerometers and rate gyroscopes. *Med. Biol. Eng. Comput.*, 39, pp. 294–302, 2001.

61. Mayagoitia, R.E., Nene, A.V., and Veltink, P.H., Accelerometer and rate gyroscope measurement of kinematics: An inexpensive alternative to optical motion analysis systems. *J. Biomech.*, 35, pp. 537–542, 2002.

62. Morris, S., A Shoe-Integrated Sensor System for Wireless Gait Analysis and Real-Time Therapeutic Feedback. Massachusetts Institute of Technology, 2004.

63. Rocon, E., Ruiz, A.F., Pons, J.L., Belda-Lois, J.M., and Sanchez-Lacuesta, J.J., Rehabilitation robotics: A wearable exo-skeleton for tremor assessment and suppression, *Proc. 2005 IEEE Int. Conf. Robot. Autom.*, IEEE, pp. 2271–2276, 2005.

64. Mathie, M.J., Coster, A.C.F., Lovell, N.H., and Celler, B.G., Accelerometry: Providing an integrated, practical method for long-term, ambulatory monitoring of human movement. *Physiol. Meas.*, 25, pp. R1–R20, 2004.

65. Najafi, B., Aminian, K., Paraschiv-Ionescu, A., Loew, F. et al., Ambulatory system for human motion analysis using a kinematic sensor: Monitoring of daily physical activity in the elderly. *IEEE Trans. Biomed. Eng.*, 50, pp. 711–723, 2003.

66. Chen, K.Y., and Bassett, D.R., The technology of accelerometry-based activity monitors: Current and future. *Med. Sci. Sports Exerc.*, 37, S490–S500, 2005.

67. Hogue, A., *MARVIN: A Mobile Automatic Realtime Visual and INertial Tracking System*, Master's Thesis, York University, 2003.

68. Latré, B., Braem, B., Moerman, I., Blondia, C., and Demeester, P., A survey on wireless body area networks. *Wirel. Networks*, 17, pp. 1–18, 2010.

69. Gooding, J., Hackmann, L., Claus, T., and Disselhorst-Klug, C., A novel portable system for gait analysis and rehabilitation. *Biomed. Tech. (Berlin)*, vol. 58, p. 2, 2013.

70. Kavanagh, J.J., Morrison, S., James, D.A., and Barrett, R., Reliability of segmental accelerations measured using a new wireless gait analysis system. *J. Biomech.*, 39, pp. 2863–2872, 2006.

71. Stroiescu, F., Daly, K., and Kuris, B., Event detection in an assisted living environment. *Conf. Proc. IEEE Eng. Med. Biol. Soc.*, 2011, pp. 7581–7584, 2011.

72. Baraka, A., A WBAN for human movement kinematics and ECG measurements. *E-Health Telecommun. Syst. Networks*, 1, pp. 19–25, 2012.

73. Jovanov, E., Milenkovic, A., Otto, C., and de Groen, P.C., A wireless body area network of intelligent motion sensors for computer assisted physical rehabilitation. *J. Neuroeng. Rehabil.*, 2, 6, 2005.

74. Ylisaukko-oja, A., Vildjiounaite, E., and Mantyjarvi, J., Five-point acceleration sensing wireless body area network—Design and practical experiences. *Eighth Int. Symp. Wearable Comput.*, vol. 1, IEEE, n.d., pp. 184–185.

75. Shadow, R.C., http://www.shadowrobot.com/products/air-muscles/. *Shad. 30mm Air Muscle—Specif.*, 2011, Accessed on June 25, 2014.

76. Vanderborght, B., http://lucy.vub.ac.be/gendes/actuators/muscles.htm. *Vrije Unversiteit Brusse*, 2005, Accessed on July 10, 2014.

77. Miles, D., DARPA's Cutting-Edge Programs Revolutionize Prosthetics. *Am. Forces Press Serv.*, 2006.

78. Adee, S., Dean Kamen's Luke Arm Prosthesis Readies for Clinical Trials. *IEEE Spectr.*, 2008, Available at http://spectrum.ieee.org/biomedical/bionics/dean-kamens-luke-arm-prosthesis-readies-for-clinical-trials, Accessed on June 24, 2015.

79. Adee, S., Reengineering the Prosthetic-Arm Socket. *IEEE Spectr.*, 2008, Available at http://spectrum.ieee.org/biomedical/devices/reengineering-the-prostheticarm-socket, Accessed on June 28, 2015.

80. Kuiken, T., Targeted reinnervation for improved prosthetic function. *Phys. Med. Rehabil. Clin. N. Am.*, 17, pp. 1–13, 2006.

81. Bionic Arm Controlled by Patient's Own Thoughts, *Singularity Hub*, 2009, Available at http://singularityhub.com/2009/02/13/bionic-arm-controlled-by-patients-own-thoughts/, Accessed on July 28, 2015.

82. Filipe, J., and Luís, S., Perceptual Feedback of Grasping Touch to Prosthesis Hand User. Universidade do Porto, 2010.

83. Guizzo, E., Dean Kamen's "Luke Arm" Prosthesis Receives FDA Approval. *IEEE Spectr.*, 2014, Available at http://spectrum.ieee.org/automaton/biomedical/bionics/dean-kamenluke-arm-prosthesis-receives-fda-approval, Accessed on June 24, 2015.

84. Adee, S., A "Manhattan Project" for the Next Generation of Bionic Arms. *IEEE Spectr.*, 2008, Available at http://spectrum.ieee.org/biomedical/bionics/a-manhattan-project-for-the-next-generation-of-bionic-arms, Accessed on June 28, 2015.

85. Aeyels, B., Peeraer, L., Vander Sloten, J., and Van der Perre, G., Development of an above-knee prosthesis equipped with a microcomputer-controlled knee joint: First test results. *J. Biomed. Eng.*, 14, pp. 199–202, 1992.

26

Biopotentials and Electrophysiology Measurements

Nitish V. Thakor

CONTENTS

26.1 Introduction

This chapter reviews the origins, principles, and designs of instrumentation used in biopotential measurements, in particular for ECG, EEG, EMG, and electrooculogram (EOG). These biopotentials represent the activity of their respective organs: the heart, the brain, muscles, and eyes. These biopotentials are acquired with the help of specialized electrodes that interface with the organ or the body and transduce low-noise, artifact-free signals. The basic design of a biopotential amplifier consists of an instrumentation amplifier. The amplifier should possess several characteristics, including high amplification, input impedance, and the ability to reject electrical interference, all of which are needed for the measurement of these biopotentials. Ancillary useful circuits are filters for attenuating electric interference, electrical isolation, and defibrillation shock protection. Practical considerations in biopotential measurement involve electrode placement and skin preparation, shielding from interference, and other good measurement practices.

26.2 The Origins of Biopotentials

Many organs in the human body, such as the heart, the brain, muscles, and eyes, manifest their functions through electrical activities [1]. The heart, for example, produces a signal called an electrocardiogram (Figure 26.1a). The brain produces a signal called an electroencephalogram (Figure 26.1b). The activity of muscles, such as contraction and relaxation, produces an electromyogram (Figure 26.1c). Eye movement results in a signal called an electrooculogram (Figure 26.1d), and the retina within the eyes produces an electroretinogram (ERG). Measurements of these and other electric signals from the body can provide vital clues as to normal or pathological functions of the organs. For example, abnormal heartbeats or arrhythmias can be readily diagnosed from an ECG. Neurologists interpret EEG signals to identify epileptic seizure events. EMG signals can be helpful in assessing muscle function as well as neuromuscular disorders. EOG signals are used in the diagnosis of disorders of eye movement and balance disorders.

The origins of these biopotentials can be traced to the electric activity at the cellular level [2]. The electric potential across a cell membrane is the result of different ionic concentrations that exist inside and outside the cell. The electrochemical concentration gradient across a semipermeable membrane results in the Nernst potential. The cell membrane separates high concentrations of potassium ion and low concentrations of sodium ions (along with other ions such as calcium in less significant proportions) inside a cell and just the opposite outside a cell. This difference in ionic concentration across the cell membrane produces the resting potential [3]. Some of the cells in the body are excitable and produce what is called an action potential, which results from a rapid flux of ions across the cell membrane in response to an electric stimulation or transient change in the electric gradient of the cell [4]. The electric excitation of cells generates currents in the surrounding volume conductor, manifesting itself as potentials on the body.

Figure 26.2 illustrates the continuum of electrophysiological signals from (a) a heart cell, (b) the myocardium (the heart muscle), and (c) the body surface. Each cell in the heart produces a characteristic action potential [4]. The activity of cells in the sinoatrial node of the heart produces an excitation that propagates from the atria to the ventricles through

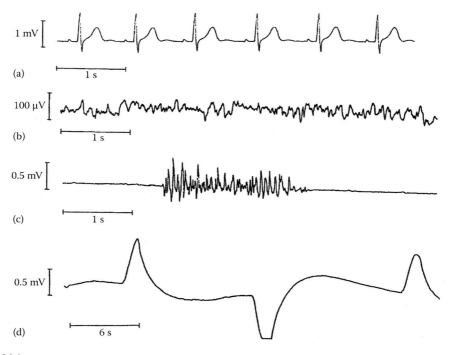

FIGURE 26.1
Sample waveforms: (a) ECG, normal sinus rhythm; (b) EEG, normal patient with open eyes; (c) EMG, flexion of biceps muscles; and (d) EOG, movement of eyes from left to right.

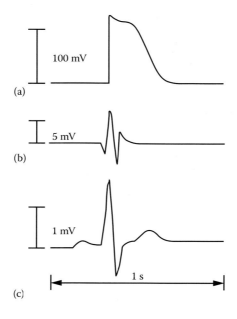

FIGURE 26.2
Schematic showing origins of biopotentials: (a) an action potential from a heart cell (recorded using a micro-electrode); (b) the electrocardiogram from the heart surface (recorded using an endocardial catheter); and (c) the ECG signal at the chest (recorded using surface electrodes).

well-defined pathways and eventually throughout the heart; this electric excitation produces a synchronous contraction of the heart muscle [5]. The associated biopotential is the ECG. Electric excitation of a neuron produces an action potential that travels down its dendrites and axon [4]; activity of a massive number of neurons and their interactions within the cortical mantle result in the EEG signal [6]. The excitation of neurons transmitted via a nerve to a neuromuscular junction produces stimulation of muscle fibers. Constitutive elements of muscle fibers are the single motor units, and their electric activity is called a single-motor-unit potential [7]. The electric activity of large numbers of single-motor-unit potentials from groups of muscle fibers manifests on the body surface as the EMG. Contraction and relaxation of muscles is accompanied by proportionate EMG signals. The retina of the eye is a multilayered and rather regularly structured organ containing cells called rods and cones, cells that sense light and color. The motion of the eyeballs inside the conductive contents of the skull alters the electric potentials. Placing the electrode in the vicinity of the eyes (on either side of the eyes on the temples or above and below the eyes) picks up the potentials associated with eye movements called EOGs. Thus, it is clear that biopotentials at the cellular level play an integral role in the function of various vital organs.

26.3 Biopotentials

Biopotentials from organs are diverse. Table 26.1 lists some of these biopotentials, their representative clinical applications, and their key measurement indexes and associated sensors. Note that all acquisitions are made with the aid of specialized electrodes in which the actual design may be customized for specific needs. The most noteworthy features of biopotentials are [1,8] the following:

- Small amplitudes (10 μV to 10 mV)
- Low-frequency range of signals (from direct current [DC] to several hundred hertz)

TABLE 26.1

Biopotentials, Specifications, and Applications

Source	Amplitude (mV)	Bandwidth (Hz)	Sensor (Electrodes)	Measurement Error Source	Selected Applications
ECG	1–5	0.05–100	Ag–AgCl disposable	Motion artifact, 50/60 Hz power-line interference	Diagnosis of ischemia, arrhythmia, conduction defects
EEG	0.001–0.001	0.5–40	Gold-plated or Ag–AgCl reusable	Thermal (Johnson) RF noise, 50/60 Hz	Sleep studies, seizure detection, cortical mapping
EMG	1–10	20–2000	Ag or carbon, stainless steel, needle	50/60 Hz, RF	Muscle function, neuromuscular disease, prosthesis
EOG	0.01–0.1	DC–10	Ag–AgCl	Skin potential motion	Eye position, sleep state, vestibulo-ocular reflex

The most noteworthy problems of such acquisitions are the following:

- Presence of biological interference (from skin, electrodes, motion, etc.)
- Noise from environmental sources (power line, radio frequency, electromagnetic, etc.)

These signal-acquisition challenges and problems for each of the biopotentials are considered in detail in the following.

26.3.1 Electrocardiogram

ECG signals are acquired by placing electrodes directly on the torso, arms, and legs (Figure 26.3a). The activity on the body surface is known to reflect the activity of the heart muscle underneath and in its proximity. A clinically accepted lead system has been devised and is called the 12-lead system [9–10]. It comprises a combination of electrodes taking measurements from different regions' designated limb leads, the precordial leads, and the chest leads. Limb leads derive signals from electrodes on the limbs and are designated as leads I, II, and III. Precordial leads are designated aVR, aVL, and aVF and are derived by combining signals from the limb leads. The remaining six leads, V1 to V6, are chest leads. Together, ECGs from these various leads help define the nature of the activity on a specific part of the heart muscle, for example, ischemia (impaired oxygen supply to the muscle) or infarction (damage to the muscle) on the left side of the chest may be noticeable in lead III.

The ECG signals at the surface of the body are small in amplitude, which make the measurements susceptible to artifacts [11], generated by the relative motion of the electrode and the skin as well as by the activity of the nearby muscles. An important consideration in good ECG signal acquisition is the use of high-quality electrodes [12]. Electrodes made out of silver coated with silver chloride or of sintered Ag–AgCl material are recommended. An electrolytic gel is used to enhance conduction between the skin and the electrode metal. Artifacts at the electrode–skin contact as well as electromagnetic interference from all sources must be minimized [13]. Since ECG instruments are often used in critical care environments, they must be electrically isolated for safety [14] and protected from the high voltages generated by defibrillators [15].

ECG biopotential amplifiers find use in many monitoring instruments, pacemakers, and defibrillators [16]. ECG signal acquisition is also useful in many clinical applications, including diagnosis of arrhythmias, ischemia, or heart failure.

26.3.2 Electroencephalogram

EEG signals are characterized by their extremely small amplitudes (in the microvolt range). Gold-plated electrodes are placed very securely on the scalp to make a very low-resistance contact. A clinically accepted lead system [17], which includes several electrodes placed uniformly around the head, is called the 10–20 lead system (Figure 26.3b). This comprehensive lead system allows localization of diagnostic features, such as seizure spikes, in the vicinity of the electrode [18].

EEG signals are difficult to interpret since they represent the comprehensive activity of billions of neurons transmitted via the brain tissues, fluids, and scalp [18]. Nevertheless, certain features can be interpreted. In the waveform itself, it is possible to see interictal seizure spikes or a full seizure (such as a petit mal or a grand mal epilepsy) [18]. Analysis of the frequency spectrum of the EEG can reveal changes in the signal power at different

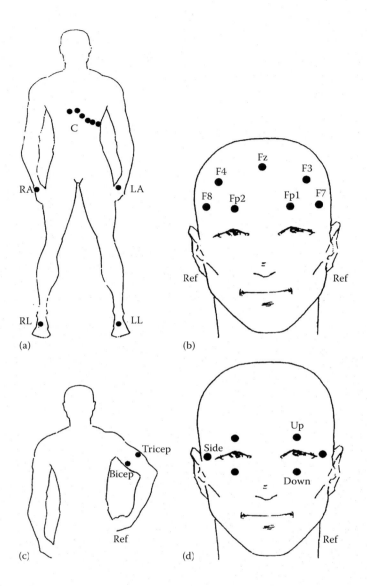

FIGURE 26.3
Schematics showing how biopotential signals are recorded from the human body. (a) ECG: 12-lead ECG is recorded using a right arm (RA), a left arm (LA), a left leg (LL), a right leg (RL), and six chest (C) electrodes; (b) EEG: selected electrode locations from the standard 10–20 EEG lead system with ears used as reference; (c) EMG: recording electrodes on the biceps and triceps with an independent reference; and (d) EOG: electrodes above or below (up–down) and the sides of the eyes along with an independent reference.

frequencies being produced during various stages of sleep, as a result of anesthetic effects and sometimes as a result of brain injury [17].

Practical problems and challenges associated with EEG signal recordings arise from physiological, environmental, and electronic noise sources. Physiological sources of interference are motion artifact, muscle noise, eye motion or blink artifact, and sometimes even heartbeat signals. Electrical interference arises from the usual sources: 60 Hz power lines, RFs, and electrically or magnetically induced interference. Moreover, the electronic components in the amplifier also contribute noise. Good design and measuring techniques can mitigate the effects of such noise and interference.

26.3.3 Electromyogram

Muscle fibers generate electric activity whenever muscles are active [19]. EMG signals are recorded by placing electrodes close to the muscle group (Figure 26.3c). For example, a pair of electrodes placed on the biceps and another pair placed on the triceps can capture the EMG signals generated when these muscles contract. EMG signals recorded in this manner have been shown to give a rough indication of the force generated by the muscle group [8]. Electrodes used for such applications should be small and securely attached and should provide recordings free of artifacts. Either silver–silver chloride or gold-plated electrodes perform quite well, although inexpensive stainless-steel electrodes may also suffice.

Since the frequency range of EMG signals is higher than that of ECG and EEG signals and since the signals are of comparable or larger amplitudes, the problem of motion artifact and other interference is relatively less severe. Filtering can reduce the artifact and interference; for example, setting the bandwidth to above 20 Hz can greatly reduce the skin potentials and motion artifacts.

Recording activity directly from the muscle fibers themselves can be clinically valuable in identifying neuromuscular disorders [19]. Therefore, invasive electrodes are needed to access the muscle fibers or the neuromuscular junction. Fine-needle electrodes or thin stainless-steel wires are inserted or implanted to obtain local recording from the fibers or neuromuscular junctions [7].

26.3.4 Electrooculogram

Electric potentials are generated as a result of movement of the eyeballs within the conductive environment of the skull. The generation of EOG signals can be understood by envisaging dipoles (indicating separated positive and negative potential sources) located in the eyeballs. Electrodes placed on either side of the eyes or above and below them pick up the potentials generated by the motion of the eyeball (Figure 26.3d). This potential varies approximately in proportion to the movement of the eyeballs; hence, EOG is sometimes used to study eye positions or disorders of eye movement and balance (a reflex called vestibulo-ocular reflex affects the nystagmus of the eye). Similarly, saccades inherent in eye motion as well as blinking of the eyelids can produce changes in the EOG signal.

The EOG signal is small (10 to 100 µV) and has low frequencies (DC to 10 Hz) [8]. Hence, an amplifier with a high gain and good low-frequency response and DC stability is desirable. Additionally, the electrode–gel combination should be such that it produces low levels of junction potential, motion artifacts, and drift in the DC signal [20]. Practical problems associated with DC drift, motion artifacts, and securing electrodes in the vicinity of the eyes make their long-term use problematic. Nevertheless, EOG signals can be useful clinically in acute studies of human disorders; therefore, careful acquisition of the signal followed by appropriate analysis is used to interpret the EOG potentials.

Other biopotential-recording techniques follow similar principles of measurements. The electrode design should be specifically adapted to the source of the signal. A thorough effort is required to minimize the noise and interference by improving electrode design and placement and optimizing the amplifier circuit. Good electrode attachment along with selective filtering at the amplifier can help obtain relatively noise-free recording. The design principles and practical considerations are described next.

26.4 The Principles of Biopotential Measurements

The unifying principles of biopotential recordings involve

- Electrode design and its attachment suited to the application;
- Amplifier circuit design for suitable amplification of the signal and rejection of noise and interference; and
- Good measurement practices to mitigate artifacts, noise, and interference.

26.5 Electrodes for Biopotential Recordings

Electrodes for biopotential recordings are designed to obtain the signal of interest selectively while reducing the potential to pick up artifact. The design should be pragmatic to reduce cost and allow for good manufacturing and reliable long-term use. These practical considerations determine whether high-quality but reusable electrodes made of silver or gold or cheaper disposable electrodes are used [20].

26.5.1 Silver–Silver Chloride Electrodes

The classic, high-quality electrode design consists of a highly conductive metal (silver) interfaced to its salt (silver chloride) and connected via an electrolytic gel to the human body [21]. Silver–silver chloride–based electrode design is known to produce the lowest and most stable junction potentials [1,20]. Junction potentials are the result of the dissimilar electrolytic interfaces and are a serious source of electrode-based motion artifacts. Therefore, additionally, an electrolytic gel typically based on sodium or potassium chloride is applied to the electrode. A gel concentration on the order of 0.1 M (molar concentration) results in a good conductivity and low junction potential without causing skin irritation.

Reusable silver–silver chloride electrodes (Figure 26.4a) are made of silver disks coated electrolytically by silver chloride [1], or, alternatively, particles of silver and silver chloride are sintered together to form the metallic structure of the electrode. The gel is typically soaked into a foam pad or is applied directly in a pocket produced by the electrode housing. The electrode is secured to the skin by means of nonallergenic adhesive tape. The electrode is connected to the external instrumentation typically via a snap-on connector. Such electrodes are well suited for acute studies or basic research investigations.

Disposable electrodes are made similarly, although the use of silver may be minimized (for example, the snap-on button itself may be silver coated and chlorided). To allow for a secure attachment, a large foam pad attaches the electrode body with adhesive coating on one side (Figure 26.4b). Such electrodes are particularly suited for ambulatory or long-term use.

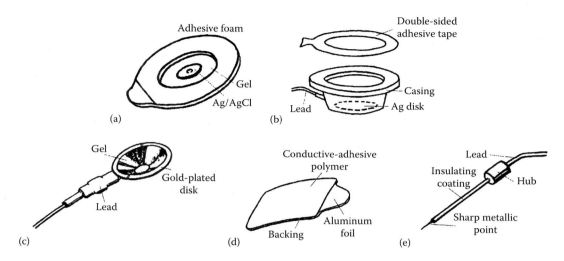

FIGURE 26.4

Examples of electrodes used in biopotential recordings: (a) disposable Ag–AgCl electrode, (b) reusable Ag–AgCl disk electrode, (c) gold disk electrode, (d) disposable conductive polymer electrode, and (e) needle electrode.

26.5.2 Gold Electrodes

Gold-plated electrodes (Figure 26.4c), which have the advantages of high conductivity and inertness desirable in reusable electrodes, are commonly used in EEG recordings [1]. Small reusable electrodes are designed so that they can be securely attached to the scalp. The electrode body is also shaped to make a recessed space for electrolytic gel, which can be applied through a hole in the electrode body [18]. The electrodes are attached in hair-free areas by use of a strong adhesive such as colloidon or securely attached with elastic bandages or wire mesh. Similar electrodes may also be used for recording EMG, especially when a great deal of motion is expected. Disadvantages of using gold electrodes over silver–silver chloride electrodes include greater expense, higher junction potentials, and greater susceptibility to motion artifacts [20]. On the other hand, gold electrodes maintain low impedance, are inert and reusable, and are good for short-term recordings as long as a highly conductive gel is applied and they are attached securely.

26.5.3 Conductive Polymer Electrodes

It is often convenient to construct an electrode out of a material that is simultaneously conductive and adhesive [20]. Certain polymeric materials have adhesive properties and by attaching monovalent metal ions can be made conductive. The polymer is attached to a metallic backing made of silver or aluminum foil, which allows electric contact to external instrumentation (Figure 26.4d). This electrode does not need additional adhesive or electrolytic gel and, hence, can be immediately and conveniently used. The conductive polymeric electrode performs adequately as long as its relatively higher resistivity (over metallic electrodes) and greater likelihood of generating artifacts are acceptable. The higher resistivity of the polymer makes these electrodes unsuitable for low-noise measurement. The polymer does not attach as effectively to the skin as does the conventional adhesive on disposable ECG electrodes built with a foam base; furthermore, the potentials generated at the electrode–skin interface are more readily disturbed by motion. Nevertheless, when the signal level is high and when restricting the subject movement minimizes artifact, the polymeric electrode offers a relatively inexpensive solution to biopotential recording.

26.5.4 Metal or Carbon Electrodes

Although other metals such as stainless-steel or brass electrodes [21] are used rather infrequently now because high-quality noble-metal electrodes or low-cost carbon or polymeric electrodes are so readily available, historically these metallic electrodes were used in laboratory or clinical settings because of their sturdy construction and reusability. Electrode gel is applied to the metal electrode which is fastened to the body by means of a rubber band. These electrodes have the potential for producing very high levels of artifact and are bulky and awkward to use but do offer the advantage of being reusable and tend to be inexpensive. Carbon or carbon-impregnated polymer electrodes are also used occasionally (although they are mainly used as electrical stimulation electrodes) [20]. These electrodes have a much higher resistivity and are noisier and more susceptible to artifacts, but they are inexpensive, flexible, and reusable and, thus, are chosen for applications such as electric stimulation or impedance plethysmography. For these applications, gel is usually not applied and the electrodes are used in "dry" form for easy attachment and removal.

26.5.5 Needle Electrodes

Needle electrodes (Figure 26.4e) comprise a small class of invasive electrodes, used when it is absolutely essential to record from the organ itself. The most common application is in recording from muscles or muscle fibers [8]. A metallic, typically steel, wire is delivered via a needle inserted at the site of the muscle fiber. The wire is hooked and, hence, fastens to the muscle fiber, even as the needle is removed. Small signals such as motor-unit potentials can be recorded in this manner [7]. For research applications, similar needle or wire electrodes are sometimes connected directly to the heart muscle. Since such electrodes are invasive, their use is limited to only highly specialized and supervised clinical or research applications.

26.6 The Biopotential Amplifier

Biopotentials exhibit small amplitudes and low frequencies [22]. Moreover, biopotential measurements are corrupted by environmental and biological sources of interference. Therefore, the essential, although not exhaustive, design considerations include proper amplification and bandwidth, high input impedance, low noise, and stability against temperature and voltage fluctuations. The key design component of all biopotential amplifiers is the instrumentation amplifier [21]. However, each biopotential-acquisition instrument has a somewhat differing set of characteristics, necessitating some specialization in the design of the instrumentation amplifier. Table 26.2 summarizes the circuit specialization needed in various biopotential amplifiers, with the ECG amplifier used as the basic design.

26.6.1 The Instrumentation Amplifier

The instrumentation amplifier is a circuit configuration that potentially combines the best features desirable for biopotential measurements [8], namely, high differential gain, low

TABLE 26.2

Distinguishing Features and Design Consideration for Biopotentials

Biopotential	Distinguishing Feature	Exclusive Amplifier Design Consideration	Additional Features Desired
ECG[a]	1 mV signal, 0.05–100 Hz BW	Moderate gain, BW, noise, CMRR, input R	Electrical safety, isolation, defibrillation protection
EEG	Very small signal (microvolts)	High gain, very low noise, filtering	Safety, isolation, low electrode–skin resistance
EMG	Higher BW	Gain and BW of op-amps	Postacquisition data processing
EOG	Lower frequencies, small signal	DC and low drift	Electrode–skin junction potential, artifact reduction

Note: BW: bandwidth; CMRR: common-mode rejection ratio; op-amps: operational amplifiers.
[a] The ECG signal acquisition is considered the standard against which the other acquisitions are compared.

common-mode gain, high CMRR, and high input resistance [23]. Figure 26.5 shows the design of the basic instrumentation amplifier. The basic circuit design principles have been described elsewhere [23–24]. The instrumentation amplifier is constructed from operational amplifiers, or op-amps, which have many of the desirable features listed above [25]. The front end of the amplifier has two op-amps, which consists of two noninverting amplifiers that have been coupled together by a common resistor R_1. The gain of the first stage is $(1 + 2R_2/R_1)$. The second stage is a conventional differential amplifier with gain of $-(R_4/R_3)$. This design results in the desired differential gain distributed over two stages of the amplifier. It also achieves a very high input resistance as a result of the noninverting amplifier front end. It exhibits a very high CMRR as a result of the differential first stage

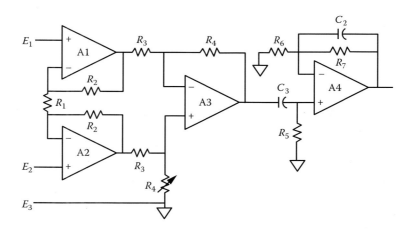

FIGURE 26.5
The instrumentation amplifier. This amplifier has a very high input impedance, high CMRR, and a differential gain set by the resistors in the two amplifier stages. The gain of the first stage (amplifiers A1 and A2) is $1 + 2R_2/R_1$; the second stage (amplifier A3), R_4/R_3; and the third stage (amplifier A4), $1 + R_7/R_6$. The lower-corner frequency is $1/(2\pi R_5 C_1)$ and the upper-corner frequency is $1/(2\pi R_7 C_2)$. The variable resistor R is adjusted to maximize the CMRR. Electrodes E_1 and E_2 are the recording electrodes, while E_3 is the reference or the ground electrode.

followed by a second-stage differential amplifier. The CMRR is enhanced by adjusting one of the matching resistors and by selecting high-CMRR op-amps. This instrumentation amplifier is a key design component universal to many biosensor interfaces and almost all biopotential instruments [22].

26.6.2 The Electrocardiogram Amplifier

The ECG amplifier can readily be designed using the instrumentation amplifier as the principal building block. Active filters with a lower-corner frequency of 0.05 Hz and an upper-corner frequency of 100 Hz are also typically added [8].

ECG amplifiers are needed in many applications, such as monitoring in cardiac intensive care units, where safety and protection are of paramount importance. Because the possibility of a direct or low-resistance access to the heart via catheters or intravenous lines exists in such settings, very small electric leakage currents can be fatal. Consequently, leakage from the amplifier is required to be below the safety standard limit of 10 μA [14]. Additionally, safety of the patient is achieved by providing electrical isolation from the power line and the earth ground, which prevents passage of leakage current from the instrument to the patient under normal conditions or under reasonable failure conditions. Electrical isolation is achieved by using transformer or optical coupling components [9], although it is important to remember that any such design should preserve the bandwidth and linearity of the amplifier. ECG amplifiers are also likely to be operated in circumstances where defibrillators might be used; thus, the amplifier circuit must be protected against the high defibrillation voltages and must be augmented by circuit components such as current-limiting resistors, voltage-limiting diodes, and spark gaps [15].

26.6.3 The Electroencephalogram Amplifier

The distinguishing feature of an EEG amplifier is that it must amplify very small signals [8]. The amplifier gain must be suitably enhanced to deal with microvolt or lower levels of signals. Furthermore, all components of the amplifier must have a very low thermal noise and in particular low electronic (voltage and current) noise at the front end of the amplifier. EEG amplifiers used in clinical applications again must be electrically isolated and protected against high defibrillation voltages, similar to the ECG amplifier.

26.6.4 The Electromyogram Amplifier

EMG amplifiers are often used in the investigation of muscle performance and neuromuscular diseases and in building certain powered or smart prostheses. In such applications, slightly enhanced amplifier bandwidth suffices. In addition, postprocessing circuits are almost always needed. For example, a rectified and integrated EMG signal has been shown to give a rough indication of the muscle activity, approximately related to the force being generated at the location of the EMG electrode [8].

26.6.5 The Electrooculogram Amplifier

The EOG signal is small in amplitude and consists of very low frequencies. Therefore, an EOG amplifier must not only have a high gain, but also a very good low frequency, or even near DC, response. This frequency response also makes the amplifier potentially susceptible to shifts in the junction potential at the skin–electrode interface and to drift in the electronic circuit characteristics. In addition to using good electrodes (Ag–AgCl) and gel (high conductivity), some type of active DC or drift cancellation or correction circuit design may be necessary.

26.7 Circuit Enhancements

The basic biopotential amplifier described above, along with the specific design considerations for each biopotential, can yield a signal acquisition of acceptable quality in most laboratory settings. In practice, however, further enhancements are always necessary to achieve acceptable clinical performance in novel applications. These enhancements include circuits for reducing electric interference, filtering noise, reduction of artifacts, electrical isolation of the amplifier, and electrical protection of the circuit against defibrillation shocks [9].

26.7.1 Electrical Interference Reduction

Environmental electric interference is always present, especially in urban hospital environments. It is desirable to eliminate interference before it enters the amplifier, for example, by proper shielding of the subject, leads, and the instrument and by grounding the subject and the instrument. Sources of interference include induced signals from power lines and electric wiring; RF from transmitters, electric motors, and other appliances; and magnetically induced currents in lead wires [13]. Interference induced on the body common to the biopotential sensing electrodes is called the common-mode interference (as distinguished from the biopotential that is differential to the sensing electrodes). If the induced current is i_d and the resistance to ground is R_0, then the common-mode interference potential is $V_c = i_d R_0$. The common-mode interference is principally rejected by a differential or instrumentation amplifier with a high CMRR. Further improvement is possible by use of the "driven right leg circuit." The right leg lead, by standard convention, is used as the ground or the circuit reference. The driven right leg circuit employs the clever idea of negative feedback of the common-mode signal into this lead. The common-mode signal is sensed from the first stage of the instrumentation amplifier, amplified and inverted, and fed back into the right leg lead (Figure 26.6a). At this stage the common-mode signal is reduced to $(i_d R_0)/(1 + 2R_2/R_1)$. Thus, the common-mode interference is greatly reduced at its source. The driven right leg circuit along with a high CMRR of the amplifier and filtering permit very high-quality biopotential measurements.

26.7.2 Filtering

After following the precautions described above, filtering at the front end of the amplifier and limiting the bandwidth of the biopotential amplifier can further help to reduce the interference (Figure 26.6b). Small inductors or ferrite beads in the lead wires help to block very high-frequency electromagnetic interference. Small capacitors between each electrode lead and ground filter the RF interference. Bandwidth limitation can be imposed at each stage of the amplifier. Because DC potentials arising at the electrode–skin interface must be blocked well before the biopotential is amplified greatly (otherwise, the amplifier

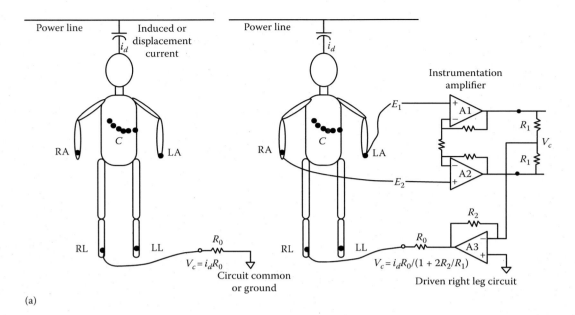

(a)

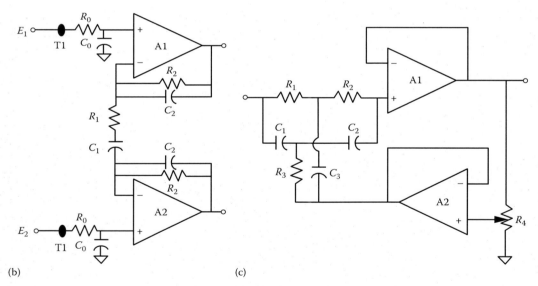

(b)   (c)

FIGURE 26.6

Circuit enhancements for biopotential measurements. (a) *Left:* Electric interference induced by the displacement current i_d from the power line; this current flows into the ground electrode lead generating common-mode voltage V_c. *Right:* A driven right leg circuit using negative feedback into the right leg electrode to reduce the effective common-mode voltage. (b) Amplifier front-end filters: T1—RF choke; R_0 and C_0—RF filter; R_1 and C_1—high-pass filter; and R_2 and C_2—low-pass filter. (c) Notch filter for power-line interference (50 or 60 Hz): twin T notch filter in which notch frequency is governed by R_1, R_2, R_3, C_1, C_2, and C_3 and notch tuning by R_4.   (*Continued*)

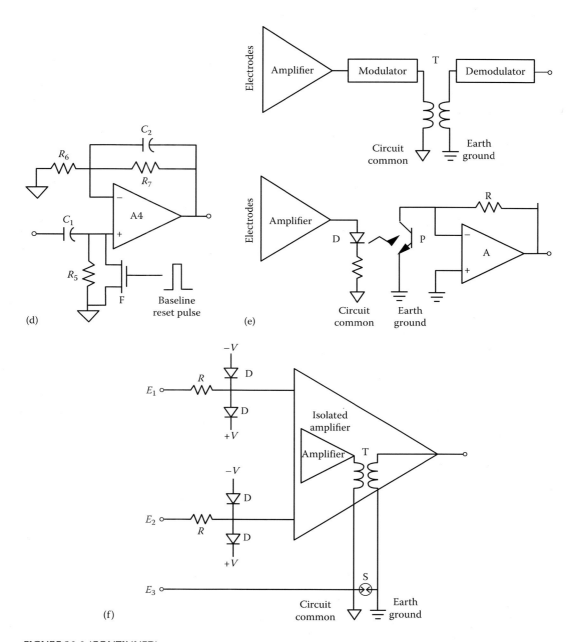

FIGURE 26.6 (CONTINUED)
Circuit enhancements for biopotential measurements. (d) Baseline restoration circuit: the high-pass filter capacitor C_1 is discharged by field-effect transistor F when activated manually or automatically by a baseline restoration pulse. (e) Electrical isolation: transformer coupled using the transformer T (top) or optical using the diode D and the photodetector P (bottom). Note that the isolator separates circuit common on the amplifier side from the earth ground on the output side. (f) Electrical protection circuit: resistance R limits the current, reverse-biased diodes D limit the input voltage, and the spark gap S protects against defibrillation pulse-related breakdown of the isolation transformer T.

could saturate), use of high-pass filtering in the early stages of amplification is recommended. Low-pass filtering at several stages of amplification is recommended to attenuate residual RF interference as well as muscle signal interference. Power-line interference at 50 or 60 Hz and their harmonics clearly poses the biggest problem in biopotential measurement [11,13]. Sometimes it may be desirable to provide a 50 or 60 Hz notch filter to remove the power-line interference (Figure 26.6c), an option that is often available with low-level signal (EEG and EOG) measuring instruments. The risk of a distorted biopotential signal arises when a notch filter is used and this may affect diagnosis. Filtering should, therefore, be used selectively.

26.7.3 Artifact Reduction

One principal source of artifact is the potential arising at the electrode–skin interface [11]. Slow changes in the baseline can arise due to changes in the junction potential at this interface and, in some instances, can cause a temporary saturation of the amplifier [9]. This event is detected manually or automatically (by quickly discharging the high-pass capacitor in the amplifier to restore the baseline; Figure 26.6d). Movement of the subject or disturbance of the electrode can produce motion artifacts [11], which can be reduced by filtering the signal, but as suggested above, such filtering, typically high pass, can severely distort the biopotential being measured. Alternatively, computerized processing may be necessary to identify an artifact and delete it from display and processing. Of note, a biopotential source could be the desired one in one case, but an unwanted artifact in another case. For example, EOG signal resulting from blinking of eyes can produce a rather significant artifact in EEG recordings. Similarly, EMG signals become unwanted artifacts in all other non-EMG biopotential measurements. ECG monitoring must especially account for EMG artifact for high-fidelity recording. Another example is the pacemaker pulse. Since a pacemaker pulse can be detected and amplified as a short (about 2 ms) pulse preceding a QRS complex, it can be mistakenly interpreted as a heartbeat by some circuits for automatically determining heart rate. Special circuits must be designed to identify and delete this artifact [9].

26.7.4 Electrical Isolation

Electrical isolation limits the possibility of the passage of any leakage current from the instrument in use to the patient [22]. Conversely, patient safety must be ensured by electrical isolation to reduce the prospect of leakage of current from any other sensor or instrument attached to the patient to the earth ground of the instrument being tested [8]. Passage of leakage current through the patient could be harmful or even fatal if this current were to leak to the heart via a catheter or an intravenous line. Electrical isolation can be done electrically by inserting a transformer in the signal path or optically by introducing an optical coupler (Figure 26.6e). Since the primary and the secondary of the transformer remain electrically isolated, no direct path to ground can exist. One problem with this approach is that the transformer is inherently an alternating current (AC) high-frequency device. Therefore, a suitable solution is to modulate the biopotential signal by using a high-frequency carrier preferred by the transformer. An alternative solution is to use optical isolation. The electric signal from the amplifier is first converted to light by a light-emitting diode (LED). This optical signal is modulated in proportion to the electric signal and transmitted to the detector. A photodetector (photodiode or a phototransistor) then picks up the light and converts it into an electric signal, which is then demodulated to recover the original signal. The optical signal is typically pulse code modulated to circumvent the inherent nonlinearity of the LED–phototransistor combination.

26.7.5 Defibrillation Protection

Biopotential-measuring instruments can encounter very high voltages, such as those from electric defibrillators that can damage the instrument [9]. For example, electric shocks on the order of 1500 to 5000 V may be produced by an external defibrillator [1]. Other high-voltage sources are electrocautery (used in surgery) and power lines (inadvertent short circuits in the instrument). Therefore, the front end of the biopotential instrument must be designed to withstand these high voltages (Figure 26.6f). Use of resistors in the input leads can limit the current in the lead and the instrument. Protection against high voltages is achieved by the use of diodes or Zener diodes. These components conduct at 0.7 V (diode conduction voltage) or 10 to 15 V (depending on the Zener diode breakdown voltage), thus protecting the sensitive amplifier components. Since it is more likely that protection against higher voltages will be needed, low-pressure gas discharge tubes such as neon lamps are also used. They break down at voltages on the order of 100 V, providing an alternative path to ground for the high voltages. As a final line of protection, the isolation components (optical isolator or transformer) must be protected by a spark gap that activates at several thousand volts. The spark gap ensures that the defibrillation pulse does not breach the isolation.

26.8 Measurement Practices

Biopotential measurements are made feasible, first of all, by good amplifier designs. High-quality biopotential measurements require use of good electrodes and their proper application on the patient, along with good laboratory or clinical practices. These practices are summarized below.

26.8.1 Electrode Use

Various electrodes best suited for each biopotential measurement were described earlier. First, different electrodes by virtue of their design offer distinguishing features: more secure (use of strong but less irritating adhesives), more conductive (use of noble metals such as silver and gold), less prone to artifact (use of low–junction potential materials such as Ag–AgCl). Electrode gel can be of considerable importance in maintaining a high-quality interface between the electrode metal and the skin. High-conductivity gels, in general, help reduce the junction potentials along with the resistance (they tend, however, to be allergenic or irritating; hence, a practical compromise in terms of electrolyte concentration must be found) [20]. Movement of the electrode with respect to the electrode gel and the skin is a potential source of artifact (Figure 26.7a). Such movements can change the electrode junction to skin potentials, producing motion artifacts [21]. Placement above bony structures where there is less muscle mass can reduce unwanted motion artifact and EMG interference (Figure 26.7b). Electrodes must be securely attached, for example, with stress loops secured away from the electrode site, so that motion artifact can be reduced. In certain instances, the electrodes may be essentially glued to skin, as in the case of EEG measurements.

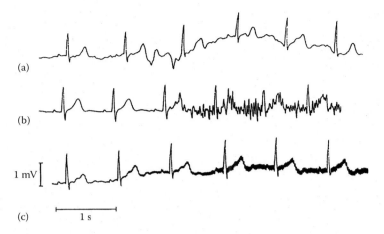

FIGURE 26.7
Examples of electric interference in biopotential recordings: (a) ECG signal with baseline changes and motion artifacts, (b) muscle signal interference, and (c) electromagnetic interference (60 Hz power line and RF).

26.8.2 Skin Preparation

The potentials existing at the skin surface, attributable to potentials at the membranes of cells in the epidermal layers of the skin, can result in a large DC potential (which can be a significant problem in EOG measurements). Any disturbance of the skin by motion, touching, or deformation can cause this potential to change and result in motion artifacts (Figure 26.7a). Sweat glands in the epidermis can also contribute varying extents of skin resistance and skin potential. Such potentials and artifacts can be reduced by abrading the epidermal skin. A mild abrasion by sandpaper or its equivalent can significantly reduce skin resistance and skin potential and, thereby, reduce artifact [26]. A less traumatic, but somewhat less effective approach, is to use an alcohol swab or similar skin-cleansing solution to wet and clean the skin surface to remove debris, oils, and damaged or dead epidermal cells. Sometimes, as with EEG measurements where very low signals are recorded and very low noise is permitted, skin resistance must be significantly lowered, perhaps to below 2 kΩ [18]. Obviously, reduced motion or muscle activity while measurement is carried out also helps.

26.8.3 Reduction of Environmental Interference

Electromagnetic interference radiated from the power lines, RF interference from machines, induced magnetic field in the leads, and electric currents induced on to the body are all potential sources of environmental interference (Figure 26.7c). Shielding of the amplifier along with the electrode and the lead, and in certain extreme conditions, shielding of the subject (for example, when taking magnetic field measurements from the body) can greatly help reduce the signals picked up by or induced into the amplifier. The electrode leads can be shielded or, at the very least, twisted together to reduce induced electromagnetic interference.

The amplifier circuit should also have extensive filtering of unwanted electromagnetic interference. To eliminate RF interference, filter capacitors should be used in the front end of the amplifier as well as at various stages of the amplifier. Very high frequencies can be blocked by the use of a choke or an inductor at the input leads. The effect of electrostatic interference can be minimized or eliminated by grounding the instrument.

Electric interference in the environment induces current into the body, which is then picked up by the biopotential amplifier as a common-mode voltage [27]. The CMRR property of the amplifier is essential for reduction of the common-mode voltage [25]. Finally, the driven right leg design [27], described earlier, can be optionally used to reduce further the common-mode voltage and the effective interference.

26.9 Conclusions

Biopotential acquisition is a well-developed science, and acceptable engineering design solutions do exist. It is apparent that each biopotential source presents its own distinct challenge in terms of electrode interface, amplifier design, pre- or postprocessing, and practical implementation and usage. ECG signals can be best acquired using Ag–AgCl electrodes, although good experimental/clinical practice is needed to reduce biological and environmental interference. Further circuit protection and isolation are necessary in clinical usage. EEG signals are distinguishable by their very low amplitude; hence, EEG electrodes must be securely attached via a very small electrode–skin resistance and the amplifier must exhibit exceptionally low noise. For EMG acquisition, electrodes are needed that can be attached for long periods to the muscle groups under study. The EMG signal inevitably needs postprocessing, such as integration, to derive a measure of muscle activity. EOG signals have small amplitudes and are characterized by DC or low frequencies. Skin–electrode potentials and DC drift of the amplifier are, therefore, important considerations.

These biopotential measurement principles are applicable to a variety of conventional as well as emerging applications. For example, although ECG acquisition is used mainly in cardiac monitors, it is also of interest and importance in implantable pacemakers and defibrillators. EEG acquisition is useful in the detection of seizure spikes and study of sleep patterns and it may also be used to identify cortical dysfunction after trauma or stroke. EMG acquisition is used in diagnosing neuromuscular diseases. Interesting attempts have been made to use EMG for controlling prostheses. EOG has been helpful in diagnosing vestibulo-ocular disorders and also has been studied as a way of operating communication devices (pointing) used by quadriplegics. The measurement and instrumentation principles described in this chapter would be applicable, with some modifications, to these emergent applications.

References

1. L. A. Geddes and L. E. Baker, *Principles of Applied Biomedical Instrumentation*, Third ed., New York: Wiley, 1989.
2. R. Plonsey, *Bioelectric Phenomena*, New York: McGraw-Hill, 1969.
3. R. Plonsey and R. C. Barr, *Bioelectricity*, New York: Plenum, 1988.
4. R. C. Barr, "Basic electrophysiology," in *The Biomedical Engineering Handbook*, J. Bronzino, Ed., Boca Raton: CRC Press, pp. 101–118, 1995.

5. D. Durrer et al., "Total excitation of the isolated human heart," *Circulation*, 41, 899–912, 1970.

6. P. L. Nunez, *Electric Fields of the Brain*, New York: Oxford University Press, p. 484, 1981.

7. K. A. Henneberg, "Principles of electromyography," in *The Biomedical Engineering Handbook*, J. D. Bronzino, Ed., Boca Raton: CRC Press, pp. 191–200, 1995.

8. J. G. Webster, Ed., *Medical Instrumentation: Application and Design*, Fourth ed., Hoboken: Wiley, 2010.

9. N. V. Thakor, "Electrocardiographic monitors," in *Encyclopedia of Medical Devices and Instrumentation*, J. G. Webster, Ed., New York: Wiley, pp. 1002–1017, 1988.

10. H. V. Pipberger et al., "Recommendations for standardization of leads and specifications for instruments in electrocardiography and vector cardiography," *Circulation*, 52, 11–31, 1975.

11. J. G. Webster, "Reducing motion artifacts and interference in biopotential recording," *IEEE Trans. Biomed. Eng.*, 31, 823–826, 1984.

12. N. V. Thakor and J. G. Webster, "Electrode studies for the long-term ambulatory ECG," *Med. Biol. Eng. Comput.*, 23, 116–121, 1985.

13. J. C. Huhta and J. G. Webster, "60-Hz interference in electrocardiography," *IEEE Trans. Biomed. Eng.*, 20, 91–101, 1973.

14. Anonymous, "American National Standard Safe Current Limits for Electromedical Apparatus," *ANSI/AAMI*, vol. SCL 12/78, 1978.

15. Anonymous, "American National Standard for Diagnostic Electrocardiographic Devices," *ANSI/AAMI*, vol. EC11-1982, 1984.

16. N. V. Thakor, "From Holter monitors to automatic defibrillators: developments in ambulatory arrhythmia monitoring," *IEEE Trans. Biomed. Eng.*, 31, 770–778, 1984.

17. A. S. Gevins and M. J. Aminoff, "Electroencephalography: Brain electrical activity," in *Encyclopedia of Medical Devices and Instrumentation*, J. G. Webster, Ed., New York: Wiley, pp. 1084–1107, 1988.

18. E. Niedermeyer and F. Lopes da Silva, *Electroencephalography*, Baltimore: Urban & Schwarzenberg, 1987.

19. C. J. De Luca, "Electromyography," in *Encyclopedia of Medical Devices and Instrumentation*, J. G. Webster, Ed., New York: Wiley, pp. 1111–1120, 1988.

20. H. Carim, "Bioelectrodes," in *Encyclopedia of Medical Devices and Instrumentation*, J. G. Webster, Ed., New York: Wiley, pp. 195–226, 1988.

21. M. R. Neuman, "Biopotential electrodes," in *Medical Instrumentation: Application and Design*, Fourth ed., J. G. Webster, Ed., Hoboken: Wiley, 2010.

22. J. H. Nagle, "Biopotential amplifiers," in *The Biomedical Engineering Handbook*, J. D. Bronzino, Ed., Boca Raton: CRC Press, pp. 1185–1195, 1995.

23. S. Franco, *Design with Operational Amplifiers*, New York: McGraw-Hill, 1988.

24. W. J. Jung, *IC Op Amp Cookbook*, Third ed., Indianapolis: Howard W. Sams, 1986.

25. P. Horowitz and W. Hill, *The Art of Electronics*, Second ed., Cambridge, England: Cambridge University Press, 1989.

26. H. W. Tam and J. G. Webster, "Minimizing electrode motion artifact by skin abrasion," *IEEE Trans. Biomed. Eng.*, 24, 134–139, 1977.

27. M. R. Neuman, "Biopotential amplifiers," in *Medical Instrumentation: Application and Design*, Fourth ed., J. G. Webster, Ed., Hoboken: Wiley, 2010.

27

Sensor Signal Conditioning for Biomedical Instrumentation

Tomas E. Ward

CONTENTS

27.1 Introduction

Many connected health applications as with telemedical systems rely on the acquisition and transmission of physiological measurement in a reliable and robust fashion. This is achieved by sensor front ends, which are increasingly found in both wearable and ambient forms. These embodiments require nonintrusive, low-power, safe, and reliable operation in the face of activities of daily living, which place the technology in environments well outside the regulated conditions of the research laboratory (Sweeney et al., 2012a). Furthermore, as digital communication systems provide ever-improving performance in terms of information rate, reliability, power consumption, and physical size, there is increasing emphasis on the sensor and sensor-side processing to deliver high-fidelity representations of the measurand under consideration. Taken together, these push demand toward ever more sophisticated sensor processing yet in physical forms which are smaller, faster, and cheaper. Computational approaches to sensor output processing, especially with respect to the processing of physiological artifacts, are driving the modern tendency toward capturing the output of the sensor in digital form as early as possible in the signal pipeline. In this scenario, only essential filtering and preprocessing steps are performed in the analog domain. Notwithstanding this tendency, there is considerable scope to condition the sensor output to contribute significantly to the overall pipeline signal-to-noise ratio (SNR) figure. It is tempting, due to the ever-increasing capability of modern computational methods to clean up a noise-corrupted sensor signal, to be complacent in regard to noise reduction at the sensor side but it is important to note that there is no substitute for good-quality, clean data (Kappenman and Luck, 2013). The introduction of noise occurs easily and through many mechanisms—removing it is always difficult so it is best to minimize its introduction in the first place (Webster, 1998). Appropriate shielding, sensor design, wiring, and grounding practice has very significant impact in terms of reducing noise onto the sensor reading, while judicial use of basic filtering and amplification can subsequently improve the SNR at an early stage (Metting van Rijn et al., 1990). The level of acceptable signal-to-noise ratio on a sensor reading is application specific; therefore, the SNR over the complete sensor-processing pipeline must be considered during design to develop an appropriate approach in line with the engineering constraints.

For example, single-trial event-related potential (ERP) measurement via noninvasive EEG for the purposes of brain–computer interfacing presents very poor SNR, and consequently, analog design engineers use considerable ingenuity in devising circuitry which can deliver an acceptable SNR for subsequent ERP interpretation (Metting van Rijn et al., 1990, 1991). In contrast, the use of a thermistor as a sensor as part of a system for measuring respiratory rate (such a thermistor is placed near the nostrils and/or integrated into a breathing mask) is characterized by robust SNR and only relatively simple circuit techniques are required to produce a signal suitable for basic respiratory rate analysis (Bronzino, 1995).

Regardless of the precise nature of the application or sensor modality, an abstraction of the sensor-side processing stages is easily to visualize. The canonical signal-processing pipeline under consideration in this chapter is illustrated in Figure 27.1. Such a pipeline serves the purpose of optimizing SNR for data acquisition and subsequent processing for

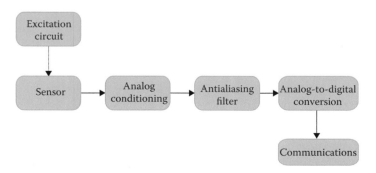

FIGURE 27.1
High-level view of the sensor-processing pipeline.

a specific application. It is worth remarking that if there are potentially many applications that may require the output of this particular sensor, then the SNR budget (end-to-end) needs to be considered for the most demanding case.

The purpose of this chapter then is to highlight the signal-conditioning components in Figure 27.1, including some practical advice on implementation. This signal-conditioning pipeline can range from the very simple—such as a signal buffer—to very complex analog systems. In this chapter, we give a concise overview of the possibilities with an emphasis on basic analog building blocks that can then be used to construct the range of processing functions most commonly encountered. We start on the left-hand side with a short introduction to sensors before traversing the various stages that comprise the pipeline.

27.2 Sensors

A sensor is a device that measures a physical quantity and converts it into a form that can ultimately be interpreted by an observer. It is natural to think of the observer as a person but it can just as easily be any subsequent system whose state is influenced by the sensor output—any measurement instrument can be interpreted in this way—so too are more complex systems, including software systems—from agents to expert systems which rely on sensor output to produce useful output. Most sensors act as transducers; i.e., energy is converted from one form into another. For the purposes of connected health applications as discussed in this chapter, whatever the physical quantity of interest is, it must ultimately be converted into an electrical form of energy and, more specifically, it is desired that changes in the physical quantity are mapped onto changes in voltage as commodity digitization systems (analog-to-digital convertors) are overwhelmingly voltage-based. For some sensors, this conversion to electrical forms of energy is inherent to the physics of the sensing phenomenon (e.g., the use of electrodes to measure electrical activity in the body); while for others, subsequent processing, usually within the sensor package, is required to produce this conversion effectively. As an example of the latter, a digital stethoscope requires an appropriate sensor to convert acoustic energy into an electrical signal. Such a sensor comprises a diaphragm that provides the acoustic amplification and which, in the past, constituted the sensor because an observer—in this case, a person—could now listen to the suitably amplified sound either directly or through an acoustic tube attached to the ear (the classic stethoscope). A digital stethoscope augments the sensing component through the addition of a transducer that converts the acoustic energy to electrical energy—the transducer is typically, in this case, a condenser microphone that converts acoustic energy through changes in capacitance to changes in voltage. This voltage signal can be digitized and stored electronically and used for interpretation either directly by a human expert or is subsequently processed as part of a more sophisticated system; e.g., acoustic signals of a cardiac origin can be analyzed automatically for specific abnormal patterns (Wang et al., 2009). While it is important to appreciate the fundamental aspects of sensing as discussed above, it should be noted that modern sensors are designed and packaged such that their output is amenable for processing directly with electrical circuits and increasingly directly with digital systems (Fraden, 2010). We do not consider the latter in this chapter and will instead focus on sensors that produce an analog voltage that can be directly related to the measurand under consideration.

Sensors for physiological measurement come in many shapes, sizes, types, and configurations. There are journals and texts devoted to the topic and it is a continuously active focus of research and development, especially in the healthcare domain. In the realm of health, the most common measurements of interest relate to the primary vital signs—temperature, blood pressure, pulse, respiratory rate along with electrical activities of the heart and of the brain, and arterial oxygenation saturation levels (Humphreys et al., 2007). Many of the above measurements are obtained using self-contained systems complete with digital output and, increasingly, web connectivity (Carlos et al., 2011). For the purposes of this chapter, we will focus our attention on sensor systems that require further analog processing such as bespoke EEG, EMG, ECG, and biophotonic sensor systems. Returning now to the basic sensor-processing pipeline as illustrated in Figure 27.1, we next examine the signal-conditioning stage. This stage is the primary focus of this chapter and considerable attention will be devoted to approaches commonly used here.

27.3 Signal Conditioning

In the context of sensors, signal conditioning refers to the process by which the output of the sensor package, which as discussed earlier will be assumed to be a voltage signal, is subsequently processed to enhance its utility for the intended application. For many applications of sensors, this stage is concerned with enhancing SNR—this is the utilitarian function required from the signal-conditioning stage and is achieved primarily through amplification and filtering. More sophisticated processing such as feature extraction can also be carried out here, but the contemporary approach is to reserve these processing steps for the digital domain and, more specifically, for software-driven processing routines.

The analog circuitry in modern data-acquisition systems is consequently concerned with amplification and the systems used are built using the basic analog amplifier building block—the operational amplifier (op-amp). In this chapter, we will highlight the utility and versatility of the operational amplifier in this regard.

27.3.1 The Operational Amplifier

Many sensors produce low-level signals that are not suitable for directly applying to an analog-to-digital conversion (ADC) process. Consequently, signal conditioning is required and the basic signal-conditioning building block in the analog domain is the operational amplifier (Horowitz, 1989). The op-amp is a high-gain voltage amplifier with differential input and typically a single output. The open-loop gain between the differential input and the output is very high such that open-loop operation drives the amplifier into saturation; i.e., the output voltage tries to swing in excess of the supply voltage. While the op-amp has some utility in this mode, for example, when configured as a comparator, most utility is obtained through operation in a closed-loop mode in which some of the output is fed back to the input. Most op-amp circuits can be understood using a few simple rules and basic circuit theory, although the device itself internally is complex (as shown in Figure 27.2). Configured with external feedback networks, operational amplifiers exhibit great flexibility in analog computation and can perform an incredible number of useful signal-conditioning functions.

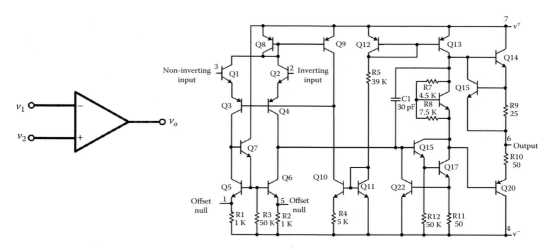

FIGURE 27.2
The left-hand side of this figure shows a well-understood standard symbol for the operational amplifier showing input (v_1 and v_2) and output voltages (v_o). The right-hand side shows a component-level diagram of the archetypal 741 operational amplifier. (Courtesy of Texas Instruments.)

The rules for understanding basic op-amp behavior are as follows:

- **Rule 1:** When the op-amp output is in linear range (for example, when there is negative feedback between output and negative input terminal), the two input terminals are at the same voltage.
- **Rule 2:** There is no current flow into either input terminal.

Rule 1 applies once there is negative feedback—i.e., connection from v_1 to v_2.
Rule 2 follows from the observation that as the input impedance is infinite (or even if not, $v_1 - v_2 = 0$), there is no input current by Ohm's law.
Application of these relationships can be used to understand the function of important operational amplifier-based signal-conditioning building blocks.

27.3.2 Signal Amplification with Operational Amplifiers

Figure 27.3 shows the two basic amplification stages that are most commonly realized with an operational amplifier device. Figure 27.3a demonstrates a unipolar design that provides a gain (and a signal inversion) that is set by the ratio of the feedback and input resistances.

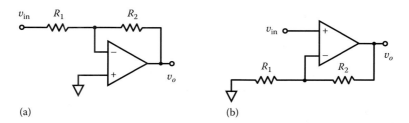

(a) (b)

FIGURE 27.3
Basic signal amplification using (a) inverting and (b) noninverting op-amp designs.

Figure 27.3b demonstrates a noninverting configuration—note how the gain is related to the externally connected components.

Application of the basic rules of op-amp analysis in the context of these circuits facilitates an understanding of function. For example, the noninverting amplifier in Figure 27.3b can be understood in the following way: As the difference between the input terminals of the amplifier can be considered zero in this situation, the voltage at the junction of R_1 and R_2 must be v_{in}. Then, through application of the potential divider relationship, we can say that

$$v_o \frac{R_1}{R_1 + R_2} = v_{in}, \tag{27.1}$$

$$\frac{v_o}{v_i} = 1 + \frac{R_2}{R_1}. \tag{27.2}$$

Thus, it is clear that the gain of this amplifier is parameterized by the resistor values R_1 and R_2. This amplifier has the virtues of positive gain and very high input impedance (essentially that of the op-amp), especially op-amps whose inputs are built with field-effect transistor (FET) technology. Such an amplifier is ideal for situations which require conditioning of output from sensors with high output impedance. The same process can be applied to the inverting amplifier:

$$i = \frac{v_i}{R_1}, \tag{27.3}$$

$$v_o = -iR_2 = -v_i \frac{R_2}{R_1}, \tag{27.4}$$

$$\frac{v_o}{v_i} = -\frac{R_2}{R_1}. \tag{27.5}$$

Such an amplifier is simple to realize; however, it should be noted that the input resistance is effectively R_1, and consequently, there is loading of the signal source that is problematic for sensors characterized by high output impedance. It is worth noting that Equations 27.3 and 27.4 demonstrate that this amplifier can also be considered as a voltage-controlled current source. Such a device is useful for some sensor solutions as part of an excitation circuit (as suggested in Figure 27.1). For example, this circuit can form the basis for a LED driver, which is useful when making optical physiological measurements such as in photoplethysmography, pulse oximetry, galvanic skin response, and even brain–computer interfacing (Soraghan et al., 2009). We will also see such a device utilized as part of an optical isolation amplifier later.

The same basic circuit can be extended further. Through use of Kirchoff's current law at the negative input, a number of voltage signals can undergo a weighted summation process as shown in Figure 27.4.

As v_- is the same as v_+ which is at ground, then we can describe the current at the node represented by the negative terminal of the op-amp as

$$\frac{v_1}{R_1} + \frac{v_2}{R_2} + \frac{v_3}{R_3} = -\frac{v_o}{R_f}, \tag{27.6}$$

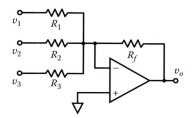

FIGURE 27.4
A summing amplifier.

and

$$v_o = -R_f \left(\frac{v_1}{R_1} + \frac{v_2}{R_2} + \frac{v_3}{R_3} \right). \tag{27.7}$$

This operation is useful, for example, in the removal of offsets where through the use of a variable resistor, an adjustable voltage level can be subtracted from the other inputs. We will see a specific example of the use of this op-amp circuit later in this chapter for a piezo-electric force-transducer application.

Returning to the noninverting amplifier again, if we set $R_i = inf$ and $R_f = 0$, we have the voltage follower circuit or *buffer* with $v_o = v_{in}$. This is a very useful circuit as it presents very high input impedance (especially FET implementations, as already mentioned). Such circuits (Figure 27.5) can be used to isolate the source from downstream circuit operation and loading effects. We will see it used in the context of filtering later in this chapter.

Now, if we replace R_2 with a capacitance C, we have a frequency-dependent component that allows, via Kirchoff's current law at the inverting input, to state that

$$\frac{v_i}{R_1} = -C \frac{dv_o}{dt}. \tag{27.8}$$

Such a circuit, therefore, performs an integrating function. The configuration is shown in Figure 27.6.

If we swap R_i and C, we yield the circuit in Figure 27.7, which can be considered a high-pass filter performing a differentiation function:

$$v_o = -\frac{1}{R_1 C} \int v_i \, dt + C. \tag{27.9}$$

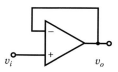

FIGURE 27.5
The operational amplifier providing a buffer function.

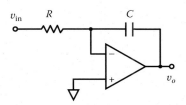

FIGURE 27.6
An operational amplifier circuit for performing the mathematical operation of integration.

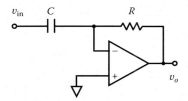

FIGURE 27.7
An operational amplifier circuit for performing the mathematical operation of differentiation.

$$C\frac{dv_i}{dt} = -\frac{v_o}{R_f};$$ (27.10)

hence,

$$v_o = -R_f C\frac{dv_i}{dt}.$$ (27.11)

One final basic operational amplifier building block that is useful to present here is the differential amplifier; it is illustrated in Figure 27.8.

Through applications of voltage division and Kirchoff's current law, a straightforward algebraic manipulation reveals amplification of the difference voltage ($v_2 - v_1$) as follows:

$$v_+ = v_2 \frac{R_2}{R_1 + R_2},$$ (27.12)

$$v_- = v_+.$$ (27.13)

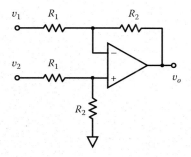

FIGURE 27.8
Basic differential amplifier design using an op-amp.

The current at the upper node in the figure can be described as follows:

$$i = \frac{v_1 - v_-}{R_1} = \frac{v_+ - v_o}{R_2}.$$ (27.14)

Substituting for v_+ and rearranging yields

$$\frac{v_o}{v_2 - v_1} = \frac{R_2}{R_1}.$$ (27.15)

For difference signals, i.e., where $v_1 \neq v_2$, we denote the gain above as differential gain G_d. Notice that for common-mode signals, i.e., $v_1 - v_2 = 0$, the gain is zero. This figure is called the common-mode gain, which we denote G_c. Real difference amplifiers do not exhibit $G_c = 0$ as in practice this is just not possible. Consequently, a figure of merit that can be used to quantify the performance of such amplifiers is the common-mode rejection ratio (CMRR), which is defined as follows:

$$CMRR = \frac{G_d}{G_c}.$$ (27.16)

The ability to measure the difference between two voltages is particularly useful in the context of measuring electrical potentials in the body. The electrical potential measured at a single location on the human body with respect to a distant reference is termed *unipolar measurement* and can be achieved through the use of amplifiers such as those already described. Commonly, there are significant common-mode signals that can be regarded as noise (particularly mains interference). These can saturate a high-gain amplifier. Consequently in situations where this occurs, differential measurements are made (also referred to as bipolar measurements) instead using the differential amplifier concept.

With the basic building blocks described above, it is possible to perform analog processing that may in some cases substitute successfully for digital computation further along the signal chain. The following example illustrates the potential utility in a force-measurement application.

27.3.2.1 Example: Piezoelectric Transducer Compensation

Piezoelectric crystals can be used as the basis for force measurement. An equivalent circuit is shown in Figure 27.9, which captures the electrical characteristics of this material under an applied force.

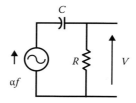

FIGURE 27.9
An electrical equivalent circuit for a piezoelectric transducer.

In an approach suggested by Ward and de Paor (1996), an application of Kirchoff's current law in this circuit yields

$$C \frac{d}{dt}(\alpha f - v) = \frac{v}{R}, \qquad (27.17)$$

assuming no current into the amplifier to which the output is connected.
Then

$$RC \frac{dv}{dt} + v = RC\alpha \frac{df}{dt}. \qquad (27.18)$$

We can use the Laplace transform to yield an algebraic expression as follows:

$$v(s) = \frac{RC\alpha s}{1 + RCs} F(s). \qquad (27.19)$$

This equation is the transfer function that relates the input, in this case an applied force $f(t)$, to the voltage across the input of the amplifier. The effect of this is a filtering process that distorts the original signal. To appreciate this, let us consider a step input to such a transducer and calculate the expected output signal:

$$f(t) = F_m \quad \text{for } t \geq 0,$$
$$f(t) = 0 \quad \text{for } t < 0. \qquad (27.20)$$

Therefore, our force-controlled voltage source is described by

$$v_i(t) = \alpha F_m \quad \text{for } t \geq 0,$$
$$v_i(t) = 0 \quad \text{for } t < 0, \qquad (27.21)$$

the Laplace transform of which is

$$v_i(s) = \alpha F_m \frac{1}{s}. \qquad (27.22)$$

Substituting this into the relationship above yields the following output for the voltage across the input of the amplifier in the s domain:

$$v(s) = \frac{\alpha F_m}{\frac{1}{RC} + s}. \qquad (27.23)$$

Using the well-known Laplace transform pair,

$$\frac{1}{s+a} \Leftrightarrow e^{-at}, \qquad (27.24)$$

We can recover a time-domain representation of the response to a step (Figure 27.10) as input then as

$$v(t) = \alpha F_m e^{-t/RC} \tag{27.25}$$

Clearly, static loads are not faithfully represented in this particular piezoelectric transducer. It is possible to compensate for such dynamics by using an appropriate conditioning circuit. If we rearrange Equation 27.19, we can relate the desired force measurement in terms of the voltage produced at the amplifier input as follows:

$$\Rightarrow \alpha F(s) = \left(1 + \frac{1}{RCs}\right) v(s), \tag{27.26}$$

Taking the inverse Laplace transform gives us the following relationship, which describes the analog computational operations required to recover $f(t)$:

$$f(t) \propto v + \frac{1}{RC} \int_0^t v(t) \, dt. \tag{27.27}$$

From an examination of the structure, an appropriate circuit could be designed with the op-amp building blocks already described; for example, a buffer, an integrator, and a summing amplifier can together perform the necessary analog computation.

It should be clear from the discussion so far that op-amps are versatile electronic devices that can be deployed in many ways to implement a wide range of functions in the analog domain. For a deeper exposition of the function of op-amps circuits for measurement and sensor conditioning including a large array of ingenious designs, the reader is referred to textbooks on the topic (Northrop, 2014; Webster, 1998) and application notes from manufacturers available online. In the next subsection, we will proceed to examine more sophisticated electronic conditioning circuits that are ubiquitous in most sensor-processing pipelines. In so doing, we can utilize our appreciation of op-amp circuits to understand how these systems are implemented and indeed how they work.

27.3.3 The Instrumentation Amplifier

The difference amplifier shown in Figure 27.8 is a basic amplifier block that can be extended to measure very small signals differentially in the body. Such measurements are the basis behind common connected health signal–acquisition targets such as the electrical activity of the heart (EKG/ECG), those of the muscles (EMG), and that

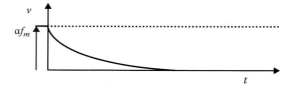

FIGURE 27.10
Step response of the piezoelectric crystal model.

of the brain (EEG). All of these signals are characterized by very low signal levels, high common-mode noise levels, and relatively high output impedances. An instrumentation amplifier is a high-performance version of the differential amplifier we have already examined. It is characterized by high common-mode rejection ratio, high input impedance, low drift, low noise, and high open-loop gain. It is ideally suited to measurement of the aforementioned signals.

The basic instrumentation amplifier design arises from a more detailed assessment of the shortcomings of the basic differential amplifier design. One of the primary problems of the basic differential amplifier design is the input impedance, which is relatively low and leads to loading of the signal source. Consequently, difference amplifiers are more commonly seen in biosignal applications where the source has low impedance, such as when measuring kinetic signals through the use of, for example, strain gauges configured as part of a Wheatstone bridge (Northrop, 2014). Given the basic operational amplifier building blocks we have seen, a natural solution to this problem is to buffer each input to a differential amplifier stage with the voltage follower circuit in Figure 27.5. FET-input operational amplifiers configured as such buffers provide very high input impedance, facilitating excellent transfer of voltage signal from the source to the input of the system. A further step is to add gain with appropriate resistors at this stage of the circuit, in effect replacing the voltage followers with noninverting amplifier stages. This leads to the canonical form of the instrumentation amplifier in Figure 27.11 (Webster, 1998).

The analysis of this circuit is somewhat more complex than the previous circuits introduced in this chapter; however, using the same basic procedure as before, one can arrive at an understanding of circuit performance. For the purposes of this exposition, we will focus on deriving CMRR, which is one of the most important performance metrics.

Thus, if we consider common-mode gain, first we begin by considering the case where $v_1 = v_2$, which is the case for common-mode signal. As the voltages between the terminals of the operational amplifiers are zero, we can then infer that the voltage across the resistor R_1 is given by the difference between v_1 and v_2, which is zero in the common-mode case. As a result, the current flowing through R_1 must be zero. As there is no current flow into the terminals of the op-amp, then this means that the currents through labeled R_2 must also be equal to zero. As a result,

$$v_3 = v_1 \tag{27.28}$$

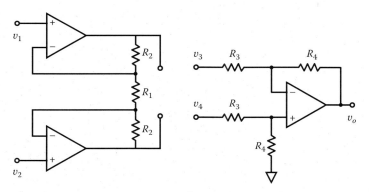

FIGURE 27.11
Basic instrumentation amplifier design illustrating the two-stage process.

and

$$v_4 = v_2. \tag{27.29}$$

So we can say that for the front end of the operational amplifier, the common-mode gain $G_c = 1$.

If we consider differential signals, i.e., $v_1 \neq v_2$, then the analysis becomes a little more involved. As before, the current flow through R_1 is determined by the difference between v_1 and v_2. This same current now flows through the resistors denoted R_2. From Ohm's law we can then see that

$$v_3 - v_4 = i(R_1 + R_2 + R_2) \tag{27.30}$$

and that

$$v_1 - v_2 = iR_1. \tag{27.31}$$

This means that the differential gain is as follows:

$$G_d = \frac{v_3 - v_4}{v_1 - v_2} = \frac{(R_1 + 2R_2)}{R_1} \tag{27.32}$$

and that the common-mode rejection ratio is

$$\text{CMRR} = \frac{G_d}{G_c} = \frac{(R_1 + 2R_2)}{R_1}. \tag{27.33}$$

Finally, the second stage, which is the differential amplifier design considered earlier, provides additional gain determined by the ratio of the resistors R_4 and R_3, so we can describe the overall gain as

$$G = \left(1 + \frac{2R_2}{R_1}\right)\left(\frac{R_4}{R_3}\right). \tag{27.34}$$

It is apparent from the above that this three–op-amp design has high input impedance and high common-mode rejection ratio, making it a very useful building block for biopotential amplifiers. While functional instrumentation amplifiers can be built using standard operational amplifier building blocks, usually more specialized operational amplifier devices are chosen in order to get the best performance match with the design requirements. For example, Figure 27.12 shows a three-device design suggested by the corporation Analog Devices for a high-speed, low-drift, and low-offset design suitable for even high-impedance sources.

The design utilizes the AD8271—a specialized difference amplifier with laser-trimmed matching thin-film resistors ensuring stable CMRR across a wide frequency range. The input or buffering amplifier ADA4627-1 has junction gate field-effect transistor (JFET) inputs which are characterized by very high input impedance and very low bias currents, making it useful when sources are themselves high impedance and when high gains are

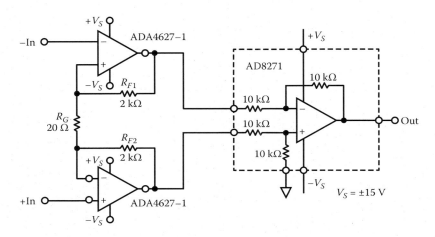

FIGURE 27.12
A three-device design by Analog Devices. (Courtesy of Analog Devices, Norwood, Massachusetts, 2014.)

required. It should be noted that a complete (and practical) design would include a decoupling capacitor network.

Even more conveniently, instrumentation amplifiers are also available from a range of semiconductor companies as a single device—a monolithic instrumentation amplifier. There are many such products available from many producers such as the AD624 made by Analog Devices, or as shown in Figure 27.13, the INA121 made by Texas Instruments. This particular device has FET inputs, is available in surface mount and dual in-line packages, and is a versatile, easy-to-use amplifier with wide applicability.

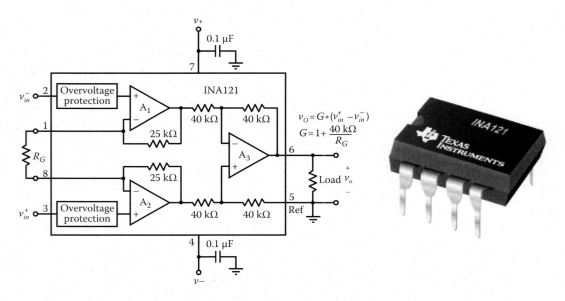

FIGURE 27.13
The Texas Instruments INA121 Instrumentation Amplifier showing high-level subsystem diagram and dual in-line package. (Courtesy of Texas Instruments.)

Usually data sheets for such devices will express CMRR through the related term *common-mode rejection* (CMR), which is understood as follows:

$$CMR = 20 \log_{10} (CMRR). \tag{27.35}$$

CMR for good-quality monolithic instrumentation amplifiers exceed 100 dB. The development of the instrumentation amplifier circuit on the same die facilitates the use of laser-trimmed resistors and better matching components throughout, which gives greater resistance to temperature variation. Such developments are leading to very high-accuracy devices that are being increasingly refined to meet modern healthcare technology demands. The presentation here gives only a brief introduction and interested readers should consult application notes from major semiconductor companies for solutions specific to their needs.

27.4 The Analog-to-Digital Conversion Process

Now that we have examined how sensor signal conditioning can be achieved through the use of analog computation blocks, the next step along the data-acquisition pipeline is the analog filtering stage. However, it is difficult to appreciate the constraints of the antialiasing filtering requirements without first examining the process of digitization of analog signals. Through an understanding of this process, the requirements and demands of the filtering stage are easier to appreciate. So what is the digitization process? Simply put, this is a conversion of a continuous signal to a discrete set of finite precision numbers, which retains to an acceptable level all the information required for the intended application. This conversion process is generally described as consisting of two processes—*sampling* and *quantization*. How these two processes are performed impacts on the fidelity of the information retained in the digital signal produced. Before we delve into detail on this topic, we first describe the basic ADC block diagram shown in Figure 27.14.

An ADC system consists of a sample–and-hold stage, which discretizes the signal in terms of time (sampling) and produces an output which can be processed further to produce an amplitude-discretized signal (quantization). The sample-and-hold unit is usually a switch, driven by a clock which allows a capacitor to charge up/down when connected by the switch, to the level of the input voltage (Figure 27.14). This is the sample step. During the hold step, the capacitor will inevitably discharge; however, the following buffer is designed

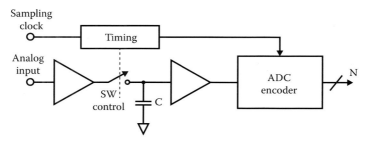

FIGURE 27.14
Block diagram of key elements of the ADC processing illustrating sampling and quantization blocks.

to have sufficiently high input impedance such that it discharges by a tolerable amount (< 1 least significant bit [LSB]) during this time. During this hold time, the encoder produces an N-bit representation of the sampled signal resulting in an amplitude discretization.

27.4.1 The Sampling Process

Figure 27.15 (top panel) demonstrates the sampling process in terms of the discrete moments in time during which the signal amplitude is captured. The temporal discretization is called sampling and we consider only the case of regular sampling in this chapter. Given such a process of regular sampling, the key question is, how often should we sample? This is an interesting question and the answer should be understood in terms of what is considered correct sampling. Properly performed sampling should allow the unambiguous reconstruction of the original continuous waveform from the samples. There should be a one-to-one mapping between the set of samples produced and the original analog signal in the continuous domain. Aliasing describes the phenomenon in which this condition is not met and it arises when the sampling rate is less than twice the highest frequency contained in the analog signal. Figure 27.16 illustrates this situation in which we have an original signal (the thicker line), which takes the form of a sinusoidal oscillation at frequency f_1 and a set of regular samples taken at a rate f_s, where f_s is just less than f_1. It is clear from the interpolation shown in with the thin line that this set of samples can describe two sinusoids—the original and a second aliased sinusoid with apparent frequency $f_s - f_1$. To avoid this we must adhere to the Nyquist sampling theorem (Unser, 2000), which states that in order to have unambiguous accurate capture of a signal through sampling then the

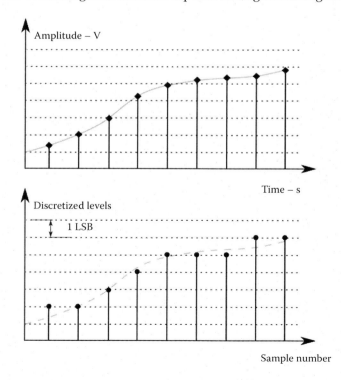

FIGURE 27.15
Quantization process.

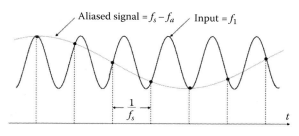

Note: f_a is slightly less than f_s

FIGURE 27.16
Aliasing in an improperly sampled signal.

sampling frequency must be at least twice the highest frequency contained in the signal of interest.

This sampling process once properly executed leads to a set of samples which have discretized the time dimension of the original signal; however, the amplitude of the signals produced is still taken from a continuous domain. As digital computers operate with finite precision, a further process of amplitude discretization is required. As we will see next, this quantization process does not come without cost in terms of signal fidelity.

27.4.2 The Quantization Process

The process of encoding the output from the sample-and-hold stage is a process that adds uncertainty and, hence, noise to the ADC process. To understand why this is so, we need to consider that the encoder will take the amplitude-continuous instantaneous sample and assign it to the nearest integer that can be represented by the N-bit ADC encoder. This quantization process is illustrated in Figure 27.15. It is clear from the lower panel of Figure 27.15, especially for the latter half of the signal, that distinctly different amplitude levels have been mapped onto a smaller set of values, resulting in the introduction of uncertainty into the process. This information loss constitutes a type of noise and is called quantization noise. This noise is distributed uniformly in the range of ±0.5 LSB, where LSB stands for the smallest quantization step in the encoder. This LSB can be directly related to a voltage (it will depend on the span of the ADC); therefore, we can quantify the noise introduced as $1/\sqrt{12}$ times the quantization step q. It is often computed simply as ~0.29 LSB and is independent of all other factors (in most cases).

The impact of this noise source depends on the application and, of course, the noise already present. If we consider an analog signal, say, for example, an EEG signal which after amplification is 100 mV peak and contains noise of 10 mV root-mean-square (RMS) (arising from the instrumentation, environmental, and physiological sources). If our ADC has a span of 10 V, then if we choose to digitize at 8 bits, we will have $q = 1/256$ V and introduced quantization noise of about 11.3 mV. If instead we choose a 24-bit ADC as per the Texas Instruments ADS1299 discussed later, we will introduce noise of 0.172 μV.

For uncorrelated noise sources, we add these noise components in quadrature (Moore and McCabe, 1999) to get the total power in the measurement. So in the case of the 8-bit ADC, the total noise on our EEG signal is now $\sqrt{(0.010)^2 + (0.0113)^2} = 15.1$ mV RMS, an increase of over 50% in the noise on our signal, while for the 24-bit case, we have just $\sqrt{(0.010)^2 + (1.72 \times 10^{-7})^2} \approx 10$ mV; i.e., there is negligible additional noise in this case. It is clear then the decision on the number of quantization levels has a direct bearing on the amount of noise introduced; however, the impact of this noise must be judged with respect to the noise already on the signal and the amount of noise which can be tolerated in the application.

After these digitization issues have been thought through for a specific application, an important and sometimes neglected aspect of the ADC process is the prefiltering of the signals with an appropriate antialiasing filter. This topic is considered next and should be considered in tandem with the sampling requirements.

27.4.3 Antialiasing Filters

A filter is, in the context of this discussion, an electrical circuit which alters the spectral composition of the signal upon which it operates. Filters are usually used to attenuate unwanted spectral ranges, although they can also be used to increase signal power in specific bands. Sometimes both are required. Filters can take the form of single passive circuits containing few elements to complex systems requiring sophisticated analysis to model fully.

Analog filters are a critical component of the data-acquisition pipeline. While they can be used at any point in the pipeline to enhance the signals of interest and suppress artifact— for example, a common use is a notch filter for removing the effects of 50 Hz (or 60 Hz) mains interference (Sweeney et al., 2012b)—in this chapter we focus primarily on their utility in ensuring that we achieve proper sampling as defined in Subsection 27.4.1. The reader will recall that in order to prevent aliasing, we must ensure that the signal to be sampled contains no components higher than half the sampling rate. To ensure that this is indeed the case, we use a filter with a low-pass cutoff suitably chosen such that the signal has no significant components above this special frequency. Such a filter is called an antialiasing filter and it is placed right before the ADC unit as shown in Figure 27.1.

Given that the requirement of an antialiasing filter is to meet stop-band attenuation levels at a frequency corresponding to half the sample rate, then there are many filter designs which can meet such a criterion. In this subsection, we give a brief introduction to some practical ideas here, beginning with the simplest filter possible—the passive RC first-order filter shown in Figure 27.17.

The filtering behavior of the passive RC low-pass filter can be understood easily by using the potential divider approach and Laplace equivalents in the s domain for the circuit components. Thus, using the potential divider approach and rearranging, we have

$$\frac{v_o}{v_i} = \frac{\frac{1}{sC}}{\frac{1}{sC} + R} = \frac{1}{1 + sRC}. \tag{27.36}$$

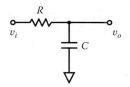

FIGURE 27.17
A first-order passive RC low-pass filter.

If we next consider the magnitude of the response, as a function of frequency, we calculate

$$\left|\frac{v_o}{v_i}\right| = \left|\frac{1}{1+sRC}\right| = \left|\frac{1}{1+j\omega RC}\right| = \frac{1}{\sqrt{1+(\omega RC)^2}}. \tag{27.37}$$

From this we can see that at DC when $\omega = 0$ that the gain is 1 and that as ω tends larger, the gain increasingly diminishes, demonstrating that this is indeed a low-pass filter. The cutoff frequency, which is defined as the half-power frequency, can then be derived as

$$\left|\frac{v_o}{v_i}\right|^2 = \frac{1}{2} = \frac{1}{1+(\omega RC)^2}, \tag{27.38}$$

which is equivalent for $\omega = 1/RC$ rad/s or $f = 1/2\pi RC$ in hertz.

This is a first-order passive low-pass filter with a roll-off of 20 *dB/decade*. This basic filtering unit can be cascaded to create higher-order and, hence, steeper roll-off designs, although to prevent loading effects, the buffer amplifier, as shown in Figure 27.5, should be used to isolate the stages. A second-order implementation of this is shown in Figure 27.18.

The preceding filters are pretty good for many antialiasing filtering needs; however, it is common to utilize filter designs that allow good trade-offs between filter characteristics that are specific to various application requirements. To better capture these ideas, we next examine a number of key analog filter design parameters (as in Figure 27.19) that we will refer to when considering these standard designs.

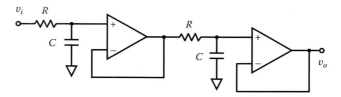

FIGURE 27.18
A second-order passive RC low-pass filter.

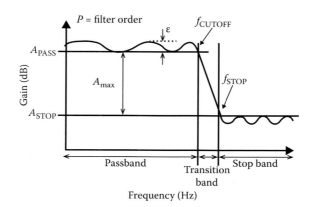

FIGURE 27.19
Analog filter design parameters.

Figure 27.19 shows the key parameters of a low-pass filter that are relevant to the design of antialiasing solutions. The cutoff frequency f_c referred to earlier is indicated as a corner frequency in the top right of the plot. The passband refers to the frequency range from DC to f_c. The stop-band frequency f_s is the point at which the minimum attenuation is first reached for the filter. The magnitude of the gain in the passband is referred to here as A_p, while the gain in the stop band is referred to as A_s. Both are usually expressed in decibels. Some filter designs have a ripple in either or both of the passband and the stop band. The magnitude of this ripple is represented here as ε. In filters with ripple, A_p and A_s refer to the minimum and maximum gains in their respective bands. An important design parameter is A_{max}, which represents the dynamic range or difference between the passband and stop-band gains (i.e., $A_{max} = A_p - A_s$). Finally, as the filter response extends beyond the cutoff frequency, it falls through what is referred to as the *transition band* to the stop-band region. The bandwidth of the transition band is determined by the precise filter design and the order (P) of the filter. As a general rough and conservative guide, the final rate of power roll-off is 20 dB/decade per pole. The filter order is determined by the number of poles in the transfer function. For instance, if a filter has two poles in its transfer function, it is described as a second-order filter.

As more poles are added and, hence, higher order is achieved, the transition bandwidth reduces and will approach the ideal "brick-wall" filter characteristic. This would suggest one should use a very high-order filter; however, as Figure 27.18 illustrates, every pole involves additional circuitry with a concomitant impact on noise, offsets, power. and bulk. Usually appropriate minimal designs are chosen based on the application needs and can proceed as follows.

As a general rule of thumb, we can estimate the order of the filter required for a particular application as

$$P = \left\lceil \frac{A_{max}}{20 \log_{10}\left(\dfrac{f_s}{2f_c}\right)} \right\rceil, \tag{27.39}$$

where the brackets used indicate rounding up. This is a conservative (as it assumes that there is full signal power at and above half the sampling rate) but useful starting point in filter design. The calculation is based on the observation that the roll-off can be estimated as 20 dB/decade per pole and we can conservatively set f_{stop} as equal to $f_s/2$ so that we have maximum attenuation at the desired frequency (half the sampling rate). Figure 27.20 helps illustrate this idea.

For example, let us consider that we have an EEG signal requiring sampling at 500 Hz, which means that we need to ensure that there is negligible signal above half the sampling rate. If we consider that there are no EEG components above 100 Hz of interest and that we need A_{max} of at least 60 dB, then

$$P = \frac{60}{20 \log_{10}\left[\dfrac{500}{2(100)}\right]} = 8. \tag{27.40}$$

Thus, we should use an eighth-order filter. Of course, when one considers the use of a filter, it is appropriate to think of what filter precisely one should use. As mentioned earlier,

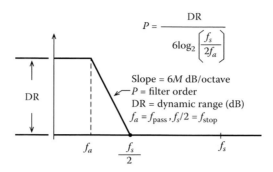

FIGURE 27.20
Filter order determination using Equation 27.39.

there are a number of standard designs that offer compromise in terms of the filter parameters described earlier. The most common filter designs are the following:

- Butterworth
- Chebyshev
- Bessel

Each of these filter designs represents a trade-off between roll-off, passband ripple, and step response. To understand why this is so, it requires an in-depth examination of the underlying mathematical models underpinning the design. This is a topic which cannot be covered here due to space constraints but interested readers can consult any standard text on analog circuits for an exposition of the principals involved (Smith, 1997). The basic trade-offs are described very succinctly below:

- *Roll-off*: Roll-off describes the width of the transition band in a filter design. The ideal filter has a transition band of zero and is thought of as a brick-wall filter. Such ideal filters are not realizable in practice, but as we have seen, real filters have a roll-off that can be improved through increasing the filter order.

  A simple rule of thumb which allows one estimation of the required order for a given filtering need has already been presented. However, it should be stressed that this is only approximate and that the precise roll-off behavior of a filter can be improved through various methods, such as, for example, introducing ripple into the filter band. The Chebyshev filter is an example of this, and of the three designs, it presents the best roll-off; i.e., for a given filter order the transition band is shortest in this design.

- *Passband ripple*: The passband of a filter should ideally be flat. As we have seen, the Chebyshev filter sacrifices a flat passband for faster roll-off. The Butterworth filter, on the other hand, has been designed to have the shortest possible transition band while still maintaining a flat passband; i.e., there is no ripple. The Bessel filter also has a flat passband; however, it has the worst roll-off of the three designs.

- *Step response*: Many filters designs exhibit ringing when subjected to an abruptly changing input such as in a step response test. The Bessel filter has been designed to minimize overshoot and ringing compared to the Butterworth and Chebyshev (worst performing in this regard) and this has come at the cost of the roll-off performance.

Now, given that these designs represent trade-offs between these filter characteristics, how should the filter choice be decided? This depends, understandably, on how these filter characteristics impact on the information bearing aspect of the signals in question. If the relevant information in a signal is embedded in terms of its spectral content, then it is important to reduce the possibilities of frequency-dependent distortion in the passband as it will lead to uncertainty in the interpretation of the information. Many EEG studies, for example, make use of the relative powers in a finite set of relatively narrow bands. For such applications, aliasing has significant impact and so aggressive roll-off filters such as Chebyshev or Butterworth (which, in addition, has flat passband) are preferred in these applications. Other signals such as the ECG, for example, are best characterized in terms of their time-domain features, such as the QRS complex. For such applications, the Bessel filter is usually the best choice as it can best retain information about such temporal features without undue distortion dues to its superior step response. For applications in which the signals contain time-varying spectral components, a popular compromise is to use a Butterworth filter. The Butterworth design is a good all-around filter.

These filter designs can be readily implemented using an op-amp circuit building block called the Sallen–Key filter—the design of which is shown in Figure 27.21 (Sallen and Key, 1955). Using this building block, which is a two-pole unit, all of the above filter designs can be implemented to a high order using suitable values for the resistors and capacitors. Increasing order is achieved simply by cascading stages and order-specific component values.

From a mathematical analysis, one can calculate the appropriate values of resistors and capacitors to achieve Sallen–Key implementations for the filter designs discussed here. However, a much more convenient way to design such a filter is to use the many design tools available online which will, upon specification of the required filter parameters, produce a detailed design including a bill of materials. Further, for many applications, single integrated circuit filter solutions, especially in the form of switched capacitor implementations, are very convenient where a clock signal is available. Their versatility, performance, and compact form factor make them an excellent choice for many. The switched capacitor filter is a mix of analog and digital components and it is from this mix that it derives its flexibility. We continue this theme of distributing function across the analog and digital domains through presenting next a brief discussion of the merits of oversampling and decimation.

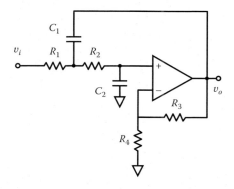

FIGURE 27.21
The Sallen–Key two-pole building block for active filter design.

27.4.4 Oversampling and Decimation

There is a tendency to reduce the amount of analog circuitry in modern connected health systems. Surfeit bandwidth and computational capacity, both local and remote, means that many operations which hitherto might have been performed using analog circuitry can now take place in the digital domain, reducing sensor node complexity and bulk. The complexity of the device impacts unit cost, failure modes, and device fixes. In terms of the last one, this is an important consideration; for example, a sensor-conditioning operation requiring replacement or update implemented in the digital domain can be upgraded via a firmware upgrade possibly implemented over the air. An analog implementation, in contrast, requires that the sensor is sent back for replacement or upgrading. This is a logistically far more complex and usually expensive task. The reduction in sensor node volume and weight associated with a reduction in circuitry has benefits in producing a smaller, and more wearable, device, which can improve user compliance, depending on the specific application. Finally, it should be noted the net reduction in sensor node bulk through removal of analog components can be even greater than accounted for by the circuitry alone, as often the associated battery requirements can be reduced to enable even greater savings. The precise nature of this particular type of saving is not guaranteed as it depends on the implementation of the processing off-loaded to the digital domain—this processing may incur additional power in terms of the digital implementation and/or transmission. While it is important then to carefully consider all aspects when deciding the analog/digital breakdown, the current trend is to reduce analog requirements through digital means. An interesting example of this is demonstrated in the area of antialiasing we have just looked at.

27.4.4.1 Oversampling

From consideration of our basic antialiasing filter design equation, Equation 27.38, we can see that one way to simplify the filter design required is to push out $f_s/2$ such that the roll-off can be less aggressive and, therefore, fewer poles are required. This necessitates sampling at rates higher than the minimum required—this would normally be twice the highest frequency of interest in the signal to be sampled. The use of a much higher sampling rate to accommodate a gentler filter roll-off is called oversampling; and given the power of modern digital signal processing (DSP), which can accommodate very high sample rates and high-throughput ADCs, it is something which can be done relatively easily for most biosignals. However, there is a lot of redundant information in the data stream produced which without compression (which requires computation), will have an impact on bandwidth needs if we are considering a connected health application. An alternative idea which also provides an opportunity to touch upon the versatility and utility of digital filtering is a technique called multisampling, which is a combination of oversampling and decimation.

27.4.4.2 Multisampling

Multirate sampling makes use of both oversampling and decimation to provide a relaxation on both the analog filtering requirements for antialiasing and the final data rate at the system output. The process works by oversampling first. Usually some multiple of the final desired data rate, which we consider, as before, as f_s is used. This relaxes the roll-off requirement by pushing out half the sampling rate used in Equation 27.40. Let us call this rate $k \cdot f_s$. Now, immediately following the ADC, we can implement a digital filter which will seek to implement the original required antialiasing function for $f_s/2$. Digital filters can accommodate very sharp cutoffs and high order relative to their analog counterparts. In addition, they are easy to implement and flexible. As a result of this operation, the resultant digital signal is highly oversampled yet contains insignificant signal above $f_s/2$. There is now considerable redundancy; however, this redundancy can be removed through the simple process of decimation. In this case we can retain every kth sample and lose the rest. Let us consider the EEG example already considered. If we push the sample rate from 500 Hz to 2 kHz, then we now need use only a third-order filter to achieve the necessary antialiasing requirements. We now have an unusable band between 100 Hz and 1 kHz. We can now design and use a digital filter with a cutoff of 100 Hz to remove these components and revert to our original desired data rate by a decimation process that retains only one out of every four samples.

The above approach highlights very well the additional benefits that can be accrued from considering the sensor-processing pipeline from both analog and digital domains simultaneously. The idea of hybrid analog and digital processing done in a tightly coupled integrated fashion is increasingly alluring for both device manufacturers, who can now provide (and sell) complete solutions, and engineers, who can benefit from smaller footprint designs and simpler design phases. We briefly examine this phenomenon next as it is a logical progression to the material presented so far and is a signpost toward future developments in sensor processing relevant to the field of connected health.

27.5 Integrated Solutions

Increasingly, semiconductor circuit providers are providing integrated systems-on-a-chip for specific application needs. A useful example which illustrates this concept very well is the ADS1299 (Texas Instruments; http://www.ti.com), shown in Figure 27.22. The ADS1299 is a low-noise, eight-channel, 24-bit, analog front end for biopotential measurement and digitization. It contains all the individual stages we have already described as well as several new ones specific to biopotential medical instrumentation applications. All of these separate stages have been finely tuned to provide a high-performance integrated solution particularly well suited for EEG/ECG measurement. Furthermore, the single-piece implementation and surface-mount form factor help contribute to a reduction in size, power requirements, and cost compared to a solution developed through the use of individual discrete stages.

Rather than producing an analog output, the ADS1299 is interfaced through the popular asynchronous serial data link protocol—*serial protocol interface* (SPI). This serial protocol supports full duplex communication, control, and data transmission and is well suited to microprocessor interfacing. The design is a hybrid of analog and digital technology with the benefits of both and the drawbacks of neither. Feature-rich devices such as the ADS1299 hold great promise for accelerating the adoption of wearable biopotential measurement and are a useful starting point for those interested in this sort of biosignal measurement.

At this stage we have touched upon, albeit in a succinct manner, many of the elements involved for sensor processing are at the predigital level. With this basic understanding, it is possible to appreciate the design of many existing systems and engage in

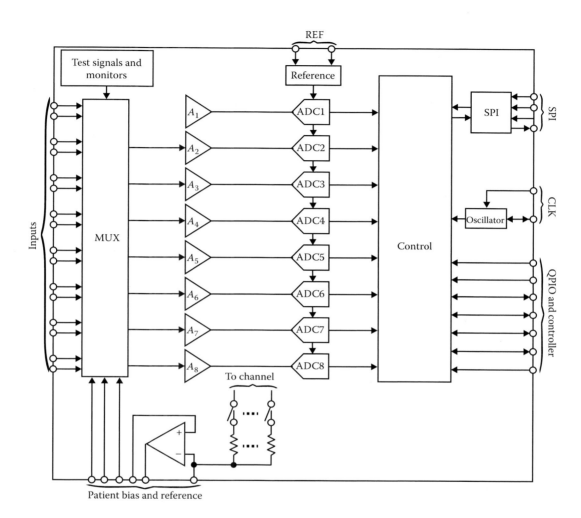

FIGURE 27.22
ADS1299 internal architecture. (Courtesy of Texas Instruments.)

the development of new applications. However, primarily because of its importance at the hardware level, we will briefly visit the user safety and the design of appropriate circuits which can help isolate human subjects from risk of shock and other hazards associated with electrical energy. A motivating factor here is that the availability and reduction in the cost of so many sensors and sensor-processing modules is leading to a huge growth in experimentation and applications of sensor processing for health and wellness applications. The ease with which people can try out these sensors may lead to overenthusiastic adoption and implementation without due care and attention to some of the more mundane yet complex aspects of the design, such as those for subject safety. Section 27.6 provides a short introduction to some of the solutions which are available and which should be integrated, when appropriate to sensor systems, which will be used with human subjects.

27.6 Isolation Circuits

For safety, it is important to protect the user from the hazards of electrical shock. While this seems obvious, it is often an afterthought, especially at the research and prototyping stage. Electrical shock can always present a safety risk with electrical circuits and it is important to consider the problem seriously. It is worth highlighting that it is current, not voltage, which is the real hazard here. Current flow in tissue can cause excessive resistive heating ($P = I^2R$ effects), leading to burns, electrochemical heating, and electrical stimulation of neuromuscular systems. In the case of electrical stimulation, there are obvious and potentially lethal dangers. For example, even 100 mA of current can lead to disruption to the delicately balanced patterns of neuromuscular interaction which govern the proper functioning of the heart. Even lower levels ~15 mA can lead to respiratory disruption, including paralysis.

Instrument front ends including sensors which may make electrical contact with the patient must be completely isolated from the mains power supply by a nonconducting barrier. This barrier must be able to withstand potential differences of several thousand volts. Isolation is accomplished by separating the input stage of the isolation amplifier from the output stage in a galvanic sense. Such a requirement necessitates that the input stage has a separate, floating power supply and a return path that is connected to the output stage of the isolation amplifier by a very high resistance and a parallel capacitance in the picofarad range. Similarly, and appropriately, high impedance isolates the input signal terminals of the front end from the output of the isolation amplifier.

27.6.1 Methods of Isolation

There are three principal means of implementing the required coupling:

- Optical
- Capacitive
- Magnetic

27.6.1.1 Capacitive Isolation Amplifiers

An example of a capacitive isolation amplifier is shown in Figure 27.23. The basic idea here is simply to modulate the input signal up into a band, and via a modulation scheme, which allows high-fidelity signal transmission across isolating capacitors. The output section then must demodulate the transmitted signal in order to recover the original biosignal. The precise implementation detail (for example, how modulation is accomplished) varies but the basic concept remains. In many designs, the power lines are isolated through an isolation transformer.

27.6.1.2 Optical Isolation Amplifiers

While implementation details again here can vary, the basic concept is straightforward and involves the conversion of electrical energy to optical energy and back again. The information in the analog signal is faithfully maintained (or at least maintained to an

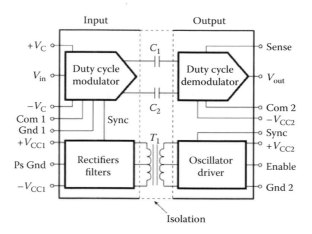

FIGURE 27.23
An example of a capacitive isolation amplifier. (Courtesy of Measurement Computing.)

acceptable level) across modalities, while the galvanic connection is eliminated in the process. The canonical implementation is that the buffered input signal is used to generate a current that drives a LED whose optical output captures changes in the original driving signal. The light produced falls on an appropriately positioned photodiode that performs the process of converting photonic energy to conventional current flow. Typically, a transimpedance amplifier is used to convert the current produced to a voltage output—a signal that should now reflect the input signal. A schematic taken from the data sheet of a popular and representative device shows such an implementation in Figure 27.24a.

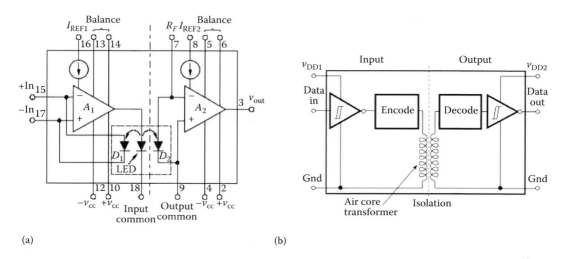

(a)   (b)

FIGURE 27.24
(a) Optical isolation amplifier example, Burr Brown ISO100 (Courtesy of Texas Instrument), and (b) digital isolation concepts. (Courtesy of Measurement Computing.)

27.6.1.3 Magnetic Isolation Amplifiers

Magnetic isolation amplifiers incorporate some of the basic ideas from both the capacitive and optical isolators. Like their optical counterparts, the basic mechanism by which isolation is achieved is through a transducer process in which energy is converted from one form into another in such a way as to eliminate galvanic connections. In a magnetic isolation amplifier, the specific means involves conversion of the voltage signal at the input to a current that is then used to generate changes in a magnetic field. Similarly as in the capacitive isolation amplifier design, this current signal is used to modulate a carrier such that there is a good coupling between the input and output stages via the bridging mechanism—in this case a coil which will either form one side of a transformer or a giant magnetoresistor bridge. In either case, the output is a voltage signal that reflects the original input voltage signal.

27.6.1.4 Digital Isolation

The techniques for achieving isolation that are already described have all operated directly with the analog biosignal. The same ideas can be used with digital signals, and in such cases, digital data are transmitted across the isolation barrier. A representative schematic is shown in Figure 27.24b.

It is clear then that there are a range of solutions available to provide subject safety when operating a sensor-processing pipeline at the hardware level. As all these solutions are available inexpensively from companies such as Texas Instruments and Maxim, to name but a few, there is no excuse for not giving this aspect of the sensor-processing system as much care and diligence as the other components we have already seen. Even with battery-operated technology, it is always important to be aware of the risks and potentials for harm with any sensor technology processing circuitry and to design accordingly.

27.7 Conclusions

This chapter has presented a whirlwind tour of the more fundamental stages in the processing of sensors for medical instrumentation purposes with an emphasis on considerations relevant to connected health and wellness applications. Such systems present challenges in terms of instrumentation size, power consumption characteristics, and ability to maximize SNR. The chapter has primarily focused on the analog domain, although discussion of the synergetic impact available through consideration of processing in both the digital and analog domains, where appropriate, has also been presented. In terms of the primary analog focus, the chapter demonstrates the versatility of analog processing using the classic building block of the operational amplifier. The basic introduction to the operational amplifier serves both the purpose of demonstrating the preprocessing possibilities in the analog domain and the laying of conceptual foundations which support better understanding of newly available, powerful, and consequently complex single monolithic integrated circuits. These modern devices will increasingly be used as the core processing elements for modern and future sensor applications. The short introduction to the sampling and quantization process is practical and usable for most applications although it does not touch on modern concepts in sampling systems such as irregular sampling (Unser, 2000) or compressive sensing (Candes and Wakin, 2008), both of which have been shown to have utility in healthcare domains. Finally, while not often considered as part of the core sensor-processing pipeline, issues of user safety from an electrical perspective are often an overlooked and poorly thought-out aspect of the hardware involved in sensor processing. Obviously in the medical device industry the opposite is the case, but given the growing number of health and wellness technology enthusiasts, from research groups in universities to "bedroom hackers," the author has thought it worth at least highlighting that there are options available and these can easily be integrated into the signal-processing pipeline already presented.

References

Bronzino, J. (1995). *The Biomedical Engineering Handbook*. Boca Raton: CRC Press; IEEE Press.

Candes, E. J., and Wakin, M. B. (2008). An introduction to compressive sampling. *IEEE Signal Processing Magazine*, 25(2) pp. 21–30, doi:10.1109/MSP.2007.914731.

Carlos, R., Coyle, S., Corcoran, B., Diamond, D., Ward, T. E., McCoy, A., and Daly, K. (2011). Web-based sensor streaming wearable for respiratory monitoring applications. In *2011 IEEE SENSORS Proceedings* (pp. 901–3). IEEE. doi:10.1109/ICSENS.2011.6127168.

Fraden, J. (2010). *Handbook of Modern Sensors: Physics, Designs, and Applications* (Google e-book) (p. 678). Springer.

Horowitz, P. (1989). *The Art of Electronics*. Cambridge [England]; New York: Cambridge University Press.

Humphreys, K., Ward, T. E., and Markham, C. (2007). Noncontact simultaneous dual wavelength photoplethysmography: A further step toward noncontact pulse oximetry. *The Review of Scientific Instruments*, 78(4), 044304(2007).

Kappenman, E. S., and Luck, S. J. (2013). *The Oxford Handbook of Event-Related Potential Components* (p. 664), Reprint Edition. New York: Oxford University Press.

Metting van Rijn, A. C., Peper, A., and Grimbergen, C. A. (1990). High-quality recording of bioelectric events—Part 1: Interference reduction, theory and practice. *Medical & Biological Engineering & Computing*, 28(5), 389–97.

Metting van Rijn, A. C., Peper, A., and Grimbergen, C. A. (1991). High-quality recording of bioelectric events—Part 2: Low-noise, low-power multichannel amplifier design. *Medical & Biological Engineering & Computing*, 29(4), 433–40.

Moore, D., and McCabe, G. (1999). *Introduction to the Practice of Statistics*. New York: Freeman.

Northrop, R. B. (2014). *Introduction to Instrumentation and Measurements*, Third Edition (p. 927). Boca Raton: CRC Press.

Sallen, R. P., and Key, E. L. (1955). A practical method of designing RC active filters. *IRE Transactions on Circuit Theory*, 2(1), 74–85. doi:10.1109/TCT.1955.6500159.

Smith, S. W. (1997). *The Scientist and Engineer's Guide to Digital Signal Processing* (p. 626), First Edition. San Diego: California Technical Publishing.

Soraghan, C. J., Markham, C., Matthews, F., and Ward, T. E. (2009). Triple wavelength LED driver for optical brain computer interfaces. *Electronics Letters*, 45(8), 392. doi:10.1049/el.2009.0214.

Sweeney, K. T., Ayaz, H., Ward, T. E., Izzetoglu, M., McLoone, S. F., and Onaral, B. (2012a). A methodology for validating artifact removal techniques for physiological signals. *IEEE Transactions on Information Technology in Biomedicine*, 16(5), 918–26. doi:10.1109/TITB.2012.2207400.

Sweeney, K. T., Ward, T. E., and McLoone, S. F. (2012b). Artifact removal in physiological signals—Practices and possibilities. *IEEE Transactions on Information Technology in Biomedicine*, 16(3), 488–500. doi:10.1109/TITB.2012.2188536.

Unser, M. (2000). Sampling—50 years after Shannon. *Proceedings of the IEEE, 88*(4), 569–87. doi:10.1109 /5.843002.

Wang, H., Chen, J., Hu, Y., Jiang, Z., and Samjin, C. (2009). Heart sound measurement and analysis system with digital stethoscope. In *2009 2nd International Conference on Biomedical Engineering and Informatics* (pp. 1–5). IEEE. doi:10.1109/BMEI.2009.5305287.

Ward, T., and de Paor, A. (1996). The design and implementation of a kinetic data extraction and analysis system for human gait. Master of Engineering Science, University College Dublin, Faculty of Architecture and Engineering.

Webster, J. (1998). *Medical Instrumentation: Application and Design*. New York: Wiley.

28

Sensor-Based Human Activity Recognition Techniques

Donghai Guan, Weiwei Yuan, and Sungyoung Lee

CONTENTS

ABSTRACT Activity recognition (AR) has become a prevalent research topic, as it provides personalized support for many diverse applications, such as healthcare and security. Numerous AR systems have been developed. In general, these systems utilize diverse sensors to obtain activity-related information, which is then used by machine learning techniques to infer the subject's ongoing activities. According to the types of sensors used, existing AR systems can be roughly divided into three categories: video sensor-based AR, which remotely observes human activity using video sensors; wearable sensor-based AR, which uses sensors attached to the subjects for activity recognition; and object usage-based AR, which uses sensors attached to objects which implicitly infer the subject's activity. We review the different types of ARs. For each type of AR, the main techniques, characteristics, strengths, and limitations are discussed and summarized. We also outline a comparative analysis of them and point out the main research challenges in AR.

28.1 Introduction

Understanding human activities and behavior has long been a research goal. In terms of daily lives, awareness of human activities can support many diverse applications. For example, in a home environment, users can be reminded to perform activities (e.g., taking medicine). In a hospital environment, a patient can be reminded to complete an exercise rehabilitation program. Employing a person to monitor other persons' activity 24 h a day is unrealistic; therefore, the automated, automatic recognition of human activities is important and necessary.

Although it is natural for humans to recognize their activities, it is not an easy task for a computer. Computers need to analyze the information gathered from sensors and infer the ongoing activity. Sensors' noises and variances make activity recognition (AR) even more complex for computers. For the recognition of a cooking activity, people might utilize various cues, such as recognizing mealtime, smelling the food being cooked, or seeing that the stove is on. Humans easily use evidence and their past experiences to infer activities. However, all the functions involved in sensing the environment, learning from past experience, and applying knowledge for inference are still a challenge for computers. The goal of activity-recognition research is to enable computers to have capabilities comparable to people's capability to recognize human activity. Only if computers can reliably recognize people's various activities can intelligent systems (such as healthcare and security systems) actually work. Moreover, research on activity recognition has a huge impact on behavioral, social, and cognitive sciences.

To build an automatic activity-recognition system, the first thing to consider is the system's sensing ability, how the system recognizes the state of the physical world. Sensing ability comes from low-level sensors (Figure 28.1). Sensed information is processed by machine learning techniques (e.g., noise filtering, feature extraction, and classification techniques) to infer the human activity (e.g., sitting, raising hand, or raising foot). Due to the variations in sensing techniques, a considerable number of activity-recognition systems have been developed.

According to the sensing techniques, the existing activity recognition can be broadly divided into three categories:

1. Video sensor-based activity recognition (VSAR): This is the classical activity-recognition approach which attempts to capture a human's activity information using appropriately placed cameras. Video analysis is the most important and challenging aspect of VSAR.

2. Wearable sensor-based activity recognition (WSAR): In this AR, physical sensors are attached to the body. This is a relatively new approach which emerged with the development of wearable computing. Motion sensors (accelerometers

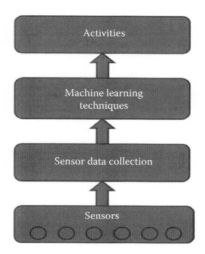

FIGURE 28.1
Work flow of activity-recognition system.

and gyroscopes) are the most common wearable sensors. They are attached to relevant parts of the body, in particular, hands and arms, to provide information about the limbs' motions. Other sensors which are usually used as complements to the motion sensors include microphones, GPS devices, and light sensors. They provide additional information about the user's environment to help improve the accuracy of activity recognition.

3. Object usage-based activity recognition (OUAR): In this AR, sensors are attached to objects. With this design, a system can recognize the objects accessed by the users and what the users do with them. Although OUAR does not explicitly monitor human activity, it is applicable, because many activities can be implicitly inferred by knowing the objects utilized. For example, if we know that the stove is on, we might infer that someone is cooking. Binary sensors and radio frequency identification (RFID)-based sensors are commonly used sensors in OUAR. They are cheap and can be ubiquitously deployed.

Of the different types of sensor-based activity-recognition systems, no one is the best. Each of them has its own characteristics and suitable applications. In this chapter, we outline the strong/weak points, main techniques, and suitable applications of each type of system, ultimately comparing them.

28.2 Video Sensor-Based Activity Recognition

VSAR is the classical activity-recognition method and is based on computer vision techniques (Figure 28.2). It is the most common among the three AR approaches. Currently, it is widely used in various domains, including industry, academia, security, and consumer agencies.

FIGURE 28.2
Video sensor-based activity recognition.

28.2.1 Video Sensor-Based Activity Recognition Applications

In this subsection, we present some typical applications of video sensor-based activity-recognition systems:

1. Surveillance systems: Surveillance systems can support many applications, such as healthcare and unusual-event alarms. Traditional surveillance systems require a human operator to constantly monitor the scene, which is tedious and inefficient. Video-based activity-recognition systems can replace or help human operators monitor anomalies and interesting activity.
2. Human–machine interaction: A human–machine interaction needs to understand and interpret human actions. Information on people's activities is particularly important in understanding intentions and purposes. For example, researchers have proposed vision-based activity recognition to develop alternatives to the traditional mouse-and-keyboard–controlled graphical user interface (GUI).
3. Sports analysis: In many kinds of sports (e.g., basketball and soccer), during a match or training, a player's performance and behavior are recorded. This has many potential usages. For example, the coach might adjust a soccer player's shooting action after analyzing the videos. Or during the match, if a penalty dispute happens, a referee can analyze the video. Due to the explosive increase in video use, it has been necessary to develop efficient indexing schemes and sports players' activity information is one of the most popular schemes for video indexing.

28.2.2 Feature Extraction in Video Sensor-Based Activity Recognition

The major steps involved in VSAR include (1) video (or image) gathering, (2) feature extraction from images, and (3) activity recognition based on the extracted features. Among these three steps, feature extraction is important and challenging. Extracting good features is the mark of a good recognition performance. The good features should generalize small variations in the images, such as personal appearance, background, viewpoint, and action execution, while being sufficient to support a robust recognition of activities.

The features in VSAR can be divided into two categories: global features and local features. Global features need the procedure of localizing a person and encoding a region of interest (ROI) and have powerful expressing abilities when rich information is encoded. However, they need accurate localization, background subtraction, and tracking.

Conversely, local features describe observations as a collection of independent patches, detecting spatiotemporal interest points and calculating the local patches around these points. Finally, the patches are integrated into a final representation. Compared with global features, local features are less sensitive to noises and do not require high-quality background subtraction or tracking; however, they require more preprocessing efforts to extract a sufficient amount of relevant interest points.

28.2.2.1 Global Features of Video Sensor-Based Activity Recognition

Global features represent the ROI of a person as a whole. The ROI is usually defined using background subtraction or tracking where common global features can be derived from silhouettes, edges, and optical flows or trajectories.

In using background subtraction segments, a scene into a background and a foreground to isolate the moving parts. The segmentation helps to evaluate the difference in pixel features between the current scene and the reference background image. Background subtraction is sensitive to environmental changes, such as illumination variations.

In general, human silhouette information is commonly employed to represent human activities [1]. In fact, human activities can be considered as temporal variations of human silhouettes. In addition to silhouette shape, motion information is also used. Optical flow is defined as the apparent motion of individual pixels on the image plane. It often serves as a good approximation of true physical motion projected onto an image plane.

Trajectories of moving objects have also been used as features to infer the activities of objects [2]. The image-plane trajectory itself is not very useful, as it is sensitive to translations, rotations, and scale changes. Alternative representations, such as trajectory velocities, trajectory speeds, and relative motion, have also been proposed.

Although the features extracted using the above methods are informative, they are sensitive to noise, partial occlusions, and variations in viewpoint. To overcome these limitations, grid-based features and space–time volumes have been utilized. Global grid-based features divide the ROI into a fixed spatial or temporal grid. Each cell in the grid describes the image observation locally; thus, partial occlusions and changes in viewpoints can be partially solved [3]. Multiple images over time can be stacked to form a three-dimensional space–time volume. A space–time volume was originally proposed by Blank et al. [4] and has since been used by other researchers.

28.2.2.2 Local Features of Video Sensor-Based Activity Recognition

Local features describe the observations as a collection of local patches, which can be samples, either densely spaced or at space–time interest points. Superior to global features, local features do not require accurate localization and background subtraction. The first type of local features is space–time interest point detectors, which are the locations in space and time where sudden changes in movement occur in the video. The locations are viewed as the most informative source for human activity recognition. Researches that use space–time interest points include those by Laptev and Lindeberg [5] and Harris and Stephens [6]. The main limitation of using these features is that the number of stable interest points is normally small. Fortunately, this limitation has been partially addressed by Dollár et al. [7] and Chomat et al. [8].

Local descriptors are the second type of local features. They summarize an image or video patch in a representation that is ideally invariant to background clutter, appearance, and occlusions and possibly to rotation and scale. Researches that have used this type of features include those by Schüldt et al. [9] and Niebles et al. [10]. Comparing sets of local descriptors is not straightforward due to the possibly of different numbers and the high dimensionality of the descriptors. Therefore, a codebook is usually generated by clustering patches and selecting the cluster centers of the closest patches as code words.

Similar to the global grid-based features, local grid-based features also exist. These features are the third type of local features and are have theoretical similarities to global grid-based features. Related works include those by Ikizler and Duygulu [11] and Zhao and Elgammal [12]. The last type of local features is a correlation between local descriptors. Grid-based representations model spatial and temporal relationships between local descriptors to some extent, but they are often redundant and contain unimportant or uninformative features. Therefore, some researchers exploit correlations between local descriptors as features.

28.2.3 Recognition Techniques in Video Sensor-Based Activity Recognition

Recognition techniques are used to infer activity information based on the aforementioned extracted features. An activity label can be predicted for each frame or sequence of frames. Four types of recognition techniques have been widely used in VSAR: nonparametric, volumetric, temporal-independent, and temporal-based techniques.

28.2.3.1 Nonparametric Techniques

Nonparametric approaches typically extract a set of features from each video frame. The features are then matched to a stored template. The template can be either 2D or 3D. When using nonparametric techniques, the typical procedures consist of motion detection and human tracking in the scene, which enables the construction of a sequence. Then, a periodicity index is computed and the periodicity sequence is segmented into individual cycles for recognition.

28.2.3.2 Volumetric Techniques

Volumetric techniques do not extract features on a frame-by-frame basis. Instead, they use a video as a 3D volume of pixel intensities and extend standard image features to a 3D case. Volumetric techniques mainly include four types of methods: spatiotemporal filtering, which filters video volume using a large filter bank; part-based approach, which considers video volume as a collection of local parts; subvolume matching, which matches videos by matching subvolumes between videos and templates; and a tensor-based approach, which is a generalized collection of matrices for multiple dimensions.

28.2.3.3 Temporal-Independent Techniques

Temporal-independent techniques neglect the information in the temporal domain. They summarize all of the frames for an observed sequence into a single representation, subsequently performing activity recognition for each frame. The main techniques in this category include k-nearest neighbors (KNN) and support vector machines (SVMs).

When using KNN, the distance measurement can affect recognition performance. Euclidean distance with global features and histograms of code words has been used [13]. Other measures, like the Mahalanobis distance, have also been utilized.

Another popular temporal-independent recognition method in VSAR is SVM. It determines a hyperplane in feature space, described by a weighted combination of support vectors. SVM has been used jointly with local features of fixed lengths, such as histograms of code words, for activity learning [14].

28.2.3.4 Temporal-Based Techniques

Temporal-based techniques fully utilize the temporal relationships of activities. They can be either generative or discriminative. Dynamic time warping (DTW) is a generative learning method. It models highly nonlinear warping functions based on the distance measure between two sequences. Hidden Markov models (HMMs) are the other common generative learning method in VSAR. They use hidden states that correspond to different phases in the performance of an action. They model state transition probabilities and observation probabilities. Although HMMs are widely used, these are limited to observations which are time independent. Discriminative models overcome this issue by modeling the conditional distribution of activity labels given the observations. Conditional random fields (CRFs) are the most commonly used discriminative models in VSAR. They use multiple overlapping features. A detailed comparison between a CRF and a HMM is given by Lafferty et al. [15].

28.3 Wearable Sensor-Based Activity Recognition

WSAR occurred with the development of wearable computing. The goal is to create personal applications that can adapt and react to the user's current context. The scope of context is diverse, and in WSAR, it refers to the current activity information.

Unlike VSAR, WSAR requires more from the sensors; therefore, it is a newer technology, dating back only to the 1990s, when hardware (including sensing, display, and computing equipment) became lightweight enough that an integrated mobile system could be "worn" by a person for an extended period of time. With the development of hardware techniques, there now exist a large number of WSAR systems that use various sensors, and recognition techniques, and are used for diverse applications.

28.3.1 Applications of Wearable Sensor-Based Activity Recognition

An important application of WSAR is in healthcare and assisted living. There also exist a number of other applications, such as industrial applications and applications for entertainment and gaming:

1. *Healthcare and assisted living*: WSAR can help elderly people live more independent lives and, thus, reduces the burden on caregivers. By detecting the activities of elderly people, we can implicitly detect potentially dangerous situations in daily life. The typical functions of these systems are to detect when people fall and to monitor vital signs.

2. *Prevention of diseases*: Some diseases can be detected from activity information prior to being diagnosed (Figure 28.3). For example, Alzheimer's can be detected

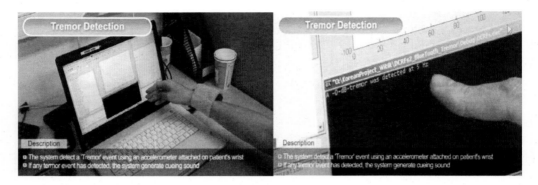

FIGURE 28.3
Tremor detection based on wearable sensor-based activity recognition.

through early symptoms related to human activity. These applications accumulate and summarize statistics about daily activities and continuous recordings of
physiological parameters. These applications can help physicians and caregivers
to estimate the physical well-being of a person.

3. *Promoting exercise*: This type of system aims to aid elderly or disabled people in
performing everyday activities, thus generating a healthier lifestyle (Figure 28.4).
With the increasing intelligence and processing power of mobile phones, many
such applications use mobile phones to remind people to perform certain activities.

28.3.2 Sensors in Wearable Sensor-Based Activity Recognition

VSAR uses only video sensors; however, WSAR involves many different kinds of sensors.
Each type of sensor has its own strengths and suitable applications.

The types of sensors used in WSAR range from relatively simple sensors with discrete
output, such as ball switches, to sensors with continuous output, such as accelerometers,
to more complex sensing methods such as audio processing. Video sensors are wearable,
with accelerometers being the most commonly used type of sensors, as they usually give
good results in terms of physical activity recognition. The most commonly used accelerometers and their vender information are summarized in Table 28.1.

FIGURE 28.4
Gait detection based on wearable sensor-based activity recognition.

TABLE 28.1

Accelerometers Used in WSAR

Sensor	Vendor
CDXL04M3 [16]	Crossbow Technologies
ADXL210E [17]	Analog Devices
LIS302DL [18]	STMicroelectronics
MMA7260Q [19]	Freescale Semiconductor
ADXL330 [20]	Analog Devices

In addition to the aforementioned sensors, there are some sensors which are less commonly used. Fiber-optic sensors are used to measure posture. Foam pressure sensors are used to measure respiration rates. Force-sensitive resistors are used to measure muscle contractions. Moreover, various kinds of physiological sensors, such as oximetry sensors, skin conductivity sensors, electrocardiographs, and body temperature sensors, have also been used for various applications. As different types of sensors can provide complementary information, using many various sensors simultaneously usually gives results that are more accurate. For example, the Ubiquitous Computing Lab of Kyung Hee University has developed the Mobile Activity Sensor Logger (MASoL). MASoL consists of various sensors which record a personal activity log. All of the sensors in the MASoL platform are embedded on one chip, to decrease user discomfort (Figures 28.5 and 28.6).

Depending on the type of activity, recognition performance can be improved by using the same type of sensor at multiple body locations, employing networks of heterogeneous sensors or integrating a variety of sensors on a single device. Combining two or more complementary types of sensor data can also help in recognizing activities, by combining motion and audio data, motion and proximity data, or motion and location data.

FIGURE 28.5
MASoL.

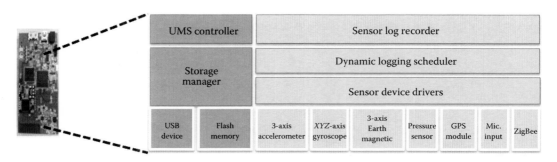

FIGURE 28.6
Key components in MASoL.

28.3.3 Recognition Techniques for Wearable Sensor-Based Activity Recognition

Quite a range of different recognition techniques have been used for WSAR. There is no best recognition technique for WSAR, as considerations include the kind of data to be processed and the types of activities to be recognized. The recognition techniques for WSAR can be classified into two types: supervised and unsupervised. Supervised learning requires labeled data. Conversely, unsupervised learning tries to directly construct models from unlabeled data, either by estimating the properties of its underlying probability density or by discovering groups of similar examples. When using labeled data, supervised learning usually outperforms unsupervised learning; therefore, supervised learning is the predominant approach for WSAR.

28.3.3.1 Supervised Recognition Techniques

There exists a wide range of algorithms and models for supervised learning. Commonly used methods in WSAR include naive Bayesian (NB), C4.5 decision trees, and nearest neighbors. HMMs are well suited for capturing temporal patterns in the data but can be difficult to train due to an abundance of parameters [21–22]. Other methods that have been applied include SVMs [23–24] and, more recently, string-matching methods. Boosting uses multiple classifiers to boost the classification performance and has also been used in WSAR.

28.3.3.2 Unsupervised Recognition Techniques

In Clarkson and Pentland's work [25], hierarchies of HMM are used to learn the locations and scenes, such as walking through a supermarket, from audio and video data in an unsupervised fashion. Graphical model-based unsupervised learning schemes have been used wherein the focus was on inferring the transportation modes (such as by bus, by car, or walking) and the destinations of the users. One work [26] combines discrete string-matching techniques with continuous HMM classifiers to discover short recurring motifs in acceleration data. They aimed to discover and model short-term motion primitives, such as those occurring during physical exercise. Other research [27] used the concept of multiple eigenspaces for the unsupervised learning of activities such as walking or juggling.

28.4 Object Usage-Based Activity Recognition

In daily life, people usually perform an activity by interacting with a series of objects. For example, bathing people may interact with a door, light, an exhaust fan, a shower faucet, etc. The strategy of OUAR is to attach sensors to these objects so that it is possible to determine the state of that object when a person interacts with it.

OUAR [28–30] can be particularly useful in domains such as cooking, which involve a relatively small number of repeated actions (chopping, pouring, spreading, etc). Object-use information can help discriminate between activities which may be similar from the standpoint of the activity alone, such as making toast and making a sandwich. Such distinctions can be important for application domains, such as health monitoring or memory aids.

28.4.1 Sensors in Object Usage-Based Activity Recognition

There are two main types of sensors used in OUAR systems: RFID-based sensors and simple binary sensors.

28.4.1.1 Radio Frequency Identification-Based Sensors

To use an RFID-based sensor, one needs RFID tags and an RFID reader. RFID tags are small and are attached to objects. Subjects need to wear a short-range RFID reader (or equipment with the RFID reader integrated). An RFID-based sensor will generate object-use events when a tagged object is manipulated during an activity. Whenever the user's hand is in close proximity to a tagged object, the reader indicates this. The tags are inexpensive and maintenance free given that they do not contain batteries. For example, by using an RFID-based sensor, it is possible to detect that a user is reading.

28.4.1.2 Binary Sensors

The other type of sensor which has received widespread acceptance in OUAR is the binary sensor. Binary sensors are usually simple and anonymous. They are unable to directly identify occupants; however, at any given time a binary value may be obtained from them. Whenever the state of a certain context (object and movement) associated with a binary sensor changes, the value of the sensor changes to "1" from "0" (indicating its static state). There are many diverse types of binary sensors, including movement detectors, contact switches, and pressure mats. They are applicable to different scenarios. For example, contact switches may be installed on the doors of cupboards, the fridge, and the microwave. Pressure mats may be discreetly installed in objects such as chairs, sofas, and beds and, in some instances, may be used to locate specific movements within rooms. They have the advantage that they do not require the person being monitored to wear or carry any new technology.

28.4.2 Recognition Algorithms

Temporal probabilistic models are most common in OUAR. Sets of probabilistic models have been proposed in OUAR: the Naive Bayesian (NB) by Tapia et al. [31], the Hidden Markov Models (HMMs) in HMMs by Van Kasteren et al. [32], and the Condition Random Field (CRF) in by Hu et al. [33]. HMMs are generative probabilistic models consisting of a hidden variable and an observable variable at each time step. For OUAR, the hidden variable is the activity performed and the observable variable is the vector of the sensor readings.

A CRF is a discriminative probabilistic model. Van Kasteren et al. [32] compared CRF with HMMs and found that CRF outperformed HMMs in all cases with respect to time slice accuracy, but HMMs achieved the overall highest accuracy. This is due to the way both models maximize their parameters. HMMs make use of a Bayesian framework in which a separate model is learned for each class. A CRF uses a single model for all classes. A comparison of HMMs and CRF was also discussed by Hu et al. [33], who found that CRF is able to easily incorporate a wide variety of computed features, which allows domain knowledge to be added to the models. They also showed that due to the independence assumptions inherent in HMMs, such computed features are not nearly as effective in improving classification accuracy. Thus, CRF's classification accuracy has shown to be consistently higher than HMM's.

28.5 Comparisons of Video Sensor-Based, Wearable Sensor-Based, and Object Usage-Based Activity Recognition

In this section, we discuss the strengths and limitations of VSAR, WSAR, and OUAR. Based on our comparisons, fusing these three approaches seems to be the most promising solution for complex activity recognition applications.

28.5.1 Video Sensor-Based Activity Recognition

VSAR is the traditional activity-recognition method. Among the three approaches, a video system with appropriately placed cameras provides the richest information and, in principle, could facilitate the most detailed analysis. However, the signal processing involved in extracting this information can be computationally intensive and, in general cases, remains an open problem. In general, VSAR often works well in a laboratory or well-controlled environment. However, it fails to achieve the same level of accuracy for real home settings due to the clutter, variable lighting, and highly varied activities that take place in natural environments. Video sensor-based activity recognition is complex to implement because it requires processing highly multidimensional data. Additionally, video information may violate the user's privacy.

28.5.2 Wearable Sensor-Based Activity Recognition

The WSAR approach tries to recognize a user's activities by employing sensors such as body-worn accelerometers and microphones to capture characteristic repetitive motions, postures, and sounds of the activities. Using these types of wearable sensors, sensing studies have successfully recognized such activities as walking, bicycling, brushing teeth, speaking and laughing, and workshop activities that have characteristic motions and/or sounds, such as sawing and drilling. An advantage of this approach is that it does not require environment-embedded sensors. That is, this approach incurs no cost in terms of money or time for embedding sensors in indoor objects and furniture. Furthermore, users can easily turn off their wearable devices when they want to preserve their privacy. Wearable sensors are capable of measuring mobility directly. They are well suited to collecting data on daily activity patterns over an extended period, as they can be integrated into clothing or worn as wearable devices. Since they are attached to the subjects they are monitoring, wearable sensors can, therefore, measure physiological parameters which may not be measurable using ambient sensors.

Although body-attached sensors are promising in identifying primitive sequences of movements, such as walking and running, it is difficult to identify goal-oriented activities (e.g., making tea and taking medicine). In addition, designing of wearable systems is complicated in size, weight, and power consumption requirements.

28.5.3 Object Usage-Based Activity Recognition

OUAR is the third type of activity-recognition approach. It infers a person's activity by analyzing the person's interactions with various objects. This approach is newer than VSAR and WSAR and was proposed to overcome the limitations of VSAR and WSAR. Many people are uncomfortable living with cameras. Moreover, people are often unwilling or forget to wear sensors. However, OUAR does have some drawbacks. For extensive and detailed recognition, OUAR requires a large number of objects to be attached to sensors. This is often either infeasible or too expensive. The cost of the sensors and sensor acceptance are pivotal issues, especially in homes.

Each of the three types of AR approaches has its own advantages and disadvantages; however, given the specific application and associated requirements of this study, there is a most suitable approach (Table 28.2).

As shown in Table 28.2, the three types of AR are complementary. For example, VSAR violates user privacy; however, WSAR and OUAR do not violate privacy. OUAR is usually limited to the home environment, whereas WSAR and VSAR can work outside home. In our ubiquitous healthcare (uHealthcare) project, we tried to combine these three approaches for our uHealthcare applications [34–35].

Our system needs to constantly monitor human-activity information to improve health conditions and determine potential health problems as early as possible; therefore, all three types of AR approaches were used together (Figure 28.7).

The cameras used for VSAR were in the lobby, the living room, and the kitchen. Due to privacy issues, they were not used in the bedroom. For WSAR, we developed the MASoL, which is a low-cost, low-energy consumption sensor device. MASoL contains 13-axis sensors, gathering many kinds of behaviors and storing them. In addition, we also used the OUAR approach with an RFID-based sensor, with RFID tags attached to the book and cup.

Combining the three AR approaches enabled our system to accurately recognize many kinds of activities. Although it is logical that the combination of three AR approaches would outperform an individual AR approach, their fusion is not trivial, as we developed a sophisticated probabilistic fusion model for this task.

The experience of building our system suggests that each AR approach has some limitations and that combining them results in a promising approach for complex recognition.

TABLE 28.2

Analysis of VSAR, WSAR, and OUAR

Category	Strengths	Limitations	Applications
VSAR	1. Video includes richest information; thus, applied range is the widest 2. Easy setup of environment	1. Violates privacy 2. Sensitive to environment (e.g., light condition and viewpoint variation)	1. Healthcare 2. Security 3. Interactive applications 4. Content-based video analysis
WSAR	1. Setup of environment not required 2. Good for applications requiring explicit motion analysis	1. Wearing sensors a burden 2. Unable to separate similar actions (e.g., making tea and making coffee)	1. Healthcare 2. Interactive applications
OUAR	1. Does not violate privacy 2. Good at recognizing goal-oriented activities	1. Recognizable activities need to be related to objects 2. Setup of environment takes more effort than for VSAR and WSAR	1. Healthcare 2. Security

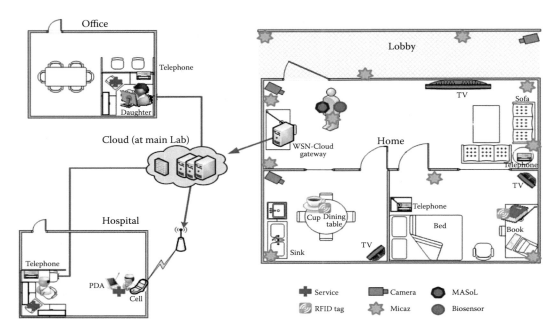

FIGURE 28.7
Activity recognition for uHealthcare application [34]. (From Guan, D., T. Ma, W. Yuan, Y.-K. Lee, and A. M. J. Sarkar, Review of sensor-based activity recognition systems, *IETE Tech. Rev.*, 28(5), 418–433, 2011. With permission.)

28.6 Challenges in Sensor-Based Activity Recognition

The significant potential of activity-recognition systems has been recognized by many researchers who have attempted different approaches, including video sensor-based, wearable sensor-based, and object usage-based approaches. In the last 20 years, many activity-recognition systems have been developed; however, activity-recognition systems still have limited functions. The main challenges include the following:

1. *Real-world conditions*: Most existing systems are designed and tested in constrained conditions. Many factors in the real world can severely limit the applicability of a system. For example, in VSAR, these factors can include noise, occlusions, and shadows. Another challenge is the robustness of system. In real-world conditions, there might be a wide variability within the same activity class. Therefore, we need to find methods which can explain and withstand the wide variability of features that are observed within the same activity class. Overall, more research needs to be done to address these practical issues in real-world conditions.

2. *Evaluation of systems*: A common problem in the activity-recognition community is the lack of annotated reference data and standardized test beds that could help researchers compare the performances of their approaches. Although there exist several benchmark data sets, most of these data sets consist of simple activities, such as walking, running, sitting, and sleeping. Very few common data sets exist

for evaluating higher-level complex activities and reasoning algorithms. A lack of evaluation standards is a challenge which makes it difficult to compare the performances of different systems.

3. *Hardware challenge*: With the development of the sensor industry, sensors have gotten stronger processing power and smaller sizes. However, the energy supply is still a problem, particularly in long-term activity recognition. Moreover, making sensors easier to use and less obtrusive also needs to be addressed.

4. *Privacy and security*: In order for activity recognition to become widely used, the users' privacy must be considered. Therefore, comprehensive privacy and security mechanisms must be developed for activity-recognition systems.

References

1. A. F. Bobick, Movement, activity and action: The role of knowledge in the perception of motion, *Philosophical Trans. Royal Soc. B: Biol. Sci.*, Vol. 352, No. 1358, pp. 1257–1265, 1997.
2. C. Cedras and M. Shah, Motion-based recognition: A survey, *Image Vision Comput.*, Vol. 13, No. 2, pp. 129–155, 1995.
3. V. Kellokumpu, G. Zhao, and M. Pietikäinen, Human activity recognition using a dynamic texture based method, in *Proc. Br. Machine Vision Conf. (BMVC '08)*, pp. 885–894, 2008.
4. M. Blank, L. Gorelick, E. Shechtman, M. Irani, and R. Basri, Actions as space-time shapes, in *Proc. Int. Conf. Computer Vision*, pp. 1395–1402, 2005.
5. I. Laptev and T. Lindeberg, Space-time interest points, in *Proc. Int. Conf. Computer Vision*, pp. 432–439, 2003.
6. C. Harris and M. Stephens, A combined corner and edge detector, in *Proc. Alvey Vision Conf.*, pp. 147–151, 1988.
7. P. Dollár, V. Rabaud, G. Cottrell, and S. Belongie, Behavior recognition via sparse spatio-temporal features, in *Proc. Int. Workshop Visual Surveillance Performance Evaluation Tracking Surveillance*, pp. 65–72, 2005.
8. O. Chomat, J. Martin, and J. L. Crowley, A probabilistic sensor for the perception and recognition of activities, in *Proc. Eur. Conf. Computer Vision*, pp. 487–503, 2000.
9. C. Schüldt, I. Laptev, and B. Caputo, Recognizing human actions: A local SVM approach, in *Proc. Int. Conf. Pattern Recognition*, pp. 32–36, 2004.
10. J. C. Niebles, H. Wang, and L. Fei-Fei, Unsupervised learning of human action categories using spatial-temporal words, *Int. J. Computer Vision*, Vol. 79, No. 3, pp. 299–318, 2008.
11. N. Ikizler and P. Duygulu, Histogram of oriented rectangles: A new pose descriptor for human action recognition, *Image Vision Computer*, Vol. 27, No. 10, pp. 1515–1526, 2009.
12. Z. Zhao and A. Elgammal, Human activity recognition from frame's spatiotemporal representation, in *Proc. Int. Conf. Pattern Recognition*, pp. 1–4, 2008.
13. D. Batra, T. Chen, and R. Sukthankar, Space-time shapelets for action recognition, in *Proc. Workshop Motion Video Computer*, pp. 1–6, 2008.
14. M. D. Rodriguez, J. Ahmed, and M. Shah, Action MACH: A spatio-temporal maximum average correlation height filter for action recognition, *Proc. Conf. Comput. Vision Pattern Recognition*, pp. 1–8, 2008.
15. J. D. Lafferty, A. McCallum, and F. C. Pereira, Conditional random fields: Probabilistic models for segmenting and labeling sequence data, in *Proc. Int. Conf. Machine Learning*, pp. 282–289, 2001.
16. N. Ravi, N. Dandekar, P. Mysore, and M. L. Littman, Activity recognition from accelerometer data, in *Proc. 17th Conference on Innovative Applications of Artificial Intelligence Conference*, Vol. 3, pp. 1541–1546, 2005.

17. L. Bao, *Physical Activity Recognition from Acceleration Data under SemiNaturalistic Conditions*, Master's thesis, Massachusetts Institute of Technology, Cambridge, 2003.
18. J. R. Kwapisz, G. M. Weiss, and S. A. Moore, Activity recognition using cell phone accelerometers, in *Proc. Fourth Int. Workshop Knowledge Discovery Sensor Data*, pp. 10–18, 2010.
19. J. B. Chong, *Activity Recognition Processing in a Self-Contained Wearable System*, Master's thesis, Virginia Polytechnic Institute and State University, Blacksburg, 2008.
20. P. Zappi, T. Stiefmeier, E. Farella, D. Roggen, L. Benini, and G. Troster, Activity recognition from on-body sensors by classifier fusion: Sensor scalability and robustness, in *Proc. ISSNIP*, pp. 281–286, 2007.
21. J. A. Ward, P. Lukowicz, G. Tröster, and T. E. Starner, Activity recognition of assembly tasks using body-worn microphones and accelerometers, *IEEE Trans. Pattern Analysis Machine Intelligence*, Vol. 28, No. 10, pp. 1553–1567, 2006.
22. J. Lester, T. Choudhury, N. Kern, G. Borriello, and B. Hannford, A hybrid discriminative/ generative approach for modeling human activities, in *Proc. IJCAI*, pp. 766–772, 2005.
23. T. Hùynh, U. Blanke, and B. Schiele, Scalable recognition of daily activities with wearable sensors, in *3rd Int. Symp. Location Context-Awareness*, pp. 50–67, 2007.
24. G. Loosli, S. Canu, and A. Rakotomamonjy, Détection des activités quotidiennes à l'aide des séparateurs à Vaste Marge, in *RJCIA*, pp. 139–152, 2003.
25. B. Clarkson and A. Pentland, Unsupervised clustering of ambulatory audio and video, in *ICASSP*, pp. 3037–3040, 1999.
26. D. Minnen, T. Starner, I. Essa, and C. Isbell, Discovering characteristic actions from on-body sensor data, in *Proc. ISWC*, pp. 11–18, 2006.
27. T. Hùynh and B. Schiele, Unsupervised discovery of structure in activity data using multiple eigenspaces, *Location Context-Awareness Second International Workshop, LoCA, Proceedings (Lecture Notes in Computer Science)* Vol. 3978, pp. 151–167, 2006.
28. J. R. Smith, K. P. Fishkin, B. Jiang, A. Mamishev, M. Philipose, A. D. Rea, S. Roy, and K. Sundara-Rajan, RFID-based techniques for human activity detection, *Commun. ACM*, Vol. 48, No. 9, pp. 39–44, 2005.
29. M. Philipose, K. P. Fishkin, M. Perkowitz, D. J. Patterson, D. Fox, H. Kautz, and D. Hahnel, Inferring activities from interactions with objects, *IEEE Pervasive Comput.*, Vol. 3, No. 4, pp. 50–57, 2004.
30. D. H. Wilson and C. Atkeson, Simultaneous tracking and activity recognition (STAR) using many anonymous, binary sensors, in *Pervasive Computing: Lecture Notes in Computer Science*, Vol. 3468, pp. 62–79, 2005.
31. E. M. Tapia, S. S. Intille, and K. Larson, Activity recognition in the home using simple and ubiquitous sensors, in *Pervasive Computing: Lecture Notes in Computer Science*, Vol. 3001, pp. 158–175, 2004.
32. T. van Kasteren, A. Noulas, G. Englebienne, and B. Krose, Accurate activity recognition in a home setting, in *Proc. UbiComp.*, pp. 1–9, 2008.
33. D. H. Hu, S. J. Pan, V. W. Zheng, N. N. Liu, and Q. Yang, Real world activity recognition with multiple goals, in *Proc. UbiComp.*, pp. 30–39, 2008.
34. Guan, D., T. Ma, W. Yuan, Y.-K. Lee, and A. M. J. Sarkar, Review of sensor-based activity recognition systems, *IETE Tech. Rev.*, 28(5), 418–433, 2011.
35. A. M. Khattak, P. T. H. Truc, L. X. Hung, L. T. Vinh, V.-H. Dang, D. Guan, Z. Pervez, M. Han, S. Lee, and Y.-K. Lee, Towards smart homes using low level sensory data, *Sensors (Basel)*, Vol. 11, No. 12, pp. 11581–11604, 2011.

29

Very Large-Scale Integration Bioinstrumentation Circuit Design and Nanopore Applications

Jungsuk Kim and William B. Dunbar

CONTENTS

29.1 Introduction

Submicron CMOS process technology has enabled the implementation of low-power, low-noise, low-cost microelectronics on a silicon chip. Using these features, biomedical very large-scale integration (VLSI) circuits have been designed and fabricated, along with the use of microelectromechanical systems (MEMs) and nanoelectromechanical systems (NEMSs) within biomedical devices. Many of the applications for these devices require sensing very weak biosignals. For instance, integrated preamplifiers, along with a micro-electrode array, are employed to amplify very small biopotential signals such as EEG, ECG, and EMG [1–2]. Recently, on-chip preamplifiers have been used to sense the small current variations that are generated when individual DNA molecules pass through a nanopore sensor.

29.1.1 Nanopore Method and Measurement

Nanopores provide an electrical DNA-sequencing method without the need for optical particle labeling or amplification of the molecular samples [3]. While other previous technologies depend on improvements of chemical, optical, or bioinformatics procedures, nanopore sensors can detect a direct electrical signal that is influenced by the nucleotides within a passing DNA strand. There are two types of nanopores, the biological nanopore [4] and the solid-state nanopore [5]. Figure 29.1a displays a biological nanopore which is

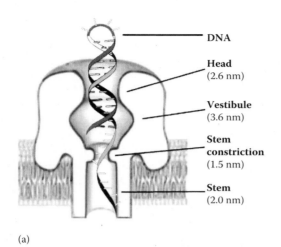

DNA

Head
(2.6 nm)

Vestibule
(3.6 nm)

Stem
constriction
(1.5 nm)

Stem
(2.0 nm)

(a)

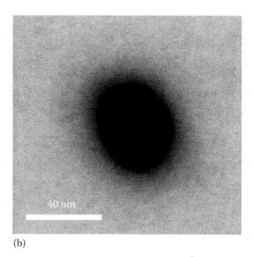

40 nm

(b)

FIGURE 29.1
(a) Schematic cross section of the α-hemolysin nanopore inserted into a lipid bilayer with 1.5 nm limiting diameter and (b) close-up image of a 40 nm solid-state nanopore fabricated in a silicon dioxide membrane.

composed of a single α-hemolysin (α-HL) protein channel in a lipid bilayer. The cis opening of the pore, termed *head* in the figure, is approximately 2.6 nm in diameter and opens to 3.6 nm at the vestibule. At the stem constriction, the opening pore gets narrow again to 1.5 nm and the trans opening diameter is expanded to 2 nm [6]. When single-stranded DNA (ssDNA) passes though the α-HL protein channel, its nucleotide bases—cytosine, guanine, adenine, and thymine—instantaneously block the channel at the stem constriction aperture. This results in a variation in the ionic current which flows through the channel. The current's amplitude and duration can change in accordance with the nucleotide bases in the smallest portion of the channel. Using the current signal variations to identify the DNA bases is known as nanopore sequencing technology.

The nanoscale pore can be also manufactured by mechanically drilling or chemically etching in synthetic materials, for example, silicon dioxide (SiO_2) [5] or glass [7]. This solid-state nanopore offers a more durable and scalable platform than the biological nanopore. Moreover, a larger voltage and temperature range can be achieved than with biological pores. Since the diameter of solid-state pores can be varied, they also have more applications beyond sequencing [3,5]. Figure 29.1b portrays a solid-state nanopore fabricated in thin SiO_2 layer. Other forms of solid-state nanopores include graphene, solid-state/biopore hybrids [8], and solid-state pores with tunneling currents [9].

To measure the minute ionic current change on the nanopore sensors, the patch-clamp technique is applied; an example configuration is illustrated in Figure 29.2. The patch-clamp technique has been widely utilized in electrophysiology and life sciences to study the characteristics of ion channels in cellular membranes [10]. Because of the high gain of the amplifier A1, the inverting input, connected to the recording electrode (E_{REC}), follows the command voltage (V_{CMD}) applied to the noninverting input. This gives rise to an electrical field between the cis and trans reservoirs that are filled with a buffered ionic (KCl) solution. Since DNA is negatively charged by virtue of the phosphate groups in the backbone, a DNA near the pore starts moving from the cis chamber through the pore and into the trans side when a trans-positive voltage is applied. While it traverses the pore, the ionic currents carried by potassium (K^+) and chloride (Cl^-) ions momentarily decrease in

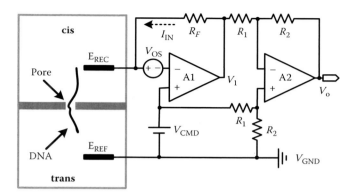

FIGURE 29.2
Conventional patch-clamp amplifier setup for detecting DNA passage through the nanopore.

the range of tens to hundreds of picoamperes. Experimentally, it has been reported that the ionic current variation is on average 120 pA when 11 kilo–base pair double-stranded DNA molecules pass through the solid-state nanopore with a diameter of 10 nm at V_{CMD} of 120 mV [5]. This amplitude varies with the diameter of the nanopore, length of DNA strand, temperature, and pH and ion concentration of the buffer.

29.1.2 Design Requirements

To accurately monitor DNA translocation events through the pore, the patch-clamp amplifier should accomplish a low input-referred noise current. From the above example, the input-referred RMS noise current of the amplifier has to be smaller than 12 pA_{RMS} to achieve the signal-to-noise ratio (SNR) of 20 dB necessary for detecting the presence of DNA in a nanopore. To sequence the DNA strand as it passes through the pore, considerably lower noise is required, and this can be achieved by additional low-pass filtering before sampling the current measurement. In addition to the low-noise performance, the patch-clamp amplifier needs a compensation technique for input parasitic capacitance. In particular, inverting-input parasitic capacitances of the nanopore, E_{REC}, and the large feedback resistor must be cancelled out to prevent the head stage saturation and data loss.

To actively control DNA motion in a nanopore, e.g., to quantify protein-DNA binding kinetics [11], we have to change V_{CMD} stepwise during sensing. When V_{CMD} is varied to control DNA movement during the nanopore sensing, the shift of the inverting input is delayed due to the parasitic capacitances. This transient difference is instantaneously amplified through the high-gain head stage so that its output becomes saturated, or significantly perturbed, until the capacitances are charged up and the inverting input once again is equal to V_{CMD}. During this interval, termed as the *dead time*, all incoming data become lost [12].

Over the past few years, research groups have developed low-noise CMOS patch-clamp amplifiers in order to detect the ionic current variations through a cellular membrane channel or a nanopore. For instance, using silicon-on-sapphire CMOS technology, Dr. Culurciello's group reduced electronic coupling noise present in sensitive current-recording equipment such as patch-clamp amplifiers [13–15]. Rosenstein et al. developed a low-noise preamplifier that was part of the reduced high-frequency noise nanopore setup [16]. Specifically, they formed a solid-state nanopore on the CMOS amplifier chip by using a postfabrication process, resulting in a reduction of input parasitic capacitance and nanopore capacitance, which, in turn, reduced the high-frequency noise and permitted higher bandwidth detection of DNA events. In this chapter, we will investigate VLSI bioinstrumentation circuit design techniques that accomplish low-noise, low-input offset voltage, and dead-time cancellation for nanopore applications, and present experimental results using an α-hemolysin nanopore sensor. Finally, we will close this chapter with future research goals for direct nanopore sequencing.

29.2 Very Large-Scale Integration Bioinstrumentation Circuit Design

This section is composed of four subsections with the following topics: (1) noise analysis, (2) low-noise core-amplifier design, (3) dead-time compensation technique, and (4) input offset voltage cancellation. Considering the four design issues, we will study the VLSI bioinstrumentation design method, analysis, and implementation.

29.2.1 Noise Analysis

In this subsection, we will discuss noise sources that arise from the nanopore itself, E_{REC}, and the head stage during nanopore sensing. For the noise analysis, we use the simplified electrical models shown in Figure 29.3. Here, R_N and C_N model the resistance and capacitance of the nanopore, respectively. These vary with nanopore diameter, thickness, and material. E_{REC} is modeled by the series resistance R_E and a double-layer capacitance C_E, where R_E is much smaller than R_N. Note that in this chapter, we pay attention to the resistive-feedback transimpedance amplifier (TIA) due to its simple hardware structure and robust reliability, rather than a capacitive-feedback TIA, which requires a disruptive periodic reset [12,15]. Readers who are interested in the capacitive-feedback TIA can refer to the paper of Kim et al. [17].

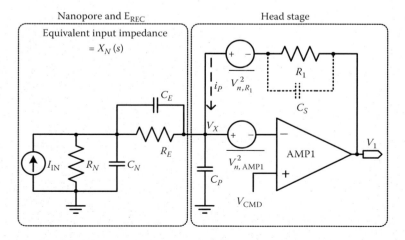

FIGURE 29.3
Simplified electrical model of the nanopore, E_{REC}, and head stage in the steady-state mode where V_{CMD} variation causes the head-stage saturation.

In terms of this equivalent circuit of the nanopore and E_{REC}, the thermal noise current [17] is

$$S_N(f) = \frac{4kT}{\text{Re}[X_N(f)]}.$$ (29.1)

Here, k and T are the Boltzmann constant and absolute temperature, respectively and Re stands for real part of complex component. $X_N(f)$ is the impedance of the equivalent circuit; it is expressed as

$$X_N(f) = \frac{(R_E + R_N) + j2\pi f \cdot (R_N R_E C_N + R_N R_E C_E)}{1 - (2\pi f)^2 \cdot (C_E C_N R_E R_N) + j2\pi f \cdot (C_N R_N + C_E R_E)}.$$ (29.2)

Because DNA translocation events have main power spectral densities with a bandwidth of ~10 kHz and R_E is also negligible compared to R_N, $X_N(f)$ is approximated to $R_N/(1 + j2\pi f \cdot C_N R_N)$. As a result, at low frequencies, $S_N(f)$ can be simplified to $4kT/R_N$, which is proportional to the conductance of the nanopore. Nanopore flicker noise is given by $(\alpha \times I^2)/(N_C \times f)$. Here, I, α, and N_C are the direct current, the Hooge parameter, and the number of charge carriers [18], respectively, all of which are associated with the KCl buffer concentration and nanopore material. To increase R_N, a nanopore with a narrow diameter should be adopted. For instance, the α-hemolysin pore with a limiting diameter of 1.5 nm has a resistance of 3 GΩ in 0.3 M KCl or 1 GΩ in 1 M KCl.

Next, the input-referred noise current of the head stage is considered. This makes a critical impact on the background noise principally due to the flicker noise of its input-pair transistors, which dominates at low frequencies. In Figure 29.3, $V_{n,\text{AMP1}}^2$ is the input-referred noise voltage of the core amplifier AMP1 and $V_{n,R1}^2$ is the thermal noise of R_1, both of which are uncorrelated. The stages' other noise sources can be neglected because of the relatively high gain of the head stage. Applying the noise-analysis method by Razavi [19], the input-referred noise current of the head stage $S_H(s)$ is calculated in the following.

In Figure 29.3, the node of V_X and current i_P have the relationship described as

$$V_X = V_{n,\text{AMP1}} - \frac{V_{n,\text{out}}}{A(s)},$$

$$V_X = \left[X_N(s) \Big\| \frac{1}{sC_P} \right] \times i_P,$$

$$V_X - V_{n,R_1} + i_P \times R_1 = V_{n,\text{out}}.$$

Accordingly, $V_{n,\text{out}}(s)$ can be rewritten as

$$V_{n,\text{out}}(s) = \left\{ \frac{A(s) + A(s) \cdot [sC_P + 1/X_N(s)] \cdot R_1}{1 + A(s) + [sC_P + 1/X_N(s)] \cdot R_1} \right\} \cdot V_{n,\text{AMP1}}(s)$$

$$- \left\{ \frac{A(s)}{1 + A(s) + [sC_P + 1/X_N(s)] \cdot R_1} \right\} \cdot V_{n,R_1}(s).$$

Since the input-referred noise current $S_H(s)$ is defined by $[V_{n,\,out}(s)/R_1]^2$, $S_H(s)$ is stated as

$$
S_H(s) = \left\{ \frac{A(s) + A(s) \cdot [sC_P + 1/X_N(s)] \cdot R_1}{1 + A(s) + [sC_P + 1/X_N(s)] \cdot R_1} \right\}^2 \cdot \frac{V_{n,\,AMP1}^2(s)}{R_1^2}
$$
$$
+ \left\{ \frac{A(s)}{1 + A(s) + [sC_P + 1/X_N(s)] \cdot R_1} \right\}^2 \cdot \frac{V_{n,\,R_1}^2(s)}{R_1^2}. \tag{29.3}
$$

Here, C_P is a parasitic capacitance stemming from the combination of the coaxial cable, Bayonet Neill–Concelman (BNC) connector, printed circuit board (PCB) trace, lead frame, bonding pad, electrostatic discharge (ESD) cell, and input-pair transistors of the core amplifier. $A(s)$ is the amplifier's transfer function, $A_0/(1 + s/\omega_0)$, where A_0 and ω_0 are its open-loop gain and bandwidth, respectively.

For simplicity, it is assumed that ω_0 is much higher than the 10 kHz bandwidth of interest. Therefore, $1/A(s)$ can be neglected, leading to the simplified expression

$$
S_H(f) \approx \left(\frac{1}{R_1} + \frac{1}{R_N} + 2\pi f \cdot C_T \right)^2 \overline{V_{n,\,AMP1}^2(f)} + \frac{4kT}{R_1}. \tag{29.4}
$$

Here, C_T denotes $C_P + C_N$, and $1/R_N$ can be ignored since $R_N \gg R_1$. The input-referred noise in Equation 29.4 can be reduced by increasing R_1 at the cost of larger area and a smaller bandwidth. To further diminish the noise, the both C_T and $V_{n,\,AMP1}^2$ should be minimized.

From Equations 29.1 and 29.4, the background noise during nanopore sensing is

$$
S_B(f) \approx \left(\frac{1}{R_1} + 2\pi f \cdot C_T \right)^2 \overline{V_{n,\,AMP1}^2(f)} + \frac{4kT}{R_N} + \frac{4kT}{R_1} + \frac{\alpha I^2}{N_C f}. \tag{29.5}
$$

Here, the nanopore thermal noise term $4kT/R_N$ ($= 0.1656 \times 10^{-28}$ A^2/Hz) can be neglected because it is much smaller than $4kT/R_1$ ($= 8.28 \times 10^{-28}$ A^2/Hz) [20]. The flicker noise can be roughly calculated, using $I = 180$ pA, $\alpha = 2 \times 10^{-3}$, and $N_C = 2200$ in 1 M KCl [21], which results in 5.82 fA$_{RMS}$ in a bandwidth of 10 kHz. It means that the head-stage noise in Equation 29.4 is the most prominent. In order to lessen the overall background noise, a low-noise core-amplifier should be designed for the head stage. In addition to the head-stage noise, the zero term in Equation 29.5 which is composed of C_T and R_1 induces a significant high-frequency noise. Thus, to achieve high-resolution translocation measurement, we also need to reduce the input parasitic capacitance C_T. For this purpose, Rosenstein et al. proposed an integration nanosystem where the CMOS bioinstrumentation is embedded onto the solid-state nanopore device, resulting in diminished input parasitics [16]. In the next subsections, we are going to discuss circuit techniques for reducing the head-stage noise and compensating for the input parasitics.

29.2.2 Low-Noise Core-Amplifier Design

A core amplifier for nanopore applications should meet the following design requirements: (1) low noise, (2) low power, and (3) wide and symmetric input common-mode range (ICMR), and (4) hardware simplicity. Empirically, nanopore applications require an

ICMR of ±500 mV or less to actively control and detect DNA motion in the pore. If V_{CMD} exceeds this range, the DNA will pass through the nanopore too fast to obtain an accurate reading. For nanopore applications, another important aim is to minimize wiring complexity to restrict opportunities for interference ingress. This issue becomes critical for future multichannel patch-clamp amplifier implementations. Therefore, a self-biased differential Bazes operational transconductance amplifier (OTA), shown in Figure 29.4a, can meet the above desired performance goals for nanopore applications. By virtue of the symmetric structure between n-channel metal–oxide–semiconductor (NMOS) and p-channel metal–oxide–semiconductor (PMOS) transistors, the Bazes OTA used in this design can meet the symmetrical ICMR requirement, without the need for a folded structure that requires biasing wires. The topology also enables a high gain because both the NMOS and PMOS contribute to the effective transconductance (g_m) and helps in reducing power consumption due to current reusing structure. As a result, this amplifier has a gain of $(g_{m2} + g_{m4}) \times (r_{o3} \| r_{o5})$, where r_o is the output resistance. Since the output equivalent noise is divided by the squared amplifier gain, the input-referred noise voltage of this OTA can be approximated as

$$\overline{V_{n,\,AMP1}^2(f)} = \underbrace{\left(\frac{1}{W_2 L_2} + \frac{1}{W_4 L_4} \right) \cdot \frac{2K}{f \cdot C_{ox}}}_{\text{Flicker noise}} + \underbrace{\frac{8kT\gamma}{g_{m2} + g_{m4}}}_{\text{Thermal noise}}. \tag{29.6}$$

Here, K is the process-dependent flicker noise constant, which is on the order of 10^{-25} V^2·F, and W and L indicate the transistor's width and length, respectively. In Equation 29.6, the flicker noise can be reduced by enlarging $W_2 L_2$ and $W_4 L_4$. In this design, PMOS and NMOS transistors with large aspect ratios of 72 μm/1.5 μm and 54 μm/1.5 μm for the input pair

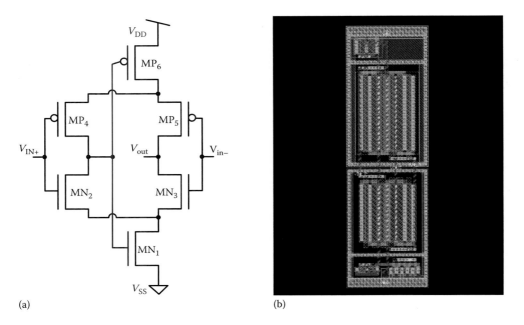

(a)   (b)

FIGURE 29.4
(a) Schematic of self-biased Bazes OTA and (b) layout of the Bazes OTA.

were chosen. This choice also reduces the portion of the input offset owing to random process mismatches.

29.2.3 Dead-Time Compensation

In this section, we will discuss a technique to compensate for the deleterious effects of input parasitic capacitances. In patch-clamp recording, two compensation functions are required to make up for the voltage drop caused by the gigaohm seal formed between a micropipette and a cell membrane, and a transient delay by the inverting-input parasitic capacitances at the head-stage input. In nanopore sensing, the voltage drop is insignificant because the gigaohm seal formation is unnecessary. In addition, other parasitic resistances due to the E_{REC} and the buffer solution are relatively small compared with the nanopore resistance. However, the transient difference between inputs still exists, typically resulting in transient output saturation.

As V_{CMD} changes in Figure 29.3, the head stage works as a noninverting amplifier with the instantaneous gain of

$$\frac{V_1}{V_{CMD}} = \left[1 + \frac{R_1 + sC_P R_1 X_N(s)}{X_N(s) + sC_S R_1 X_N(s)} \right].$$
(29.7)

Here, C_S includes the stray capacitance due to the large feedback resistor and a feedback capacitor which is employed for stability of the TIA. Assuming that R_E and C_E of E_{REC} are negligible, $X_N(s)$ can be simplified to $R_N/(1 + sC_N R_N)$. Thus, the gain is rewritten as

$$\frac{V_1}{V_{CMD}} \approx \left[1 + \frac{R_1 + s(C_N + C_P)R_1 R_N}{R_N + sC_S R_1 R_N} \right].$$
(29.8)

In the case where V_{CMD} varies rapidly, the gain becomes $[(C_N + C_P)/C_S]$. Because $C_N + C_P$ having tens of picofarads is much greater than C_S with tens of femtofarads, a small V_{CMD} variation is immediately amplified by a factor of several hundred. As a result, the head stage is easily saturated and remains unusable until the capacitances are fully charged, typically resulting in a relatively long dead time.

In order to prevent head-stage saturation and reduce the dead time, compensation techniques for injecting a roughly predicted amount of current directly to the inverting-input node through a compensation capacitor have been presented by Prakash et al. [12], Yokichi [22], and Sakmann and Neher [10] at the cost of hardware complexity. Novel compensation circuit for minimizing the complexity of and quickly precharging the inverting node has been devised by Kim et al. [20]; this circuit is illustrated in Figure 29.5. In this design, a compensation switch M_1 is added in parallel with R_1. This switch is turned on 20 ns before changes in V_{CMD} are initiated and maintained in that state for the 8 µs which is required to complete the shift of V_{CMD}. This time is variable and designed to be dependent on the rise and fall times of V_{CMD}. The transition of the head stage to a unity-gain buffer causes the impedance at the inverting node to become very low; thus, the input rapidly follows the desired V_{CMD} variation without delay and saturation. That is, R_1 in Equation 29.8 becomes zero, resulting in $V_1 = V_{CMD}$. This new compensation circuit needs to avoid undesirable glitching caused by switching. Thus, a track-and-hold (T/H) circuit is required, where the T/H circuit stays in the hold mode during the precharge operation (this T/H circuit is

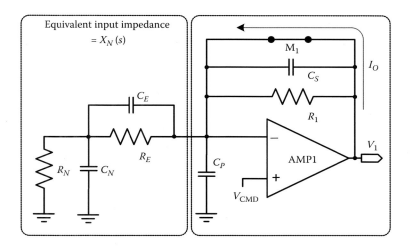

FIGURE 29.5
Transient mode in which the compensation switch M_1 is turned on and, thus, the inverting input follows V_{CMD} variation without delay and saturation.

placed after the second stage as shown in Figure 29.7). After the 8 µs transient, M_1 is turned off and the T/H circuit returns to the track mode with a unity gain. In this system, the compensation pulse that activates the switch M_1 is generated in synchronization with V_{CMD}, the variation of which both are governed by a control unit, i.e., field-programmable gate array (FPGA). The simulated result for this compensation operation is shown in Figure 29.6.

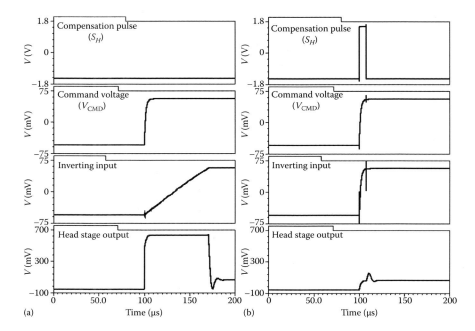

FIGURE 29.6
Simulated results of the compensation switch: (a) compensation pulse is deactivated and (b) the pulse is activated and but its falling edge results in glitch and overshoot at the head-stage output.

29.2.4 Input Offset Voltage Cancellation

It is also essential to minimize the input offset voltage of the head stage. Even though a high-gain amplifier is employed for the multistage, some offset induced by the first stage always exists because of random process mismatch and systematic variations. This offset is amplified by the second-stage voltage-gain difference amplifier (VDGA), shown in Figure 29.2, thus limiting the output dynamic range. The random offset, mainly caused by random fluctuation of process model parameters during the process, can be reduced by enlarging the input-pair transistors. Because large input-pair transistors were already used to suppress the flicker noise, the random offset is effectively suppressed in this design. The remaining offset arises from systematic causes such as layout differences, CMOS process, and thermal gradients. In addition to the intrinsic systematic offset, the compensation switch in this design leads to an offset due to charge injection. To minimize the deleterious effect of the input offset voltage, a stabilized chopper or an autozeroing amplifier, which operates based on the chopper sampling technique, has been used by Enz and Temes [23]. However, this amplifier suffers from the large ripple voltages at the input node, generated by chopping, which would result in V_{CMD} fluctuation during nanopore sensing. Another alternative is to use the instrumentation amplifier topology shown in Figure 29.7 to minimize the input offsets, without the need for chopping. In the track mode, the nodes of V_1, V_2, V_3, and V_4 have the quiescent (Q) point biases of $(V_{CMD} + V_{OS1})$, $(V_{CMD} + V_{OS2})$, $(V_{CMD} + V_{OS2}) - (V_{OS1} - V_{OS2}) \times (R_4/R_3)$, and $(V_{CMD} + V_{OS2})$. As a result, the output V_O has the offset of $(V_{OS1} - V_{OS2}) \times (R_4/R_3)$. To suppress the systematic offsets, all amplifiers AMP1–AMP5 are composed of an identical structure, shown in Figure 29.4b, which is designed using the common centroid layout method and dummy transistor. Furthermore, the differential structure of the instrumentation amplifier helps to minimize the residual systematic offset because it can make V_{OS1} equal to V_{OS2}.

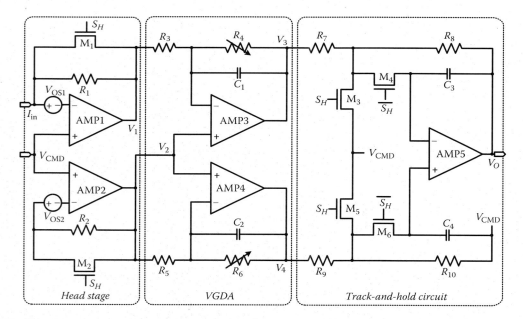

FIGURE 29.7
Schematics of the amplification block composed of three stages: (1) head stage, (2) voltage-gain difference amplifier, and (3) track-and-hold circuit. Here, an instrumentation amplifier topology, AMP1 and AMP2, and compensation switches M_1 and M_2 are adopted for the head stage.

In more detail, the architecture of the VLSI bioinstrumentation circuit shown in Figure 29.7 consists of three stages: (1) head stage, (2) VGDA, and (3) T/H circuit. For the head stage, a resistive-feedback transimpedance amplifier is utilized with an instrumentation amplifier topology that is capable of minimizing the input offset voltage as well as increasing common-mode rejection for immunity to interference. Here, compensation switches M_1 and M_2 are located in parallel with the feedback resistors R_1 and R_2 to compensate for the effects induced by the inverting-input parasitic capacitances. The inverting input of the core amplifier AMP1 is connected to E_{REC}, through which a minute signal current is fed while the noninverting inputs of AMP1 and AMP 2 are driven with the adjustable V_{CMD}. The counterpart, AMP2, plays a role in canceling out systematic input offsets, principally arising from change injection by the compensation switches.

For the second-stage VGDA, a resistive-feedback inverting amplifier with a gain of $-R_4/R_3$, where R_4 is controllable, is chosen rather than a capacitive-feedback amplifier in order to allow measurement of the DC information provided by V_{CMD}. The observed DC variations at the output V_O can be used to estimate the nanopore conductance. By adding capacitors C_1 and C_2 to the feedback resistors of R_4 and R_6 in parallel, we form a first-order LPF to remove high-frequency noise injected by the head stage. This stage also provides programmable gain to accommodate the dynamic range of the signal by adjusting R_4 and R_6. Here, we are able to discretely select one of gains 0, 14, 23.5, or 32 dB by setting R_4 to 100 kΩ, 400 kΩ, 1 MΩ, or 2.5 MΩ with respect to R_3 at 100 kΩ, which is controlled by a 2-bit thermodecoder.

An inverting T/H circuit with clock-feedthrough cancellation is employed for the third stage. This T/H circuit is used to avoid glitch injection during the dynamic compensation interval rather than to provide an additional gain. By setting both of R_7 and R_8 to 100 kΩ, in the track mode this stage operates as a unity-gain difference amplifier while the captured data are stored on C_3 (= 2 pF) in the hold mode. By virtue of the differential structure of the T/H circuit, charge injection and clock feedthrough by the clock switching on M_3 and M_4 are cancelled.

29.3 Implementation and Experimental Results

This low-noise VLSI bioinstrumentation circuit was fabricated in a 0.35 μm CMOS process and tested with an α-hemolysin nanopore with a diameter of 1.5 nm. Figure 29.8a displays a micrograph of this prototype chip occupying an active-die area of 0.3038 mm² [20]. The experimental setup for verifying the chip performance is shown in Figure 29.8b. Although not illustrated in Figure 29.7, an off-chip fifth Bessel low-pass filter (LPF) with a bandwidth of 10 KHz is used after the unity-gain buffer on the chip to further restrict the bandwidth and reduce the noise while maintaining desired dynamics. To ward off 60 Hz noise induced from ground, two 1.5 V coin batteries are adopted for power. The middle tap between two batteries is used as a chip ground and is also connected to the trans chamber through the reference electrode (E_{REF}). This PCB is placed along with the biological nanopore sensor in a Faraday cage to further shield 60 Hz noise injection.

Employing the VLSI bioinstrumentation, we measured the open-channel current on the α-hemolysin nanopore and compared the result with commercial benchtop amplifier, the Axopatch 200B. The conductance of the pore was determined from the measurement of

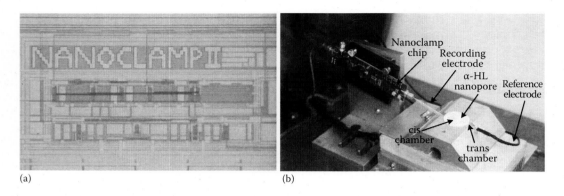

(a)   (b)

FIGURE 29.8
(a) Micrograph of VLSI nanoclamp chip and (b) experimental setup for the nanopore recording which is placed in a Faraday cage.

steady-state current changes as V_{CMD} was varied between two values. The conductance graphs are shown in Figure 29.9, with 1.16 and 1.12 nS measured by the nanoclamp system and the Axopatch 200B, respectively. Here, the difference of 3.6% arises primarily from evaporation of the buffered ionic solution, which causes a gradual increase in conductance during the measurement. We measured values first with the Axopatch 200B and then with the VLSI bioinstrumentation circuit.

In order to measure the ultralow current attenuation associated with DNA molecules that pass through the nanopore, single-stranded DNAs with 40-mer length (1 M) were added to the solution in the cis chamber which was filled with 1 M KCl at pH 8 and 23°C. By setting the command voltage to –200 mV (if the V_{CMD} exceeds –200 mV, the formation

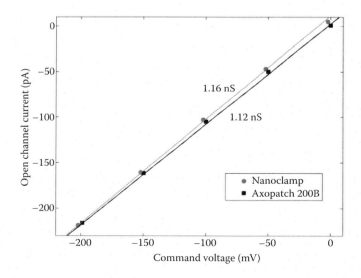

FIGURE 29.9
Measured nanopore conductance graphs using a VLSI nanoclamp (light grey) and the commercial Axopatch 200B (dark line).

on the biological pore will be dissolved), translocation events of DNA traversing the nanopore were induced, as displayed in Figure 29.10a. Representative events are shown in Figure 29.10b.

Figure 29.11 illustrates the VLSI bioinstrumentation circuit noise and background noise spectral densities during nanopore sensing. Here, the solid line indicates the input-referred noise current of just the bioinstrumentation, while the dashed line shows the background noise current including the bioinstrumentation, nanopore, E_{REC}, and ionic solution. Thus, nanopore noise can be roughly estimated by subtracting the bioinstrumentation noise from the background noise. In the high-frequency regime of around 10 KHz, above where the nanopore and bioinstrumentation noises are filtered by the off-chip fifth-order Bessel LPF, the background noise increase results from the effects of the thermal noise of the self-biased OTA, the large nanopore, and shaping due to the input parasitic capacitances C_T. In the low-frequency regime, the nanopore has a flicker noise, although it is lower than for the VLSI bioinstrumentation, which stems from nanopore surface charge fluctuations due to charge traps, and possibly the presence of nanobubbles fluctuating within the pore. The measured bioinstrumentation noise and the background noise are 4.21 and 5.87 pA$_{RMS}$ within a 10 KHz bandwidth, respectively. This noise level will be further reduced to sequence DNA in the future.

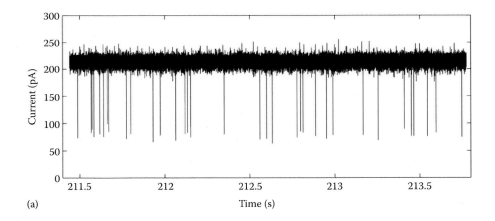

(a)

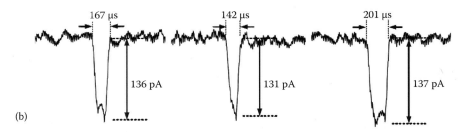

(b)

FIGURE 29.10
(a) Translocation events measured with the nanoclamp, where the α-hemolysin protein nanopore was used. Most of the signal at 200 mV$_{CMD}$ is at the open-channel of ~210 pA, with brief attenuations at 70–80 pA showing individual DNA capture and translocation events. (b) Inset representative blockades induced by 40-mer single-strand DNAs passing through the α-hemolysin nanopore.

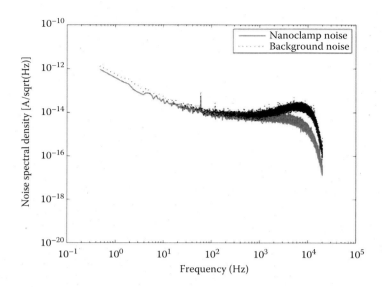

FIGURE 29.11
Measured input-noise spectral densities of VLSI nanoclamp and background in the nanopore sensing.

29.4 Scalability and Multichannel Implementation

A primary goal in nanopore research is to directly analyze DNA (e.g., for sequencing) and to miniaturize the technology into a portable (ideally, handheld) diagnostic platform. Related to this, we consider the scalability of the circuitry for a handheld nanopore array [24], which is associated with miniaturization of a DNA-analysis technology. In particular, we assess how many channels can be functionalized on a device based on existing fabrication methods and relevant patch-clamp amplifier technology; we do not consider the scalability of the nanopore sensors of the fluidics portion of the technology. A multichannel nanopore array on the small footprint device requires a multichannel patch-clamp amplifier integrated on a single chip, and each amplifier should be connected with an individual nanopore in order to prevent signal interference between nanopores. Thus, the number of patch-clamp amplifiers should be the same as the number of nanopores. There is also analog signal acquisition from each nanopore through the patch-clamp amplifiers, and each analog signal must be digitized using an analog-to-digital converter. The multichannel patch-clamp amplifier should have two main features: miniaturization and high channel density. For the purpose of establishing a geometric size constraint, within which the circuitry should be housed, we consider the iPhone as our ideal housing dimension ($5.86 \times 11.52 \times 0.93$ cm^3). Ideally then, several hundreds or thousands of patch-clamp amplifiers have to fit in an iPhone. A patch-clamp amplifier from Tecella supports up to 384 channels (according to their company web page), but its size ($19.8 \times 23.6 \times 35.6$ cm^3) is too large for our thought exercise here. CMOS process technology enables space minimization, thus helping to increase channel density on a limited area. According to our previous work [24], Figure 29.12 illustrates a graph showing the trade-off between die cost and die yield when varying the number of patch-clamp amplifiers on a die. Although we could integrate the maximum of 1500 patch-clamp amplifiers on a die area of 600 mm^2, the yield would be poor, resulting in increased

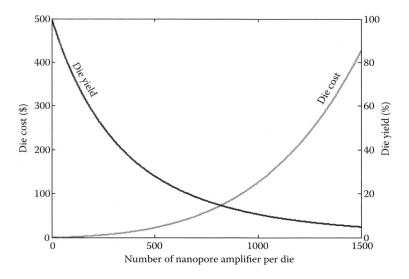

FIGURE 29.12
Trade-off curve between die yield and cost of varying the number of patch-clamp amplifiers on the die. The optimal value is approximately 820 patch-clamp amplifiers on one die, resulting in a cost of roughly $70.

cost. Considering the trade-off in cost and yield, the curve suggests an optimal choice of integrating 820 patch-clamp amplifiers on one die, corresponding to a cost of approximately $70. If other costs related to circuitry are considered, such as testing verification procedures and chip-packaging processes, the chip cost would be increased up to a few hundred dollars. Based on our thought experiment, we reason that 800–1000 channels are reasonable for a handheld device, with costs (excluding usage and microfluidics fabrication) well below $1000 per instrument. As stated, it is essential that each measured current in the array have as low noise as possible, especially in nanopore sequencing, so that robust base-calling algorithms can be employed and error probabilities can be assigned to the sequences [25].

When we implement several hundred amplifiers, pad size and location become critical on the chip because the number of pads can limit the chip area. For example, if the 820 amplifiers are integrated on the chip, at least 820 pads would be required to access the nanopore array and 205 pads are placed on each side. Let us assume that one pad area is approximately $50 \times 50 \ \mu m^2$ and the pitch between pads is also 50 μm. In this case, 205 pads occupy 20.5 mm in length; thus, the minimum chip size would be $20.5 \times 20.5 \ mm^2$. If we consider other functional blocks' inputs and outputs, the number of pads will be raised. Naturally, this affects the number of amplifiers that can be fitted on the chip. To save the active area, we can apply an area-efficient pad technique [26] that is widely used for high–pin count chips. Figure 29.13a displays a conventional cross-sectional structure of pad, ESD, and amplifier. Because the amplifier is located in the center of the chip, the amplifier is connected to the pad and ESD through the long metal line. Here, the pad and ESD are placed in the peripherals of the chip as shown in the top view. Figure 29.13b illustrates the proposed cross-sectional structure of pad, ESD, and amplifier where ESD and amplifier are placed under the pad at the center of the chip. This technique will be helpful in shrinking the chip size. In this case, a flip-chip bonding can be adopted to connect the chip to the multinanopore array on a PCB. Employing this technique, a conceptual architecture of the multichannel electronics is displayed in Figure 29.14 and presently this multichannel nanopore-sensing chip is under development.

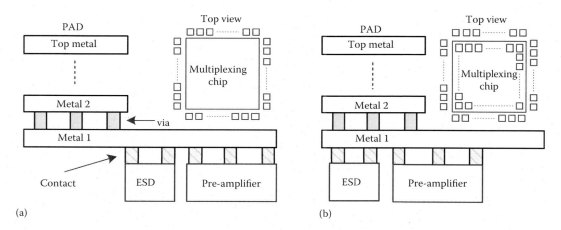

FIGURE 29.13
(a) Conventional cross-sectional structure of pad, ESD, and amplifier and (b) proposed cross-sectional structure of pad, ESD, and amplifier.

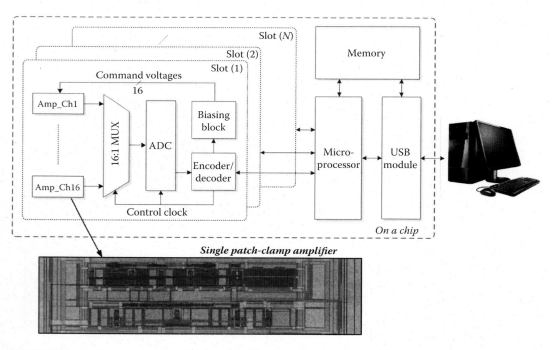

FIGURE 29.14
Conceptual system architecture for a future handheld VLSI bioinstrumentation device.

References

1. R.R. Harrison et al., "A low-power integrated circuit for a wireless 100-electrode neural recording system," *IEEE J. Solid-State Circuits*, vol. 42, no. 1, pp. 123–133, Jan. 2007.
2. K.D. Wise, D.J. Anderson, J.F. Hetke, D.R. Kipek, and K. Najafi, "Wireless implantable microsystems: High-density electronic interfaces to the nervous system," *Proc. IEEE*, vol. 92, no. 1, pp. 76–97, Jan. 2004.
3. B.M. Venkatesan and R. Bashir, "Nanopore sensors for nucleic acid analysis," *Nature Nanotechnol.*, vol. 6, 10.1038, Sep. 2011.
4. D. Branton et al., "The potential and challenges of nanopore sequencing," *Nature Biotechnol.*, vol. 26, no. 10, pp. 1146–1153, Oct. 2008.
5. C. Dekker, "Solid-state nanopores," *Nature Nanotechnol.*, vol. 2, pp. 209–215, Mar. 2007.
6. L. Song, M.R. Hobaugh, C. Shustak, S. Cheley, H.-G. Bayley, and J.E. Gouaux, "Structure of staphylococcal alpha-hemolysin, a heptameric transmembrane pore," *Science*, vol. 274, no. 5294, pp. 1859–1866, Dec. 1996.
7. E.N. Ervin, R. Kawano, R. White, and H.S. White, "Simultaneous alternating and direct current readout of protein ion channel blocking events using glass nanopore membranes," *Anal. Chem.*, vol. 60, no. 6, pp. 2069–2076, Jan. 2008.
8. A.R. Hall, A. Scott, D. Rotem, K.K. Mehta, H. Bayley, and C. Dekker, "Hybrid pore formation by directed insertion of α-hemolysin into solid-state nanopores," *Nature Nanotechnol.*, vol. 5, no. 12, pp. 874–877, Dec. 2010.
9. A.P. Ivanov, E. Instuli, C.M. McGilvery, G. Baldwin, D.W. McComb, T. Albrecht, and J.B. Edel, "DNA tunneling detector embedded in a nanopore," *Nano Lett.*, vol. 11, no. 1, pp. 279–285, Jan. 2011.
10. B. Sakmann and E. Neher, *Single-channel recording*, Second ed., Springer, New York, 1995.
11. H. Wang, N. Hurt, and W. Dunbar, "Measuring and modeling the kinetics of individual DNA–DNA polymerase complexes on a nanopore," *ACS Nano*, vol. 7, no. 5, pp. 3876–3886, 2013.
12. J. Prakash, J.J. Paulos, and D.N. Jensen, "A monolithic patch-clamping amplifier with capacitive feedback," *J. Neurosci. Methods*, vol. 27, no. 2, pp. 165–172, Mar. 1989.
13. P. Weerakoon, E. Culurciello, Y. Yang, J. Santos-Sacchi, P. Kindlmann, and F. Sigworth, "Patch-clamp amplifiers on a chip," *J. Neurosci. Methods*, vol. 192, no. 2, pp. 187–194, Jul. 2010.
14. P. Weerakoon, E. Culurciello, K. Klemic, and F. Sigworth, "An integrated patch-clamp potentiostat with electrode compensation," *IEEE Trans. Biomed. Circuits Systems*, vol. 3, no. 2, pp. 117–125, Apr. 2009.
15. F. Laiwalla, K. Klemic, F.J. Sigworth, and E. Culurciello, "An integrated patch-clamp amplifier in silicon-on-sapphire CMOS," *IEEE Trans. Circ. Syst. I: Regular Papers*, vol. 53, no. 11, pp. 2364–2370, Nov. 2006.
16. J. Rosenstein, M. Wanunu, C. Merchant, M. Drndic, and K. Shepard, "Integrated nanopore sensing platform with sub-microsecond temporal resolution," *Nature Method*, vol. 9, pp. 487–492, 2012.
17. D. Kim, B. Goldstein, W. Tang, F. Sigworth, and E. Culurciello, "Noise analysis and performance comparison of low current measurement systems for biomedical applications," *IEEE Trans. Biomed. Circuit Systems*, vol. 7, no. 1, pp. 52–62, Feb. 2013.
18. F.N. Hooge, T.G.M. Kleinpenning, and L.K.J. Vandamme, "Experimental studies on $1/f$ noise," *Rep. Prog. Phys.*, vol. 44, no. 5, pp. 479–532, May 1981.
19. B. Razavi, *Design of integrated circuits for optical communications*, McGraw-Hill Science, New York, Sep. 2002.
20. J. Kim, R. Maitra, K. Pedrotti, and W.B. Dunbar, "A patch-clamp ASIC for nanopore-based DNA analysis," *IEEE Trans. Biomed. Circuit Systems*, vol. 7, no. 3, pp. 285–295, Jun. 2013.

21. R.M.M. Smeets, U.F. Keyser, N.H. Dekker, and C. Dekker, "Noise in solid-state nanopores," *Proc. Natl. Acad. Sci. USA*, vol. 105, no. 2, pp. 417–421, Jan. 2008.

22. J.T. Yokichi, "Subsystems and methods for use in patch clamp systems," *US Patent No. 7741829 B2*, Jun. 2010.

23. C.C. Enz and G.C. Temes, "Circuit techniques for reducing the effects of op-amp imperfections: Autozeroing, correlated double sampling, and chopper stabilization," *Proc. IEEE*, vol. 84, no. 11, pp. 1584–1996, Nov. 1996.

24. R. Maitra, J. Kim, and W.B. Dunbar, "Recent advances in nanopore sequencing," *Electrophoresis*, vol. 33, no. 23, pp. 3418–3428, Dec. 2012.

25. C.R. O'Donnell, H. Wang, and W.B. Dunbar, "Error analysis of idealized nanopore sequencing," *Electrophoresis*, vol. 34, pp. 2137–2144, May 2013.

26. L. Luh, J. Chroma, and J. Draper, "Area-efficient area pad design for high pin-count chips," *Proc. IEEE Ninth Great Lakes Symp: VLSI*, pp. 78–81, Mar. 1999.

30

Wireless Electrical Impedance Tomography: LabVIEW-Based Automatic Electrode Switching

Tushar Kanti Bera

CONTENTS

ABSTRACT   Electrical impedance tomography (EIT) is a computed tomographic method which reconstructs the spatial distribution of electrical conductivity within a closed domain from the boundary potentials developed by a constant-current injection. A low-frequency low-magnitude constant sinusoidal current is injected to the closed domain under test through the current electrodes attached to the domain boundary and the conductivity distribution is reconstructed from the surface potentials collected from the voltage electrodes by using an image-reconstruction algorithm in a personal computer. Modern EIT systems generate the digital data either by using a microcontroller with a software like MATLAB® or C programs (or any other software) or by modern data-acquisition systems (DASs) using signal generation cards with appropriate data-acquisition software, such as Laboratory Virtual Instrumentation Engineering Workbench (LabVIEW). In this chapter we discuss digital data generation, analog data acquisition, image reconstruction, and image analysis of a wired EIT system. The electrode switching technique of a wireless EIT system is also detailed. The DAS design aspect and operating principle and the experimental results are discussed. Moreover, EIT hardware and software and the practical experimentations are highlighted. In the sections, the theory of the system components (both software and hardware), system design procedure (both software and hardware), system operation methodology, and the practical experimentation are presented. The performance of the developed system, system efficiency, system reliability, and the system advantages, limitations, and precautions are detailed. The chapter concludes with overall system performance, system efficiency, system reliability, the results obtained in experiments, and the system advantages and limitations as well as with the present challenges and the future aspects.

KEY WORDS: *wireless electrical impedance tomography, electrode, electric switching, LabVIEW, electrode switching module.*

30.1   Introduction

Electrical impedance tomography (EIT) [1–5] is a computed tomographic method [6–7] which reconstructs the spatial distribution of electrical conductivity within a closed domain (Ω) from the boundary potentials developed by a constant-current injection (Figure 30.1). In EIT, a low-frequency low-magnitude constant sinusoidal current is injected to the closed domain under test (DUT) through the current electrodes attached to the domain boundary ($\partial\Omega$) and the conductivity distribution is reconstructed from the surface potentials collected from the voltage electrodes (Figure 30.1) by using an image-reconstruction algorithm [8–12] in a PC. EIT is being studied in different areas of science and technology as an image-reconstruction modality due to its several advantages [3–4,13–14] over other computed tomographic techniques. Being a noninvasive, nonradiating, nonionizing, and inexpensive methodology, EIT has been extensively researched in clinical diagnosis [15],

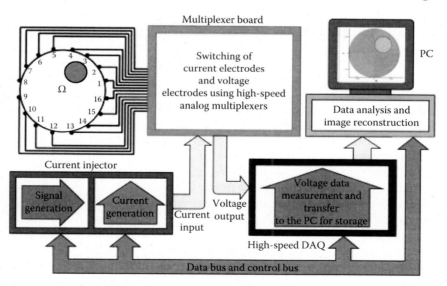

FIGURE 30.1
EIT system schematic.

biomedical engineering [16], biotechnology [17], industrial engineering [18], chemical engineering [19], oceanography [20], civil engineering [21], and so on.

In EIT, the surface electrodes are the interfacing sensors [22] and used for current injection and the voltage measurements. Boundary data are collected by switching the surface electrodes in a particular fashion depending on the current-injection protocol [1,23–25]. Analog multiplexers (MUXs) [26] are used for electrode switching of the hardware in both wired and wireless EIT systems. Generally, the current injection and voltage measurement are performed using four-probe or electrode methods [27]. In the four-electrode method, for example, the hardware needs four analog multiplexers. Therefore, the four electrode method-based data collection procedure for a 16-electrode EIT system [28–42] requires an automatic electrode switching module (ESM) developed with 16:1 analog MUXs. The sets of 16-bit parallel digital data are generally required to operate four 16:1 analog multiplexers switching at the same time. Bera and Nagaraju [43–45] studied the surface electrode switching of a 16-electrode EIT system using 16-bit parallel digital data obtained from 8-bit parallel digital data. In this study, a USB-based data-acquisition system (DAS) generated 8-bit parallel data and a binary adder circuit (BAC) was used to convert these to 16-bit parallel digital data. The sets of 8-bit parallel digital bits were generated using a National Instruments (NI) (United States) USB 6251 card in LabVIEW-based software and sent to the BAC, which converted it to the sets of 16-bit parallel data. The BAC fed the 16-bit parallel data sets to the multiplexers in ESM and operated the multiplexers to switch the surface electrodes in a particular sequence required for successful data acquisition (DAQ).

A wireless data-acquisition system (WL-DAS) provides some unique advantages; hence, WL-DAS is sometimes preferred in medical EIT systems for data acquisition as well as electric isolation property for maximum patient safety. In EIT, for both wired and wireless systems (Figure 30.2), surface electrodes are required to be switched in a particular fashion depending on the current-injection protocol applied. Analog multiplexers or other electronic switches are used to switch the electrodes in modern digital EIT hardware [46–49]. A greater number of electrodes in an EIT system require a greater number of switching points in a multiplexer; hence, more digital bits are needed to operate the corresponding multiplexers. In wireless EIT systems, the required parallel data are generated in EIT instrumentation and transmitted from the hardwired instrumentation to be received at the phantom side.

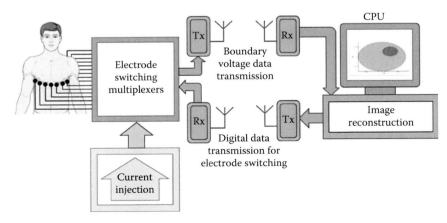

FIGURE 30.2
A wireless digital data-transmission scheme developed with RF-based transmitter (Tx) and receiver (Rx) modules for a 16-electrode EIT.

30.2 Electrical Impedance Tomography Wireless Instrumentation

Like a wired EIT system, an analog instrumentation is required in EIT to inject a constant current to the phantom boundary and measure the boundary potentials developed at the surface electrodes. Modern wireless EIT instrumentation can be developed with a constant-current injector (CCI), automatic ESM, signal-conditioning circuits (SCCs), and a wireless DAS which may consist a wireless digital data transmission system (WL-DDTS) for electrode switching and wireless analog data transmission system (WL-ADTS) for analog signal transfer.

30.2.1 Constant-Current Injector

A CCI is generally developed with a voltage-controlled oscillator (VCO) and voltage-to-current converter, which may be a voltage-controlled current source (VCCS). The VCO [43] consisted of a variable frequency function generator. Bera and Nagaraju [43–45] reported a VCO which was developed with MAX038 (Figure 30.3), a high-frequency, precision function generator which can produce accurate triangle, sawtooth, sine, square, and pulse waveforms with a minimum number of external components [50].

In the electrode switching studies, the VCCS was developed with a modified Howland constant-current generator (Figure 30.4) and it was fed by the VCO. The VCCS had two operational amplifier AD811 ICs (Analog Devices Inc., United States) [51]. The AD811 is a high-speed wideband current feedback operational amplifier which is suitably used in modified Howland current source with a feedback resistor [50].

30.2.2 Electrode Switching in Electrical Impedance Tomography

An EIT system needs an automatic ESM. An electrode switching module may be developed with analog multiplexers or any other automatic electronic switching devices, like relays.

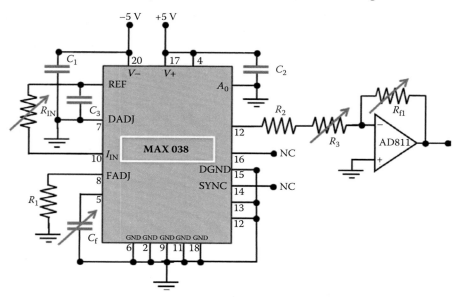

FIGURE 30.3
VCO with MAX038 IC and inverting amplifier.

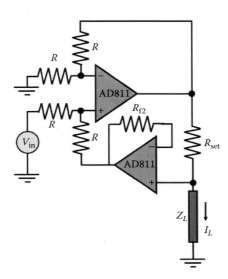

FIGURE 30.4
Modified Howland-based VCCS using AD811 ICs.

The electrodes can also be switched with any other electronic switching devices, like low resistive electronic switches such as dual in-line package (DIP) switches [5].

The number of the channels of the multiplexers depends on the number of the electrodes in the system. A system with N surface electrodes needs N:1 analog multiplexers. In principle, for better image resolution, a greater number of surface electrodes are required; hence, high-resolution systems need advanced multiplexers with a large number of channels. For a 2^N-electrode system, $4 \times N$ parallel digital data are required to operate all the four (2^N):1 analog multiplexers switching at the same time.

Generally, an EIT system generates the digital data either by using a microcontroller with a software like MATLAB or C programs or by data-acquisition and/or signal-generation cards backed up by data-acquisition software like LabVIEW. A LabVIEW-based data-generation system is found to be fast, easy to control, easy to program, user friendly, and compact [52–53].

30.2.3 Electrode Switching Module

In EIT, the DUT is surrounded by the surface electrodes and a constant current is injected to the domain through all the possible combinations of current electrodes (E_I) and the voltage data are collected from the voltage electrodes (E_V). The current injection through a particular current electrode pair and corresponding voltage data collection from all the possible voltage electrode pairs is known as a current projection (P_N). In an N-electrode EIT system, the current signal is injected through two electrodes and the differential potential data are collected across all the possible electrode pairs made up of the rest of the voltage electrodes. Hence, there are N different current projections each of which yield $N - 3$ differential voltage data.

An ESM is used to switch the surface electrodes in a particular sequence controlled by the PC. A PC-based automatic ESM [43] is illustrated in Figure 30.5 with high-speed CMOS analog MUXs. The current signal is injected to the DUT through four 16-channel CMOS analog multiplexers ICs (CD4067BE) for the 16-electrode EIT system. The current signal is injected to the DUT through two 16:1 MUXs (MUX-I_1 and MUX-I_2) called current electrode switching multiplexers (CESM-MUXs) and the voltage data are collected through two other MUXs (MUX-V_1 and MUX-V_2) called voltage electrode switching multiplexers (VESM-MUXs).

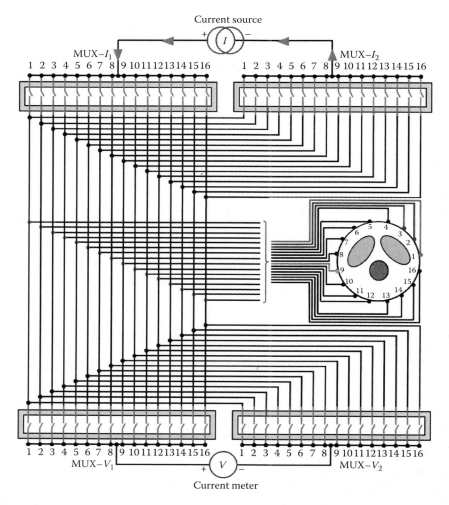

FIGURE 30.5

ESM and the electrode connections of a 16-electrode EIT system (current source is denoted by I and the voltage source is denoted by V).

30.2.4 WL-DDTS with Radio-Frequency Transmitter and Receiver

The wireless EIT data-transmission scheme (WL-DDTS) generates an 8-bit parallel data set to be converted into the serial data by an encoder [45]. The encoded serial data are then transmitted through the wireless transmission module (Figure 30.6a) and received by the receiver module [45].

The wireless RF transmission system developed using an amplitude shift keying (ASK)-based transmitter/receiver (Tx/Rx) pair operating at 433 MHz [45] is shown in Figure 30.6a. The scheme contains the 8-bit decoder/encoder ICs. Eight-bit parallel digital data are generated in DAQ card and fed to the encoder IC (HT640) in the transmission module. The encoder IC converts the 8-bit parallel data into 8-bit serial data and then feeds the 8-bit serial data to the transmitter block, which transmits the serial data to the receiver. At the receiving end, the serial data are received and supply the 8-bit data to the decoder IC (HT648L), which converts the serial data into parallel data. As the 16-electrode EIT system

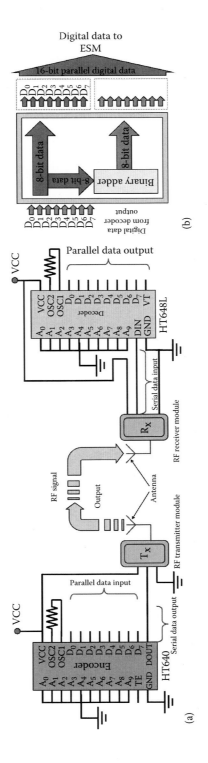

FIGURE 30.6

Transmission of 8-bit data in wireless EIT: (a) RF Tx/Rx module and (b) schematic of the binary adder circuit. (From Bera T. K. and Nagaraju J., Switching of a sixteen electrode array for wireless EIT system using a RF-based 8-bit digital data transmission technique, *ObCom 2011, Part I, CCIS 269*, pp. 202–211, 2012. With permission.)

requires 16-bit parallel digital data for electrode switching, the 8-bit parallel digital data received from the wireless transmission module are converted to the 16-bit parallel digital data by using the binary adder circuits (Figure 30.6b) [45].

By using 16-bit encoder/decoder ICs, the direct transmission of 16-bit data is possible (Figure 30.7) and this eliminates the binary adder circuit required for converting 8-bit data to 16-bit data. A 16-bit data-transmission scheme uses 16-bit encoder/decoder ICs (e.g., GL116 [Glolab Corporation, United States]). The transmitter module transmits the 16-bit serial data for the receiver to be converted to 16-bit parallel data.

In both wired and wireless EIT, the electronic instrumentation generates parallel digital data required for multiplexers to switch the current and voltage electrodes. The parallel bits are generated by DAS and fed to the ESM connected to the surface electrodes. For an N-electrode system, ESM should be capable of switching the N number of electrodes independently for current injection and voltage measurement using digital data, which are essentially required to be studied and tested before using it for MUX operation to ensure to feed the accurate digital data for correct electrode switching.

30.3   Digital Logic for Electrode Switching in LabVIEW-Based Algorithms

The electrode switching program can be written in LabVIEW or another software to control the NI DAQ card which sequentially generated data bits required for the multiplexer operation. The LabVIEW program, called the DAS program, consisted of two "for" loops (each of which were used to generate the parallel digital bits), one for current electrode switching and the other for voltage electrode switching. Hence, the 16-bit parallel data were generated in LabVIEW using two for loops. Further, the 16-bit parallel data are divided in four sets of 4-bit parallel data. These four sets of parallel data are then fed to four multiplexers in different ways depending upon the logic used in the DAS program.

By using the DAS, different current-injection protocols which injected the current in different electrode combinations (neighboring, opposite, reference, etc.) were developed [23]. The potentials of all the voltage electrodes were measured with respect to the ground point. However, in some cases, instead of 16-bit parallel data, sets of 8-bit parallel data were generated and then converted to 16-bit parallel data.

To evaluate the performance of the DAS and to evaluate the LabVIEW program operations, a digital signal-testing module (DSTM) [45] was developed by using eight low-power LEDs and eight high-precision resistors (1 kΩ, 1% tolerance) as shown in Figure 30.8. All the anodes of the LEDs were connected in series with the eight high-precision resistors and the cathodes were shorted and connected to the ground point of the electronic circuit. A time delay is programmed on to the electrode switching and the outputs of the eight I/Os of the DAQ card are fed to the DSTM. To assess the DAS operation and its digital data generation, the boundary potential data were collected by a USB-based data-acquisition system.

30.4   Electrode Protocols and Data Generation

This section discusses the current-injection strategies and data-collection techniques used in different current-injection methods, namely, the neighboring and opposite methods.

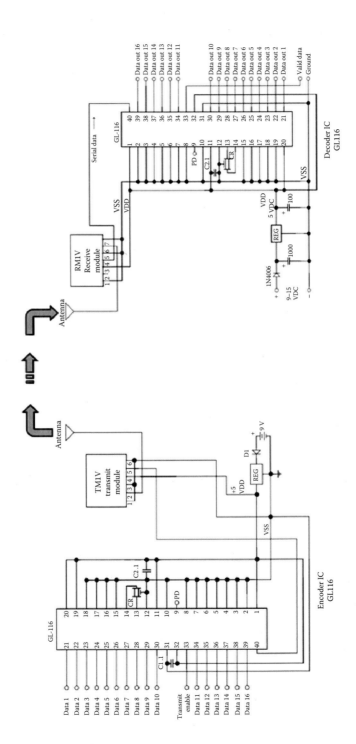

FIGURE 30.7

Transmission of 16-bit data in wireless EIT: (a) RF Tx/Rx module and (b) schematic of the binary adder circuit.

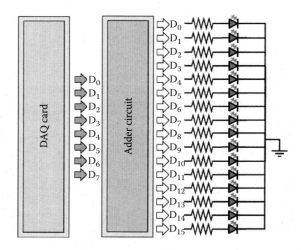

FIGURE 30.8
DSTM.

30.4.1 Neighboring Method

The neighboring or adjacent current-injection method was suggested by Brown and Segar [54]. In this method, the current is applied through neighboring or adjacent electrodes and the voltage is measured successively from all other adjacent electrode pairs excluding the pairs containing one or both current electrodes. The neighboring method applied for a 16-electrode EIT system with a circular domain under test (within a cylindrical volume conductor) is shown in Figure 30.9. A complete scan on a 16-electrode EIT system yields $16 \times 13 = 208$ voltage measurements, but due to the reciprocity [25], 104 data out of

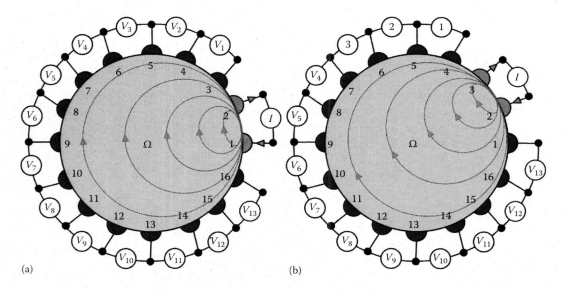

FIGURE 30.9
Boundary potential data-collection schematic in the neighboring current method for a 16-electrode EIT system with a circular domain (a) projection 1 (P_1) and (b) projection 2 (P_2).

TABLE 30.1

16 Bits of Parallel Digital Data (Converted from 8-Bit to 16-Bit Data in BAC) Required to Operate Four 16:1 Multiplexers in 16 Current Projections in the Neighboring Method in 16-Electrode EIT

Current Projection	MUX-I$_1$ (E$_1$) $D_{15}D_{14}D_{13}D_{12}$	MUX-I$_2$ (E$_2$) $D_{11}D_{10}D_9D_8$	MUX-V$_1$ (E$_3$) $D_7D_6D_5D_4$	MUX-V$_2$ (E$_4$) $D_3D_2D_1D_0$
P$_1$	0000	0001	0010	0011
			0011	0100
			0100	0101
			–	–
			1110	1111
P$_2$	0001	0010	0011	0100
			0100	0101
			0101	0110
			–	–
			1111	0000
P$_3$	0010	0011	0100	0101
			0101	0110
			0110	0111
			–	–
			0000	0001
P$_4$	0011	0100		
P$_5$	0100	0101		
P$_6$	0101	0110		
P$_7$	0110	0111		
P$_8$	0111	1000		
P$_9$	1000	1001		
P$_{10}$	1001	1010		
P$_{11}$	1010	1011		
P$_{12}$	1011	1100		
P$_{13}$	1100	1101		
P$_{14}$	1101	1110		
P$_{15}$	1110	1111	0000	0001
			0001	0010
			0010	0011
			–	–
			1100	1101
P$_{16}$	1111	0000	0001	0010
			0010	0011
			0011	0100
			–	–
			1101	1110

the 208 differential voltage measurements are independent of each other. The electrode switching program is written with two for loops as shown in Table 30.1.

30.4.2 Opposite Method

Hua et al. [55] proposed a method in which the current is injected through two diametrically opposed electrodes (as the name of the method suggests) and the differential

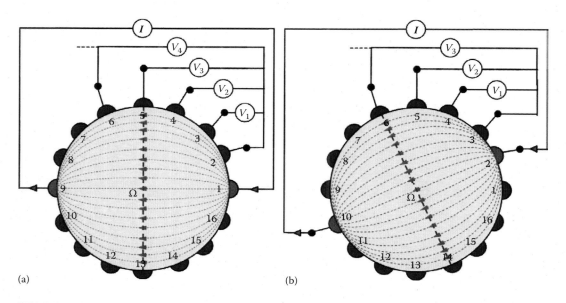

(a)   (b)

FIGURE 30.10
Boundary potential data-collection schematic in opposite current method for a 16-electrode EIT system with a circular domain (a) P_1 and (b) P_2.

potentials are measured on the voltage electrodes with respect to the reference electrode adjacent to the current-injecting electrode (Figure 30.10).

Thirteen differential potentials are measured successively from the 13 electrode pairs E_2-E_3, E_2-E_4, ... , and E_2-E_{16}, considering E_2 as the reference, as in Figure 30.10a. The injected current distribution is more uniform [1] in the opposite method; therefore, it has a good sensitivity [1]. Similarly, as shown in Figure 30.10b, P_2 injects the current signal through E_2 and E_{10} and measures 13 differential voltage (V_d) data (V_1, V_2, V_3, ... , V_{13}) successively from the 13 electrode pairs E_3-E_4, E_3-E_5, ... , and E_3-E_1. As in the previous case, the opposite current-injection method yields 16 current projections in a 16-electrode EIT system, and each of the current projections yields 13 differential voltage data producing $16 \times 13 = 208$ voltage measurements. But, due to the reciprocity [25], among the 208 differential voltage measurements, only 104 data are independent of each other.

The electrode switching program is written with two for loops as shown in Table 30.2.

30.5 Wireless Experimental Data Collection and Image Reconstruction

A number of EIT phantoms were developed for experiments, as shown in Figure 30.11a and b by using a 150 mm diameter shallow glass tank. Sixteen rectangle-shaped (34×10 mm^2) stainless steel electrodes [43] were equally spaced in the tank. The electrodes are cut from a high-quality (type 304) thin stainless-steel sheet to avoid the localized pitting corrosion leading to possible creation of small holes [43]. Electrodes were fixed over the inner wall of the tank by using insulated steel clips connected to the hardware through the steel crocodile clips (Figure 30.11a). Equal-length copper wires were used for connections with low resistivity. A stainless-steel common-mode electrode (CME) [43] was placed at the

TABLE 30.2

16 Bits of Parallel Digital Data (Converted from 8-Bit to 16-Bit Data in BAC) Required to Operate Four 16:1 Multiplexers in 16 Current Projections in the Opposite Method in 16-Electrode EIT

Current Projection	MUX-I$_1$ (E$_1$) $D_{15}D_{14}D_{13}D_{12}$	MUX-I$_2$ (E$_2$) $D_{11}D_{10}D_9D_8$	MUX-V$_1$ (E$_3$) $D_7D_6D_5D_4$	MUX-V$_2$ (E$_4$) $D_3D_2D_1D_0$
P$_1$	0000	1000	0001	0010
			0001	0011
			0001	0100
			–	–
			0001	1111
P$_2$	0001	1001	0010	0011
			0010	0100
			0010	0101
			–	
			0010	0000
P$_3$	0010	1010	0011	0100
			0011	0101
			0011	0110
			–	–
			0011	0001
P$_4$	0011	1011		
P$_5$	0100	1100		
P$_6$	0101	1101		
P$_7$	0110	1110		
P$_8$	0111	1111		
P$_9$	1000	0000		
P$_{10}$	1001	0001		
P$_{11}$	1010	0010		
P$_{12}$	1011	0011		
P$_{13}$	1100	0100		
P$_{14}$	1101	0101		
P$_{15}$	1110	0110	1111	0000
			1111	0001
			1111	0010
			–	–
			1111	1101
P$_{16}$	1111	0111	0000	0001
			0000	0010
			0000	0011
			–	–
			0000	1110

phantom center for reducing the common-mode errors [43] of the electronic circuit. The phantom tank was filled with a 0.9% (weight/volume) NaCl solution.

To collect data, the DAS program was written in LabVIEW [56] for interfacing the DAQ with the PC. The software was written to operate and control the EIT system hardware for an automatic current injection and voltage measurement through the ESM. LabVIEW is system design software which can be used as a system design platform and development environment for a visual programming language from National Instruments [51].

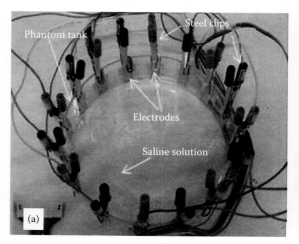 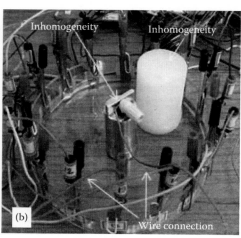

FIGURE 30.11
Reconfigurable practical phantom: (a) glass tank with surface electrodes and (b) phantom tank filled with NaCl solution, nylon inhomogeneity near electrode 7, and stainless-steel CME.

The DAS program was written in such a way that the DAS can be used for different current-injection protocols (neighboring, opposite, etc.) in EIT [43]. The DAS simultaneously generated the digital data to operate the current and voltage electrodes and acquired the developed voltage signals from the practical phantom or subject under test (SUT). The parallel digital bits were generated in NI USB 6251 card and obtained from the digital I/O ports of the card. The outputs of the I/O ports were fed to the selector channels of the multiplexers in ESM. The analog voltage signals generated on the voltage electrode were fed to the analog I/O port through the ESM and simultaneously measured.

Using the DAS, a 1 mA, 50 kHz sinusoidal constant-current signal was injected to the phantom boundary and the electrode potentials were collected using a particular current-injection protocol. A constant-current signal was injected to the phantom boundary without inhomogeneity as well as with inhomogeneity and the entire electrode potentials were collected using opposite current-injection protocol. Nylon cylinders were put as the inhomogeneity near different electrode positions and the current signal was injected (Figure 30.10b). The current flux generated due to the constant current-injection produces the potentials in the phantom domain (Ω). The boundary potentials developed (V_1 through V_{16}) at the surface electrodes (E_1 through E_{16}, respectively) attached to the phantom boundary were collected using opposite current-injection protocol and sent to the PC for the analysis and image reconstruction.

Although, generally, the boundary potentials are not measured on current electrodes for contact impedance problem [25], the voltage measurements were conducted on current electrodes to obtain the greatest sensitivity to the resistivity changes in the domain [57]. They measured the RMS values of boundary potentials on all the electrodes with respect to the analog ground of the electronic circuits. As the potential data were collected from all the electrodes, a complete scan over the 16-electrode system yielded 256 potential data from 16 current projections. The complete voltage data set containing 256 data for a 16 current projections for the 16-electrode EIT system were saved as a .txt file in PC [23,43–45] for computation. Boundary data were collected for the practical phantom with different inhomogeneity configurations and the resistivity images were reconstructed. Resistivity images were reconstructed from the boundary data using the Electrical Impedance Tomography and Diffuse Optical Tomography Reconstruction Software (EIDORS) [58].

It was observed that the potential profiles of all the current projections were symmetric for homogeneous medium with circular domain phantoms, which proved the stability of the system. It is also noticed that for the opposite method, only the first 128 voltages measured for the first eight projections were sufficient to reconstruct the inhomogeneities; hence, the first 128 voltages were measured and saved in the PC for imaging study. This reduced the data-acquisition time to 50% of the full data-acquisition time of the 16-electrode EIT system. Results also showed that the boundary potentials developed for homogeneous medium (NaCl solution only) were found suitable for efficient impedance imaging.

30.5.1 Advantages and Disadvantages

The DAS was developed with an NI USB 6251 data-acquisition card and controlled by the LabVIEW software. A user-friendly virtual instrumentation front panel developed in LabVIEW establishes the excellent graphical user interface (GUI). The control program was written in block diagram in LabVIEW was found easy to operate and modify for the other EIT systems. LabVIEW and its graphical development platform and the modularity extended by the PCI (peripheral component interconnect) eXtensions for instrumentation (PXI) architecture with the tools required to create and deploy measurement and control systems through unprecedented hardware integration [54]. The LabVIEW-based GUI was found suitable to fully control the EIT system for current injection and the boundary data acquisition. Boundary data for first eight current projections were found sufficient to produce impedance image; hence, the other eight current projections were not collected, which reduced the data-acquisition time to 50% of its total data-acquisition time (Table 30.3). The WL-DDTS with DAQ card is found with higher cost compared to the other microcontroller based systems. The EIT systems with higher number of electrodes need DAS with higher number of digital I/Os.

30.6 Mathematical Approach and Electrode Models

The conductivity of the domain under test is reconstructed in inverse solver by using the measured boundary potentials developed for a constant-current injection and the calculated boundary potentials obtained from the solution of the EIT governing equation solved by forward solver. The governing equation of EIT represents the relation between the conductivity of the DUT and the developed potential for a constant-current injection. If a low-frequency sinusoidal constant current applied to a homogeneous and isotropic medium (DUT) with low magnetic permeability (biological tissue) and electrical conductivity (σ) develops the electrical potential (ϕ) at all the points within DUT, the EIT governing equation can be represented as [1–2]:

$$\nabla \cdot \sigma \nabla \phi = 0, \tag{30.1}$$

where ∇ is the gradient operator in the same system.

This nonlinear partial differential equation represents the electrodynamics of impedance imaging. Boundary conditions [1–2] are applied to restrict the solutions of this nonlinear partial differential equation.

TABLE 30.3

16 Bits of Parallel Digital Data (Converted from 8-Bit to 16-Bit Data in BAC) Required to Operate Four 16:1 Multiplexers in 16 Current Projections in the Opposite Method in a 16-Electrode EIT

Current Projection	MUX-I_1 $D_{15}D_{14}D_{13}D_{12}$	MUX-I_2 $D_{11}D_{10}D_9D_8$	MUX-V_1 $D_7D_6D_5D_4$	MUX-V_2 $D_3D_2D_1D_0$
P_1	0000	1000	0000	1000
			0001	1000
			0010	1000
			–	–
			1111	1000
P_2	0001	1001	0000	1001
			0001	1001
			0010	1001
			–	–
			1111	1001
P_3	0010	1010	0000	1010
			0001	1010
			0010	1010
			–	–
			1111	1010
P_4	0011	1011	0000	1011
			0001	1011
			0010	1011
			–	–
			1111	1011
P_5	0100	1100		
P_6	0101	1101		
P_7	0110	1110		
P_8	0111	1111		
P_9	1000	0000		
P_{10}	1001	0001		
P_{11}	1010	0010		
P_{12}	1011	0011		
P_{13}	1100	0100		
P_{14}	1101	0101	0000	0101
			0001	
			0010	
			–	
			1111	
P_{15}	1110	0110	0000	0110
			0001	0110
			0010	0110
			–	–
			1111	0110
P_{16}	1111	0111	0000	0111
			0001	0111
			0010	0111
			–	–
			1111	0111

As Equation 30.1 has an infinite number of solutions to constrain its solution space and to obtain a reasonable physical model for EIT, the boundary conditions are required to be implemented. The boundary conditions are the sets of conditional information which are evoked from the electrode models of current injection and voltage measurement; these are briefly discussed below [59].

In EIT, the electrical conductivity or resistivity of a conducting domain Ω is reconstructed from the surface potential developed by a current signal injected at the domain boundary $\partial\Omega$. An electric power signal (current or voltage) is applied to the boundary of the subject under test through different pairs of the surface electrodes (yielding different current projections) attached to the SUT and the signal generated at the boundary (voltage or current) are collected (Figure 30.12) from the voltage electrodes for all current projections to obtain a complete scan (around the entire boundary or 360° angular orientation).

Considering a closed domain Ω as shown in Figure 30.12, let us assume that a small cylindrical volume element (τ) is placed on the surface of the domain such that the top and bottom of the elemental cylinder are almost parallel with the boundary ($\partial\Omega$). Now, the electric field (E) is defined as

$$E = -\nabla\phi. \tag{30.2}$$

Considering the source current density (J_s) inside the object as zero for low-frequency range in EIT, we have the boundary condition as

$$\sigma\frac{\partial\phi}{\partial n} = I, \tag{30.3}$$

where I is the negative normal component of the injected current density J_s and it is referred to as the injected current in EIT [59].

30.6.1 Continuum Model

The continuum model is an electrode model which assumes that there are no electrodes and injected current I is a continuous function satisfying

$$I(\zeta) = C \cos (k\zeta), \tag{30.4}$$

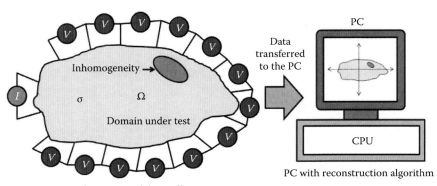

Boundary potential data collection

PC with reconstruction algorithm

FIGURE 30.12
EIT schematic.

where C is constant [59]. The experimental studies it has been shown that this model over-estimates the resistivities as much as 25% [59] because in this model the effects of the electrodes are totally ignored.

30.6.2 Gap Model

The gap model in EIT with L number of electrodes assumes that the injected current I is given by [59]

$$I = \begin{cases} \dfrac{I_l}{|E_l|} & \text{for } x \in E_l, \ l=1,2,\ldots,L, \\[2ex] 0 & \text{for } x \in \partial\Omega \Big/ \bigcup_{l=1}^{L} E_l, \end{cases} \tag{30.5}$$

where $|E_l|$ denotes the area of the electrode E and I is injected current into the lth electrode.

Although the gap model provides only a slight improvement over the continuum model, it still overestimates the resistivities. It is important to note that both the continuum and the gap models ignore not only the shunting effect of the electrodes and but also the contact impedances [1,59] which are produced by the electrochemical effect of the electrode–electrolyte interface [1].

30.6.3 Shunt Model

The shunt model [59] is proposed by taking into account the shunting effect of the electrode; that is, the potential on the electrode is constant. In addition, in the shunt model, the boundary condition represented by the continuum (Equation 30.3) model is modified with a more reliable condition given by

$$\int_{E_l} \sigma \frac{\partial \phi}{\partial n} dS = I_l, \qquad x \in E_l, \ l=1,2,\ldots,L. \tag{30.6}$$

In the shunt model, the shunting effect is taken into account by considering an extra condition given by

$$\phi = V_l, \qquad x \in E_l, l = 1, 2,\ldots, L, \tag{30.7}$$

where V is the measured voltage on the lth electrode.

Unfortunately, the shunt model underestimates the resistivities since the contact impedances problems are still ignored [59].

30.6.4 Complete Electrode Model

To obtain better EIT results in a real subject, the complete electrode model is proposed. The complete electrode model works by taking into account both the shunting effect of the electrodes and the contact impedances for the electrode–electrolyte interface [59]. The complete electrode model consists of Equation 30.1 and the boundary conditions as follows:

$$\phi + Z_l \sigma \frac{\partial \phi}{\partial n} = V_l, \qquad x \in E_l, \ l = 1, 2, \ldots, L, \tag{30.8}$$

$$\int_{E_l} \sigma \frac{\partial \phi}{\partial n} dS = I_l, \qquad x \in E_l, \ l = 1, 2, \ldots, L, \tag{30.9}$$

$$\sigma \frac{\partial \phi}{\partial n} = 0, \qquad x \in \partial\Omega \Big/ \bigcup_{l=1}^{L} E_l, \tag{30.10}$$

where Z_l is the effective contact impedance at the electrode–electrolyte interface.

In complete electrode model, in addition to the above boundary conditions, the following two extra boundary conditions for the injected current and measured voltages are also incorporated to ensure existence and uniqueness of the result in EIT. These two conditions for the injected current and measured voltages are

$$\sum_{l=1}^{L} I_l = 0, \tag{30.11}$$

$$\sum_{l=1}^{L} V_l = 0. \tag{30.12}$$

The complete electrode model has been shown to predict the measured voltages at the precision of the measurement system.

30.7 Results and Discussion of Results

Using the developed instrumentation, a 1 mA, 50 kHz sinusoidal constant-amplitude current signal is injected for opposite [44] and common-ground [44] methods of current injection [44] and the boundary potentials are collected. The RF transmission technology-based [60] digital data–transmission technique has been studied with a wireless digital data transmission system (WL-DDTS) to evaluate the surface electrodes switching of a 16-electrode EIT system. To study the performance of the WL-DDTS, the data are collected for different inhomogeneity configurations obtained from a shallow saline phantoms containing single or multiple inhomogeneities and the corresponding boundary data profiles are analyzed. The absolute resistivity images are also reconstructed from the saline phantom data collected with the WL-DDTS and the images are studied to analyze the performance of WL-DDTS.

Results show that the boundary potential profiles obtained for homogeneous medium as well as for inhomogeneous medium with different inhomogeneity configurations are of similar fashion. The boundary data profiles are also found with good correlations between the WL-DDTS and wired digital data–transmission system (W-DDTS) for both opposite and common-ground current-injection methods. It is observed that for both the opposite and common-ground methods, the boundary potential data obtained from a homogeneous medium (NaCl solution) phantom with the electrode switching by WL-DDTS, the boundary potential profiles are found similar to those obtained for a W-DDTS system.

The stability of the system with WL-DDTS is studied with symmetric electrode pair (SEP) analysis [44] and it is compared with stability of the W-DDTS–based system. The SEP analysis shows that the profiles of the boundary obtained for all the current projections in opposite [44] and common-ground [44] methods are found symmetric, which proved the stability of the system [44]. The potential of the SEPs (Figure 30.13a and b) shows that the experimental phantom is almost symmetric with respect to the axis of symmetry of the current flux line.

Circular inhomogeneities (35 mm) made up of polypropylene (PP) cylinders are placed at different electrode positions (Figures 30.13a, 30.14a, 30.15a, and 30.16a) and the RMS potentials are measured on the surface electrodes with opposite and common-ground methods by using W-DDTS based DAS. Boundary data profiles of the phantoms shown in Figures 30.14b, 30.15b, and 30.16b. For all the phantoms, it is observed that the maximum potential (V_{ei}) of the boundary data for a phantom with inhomogeneity placed near the Nth electrode occurs in the $\{[(N-1) \times 16] + N\}$th data point in the boundary potential matrix $[\mathbf{V_m}]$. The maximum potential V_{ei} point in the $[\mathbf{V_m}]$ matrix is the potential of the electrode near the inhomogeneity when the positive terminal of the current source is connected to the same electrode because of the highest voltage drop occurring across the current path along the inhomogeneity and the electrode, due to the high inhomogeneity impedance (Z_{inhom}) and the high electrode contact impedance (Z_{ec}) [44]. For opposite current-injection protocol with CME at phantom center or for common-ground current-injection protocol, the boundary potential curves with an inhomogeneity placed near the electrode E_L ($L = 1$, $2, 3, \ldots, 16$) show that the highest voltage peak is appeared at the Lth projection [44]; hence, by studying the boundary data, the position of inhomogeneity can be found.

For example, the boundary potentials obtained for a phantom with a single inhomogeneity at electrode 3 (Figure 30.14a) shows that the high voltage peak appeared at the third

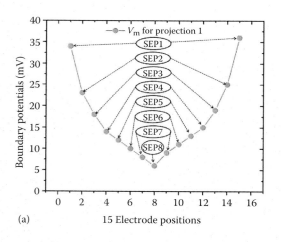

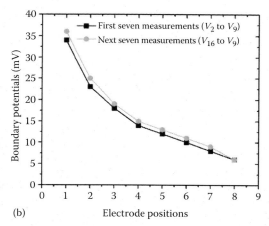

(a) 15 Electrode positions

(b) Electrode positions

FIGURE 30.13

(a) Homogeneous boundary potentials for opposite method (SEP electrode points are shown); (b) SEP electrode potentials for the first seven measurements (V_2 to V_9) and the next seven measurements (V_{16} to V_9) from a homogeneous medium. (Reprinted from *Measurement*, 45, Bera, T. K., and J. Nagaraju, Surface electrode switching of a 16-electrode wireless EIT system using RF-based digital data transmission scheme with 8 channel encoder/decoder ICs, 541–555, Copyright (2013), with permission from Elsevier.)

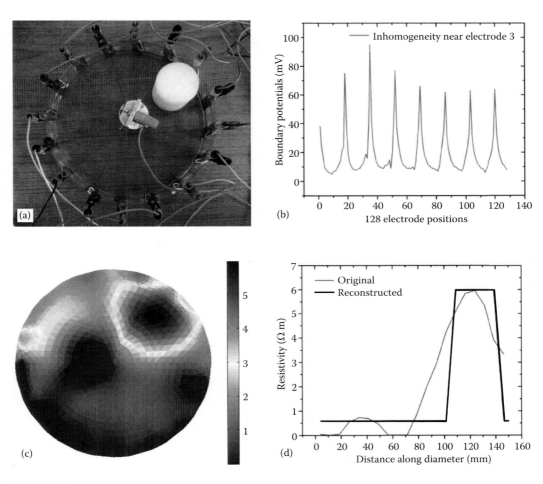

FIGURE 30.14
(a) Phantom 1 with PP cylinder at electrode 3 and its (b) boundary potentials, (c) resistivity image, and (d) diametric resistivity plot (DRP). (Reprinted from *Measurement*, 45, Bera, T. K., and J. Nagaraju, Surface electrode switching of a 16-electrode wireless EIT system using RF-based digital data transmission scheme with 8 channel encoder/decoder ICs, 541–555, Copyright (2013), with permission from Elsevier.)

projection (Figure 30.14b), which clearly indicates the presence of inhomogeneity at the third electrode position. In the potential curves with inhomogeneity at electrode 5 (Figure 30.15a), it is also observed that the high voltage peak appears at the fifth projection (Figure 30.15b), which clearly indicates the presence of inhomogeneity at the fifth electrode position. Similarly, the high voltage peak appeared at the seventh projection for the boundary data (Figure 30.16b) obtained with inhomogeneity at electrode 7 (Figure 30.16a) indicates the inhomogeneity position [44].

The resistivity images reconstructed from the boundary data collected by WL-DDGS are also studied to analyze the WL-DDGS performance. Three phantom configurations, phantoms 1, 2, and 3, are developed with single circular inhomogeneity near electrodes 3, 5, and 7. Boundary data profiles of phantoms 1, 2, and 3 were collected with 1 mA, 50 kHz sinusoidal current injection. Absolute resistivity images, reconstructed in EIDORS are shown in Figures 30.14c, 30.15c, and 30.16c. It is observed that for all the phantoms the

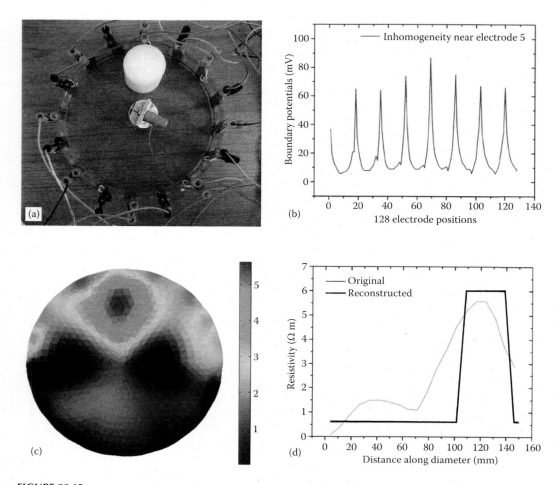

FIGURE 30.15
(a) Phantom 2 with PP cylinder at electrode 5 and its (b) boundary potentials, (c) resistivity image, and (d) DRP. (Reprinted from *Measurement*, 45, Bera, T. K., and J. Nagaraju, Surface electrode switching of a 16-electrode wireless EIT system using RF-based digital data transmission scheme with 8 channel encoder/decoder ICs, 541–555, Copyright (2013), with permission from Elsevier.)

inhomogeneities are successfully reconstructed with almost its proper shape and position. The average background resistivity (neglecting the background error near inhomogeneity) in the reconstructed images of phantoms 1, 2, and 3 are found to be 0.62, 0.78, and 0.89 Ω m. respectively. The resistivity of the background (0.9% NaCl solution) is measured as 0.61 Ω m in impedance analyzer [28,44].

As the resistivity of nylon inhomogeneity is very high compared to the background (NaCl solution) resistivity, it is very difficult to reconstruct the resistivity images with actual nylon resistivity. However, the resistivity of the inhomogeneity is found to be almost 10 times more than the bathing solution resistivity for all the phantom configurations. The average inhomogeneity resistivities (IR_{Mean}) reconstructed for phantoms 1, 2, and 3 are found to be 5.09, 4.95, and 5.28 Ω m, respectively, and the maximum inhomogeneity resistivities (IR_{Max}) are found to be 5.97, 5.61, and 6.46 Ω m. Result also show that the

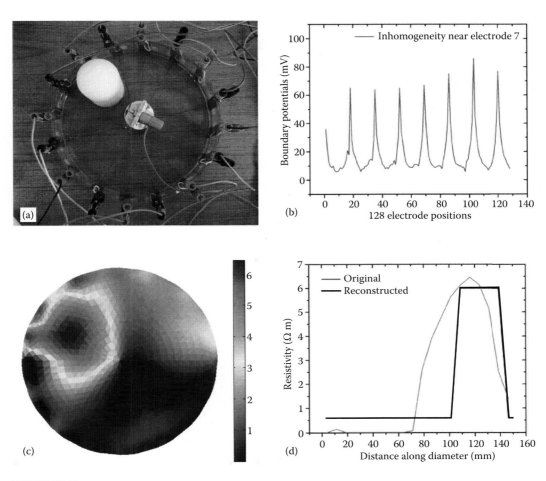

FIGURE 30.16
(a) Phantom 3 with PP cylinder at electrode 7 and its (b) boundary potentials, (c) resistivity image, and (d) DRP. (Reprinted from *Measurement*, 45, Bera, T. K., and J. Nagaraju, Surface electrode switching of a 16-electrode wireless EIT system using RF-based digital data transmission scheme with 8 channel encoder/decoder ICs, 541–555, Copyright (2013), with permission from Elsevier.)

contrast-to-noise ratios (CNRs) of the reconstructed images obtained for phantoms 1, 2, and 3 are 3.14, 3.24, and 2.53, respectively.

The percentages of contrast recovery (PCRs) of the reconstructed images obtained for phantoms 1, 2, and 3 are 70.01%, 65.84%, and 69.00%, respectively. The coefficients of contrast (COCs) of the reconstructed image for phantoms 1, 2, and 3 are 3.88, 3.55, and 3.39, respectively. Hence, it is observed that for the absolute resistivity imaging study with WL-DDGS–based DAS with common-ground method, the average CNR, the average PCR, and the average COC in the reconstructed images for inhomogeneities near all the electrodes are 2.97, 68.28%, and 3.61, respectively [44]. As the proper image reconstructions in the standard software reflect the accurate and efficient boundary data collection, it is concluded that WL-DDGS works efficiently.

The DRPs [5,11–12] of the resistivity images of phantoms 1, 2, and 3 are shown in Figures 30.14d, 30.15d, and 30.16d, respectively. It is observed that for the resistivity imaging study with common-ground method with WL-DDGS–based DAS, all the DRPs almost follow the DRP of the original object and the DRPs show that the centers of the reconstructed images are found at their original positions (centers of the original objects) except the image center of inhomogeneity at electrode 7 [44].

30.8   Conclusions

The LabVIEW program was found very suitable for both the wired and wireless EIT data acquisition. It was fast, user friendly, accurate, and easy to use. The DAS efficiently generated 8-bit parallel digital data, which were then converted to 16-bit parallel digital data suitable for operating the ESM. It was observed that the 8-bit and 16-bit parallel digital data were generated in LabVIEW. The first for loop in the LabVIEW program sequentially generated 16 sets of 4-bit parallel digital data changing from 0000 to 1111 and all the 16 sets of the data were fed to the MUX-I$_1$. Results show that the LabVIEW program along with the DAS generated the 8-bit parallel digital data, which were tested by 8-bit DSTM. All the sets of 8-bit digital data were found suitable and sequentially correct for electrode switching.

The surface electrodes of a 16-electrode EIT system can be switched with either an 8-bit digital data–transmission scheme or a 16-bit digital data–transmission scheme. The 8-bit digital data–transmission scheme needed a binary adder circuit to convert the 8-bit digital data into 16-bit digital data required for electrode switching in a 16-electrode system. The wireless digital data generation system (WL-DDGS) based on surface electrode switching of 16-electrode wireless EIT an radio frequency digital data–transmission technique operating with eight-channel encoder/decoder ICs. The ICs suitably interfaced the developed transmitter/receiver module with the USB-based DAQ system through the analog multiplexer-based electrode switching module. Eight-bit parallel digital data generated by DAQ card were sent to the transmission module to transmit the digital data bits to the receiver. The receiver module fed the serial data to the decoder IC and the decoder IC outputs were sent to the binary adder circuits for 8-bit to 16-bit conversion by adding the specified data set as per the current-injection patterns. The 16-bit data obtained at the adder outputs were fed to the multiplexers of the ESM for surface electrode switching. Results demonstrate that the wireless module successfully transmits and receives the parallel data and all the electrodes were switched as per the system requirements. Furthermore, for remote area, the wireless EIT system can acquire the data from a long distance with a suitably designed wireless data-transmission system with higher transmission range. In a hospital, an operator or a doctor sitting in a central control room can acquire the EIT data of patient in ICU or any other place.

References

1. Webster J. G., *Electrical Impedance Tomography* (*Adam Hilger Series of Biomedical Engineering*), Adam Hilger, New York, 1990.
2. Holder D. S., *Electrical Impedance Tomography: Methods, History and Applications (Series in Medical Physics and Biomedical Engineering)*, First ed., Institute of Physics Publishing Ltd., London, 2005.
3. Denyer C. W. L., Electronics for Real-Time and Three-Dimensional Electrical Impedance Tomographs, PhD Thesis, Oxford Brookes University, January 1996.

4. Metherall P., Three Dimensional Electrical Impedance Tomography of the Human Thorax, PhD Thesis, University of Sheffield, 1998.

5. Bera T. K. and Nagaraju J., Resistivity imaging of a reconfigurable phantom with circular inhomogeneities in 2D-electrical impedance tomography, *Measurement*, vol. 44, no. 3, pp. 518–526, doi:10.1016/j.measurement.2010.11.015, March 2011.

6. Bushberg J. T., Seibert J. A., Leidholdt Jr. E. M., and Boone J. M., *The Essential Physics of Medical Imaging*, Second ed., Lippincott Williams & Wilkins, Philadelphia, 2001.

7. Avinash C. K. and Slaney M., *Principles of Computerized Tomographic Imaging*, IEEE Press, Hoboken, 1999.

8. Yorkey T. I., Webster J. G., and Tompkins W. J., Comparing reconstruction algorithms for electrical impedance tomography, *IEEE Trans. Biomed. Eng.*, vol. 34, pp 843–852, 1987.

9. Lionheart W. R. B., EIT reconstruction algorithms: Pitfalls, challenges, and recent development, Review Article, *Physiol. Meas.*, vol. 25, pp. 125–142, 2004.

10. Breckon W. R., Image Reconstruction in Electrical Impedance Tomography, PhD Thesis, Oxford Polytechnic, 1990.

11. Bera T. K., Biswas S. K., Rajan K., and Nagaraju J., Improving image quality in electrical impedance tomography (EIT) using projection error propagation-based regularization (PEPR) technique: A simulation study, *J. Elec. Bioimpedance*, vol. 2, pp. 2–12, doi:10.5617/jeb.158, 2011.

12. Bera T. K., Biswas S. K., Rajan K., and Nagaraju J., Improving conductivity image quality using block matrix-based multiple regularization (BMMR) technique in EIT: A simulation study, *J. Elec. Bioimpedance*, vol. 2, pp. 33–47, doi:10.5617/jeb.170, 2011.

13. Bayford R. H., Bioimpedance tomography (electrical impedance tomography), *Annu. Rev. Biomed. Eng.*, vol. 8, pp. 63–91, 2006.

14. Riera J., Riu O. J., Casan P., and Masclans J. R., Electrical impedance tomography in acute lung injury, Review, *Med Intensiva*, vol. 35, no. 8, pp. 509–517, 2011.

15. Holder D. S., *Clinical and Physiological Applications of Electrical Impedance Tomography*, First ed., Taylor & Francis, London, July 1, 1993.

16. Brown B. H., Medical impedance tomography and process impedance tomography: A brief review, *Meas. Sci. Technol.*, vol. 12, pp. 991–996, August 2001.

17. Linderholm P. et al., Cell culture imaging using microimpedance tomography, *IEEE Trans. Biomed. Eng.*, vol. 55, no. 1, pp. 138–146, 2008.

18. Dickin F. and Wang M., Electrical resistance tomography for process applications, *Meas. Sci. Technol.*, vol. 7, p. 247, 1996.

19. Damasceno V. M. and Fratta D., Chemical diffusion detection in a porous media using electrical resistance tomography, *ASCE Geotechnical Special Publication (GSP) 149: Site and Geomaterial Characterization*, pp. 174–181, 2006.

20. Ingham M., Pringle D., and Eicken H., Cross-borehole resistivity tomography of sea ice, *Cold Reg. Sci. Technol.*, vol. 52, pp. 263–277, 2008.

21. Karhunen K., Seppänen A., Lehikoinen A., Monteiro P. J. M., and Kaipio J. P., Electrical resistance tomography imaging of concrete, *Cement Concrete Res.*, vol. 40, pp. 137–145, doi:10.1016/j.cemconres.2009.08.023, 2010.

22. Webster J. G., *Measurement, Instrumentation, and Sensors Handbook*, CRC Press, Boca Raton, 1999.

23. Bera T. K. and Nagaraju J., Studying the resistivity imaging of chicken tissue phantoms with different current patterns in electrical impedance tomography (EIT), *Measurement*, vol. 45, pp. 663–682, doi:10.1016/j.measurement.2012.01.002, 2012.

24. Bera T. K. and Nagaraju J., Studying the 2D resistivity reconstruction of stainless steel electrode phantoms using different current patterns of electrical impedance tomography (EIT), Biomedical Engineering, *Proceedings of the International Conference on Biomedical Engineering 2011 (ICBME-2011)*, Narosa Publishing House, New Delhi, pp. 163–169, 2011.

25. Malmivuo J. and Plonsey R., *Bioelectromagnetism: Principles and Applications of Bioelectric and Biomagnetic Fields*, Chap. 26, Sec. 26.2.1, Oxford University Press, New York, 1995.

26. Mano M., *Digital Design*, Third ed., Prentice Hall, Upper Saddle River, p. 173 (Chap. 5, Secs. 5–6), 2001.

27. Bera T. K. and Nagaraju J., A chicken tissue phantom for studying an electrical impedance tomography (EIT) system suitable for clinical imaging, *Sensing and Imaging*, vol. 12, nos. 3–4, pp. 95–116, doi:10.1007/s11220-011-0063-4, 2011.

28. Griffiths H., Zhang Z., and Watts M., A constant-perturbation saline phantom for electrical impedance tomography, *Phys. Med. Biol.*, vol. 34, no. 8, pp. 1063–1071, 1989.

29. Bera T. K. and Nagaraju J., A reconfigurable practical phantom for studying the 2-D electrical impedance tomography (EIT) using a FEM based forward solver, *10th International Conference on Biomedical Applications of Electrical Impedance Tomography (EIT 2009)*, School of Mathematics, The University of Manchester, United Kingdom, June 16–19, 2009.

30. Hahn G., Just A., Dittmar J., and Hellige G., Systematic errors of EIT systems determined by easily-scalable resistive phantoms, *Physiol. Meas.*, vol. 29, pp. S163–S172, 2008.

31. Bera T. K. and Nagaraju J., A study of practical biological phantoms with simple instrumentation for electrical impedance tomography (EIT), *Proceedings of IEEE International Instrumentation and Measurement Technology Conference (I²MTC 2009)*, Singapore, May 5–7, 2009, pp. 511–516. doi:10.1109/IMTC.2009.5168503, 2009.

32. Holder D. S., Hanquan Y., and Rao A., Some practical biological phantoms for calibrating multi-frequency electrical impedance tomography, *Physiol. Meas.*, vol. 17, pp. A167–A177, 1996.

33. Bera T. K. and Nagaraju J., A simple instrumentation calibration technique for electrical impedance tomography (EIT) using a 16-electrode phantom, *Proceedings of the Fifth Annual IEEE Conference on Automation Science and Engineering (IEEE CASE 2009)*, Bangalore, India, August 22–25, 2009, pp. 347–352, doi:10.1109/COASE.2009.5234117, 2009.

34. Thomas D. C., Siddall-Allum J. N., Sutherland I. A., and Beard R. W., Correction of the non-uniform spatial sensitivity of electrical impedance tomography images, *Physiol. Meas.*, vol. 15, pp. A147–A152, 1994.

35. Bera T. K., Biswas S. K., Rajan K., and Nagaraju J., Image reconstruction in electrical impedance tomography (EIT) with projection error propagation-based regularization (PEPR): A practical phantom study, *Lecture Notes in Computer Science*, vol. 7135/2012, pp. 95–105, doi:10.1007/978-3-642-29280-4_112011, 2012.

36. Hartinger A. E., Gagnon H., and Guardo R., Accounting for hardware imperfections in EIT image reconstruction algorithms, *Physiol. Meas.*, vol. 28, pp. S13–S27, 2007.

37. Bera T. K. and Nagaraju J., A gold sensors array for imaging the real tissue phantom in electrical impedance tomography, International Conference on Materials Science and Engineering 2012 (ICMST 2012), Kerala, India, 2012.

38. Bera T. K. and Nagaraju J., Studying the boundary data profile of a practical phantom for medical electrical impedance tomography with different electrode geometries, *Proceedings of the World Congress on Medical Physics and Biomedical Engineering—2009*, Munich, Germany, Dössel O. and Schlegel W.C. (Eds.), IFMBE Proceedings, vol. 25/II, pp. 925–929, doi:10.1007/978-3-642-03879-2_258, September 2009.

39. Hahn G., Just A., and Hellige G., Determination of the dynamic measurement error of EIT systems, *ICEBI, IFMBE Proceedings*, vol. 17, pp. 320–323, 2007.

40. Bera T. K. and Nagaraju J., Gold electrode sensors for electrical impedance tomography (EIT) studies, *IEEE Sensors Application Symposium (IEEE SAS 2011)*, February 22–24, 2011, United States, pp. 24–28. doi:10.1109/SAS.2011.5739810, 2011.

41. Bera T. K., Biswas S. K., Rajan K., and Nagaraju J., Improving the image reconstruction in electrical impedance tomography (EIT) with block matrix-based multiple regularization (BMMR): A practical phantom study, *IEEE World Congress on Information and Communication Technologies 2011(WICT-2011)*, Mumbai, India, pp. 1346–1351, 2011.

42. Bera T. K. and Nagaraju J., A multifrequency electrical impedance tomography (EIT) system for biomedical imaging, *IEEE International Conference on Signal Processing and Communications (SPCOM 2012)*, Bangalore, India, pp. 1–5, 2012.

43. Bera T. K. and J. Nagaraju J., Switching of the surface electrodes array in a 16-electrode EIT system using 8-bit parallel digital data, *IEEE World Congress on Information and Communication Technologies 2011(WICT-2011)*, Mumbai, India, pp. 1288–1293, 2001.

44. Bera T. K. and Nagaraju J., Surface electrode switching of a 16-electrode wireless EIT system using RF-based digital data transmission scheme with 8 channel encoder/decoder ICs, *Measurement*, vol. 45, pp. 541–555, doi:10.1016/j.measurement.2011.10.012, 2013.

45. Bera T. K. and Nagaraju J., Switching of a sixteen electrode array for wireless EIT system using a RF-based 8-bit digital data transmission technique, *ObCom 2011, Part I, CCIS 269*, pp. 202–211, 2012.

46. Bera T. K. and Nagaraju J., A LabVIEW based multifunction multifrequency electrical impedance tomography (MfMf-EIT) instrumentation for flexible and versatile impedance imaging, *15th International Conference on Electrical Bio-Impedance (ICEBI) and 14th Conference on Electrical Impedance Tomography (EIT)*, Germany, April 22–25, 2013, p. 216, https://app.box.com /s/7j76wvclfss3cpljgfs9, Accessed on September 1, 2014.

47. Bera T. K. and Nagaraju J., A battery based multifrequency electrical impedance tomography (BbMf-EIT) system for impedance imaging of human anatomy, *15th International Conference on Electrical Bio-Impedance (ICEBI) and 14th Conference on Electrical Impedance Tomography (EIT)*, Germany, April 22–25, 2013, p. 217, https://app.box.com/s/7j76wvclfss3cpljgfs9, Accessed on September 1, 2014.

48. Bera T. K., Saikia M., and Nagaraju J., A battery-based constant current source (Bb-CCS) for biomedical applications, *2013 International Conference on Computing, Communication and Networking Technologies (ICCCNT 2013)*, Tamil Nadu, India, pp. 1–5, 2013, http://ieeexplore.ieee .org/stamp/stamp.jsp?tp=&arnumber=6726810, Accessed on 1st September, 2014.

49. Bera T. K. and Nagaraju J., Practical phantom studies with a battery based electrical impedance tomography system, *1st International & 16th National Conference on Machines and Mechanisms (iNaCoMM 2013)*, Indian Institute of Technology Roorkee, Roorkee, India, 2013, pp. 1040–1043, http://www.nacomm2013.org/Papers/149-inacomm2013_submission_363.pdf, Accessed on September 1, 2014.

50. Data Sheet, MAX038-high-frequency waveform generator, Maxim Integrated Products, Inc., California.

51. Data Sheet, AD811—High performance video op amp, Analog Devices, Inc., Massachusetts.

52. Wei H. Y. and Soleimani M., Hardware and software design for a National Instrument-based magnetic induction tomography system for prospective biomedical applications, *Physiol. Meas.*, vol. 33, p. 863, doi:10.1088/0967-3334/33/5/863, 2012.

53. Bera T. K. and Nagaraju J., A LabVIEW based electrical impedance tomography (EIT) system for radiation free medical imaging, Limited Edition Awards Book of Top 50 Case Studies, National Instruments Graphical System Design Achievement Awards 2011 (NI GSDAA 2011), National Instruments, http://sine.ni.com/cs/app/doc/p/id/cs-14779 [Published Online], Accessed on September 1, 2014.

54. Brown B. H. and Segar A. D., The Sheffield data collection system, *Clin. Phys. Physiol. Measur.*, vol. 8 (suppl. A), pp. 91–97, 1987.

55. Hua P., Webster J. G., and Tompkins W. J., Effect of the measurement method on noise handling and image quality of EIT imaging, *Proceedings of Ninth International Conference of the IEEE Engineering in Medicine and Biology Society*, vol. 2, Institute of Electrical and Electronics Engineers, New York, pp. 1429–30, 1987.

56. Travis J. and Kring J., *LabVIEW for Everyone: Graphical Programming Made Easy and Fun*, Third ed., Prentice Hall, Upper Saddle River, 2006.

57. Cheng K. S., Simske S. J., Isaacson D., Newell J. C., and Gisser D. G., Errors due to measuring voltage on current-carrying electrodes in electric current computed tomography, *IEEE Trans. Biomed. Eng.*, vol. 37, no. 60, pp. 60–65, 1990.

58. Vauhkonen M., Lionheart W. R. B., Heikkinen L. M., Vauhkonen P. J., and Kaipio J. P., A MATLAB package for the EIDORS project to reconstruct two dimensional EIT images, *Physiol. Meas.*, vol. 22, pp. 107–111, 2001.

59. Vauhkonen M., Electrical Impedance Tomography and Prior Information, PhD Thesis, *Kuopio University Publications C: Natural and Environmental Sciences*, p. 110, 1997.

60. Voldman S. H., *ESD: RF Technology and Circuits*, John Wiley & Sons Ltd., Chichester, 2006.

Index

Page numbers followed by f and t indicate figures and tables, respectively.